PLASTICS AND THE ENVIRONMENT

PLASTICS AND THE ENVIRONMENT

Edited by

ANTHONY L. ANDRADY
Research Triangle Institute
Research Triangle Park, NC

A JOHN WILEY & SONS PUBLICATION

Copyright © 2003 by John Wiley & Sons, Inc. All rights reserved.

Published by John Wiley & Sons, Inc., Hoboken, New Jersey.
Published simultaneously in Canada.

No part of this publication may be reproduced, stored in a retrieval system, or transmitted in any form or by any means, electronic, mechanical, photocopying, recording, scanning, or otherwise, except as permitted under Section 107 or 108 of the 1976 United States Copyright Act, without either the prior written permission of the Publisher, or authorization through payment of the appropriate per-copy fee to the Copyright Clearance Center, Inc., 222 Rosewood Drive, Danvers, MA 01923, 978-750-8400, fax 978-750-4470, or on the web at www.copyright.com. Requests to the Publisher for permission should be addressed to the Permissions Department, John Wiley & Sons, Inc., 111 River Street, Hoboken, NJ 07030, (201) 748-6011, fax (201) 748-6008, e-mail: permreq@wiley.com.

Limit of Liability/Disclaimer of Warranty: While the publisher and author have used their best efforts in preparing this book, they make no representations or warranties with respect to the accuracy or completeness of the contents of this book and specifically disclaim any implied warranties of merchantability or fitness for a particular purpose. No warranty may be created or extended by sales representatives or written sales materials. The advice and strategies contained herein may not be suitable for your situation. You should consult with a professional where appropriate. Neither the publisher nor author shall be liable for any loss of profit or any other commercial damages, including but not limited to special, incidental, consequential, or other damages.

For general information on our other products and services please contact our Customer Care Department within the U.S. at 877-762-2974, outside the U.S. at 317-572-3993 or fax 317-572-4002.

Wiley also publishes its books in a variety of electronic formats. Some content that appears in print, however, may not be available in electronic format.

Library of Congress Cataloging-in-Publication Data:

Andrady, A. L. (Anthony L.)
   Plastics and the environment / Anthony L. Andrady.
     p. cm.
   ISBN 0-471-09520-6 (cloth : acid-free paper)
   1. Plastics–Environmental aspects. I. Title.

TP1120 .A54 2003
668.4′028′6–dc21

Printed in the United States of America

10 9 8 7 6 5 4 3 2 1

To my parents
Vincent and Rita Andrady

CONTENTS

Preface	xiii
Acknowledgments	xix
Contributors	xxi

PART 1 — 1

1 AN ENVIRONMENTAL PRIMER — 3
Anthony L. Andrady

1.1	Introduction	8
1.2	The Big Picture — Earth and its Environment	11
1.3	The Small Picture — Business Enterprises	18
1.4	Valuation of Environmental Resources	25
1.5	Stewardship of the Environment	27
1.6	Environmental Issues Related to the Plastics Industry: Global Concerns	32
1.7	Environmental Issues Related to the Polymer Industry: Local and Regional Concerns	51
1.8	Present Treatment	56
	Appendix A: Global Warming	62
	Appendix B: Depletion of Stratospheric Ozone	69

2 COMMON PLASTICS MATERIALS — 77
Anthony L. Andrady

- 2.1 Common Thermoplastics — 79
- 2.2 Polyethylenes — 83
- 2.3 Polypropylene — 93
- 2.4 Poly(vinyl Chloride) — 96
- 2.5 Polystyrene — 101
- 2.6 Poly(ethylene Terephthalate) — 103
- 2.7 From Resins to Thermoplastic Products — 105
- 2.8 Polyurethane and other Polymeric Foams — 113

3 POLYMERS AND ENERGY — 123
Ian Boustead

- 3.1 Introduction — 123
- 3.2 Contributions to Energy — 124
- 3.3 Feedstock — 125
- 3.4 Transport — 125
- 3.5 Representations of Energy — 126
- 3.6 Gross and Net Calorific Values — 127
- 3.7 Typical Gross Energies for Polymer Production — 128
- 3.8 Conversion Processes — 132
- 3.9 Energy and Polymer Recycling — 133

PART 2 — 137

4 PLASTICS IN PACKAGING — 139
Susan E. M. Selke

- 4.1 Introduction — 139
- 4.2 Packaging Functions — 142
- 4.3 Advantages of Plastics in Packaging Applications — 143
- 4.4 Types of Plastics Packaging — 145
- 4.5 Common Packaging Plastics — 148
- 4.6 Biodegradable Plastics — 158
- 4.7 Source Reduction — 160
- 4.8 Reuse — 162
- 4.9 Recycling — 163

	4.10	Energy and Environmental Assessments	166
	4.11	Legislation and Regulation	177

5 PLASTICS IN AGRICULTURE — 185
Ikram Hussain and Halim Hamid

	5.1	Introduction	185
	5.2	Greenhouse Films	187
	5.3	Weather Parameters that Affect Plastics Lifetime	190
	5.4	Stabilization of Plastics	192
	5.5	Mulch Films	197
	5.6	Plastics in Silage	199
	5.7	Disposal of Waste Plastic Films	199
	5.8	Drip Irrigation Systems	201
		Appendix: Structure of Selected UV-Stabilizers	207

6 COATING — 211
Loren W. Hill

	6.1	Environment Cost and Benefits Coatings	211
	6.2	Coatings and the Energy Crisis	212
	6.3	Coatings and the Material Crisis	212
	6.4	Historical Development of the Coatings Industry	213
	6.5	Binders and Solvents Used in Modern Coatings	215
	6.6	Effect of Government Regulations on the Coatings Industry	220
	6.7	Coatings Industry Responses to Regulation	228

7 WASTES FROM TEXTILE PROCESSING — 243
Brent Smith

	7.1	Introduction	243
	7.2	Yarn Formation	266
	7.3	Yarn Preparation	267
	7.4	Fabric Formation	269
	7.5	Textile Wet Processing Chemicals	271
	7.6	Substrate Preparation (Pretreatment)	276
	7.7	Coloration: Dyeing and Printing	283
	7.8	Finishing	299
	7.9	Product Fabrication	303
	7.10	Conclusion	304

PART 3   311

8 ENVIRONMENTAL EFFECTS ON POLYMERIC MATERIALS   313
Norma D. Searle

    8.1 Introduction   313
    8.2 Weather Factors and Their Effects on Polymeric Materials   314
    8.3 Environmental Stability and Degradation Mechanisms of Polymeric Materials   320
    8.4 Stabilization of Polymeric Materials Against Environmental Effects   330
    8.5 Environmental Weathering Tests   339
    8.6 Laboratory-Accelerated Weathering Tests   342
    8.7 Laboratory-Accelerated versus Environmental Weathering Tests   351

9 BIODEGRADABLE POLYMERS   359
Stephen P. McCarthy

    9.1 Introduction   359
    9.2 Justification   360
    9.3 Requirements for Design and Manufacture   361
    9.4 Biodegradable Polymers from Renewable Resources   361
    9.5 Biodegradable Polymers from Petroleum-Derived Products   367
    9.6 Future Developments   368

10 PLASTICS IN THE MARINE ENVIRONMENT   379
Murray R. Gregory and Anthony L. Andrady

    10.1 Introduction   379
    10.2 Plastic Litter and other Marine Debris   381
    10.3 Biological and Environmental Impacts   385
    10.4 Degradation of Plastics at Sea   389
    10.5 Photodegradable Plastics as a Mitigation Strategy   394
    10.6 Conclusions   397

11 FLAMMABILITY OF POLYMERS   403
Archibald Tewarson

    11.1 Introduction   403
    11.2 Heat Exposure of a Polymer   406

	11.3	Release of Polymer Vapors	407
	11.4	Polymer Melting	408
	11.5	Polymer Vaporization/Decomposition	417
	11.6	Ignition of Polymer Vapors	417
	11.7	Combustion of Polymer Vapors	435
	11.8	Fire Propagation	461
		Nomenclature	486

12 BIODEGRADABLE WATER-SOLUBLE POLYMERS   491
Graham Swift

	12.1	Introduction	491
	12.2	Definitions	493
	12.3	Opportunities for Biodegradable Water-Soluble Polymers	495
	12.4	Test Methods for Biodegradable Water-Soluble Polymers	497
	12.5	Synthesis of Biodegradable Water-Soluble Polymers	499
	12.6	Modified Natural Polymers	509
	12.7	Conclusions	513

PART 4   521

13 POLYMERS, POLYMER RECYCLING, AND SUSTAINABILITY   523
Johannes Brandrup

	13.1	Introduction: What is Sustainability	523
	13.2	Polymers and Sustainability	524
	13.3	Polymer Recycling and Sustainability	530
	13.4	Conclusions	560

14 PLASTICS RECYCLING   563
Michael M. Fisher

	14.1	Introduction	563
	14.2	Basic Plastics Recycling Definitions and Nomenclature	564
	14.3	Polymer Recovery, Recycling, Resource Conservation, and Integrated Resource Management as Global Concepts	566
	14.4	Early History of Plastics Recycling (Pre-1990)	567
	14.5	Polymer Recycling Statistics	569
	14.6	Marking of Plastic Packages and Products	574
	14.7	Collection of Plastics for Recycling	577

	14.8 Overview of Plastics Recycling Technology	583
	14.9 Looking Ahead	617
15	**THERMAL DESTRUCTION OF WASTES AND PLASTICS**	**629**
	Ashwani K. Gupta and David G. Lilley	
	15.1 Introduction	629
	15.2 Magnitude of the General Incineration Problem	630
	15.3 Sources, Disposal, and Recycling of Plastic and other Wastes	634
	15.4 Incineration of Cellulose and Surrogate Solid Wastes	641
	15.5 Thermal Destruction of Plastic and Nonplastic Solid Waste	655
	15.6 Thermal Destruction of Polypropylene, Polystyrene, Polyethylene, and Polyvinyl Chloride	673
	15.7 Prospects, Challenges, and Opportunities for Energy Recovery from Wastes	693
	15.8 Closure	694
16	**RECYCLING OF CARPET AND TEXTILE FIBERS**	**697**
	Youjiang Wang, Yi Zhang, Malcolm B. Polk, Satish Kumar, and John D. Muzzy	
	16.1 Introduction	697
	16.2 Textile Waste and Recycling	698
	16.3 Carpet Waste and Composition	699
	16.4 Fiber Recycling Technologies	701
	16.5 Summary	721
17	**POLYMERS IN AUTOMOBILE APPLICATIONS**	**727**
	Wendy Lange	
	17.1 Introduction	727
	17.2 General Environmental Impacts	730
	17.3 Using Polymers in Vehicles	731
	17.4 Summary	744
Index		747

PREFACE

ON THE ENVIRONMENT...

Ironically the most significant barrier to developing and implementing a viable environmental strategy appears to be the environmental debate itself. It has long left its home territory of science and ventured into the less predictable public arena of green groups, corporate interests, and public watchdogs. Politicization of environmental issues has lead to a degree of polarization among various interests that seriously interferes with objective scientific evaluation of the key issues. Yet it is precisely that type of critical analysis that is crucial to preserve the global environment. It is certainly not a case of merely doing something "for the environment" to make us feel better for having been sympathetic to the plight of the ecosystem but of consistently contributing to what is really needed to protect the biosphere.

The key environmental caution articulated by Malthus back in 1798 certainly remains valid today. Critics and skeptic environmentalists have repeatedly pointed out that Malthusian predictions have never materialized as yet. The population has indeed increased as predicted, with the overall affluence improved globally, but quite contrary to Malthus's expectation, we have more than enough food to feed the masses. But also, we have so far been running ahead of Malthus in a race where our advantage has been technical innovation. Technology has certainly not run out of steam, and the promise of coming days of new cheaper energy sources and alternate abundant (even perhaps) renewable materials are certainly credible from a technical vantage point. The crucial issue, however, is one of timing. Can humanity maintain the advantage it presently enjoys, approximately matching the rate of depletion of the current set of fossil fuel energy and material resources given our rate of progress in developing the next set of replacement resources and technologies? Specifically, can the petroleum energy reserves (now known and yet to be discovered) last long enough for us to develop, for instance, practical cost-effective fusion energy or to make the breakthrough in efficient harvesting of solar power? Can we stretch our strategic metal resources far enough until better manufacturing technologies no longer dependent on these can be discovered and developed? Can these goals be achieved at the rate needed without causing global environmental ills such as climate forcing? That is the relevant articulation of Malthus's caution.

A reasonable answer to these questions is that we do not know for sure but are justifiably optimistic. Our estimates of the longevity of available natural resources crucial to life as we know it are woefully inadequate because we do not know the full extent of reserves or the rate of depletion reliably. We know even less about the rate of relevant technology breakthroughs in the future. Therefore, we cannot know which of these is the faster process, but we should be convinced of the need to run ahead of Malthus at all future times. Most would agree with this conclusion but differ on their interpretation of what "running ahead" means in practical terms. Two schools of thought, the progeny of the aforementioned polarization of the community, are evident in the literature.

Some, including scientists, have adopted a somewhat narrow vision of the task at hand. In this vision, achieving economic development and global affluence is counterbalanced by environmental damage and loss of resources. The logic of the position assumes that humanity draws upon a constant resource pool and the two activities are therefore mutually exclusive. Thus, slowing down development (and therefore our economic well-being) becomes synonymous with environmental preservation. Implicit in the argument is an unreasonable and premature rejection of the promise of technology.

Others propose that we should live as we have been living, continuing to boost the energy intensity of our life-styles, focusing on economic development and the social well-being of the global citizenry. It is, they argue with some credibility, the economic development that invariably fuels and facilitates high levels of environmental stewardship. We, the hare, are so far ahead that the Malthusian tortoise can hardly hope to catch up with us. The enormous power of technology will save us each and every time, substituting resources, cleaning up the environment, serving its master unfailingly (until perhaps humanity faces entropic death of the system at the very end of our days!).

The consumer, using the powerful tools of his or her voting right in a democracy and his or her enormous collective pocketbook, endorses one or the other of these groups. Issues relating to the environment, however, are complex enough to lead objective unbiased scientists to debate for years with each other. The lack of time, resources, and inclination of the general public to study environmental issues and come to educated conclusions defers this important task to governments, educators, and activists. The public therefore tends to vote on perceptions, listening to the loudest voice with articulate claims repeated most often.

Both camps are wrong in their rather extreme interpretation of what the "race against Malthus" really entails. Truth of the matter is that we cannot guess the timing of needed future innovations. Prudence therefore dictates that we consume at a rate to maximize the chance of such technologies emerging in a relevant time scale. For instance, shale oil reserves do exist, and even with present-day technology is likely to be extracted to provide oil at less than twice the today's cost of oil. To therefore argue that we are free of any energy problems in the short or the long term is simplistic. Environmental cost of producing and using shale-derived oil is an integral part of this solution and needs to be considered as well and be reflected in the cost of that oil. In studying scarcity, no commodity can be considered in isolation; the sociopolitical impact of using the resource and how it impacts availability of other commodities in an international scenario of anarchic states need to be taken into account as well. A particularly dangerous line of reasoning, which appears on the surface to be somewhat persuasive, is that if the "cost" of (foreseeable and known) damage to the environmental is less than the (estimated) cost of repairing the damage after the fact, then it is best to ignore the problem. This logic reflects a lack of appreciation of the interrelated nature of the environment and how damage to any single facet of the environment can have far-reaching effects on others and the global ecosystem as a whole.

It is equally simplistic to argue that since economic growth means using resources, we should slow down growth to conserve the resources. It is after all the communities with more disposable income that are able to practice environmental stewardship at the needed level. Innovations that would save the day are likely to come from industry looking for better ways to produce and market energy.

An appropriate balanced strategy would be a two-pronged one: (a) conserve resources through waste prevention, reuse, and recycling of resources where the energy savings and externality considerations warrant it, and (b) provide incentives and otherwise facilitate innovation pertinent to key environmental issues within academia as well as private industry. The present course we are on likely overemphasizes the former and is inadequate in the latter. We rely heavily on the free-market mechanism for stewardship of the commons and for timely innovations. Free markets are well known to operate poorly when it comes to protecting the commons. They do well in promoting target innovations that can help the environment, particularly with incentives and rewards to promote the activity. However, being market driven, their innovative effort will not always be equitably distributed over the landscape of urgently needed environmental innovations. Regulatory mechanisms alone do not elicit corporate environmental interest and fall short of exploiting the full capability of the technical prowess of the establishment to benefit the environment.

...AND ON PLASTICS...

Plastics, as a class of material, is a truly exceptional one in that within a short span of less than a single lifetime it has pervaded nearly all aspects of modern life in all parts of the civilized world. Examples of successful replacement of conventional materials by plastics are far too numerous to list. What is important to note, however, is that nearly all of these substitutions survived in the marketplace and often continue to increase their market share in the relevant sectors. These obviously provide good value for the money because successful applications of plastics deliver performance comparable to (or better than) the materials they replaced but at a lower cost. A valid argument might be made that the market "cost" of plastics seriously underestimates the "true" cost, which reflects the use of common resources and externalities associated with their production. The same, however, holds true for competing materials as well. The available (albeit incomplete) data suggest that even a comparison based on the true cost of materials would find plastics to be an exceptional value.

Making plastics, of course, uses fossil fuel resources and invariably creates emissions; using plastics and especially disposing of postconsumer plastic waste has an associated environmental cost. This is by no means a phenomenon unique to plastics. It is common to all manufacturing and service industries. The cost is routinely paid globally, for instance, in using oil for transportation. Transportation is critical for the functioning of society, and the cost indeed is a reasonable and politically acceptable one. The question then is: Do the benefits provided by the use of plastics justify the environmental costs associated with their use?

Perhaps the best strategy in addressing the issue of plastics and the environment is to develop and present to the consuming public a balanced defendable cost–benefit study. Quantifying all the pertinent costs and both direct as well as indirect benefits of any material is not a straightforward undertaking. For instance, the health benefits of clothing, the protective value of surface coatings, the waste prevention by food-packaging plastics, and the quality of life enhancement by numerous plastic medical products need to be counted in a comprehensive assessment. This would avoid the qualitative discussions relating to isolated environmental impacts relating to plastics.

But, continuing improvements by the plastics industry to use less of the resources and to develop lower-polluting technologies are important because of their obvious environmental merit. Also, these efforts communicate the industry position on the environmental stewardship unequivocally to the user community.

The present effort attempts to facilitate this latter process by presenting in a single volume, technical discussions on various aspects of plastics of relevance to the environment. The first section of the book is intended to be an introduction to the key subject matter of plastics and environment, for those needing such an introduction. The second section explores several major applications of plastics with environmental implications: packaging, paints and coatings, textiles, and agricultural film use. The next section explores the behavior of plastics in some of the environments in which they are typically used, such as outdoors, in biotic environments, or in fires. The final section consists of chapters on recycling and thermal treatment of plastics waste.

ANTHONY L. ANDRADY

Research Triangle Park, NC

ACKNOWLEDGMENTS

I am very grateful to all the scientists whose valuable contributions have made this volume a reality. If not for their dedication and generosity the wide subject area covered here would have been simply impractical. I would like to acknowledge the publishers and authors who granted me the permission to reproduce or use some of their material throughout the book, and numerous colleagues who enhanced the scientific quality of the book through their discussion and critique of this work. The Research Triangle Institute deserves special thanks for its encouragement and support toward this effort. Finally, I am particularly grateful to my wife Lalitha and my sons (Anthony and Gerald) for their patience and understanding over the innumerable evenings and weekends I had to devote to preparing this manuscript.

CONTRIBUTORS

Anthony L. Andrady, Engineering and Environmental Technology, Research Triangle Institute, Research Triangle Park, Durham, North Carolina 27709, USA

Ian Boustead, Boustead Consulting Ltd, Black Cottage, West Grinstead, Horsham, West Sussex, RH13 8GH, United Kingdom

Johannes Brandrup, Am Allersberg 6,65191, Wiesbaden, Germany

Michael M. Fisher, Director of Technology, American Plastics Council, 1300 Wilson Blvd., Suite 800, Arlington, Virginia 22209, USA

Murray R. Gregory, Department of Geology, The University of Auckland, Chemistry Building Level One, 23 Symonds St., Auckland, New Zealand

Ashwani K. Gupta, Department of Mechanical Engineering, University of Maryland, College Park, Maryland 20742, USA

Halim Hamid, Center for Refining and Petrochemicals Research Institute, King Fahd University of Petroleum and Minerals, Dhahran 31261, Saudi Arabia

Loren W. Hill, Coating Consultant, 9 Bellows Road, Wilbraham, Massachusetts 01095, USA

Ikram Hussain, Center for Refining and Petrochemicals Research Institute, King Fahd University of Petroleum and Minerals, Dhahran 31261, Saudi Arabia

Satish Kumar, School of Textile and Fiber Engineering, 801 Ferst Dr., Georgia Institute of Technology, Atlanta, Georgia 30332, USA

Wendy Lange, Energy and Environment Technical Integration Engineer, General Motors Corporation, 7000 Chicago Rd., Warren, Michigan 48090, USA

David G. Lilley, School of Mechanical & Aerospace Engineering, Oklahoma State University, Stillwater, Oklahoma 74078, USA

Stephen P. McCarthy, Department of Plastics Engineering, University of Massachusetts at Lowell, Lowell, Massachusetts 01854, USA

John D. Muzzy, School of Chemical Engineering, Georgia Institute of Technology, Atlanta, Georgia 30332, USA

Malcolm B. Polk, School of Textile and Fiber Engineering, 801 Ferst Dr., Georgia Institute of Technology, Atlanta, Georgia 30332, USA

Norma D. Searle, Consultant, 114 Ventnor F, Deerfield Beach, Florida 33442, USA

Susan E. M. Selke, School of Packaging, Michigan State University, East Lansing, Michigan 48824, USA

Brent Smith, Cone Mills Professor of Textile Chemistry, North Carolina State University College of Textiles, Raleigh, North Carolina 27695, USA

Graham Swift, GS Polymer Consultants, 10378 Eastchurch, Chapel Hill, North Carolina 27517, USA

Archibald Tewarson, FM Global Research, 1151 Boston-Providence Turnpike, Room 314, Norwood, Massachusetts 02062, USA

Youjiang Wang, School of Textile and Fiber Engineering, 801 Ferst Dr., Georgia Institute of Technology, Atlanta, Georgia 30332, USA

Yi Zhang, DuPont Teijin Films, Hopewell, VA 23860

To waste, to destroy, our natural resources, to skin and exhaust the land instead of using it so as to increase its usefulness, will result in undermining in the days of our children the very prosperity which we ought by right to hand down to them amplified and developed.

—Theodore Roosevelt
(circa 1913)

—I want to say one word to you.
—Yes Sir.
—Are you listening?
—Yes Sir, I am.
—PLASTICS!

—The Graduate
(1967)

PART 1

CHAPTER 1

AN ENVIRONMENTAL PRIMER

ANTHONY L. ANDRADY
Engineering and Environmental Technology, Research Triangle Institute

Even a cursory study of the global environment leads to the inescapable and gloomy conclusion that humankind is fast approaching a crisis. A combination of factors has lead the most successful species ever to inhabit Earth to reach a rate of growth that can no longer be sustained by the resource base available on the planet. The world population is increasing and by latest estimates will reach 10.4 billion in year 2100. The food requirements to support this large a population and their activities come for the most part, directly or indirectly, from daily insolation of Earth by the sun. Primarily, this is the solar energy trapped by plants through photosynthesis and transferred up to various levels of the food chain.

Apart from the energy derived from food, our modern life-style demands additional sources of energy primarily for transportation and industrial production of goods. Unlike other species that grow within a diurnal solar energy budget, humans have developed technologies to access Earth's fossil fuel reserves, particularly petroleum resources, to supplement the meager budget of daily solar insolation. Deposits of fossil fuels created over millions of years via collection and storage of solar energy by plants have served humans very well since the days of the industrial revolution. The problem is that our modern life-style demands that we draw heavily upon these limited and diminishing reserves, at a rate very much faster than they can be replaced by the natural geochemical cycles. Even worse, in using these to produce the goods and services to keep society supplied, we not only spend these energy reserves inefficiently but also pollute the environment and create enormous amounts of waste in the process. The problem of waste generation and industrial pollution, once a local or at best a regional concern has now turned into a growing transboundary problem. In terms of impact it is no longer merely a question of pollution causing illnesses, birth defects, urban smog, and Love Canals that can be cordoned off. Today,

---

Plastics and the Environment, Edited by Anthony L. Andrady.
ISBN 0-471-09520-6 © 2003 John Wiley & Sons, Inc.

human activity is beginning to threaten the very fabric of the biosphere. The global ecosystem is already showing signs of being compromised by radiative forcing from greenhouse gases in the atmosphere, acidification of water resources, and loss of biodiversity. This is the first time in the history of the planet that human activity has resulted in such intense and perhaps irreparable damage to the environment.

Surprisingly, the amount of attention paid to this global threat has been quite disproportionate to its potential impact. While it has certainly prompted some discussion in academia as well as in industry, few concerted plans of action have emerged from the rhetoric. In the United States, national commitment to research the problem and options for its mitigation has been at best a limited one. At the international level, the progress in some fronts, such as in the Montreal Protocol on limiting the use of ozone-depleting chemicals, have proven that collective activity on global problems is indeed possible. Yet, the momentum on other crucial environmental issues such as the reduction of carbon dioxide emissions or the preservation of biodiversity has been disappointing.

What has all this got to do with polymers? A major goal of this work is to assess the extent to which the production and use of plastics[1] contribute to global environmental ills. The plastics industry is healthy with an annual growth rate of 4–5% between 1996 and 1999. The world resin production in the year 2000 was approximately 135 million tons, with North American capacity accounting for about a third of it. Western Europe and Asia are also major manufacturing regions each accounting for 20–23% of the total production. This situation, however, is likely to change with developing countries turning to increased production and utilization of plastics in the future. Plastics are rapidly becoming cost competitive with most traditional materials of construction and will continue to be increasingly popular in developing countries.

Presently, the use of plastics in consumer goods is relatively more prevalent in the developed countries. In fact the per capita consumption of plastics worldwide correlates surprisingly well with the per capita gross national product (GNP) of the country [1] (see Fig. 1.1). In affluent countries such as Japan or in western Europe plastics consumption can be as high as 200 lb/person-year as opposed to less than 10 lb/person-year in the least developed countries. However, on a volume basis, the future consumption of plastics is likely to increase the fastest in developing nations, particularly in China, India, and some countries in Latin America. Even a very small increase in the per capita consumption of plastics in highly populated countries translates into a very large increase in the volume of polymer usage. Among the healthiest markets for plastics in most countries is packaging applications, which is growing at a steady pace, and is consistently replacing conventional materials such as glass or metal. Increasing global affluence has encouraged the increased use of disposable plastic

---

[1] Polymers include plastics as well as elastomers, although the plastics segment of the industry is relatively much larger. This discussion focuses primarily on plastics, as most of the common and visible environmental concerns pertain to the plastics industry.

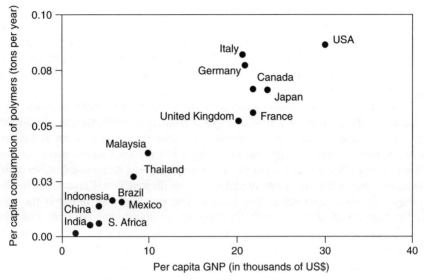

Figure 1.1. Dependence of annual per capita consumption of plastics on the GNP of a country.

packaging products in the developing world as well. Competitive wages in these countries attract polymer-based industries, particularly the low-technology processing operations, further encouraging the use of plastics in the developing world. The large disparity in labor rates[2] for plastics industry workers in the United States and in places as close as Mexico has remained unchanged in the past decade. Processing operations such as fiber spinning, film blowing, pipe or conduit extrusion, and injection-molding are increasingly being carried out in the developing countries with low labor costs. Not surprisingly, plastics processing equipment is also increasingly being manufactured in developing countries as well as in the West. For instance, in 1993 the number of injection-molding machines manufactured in Asia exceeded that manufactured in the United States

---

[2] The average hourly compensation for production workers in the plastics industry (SIC 308) in the United States during 1998 was US$15.39. The same rate in U.S. dollars for some selected countries follows:

Country	Hourly wage in US$	Country	Hourly wage in US$
Mexico	$1.47	Germany	$22.27
Taiwan	$4.98	Japan	$17.34
Hong Kong (China)	$5.75	Canada	$12.70

Source: Bureau of Labor Statistics, Washington, DC, 1997. Web site: *http://stats.bls.gov/fls/flshcind.htm*.

and Germany combined [2]. Also, the considerably less stringent environmental regulations in developing regions tend to lower operational costs for chemical industries relative to that in the United States or western Europe. The global trend is clear; the plastics industry will continue to prosper worldwide with increasing quantities being manufactured, processed, and used in the developing nations.

Most discussions on plastics and the environment center on the potential contribution of the production use and disposal plastics to environmental pollution. Plastics in common with other man-made (or man-processed) materials such as paper or glass require energy and raw materials for their manufacture. As with other materials, the production of plastics does use fossil fuel, contribute to emissions, and leaves behind waste materials to be disposed of. Plastics are hardly a unique class of materials in this regard. The issue at hand is not if the production and use of plastics results in a significant environmental disruption, but whether the environmental cost is ultimately justified in terms of economical, environmental, and other societal benefits gained through the use of plastics. The pertinent question to pose is whether the same degree of benefits offered by plastics (in a given application such as packaging) might be delivered by an alternate material that has a lower environmental penalty associated with its use. For instance, over half of the crude oil production is used in the area of transportation,[3] and this application does result in considerable resource depletion as well as environmental damage. But, no convenient, cost-effective, less polluting alternative for replacing gasoline-driven mass transportation is presently available. At a societal level the benefits of an efficient transportation system evidently justifies the substantial environmental "footprint" associated with transportation. The relative small expenditure of precious fossil fuel reserves and the potential industrial emissions in the plastics industry (or any other industry) needs to be similarly evaluated. What are the benefits and environmental costs of using plastics? The impressive record of plastics in the marketplace is compelling evidence that plastics provide good value for the money. A class of materials only about half a century old, plastics have already pervaded most applications ranging from medical implants to aircraft parts. Novel uses for plastics continue to emerge on a regular basis and all indications point to their continued success.

With increased use of plastics in consumer applications environmental concerns relating to these materials are beginning to be raised. For instance, a particularly visible plastics-related environmental problem is that of municipal solid waste disposal. With about 30% of the plastics production used in packaging, it is not surprising to find a significant and growing fraction of plastics in the municipal solid waste and in urban litter streams. Consumer awareness and sensitivity to the environmental impact of solid waste has never been at a higher

---

[3] Automobile use is the main drain on fossil fuel reserves, and in the United States ownership level has reached 0.77 automobiles per licensed driver by 1999. About 15% of the greenhouse gas emissions in the United States are attributed to automobiles [60].

level.[4] Today, the majority of American consumers go so far as to claim that corporate environmental reputation directly affects their brand loyalty and product choice in the marketplace. For a while, the "plastic or paper" issue was heatedly debated and even influenced consumer preferences for bagging at supermarket checkout counters. Similar concerns on Styrofoam packaging in the waste stream so strongly affected consumer behavior that major fast-food chains switched over from Styrofoam sandwich boxes to paper products in their disposable fast-food boxes. Published reports [3] of persistent nonbiodegradable Styrofoam packaging waste in old landfills played a role in crafting this consumer preference. International concerted response to the recognition of plastics as a significant xenobiotic pollutant of the oceans resulted in the U.S. ratification of an international maritime agreement (MARPOL Annex V)[5] in 1987 that severely limited the discharge of plastics waste into the sea. The Annex V of MARPOL implemented in the U.S. Congress via Public Law 100–220 lead to military vessels complying with these restrictions. Numerous related concerns in the United States and abroad can be added to the list, which will continue to grow for some time. The most recent major piece of legislation affecting polymer use is perhaps the control of volatile organic compounds (VOCs) in polymer-based coating formulations, promulgated in 1999 (under the 1990 Amendments to the Clean Air Act). These VOC restrictions will continue to be reviewed, and additional controls on the release of hazardous air pollutants are expected to further limit the coatings (and adhesive) formulation industries in the near future.

For the present purpose, it is convenient to view the polymer industry in segments. Technology that synthesizes the resin, forms it into prils, and transports it to the point of use is the manufacturing segment of the industry. Processing of the resin into an intermediate product (such as a masterbatch) intended for further processing or a final product for consumer use is a second segment. The most visible segment is where the consumer uses the plastic product and the product is either disposed of or recycled at the end of its useful life (in some cases as short as a few minutes). Each of these stages in the life cycle of a plastic product has associated environmental concerns. These need to be examined to determine if they are real or perceived issues and which are significant in quantitative terms. Ideally, the issues must be considered in the broader context of the total global environment and in relation to the long-term strategies for sustainable growth of human societies.

---

[4] The year 2000 survey of consumer attitudes reported in the *Green Gauge Report* by Roper-Starch Worldwide (New York) shows 49% of the consumers polled were concerned enough about the environment to pay a small premium for environmentally compatible consumer products.

[5] Three international treaties comprise MARPOL: (a) the Convention for the Prevention of Marine Pollution by Dumping from Ships and Aircraft, adopted at Oslo on February 15, 1972; (b) the International Convention for the Prevention of Pollution from Ships, 1973, adopted at London on November 2, 1973, and (c) the Protocol of 1978 relating to the International Convention for the Prevention of Pollution from Ships, 1973, adopted in London on February 17, 1973.

8 AN ENVIRONMENTAL PRIMER

1.1. INTRODUCTION

Figure 1.2 introduces the root cause of environmental problems. As shown in the figure, Earth's population and the global use of fossil fuel energy are both growing at an alarming rate. The different scenarios referred to in the figure are based on slightly different assumptions in the population and energy consumption models used in the projections. On August 12, 1999, the world population reached 6 billion, and according to conservative estimates will reach around 9.8 billion by 2050 (with a variation of approximately 2 billion about this estimate). This increase has been the result of dramatic reductions in the death rates in all parts of the world; the birthrate is also presently on the decline in all parts of the world. The world population is expected to finally begin flattening out about the year 2100, assuming the developing nations reduce their rates of population growth as anticipated, in the future. Presently, however, the population is increasing exponentially at the rate of approximately 1.3% annually or by about 80 million people a year. That is about the total population of Germany or Mexico being added to the global citizenry each year! In one human life span (1930 to present) the population grew from 2 billion to the present 6.5 billion.[6] The implications of this high rate of growth are underlined by the fact that most of this increase takes place within the developing regions of the world. Also, the life expectancy in all parts of the globe is slowly increasing, reaching an estimated average of

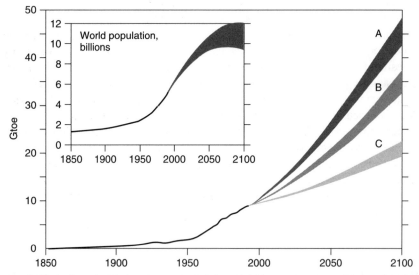

Figure 1.2. Historical development of global use of primary energy (in Gtoe) including projected trends to 2100. Insert shows the projected population growth for same period (courtesy of World Energy Council, London, England).

---

[6] A world population of about 2 billion in 1927 doubled to 4 billion in the following 170 years. It is expected to double again to reach about 8 billion in 2028 (within a mere 54 years).

about 77 years by 2050. The global per capita energy consumption, however, has remained approximately the same in the recent past, and the demand for energy is at this time keeping pace with the increase in population. This projected population increase by 2050 should translate into at least a 75% increase in the future global demand for energy.

In 1798, Thomas Malthus [4] predicted a limit to growth in world population due to food production falling short of the demand by the rapidly growing humankind. The Malthusian prediction has not as yet materialized thanks to dramatically increased food production over the years. Green technology has until now prevailed and the food production has generally kept pace with the growth in world population. World grain production, for instance, has increased with the population growth, with the per capita production remaining around 300–344 kg from 1970 until the present time.[7] Experts agree that today the world has more than enough food[8] to feed the population, and hunger is merely a result of difficulties in distribution and allocation rather than unavailability of food. About a billion people suffering from hunger in today's world are mainly in Asia and Africa [5]. (Interestingly, about the same number, but primarily in the West, suffers from obesity due to overconsumption.) Until about the year 2050 the demand for additional food will probably be met in most regions of the world (except perhaps in Africa, which will require a trebling of the food energy to sustain its population even at the medium variant of UN projections). Longer term supply of food relies to a great extent on future technological advances achieved in a timely manner, particularly on genetic engineering that would provide higher-yielding strains of cultivars that are also resistant to pests, better irrigation engineering, improved fertilizers, and more effective pesticides. Because of the impressive past performance, the Food and Agriculture Organization (FAO) is confident of no imminent scarcity of food supplies despite the rapid population growth. While the world has enough unused land capacity that might be put to agricultural use, the ability of the ecosystem to sustain intense food production in the future is not clear. In the past 25 years or so, the increase in food production was achieved not by increasing the area of total arable cropland but by more extensive use of fertilizers to increase crop yield. For instance, when the population doubled between 1960 and the year 2000, the supply of food comfortably kept pace with the increase, although the increase in farmland was only 12%. But the use of fertilizer jumped eight times to 80 million tons in the year 2000 [6]! Intensive agriculture is, of course, not without associated environmental damage. A common consequence of increasing fertilizer use in agriculture is the increased runoff entering large bodies of water, polluting them. Fertilizer-induced hypoxia

---

[7] The soybean production on a per capita basis has nearly doubled over the same period to 23 kg. The per capita meat and fish production over the same period has either stayed the same or slightly increased (Worldwatch Institute).

[8] It is adequate to provide the population with a nutritionally acceptable, but mainly vegetarian, diet. If the populace were given the free choice of the types of food they would like to consume, including some animal proteins, this may not be entirely true.

is already apparent in some regions of the northern Gulf of Mexico (cutting off oxygen particularly to the benthic organisms). Eutrophication due to excess phosphorous runoff from erosion of agricultural land threatens the health of freshwater ecosystems. A related concern is the environmental cost of manufacturing phosphate fertilizer itself; a by-product in the production of each ton of fertilizer is about 5 tons of phosphogypsum. Because of its trace radium content,[9] this by-product is not used to any significant extent [7].

The burgeoning world population has to be supplied with goods and services as well as food, and the demand for goods has risen over a 100-fold since 1900. In the mid-1990s it took about 310 EJ (exajoules[10]) of energy, an all-time high, to support the global human activities. On average, each person on Earth consumes about the equivalent of a gallon of oil per day. A vast majority of Earth's population uses considerably less than this, while those in North America (and western Europe) typically use considerably more. The demand for energy has consistently increased year to year and is likely to continue to do so in the future, particularly in the developing economies. The fastest growth in fossil fuel consumption in recent years has been in North America.

Compounding the problem of uncontrolled population growth is the trend toward urbanization. Primarily because of the practical advantages of centralization of business and industry, cities have been a particularly successful phenomenon. Two clear trends are discernible; accelerated urbanization across the globe, and the increasing numbers of megacities with immense population densities. Better economic prospects and social opportunities attract large numbers to live under crowded, sometimes impoverished, conditions in megacities; there are presently about 20 cities with over 10 million inhabitants. While the most common type of human settlement still remains the rural village or town with less than about 10,000 inhabitants, over a half of the world's population by the year 2050 is expected to live in large cities. Cities such as New York and Mexico City already hold more people than the total population of small countries such as Sri Lanka, and those with populations of over 25 million will be common by the end of this century. (Good candidates include São Paulo, Tokyo, Mexico City, Calcutta, and Bombay.) Urban areas supporting a high population density must invariably suffer high levels of environmental damage and scarcity of natural resources as well. Any disproportionate expansion of a single species in the ecosystem must invariably be at the expense of others. Large-scale deforestation, loss in biodiversity, overfishing of the oceans, and increasing carbon dioxide loads released into the atmosphere are all indicators of this trade-off.

This book is more about polymers than environmental science. To achieve its goals fully, however, some minimal amount of environmental knowledge will

---

[9] Phosphogypsum is chemically similar to natural gypsum but because of its radium content is unused and stockpiled. The production during the last 50 years is 700 Mt in Florida alone (USGS Fact Sheet FS 155-99 September 1999).

[10] An exajoule is 10^{18} joules or one quadrillion Btus of energy. One exajoule is approximately equal to the energy derived from 163 million barrels of crude oil.

obviously be needed. To those as yet uninitiated, the summary discussion below will serve as a convenient starting point into the study of the environment. Several excellent sources are suggested for those interested in detailed information and a more complete discussion of the environmental arguments. This is followed by a collection of chapters on how plastics are related to specific environmental issues of interest. Those already familiar with environmental issues may wish to skip to these subsequent chapters.

1.2. THE BIG PICTURE — EARTH AND ITS ENVIRONMENT

Earth is not a thermodynamically closed system; all bio-geo-chemical cycles within its boundaries are driven by the continuous influx of solar radiation. The complex ecological systems in the biosphere as well as nature's recycling processes for the key elements such as carbon are all fueled by this single source of energy. The primary anabolic activity in the biosphere uses solar energy to create products such as carbohydrates via photosynthesis, and dissipates low-grade thermal energy (energy that is difficult to collect and put to use) in the process. The second law of thermodynamics requires that this conversion of photon energy into other forms be less than 100% efficient and some release of thermal energy is inevitable.

Nature is able to methodically pass on food products throughout the ecosystem employing a complicated web of nutritionally interdependent life forms. This allows, for instance, the phytoplankton in a marine ecosystem to be consumed by zooplanktons that are in turn consumed by fish, and these in turn by birds, and so on. Depending on how complicated the ecosystem of interest is, it is possible to discern linear food chains or complicated food webs along which photosynthetically trapped energy flows through the system to sustain higher organisms. Dead organic matter (leaves or animal carcasses) are a valuable carbon source to soil microbiota and are readily biodegraded and their constituent elements returned promptly to their respective bio-geo-chemical cycles. The cycles work so well because the occupants of various niches in the ecosystem have evolved and modified themselves to best fit those individual roles. This makes it possible for nature to efficiently create low-entropy materials using solar energy, to sustain a complex web of life, and to effectively reuse the same elemental building blocks over and over again.

1.2.1. The Energy Crisis

Most human endeavors demand large amounts of energy to be spent in a short time scale (i.e., are energy-intensive). This energy is presently available largely from fossil fuels or from a secondary source derived from them. Since the industrial revolution, we have invented machinery allowing a single person to extract and control far more energy than he or she could individually generate or even consume. (The human body can barely produce about 100 W of energy.) From then on we have been drilling out millions of barrels of oil a day, tapping natural gas reserves, and mining increasingly large amounts of coal. This is a consequence of both the population increases and societal change tending toward increasingly energy-intensive life-styles. These fossil fuel reserves built up over millions of years are of course nonrenewable within the short time scale of their use. The coal reserves are fossilized land plants that lived 300 million or more years ago, while oil is formed from the phytoplankton sediments deposited 100 million years ago in the bottom of prehistoric oceans. These fossil fuels are currently being depleted at a rate that is 100,000 times faster than they are formed and with alarming inefficiency. The conversion efficiency of the extracted fossil fuel into useful heat, electricity, or mechanical energy to power the automobiles are ridiculously low.

The world consumption of fossil fuels was around 7600 million tons of oil equivalent in 1999. The United States with only about 5% of the world's population accounts for about 21% of the global use of fossil fuel (but also accounts for 26% of the global gross domestic product, GDP[11]). High energy intensities are generally associated with high standards of living. The disparity in living standards in the developing and developed nations is appallingly large and continues to widen. Yet, achieving parity in living standards worldwide is prohibitively expensive in terms of both energy and material requirements. The aggregate area of ecologically productive land needed to provide the entire population of Earth with the present North American life-style is estimated to take two additional Earths [8]! The present paradigm of development is clearly not sustainable in the long term.

There is little doubt that a major shift toward sustainable growth[12] is inevitable in all parts of the world and will likely include changes in technology as well as in life-styles. Defining "sustainable development" has not been easy; the UN World Commission on Environment and Development offers a working definition as, "meeting the needs of the present without compromising the ability of future generations to meet their own needs." This implies a long-term perspective and even an appreciation of the demands by future generations on the limited energy and material resources. Sustainability can only be achieved when environmental concerns become an integral part of the economic process and is paid the same close scrutiny devoted to economic competitiveness and profitability. The difficulty in its implementation lies in the fact that the major beneficiaries of the altruism demanded of the present generation in adopting these goals are the as-yet nonexistent future generations. Strictly speaking, any sustainable functioning demands that fossil energy be used only at the rate it can be replenished, maintaining the net pool of potential reserves at about the same level. This is clearly an impractical goal. In practical terms it translates into using fossil fuel reserves more efficiently, only for the more worthwhile of the human

---

[11] The gross domestic product of a country is the sum of goods and services produced in the country or, in this instance, in the whole world.

[12] Growth on Earth is presently driven by the finite reserves of fossil energy and materials. As such, growth cannot be sustainable in the long run; strictly speaking the term "sustainable growth" is a misnomer.

endeavors and that too at only a minimal rate. Accelerating technological breakthroughs that improve energy efficiency and deliver alternative energy options should be a crucial component of any long-term strategy on ensuring energy for the future.

In the developing countries in Africa or Latin America it is not uncommon for renewable resources to meet as much as 60–70% of the demand for energy. Worldwide, however, the renewable sources accounted for only 18% of the energy consumed in 1995. Renewed interest in non-fossil-fuel technologies has encouraged the development of modern, biomass-based plants as well as innovative solar or wind technologies. Given the present scale of their development, the energy produced by these will probably remain a small fraction of that from conventional renewable resources (such as straw, firewood, or small-scale hydroelectric plants) in the foreseeable future.

The gloomy scenario of possible energy and raw material limitations applies only to the long-term rather than to the immediate future. Even if we assume that the remaining global oil potential of about 2 trillion barrels[13] was all that was available, this still allows about 100 years of use at present rates of energy consumption. Assuming that technologies to recover estimated but unproven oil reserves and including coal[14] and natural gas reserves become available, this time frame can be as long as a few centuries.[15] The estimated "reserves" are continually upgraded, as new information becomes available, making the numerical values of reserves woeful underestimates. Potential fuel reserves in the United States that are more difficult to access—such as the western shale, eastern shale, and tar sands—can and will be exploited as the rising price of crude oil makes these explorations cost competitive. A reasonable window of time is therefore available for this transition of the global economy from uncontrolled energy-driven growth to a sustainable mode of development[16]. How wide this window is depends on the optimism of the analyst; the most optimistic expect 1000 years worth of oil, and about 5 millennia worth of oil from the exploitation of shale oil reserves [9]. Figure 1.3 shows the past and future course of world reliance on different primary energy resources. The heavy line starting at the bottom of the graph shows the evolution of energy use from 1850 to the present time. The set of lines starting in the 1990s and trending downward shows the expected paths of energy resource use in the future. The labels A, A1, B, C1, and C2 refer

---

[13] Estimate of the proven reserves of conventional oil in the world is about 140,200 million tons at this time. However, this estimate increases with time as additional reserves are discovered and because of advances in recovery techniques that makes more of the reserve exploitable.

[14] Estimated coal reserves in the world at the end of 1998 amounted to 984,211 million tons, sufficient to last for the next 200 years. The natural gas supplies in the United States alone is estimated to be sufficient for 60–120 years of consumption at the present rate.

[15] The Department of Energy estimates of U.S. crude reserves accessible with existing technology is about 99 billion barrels. Using a higher cost of $27/barrel that allows the use of advanced technologies to be used in extraction, the reserves can be as high as 575 billion barrels [61].

[16] At the UN Conference on Environment and Development in Rio de Janeiro in 1992 an agenda for sustainable development, the Agenda 21, was laid out.

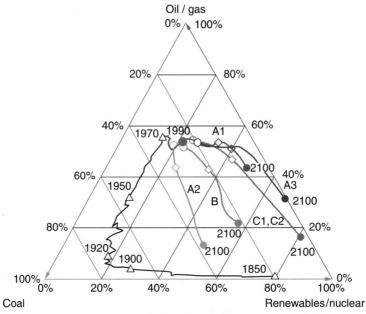

Figure 1.3. Evolution of the world reliance on different sources of primary energy from 1850 to 1990s (heavy line). Other lines originating in the center of the figure show possible alternate paths of development under different scenarios (courtesy of World Energy Council, London, England).

Table 1.1 Percentage of Global Energy Production from Different Sources in 1998 [10]

Source		Source	
Oil	37.5	Hydroelectric	6.6
Coal	21.8	Biomass	6.4
Natural Gas	21.1	Geothermal	0.12
Nuclear	6.0	Solar + Wind	0.05

to different scenarios that assume different combinations of factors such as the population growth, the gross world product (GWP), energy demand in the future, technology cost and dynamics, and environmental taxes. Table 1.1 allocates the world production among various energy sources. The argument that nonrenewable fossil fuel resources on Earth will run out in a few generations cannot be dismissed as a mere doomsday prediction. Alternate energy technologies that can potentially replace fossil fuels are emerging, but only very slowly. Hydropower, which currently accounts for less than 10% of commercial energy worldwide (and accounting for only 9–10% of the electricity generated in the United States in 1995), is a highly efficient (∼85%) clean technology. But, it is unlikely to ever supply more than about 10 percent of the global demand.[17] The best locations worldwide for siting the hydroelectric plants have already been exploited. Factors such as the high up-front cost of new plants as well as the impact of the plants on fish and wildlife, sediment buildup, and the impact on watershed characteristics discourage further expansion of this technology. Existing hydroelectric power projects, however, produce the most cost-effective power in the United States today.

The enormous amount of solar radiation that reaches Earth's surface is more than adequate for our needs only if it could be effectively captured and stored. Depending on the latitude, 100–300 Wm^2 of solar radiation reach Earth's surface. This is a quite substantial quantity in that, at least in theory, a desert stretch of a few hundred square miles equipped with the inefficient solar cells of today would collect enough energy for the present global needs! In practice, however, cost remains a problem. Photovoltaic systems, the second fastest growing renewable energy source in the world, relies on collection plates and are therefore land intensive — requiring around 5–10 acres of land per megawatt of installed capacity due to the low (~15%) efficiency of the collectors. The up-front costs of installing these collector plates are also quite high. Dramatic reductions in cost have been achieved over the past decade, but solar-generated electricity still costs several times more per kilowatt compared to the conventional product. The thermal solar plants[18] (such as the 10-MW Solar Two demonstration plant that came on-line in 1996) also have high up-front costs.

Wind power, while a promising source in the United States (as well as in countries such as Germany, Netherlands, and Denmark), has been slow in developing. It is a nonpolluting, free source[19] of power, but the viability of the technology in the absence of government subsidies is not clear at this time. Low efficiency, high variability of production (as the availability of wind itself varies) and the danger posed to wild birds are the disadvantages of the technology. Even Denmark, the clear leader in this technology, produces less than 10% of its domestic power from this source. Globally wind power is the fastest growing renewable energy supply, based on 1990–1998 data [11].

The biomass-derived fuels including not only wood waste, but also peat wood, wood sludge, liquors, railroad ties, pitch, municipal solid waste, straw, landfill gases, fish oils, and other waste products, are important sources of energy in many parts of the world. It is also an increasingly scarce source as more and more forests worldwide are used for agriculture. It is difficult to envision the

---

[17] Although its share globally is small, individual countries rely heavily on hydroelectric power. Hydroelectric power accounts for more than half the energy produced in 63 countries!

[18] Thermal solar energy plants rely on sunlight concentrated using parabolic surfaces to operate a heat engine that converts the power into electricity. Photovoltaic technology directly converts the solar energy into electricity using solar panels (no moving parts).

[19] Note, however, that wind plants also have high up-front costs though their maintenance costs are relatively low. Wind, though a free resource, is an erratic unreliable source of power. Consequently, the plants tend to operate at low capacity levels, around 23% in 1994. Fossil fuel and nuclear power plants typically operate at 75–90% of capacity [62].

prospects of large-scale utilization of biomass as a source of energy in the United States. Other renewable sources of energy such as tidal power, geothermal energy (not strictly a renewable resource), ocean temperature gradient energy, and wave power have been even slower in developing.

Nuclear fission and/or fusion are likely to one day economically supply nearly all the global energy needs.[20] Presently the nuclear fission plants in the United States generate electricity at a cost lower than most other sources and is one of the cleanest technologies used for the purpose. It does not produce carbon dioxide in the process and when properly operated is virtually pollution free. The energy in a single gram of uranium-235 is equivalent to that in 2–3 tons of coal! In 1995, about 23% of the nation's electricity, amounting to 673 billion kilowatt hours, was supplied by nuclear fission power plants. In some European countries this percentage is much higher (e.g., 78% in France), but globally it currently accounts for only about 6% of primary energy. In the 1990s the global nuclear capacity grew by a mere 5% [12]. Safety concerns and the fact that reactors produce radioactive waste material that needs to be disposed of are the main drawbacks of this key source. With improved technology, cheaper and safer nuclear power plants will be a reality provided the issue of spent fuel disposal is resolved. Despite the well-known disasters at Three Mile Island and in Chernobyl, the average number of fatalities associated with producing energy is much lower for nuclear plants compared to that for coal, oil, or even the renewable sources. This is particularly true of plants operated in the United States. The problem with nuclear fission is that of disposing nuclear waste. A safe repository for highly radioactive waste storage is needed if growth of nuclear energy is to be a reality in the United States. (Perhaps the one presently under planning for Yucca Mountain, Nevada, will finally address this shortcoming, paving the way to more nuclear energy in the future for the United States). Given the known reserves of uranium, the fissionable material in nuclear power plants, we are likely to enjoy a steady supply long after oil, gas, and coal supplies have all been exhausted. This would allow a reliable energy source well into the next century before the fusion processes have been developed.

Research on nuclear fusion, regrettably and incredibly ignored in recent federal funding in the United States, may hold the eventual key to clean abundant energy in the distant future. The very real prospects of relatively clean abundant energy from this future technology should spur on its development. A promising and practical fusion reaction uses deuterium and tritium (at about a 100×10^6 °C) contained perhaps in a "magnetic field container." The reaction yields 17 MeV of energy (per mole of each reactant) that is carried by neutrons produced during the reaction. Given the high energy associated with the reaction and flammable materials (particularly the lithium used to make tritium) used in the process, fusion reactors are likely to be at least as hazardous as the fission reactors.

---

[20] Nuclear fission requires mining and purification to generate the reactor-grade fissionable uranium-235. It also yields spent fuel that needs to be disposed of safely. About 50,000 Mt of nuclear waste lie in pools as spent fuel at the 109 operating and 20 closed power plants around the country.

1.2.2. Materials Crisis

Even with energy expenditures stringently controlled, or an abundant clean source of energy developed, the prospects of achieving sustainability in the long term is seriously threatened by the rapid depletion of essential material and mineral resources. The manufacturing of goods invariably involve using concentrated pools of naturally occurring raw material resources leading to their eventual dissipation in the environment. For instance, the manufacture of plastics uses up a fraction of the crude oil reserves or natural gas that is invariably dissipated as postconsumer plastic waste in the municipal solid waste (MSW) stream or as gaseous incineration products in the atmosphere. Clearly, this unidirectional materials flow is viable only as long as the intact reserves are available. All estimates published for the global "reserves" of these nonfuel resources, of course, are increased frequently as new discoveries of reserves are made (as with the oil). The quantitative information is therefore easily outdated.

The estimated value of domestic (nonfuel) mineral raw materials mined in the United States is $40 billion (net imports into the United States amount to $29 billion)! Nine of these minerals have an annual production value of over $1 billion at the present time. These are mainly commodity construction materials such as sand and gravel but also include key metals such as gold, copper, iron, as well as phosphate rock. For some important minerals the United States depends entirely[21] on exports and conscientious exploitation of the ore resources in foreign countries will be crucial to domestic economic development. Ores of precious and rare-earth metals, for instance, cannot be replenished in a practical time scale and their applications typically do not allow these to be recycled effectively. An energy-rich material-poor world is as bleak a prospect as one with no future energy options.

A reasonable, albeit incomplete, measure of future availability of critical raw materials is the ratio of known world resources for a given material to the present annual consumption of that material worldwide [13]. Adequacy of global reserves has been studied using this ratio (which is an estimate of the number of years for which the reserves will last) and the findings divide the mineral reserves into several broad classes. The following longevity of the supplies was estimated assuming the consumption rates in 1997 and the reserves known in 1998 (according to the U.S. Geological Survey database):

<100-year supply	Copper, gold, silver, zinc, sulfur, tin, molybdenum
100- to 200-year supply	Nickel, boron, asbestos
>200-year supply	Aluminum, iron, cobalt, chromium

These estimates should not, of course, be taken to mean absolute depletions in the time scale indicated (i.e., for instance, that there will be no more copper

---

[21] These include alumina, natural graphite, manganese, fluorspar, quartz crystal, strontium, thallium, thorium, and yttrium. These come from places such as Canada, China, Mexico, Germany, and South Africa.

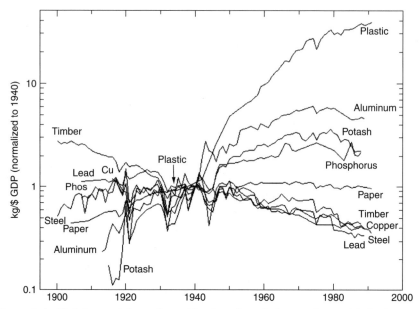

Figure 1.4. Global materials use intensities for selected classes of materials in the recent past showing increased reliance on plastics. [Source: I. K. Wernick and J. H. Ausbel, *Dardalus*, **125**(3), 171 (1996).]

ore after 2100!). It is more a guide on the relative scarcity of the materials. Exploring substitute materials that might be used in place of the more rapidly depleting critical materials needs to be an integral component of any plan toward sustainable growth.

Beginning with the post–World War II construction boom the U.S. consumption of all materials including plastics has significantly grown in volume. Data from the U.S. Geological Survey indicates an increasing trend for materials in the United States (and in the world as a whole) during 1970–1995. The oil crisis in mid-1970s and the economic recession in early 1980s had little effect on the patterns of consumption. Intensity of material use in the United States in the past decade shows that plastics use grew at a relatively faster rate compared to that of conventional materials. Figure 1.4 shows the trends in the annual U.S. consumption of key materials divided by the gross domestic product (GDP) in constant 1987 dollars (and normalized for the base year 1940). While conventional materials used in infrastructure improvement, such as steel, copper, lead, and lumber, gradually became relatively less important to the economy, the use of light-weight materials such as aluminum and plastics gained ground vigorously since World War II.

1.3. THE SMALL PICTURE – BUSINESS ENTERPRISES

The economy is based on the production of goods and services by businesses motivated by profit. As with the case of an ecosystem, a business operates as an

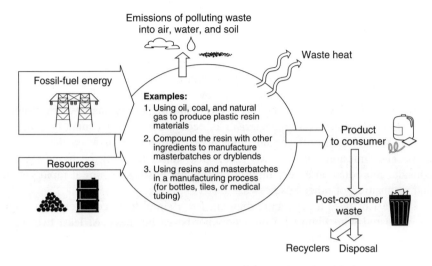

Figure 1.5. Schematic representation of a plastics-related industry.

open system consuming energy inefficiently to deliver useful goods or services. In the process, it generates by-products and waste products as ecosystems do. To perform its function, the business entity relies on a supply of raw materials and fossil fuel energy. For convenience, a business enterprise might be viewed as a black box with energy and material inputs and product and waste outputs (see Fig. 1.5). This simple model applies to the polymer industry as well. For instance, petroleum-derived monomer raw materials and an adequate source of energy are used in resin manufacture. The process invariably yields some amount of wastewater as well as emissions discharged into the atmosphere. However, the simple representation in Figure 1.5 sometimes used in the literature [14, 15] suffers from several significant limitations. Most importantly, it depicts environmental interactions solely in terms of fluxes (of energy or of materials), ignoring the many impacts of the operation not easily quantified. For instance, the aesthetic intrusion of a landfill in an urban location or the anxiety of residents downstream from a chemical processing plant are important social factors that elude description in a simple flow model. Even the quantifiable fluxes are handled rather formally with no distinction being made between a continuous, low-level discharge of a pollutant into a waterway over a period of time and the sporadic discharge of an equivalent quantity of the same pollutant in a single burst. The two modes of discharge have very different ecological implications. For the present limited purpose of comparing the ecological and business systems, however, this simple model is adequate.

The business system depicted in the figure uses up fossil fuels and dissipates the once-concentrated veins of raw-material resources in its quest to generate marketable products. In this regard it is not all that different from a living system that uses the same solar energy either directly or indirectly. Interestingly, relationships analogous to those in the biosphere, such as competition, symbiosis, parasitism, or commensalism are at times found in business arena as well. The term *industrial ecology* has been used to emphasize this analogy [16] between living and industrial systems. The ideal of industrial ecology is a set of interlocking business operations (embedded within and interfacing with the natural ecosystem). The waste collected and recycled from one operation then serves as the raw material for another. However, significant differences apparent between the two systems make this a less than satisfactory analogy. The sophisticated interdependencies characteristic of ecosystems simply cannot be achieved in commerce because of certain fundamental differences between the two systems. The "missions" of the two "ecologies" differ as well; the living systems consider reproduction a very high priority while businesses aim at rapid growth and acquisition of other businesses.

Figure 1.6 gives a schematic diagram for a simplified ecosystem and might be compared with Figure 1.5 on a polymer-based business. At least two major differences between the biosphere and industrial ecology are readily apparent:

1. A crucial difference is that natural ecosystems are energy constrained and evolve strictly within a diurnal budget of solar energy while the industrial systems operate on a user-controlled supply of fossil fuel reserves. The sum total of processes in nature draw energy from the 5×10^{22} j of solar energy reaching Earth's surface on a daily basis.[22] The supply of energy (fossil-fuel-derived energy) into a business system is not constrained and

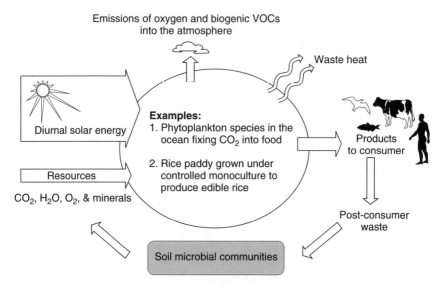

Figure 1.6. Schematic representation of an ecosystem.

---

[22] By comparison, the amount of energy derived from burning a gallon of gasoline is about 10^8 J.

THE SMALL PICTURE – BUSINESS ENTERPRISES   21

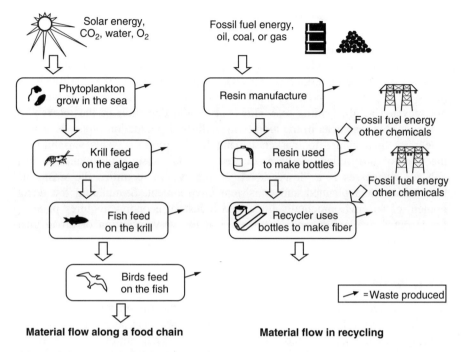

Figure 1.7. Comparison of materials flow in natural and industrial ecological systems.

is determined by the market demand for its products. Natural ecosystems are guaranteed their diurnal solar energy ration in the foreseeable future, whereas the energy-intensive industrial systems could easily face a fossil fuel energy crisis in the coming centuries. These differences are illustrated in Figure 1.7

2. Ecosystems are built upon a hierarchy of species that constitutes a food web. The solar energy initially fixed by photosynthetic primary producers is transferred in steps along the biosphere to grazing herbivores and finally to meat-eating higher animals including humans. Energy trapped in food is essentially passed along the food chain. No net energy input is needed by the herbivores to extract the energy from plants or by the carnivores to survive on herbivores or the microbial communities deriving to derive energy from dead plant or animal tissue.

The polyester bottles manufactured by industry A and disposed of as postconsumer waste by its customer can be regarded as a valuable raw material for business B. Business B recycles the shredded bottles into fiberfill for jackets or plastic tape. The recycler (business B), however, does not derive any energy from the waste polyester; it rather uses additional thermal energy again derived from fossil fuel to convert the waste plastic into fiberfill. Generally, however, energy savings do accrue in recycling because it takes less energy to reprocess already synthesized polymer compared to making new resin and processing it. It is the material resources that are passed along between the industries at different "trophic" levels. There is no flow of energy from level to level as is typical of natural food chains.

The exceptional recycling effectiveness of nature's bio-geo-chemical cycles to conserve and reuse material resources has no parallel even in the best integrated industrial systems. In industry, the waste is generally either poorly recycled or not recycled at all, leading to a slow but certain depletion of the reserves of nonrenewable resources in the long term. Industrial production proceeds at variable rates in response to market demand for its products. This variability in the rates of production of the associated material waste often makes it difficult for businesses at a "lower level" in the system to utilize the waste and by-products in a planned and consistent environment. Stimulating the development of an environmentally integrated industrial economy requires planning for resource availability and waste reuse at the very inception of broad business concepts.

1.3.1. Waste Recycling and Pollution Control

Reuse and recycling extends the useful life of the raw-material resources in instances where a market exists for the recycled products, and it is economical to carry out collection and reprocessing. From the standpoint of resource conservation, any form of recycling is obviously desirable and should be promoted. In practice, however, the merits of resource conservation must be weighed against the cost of fossil fuel energy used in recycling as well as the impact of waste from the process on the environment. In general, for recycling to be attractive, the energy cost of reprocessing must be relatively low compared to that of using virgin material:

$$E_{vir} + E_{dis} > E_{coll} + E_{recyc}$$

where E_{vir} is the energy needed to produce the plastic material starting from the naturally occurring raw-material resource, E_{dis} is the energy cost of alternate disposal (e.g., incineration), E_{coll} is the cost of collection and sorting of the plastic waste stream. As E_{dis} and E_{coll} are generally relatively smaller, the justification for recycling from energy considerations alone often depends on the magnitude of the ratio (E_{vir}/E_{recyc}), that in turn depends on the particular method of recycling involved [17]. There are particular instances where E_{coll} might be disproportionately high, as in the case of plastic litter on beaches or in the marine environment. Waste items such as fishing gear fragments, rope, and Styrofoam bait box pieces do not warrant recycling. However, their collection and disposal accrues some very significant environmental benefits.

The relative amount of pollution associated with the production of virgin material relative to that of the recycling process is also an important consideration. Generally, the emissions load on the environment associated with the recycling or reuse of a plastic material is significantly less than that of using virgin material for the same product. Selecting a mode of recycling must be based on the relative costs of operations, the true energy costs, as well as the pollution potential of the various processes. The value of recycling is not always understood or appreciated by all scientists and economists.[23] This lack of understanding is behind some of the poorly thought out green legislation that often has little effect on or even be detrimental to the environment. In promoting recycling, the primary consideration should be the existence of viable markets for the recycled products and their cost competitiveness. In cases where recycling is not feasible or is too expensive in terms of energy requirements, it is important to develop different technologies that would slow down or eliminate the use of critical resources.

Titanium dioxide (rutile) might be considered as an example of a depleting raw material. It is used as an opacifier in extruded profiles of rigid polyvinyl chloride (PVC) as well as a white pigment in coating formulations Two processes are used to manufacture the pigment; the sulfate process that uses ilmenite ore or the chloride process that uses the less abundant natural rutile. There is, of course, no convenient way to recover and concentrate the titania from discarded plastic products. Over a period of time, these polymer processing operations will remove and dissipate (either into air, water, or soil) the ilmenite ore and natural rutile, contributing to their eventual depletion. Whether this represents a serious resource problem depends on the total demand for the raw material in relation to its supply.[24] Resource conservation is generally an environmentally friendly endeavor that often calls for technological solutions. Can less of the titania, perhaps through innovative coating of the pigment, be used to achieve the same quality of PVC product? Can an alternate material serve the role of titania in these applications, broadening the available choice of opacifiers?

Creation of dangerous waste materials by business activities is certainly not a new phenomenon. The Love Canal incident in the 1970s (unfortunately involving a company associated with plastics—the Hooker Chemicals and Plastics Corporation)[25] is a reminder of the potential seriousness of the problem. In this and in numerous other cases chemical wastes were discharged into a body of water with the assumption that wastes will be readily diluted below a certain "threshold" concentration at which the pollutant posed no significant health risks. It is the same assumption behind the use of smoke stacks to control air emissions. When biological or physical mechanisms inadvertently concentrate the

---

[23] Steven E. Landsburg, for instance, in his otherwise exceptional book *The Armchair Economist* (Simon–Schuster, 1995) in the chapter entitled "Why I Am Not an Environmentalist" argues that not recycling paper would lead to increased production of paper and therefore to more forests being planted (an obviously environmentally desirable outcome). This, of course, is not a defensible argument as the production of more irgin paper will use about twice the energy, emit more air pollutants, and create more water pollution compared to recycling the same amount of paper! See, for instance, Jansen et al. [68].

[24] World reserves of titanium dioxide are estimated at 300 million tons (90% of this as Ilmenite). The world demand for TiO_2 has grown at about 3% annually and was at 414,000 Mt in 1997.

[25] From 1940 to the early 1950s the company dumped chemical wastes into Love Canal. In 1977 the health impacts from contaminated soil and water on the occupants of the housing estate built on the property was noticed. In 1978 the families affected were evacuated.

pollutants, however, the strategy does not always work. This was aptly demonstrated in the well-know case of bioaccumulation of mercury in fish in Minamata (Japan). The mercury waste discharged into the Minamata River was expected to be rapidly diluted by water to levels at which the compounds posed no health risk. Instead, the compounds concentrated in the fat tissue of fish, causing very serious mercury poisoning in the area residents who consumed the fish (the Minamata disease).[26] The bioaccumulation factor for organic mercury is particularly high (10,000–85,000). Other examples are polychlorinated biphenyls (PCBs) and herbicides such as Atrazine that have been detected at elevated levels in the Great Lakes. Alternate approaches to pollution control include sequesteration for disposal. Effectiveness of the latter approach depends on the integrity of the containment system.

A general limitation of pollution control approaches is that initially a waste material must be recognized as a hazardous pollutant prior to it being regulated. A seemingly innocuous waste material may have a serious but unknown indirect health impact. This was illustrated in the case of chlorofluorocarbons (CFCs). These inert, safe gases with very low thermal conductivity were once used popularly as a blowing agent in certain polyurethane and polystyrene foams in building insulation and in aerosol spray cans. Once discharged into the atmosphere and having reached the stratosphere, they cause the depletion of Earth's protective ozone layer (see Appendix B on Stratospheric Ozone Depletion for a detailed discussion). As the ozone layer in the upper atmosphere filters out the harmful ultraviolet (UV-C and most UV-B) radiation reaching Earth's surface, CFC emissions invariably lead to increased ultraviolet B radiation in sunlight reaching Earth. Exposure to higher UV levels is believed to cause skin carcinoma, eye damage, a lower productivity in some agricultural cultivars and damage to aquatic ecosystems [18]. The Montreal Protocol that addressed the issue at a global level came about only in 1987 and is slowly gaining international support.

There is no guarantee that the toxicity or potential health impact of a chemical will always be recognized at the time it becomes an important industrial raw material. Is it therefore reasonable to regulate emissions into the environment even in the absence of a clear and complete casual link between emissions and effects? A step in this direction is the international formulation of the *precautionary principal*,[27] which is beginning to play an increasing role in environmental

---

[26] For decades the Chisso Corporation dumped its mercury waste into the sea. By 1956, the link between mercury pollution and Minamata disease was established, but surprisingly little action was taken. The dumping continued until 1968. The 26-year ban on fishing in the affected waters was lifted recently in 1995 [63].

[27] The United Kingdom's Royal Commission on Environmental Pollution first adopted this principal. The First International Conference on the Protection of the North Sea in 1984 advocated the reduction of emissions into the oceans, of persistent, toxic waste that is liable to bioaccumulate, even in the absence of any conclusive evidence to link the waste to health effects.

A case where the principal would have paid off had it been applied is in the case of antiknock additives in gasoline. General Motors researchers back in the 1920s discovered both tetraethyl lead and ethanol to be effective antiknock additives to gasoline. The consensus finding within GM was

policy decisions (such as the Rio Declaration in 1992). This principal holds that when an emerging technology is suspected (but not proven) of environmental harm, scientific uncertainty about the scope of the harm should not necessarily preclude precautionary action [64]. This is in effect "willingness to err on the side of caution" or "giving the environment the benefit of the doubt" in situations where the science is not clear of the full impact of a material or a practice. When dealing with statistical uncertainty this amounts to minimizing type II errors (where one fails to reject a null hypothesis that is false) as opposed to type I errors. Despite the tendency to do the opposite in scientific analyses, this more conservative approach can be justified in the case of environmental decision making where it better protects the large populations of individuals [19]. In the face of scientific or decision-theoretic uncertainty as well, a strong case can be made for adopting the more conservative position, that it is best to select the course of action or the policy that avoids the worst possible consequence of all choices of actions.[28] [19, 20].

1.4. VALUATION OF ENVIRONMENTAL RESOURCES

The popular overall measures of economic well-being of a country, such as the GDP, do not include measures of environmental quality as one of its components. Even worse, environmental destruction (such as oil spills, deforestation, burning of oil wells in wartime) that result in massive cleanup efforts and medical support to those affected, are in fact counted as additions to the GDP [21]! The value of the environmental assets lost that resulted in these services being mobilized is ignored completely. Manufacturing processes typically use resources inadequately titled to any single entity. These, such as atmospheric and water resources, are a part of the territorial or even global commons and is freely accessible by anyone. Damage to the atmosphere, the deep-sea environment, layers of topsoil, or the inland water resulting from industrial processes represent a real but invisible cost. It is a cost silently borne by the biosphere and by the society at large including even future generations and is further exacerbated by the large disparity in the levels of production of commodities (that tap these resources) in different countries. Since the cost of using these resources is often excluded from the conventional accounting practices, they are sometimes referred to as *externalities*. The flawed belief is that these environmental goods are in plentiful supply and therefore have no marginal value has precluded their recognition within conventional accounting. While comprehensive accounting of

---

that ethanol was the additive of choice, but unlike tetraethyl lead it was not patentable. GM went along with the lead-based additive although its own medical committee delivered a highly cautionary report on the additive. Had the precautionary principal been followed, the U.S. environment would have been free of the lead from 7 million tons of the lead-based additive (*C&E News*, **78**(16), 29 (2000)).

[28] The other school of thought prescribes the use of Baysean approach of maximizing the expected utility in decision making under uncertainty.

all resources, including the "commons" resources, would marginally increase the market cost of products, it could also generate a stream of funding for conservation and environmental management. More importantly it would provide the consumer with a valid measure of the true cost of production of the products of interest. The practice of ignoring the value of natural resources and environmental assets also extends into the assessment of national economies[29] sometimes with very undesirable consequences.

Valuation suggests a method for converting to monetary value things (such as environmental goods) not typically sold in a marketplace. This is an inherently difficult and to some extent subjective undertaking. Environmental accounting is impossible without reliable and good valuation procedures. Early attempts at valuation of "externalities" were primarily via life-cycle analysis (LCA) [22]. Cost–benefit analysis (CBA), for instance, assumes that if environmental goods and services were assigned a monetary value then the environmental effects will be properly accounted for in decision making [23]. However, as most of the relevant externalities do not have readily identifiable markets, this is not easy to achieve and is inadequate for the purpose. A generally used methodology is the contingent valuation method (CVM) [24] where a scenario is developed setting up a hypothetical market for the environmental goods and services in question. The affected consumers are surveyed to assess their "willingness to pay" for a particular intangible commodity, obtaining an estimate of the monetary value of the goods and services.

These approaches, however, are not entirely satisfactory. In the recent past approaches such as LCA were exploited by manufacturers eager to present their products as being "greener" than those of their competitors. Placing a reasonable monetary value on environmental goods is difficult because it forces consumers to value intangible fragments of the environment taken in isolation, ignoring the overall impacts on the whole complex system. The thought processes used in valuation of an environmental asset should involve long-term considerations of the intangible benefits to the community. These are very different from those used in valuing capital goods or services intended for immediate consumption. Consumers have little experience in placing a dollar value on such intangible diffuse commodities. For instance, what is the damage of releasing an ozone-depleting substance (ODS) such as a Freon into the air versus that of releasing waste chromium into the water in an estuary? Or, what is the relative impact of a continuous low-level release of a pollutant versus the same quantity released at once into the atmosphere? Using inadequate approaches in the valuation of environmental assets is not necessarily better than using no valuation at all because of the political and economical implications of such exercises. Particularly worrisome is the manner in which these types of analyses might be used and the impact they can have on public policy and legislation.[30]

---

[29] The United Nations System of National Accounts, for instance, has done little to correct the inconsistency in its treatment of natural resources [65].

[30] A World Bank–funded project in Brazil resulted in rapid deforestation in the state of Rondonia and increased the denuded area of the state from 1.7 to 16.1% during 1978–1991] [66].

In 1993 the UN Statistics Division published the handbook *Integrated Environmental and Economic Accounting*, which is being revised by the London Group on Environmental Accounting (their report was expected to be released in 2002). This document attempts to integrate the different methods available for environmental accounting into a single framework. It proposes a series of versions or "building blocks" for the construction of the accounts and includes calculation of depletion and estimation of the maintenance costs required for sustainable use of resources.

1.5. STEWARDSHIP OF THE ENVIRONMENT

Who should be invariably held responsible for protecting the environment? The simple answer is, of course, the general public. A detailed consideration, however, points to at least three potential groups that can act as responsible environmental stewards: the industry, consumer, and the government.

1.5.1. Industry

Businesses function because they reduce the transactional cost of acquisition of goods and services by individuals. They exact a "fee" for this service and exist for the sole purpose of making a profit for their owners. Implicit in any business model within a free-market economy is the tendency to continuously improve the efficiency of its operation, for instance, by reducing the cost of raw materials needed for its processes or by exploring alternate product delivery systems. Environmental benefits are therefore assured in a business when good environmental stewardship leads to a higher profit potential.

There are numerous examples of businesses being effective and even aggressive stewards of the environment in instances where these two considerations were congruent. Such opportunities are not difficult to identify. The following examples illustrate instances where environment-friendly changes in the design of a product lead to cost savings and therefore to increased profits:

- A large manufacturer realized immediate savings in packaging and transportation costs by moving from the traditional cylindrical design to a novel rectangular one for its 32-and 48-oz plastic vegetable oil containers. The new bottle used about 30% less plastic resin (eliminating 2.5 million pounds of plastic consumption annually) and was easier to pack into boxes. The change also allowed the company to cut down on its use of corrugated boxes for shipping of this product by 1.3 million pounds annually!
- A leading automobile manufacturer switched to the use of recycled poly(ethylene terephthalate) (PET)(from soda bottles) to fabricate components such as grille reinforcements, window frames, engine covers, and trunk carpets. The use of recycled grade of resin as opposed to virgin resin not only reduced the cost of production but also provided a significant (in 1999, 7.5 million pounds) annual market for recycled PET resin.

- Over a period of two decades a leading beer manufacturer reduced the weight of its aluminum beverage cans (by 33%) and glass bottles (by 23%) using innovative improvements in the design and materials used in manufacture. This resulted in increased profits and better environmental stewardship at the same time. The cost savings from weight reduction of the cans combined with those from its aggressive aluminum-recycling program saves the company some $200 million annually.
- In 1999 The General Electric Corporation developed a superior synthetic process for its ULTEM polyetherimide that allowed the resin to be made at a saving of 25% of the direct energy used for the process. The new approach also resulted in substantial reduction of the process waste that required disposal or treatment. (A Presidential Green Chemistry Challenge Award recognized this particular innovation in 1999.)
- Until the late 1980s the nylon carpet fibers manufactured by Du Pont were made available in white to be dyed at the mills in a typical wet process. The process requires heat to maintain the dye baths at proper temperature and is polluting, as considerable amounts of water were needed to wash the fiber. A new process (solution dyed nylon) was developed where the dye was incorporated into the fiber during extrusion processing. This lead to improved color fastness, cost savings, and significant energy savings as well as reduction in emissions.

Even a simple environmental audit of an industrial operation can often uncover profitable waste reduction opportunities. A case in point is the list of 17 priority chemicals that the U.S. Environmental Protection Agency (EPA) asked the industry to voluntarily reduce. In 1992, the Chemical Manufacturers Association claimed its member companies had reduced the emission of these by 33% and expected to cut emissions down to 50% in 2 years, well ahead of the agreed upon schedule. Cadmium-based pigments used in plastic resins were included in this list. The industry changed to noncadmium colorants, thus reducing the cadmium eventually disposed of in postconsumer waste by 96%. Evidently, businesses that developed environmental innovations to address their pollution problems anticipate and often reap significant commercial benefits. These may come about in the form of saleable by-products, innovative low-polluting primary products, and cost-savings associated with cleanup or even incentives. A1992 study by INFORM (a New York–based organization) found that in 181 waste prevention activities reported by the 29 companies, each $1 spent on the activity yielded an average saving of $3.50 to the company. The average time to pay back the initial capital investment was just over 1 year. A particularly successful global example is the restrictions on the use of CFCs by the Montreal Protocol of 1987. The immediate reaction of the industry to control of CFCs was negative, but as the opportunities for profit by developing substitute compounds were realized, the industry actively supported the protocol. Major companies such as Du Pont invested significant research resources into developing profitable substitutes[31] to replace the ozone-depleting Freons. A recent book by O'Brien (*Ford & ISO 14001: The Synergy Between Preserving the Environment and Rewarding Shareholders*, McGraw-Hill, 2001) documents in great detail how Ford Company discovered that better environmental stewardship can be very profitable.

The difficulty arises when corporate profit goals and environmental stewardship do not coincide, and the very nature of business enterprise forces the selection of the former at the expense of the latter. In this situation, intervention by government can spur environmental benefits and stimulate corporate technological innovation. One such case is that of lead chromate, once widely used as a yellow pigment in traffic paints (usually solvent-borne paints based on alkyds or hydrocarbon resins). With the advent of regulations restricting the use of lead pigment on highways, new organic yellow pigments were adapted for the application. These are now almost used exclusively for yellow markings on highways. A similar change is now taking place in paint and adhesive industries as a result of the volatile organic compound (VOC) limits being imposed on these products as a result of the Clean Air Act Amendments.

There are also instances where businesses can be proactive in environmental stewardship despite an apparent loss in profits. In these situations, businesses seemingly compromise their profits by voluntary adopting compliance strategies. These policies of self-regulation at times reach even beyond what is recommended by the regulatory bodies. While what motivates such actions is not always clear, perhaps some of the following factors play a role in their decisions.

- The high level of consumer awareness of the environmental profile of its products may dictate such a move by the business. With consumer purchase preferences being significantly influenced by the environmental performance of the goods or services, businesses may compete for market share on the basis of positive environmental attributes of their products.
- Adopting high environmental standards ahead of anticipated regulatory pressure can be a strategic move by a business toward eventual greater market share.
- Good environmental practices, especially with the more visible consumer goods enhance the overall corporate image. The promotional value of good environmental stewardship may outweigh the fiscal loss of voluntary compliance over the long term.

A recent book by Hawkins et al. (*Natural Capitalism: Creating the next industrial revolution*, Little, Brown & Company, 2000) discusses the role of industry in detail.

---

[31] The substitutes that were developed, the hydrochlorofluorocarbons (HCFCs) and hydrofluorocarbons (HFCs) are transitional substitutes. Both are greenhouse gases, and HCFCs are also ozone-depleting substances although significantly less so than the CFCs they replaced.

1.5.2. Consumers

Often lacking the benefit of comprehensible scientific information, the layperson is generally unable to fully appreciate the magnitude and the implications of even the common environmental issues. Yet, the general public is increasingly sympathetic toward environmental preservation. The majority of American consumers claim that a company's environmental reputation affects their product choice.[32] Surveys have shown the willingness of the consumer to even pay a small premium to ensure the environmental compatibility of certain classes of products. While individual exceptions to this altruistic mood exists (e.g., in NIMBY — not in my backyard — concerns relating to the siting of facilities such as nuclear power plants or solid waste processing plants), broad-based support for environmental preservation is widespread in the United States, Japan, and in western Europe. Public activity in legislating a cleaner environment is evident in the dramatic proliferation in U.S. regulations during the past two decades [25]. The same is true of international treaties, agreements, and regulations, where most of them were reached after 1970.[33]

The technical nature of most environmental issues unfortunately precludes the general public from directly assessing and contributing to the decision-making process. Traditionally, the public has relied on experts, or scientists, to identify and clarify the relevant issues. Are paper bags really more environmentally friendly to use than plastic bags? Is using a ceramic cup instead of a paper cup for coffee more environmentally friendly? Do Styrofoam cups last hundreds of years in landfills? These popular questions do not have clear answers agreed-upon by all such "expert" scientists. Without such answers, however, the public cannot commit and mobilize itself toward sustainability at the level that will do the most good, at the local grass-roots level. This information vacuum is at times addressed by special-interest groups (both environmental and industry lobbyists) exploiting the situation to present questionable or biased information. The frustration of the general public in their attempts to find unambiguous answers to simple environmental issues might be attributed to two general causes.

1. The public perception of the "infallibility" of science that is expected to have all the correct answers to any technical question! A layperson correctly assumes scientific analyses and responses to be based exclusively on facts. The expectation therefore is that all scientists will agree on the assessment and the response to a given environmental issue. But scientists often disagree on even the very fundamental issues such as if global warming is really under way or not. Laypeople do not fully appreciate

---

[32] A 1990 survey reported this percentage to be as high as 77 [67].
[33] By 1995 over 170 international environmental treaties were in place. If accords and agreements are also counted, the number is around 800. However, the effectiveness of these in environmental preservation depends very heavily on the funding available for their implementation. Unfortunately, most of these initiatives did not reach their potential effectiveness due to lack of enabling resource commitments.

the multidimensionality of technical issues and the difficulties faced by scientists interpreting a limited set of pertinent data to arrive at the best possible answer.
2. Some of the decisions relating to environmental matters are essentially value judgments. For instance, the level of environmental intrusion that is justified in order to derive a certain social benefit is a particularly vexing question. The answer varies from person to person and is in any event not easy to formulate. Decisions made on the basis of an incomplete set of data, often difficult to interpret, is typical in environmental management. These are invariably affected by the degree to which the decision maker is risk averse, yielding a range of interpretations of the same information.

Unfortunately, the democratic process can be flawed when it comes to dealing with issues that include a strong technical component [26]. Inadequate understanding of environmental issues, the absence of resources, and even the commitment by the individuals to arrive at a preference on these issues can result in the process being deferred to the societal level. Voting on a poorly understood technical issue does not lead to good judgment; it only allows well-funded interest groups with different agendas to play a bigger role in influencing the public conscience [27]. The situation can only be improved by empowering the consumers at the grass-roots level by ensuring that relevant information in an easily understood format is made readily available to them.

1.5.3. Government

Federal programs have served a valuable role in environmental stewardship in the United States. Most of the long-term data collected on the quality of the environment has been under federal grant sponsorship. In addition to collection of data, federal and at times state organizations provide analysis and research leading to mitigation of the more visible problems. These might be technical solutions, as with the Superfund cleanup program or unique nontechnical solutions available to governments. The latter includes changes in taxation (tax credits), deposit fees, incentive programs for recycling, release permits, and investments in infrastructure to promote select pollution prevention technologies. Regulation is the most salient function of the agencies. Regulation works via bans imposed on certain materials (e.g., bans on the use of polystyrene foam packaging in several counties) or through stringent quality-control measures imposed on specific industries (e.g., ceiling levels of VOCs in solvent-borne paints). Perhaps the most important role of the government is in educating the general public on the implications of environmental preservation. In a democracy, the better educated consumer will invariably be the strongest proponent of environmental stewardship.

In the United States, the EPA created in 1970, is the key agency entrusted with the environment and its well-being. Much of the last 20 years of efforts in the agency has focused mainly on cleaning up existing pools of pollution. Presently, this focus is changing to the prevention of pollution. The advent of the

Pollution Prevention Act of 1990 is a key piece of legislation that underscores this change in emphasis. The initiative, for the first time, encourages source reduction, the efficient use of energy and raw materials, process modification to minimize the release of pollutants, and conservation of nonrenewable resources. Also interesting is the recognition by the EPA of an effective supplement to the command-and-control route of regulation: economic incentives. Chemical processes, including polymer-related processes, in less regulated times often paid little attention to potential profitability associated with pollution prevention. The presence of economic incentives to promote pollution prevention can strongly motivate industries anxious about future regulations that may restrict their operation, to adopt a "greener" operational ethic.

1.6. ENVIRONMENTAL ISSUES RELATED TO THE PLASTICS INDUSTRY: GLOBAL CONCERNS

Specific environmental issues related to the polymer-related industry have been raised from time to time by the general public as well as the scientific community. More often than not, the quantitative data that supports the concerns and estimates the magnitude of the impacts have not been available. Consumer interest in plastic-related environmental concerns, however, continue to increase. This has unfortunately lead to the inaccurate information and indefensible arguments in the press (and at times even in technical literature[34]). A newspaper article,[35] for instance, may question the recyclability of plastics, or speculate on the hundreds of years that polymers take to degrade in the soil environment. Surprisingly, the plastics industry has done little to educate the public and to promote an unbiased technical discussion on the environmental stature of plastics. It is crucial for the plastics industry of the future to play a much larger, more emphatic, role in public education and dissemination of unbiased technical information relating to the industry. All criticism leveled against plastics as a material, however, is not unjustified; several valid, concerns relating to plastics and the environment do exist.

Environmental issues of today are conveniently discussed within two broad classes; global issues and regional or local issues. Table 1.2 summarizes the various issues in each category that impact the polymer industry. The following discussion examines each of these environmental concerns in an effort to understand the extent to which the plastics industry potentially contributes to it. Serious environmental problems that do not involve the polymer industry directly, such as the loss of biodiversity or the increase in urban population density, are excluded from the present discussion.

---

[34] In *Plastics: The Making of a Synthetic Century* by Stephan Fenicell, Harper Business (1996), the author makes the incredible statement "plastics owe its bad rap and its bad rep to the undeniable fact that so many of its most fundamental applications are so chintzy and sleazy."

[35] San Francisco *Bay Guardian*, 02-28-1996, "Epicenter: The Plastics Inevitable," by Leighton Klein: "... so producing plastic requires the constant extraction of oil, coal, and natural gas. Repairing anything made of plastic is nearly impossible, and the notion of recycling it is a robust fallacy at best."

Table 1.2 The Major Global Environmental Issues and their Relevance to the Plastics Industry

Environmental Issue	Identified Cause	Effects	Relevance to Plastics
1. Depletion of fossil-fuel energy and raw-material resources	Use of fossil fuels as a source for energy as well as raw materials	1. Contributing to future energy crisis 2. Contributing to future shortage of critical raw materials	Polymers are synthesized for the most part from fossil fuel resources such as crude oil. But the resource consumption by the industry is relatively small (less than about 4%).
2. Global Warming	The release of greenhouse gases such as CO_2, methane, NO_x and CFC's into the atmosphere	1. Direct health impacts due to increased temperature. 2. Sea-level rise and possible displacement of populations 3. Unstable weather conditions 4. Increase in vector-borne and infectious diseases 5. Loss of agricultural productivity.	All industries, including the plastics industry release some greenhouse gases. Polymer industry does not produce a disproportionate share of the emissions.

(*continued overleaf*)

Table 1.2 (*continued*)

Environmental Issue	Identified Cause	Effects	Relevance to Plastics
3. Depletion of stratospheric ozone	Release of ozone depleting substances (ODS) such as CFCs and CHFCs into the atmosphere	Depletion of the ozone layer results in higher levels of UV-B radiation reaching the earth's surface. Following consequences are reported. 1. Higher incidence of skin and eye damage due to increased UV-B radiation 2. Other impacts on human health 3. Changes in agricultural productivity as well as marine and in fresh water ecosystems. 4. Possible changes in the bio-geochemical cycles.	Plastics industry in the US does not use any significant levels of CFCs. Worldwide the use of HCFCs in plastic foams is also being phased out.
4. Acidification of the environment	Acidic gas (NO_x and SO_x) release mainly from the burning of fossil fuels	1. Damage to freshwater ecosystems including inland fisheries. 2. Impairment of the fertility of agricultural soil due to acidic leaching. 3. Crop and forest damage by direct acidification and biotoxicity due to solubilized metals	Polymer industry uses fossil fuels, but not at a disproportionate level The incineration of PVC has been claimed to result in the release of HCl into the environment contributing to acidification. But this is considered a negligible contribution to acidification.

Table 1.2 Major Regional or Local Environmental Issues and their Relevance to the Plastics Industry

Environmental Issue	Identified Cause	Effects	Relevance to Plastics
1. Solid Waste	Municipal solidwaste (MSW) stream grows each year. The plastics waste content of MSW is small, but it is not biodegradable.	1. Lack of sufficient landfill space requiring transportation of waste over long distances. 2. Possible contamination of potable water supplies by leachate. 3. Need to explore alternative solid waste disposal strategies	Polymers represent only about 8% by weight of the MSW stream. The non biodegradability of plastics in a landfill is not particularly a disadvantage.
2. Urban litter	Disposable products, particularly plastic and paper packaging materials, are not disposed of properly by the consumers.	1. Negative aesthetic impact in urban areas. 2. Pollution of the world's oceans due to at-sea disposal of plastic waste 3. Some products pose a hazard to marine mammels, reptiles and birds. 4. Wash-up on beaches of improperly disposed medical waste.	Litter is a real problem in urban areas. Some plastic products such as non photodegradable varieties of six-pack yoke or even plastic bags are can pose a danger to marine life. Resin prills entering the ocean during transport is also claimed to create a hazard to birds who feed on them.

(*continued overleaf*)

Table 1.2 (*continued*)

Environmental Issue	Identified Cause	Effects	Relevance to Plastics
3. Smog	Volatile organic compounds (VOCs) emitted from solvents, paints and adhesives.	1. Creates urban smog which is unsightly and presents a health hazard.	Solvents used in the polymer industry and in applications such as paints and adhesive formulations can contribute significantly to the problem of urban smog. This is particularly true in regions with high solar UV levels.
4. Hazardous air pollutants	Industrial processes emit volatile chemicals. These include some of the VOCs mentioned above.	1. Negative health effects from overexposure to these chemicals	1. Solvents in paints and adhesive systems include HAPs. 2. Monomers can often be HAPs. 3. In thermoset systems or in-situ polymerization technologies, a small fraction of HAP reactants might be released to the environment. 4. HAPs (in this case dioxins) are claimed to be created in the incineration of polymers such as PVC.

- *Global Concerns* These environmental problems have adverse impacts that are felt far beyond the location where environmental mismanagement occurs. Their effects are transnational and can even threaten the very survival of global ecosystem in the long term. Issues such as global warming or the stratospheric ozone layer depletion fall into this category. Addressing problems in this category is complicated by the fact that the multiple perpetrators (or the polluters) are not always readily discernible. Resolution of these requires an international effort, nations acting in concert toward a common goal to ensure the rights of future generations.
- *Local and Regional Concerns* The more familiar environmental problems have to do with local or regional impacts of pollution such as the release of toxic chemicals into the environment, management of municipal solid wastes, or the disposal of spent nuclear fuel. The impact of these problems is generally limited to specific geographic regions and often results in short-term consequences to a well-defined population. The polluters are generally easy to identify and legislative or other remedies are available to address these types of issues.

The two categories might be further subdivided into environmental concerns with long-term and short-term implications. This two-dimensional approach allows the additional dimension of time to simple geographic categorizations of environmental impacts such as that by Hauschild and Nuim [28]. This yields an interesting matrix of environmental problems in which the cells are differentiated in terms of testability of hypotheses and by the perception of severity of the problem. Testing as used in the matrix refers to field experimental verification of hypothesis that address the causes, impacts, and mitigation of the environmental problems. Examples of each type are given in parentheses.

	Local	**Global**
Long Term	1. Testing not feasible. *(Erosion of top soil, deforestation, toxic waste disposal in inadequate facilities)*	2. Testing not feasible. *(Fossil fuel resource depletion, global warming, damage to stratospheric ozone layer)*
Short Term	3. Testing is feasible. *(Air pollution: SO_X and NO_X, smog problems, contamination of watershed resources)*	4. Limited testing is feasible. *(Loss in biodiversity, depletion of fish stocks)*

The long-term global problems in cells 1 and 2 allow little, if any, opportunities for field testing of hypotheses as to what needs to be done now to avert future problems because of the long time spans over which the phenomenon occurs. The consequences (or the damages) of these environmental problems, however, are generally postponed and shifted to future populations; future losses are generally assessed and valued at a discount relative to imminent losses to the present population. No time is available with such problems to experiment with different mitigation strategies to determine which works best. Both cells 3 and 4 allow testing and experimental verification of mitigation-related hypotheses because the phenomenon spans relatively shorter time scales. However, the likelihood of testing and experimentation is relatively smaller for cell 4 because it requires coordination of effort over geopolitical boundaries. With issues that fall within cell 4, such as overfishing or whaling, different national agendas come into play and their resolution often becomes politically complicated.

1.6.1. Depletion of NonRenewable Fossil Fuels and Raw-Material Reserves in Production of Plastics

The plastics industry, in common with other industries, utilizes fossil fuel energy and raw-material resources for its operation. The question to consider is if the magnitude of the drain on energy and raw materials by the plastics industry is justified by the benefits plastics offer society. This is essentially a value judgment that involves assessing the marginal utility derived from the use of plastic products (as opposed to substitute products) by the society at large. As shown in Figure 1.8 (based on 1988 data), the energy use by the U.S. plastics industry in relation to the other demands on energy is very small and amounted to about only 3% [29]. About half the energy used by the plastics industry is that contained in the feedstock or raw materials and the rest is energy used in processing. The enormous convenience and the energy savings offered by plastics in a bewildering range of applications easily justify this minimal expenditure of fossil fuel. Plastic products pervade nearly every aspect of modern living and include applications in textile, packaging, and building. About 60% of the feedstock as well as the process energy for manufacture of plastics were derived from relatively cleaner-burning natural gas and only 33% are from coal or petroleum sources.

The fraction of crude oil that is used in the worldwide production of resin amounts to less than 4% of the total global petroleum production [30]. A part of the low-molecular-weight gaseous fraction used for making commodity thermoplastics used to be flared in the refineries before the resin industry started utilizing it. Using this small a fraction[36] of the petroleum resource to provide a versatile material that impacts such a broad array of product applications is an impressive achievement.

An interesting example of the economic leverage afforded by plastics is in the commercial fishing industry in which the natural-fiber based fishing gear was abandoned in favor of nylon and polyolefin nets in the late 1940s. The

---

[36] In the United States only 1.5% of the oil and natural gas consumed annually is used in the production of plastics.

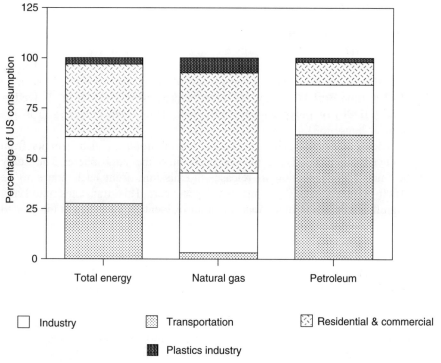

Figure 1.8. Estimated energy consumption by U.S. plastics industry relative to that by other sectors (based on Franklin, 1990) [29].

relatively small volume of plastic resin used in this application resulted in a strong, exceptionally durable fishing gear that increased the "catchability" by a factor of 2–12 times relative to the conventional natural-fiber gear [31], very significantly increasing the profitability of the industry. In transportation applications, particularly in automobile and aircraft design, the replacement of metal parts by the lighter polymeric materials contributes to energy savings far in excess of the minimal cost of raw materials involved in resin production.

Several case studies illustrate instances where the use of a relatively small amount of a plastic material leads to significant energy savings (as well as emission levels). These findings are based on comprehensive cradle-to-grave-type analysis reported for the relevent materials. Other instances of energy savings using plastics will be discussed in chapters that deal with specific areas of application.

- **Case Study** The switch over from metal or asbestos cement pipes to lighter plastic pipes in the building industry is estimated to deliver substantial savings in energy (estimated at 330 trillion Btu in the United States). The energy needed to fabricate a vinyl sewer pipe is about a third of that needed to make the same out of cast iron [32].

- **Case Study** In residential buildings, the use of plastic wrap to reduce air infiltration was recently estimated to save 12–60 million Btu annually (360–1800 million Btu over a 30-year period). Assuming the plastic reduces the infiltration by 50%, energy savings incurred from its use on all new houses in the United States between 1980 and 1997 was estimated to be 9×10^9 billion Btu! Yet, the energy cost of the plastic wrap itself was only 1–2 million Btu! The wraps are manufactured using high-density polyethylene (HDPE) or polypropylene (PP) fibers that weigh only 15–33 lb per typical house [33].

  Along with the saving in energy, the environment also benefits from reduced emission of greenhouse gases into the atmosphere. The study estimated the reduction in release of emissions from each house to be 1600–8100 lb of CO_2 equivalents per year. This translates into very significant emission reductions when all residential properties in the United States are taken into account. A similar claim might be made for the lightweight vinyl window frames that can effectively replace aluminum window frames. The heat loss through a vinyl frame is smaller than that through an aluminum frame by a factor of about 1000 [34]! Replacing selected conventional building materials with vinyl can result in very significant energy savings.

- **Case Study** Polyurethane (PU) foam is the most commonly used insulation material in the manufacture of refrigerators and freezers today. Effectiveness of the foam might be compared to that of fiberglass insulation used in the past for the same application. The energy requirement for appliances using polyurethane insulation was estimated to be only 61% of that for the appliances with fiberglass insulation. This translates into 98 million Btu over the life of a single unit (there were 140 million units in the United States alone in 1997). The total greenhouse gas emissions would be reduced by 62–64% when polyurethane was used instead of fiberglass. The energy costs to manufacture the small quantity of polyurethane foam per unit is less than 1% of the energy expended in using the appliance over its entire lifetime. But, the foam delivers disproportionately large savings in the energy cost to run the equipment over its useful lifetime. (The energy for manufacture of the appliance itself was, of course, not included in these estimates [35].)

- **Case Study** The use of plastics in automobiles leads to indirect but large savings of energy. For example, a small 1% reduction in the weight of the vehicle by replacing metal parts with plastic parts might be assumed to result in a 1% saving in fuel consumption. At a fuel consumption rate of 1 kg per 10 km and 150,000 km of driving over the lifetime of the vehicle, the savings in crude oil consumed was estimated at 180 kg of fuel. This is about 10 times the cost of production of the plastic parts (10 kg for a 1000-kg vehicle) which is about 17 kg of crude oil [36].

1.6.2. Emission of Greenhouse Gases, Particularly CO_2

Emission of greenhouse gases (see also Appendix A on Global Warming) leading to gradual but certain warming of the global environment is a particularly salient environmental problem with grave implications. Measurements made worldwide clearly indicate that average global temperatures have been on the rise since the industrial revolution. The temperature has correlated particularly well with the CO_2 concentrations in the atmosphere over that period of time, suggesting that the burning of fossil fuels played a key role in the phenomenon. In 2001, the United Nations Intergovernmental Panel on Climate Change (IPCC) concluded that most of the warming observed over the last 50 years is attributable to human activities, particularly to the burning of fossil fuels. There is some disagreement within the scientific community on related issues such as the nature of the exact correlation between CO_2 levels and the global average temperature or the effectiveness of various climatic models. But, any debate on the main causes of warming has now been put to rest and high levels of CO_2 and other greenhouse gases are well accepted to be the primary cause of global warming. The consequences of global warming are already beginning to be documented in many parts of the world. Effects predicted by IPCC include rising sea levels, increased precipitation, glacial melting, and greater extremes in weather events, such as droughts and floods. Table 1.3 lists the common greenhouse gases and their relative contribution to the global warming that is caused by human activity.

In common with all types of industry, plastics manufacturing and processing operations primarily use fossil fuel energy. The concomitant release of CO_2 and other greenhouse gases contributes in some measure to the global climate change process. The overall industrial CO_2 emissions are significant, amounting to some 23,880 million Mt worldwide in 1996 (the North American share was 5710 million Mt) A small fraction of this CO_2 emission can be undoubtedly attributed to the plastics industry. The plastics industry (as should every industry) continually makes every effort to minimize such emissions by becoming increasingly energy efficient and by developing innovative low-emitting technologies.

Table 1.3 Common Greenhouse Gases

Greenhouse Gas	Relative Warming Influence (%)[a]	Approximate Lifetime in the Atmosphere (yr)	Rate of Accumulation in the Atmosphere (%/yr)
Carbon dioxide	60	50-200	0.4
Methane	20	12	0.6
Halocarbons	14	Variable (see Table 1.8)	Variable
Nitrous oxide	6	120	0.25

[a] Based on data from the IPCC (2001) report *Climate Change 2001: The Scientific Basics*, Cambridge University Press, Cambridge, England, 2001.

Continuing efforts at reducing the environmental footprint, even when it does not translate into tangible profits, should be a guiding principal for all responsible industries in the future. See Appendix A for a detailed discussion of global warming.

1.6.3. Use of Ozone-Depleting Substances by Plastics Industry

Both ground-based and satellite measurements have unambiguously established that the stratospheric ozone concentrations have been slowly falling since the mid-1980s (see Appendix B for a discussion of Stratospheric Ozone Depletion). The ozone concentrations in the stratosphere are latitude dependent and the decline in concentrations is uneven. The Antarctic winter ozone hole (with about a 50% loss in ozone within it) in the 1990s is an extreme case of the depletion exacerbated by the polar meteorological features. A similar seasonal loss has been seen in the Arctic as well. In midlatitudes (25°N to 60°N) average total ozone was lower by 4–6% in the 1990s compared to 1980. While meteorological factors also contribute to the phenomenon of stratospheric ozone loss in midlatitudes, this depletion is believed to be mainly related to halogen-catalyzed photochemical destruction of ozone in the stratosphere.

The emission of CFCs into the atmosphere is believed to be primarily responsible for the occurrence of atmospheric chlorine. CFCs were used as physical blowing agents in the production of polyurethane and polystyrene foams used in building insulation applications and in aerosol spray cans. The excellent thermal insulation properties of the gas and its nonflammability made it a good candidate for these purposes. These gases diffuse out of the foam and into the environment over the years of product use or once the foam is destroyed. CFCs persist long enough in the environment[37] to reach the stratosphere and once there interfere with the photochemical reactions responsible for maintaining the ozone layer (see Appendix B on Ozone Layer Depletion). The loss of stratospheric ozone directly affects the UV-B content of terrestrial sunlight. The increase in the terrestrial UV-B (290–315 nm) radiation has far-reaching effects on the biosphere. Adverse impacts of increased UV-B include threats to public health, loss of yield in key agricultural crops, and possible damage to aquatic ecosystems. (On the positive side, ozone depletion has a cooling effect on the stratosphere and can offset as much as 30% of the radiative forcing due to increased greenhouse gases).

The Montreal Protocol (1987) has restricted the production and use of several classes of ozone-depleting compounds (ODS) including the original suspect, the CFCs. In July of 1992, the EPA issued its final rule implementing section 604 of the Clean Air Act Amendments of 1990, which limit the production and consumption of ozone-depleting substances. The rule requires industry to reduce production of Class I ODSs and to phase them out completely by January 1, 2000

---

[37] The fact that CFCs are denser than air does not result in their settling to ground level and away from the stratosphere. Such settling though possible in undisturbed masses of air will be short lived, and air currents will readily carry the CFCs into the stratosphere.

(2002 for methyl chloroform). The combined abundance of chlorine and bromine in the stratosphere is expected to peak out around the year 2000, and the rate of decline in ozone at midlatitudes is already apparent. Taking the anticipated reductions about in ODSs into account, the ozone layer is expected to undergo photochemical self-repair, reaching the pre-1980 levels by the about year 2050.

Ozone-Depleting Substances in Polymer Foams Expanded polystyrene and polyurethane foams are popularly used as insulation materials in the refrigeration industry as well as in the building industry. Polyurethane foams are also used as floatation aids in marine vessels. These applications typically use polyisocyanurate board stock (PIR), expanded polystyrene (XPS), and sprayed polyurethane foam (PU-S). Because of the demand for low thermal conductivity, CFC-11 and CFC-12 (mostly in XPS) was used in these foam applications. In 1988–1999, the use of CFCs in foam blowing was at an all-time high, about 200,000 Mt globally. Post–Montreal Protocol manufacturing relies on less polluting substitutes in place of these ODSs. Nearly all of the CFC blowing agents has now been replaced with HCFCs. In the transitional period following the 1995 restrictions on ODSs, HCFC 141b and HCFC 142b, with somewhat similar insulating potentials might be used. But the international agreements ultimately require the use of non-ODS blowing agents in these applications (e.g., CO_2, cyclopentanes, and certain HFCs; see Table 1.4).

A U.S. EPA study in 1993 estimated the U.S. production of flexible polyurethane foam slabstock at 600,000 tons. The 25 companies that manufactured foam at the time were estimated to emit 15,000 tons of methylene chloride (CH_2Cl_2) into the atmosphere. This amounted to 10% of the total emission of this hazardous air pollutant (HAP) into the air nationwide. The volatile emissions from the industry (estimated by industry sources in 1992) is summarized in Table 1.5 (note: no process water is used in flexible polyurethane manufacture, so only air emissions are considered.)

A majority of (HAPs) release is from the use of methylene chloride (or methyl chloroform) as an auxiliary blowing agent (ABA) in the manufacture of foam. These compounds were used for that purpose long before the popular CFC-11 chlorofluorocarbon was phased out. The primary foaming agent in polyurethane is CO_2 formed in the reaction between isocyanate and water. Auxiliary blowing agents mixed in with the reactants vaporize during the exothermic reaction of

Table 1.4 Common Classes of ODSs and Their Chemical Composition

Class	Chemical Name	Atoms	Example
CFC	Chlorofluorocarbon	C,F,Cl	CFC-11 (CCl_3F)
HCFC	Hydrochlorofluorocarbon	H,Cl,F,C	HCFC-142b ($C_2H_3F_2Cl$)
HBFC	Hydrobromofluorocarbon	H,Br,F,C	
Halon	Bromochlorodifluoromethane	H,Br,F,Cl,C	Halon 1211 (CF_2ClBr)

Table 1.5 The Annual Estimated Volatile Emissions from U.S. Flexible Polyurethane Foam Industry in 1992

Function	Emission Load (tons per year)	CH_2Cl_2	CH_3Cl_3	TDI/MDI	Naphtha	CFC-11
Slabstock						
Blowing Agent	16,968	✓	✓			
Fabrication	1401		✓			
Chemical Storage	49	✓			✓	
Rebond operation	11					
TOTAL	18,430					
Molded Foam						
Equipment cleaning	440	✓				
Blowing agent	60					✓
Mold Release	10,000				✓	
Chemical Storage	43			✓		
Other	17					
TOTAL	10,560					

Based on the USEPA report EPA 625/R-96/005 September 1996.

polyols and isocyanates, acting as an additional blowing agent and also removing excess heat that can scorch the foam. Molded foams typically do not use ABAs but integral skin foams rely on their use.

The current practice in industry has changed markedly from that in pre-Montreal Protocol days, and most of the VOCs included in the Table 1.5 have been practically eliminated. CFC-11 has been replaced by either liquid carbon dioxide or by newer classes of relatively safer blowing agents (such as HFAs) that perform equally well though at a slightly higher cost. Water-based release agents and less volatile solvents have replaced most of the naphtha-type agents that contributed to high VOC emissions. The use of high-pressure mix heads by industry also contributed to the conservation of solvents as these did not need to be flushed as with low-pressure mix heads. The subject of foams and the chemistry of blowing agents will be treated in greater detail in Chapter 2.

1.6.4. Potential Pollution of Air, Water, and Soil from Making and Using Plastic Materials

This particular concern can be local, regional, or even global depending on the nature and magnitude of the particular pollutant. For convenience, it is treated here as a global concern. Table 1.6 summarizes data on the energy consumption and emissions load associated with the manufacture of common plastics [37].

Table 1.6 Estimated Energy Costs and Emissions Associated with the Production of Plastics Materials [37]

Polymer	Energy (GJ/ton)	Emissions Air (UPA/ton)[a]	Emissions Water (UPW/ton)
PVC	53	700	3000
PE	70	265	1650
PP	73	325	3685
PS	80	255	6335
PET	84	180	8000 (estimate)
PUR	98	—	—
PC	107	180	5050
Tin plate[b]	30	3400	4600
Aluminum	223–279	9320[c]	27,000
Aluminum recycled	10	370	Negligible
Glass	9–12	109,320[d]	—

[a] UPA/ton = units of polluted water associated with the production of a ton of plastic. UPA/ton = units of polluted water associated with the production of a ton of plastic.
[b] Steel plate about 0.5 mm thick coated with a layer of tin on both sides. Most of the energy is for steel production process.
[c] Emissions include those associated with extraction of raw materials.
[d] Glass made from 43,2% cullet.

Those ranking high in each column are particularly good candidates for recycling, assuming that the energy and emissions associated with reprocessing are not excessive. Units of polluted water (UPW) and air (UPA) used in the table are measures of emissions. UPW is the number of cubic meters of water polluted up to the European drinking water standard by the production of 1 tonne of the material. UPA is a similar measure for air and gives the cubic meters of air needed to dilute the emissions to the European maximum acceptable concentration (MAC).

In terms of emissions, plastic materials compare favorably with competing materials as seen in Table 1.6. Another set of data illustrating the same compares plastics with other materials including bleached and unbleached paper. The comparison is illustrated in Figure 1.9 on the air and water pollution associated with different materials. Interestingly, in these studies, recycled aluminum (and perhaps even recycled glass) appears to be a low-energy and low-polluting competing material compared to virgin plastics. Individual estimates of energy as well as emissions vary according to the mix of energy sources used (therefore by the country of origin) as well as with the type of raw materials and the process employed. The life-cycle analysis methodology used and the assumptions therein can also affect the numbers. However, a valid approximate comparison might be made (as in Fig. 1.9) using data from a single published source.

Chapter 4 includes detailed inventories of emissions associated with the production of common thermoplastic materials. The potential for emissions, however, does exist in the synthesis, processing, and use of polymer-based products.

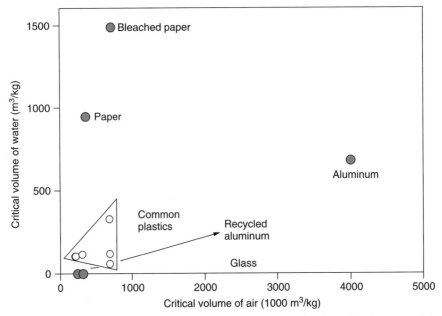

Figure 1.9. Air and water emissions associated with the production of various materials including common thermoplastics.

It is important to quantify any pollutants associated with the plastics industry, assess their impacts on the environment, and adopt pollution prevention measures to minimize the impact of such releases on the environment.

For instance, the monomer raw materials for common plastics such as polystyrene are indeed volatile organic compounds and hazardous air pollutants. Also, polyolefins manufacture in liquid slurry or solution processes use alkane solvents. Exposure to heptane, hexane, or butane solvents used in these processes can be potentially hazardous and threshold limit values (TLVs) for these chemicals are published by the ACGIF.[38] The newer installations, using gas-phase polymerization processes for polyethylene and polypropylene in fluidized beds or continuous-flow stirred-bed reactors do not employ solvents. These processes are less polluting and are more energy efficient as no energy is spent on recovering the solvent and drying the polymer. Some of the processes used in the manufacture of PVC, polystyrene (PS), as well as the many specialty polymers also involve the use of solvents. Improved engineering designs and management practices, however, have generally ensured that these solvents are used in closed systems with only minimal releases into the environment. Recovery of residual solvents from the resin product can require several steps in the manufacturing process and adds to the cost of the final product [38]. Despite increased costs, most of

---

[38] The American Conference of Governmental Industrial Hygienists, publishes the TLV in an annually revised publication. ACGIH, 1330 Kemper Meadow Drive, Cincinnati, OH 45240.

the technology for solvent removal and containment within the manufacturing environment are presently implemented.

The following examples highlight selected polymer-based industries with a particularly high potential for atmospheric emissions. These selections do not, of course, present a comprehensive coverage of all such applications.

- **Case: Paints and Adhesives** Paints and adhesives are essentially concentrated solutions (or in some instances latices) of polymers in a suitable mix of solvents.

Products, such as varnishes, lacquers, latex paints, and glues, have large amounts of solvents that are invariably volatilized into the atmosphere. In the United States, annual associated VOC load on the atmosphere is about 25 million tons [39]. In western Europe alone, 4 million tons of solvents are used annually with about a quarter of the annual total VOC (including biogenic emissions) attributed to solvent use. A conservative estimate of the VOC emissions from a single class of paints, the solvent-based paints used as U.S. highway markings, amounted to about 40 million pounds annually [40].

Film formation in paints and adhesives depends on the evaporation of these solvents into the environment. A well-known effect of VOCs in the atmosphere is the propensity for smog formation. The VOCs encourage the formation of photochemical oxidants such as ozone at ground level in environments already affected by NO_X pollution.[39] Some natural processes also contribute to smog formation (as is apparent in the morning mists of the beautiful Blue Ridge Mountains in North Carolina). Biological VOCs from vegetation (forests account for over 90% of the biological emissions such as terpenes) or from soil, combined with NO_x generated during lightning photoreact under exposure to solar ultraviolet radiation to produce this "natural" smog. High levels of ozone at ground level are particularly harmful to human health as well as to the biosphere. (High ozone levels in the "ozone layer" in the stratosphere, however, are protective and desirable.) The VOCs could be toxic or carcinogenic and may directly affect the health of populations routinely exposed to them (such as paint crews). While some of the VOCs partition into bodies of water or are deposited in the soil, most of it

---

[39] Ozone is formed photochemically in the troposphere by photolysis of NO_2 in a equilibrium process described in the following simplified scheme:

$$NO_2 + hv = NO\bullet + O\bullet$$

$$O\bullet + O_2 = O_3$$

$$NO\bullet + O_3 = NO_2 + O_2$$

The presence of VOCs in the atmosphere produce peroxy radicals that compete in and interfere with the last reaction increasing O_3 levels:

$$RO_2\bullet + NO\bullet = RO\bullet + NO_2$$

Table 1.7 Lifetimes (in years) of Some Common VOCs in Presence of Selected Reactive Species (OH radicals and Ozone) in Atmosphere [41, 42]

Compound	POCP[a]	OH Radicals	Ozone
Propane	16	13 days	>4500 years
n-Octane	12	1.7 days	>4500 years
Ethene (ethylene)	100	1.7 days	10 days
Propene (propylene)	75	7 h	1.6 days
Benzene	13	12 days	>4.5 years
Toluene	41	2.4 days	>4.5 years
m-Xylene	78	7 h	>4.5 years
Formaldehyde	49	1.5 days	>4.5 years
Acetaldehyde	33	11 h	>4.5 years
Benzaldehyde	−40	1.1 days	—
Acetone	9	65 days	—
2-Butanone	17	13 days	—
Methanol	9	15 days	>220 days
Ethanol	4	4.4 days	>50 days

[a]POCP (photochemical ozone-creating potential) is the ozone-creating potential of a compound relative to ethylene, expressed as an index with the value for ethylene being 100.

remains in the air. Three key mechanisms serve as sinks for the VOCs remaining in the atmosphere: photochemical oxidation by hydroxyl radicals, direct photolysis by ultraviolet radiation, and reaction with ozone or other reactive species such as chlorine in the atmosphere. Table 1.7 shows the lifetimes of some common VOCs in the presence of these reactive species.

Paints and coatings can be formulated as water-borne or water-based compositions with relatively small amounts of organic volatile co-solvents in their formulations. Coating systems such as powder coatings commonly employed in original equipment manufacture or the UV-cured coatings used in consumer products do not involve the evaporation of solvents. Recent amendments to the Clean Air Act of 1990 led to the adoption of ceiling values for VOCs in coatings as well as in adhesives. The paint industry has responded to the restrictions on the use of VOCs by increased reliance on powder coatings, high-solids paints, and 100% solids coatings.

- **Case: Fiber Spinning** In the production of fibers melt spinning, wet spinning or dry spinning of a polymer are employed. Dry spinning, in which a solvent-borne polymer is extruded into a fiber, is no longer popular because it requires a solvent recovery step in the manufacturing operation to be in compliance with environmental regulations. Melt spinning in which the melt is extruded through spinnerets is the most convenient and the least polluting fiber-spinning technique. Polymers such as cellulose acetate, aramids, and rayon, however, cannot be spun from the melt and wet spinning has to be used.

In wet spinning, the polymer dissolved in a solvent is spun into a nonsolvent bath where the polymer coagulates into a fiber. The fiber is washed free of the bath solvent and dried during the drawing stage. The composition of the coagulation bath changes during the process, as the wet spinning of fibers slowly uses up the chemical solution. In the case of Kevlar manufacture, for instance, the aromatic polyamide solution (or the "dope") is spun into a sulfuric acid bath. The process results in waste sulfuric acid that is not economically recoverable. The same is true of the xanthate process used in the manufacturing of cellophane or rayon. Alkaline cellulose xanthate in carbon disulfide is extruded into a bath of dilute sulfuric acid where it regenerates the cellulosic fiber or film. Economical uses need to be found for the waste from these processes to avoid their accumulation in the environment. With products such as regenerated cellulose, the advantage of using these films over those made from conventional plastics (such as polyolefins) is their ready biodegradability.

- **Case: Styrene Emissions from Composite Manufacture** A sector of the plastics industry presently under review by the EPA is that of composites. When final, the regulatory action by the agency is expected to reduce related air emissions by 14,500 tons per year (a reduction of 65% over 1997 levels).

    Unsaturated polyester thermoset systems use styrene as a reactive diluent (sometimes along with substituted styrenes such as vinyl styrene or methyl methacrylate). Some of the styrene evaporates in the process (the emission factor is estimated at about 1–3% of the styrene for composites and a little higher for continuous lamination) and poses a threat as a hazardous air pollutant. The time-weighted average (TWA) for styrene is 50 ppm (or 213 mg/m^3). In 1996 the Occupational Safety and Health Administration (OSHA) endorsed a proposal by the styrene industry to voluntarily adopt the 50-ppm exposure limit.

    The composite industry was able to achieve low emissions by a number of changes, including the increased use of low-styrene resins of lower molecular weight and higher percentage of fillers in the compounds. Also employed were suppressants to minimize the surface evaporation of styrene during the curing stage of the resin. In addition to changes in the formulation, engineering improvements in spraying technology (e.g. the use of controlled spray atomizers) and improved plant design with better airflow characteristics can also be employed to reduce emissions.

- **Case: Greener Route to Lactones Used in Synthesis of Polyols** Low-molecular-weight (generally up to about 40,000) polyols used as prepolymers in the polyurethane industry can be made by ring-opening polymerization of lactones with titanium catalysts and di- or tri-hydroxy initiators. The lactam raw material for the reaction is made by oxidation of ketones (Bayer–Villiger reaction) using a peroxycarboxylic acid. The reaction produces equimolar quantities of the lactone and reduced acid waste:

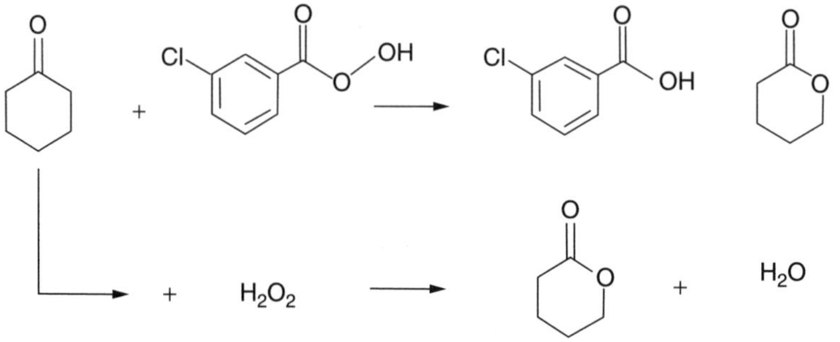

Recent research literature shows an alternative "greener" pathway to achieve the same using hydrogen peroxide as the oxidizing agent and zeolite infused with tin as the catalyst. The reaction products in this case will be the lactone and water. When adapted for commercial production, this pathway could represent a potent pollution prevention strategy for prepolymer production industries.

A 1990 EPA study on five key air pollutant emissions (CO, SO_x, NO_x, hydrocarbons, and particulates) showed the industrial sector accounted for 66.6 million tons of emissions per year, compared to 90.5 million tons by the transportation sector and 45.9 million tons by heating. It is difficult to estimate the fraction of the industrial emissions attributable to the polymer industry. Assuming the burning of fossil fuels to be the primary source of these pollutants, an approximate estimate might be made by adjusting the emission load by the ratio of energy consumed by the plastics industry to the total energy used in the United States [29]. This yields an estimate of only about 1% of the total emissions attributable to plastics industry and even this is likely to be an overestimate because the industry generally relies more on cleaner-burning natural gas than on petroleum fuels.

As with the depletion of nonrenewable resources by the polymer industry, the levels of pollution generated by the industry must be weighed against the benefits of the polymeric products to society. Some of these products could indirectly reduce the impacts of pollution in a different application or save large amounts of resources in yet other applications. (For instance, the use of polymer-based filtration systems and membrane separators can purify the effluent industrial gas streams, recycling useful products from the waste gas. A cellulose acetate fiber filter in cigarettes can reduce particulate and tar inhalation with very significant cost savings in the healthcare sector.)

1.7. ENVIRONMENTAL ISSUES RELATED TO THE POLYMER INDUSTRY: LOCAL AND REGIONAL CONCERNS

1.7.1. Acidification of the Environment

A particularly important consequence of the release of gaseous pollutants into the atmosphere is the local acidification of the environment. Rainwater reaching Earth's surface has a pH of about 5.6 (it is lower than 7 because of the dissolved carbon dioxide). In some parts of North America, Europe, and even Asia, however, acidic rainwater of pH < 5.6 is quite widespread. This is believed to be due to the presence of acidic gases, particularly oxides of sulfur (SO_x) and oxides of nitrogen (NO_x) in the atmosphere. The contribution of HCl, if any, is relatively minor. A biogenic contribution to acidification caused by emissions from volcanos or forest fires also exists, but is relatively small.

The nitric oxides, from vehicle exhaust, are insoluble in water but can be oxidized (either by peroxides or ozone) into NO_2 that can dissolve in water to yield dilute nitric acid. The NO_x emissions are relatively small amounting to only about 28 million tons, compared to the 71 million tons of sulfur emitted globally in 1994. This includes biogenic emissions, due to microbiological activity. The SO_x emissions are responsible for about 60% of the acid rain obtained worldwide [43]. A particular concern is the increased threat of acidification in the developing countries that use low grades of fuel and rely on biomass burning. For instance, the brown coal used in China has a sulfur content of 5 percent. In the United States, most of the SO_x emissions (an estimated 75%) occur from east of the Mississippi River where most industry is located. While emission of hazardous gases is generally considered a local or regional problem, the SO_x/NO_x emission can also result in impacts across national boundaries. For instance, the emissions from northern states in the United States are responsible for about half the acid deposition in southeastern Canada,[40] and those from industries in northeastern China and South Korea affect Japan.

The impacts of acidification reach well beyond the direct toxic effects of high hydrogen ion concentration on plants and animals. Decline in the quality and growth rates of forests due to acid rain is well documented in several regions of the world[41] [44]. For instance, in the Sichuan province (particularly in Omei Mountain) of China, 87% of the cedar trees are seriously affected by acidification. Scandinavian scientists in 1960s first observed reduction of fish in lakes affected by acidification. Even a moderate acidification (pH ~ 6.0) of surface waters can kill crustacean species. At pH < 5.5 loss of economically important fish species such as salmon and trout is expected. Hardy species such as the

---

[40] Under the 1991 Air Quality Accord between Canada and the United States, emissions that cause acid deposition were permanently capped in both countries (13.3 million tonnes of sulfur for the United States and 3.2 million tonnes for Canada).

[41] In Germany over 50% of the forests, particularly the fir forests in the southern part of the country, show some signs of decline. In the U.S. red Spruce is the most-affected species.

eel and the brook trout succumb at a pH $\sim$ 4.5 [45]. Fish, particularly the eggs and the fry are very sensitive sudden changes in the pH of their medium. A particularly damaging phenomenon (sometimes referred to as "acid shock") is seen in winter when acidic deposits build up in the snowpack. In spring with the melting of snowpack the large amounts of acid released in a short period of time contaminate bodies of water. An important indirect effect of acid rain is the solubilization of metal compounds that can be toxic to fish and other aquatic organisms [46]. Aluminum solubilizes and leaches out of the soil at pH < 5.5 yielding Al^{3+} and $Al(OH)_3$ that can be toxic to plants. A similar increase in concentration in the soil water is expected for other metals such as cadmium, mercury, and lead. Acidity-induced decreases in the solubility of minerals such as calcium can also be detrimental to the ecosystem. With low levels of dissolved calcium in water, invertebrates with shells (molluscs and crustaceans) can suffer poor growth. The deleterious effects of acidity on historical buildings and monuments (e.g., the Roman architecture in Italy, the Greek monuments such as the marble sculptures in Athens, and the Taj Mahal in Agra, India) are well known. International efforts are already under way to address the loss of this world cultural heritage.

The emissions of SO_x and NO_x are declining in the United States in recent years, reducing the threat of acid rain in the region. This is mainly a result of Title IV of the Clean Air Act that set a goal of reducing annual SO_2 emissions by 10 million tons below the 1980 emissions levels. The phase I effort (regulating 110 large sources, concluded in 1995) was successful, having reduced SO_x levels 40% below the required levels. It is important to continue the remedial and pollution prevention efforts globally to keep this trend going in future years.

1.7.2. Issues Related to Solid Waste Management

Municipal Solid Waste in Landfills The issue of plastics in the municipal solid waste (MSW) stream became particularly visible in the 1980s. The United States produces about 232 million tons of MSW annually (amounting to about 4.5 lb per day on a per capita basis) containing 10.7% by weight (or 25 million tons) of plastics [47]. This rate is considerably higher than that of most western countries, Canada, or Japan. Since a fraction of the waste is recycled, only about 3.5 lb per person per day is discarded in the United States according to 1994 data. Most of the plastics discarded is postconsumer packaging materials such as bags and empty containers. About 57% of the MSW was landfilled and 16% incinerated mostly in waste-to-energy plants during 1995. In the United States landfilling is likely to remain the dominant solid waste disposal strategy in the near future. Once disposed in a modern landfill, very little deterioration of the waste is expected. Even the readily biodegradable materials such as paper and yard waste fractions persist for long periods of time in landfills due to the absence of enough moisture and oxygen needed to support a biotic environment within the fill [48]. This is particularly true of food or yard waste contained in plastic bags. Not surprisingly, the plastics fraction does not break down significantly under these conditions. This lack of deterioration is not undesirable, as no polluting leachates or flammable gases are produced in any significant volumes, and the fill remains stable as the volume of compacted waste remains the same.[42]

The available landfill capacity in the United States is rapidly decreasing. An estimate in 1988 indicated that 40–45% of the then active landfills would reach capacity within 5–7 years. The generally perceived undesirability of plastics in the MSW stream stems from the argument that since plastics do not biodegrade in landfill environments they do not yield their occupied volume for additional disposal of waste [3, 49]. However, this expected additional fill volume had the plastics been biodegradable is far too small to provide any significant advantage. It is easy to confuse the relatively high volume fraction (about 18–25%) of plastics in the MSW stream with that in the landfill. Plastic packaging is compacted to a minimal volume under the weight of the fill above it and the volume of the plastics is likely to be quite small.[43] In fact, biodegradability would have the undesirable consequence of generating leachate. The whole argument rests on the assumption of imminent shortage of fill volume in the United States. The total fill volume needed even for the United States for the rest of this century is small (18 mi^2 area of land [9]) and the problem is one of siting and transportation of waste. Undoubtedly, this will slowly increase the cost of landfilling over time and promote the development of innovative alternatives.

The real issue relating to plastics waste in landfills is that it represents a valuable resource discarded well before the end of its useful lifetime (or even after only a single use). The true cost of the waste plastics can only be appreciated if all the natural resource and environmental costs invested in its production are correctly taken into account. Rather than burying these valuable raw materials, the landfills should perhaps be thought of as temporary repositories of durable plastics waste for future recycling or conversion into useful energy. Even regarding plastics in landfills as a future source of fuel is reasonable when one considers the high heat content of common plastics — 18,687 Btu/lb for polyethylene and 16,419 Btu/lb for polystyrene, compared to only about 4800 Btu/lb for mixed MSW and 6800 Btu/lb for mixed paper.

Litter Problem Plastics waste in urban litter does pose a serious and real waste management problem. Unlike paper, plastics do not degrade in the outdoor environment at a fast enough rate compared to that of littering. The result is a serious aesthetic problem in highly populated urban areas in almost any part of

---

[42] When a landfill is full, it is possible to cap it and use the surface real estate. The Santama landfill in Tokyo (Japan) 'for instance' plans to build a sports facility on the capped top of the fill after its useful life of 13 years. Other Japanese sites have made similar use of full landfills. The lack of biodegradation prevents any reduction in fill volume over time and is an important consideration when landfills are used in this manner. In fact biodegradable food or paper waste is not accepted at such landfills.

[43] The volume fraction of plastics in the MSW stream is calculated assuming volume:weight ratios ranging from 2.4 to 3.9. These estimates are very unlikely to apply to plastics compacted at the bottom of a landfill where the ratio will be closer to unity.

the world. Litter commonly encountered range from cellulose acetate filter tips from cigarettes[44] to polystyrene foam cups, and polyethylene plastic bags.

The situation is even worse in the case of the marine environment. Plastic waste, primarily packaging-related waste, finds its way into the worlds oceans from fishing, commercial, and naval vessels. The ships' waste dumped into the ocean includes plastic waste. For instance, a large passenger liner or an aircraft carrier may dispose of as many as several thousand plastic cups a day. Since the U.S. ratification of MARPOL Annex V (and compliance by the U.S. Navy), this practice has ceased. By 1999 naval surface ships were equipped with plastic waste processors for onboard melting and compaction of food-contaminated and other plastic waste. The compacted waste (a molded disk) is stored aboard the vessel for disposal at port. This is a major achievement as it prevents at-sea disposal of the 5.5 million pounds of plastic waste annually generated by the fleet. Fishing gear, however, is either routinely lost or are deliberately disposed of at sea. Commercial fishing nets such as trawls and midwater gill nets (usually made of nylon, negatively buoyant in seawater) are large and are comprised of many thousands of square meters of netting. Their loss or disposal at sea damages the fishery due to "ghost fishing." Gear made of polyolefin material (such as with trawl webbing) remains afloat for a time until heavy encrustation by foulant macroinvertebrates sink them. During this period, marine mammals, birds, fish, and turtles can be entangled or trapped by the discarded plastic waste, particularly in netting, strapping bands, and six-pack rings. Birds are even reported to feed on resin pellets spilt during transportation of resin by sea and then suffer impaired growth. Beach litter is also an important additional source of litter reaching the sea; in fact, a large majority of the plastics polluting the world's oceans include packaging materials that originate at beaches.

Littering is a behavioral problem as much as it is a technology-related issue. The standard approach of collection and recycling does not work well in the marine environment. While some efforts are made to clean beaches of plastics debris, the same is not feasible with plastics at sea. Technology can therefore hope to mitigate only a part of the problem. For example, the entrapment of animals in six-pack rings at sea might be avoided by using enhanced photodegradable six-pack rings made from (ethylene *co*-carbon monoxide) copolymer as opposed to low-density polyethylene (LDPE). (See Chapter 10) Degradable plastic bags might be used at sea as these disintegrate rapidly, avoiding ingestion by turtles (who apparently mistake the partially inflated bags for jelly fish). But, the enhanced-degradable polymer technology has not provided a generic answer to all problems associated with marine plastic waste. The ingestion of plastic pellets by birds or the introduction of microfragments of plastics into the feed of filter feeders at sea is not addressed by these technologies. The possibility of concentration of organic compounds in water in the plastic material via partitioning at

---

[44] About 4.5 trillion cigarette butts made of cellulose acetate fiber is discarded as litter each year! In addition to the fire hazard associated with carelessly discarded buts, the possibility of their leaching out the concentrated trapped chemicals into soil water has been raised.

sea, and their possible introduction into the marine food chain via ingestion by organisms, have not been adequately addressed.

Release of Hazardous Emissions from Incineration Incineration of MSW allows the recovery of energy from the flue gases formed in the process. This energy might be used to generate process steam or converted into electricity. A state-of-the-art combustion plant operates at about 22% efficiency and produces 600 kWh of electricity per ton of waste (the efficiency can be much higher for combined heat and power generation plants). Compared to the burning of coal, for instance, burning of MSW produces less acidic gases and less carbon dioxide. Incineration, however, is not always a particularly clean process, and the plastics fraction in the waste has been blamed for compounding the air pollution problem. Specifically, the production of hydrogen chloride on burning PVC, the contribution of the chlorine from PVC to the formation of dioxins,[45] and the influence of plastics on the metal content of the incinerator ash have been pointed out. The amount of data available to substantiate these claims, however, are quite small and somewhat inconsistent. For instance, incineration of feed waste containing increasing amounts of chlorine was found to have a minimal effect on the amounts of dioxin produced on incineration [50, 51]. Results such as these suggest that the chlorine content of the waste might not be the determining factor for dioxin production. Continuing studies will eventually quantify the extent to which plastics contribute to the issue.

Dehydrochlorination of PVC is a facile reaction and is expected to take place on incineration of the plastic waste [52, 53]. The corrosive fumes can potentially affect the incinerator structure itself and if allowed to escape into the atmosphere will contribute to the acidification of the environment. Analysis of flue gases from mixed MSW incinerators typically show very small amounts of HCl (<2000 mg/m^3 of gas) [54]. Either the amounts generated are quite low or the HCl formed in the incinerator undergoes reaction with other waste components or incineration products.

Particular emphasis has been placed on the emissions of polychlorinated dibenzo-p-dioxins (PCDD) and polychlorinated dibenzo-furans (PCDF) during incineration of MSW. Depending on the design, operating temperature, and the composition of waste incinerated, most incinerators may emit small amounts (ranging from a few nanograms to a few thousand nanograms per cubic meter of gas) of these highly toxic compounds. The toxicity[46] of these two classes of compounds, based on extensive animal studies, has been reviewed [55, 56] and is highly dependant on the number and position of the chlorine substituents in the molecule. Trace amounts of these already present in the MSW can be volatilized

---

[45] This is a generic name for a family of about 210 related cyclic chlorinated hydrocarbons, some of which are believed to be very toxic. Toxicity studies based on animal studies, however, show different species to vary enormously in their susceptibility to dioxins.

[46] The toxicity of these compounds is species dependent. The most toxic is 2,3,7,8-tetrachlorodibenzo-p-dioxin.

in the incinerator or these might be the result of de novo synthesis catalyzed by the fly ash at high temperatures. However, a third relevant mechanism is the possible formation of these compounds by in situ synthesis with the chlorine being derived from the hydrochlorination of PVC in the waste [57]. While there appears to be no relationship between the amount of PVC in the waste and the emission of these toxic compounds during incineration [58, 59], the issue is still being investigated.

1.8. PRESENT TREATMENT

The major global environmental issues, in spite of their urgency, are addressed inadequately and slowly because of the lack of information and the need for international efforts to mitigate them. With regional or local environmental issues, the consumer public, industry, and the government can all play an important role in their resolution. The leadership of industry can be relied upon to provide some of the solutions, particularly in the instance where the implementation of these solutions results in cost savings or increased profits to the corporate entities. In instances where this is not the case, regulatory actions can play an important positive role in motivating industry as well as the consumer toward better environmental stewardship. The most important stakeholder, the consumer, is often left out of the decision-making process because of the technical nature of the environmental issues. Efforts at educating consumers, presenting the information in a comprehensible and timely manner, will empower them to make the necessary value judgment independently.

As with other goods or services the cost of resources and impacts associated with the plastics industry must be considered a trade-off against the societal benefits plastics deliver and any cost savings enjoyed through their use. The polymer industry has been somewhat unfairly singled out in the recent years as being disproportionately responsible for various environmental concerns of the day. Some of these claims are unfounded and not supported by data, while others appear to be reasonable. However, it is clear that the polymer industry needs to be even more environmentally sensitive in the coming years. Based on the discussion here some broad overall directions the industry may take to minimize the negative impact on the environment might be identified.

1.8.1. Being Environmentally Proactive

1. The plastics industry should strive to become increasingly energy efficient in all its activities. To the extent possible, renewable energy sources and high-efficiency cleaner burning energy sources must be preferred in the manufacture and processing of plastics. The possibility of converting process waste into useful heat energy to offset some of the energy costs of processing operations should be explored[47].

---

[47] Innovative ideas such as the use of a fan-coil heat exchanger to capture the waste heat from mold cooling water in a processing plant or from a film blowing operation for heating the factory

2. The sparse fossil fuel raw materials must be used sparingly, selecting the more abundant resources in preference to rapidly depleting ones. This includes relying more on renewable raw materials in the manufacture of plastics. This possibility needs to be seriously investigated.[48] In the case of rapidly depleting (nonfuel) raw materials that the industry routinely relies on (for catalysts, stabilizers, and colorants), the use of alternate, more abundant raw materials should be considered.
3. The plastics industry must actively develop and promote strategies to extract the maximum possible useful service life from all plastics. Increasing the service life conserves raw materials as well as energy. A waste material should be viewed as an asset (a possible raw material) rather than a disposal problem, with the focus on search for processes and markets that can make a useful product out of it. Recycling should be practiced whenever the resource recovery can be achieved at substantial overall savings in fossil fuel energy with minimum pollution.
4. In resin synthesis, compounding, and processing, the industry needs to make a conscious effort to prevent or minimize environmental pollution. In instances where some pollution is inevitable, the industry should develop viable strategies to minimize damage to the environment and should quantify the magnitude of the anticipated impact. Pollution prevention should always be practiced in preference to waste treatment or clean up after the fact.
5. Chemical components in resin systems, compounds, and processing operations should be selected to minimize the toxicity (particularly to humans) if these were to be released to the environment. Particular attention should be paid to the intermediate compounds formed during photochemical, hydrolytic, or biodegradation of the plastic itself or other chemicals used in the plastic compound.
6. Applications of plastics in which the use of a relatively small amount of plastics achieve disproportionate savings in energy and resources must be actively promoted and publicized by the industry. Such "leveraging" of the value of plastics in the economy can easily justify the value of plastics as a strong competing material.
7. The polymer industry must increase its efforts to educate the consumer with unbiased, credible, scientific information on environmental issues related to plastics. Consumers (including the younger generation) should be presented with the information in a format that is easy to understand.

---

building, and the use of a heat pump to upgrade the heat energy for use in resin preheaters in processing equipment have been suggested.

[48] Examples include the polyamides synthesized from fatty acids obtained via ozoanization of vegetable oils, chlorinated or hydrochlorinated natural rubber to obtain plastic materials, and the various modifications of cellulose or chitin.

REFERENCES

1. U. Y. Kim and K. U. Kim, Research and Development Activities on Polymers for the 21st Century in Korea. Macromolecules Symposium vol. 98, 1995, p. 1261.
2. J. Stone, *Mod. Plastics.* July 11, 1993.
3. W. L. Rathje, Inside Landfills: A Preliminary Report of the Garbage Project's 1987–88 Excavation at Five Landfills, in Municipal Solid Waste Technology Conference, San Diego, CA, 1989.
4. T. R. Malthus, *An Essay on the Principle of Population; or a View of Its Past and Present Effects on Human Happiness; with an Inquiry into Our Prospects Respecting the Removal or Mitigation of the Evils which it Occasions*, J. Johnson, London, 1803.
5. B. Gardner and B. Halweil, in L. E. Brown, C. Flavin, and H. French, eds., *State of the World 2000*, Norton, New York, 2000, p. 60.
6. IFIA, *International Fertilizer Industry Association, Statistical Database*, 2000, www.firtilizer.org/stats.htm.
7. J. R. Johnson and R. Traub, *Risk Estimates for Use of Phosphogypsum*. 1996, Batelle Pacific Northwest Laboratories Institute of Phosphate Research, Bartow, FL, 1996.
8. W. Rees and M. Wackermagel, in A.-H. Jansson, ed., *Ecological Footprints and Appropriated Carrying Capacity: Measuring the Natural Capital Requirements of the Human Economy. Investing in Natural Capital*, Island Press, Washington D.C., 1994.
9. B. Lomborg, *The Skepticle Environmentalist, Measuring the Real State of the World*. Cambridge University Press, New York, 2001, p. 120.
10. P. E. Hodgson, *Nuclear Power, Energy and the Environment*, London, Imperial College Press, London, 1999, 201.
11. L. E. Brown, in L. E. Brown, C. Flavin, and H. French, eds., *State of the World 2000*, Norton, New York, 2000, p. 17.
12. N. Lennson, in L. R. Brown, M. Renner, and C. Flavin, eds., *Vital Signs 1997*, Worldwatch Institute, New York, 1997.
13. S. Kesler, *Mineral Resources Economics and the Environment*. Macmillan College Publishing, New York, 1994, p. 322.
14. T. Jackson, in S. E. Institute, ed., *Material Concerns. Pollution, Profit, and Quality of Life*, Routledge, New York, 1996.
15. R. Clift, K. Burnington, and R. E. Lofstedt, in Y. Guerrier, et al., eds., *Values and the Environment. A Social Science Perspective*, Wiley, New York, 1995.
16. R. A. Frosch and N. E. Gallopoulos, *Sci. Am.*, September, p. 142–152 1989.
17. I. Boustead, in M. B. J. Brandrup, W. Michaeli, and G. Menges, eds., *Recycling and Recovery of Plastics*, Hanser, New York, 1996, p. 73–93.
18. F. R. De Gruijl, *Eur. J. Cancer*, **35**, 1999 (2003).
19. K. Schrader-Frachette, *Risk and Rationality: Philosophical Arguments for Populist Reforms*. University of California Press, Berkley, CA, 1991.
20. R. M. Cook, *Dialec*, **36**(4), 334 (1982).
21. H. E. Daly and J. B. Cobb, *For the Common Good*, Beacon Press, Boston, 1989.
22. J. Repetto, *Sci. Am.* June, 94 (1992).
23. D. W. Pearce, ed., *Blueprint 2: Greening the Global Economy*. Earthscan, London, 1991.

24. D. W. Pearce and R. K. Turner, *Economics of Natural Resources and the Environment*. Harvester Wheatsheaf, London, 1990.
25. J. A. Cusumano, *Chemtech*. August, 482 (1992).
26. J. Parker, in Y Guerrier, et al., eds., *Enabling Morally Reflective Communities*, Wiley, New York, 1995, pp. 33–49.
27. R. J. Conner, *Economics and Decision Making for Environmental Quality*, University Press of Florida, Fernandina Beach, FL, 1974, p. 299.
28. M. Z. Hauschild and B. A. Nuim, *Selected Topics in Environmental Management*, UNESCO Series of Learning Materials in Engineering Sciences, UNESCO Press, 1993.
29. Franklin Associates, *A Comparison of Energy Consumption by the Plastics Industry to Total Energy Consumption in the United States*. Franklin Associates, Priarie Village, KS, 1990.
30. R. S. Stein and V. Wilcox, *Trends Polym. Sci.* **4**(11), 358 (1996).
31. G. Klust, *Netting Materials for Fishing Gear*, Fishing News Books, Surrey, England, 1982.
32. Franklin Associates, *Comparative Energy Evaluation of Plastics Products and Their Alternatives for the Building and Construction and Transportation Industries*. Franklin Associates, Priarie Village, KS, 1991.
33. Franklin Associates, *Plastics' Energy and Greenhouse Gas Savings Using Homewrap Applied to the Exterior of Single Family Housing in the US and Canada. A Case Study*, Franklin Associates, Priarie Village, KS, 2000, p. 20.
34. G. Gappert, *Vinyl 2020: Progress, Challenges, Prospects for the next Quarter Century*. Vinyl Institute, Morristown NJ, 1996, p. 43.
35. Franklin Associates, *Plastics' Energy and Greenhouse Gas Savings Using Refrigerator and Freezer Insulation as a Case Study*, Priarie Village, KS, 2000, p. 22.
36. L. C. E., Struik and L. A. A. Schoen, eds., *Recycling and Reuse of Polymeric Materials; The Limited Potential of Biodegradables*, 7th European Polymer Federation Symposium on Polymeric Materials (EPF 98)., T. S. S. Spychaj, ed., Wiley, Szececin, Poland, 2000, p. 1.
37. R. Jansen, M. Koster, and B. Strijtveen, *Environmentally-Friendly Packaging in the Future*. Vereniging Mileudefensie, Amsterdam, The Netherlands, 1991: F. Lox, Packaging & Ecology, Pira International, Surrey, England 1992, p. 246.
38. T. J. Cavanaugh and E. B. Neumann, *Trends in Polym. Sci.*, **3**(2), 48 (1995).
39. J. S. Young, L. Ambrose, and L. Lobo, *Stirring Up Innovation. Environmental Improvements in Paints and Coatings*. INFORM, New York, 1994, p. 10.
40. A. L. Andrady, *Pavement Marking Materials: Assessing Environment-Friendly Performance.*, in *NCHRP Report 392*, Transportation Research Board, Washington, DC, 1997, p. 46.
41. R. Atkinson, "Gas Phase Troposheric Chemistry of Organic Compounds," In *Volatile Organic Compounds in the Atmosphere*, R. E. Hester and R. M. Harrison, eds., The Royal Society of Chemistry, Herts, England, pp. 65–89.
42. J. Murlis, "Volatile Organic Compounds: The Development of UK Policy," In *Volatile Organic Compounds in the Atmosphere*, R. E. Hester and R. M. Harrison, eds., The Royal Society of Chemistry, Herts, England, pp. 125–132.
43. Longhurst, J. W. S., *Fuel*, **72**(9), 1261 (1993).

44. J. G. Irwin, *Archi. Environ. Contamin. and Toxicol.* **18**, 95 1989.
45. V. N. Bashkin, *Chem. Britain*, **37**(6), 40–42 (2001).
46. D. W. Sparling, in *Handbook of Ecotoxicology*, D. J. Hoffmann, G. A. Burton, Jr., and J. Cairnes Jr., eds., Lewis, Boca Raton, FL, 1994.
47. USEPA, *Characterization of the Municipal Solid Waste in the United States, 2000 Update*, Franklin Associates, Priarie Village, KS, 2001.
48. N. C. Vasuki, *Waste Age*, November, 165–170 (1988).
49. R. L. Rathje, et al. *Source Reduction and Landfill Myths*, in *ASTSWMO National Solid Waste Forum on Integrated Solid Management*, Lake Buena Vista, Florida, 1988.
50. F. W. Karasek, et al., *J. Chrom.*, **227**, 270 (1983).
51. W. F. Carroll, *J. Vinyl Tech.* **10**(2), 90 (1988).
52. H. Hagenmeier, in *Energy Recovery Through Waste Combustion*, A. Brown, P. Evemy, and G. L. Ferrero, eds., Elsevier Applied Science, Essex, England, 1988, 154.
53. USEPA, *Assessment of Municipal Waste Combustion Emissions Under Clean Air Act. Advance Notice of Proposed Rulemaking*, USEPA, Washington, DC, 1987.
54. G. H. Edulife, in *Waste Incineration and the Environment*, R. M. Harrison, ed., Royal Society of Chemistry, Lecthworth, Herts United Kingdom, 1994, p. 73.
55. S. A. Skene, I. C. Dewhurst, and M. Greenberg, *Human Toxicol.* **8**, 173 (1989).
56. A. W. M. Hay, *Chlorinated Dioxins and Related Compounds: Impact on the Environment*, R. W. F. O. Hutzinger, E. Merian, and F. Poochiari, eds., Pergamon, Oxford, 1982.
57. S. Marklund, *Chemosphere* **16**, 29 (1987).
58. J. R. Visalli, *J. Air Pollution Control Assoc.* **37**, 1451 (1987).
59. S. Nchida and H. Kamo, *Ind. Eng. Chem. Proc. Des. Dev.* **22**: 144 (1987).
60. S. Dunn, in L. R. Brown, M. Renner, and C. Flavin, eds. *Vital Signs 1997*, Worldwatch Institute, New York. 1997, p. 74.
61. W. L. Fisher, *An Assessment of Oil Resources Base in the U.S.*, U.S. Department of Energy, Bartlesville, OK, 1992.
62. P. Gipe, *Wind Energy Comes of Age*. Wiley, New York, 1995, p. 403.
63. K. Olsen, *San Diego Daily*, 29 July, 1997.
64. B. Hileman, *C&E News*, April 17, 2000, p. 29.
65. R. Repetto, *Sci. Am.* June, p. **94**, 266 (1992).
66. B. Rich, *Mortgaging the Earth. The World Bank. Environmental Impoverishment, and the Crisis of Development*, Beacon, Boston, 1994.
67. D. Kirkpatrick, *Fortune*, 12 (1990).
68. R. Jansen, M. Koster, and B. Strijtveen, *Environmentally-Friendly Packaging in the Future*, Vereniging Mileudefensie, Amsterdam, The Netherlands, 1991.

SUGGESTED READING

1. William P. Cunningham, Mary Ann Cunningham, and Barbara Woodworth Saigo, *Environmental Science: A Global Concern*, McGraw-Hill, December 2001.

2. G. Tyler Miller, *Environmental Science: Working with the Earth*, Brooks Cole, November 2001.
3. Bernard J. Nebel, and Richard T. Wright, *Environmental Science: Towards a Sustainable Future*, Prentice Hall, October 2001.
4. Daniel B. Botkin, and Edward A. Keller, *Environmental Science: Earth as a Living Planet*, John Wiley & Sons, July 2002.
5. Bjorn Lomborg, *The Skeptical Environmentalist: Measuring the Real State of the World*, Cambridge University Press, December 2001.
6. Daniel D. Chiras, *Environmental Science*, Brooks Cole, January 1998.

APPENDIX A: GLOBAL WARMING

The surface temperature of earth is determined by the rate at which solar radiation reaches the planet, the surface albedo, and the rate of emission of infrared radiation by the earth into the atmosphere. The earth's atmosphere contains low levels of the so called "greenhouse" gases such as carbon dioxide and water vapor, that allow the incoming solar radiation to reach the earth unhindered but trap some of the heat radiated back from earth into the atmosphere[49]. In this regard, the greenhouse gases act similar to glass in a greenhouse, trapping the long wavelength infrared radiation from escaping the earth. Figure A.1 indicates the fate of solar radiation entering the earth's atmosphere. For the most part this "greenhouse effect" is a natural process indispensable to life on earth; if not for the biogenic greenhouse gases in our atmosphere the temperature of earth will be inhospitable (colder by about 33°C). However, increasing concentrations of carbon dioxide and other greenhouse gases in the atmosphere can cause a corresponding warming of the surface temperatures on earth. The concentration of greenhouse gases, particularly CO_2 in the atmosphere has been steadily rising in the recent years mainly because of human activity. The consequent slow warming of earth measured over the recent decades (see Figure A.2) has been reliably attributed primarily to the accumulation of man-made greenhouse gases in the earth's atmosphere.

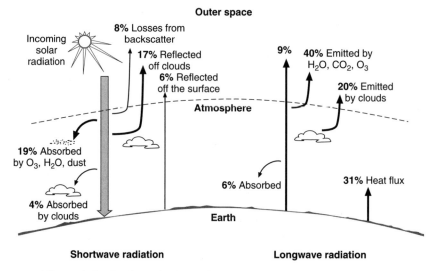

Figure A.1. The fate of solar radiation entering the earth's atmosphere.

---

[49] This is essentially the result of the large difference in the temperature of the sun (around 6000 K) and that of earth (about 286 K). Because of this large difference in temperature, the frequency of the infrared radiation emitted by the earth is much lower than that of incoming solar radiation, and is effectively absorbed by these greenhouse gases.

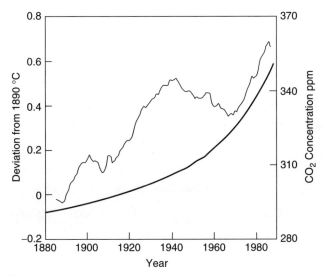

Figure A.2. Increasing trend in the mean global temperature (compared to that in 1890) and atmospheric CO_2 concentration (heavy line).

It is the burning of fossil fuels that is mostly responsible for the increased carbon dioxide levels in the present-day atmosphere to nearly 30% above that in pre-industrial times (∼1850). Today, the atmospheric CO_2 level of about 360 ppm is higher than that encountered by the earth at any time during at least the past 160,000 years [1]. Transportation and industrial uses account for most of this fossil fuel use, but CO_2 emissions also originate from cement or lime facilities, incineration plants and the flaring of refinery gases. Deforestation is a second significant source of CO_2 emissions [2]. The tropical forests were depleted at a rate as high as 12.6 million hectares per year worldwide during 1990–1995 [3]. The second key greenhouse gas is methane, released from cattle breeding operations, rice paddy cultivation, sewage treatment plants and from coal mining. Natural sources include wetland emissions, marine contributions, and termite activity. Levels of methane have more than doubled since the industrial era. A third greenhouse gas, nitrous oxide, N_2O, is released from fertilizer used in agriculture as well as from internal combustion engines. Of the total greenhouse gas emissions in the US in 1998, CO_2, CH_4 and N_2O account for 81.4, 9.9 and 6.5 percent respectively. The pivotal role of CO_2 as a greenhouse gas is apparent from the long-term correlation between the CO_2 level in the atmosphere and the average temperature on earth. Chlorofluorocarbons (CFCs) are particularly potent greenhouse gases and even at low concentrations have a very large impact in terms of global warming. Compared to a value of unity for CO_2, the global warming potential (over a period of 100 years), GWP, of $CH_4 = 23$ and that of $N_2O = 296$. The comparable GWP value for CFC-11 and CFC-12 would be 4000 and 8500! The Table A.1 below shows data on the global warming potential of common greenhouse gases and chlorofluorocarbons relative to that of carbon

Table A.1 Direct Global Warning Potentials (GWPs) of Selected Gases Relative to Carbon Dioxide

Gas		Lifetime (Years)	Global Warming Potential[a] (Time Horizon in Years)		
			20 yrs	100 yrs	500 yrs
Carbon dioxide	CO_2		1	1	1
Methane[b]	CH_4	12.0_	62	23	7
Nitrous oxide	N_2O	114_	275	296	156
Hydroflurocarbons					
HFC-23	CHF_3	260	9400	12000	10000
HFC-32	CH_2F_2	5.0	1800	550	170
HFC-41	CH_3F	2.6	330	97	30
HFC-125	CHF_2CF_3	29	5900	3400	1100
HFC-134	CHF_2CHF_2	9.6	3200	1100	330
HFC-134a	CH_2FCF_3	13.8	3300	1300	400
HFC-143	CHF_2CH_2F	3.4	1100	330	100
HFC-143a	CF_3CH_3	52	5500	4300	1600
HFC-152	CH_2FCH_2F	0.5	140	43	13
HFC-152a	CH_3CHF_2	1.4	410	120	37
HFC-161	CH_3CH_2F	0.3	40	12	4
HFC-227ea	CF_3CHFCF_3	33	5600	3500	1100
HFC-236cb	$CH_2FCF_2CF_3$	13.2	3300	1300	390
HFC-236ea	CHF_2CHFCF_3	10	3600	1200	390
HFC-236fa	$CF_3CH_2CF_3$	220	7500	9400	7100
HFC-245ca	$CH_2FCF_2CHF_2$	5.9	2100	640	200
HFC-245fa	$CHF_2CH_2CF_3$	7.2	3000	950	300
HFC-365mfc	$CH_3CH_2CF_2CH_3$	9.9	2600	890	280
HFC-43-10mee	$CF_3CHFCHFCF_2CF_3$	15	3700	1500	470
Other					
SF_6		3200	15100	22200	32400
CF_4		50000	3900	5700	8900
CH_3OCH_3		0.015	1	1	<<1
HFE-125	CF_3OCHF_2	150	12900	14900	9200
HFE-134	CHF_2OCHF_2	26.2	10500	6100	2000
HFE-143a	CH_3OCF_3	4.4	2500	750	

[a]The GWP is an index for estimating relative global warning contribution due to atmospheric emission of a kg of a particular greenhouse gas compared to emission of a kg of carbon dioxide. GWPs calculated for different time horizons show the effects of atmospheric lifetimes of different gases. Based on reports of the working Group 1, IPCC [2000].
[b]The GWPs for CH_4 includes an indirect contribution from stratospheric H_2O and O_3 production.

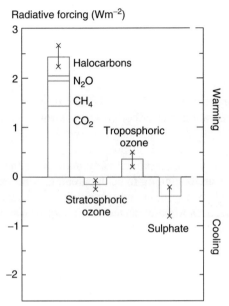

Figure A.3. The relative magnitude of radiative forcing associated with greenhouse gases, ozone in the atmosphere, and sulfates.

dioxide. The Figure A.3 is based on the recent report of the IPCC (The Intergovernmental Panel on Climate Change) where the magnitude of the radiative forcing associated with greenhouse gases, ozone in lower and upper layers of the atmosphere and sulfate aerosols (cooling) are compared. Only those factors where the level of scientific understanding is high is indicated in the figure (the original IPCC figure included other contributors such as burning of biomass or fossil fuels mineral dust and aerosols [11].

Global average temperatures have already edged up at a rate of about 0.6°C (+ or −0.2°C) over the 20th century. There is consensus among leading researchers that at least in the past 50 year period most of the observed warming of the earth is mainly the result of increased greenhouse gases in the atmosphere [4]. While a direct correlation between the rise in global average temperature and the concentration of these gases is difficult to demonstrate, the available evidence increasingly points to anthropogenic factors causing global climate change [5]. Careful measurements show a clear warming trend in the sea environment as well [6], both in the Southern and Northern Hemispheres. Unless carbon emissions into the atmosphere are drastically controlled, global warming is likely to continue unabated in to the future years. According to the best estimates, the average global temperatures are expected to rise by 1.4–5.8°C by 2100. Therefore nearly all land areas are likely to warm above the present global average, with more extreme hot days and heat waves and fewer cold days becoming increasingly common. Assumptions made in the models used to estimate these introduce a certain degree of uncertainty as to the exact number of degrees of warming expected and how such warming might be geographically distributed. However, at the present time there is worldwide consensus that global warming is indeed well underway, that anthropogenic greenhouse emissions[50] are primarily responsible for it, and on the likely consequences of this phenomenon. Even if the estimates are only partly correct, the future generations will face the unpleasant prospect of living on earth that will be the warmest ever since human beings evolved[51]. It is important to appreciate that despite the few degrees increase anticipated in global average temperature, the effect on the environment and biosphere could be devastating. After all, during the last Ice Age the global average temperature was less than 6°C lower than it is at present!

As the global warming process is already under way the associated environmental consequences are beginning to be apparent. Careful extrapolation of these effects to higher levels of warming and other studies of likely impacts, yield possible global consequences to be expected at a possibly warmer future time. The more serious of these effects are likely to be regional where extreme weather events such as drought heat waves and heavy rainfall are expected to increase over the next 100 years in affected parts of the world. Other regions will actually be cooler than at the present time. Some particularly serious consequences of global warming are as follows.

Rise in Sea Levels

Melting of the polar ice caps in warm weather as well as the volume expansion of the oceans is believed to cause the rise in sea level consequent to global warming. The process seems to be already underway. The arctic sea ice is decreasing at the rate of about 2.8 percent per year since mid 1970s. Antarctic ice shelves are also reported to be melting at an accelerated rate already, with a high retreat rate of 3000 sq. km per year [7]. The average rise in sea level during the 20th century has been estimated at 0.1 to 0.2 meters. A number of low-lying nations including many island states would be severely affected by the rise in sea level, possibly resulting in large-scale migration or resettlement of refugee populations[52]. Bangladesh, Nile Delta area, the Maldives, and the Netherlands are particularly likely candidates. The impact of sea level rise on developing nations without resources to implement adaptive measures is particularly serious.

In the US, the Everglades area in southern Florida is likely to be affected. A closely related concern is the possible intrusion of saline water into bodies

---

[50] The anthropogenic emissions are small (about 5 percent of the total) compared to that generated by nature. But it is this part of the load amounting to over 400,000 tons of CO_2 annually that has primarily affected the natural equilibrium of the carbon cycle on earth and resulted in CO_2 increase in the atmosphere.

[51] In the 1990s the record for global mean surface temperature was broken four times (1990, 1995, 1997 and 1998) making it the warmest decade on record for Northern Hemisphere in the millennium!

[52] In Bangladesh over 21 million of the population live in areas likely to be affected by a 1-meter rise in sa level. The figure for China is about 70 million!

http://www.climatechangesolutions.com/english/science/gimpact.htm

of fresh water. This would have serious implications on affected ecosystems, agriculture and drinking water supplies.

Public Health Impacts

Heat stress in the northern latitudes where the populations are not acclimatized would be the primary health impact of warming [8]. Serious illness and increased mortality is indicated in some regions (including some urban areas in the US) with a high degree of air pollution. Asthmatics and those with respiratory difficulty will be particularly affected by heat stress. An equally serious concern will be the prevalence of vector-borne disease [9] likely to be encouraged by global warming. These include malaria, schistosomiasis, dengue fever, yellow fever, and the African sleeping sickness. The reach of malaria, for instance, could extend from about 45% of the world population at the present time to as much as 60%, as a result of the projected warming. Developing countries, some of them unfortunately in low-lying areas and also susceptible to sea-level rise, will be particularly hard hit by this public health crisis. The major indirect health effects of global warming are likely to be due to overcrowding and migration of large masses of people, and impacts from malnutrition or lack of water from droughts and consequent disruption in agriculture.

Ecosystems

Warming would cause major shifts in the areas of natural ecosystem types such as temperate or boreal forest, prairie and tundra. At the IPCC projected levels of warming the boundaries of biomes could shift towards the poles by several hundred kilometers.

A particularly sensitive indicator of warming might be coral in the marine ecosystem. Corals undergo bleaching when the reef temperatures rise by even a $1-2°C$ over their normal maximum of $28-29°C$. Massive bleaching of corals has been reported [10] from many parts of the world including the Caribbean and the Philippines. Taken along with the rate of retreat of the western alpine glaciers, these phenomena can be very sensitive indirect indicators global warming.

Effect on Crops

Increased CO_2 levels and a warmer climate can have a positive effect on plant productivity in some in some mid- to high latitudes regions of the world. Because the impact depends on the latitude, the effect of low to moderate global warming on crops can be very different for developed versus the developing nations. (Industrialized countries that are able to quickly adapt to the climatic change would be relatively little affected, and are likely to enjoy increased crop yield.) This is likely a result of increased metabolism of cultivars as well as changes in life cycles of common pest species due to warmer conditions. But a decrease in productivity, unpredictable climates, and altered farming cycles can adversely impact food production in other regions. Indirect impact of climate change on pollinating insects is also a concern, With about 80% of the about 1300 crop species cultivated worldwide being insect-pollinated, any change in the insect ecology can negatively affect agriculture and therefore the worlds food supply.

A large volume of information has been published on the climate forcing and global warming. The scope of this volume does not permit a comprehensive discussion of this interesting topic. The various reports of the IPCC (2001) [11] will provide a good starting point for readers interested in a more detailed study of global warming. The Panel in their latest report identified the most effective short-term mitigation strategy to be limiting the emission of greenhouse gases. Stabilization of atmospheric CO_2 concentrations at even 450 ppm would require the global anthropogenic CO_2 emissions to drop below year 1990 levels within a few decades, and continue to decrease steadily thereafter to a small fraction of current emissions.

REFERENCES

1. C. Bright, *State of the World 1997*, W. W. Norton and Company, New York, 1997.
2. J. T. Houghton, G. J. Jenkins, and J. J. Ephraumus, eds., *Climatic Change: The IPCC Scientific Assessment*, Cambridge University Press, Cambridge, 1990.
3. F. a. A. Organization, *State of the World Forcasts*, FAO, ROME, Rome, Italy, 1997.
4. T. R. Karl, W. K. Knight, and B. Baker, *Geophysical Research Letters*, **27**, 719, 2000.
5. C. Rapley, *Polar Research*, **18**, 117, 1999.
6. S. Levitus, J. I. Antonov, T. B. Boyer, and C. Stephens, *Science*, **287**, 2225, 2000.
7. L. E. Brown (2000) *Challenges of the New Century*, ed. L. E. Brown, C. Flavin, and H. French. New York, Norton and Company Inc., pp. 6.
8. L. S. Kalkstein (1992) "*Impacts of global warming on human health: Heat stress related mortality*," ed. K. Majmdar, L. S. Kalkstein, B. Yarnal, E. W. Miller, and M. Rosenfeld, The Pennsylvania Academy of Sciences, pp. 371–383.
9. J. D. Longstreth, *Environmental Health Perspectives*, **96**, 1991.
10. R. B. Aronson, W. F. Precht, I. G. Macintyre, and T. J. T. Murdoch, *Nature*, **405**, 36, 2000.
11. The World Meteorological Organization (WMO) and the United Nations Environment Programme (UNEP) established intergovernmental Panel on Climate Change in 1988. The role of the IPCC is the assessment of the scientific, technical and socio-economic information relevant to the understanding of the risk of human-induced climate change. Its the Third Assessment Report, "Climate Change 2001" has been published in English by the Cambridge University Press. The reports are also available for downloading at the IPCC website: *www.ipcc.ch*. Climate Change 2001: The Scientific Basis, Climate Change 2001: Impacts, Adaptation & Vulnerability, Climate Change 2001: Mitigation, Climate Change 2001: Synthesis Report, are available at this website. Other interesting publications "IPCC Special Report on The Regional Impacts of Climate Change. An Assessment of Vulnerability" and the "Special Report on Emissions Scenarios."

APPENDIX B: DEPLETION OF STRATOSPHERIC OZONE

More than 90% of the ozone in the atmosphere is concentrated in the stratosphere as a layer extending from about of 5–11 miles above the earth's surface up to about 30 miles. Unlike the tropospheric ozone content in the lower layers of the atmosphere that gives rise to urban smog and air pollution, this stratospheric "ozone layer" serves a crucial purpose. Although the concentration of ozone in this layer is quite low[53] it functions as an effective shield against the harmful ultraviolet (UV) radiation in extra-terrestrial solar radiation from reaching the earth's surface. The stratospheric ozone layer completely filters out the UV-C ($\lambda < 290$ nm), and most of the UV-B radiation ($\lambda = 290$ nm to 315 nm), both particularly damaging wavelengths to living organisms, from the solar radiation passing through it. It has little impact on wavelengths longer than about 315 nm including UV-A radiation. The evolution of life on earth was only made possible by the effective shielding action afforded by the ozone layer. The ozone concentration in stratosphere varies strongly with latitude over the globe with the highest values in mid latitudes and at the poles. Lowest ozone levels are found in the tropical regions. Air currents cause a longitudinal variation in ozone levels as well.

Ozone is produced by the photolysis of oxygen (by $\lambda < 242$ nm UV radiation) and is readily reconverted to oxygen by the Chapman reaction of ozone with atomic oxygen. A simplified scheme of the photochemical equilibrium process for ozone formation is as follows. The concentration of ozone in the stratosphere is determined by the rates of chemical reactions that produce and destroy ozone.

$$O_2 + h\nu \longrightarrow 2O$$
$$O + O_2 \longrightarrow O_3$$
$$O_3 + h\nu \longrightarrow O_2 + O$$
$$O + O_3 \longrightarrow 2O_2$$

Reactive species such as NO_x molecules, chlorine radicals, and hydrogen radicals present in the stratosphere can also cause the catalytic conversion of ozone to oxygen, in competition with the last reaction shown above. Reactions such as the following, therefore interfere with the equilibrium photochemical processes, reducing the concentration of ozone in the stratosphere. The net reaction is one where ozone is converted to oxygen and catalytic Cl species survives for further reaction. Particularly effective in this regard are halogen monoxides (ClO and BrO).

$$CCl_2F_2 + UV \longrightarrow CClF_2 + Cl$$
$$Cl + O_3 \longrightarrow ClO + O_2$$
$$ClO + O \longrightarrow Cl + O_2$$

---

[53] If the ozone layer over the US were compressed into a layer of temperature 0°C at 1 atmosphere pressure, it would have a thickness of only 3 mm. As a 0.01 mm thickness of an ozone layer is defined to be 1 Dobson Unit (DU), the average thickness of the ozone layer over the US is about 300 DU.

Table B.1 Some Common Ozone Depletion Substances Controlled Under the Montreal Protocol

Name	Chemical Name	Lifetime (yrs)	ODS	GWP
CFC-11	Trichlorofluoromethane	45	1.0	4000
CFC-12	Dichlorofluoromethane	100	1.0	8500
CFC-113	1,1,2-trichlorotrifluoromethane	85	.8	5000
CFC-114	dichlorotetrafluoromethane	300	1.0	9300
CFC-115	Monochloropentafluoroethane	1700	.06	9300
Halon 1211	Bromochlorodifluoromethane	11	3.0	1300
Halon 1301	Bromotrifluoromethane	65	10.0	5600
Halon 2402	Dibromotetrafluoroethane	6	—	—
CFC-13	Chlorotrifluoromethane	640	1.0	11700
CFC-111	Pentachlorofluoroethane	—	1.0	—
CFC-112	Tetrachlorodifluoroethane	—	1.0	—
CFC-211	Heptachlorofluoropropane	—	1.0	—
CFC-213	Hexachlorodifluoropropane	—	1.0	—
CFC-214	Tetrachlorotetrafluoropropane	—	1.0	—
CFC-215	Trichloropentafluoropropane	—	1.0	—
CFC-216	Dichlorohexafluoropropane	—	1.0	—
CFC-217	Chloroheptafluoropropane	—	1.0	—
CCl4	Carbon tetrachloride	35	1.1	1400
CHCl3	1,1,1-trichloroethane	4.8	.1	110
CH3Br	Bromomethane	—	.7	5
HCFC-22	Monochlorodifluoromethane	11.8	.055	1700
HCFC-123	Dichlorotrifluoroethane	1.4	.02	93
HCFC-124	Monochlorotetrafluoroethane	6.1	.022	480
HCFC-141b	Dichlorofluoroethane	9.2	.11	630
HCFC-142b	Monochlorodifluoroethane	18.5	.065	2000
HCFC-225ca	Dichloropentafluoropropane	2.1	.025	180
HCFC-225cb	Dichloropentafluoropropane	6.2	.033	620

The magnitude of the interference depends on how much of these catalytic molecules are present in the stratosphere and their lifetimes. A single atom of chlorine can destroy thousands of ozone molecules during its lifetime in the stratosphere. The Table B.1 below summarizes the lifetimes and the ozone depletion potential (ODP)[54] of common ozone-depleting chemicals. Many of the ODSs also happen to be greenhouse gases and their global warming potentials (GWP) are indicated in the table. Chlorine-containing compounds that play this

---

[54] The Ozone Depletion Potential (ODP) of a compound is defined as the ratio of the total amount of ozone destroyed by a given amount of the compound to the amount of ozone destroyed by the same mass of CFC-11.

catalytic role particularly well include the chlorofluorocarbons (CFCs), carbon tetrachloride, halons, methyl chloroform, methyl bromide, and hydrochlorofluorocarbons (HCFCs). These, particularly the CFCs, were once extensively used in refrigeration (~32%), cleaning (~20%), propellant (~18%), and foam blowing applications (~28%) (their use has since been severely restricted by the Montreal Protocol). Bromine, though more efficient at destroying ozone relative to chlorine, is fortunately present at a much lower concentration in the stratosphere. However, the effect of bromine compounds (mostly methyl bromide emissions from natural sources) can be quite significant in determining the ozone concentrations at mid latitudes.

The primary consequence of the loss of stratospheric ozone is that more of the extraterrestrial UV-B radiation (290 nm to 315 nm in wavelength) reaches the Earth. A wide range of measurements has established this relationship between ozone levels and the UV radiation reaching the Earth's surface. In 1994 the UV-B levels reaching earth's surface were found to be 8–10% higher than those found 15 years earlier at 45°N and °S, with even higher in the polar regions. The increase in UV-B across the Earth is not uniform; there is negligible ozone depletion in the tropics (20°N to 20°S of the equator) and the maximum reported depletions in the "ozone hole" at the pole[55]. Antarctica has unique weather conditions[56] such as polar winds and polar stratospheric clouds that create particularly effective conditions for halogen-catalyzed ozone depletion. This lead to the appearance of a seasonal "ozone hole" first observed in 1985, where over half the overhead ozone within it was depleted. Recently reported increases in erythemal UV radiation from 1970 values was about 7% in Northern mid latitudes in spring, 6% in southern mid-latitudes, 130% in Antarctic spring, and 22% in Arctic Spring [1]. These values are probably close to the highest possible increase in UV radiation to be experienced on earth's surface, as the best models suggest the concentration of ozone-depleting substances in the stratosphere to have peaked around 2000.

The increase in UV-B anticipated as a result of ozone depletion is relatively small compared to the natural variations in UV-B in most places on earth[57]. However, unlike with natural variations, ozone-depletion consistently increases the UV-B levels. Changes that rely on cumulative exposure (such as certain types of skin cancer) will therefore be affected by any increase in UV-B despite the

---

[55] The most severe depletion of ozone in the ozone hole was observed in the 2000–20001 season in the Antarctica. The area of the hole was over 28 million sq.m. and the ozone levels were about 40 percent of the normally expected level!

[56] The lower stratosphere at the South Pole is the coldest spot on Earth! The polar clouds with minute ice crystals that form in the region effectively facilitate the breakdown halogen-containing molecules into reactive forms and into halogen radicals.

[57] The increased in UV-B radiation due to ozone depletion at mid latitudes is well within that obtained by moving several hundred kilometers towards the equator. At the population level (as opposed to individual level) an equivalent move would still be expected to result in increased health and other risks.

natural variability. Aerosol and pollutant build up in the atmosphere in urban areas, can generally decrease the UV-B levels, but at a high environmental cost.

In 1974 Molina and Rowland proposed the catalytic pathway for ozone depletion by stable halogen-containing chemicals such as the CFCs. The Montreal Protocol on Substances that Deplete the Ozone Layer established in 1987 sought to restrict the production and emission of chlorofluorocarbons and related compounds by signatory countries. With mounting evidence of damage to the ozone layer the protocol was soon amended (by the 1990 London Amendments) to call for a ban on the production of the most damaging of the ozone-depleting compounds by 2000 (2010 for developing countries). Subsequent amendments revised the year to 1996 for developed countries (Copenhagen Amendments in 1992), and added further restrictions to strengthen the protocol. These corrective measures have drastically reduced the input of ODSs into the atmosphere allowing the photochemical processes to reestablish the integrity off the ozone layer. Recent measurements made show a decrease in the abundance of ozone-depleting gases in the atmosphere. This decrease is almost entirely due to the outstanding success of the Montreal Protocol and associated Amendments. It is likely that the levels of ODSs will remain at near-peak levels for a few decades and then decrease significantly. Based on the present rate of recovery, the ozone layer is expected to return to normal levels (pre-1980 levels) by the middle of the next century.

The increase in UV-B radiation content in terrestrial solar radiation is undesirable and can lead to severe disruptions in different areas of human activity. These impacts and their significance along with the areas of uncertainty associated with each, are discussed in detail in a special recent issue of the Journal of Photochemistry and Photobiology (volume 46 B: Biology, October 1998). The anticipated effects of ozone depletion such as the increase in the incidence of a given disease, is based on two quantities; the radiation amplification factor, RAF, and the biological amplification factor BAF, for the particular effect. RAF is the ratio of the (percentage increase in the UV effective radiation) to the (percentage decrease in stratospheric ozone concentration). BAF is the ratio of the (percentage increase in the incidence of the disease or effect) to the (percentage increase in the UV effective radiation). The product of these two factors allows the estimation of the overall percentage increase in incidence of the disease per percentage ozone depletion. Two particularly important areas of impact[58], however, need to be pointed out.

a) **Human and Animal Health Impacts:** Several major organ systems in human beings, particularly components of the eye, the outer layers of the skin, and cells from the immune system (Langerhan's cells) that are affected via exposure through skin, are routinely exposed to solar radiation.

---

[58] A particularly informative web site that discusses the impacts of increased UV-B in sunlight is the Columbia University's Center for International Earth Science Information Network (CIESIN) http://www.ciesin.org/TG/HH/ozhlthhm.html

Eye: Absorption of UV radiation by the cornea (that absorbs most of the UV-B) and by the lens the UV radiation rarely reaches the retina of the eye. However, acute short-term exposure of the eye to UV-B radiation can lead to increased incidence of eye decease ranging from simple "snow blindness" (photokeratitis) to squamous cell carcinoma (SCC). Strongest correlation between eye diseases and exposure to solar radiation is observed for the condition pterigyium and climatic droplet keratopathy. With SCC, a recent epidemiological study found a strong relationship between the incidence of the disease and the latitude of exposure [2] (with 40–50 percent increase in incidence per 10° change in latitude).

Exposure can also lead to certain types of cataract formation increasing the risk by up to 3-fold [3]. Epidemiological data linking the cataracts and exposure to solar radiation has been reviewed [4]. Cataract is the leading cause of blindness worldwide requiring surgical removal of the affected lens, involving very significant public health costs worldwide.

Skin: Sunburn is the most common result of exposure to solar UV radiation. UV-B is well known to be about 1000 times more efficient than UV-A in causing sunburn. Even with the most sensitive skin types with no pigmentation sunscreens and controlled exposure offer a convenient means of controlling the condition.

Skin cancer is the most common form of cancer among fair-skinned people in the US and the incidence rates of common forms of skin cancer increases with the dose of UV-B radiation. The incidence of NMSC (the predominant form of which is the non-lethal basal cell carcinoma) in US correlates well with available UV-B levels. A relatively higher level of risk exists for fair-skinned populations [5] exposed to increased solar UV-B contents. The role of UV-B in promoting the more serious malignant melanoma remains unresolved. Factors such as genetics, avoidance behavior and diet can modify the impacts of exposure in both humans and in animal species.

One clear benefit of increased UV-B exposure of the human skin, however, is the increased production of vitamin D!

Immune reactions: The role of UV radiation in modulating immune reactions in humans is well established and it is likely that increased UV-B radiation will impact the incidence of infectious diseases[6]. Risk factors for certain cancers (skin cancer and non-Hodgkin's lymphoma) include immunosuppression. UV-induced immunosuppression is therefore likely to promote such cancers, but no epidemiological supporting data are available at the present time.

b) Food Web.

Crop species: Increases in solar UV-B radiation affect crop plants (as well as other plants) in several different ways including direct damage, alteration of patterns of gene activity, and changes in their physiological cycles. These changes can significantly impact crop yields, with the direction of the response depending on the species and even varieties of the same species. This is particularly true of economical important crops[59] such as soybean [7] and rice [8] where different varieties were found to either increase or decrease in yield as a result of increased UV-B content in sunlight. About half of all species investigated, however, show some effect to increased UV-B radiation, with some adapting fairly quickly and others very slowly to the exposure. Selective breeding of the UV-resistant varieties might be considered to avoid loss in crop yield under such exposure conditions. Changes in UV-B levels also affect the interaction between species in a community and can affect the susceptibility of species to pathogens and insects.

Marine environment: Depending on the clarity of water UV-B radiation could penetrate 5–20 m into the water column, and has a direct influence on the phyto- and zooplankton species in the worlds ocean. This potentially impacts the capacity for primary production and carbon dioxide fixation in these systems, [9]. UV radiation affects the growth, reproduction and the physiology of phytoplanktons (including their photosynthetic pigment content) [10] [11]. A particular concern is the effect of increased UV-B on the marine species that constitute the base of the food pyramid accounting for almost a third of the global productivity. If the species are able to sense and react to the radiation by moving away to deeper waters, their photosynthetic production will tend to decrease[60]. As with crop species the UV-B tolerance case of marine species also varies widely.

The main threats of UV-B to zooplankton species is the reduction of their food supply, the phytoplankton, and stresses at various states of developmental stages of their life cycle, possibly causing mass mortalities [13]. While planktons seem to survive well in tropical regions with already high UV-B levels, species adapted to low UV levels in the Polar Regions might be particularly susceptible to any increase in UV-B. An important consequence of any significant reduction in the plankton levels is its adverse impact on the world's fisheries. Reduced fish stocks can potentially alter the diet of a large part of the population; fish constitute about 40 percent of the dietary protein intake in Asia (where a majority of the global population is found in).

REFERENCES

1. Madronich, S., et al., *Changes in biologically active ultraviolet radiation reaching the Earth's surface*. Journal of Photochemistry and Photobiology: B Biology, **46**, 5 (1998).

---

[59] Only about 15 species of plants provide a great majority of the calories consumed and 75 percent of the world's protein consumption. In addition to the key cereals, sugar cane, sugar beet, sweet potato, cassava and soybean fall into this category. Clearly it is important to establish the effect of UV-B on the productivity of these species under different growing conditions.

[60] A 25 percent depletion of the ozone layer and the ensuing increase in UV-B reaching the surface of oceans was estimated to reduce phytoplankton productivity by as much as 35% [12].

2. Newton, R., et al., *Lancet*, **347**, 1450 (1996).
3. West, S. K., et al., *JAMA*, **280**, 714 (1998).
4. Dolin, P. J., *Br. J. Opthalmol.*, **78**, 178 (1994).
5. De Gruijl, F. R., *Skin Cancer and Solar UV Radiation. Eur. J. Cancer*, **35**, 2003 (1999): S. Madronich and F. R. de Gruijl, *Nature*, **366**, 23, (1993). Van der Leun, J. C. and F. R. de Gruijl, Influences of ozone depletion on human and animal health. Chapter 4 in *UV-B radiation and ozone depletion: Effects on humans, animals, plants, microorganisms, and materials*, ed. M. Tevini, 95–123. Ann Arbor: Lewis Publishers, (1993).
6. Robinson, J. K., *Archives of Dermatology* **126**, 477, (1990).
7. Teramura, A. and N. S. Murali, *Environmental Exp. Botany*, **26**, 89 (1986).
8. Nouchi, I., K. Kobayashi, and H. Y. Kim, *J. Agric. Meteorol.*, **52**, 867 (1997).
9. Häder, D.-P., *Ecosystems, Evolution and Ultraviolet Radiation*, ed. C. S. Cockell and A. R. Blaustein, New York: Springer-Verlag. (2001). p150.
10. Gieskes, W. W. C. and A. G. J. Buma, *Plant Ecology*, **128**, 16 (1997).
11. Hermann, H., et al., *J. Photochem. Photobiol.*, **34**, 21 (1996).: J. J. Cullen and P. J. Neale, *Photosynthesis Research*, **39**, 303 (1994).
12. Smith, R. C., Baker, K. S., Holm-Hanson, O. and Olson, R., *Photochem. Photobiol.* **31**, 585 (1980).
13. Chapman, J. and Hardy, J. T. (1988). Effects of middle ultraviolet radiation on marine fishes *Final Report Oregon State University US EPA Coop. Argrmt.* CR-812688-02-0.

CHAPTER 2

COMMON PLASTICS MATERIALS

ANTHONY L. ANDRADY

Engineering and Environmental Technology, Research Triangle Institute

The first totally man-made polymer to be synthesized was the phenol formaldehyde resin (called Bakelite at the time) made by Leo Baekeland in his garage in Yonkers, New York, back in 1907.[1] It was an immediate success not only as a replacement for shellac in electrical wiring (the primary reason for its invention) but also in numerous consumer uses including the body of the old black dial telephones and in early electrical fittings. Since that time, plastics have grown rapidly and have now become an indispensable part of everyday life. The exponential growth of plastics and rubber use, essentially over a short period of half a century, is a testimony to the versatility, high performance, and cost effectiveness of polymers as a class of materials.

Polymers derive their exceptional properties from an unusual molecular architecture that is unique to polymeric materials, consisting of long chainlike macromolecules. While both plastics as well as elastomers (rubber-like materials) are included in polymers, discussions on environment-related issues have mostly centered around plastics because of their high visibility in packaging and building applications. But elastomers are used in high-volume applications as well, particularly in automobile tires where the postconsumer product has to be effectively disposed of or recycled. With the rapid growth in automobiles, there is a consistent high demand for tires, which have a relatively short lifetime (compared to that of the automobile itself). The environmental concerns on elastomers are outside the scope of this discussion but are certainly serious enough to warrant a comprehensive study. These include the health impacts of rubber

---

[1] The first plastic material ever made was probably cellulose nitrate synthesized in 1862 by Alexander Parkes. A second cellulose-based polymer, cellophane, was invented in early 1900s by a Swiss chemist (Edwin Brandenberger). Cellophane found many applications and is still manufactured in the U.S.

Plastics and the Environment, Edited by Anthony L. Andrady.
ISBN 0-471-09520-6 © 2003 John Wiley & Sons, Inc.

compounding chemicals, the inhalation of particulate rubber that enters the atmosphere through breakdown of tires in use, and problems related to disposal of postconsumer tires.

Many of the common thermoplastics[2] used today, however, were developed after the 1930s; and a few of these even emerged after World War II. Among the first to be synthesized were the vinyl plastics derived from ethylene. The Du Pont Company and I.G. Farben (Germany) filed the key patents on the synthesis of vinyl chloride copolymers back in 1928. The homopolymer poly(vinyl chloride) (PVC) was already known but not recognized as being particularly useful at the time, although Semon in B.F Goodrich found the plasticized resin to be somewhat useful in the waterproofing of fabrics. Vinyl polymers and their plasticized compounds were being commercially produced in the United States even prior to World War II. But the now common rigid PVC used in building was a postwar development that rapidly grew in volume to a point that by the early 1970s the demand for vinyl resin was close to that for polyethylene!

Polyethylene, the plastic used in highest volume worldwide, was discovered at Imperial Chemical Industries (ICI) research laboratories in 1933, but the company filed the patent on the relevant polymerization process only in 1936. This high-pressure polymerization route was exclusively used to commercially produce low-density polyethylene (LDPE) for nearly two decades until the low-pressure processes for high-density polyethylene (HDPE) were developed in 1954. Linear low-density copolymers of ethylene (LLDP), intermediate in structure and properties between the HDPE and LDPE, followed even more recently in the 1970s. In the last decade yet another new class of polyethylene based on novel metallocene catalysts has been developed. Polypropylene manufacture started relatively late in the 1950s only after the stereospecific Ziegler–Natta catalysts that yielded high-molecular-weight propylene polymers became available. While a range of copolymers of ethylene is also commercially available, the homopolymer of propylene enjoys the highest volume of use. Polyethylene, polypropylene (and their common copolymers) are together referred to as polyolefins.

Several other common thermoplastics emerged about the same time as LDPE in 1930s. Polystyrene, for instance, was first produced in 1930 and by 1934 plants were in operation producing the commercial resin in both Germany and the United States. Poly(methylmethacrylate) (PMMA) was developed by ICI about the same period. Carothers's discovery of nylons (introduced in 1939 at the World's Fair in New York) yielded a material that particularly served the allied war effort. Nylon was used extensively in tire reinforcement, parachute fabric, as well as in everyday products such as toothbrushes and women's stockings. Engineering thermoplastics such as polycarbonate by comparison are a more recent development, with commercialization by General Electric Company around 1958.

---

[2] The term '*thermoplastics*' refers to plastic materials that can be formed into different shapes by the application of heat and pressure, over and over again. These are therefore easily recyclable into other products by remelting and processing into a different shape. The other group of plastics, the *thermosets*, (including epoxy and polyurethane material) are crosslinked on curing and will not soften on heating to allow these to be formed into a different shape.

The millions of metric tons of polymer resins manufactured annually worldwide are predominantly derived from petroleum and natural gas feedstock, but other raw materials such as coal or even biomass might also be used for the purpose. In regions of the world where natural gas is not readily available, petroleum or coal tar is in fact used exclusively as feedstock. About half the polyolefins produced in the United States today is based on petroleum, the remainder being derived from natural gas. The crude oil is distilled to separate out the lighter components such as gases, gasoline, and kerosene fractions. Cracking is the process of catalytically converting the heavier components (or "residues" from this distillation) of crude oil into lighter more useful components. About 45% of the crude oil reaching a refinery is converted to gasoline.

Ethylene from cracking of the alkane gas mixtures or the naphtha fraction can be directly polymerized or converted into useful monomers. (Alternatively, the ethane fraction in natural gas can also be converted to ethylene for that purpose). These include ethylene oxide (which in turn can be used to make ethylene glycol), vinyl acetate, and vinyl chloride. The same is true of the propylene fraction, which can be converted into vinyl chloride and to ethyl benzene (used to make styrene). The catalytic reformate has a high aromatic fraction, usually referred to as BTX because it is rich in benzene, toluene, and xylene, that provides key raw materials for the synthesis of aromatic polymers. These include p-xylene for polyesters, o-xylene for phthalic anhydride, and benzene for the manufacture of styrene and polystyrene. When coal is used as the feedstock, it can be converted into water gas (carbon monoxide and hydrogen), which can in turn be used as a raw material in monomer synthesis. Alternatively, acetylene derived from the coal via the carbide route can also be used to synthesize the monomers. Commonly used feedstock and a simplified diagram of the possible conversion routes to the common plastics are shown in Figure 2.1.

2.1. COMMON THERMOPLASTICS

The global production of plastics has grown exponentially over recent years and therefore correlates quite well with the population growth. The world production of resins, which was insignificant a mere half a century ago, now stands at about 132 million metric tons annually. Figure 2.2 illustrates this exponential growth in production and shows how it correlates with the increase in global population. If the linear relationship shown in the figure holds in the future, it might be used to estimate the approximate future demand for resins worldwide. For instance assuming a global population of about 10 billion around the year 2100, when it is expected to level off, the extrapolation suggests a annual global resin production close to 500 million tons annually.

In the United States the demand for commodity thermoplastics continually grew in the 1990s (except for a brief pullback in 1995), and, during the recent period of 1995–2000, the consumption of plastics increased at a compound growth rate of 1–5%, depending on the application area. Packaging applications are of particular interest from an environmental standpoint (because of the high

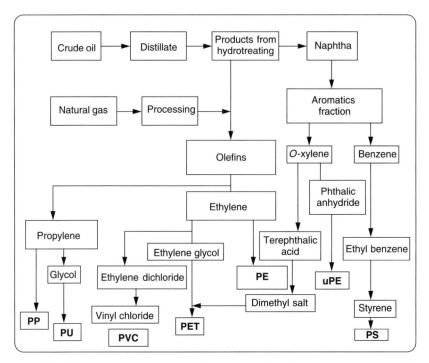

Figure 2.1. Common polymers derived from crude oil and natural gas raw materials.

visibility of packaging in municipal waste and in urban litter) and account for about a quarter of the U.S. demand for thermoplastics.[3] The highest domestic growth rates for plastics in recent years has been in bottle and container applications of HDPE, fiber and closure applications of PP, and in PVC.

As pointed out in Chapter 1, the per capita consumption of plastics in a country correlates well with its affluence. However, large countries in the developing world are rapidly growing key markets for plastics despite their relatively low per capita income, simply because of the high populations. Asian countries, particularly China (including Hong Kong), for instance, are clearly poised to become a major player in the international plastics business of the future. Asia processes about the same volume of plastics as western Europe (but less than the 34% processed in North America). About half of the Asian processing activity is in China. The annual growth rate of the plastics sector in China has been over 10% throughout the last decade [1]! Not surprisingly, China is also a leading producer of plastic processing machinery in the world. Also, about 20% of the production capacity is estimated to be in need of replacement and will result in significant

---

[3] In recent years the U.S. packaging market for thermoplastics was approximately divided up between resin types as follows: HDPE, 31%; LDPE, 12%; LLDPE, 16%; polypropylene (PP), 14%; poly(ethylene terephthalate) (PET), 15%. In the year 2000 the packaging sector in the United States used over 21 million tons of resin.

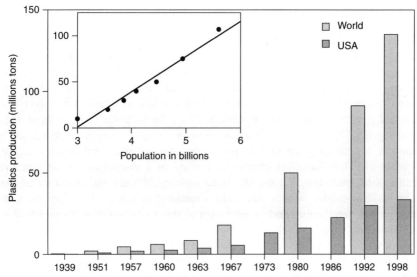

Figure 2.2. World and U.S. production of thermoplastic resins. Most of the data is in long tons with later data in tonnes. Insert shows the relationship between world resin production and the population.

technical upgrades and capacity in the sector.[4] The industry sector is particularly strong in the Pearl River Delta region of Hong Kong, which produces processing machinery as well. China is the fifth largest resin producer in the world, using only a small fraction of the resin domestically (per capita plastics use in China is less than 7 kg per year). Table 2.1 summarizes the plastic industry in different

Table 2.1 Data on Plastics Processing Industry in Asian Regions (Other Than China)

Country	Population (millions)	GDP (Billion $US)	Demand for Plastics (million tonnes)	Employed by Industry
Hong Kong[a]	27.2	220	7	1 million +
Malaysia	21	215	1	85,000
Taiwan	22	362	3	150,000
South Korea	47	580	4[b]	100,000
Thailand	60	369	2[b]	—
Japan	127	3913	9	458,000
Singapore	4	92	0.3	13,000

[a] Data for Hong Kong has been separately compiled although it is a part of China.
[b] 1997 data. Table was compiled from data in reference [1].

---

[4] Recent admission of China into the World Trade Organization (WTO) can also favor the processing industry as the import tariffs on advanced equipment are gradually reduced and export subsidies the industry operates under are removed.

Asian countries; the data is approximate [1] but adequate to yield some idea of the magnitude of the industry.

A comprehensive introduction to common polymers and their manufacture within a single chapter is impractical and is not the present objective. It is only those plastics that are used in high enough volume in common applications that interact significantly and visibly with the environment. This is particularly true of solid-waste related issues where attention has often focused on polystyrene foam and polyolefin packaging materials. Also, the magnitude of environmental impacts generally increases with the worldwide production volume of a material (although the intensity of such impacts may change with individual materials in question). Therefore, this chapter is limited to a discussion of the common thermoplastic materials that are produced in large volume and therefore of particular environmental significance. The treatment is intended to introduce the reader to the manufacture and general uses of these plastics. For the present purpose "common" plastics include the high-volume commodity resins polyethylene, polypropylene, poly(vinyl chloride), polystyrene, and thermoplastic polyester. Also included in the discussion are polymeric foams that are a visible and environmentally interesting group of products because of their link to the use of chemicals that are stratospheric ozone-depleting substances (ODS). The five most commonly used thermoplastics in the U.S. and their consumption levels are shown in Table 2.2.

A fundamental question on plastic materials is their true cost in terms of nonrenewable resource use and energy use, as well as the emission loads associated with their manufacture. This will be discussed in detail in Chapter 3. The energy estimates shown in Table 2.3 were taken from a recent U.S. study carried out by Franklin Associates for the Society of Plastics Industry [2, 3]. The total energy per 1000 lb of plastic products includes the energy content of raw materials (petroleum and natural gas resources), the energy used in manufacture of the resin, processing energy used to fabricate the product, as well as the associated transportation energy. The margin of error in the estimates is about 10%. Plastics compare well with competing materials such as glass, metal, and wood

Table 2.2 U.S. Resin Production in 2000 and Change from Previous Year

Plastic Resin	Year 2000 Production (in millions lb)	Percent Change from 1999
HDPE	13,968	+0.8
LLDPE	7,951	−1.9
LDPE	7,575	−1.6
PP	15,739	+1.6
PVC	14,442	−3.2
PET	7,029	−4.4
PS	6,844	−3.3

Table 2.3 Energy Expenditure in Producing 1000 lb of Plastic Product

Product	Raw Materials (%)	Processing (%)	Transportation (%)	Total (1000 Btu)
HDPE	61.6	36.2	2.2	37,650
LDPE	56.4	41.6	2.0	41,136
LLDPE	61.6	36.2	2.2	37,650
PP	60.5	37.4	2.1	38,353
PVC	36.0	60.8	3.2	30,539
PET	50.4	46.6	3.0	38234
PS	61.8	34.7	3.5	39,796
PS—E*	59.8	36.9	3.5	43,101
HIPS	63.7	34.2	2.1	40,333

Based on data from references [2, 3].
*Expanded polystyrene foam products.

in terms of both the performance in selected applications and in environmental compatibility.

2.2. POLYETHYLENES

Polyethylenes, the most widely used class of plastics in the world, include several copolymers of ethylene in addition to the homopolymer.[5] The polyethylene homopolymer has the simplest chemical structure of any polymer.

$$\sim\!\!-CH_2-CH_2-CH_2-CH_2-CH_2-CH_2-CH_2-CH_2-\!\!\sim$$

The commercially available resins, however, have far more complicated structures with branched chains and semicrystalline morphologies not indicated in this simple representation. Depending on their copolymer composition and the polymerization process used, commercial polyethylenes display a wide range of average molecular weights, molecular weight distributions (polydispersity), and chain branching in the resin. These molecular parameters affect the ability of the macromolecules to pack closely into a dense matrix and also control the extent of crystallinity in the material. Because of their semicrystalline nature, polyethylenes do not display their theoretical density of 1.00 g/cm^3 (or the theoretically expected melting point of about 135°C) but show a surprisingly wide range of physical properties, Based on these, particularly the bulk density, the resins are divided into three basic types.

- Low-density polyethylenes (LDPE)
- High-density polyethylenes (HDPE)
- Linear low-density polyethylene (LLDPE)

---

[5] The first synthesis of polyethylene was from diazomethane, but its commercial preparation in the 1930s was via the polymerization of the ethylene monomer.

84 COMMON PLASTICS MATERIALS

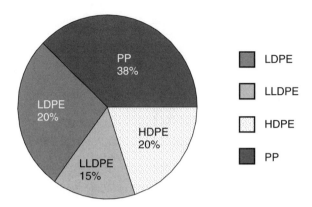

Figure 2.3. Market share of polyolefins in 2000.

The approximate U.S. market share for each type of polyethylene (and for polypropylene) is indicated in Figure 2.3

High-density polyethylene has the simplest structure and is essentially made of long virtually unbranched chains of polymer (somewhat representative of the simple structure shown above). These chains are able to align and pack easily; HDPE therefore has the highest degree of crystallinity in a polyethylene. Its molecular weight is high enough (and the chain branching minimal) to obtain a degree of crystallinity as high as 70–95% (and a correspondingly high density in the range of 0.941–0.965 g/cm^3). Low-density polyethylene on the other hand has extensive chain branching in its structure. Both long- and short-chain branching are usually present, and this results in a comparatively lower material density of 0.910–0.930 g/cm^3 and a crystallinity of only 40–60%. LDPE has a melting temperature range of 110–115°C. The amount of crystallinity and the melt temperature of the resin can even be further reduced by incorporating a small amount of a suitable co-monomer. When the branches on polymer chain are mostly short chains, a linear low-density polyethylene with a density range of 0.915–0.940 g/cm^3 and a higher degree of crystallinity of 40–60% is obtained. Figure 2.4 shows the general relationship between the crystallinity, density, and the extent of chain branching in polyethylenes, and Figure 2.5 shows a schematic of the nature of chain branching in the three varieties of polyethylene.

Since its introduction in 1968 the LLDPE resin has been extensively used in packaging films, particularly in products such as grocery bags and garbage sacks where high clarity is particularly not important. LLDPE with short branches yield exceptional strength and toughness; LDPE packaging film can often be replaced with an LLDPE film of only about a third of the thickness. Given the cost effectiveness of LLDPE, it is likely to be used increasingly (mainly at the expense of LDPE) in future packaging applications.

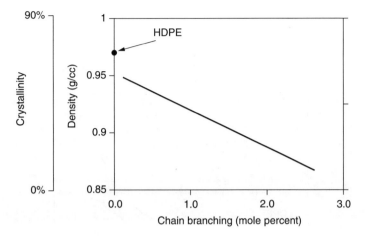

Figure 2.4. Relationship between the crystallinity, density, and extent of chain branching.

Already, a second generation of LLDPE resins based on metallocene catalysts (sometimes referred to as mLLDPE resins) is becoming available. This new catalyst system allows the production of LLDPE as well as very low-density polyethylene (VLDPE) resins of exceptionally narrow molecular weight distribution, with controlled amounts of short-chain branching. The microstructure and therefore the mechanical properties of mLLDPE resins differ substantially from those of LLDPE made using conventional Zieglar–Natta type of catalysts. The polydispersity of the mLLDPE resins approach the most probable value of 2.0 (compared to the value of 20–50 typically obtained for LDPE and 4–15 for HDPE). The narrower distribution of molecular weights in a resin generally leads to better mechanical properties such as increased toughness or hardness and better low-temperature impact strength. However, the same features also makes it difficult to melt and process the resin. Introducing low levels of certain co-monomers as well as controlled amounts of chain branching can often alleviate some of these difficulties.

The amount of chain branching in polyethylene is an important determinant of the properties of the resin. For instance, important material properties such as the hardness, or the initial tensile modulus, decrease almost linearly as the extent of chain branching increases. The presence of branch points (as well as co-monomers) in the chain reduces the likelihood of crystallite formation at those points, reducing the overall crystallinity as well as density of the resin. The different grades of polyethylenes and its common copolymers used in popular applications and their general characteristics are listed in Table 2.4.

Recent market studies indicate very good prospects for polyolefins in general (and polypropylene in particular) in the near term. Of the polyethylenes LLDPE is presently growing at the fastest rate while HDPE is continuing to compete

86 COMMON PLASTICS MATERIALS

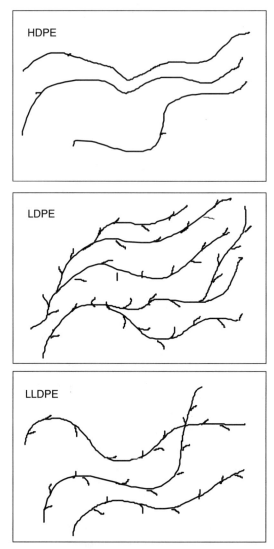

Figure 2.5. Schematic representation of levels of chain branching encountered in different types of polyethylenes.

effectively with vinyl in the pipe market [4]. It is also used extensively in bottles, containers, drums, and houseware applications where its high modulus and very low permeability is particularly desirable. The high-density variety has the highest stiffness among polyethylenes but molded parts and thick films are opaque compared to the translucent LDPE products. The latter grade therefore finds a considerable market in films for packaging (including plastic-coated paperboard packages) and agricultural use as mulch films, greenhouse films and irrigation pipes. These uses represent about two thirds of the U.S. market for

Table 2.4 Different Types of Polyethylenes Used in Packaging Applications and Their Properties

Type	Comonomer	Density[a]	Melt Index	P[b]	Applications
LDPE	None	0.919–0.923	0.2–0.8	B	Heavy duty sacs
		0.922–0.925	1.5–2.0	B	Bread, bakery and general-purpose bags
		0.930–0.935	1.5–2.0	B/C	Overwrap
		0.918–0.924	1.2–2.0	C	General-purpose packaging
		0.917–0.930	5.0–10.0	C	Extrusion coating
LDPE	2–5% VA[c]	0.925–0.930	1.5–2.0	B	Frozen food
	3–5% VA	0.923–0.927	0.2–0.4	B	Ice bags
	7% VA	0.927–0.945	0.2–0.4	B	Liquid packaging and extrusion coating
	15–18% VA	0.927–0.940	2.0–6.0	B	Disposable gloves, card stock, extrusion coating
	None	0.924–0.926	6–10	IM	Bottles and closures
		0.914–0.918	15–25	IM,BM	Bottles for pharmaceuticals and aseptic packaging
LLDPE	Butene	0.917–0.922	0.8–2.5	B	General-purpose packaging
	Hexene, octene	0.912–0.919	2.0–4.0	C	Stretch wrap
		0.928–0.935	2.0–5.0	C	Bread, bakery and overwrap
		0.924–0.928	0.7–1.5	B	Grocery sacks
		0.917–0.923	0.8–1.5	B	Ice bags
	None	0.918–0.922	0.8–1.2	IM,BM	Drum liners
		0.920–0.926	12–30	IM	Industrial containers
		0.926–0.935	50–150	IM	Closures
HDPE	None	0.960	0.35	BM	Milk jugs, bottles, pails, and containers
		0.940	0.2	B	General purpose
		0.960	30	IM	Food containers

[a] Density is in g/cm.
[b] P = processing technique; B, blown; C, cast; BM, blow molding; IM, injection molding.
Based on references [6, 33].
[c] VA = vinyl acetate.

LDPE. The trash-bag market is presently its largest single market, but LLDPE is gaining market share in this application because of its superior strength and tear/puncture resistance. The use of LLDPE in grocery bags has been an important development. Unlike with LDPE the thinner and stronger LLDPE bags are cost competitive with paper bags and can be manufactured with handles [5]. This has lead to their recent growth in that market segment (which also includes products such as garment bags and ice sacks) as well as in the multiwall paper sacks market.

Blow-molded products account for over 40% of the HDPE consumption with the milk and juice jugs being the primary packaging product. Only about 30% of the resin is used in injection-molded products such as food containers and crates, while film applications account for about 7% of the HDPE consumption [6].

2.2.1. Manufacture of Polyethylenes

Ethylene for the manufacture of polyethylene is derived from cracking various components of petroleum oil such as the gasoline fraction, gas oil, or from hydrocarbons such as ethane. While petroleum remains the predominant source of the monomer at the present time, it can also be produced using biomass. In fact ethylene has been commercially derived from molasses, a by-product of sugar cane industry, via the dehydration of ethanol.

The polymerization of ethylene might be carried out in solution or in a slurry process. But these processes are complicated by the need for a separation step to isolate the resin product from solution. The newer installations favor the gas-phase process that can produce both the low- and high-density resins. Older plants lack this versatility and are able to produce only either the high-density or the low-density type of polyethylene.

In the older process, LDPE resin was produced under high pressure (15,000–22,500 psi at 100°C–300°C) in stirred autoclave or tubular-type reactors, where the liquefied ethylene gas is polymerized via a free radical reaction initiated by peroxide or by oxygen.

$$CH_2=CH_2 \longrightarrow -[-CH_2-CH_2-]_n-$$

The reaction is highly exothermic (22 kcal/mol) and therefore requires careful control of the temperature, especially in autoclave reactors. The product generally has a high level of long chain branching from chain transfer to polymer. Short-chain branches are methyl or alkyl groups formed by the active growing chain end abstracting a hydrogen atom from another part of the chain via "back-biting" reactions. As illustrated below, the back-biting process involves hydrogen abstraction by the growing macroradical chain to yield a short (four-carbon) branch on the main chain. Such transfer of the macroradical can also occur with a separate polymer chain in the mix, yielding to longer branches typically found in high-pressure polyethylenes. The tubular reactors yielded a product with a relatively higher amount of chain branching compared to the resin made in autoclave reactors. The molecular weight of the product is determined by the reaction conditions employed:

$$\sim CH_2-CH_2\overset{CH_2}{\underset{\bullet CH_2}{\diagdown}}\overset{}{\underset{CH_2}{\diagup}}CH_2 \quad \longrightarrow \quad \sim CH_2-\overset{\bullet}{C}H\overset{CH_2}{\underset{CH_3}{\diagdown}}\overset{}{\underset{}{\diagup}}CH_2$$

$$\sim CH_2-CH-CH_2-\overset{\bullet}{C}H_2$$
$$\underset{\underset{CH_3}{|}}{\underset{(CH_2)_3}{|}}$$

The high-pressure process relied on large and complex plants that required careful process control. Therefore, the discovery in 1953 of the appropriate catalysts that allowed the process to be carried under low pressure ($\sim$500 psi) was welcomed by the industry [7]. Three types of catalysts were developed about that time: the Ziegler-type catalysts typically obtained by reacting alkyl aluminum compounds with titanium chloride[6]; metal oxide catalyst systems, developed by Phillips Petroleum in the United States, typically made of chromium oxide supported on a silicaceous carrier [8]) and a different type of oxide catalyst developed by Standard Oil Company. The first plants based on the Ziegler catalyst went on line in Germany by 1955 and a plant based on the Phillips catalyst in Texas opened in 1957. The third catalyst system developed much slower and was picked up by the Japanese plastics industry in a plant opened in 1961.

The first LLDPE resin was marketed by Du Pont in 1960 followed shortly thereafter by Union Carbide. The polymerization was effected by organo-metallic catalysts (Ziegler–Natta catalysts) and carried out at a relatively low pressure. Moderate levels of α-olefin co-monomers are used to control the density of the resin product. The inclusion of suitable co-monomers in the polymer chains disrupts crystallinity and reduces the density and rigidity of the resin. Despite its superior properties the LLDPE leaves much to be desired at the molecular level. The co-monomer is not homogeneously distributed across the different molecular weights and the molecular weight distribution is rather broad (narrower distributions are believed to yield better properties). Efforts to produce a well-defined narrow molecular weight distribution recently led to the development of yet another catalyst system, a class of single-site catalysts called metallocenes. The catalyst is an organometallic compound of a group IV metal atom attached to two cyclopetadienyl groups and two alkyl halides (or to methyl alumoxane moieties).

---

[6] $R_2AlCl + TiCl_4 \longrightarrow RAlCl_2 + RTiCl_3$

Gas-phase polymerization represents an important advance in the manufacturing technology for polyolefins. In the Unipol (gas-phase) process ethylene and any comonomers (usually other olefins such as oct-1-ene, hex-1-ene) are fed continuously into a fluidized-bed reactor at a pressure of about 0.7–2.0. MPa and at a temperature of less than 100°C. The catalyst is added directly into the bed. The gas-phase process is not constrained by problems associated with pumping or handling viscous polymer solutions and the solubility of the resin product. The solid polyethylene is directly removed from the reactor with any residual monomer being purged and returned to the bed. The gas-phase reactors are able to take advantage of the new metallocene catalysts with little engineering modification. A schematic diagram of a Unipol-type reactor is shown in Figure 2.6.

Most importantly the development of gas-phase processes allowed the energy-intensive high-pressure polymerization process to be replaced by the far more energy-efficient low-pressure process. The gas-phase processes offer the greatest versatility of products in terms of resin density and melt index of polyethylenes. From an environmental standpoint the gas-phase reaction is of particular interest and offers several advantages over the conventional technology as recently summarized by Joyce [9].

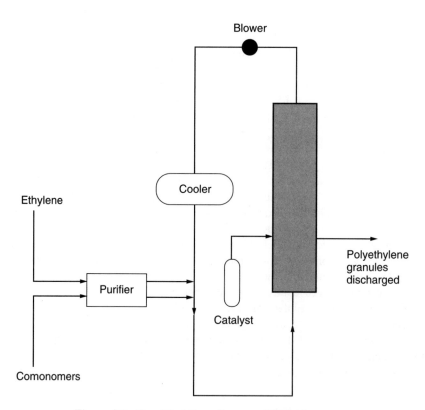

Figure 2.6. Simplified flow diagram of Unipol process.

1. The gas-phase polyethylene plants[7] introduced by Union Carbide in 1968 require only about half the capital investment of the high-pressure plants and far fewer workers to operate them. The wastage of ethylene during the polymerization process is also lower in the gas-phase process, reducing cost of the product and conserving the monomer resources.
2. The energy costs of solvent recovery associated with the older processes are saved in the gas-phase process. By the mid-1970s the process refinements had reduced the energy costs of polyethylene manufacture by about a half, compared to that of the conventional high-pressure processes. Figure 2.7, based on reference [9], illustrates the record of the polyethylene industry over the years in energy conservation via process improvement.
3. A major advantage in moving away from the solvent and slurry processes is the elimination of the emission problems associated with the use of hexane and butane solvent (or suspension agents in the case of slurries). Minimum release of volatile organic compounds (VOCs) into the atmosphere is desirable during manufacture.
4. The number of workers needed to operate a modern polyethylene plant is minimal. These labor savings do translate into very real energy and resource savings, particularly in a manufacturing setting. As opposed to the 84 individuals needed to produce a million pounds of the resin in the days the industry started in the 1940s, only a single operator in a modern plant can produce the same volume of even a better grade of resin, today.

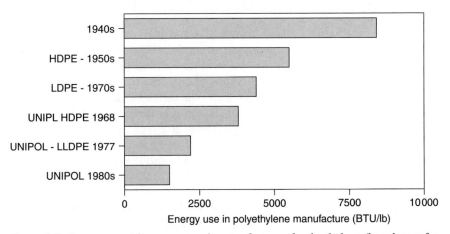

Figure 2.7. Improvement in energy use in manufacture of polyethylene (based on reference [9]).

---

[7] The Unipol process was first introduced by Union Carbide in its Seadrift, Texas, facility in 1968. Other companies such as Amoco and British Petroleum developed the technology further. Today over a 100 reactors are in operation (or in construction) worldwide with an annual capacity of 19 billion pounds of HDPE or LLDPE resin.

COMMON PLASTICS MATERIALS

Table 2.5 Low-Pressure Production Processes for Polyethylene [9]

Company	Process Type	Reactor	Solvent/ Diluent	Residence time (h)	Conversion (%)	Polydispersity Index (Typically)
Solvay	Slurry	Loop	C_6H_{14}	2.5	97	Broad
Philips	Slurry	Loop	C_4H_8	1.5–2.0	98	Broad
Hoechst	Slurry	CSTR[a]	C_6H_{14}	2.5	95–98	Broad
Stanicarbon	Solution	CSTR[a]	C_6H_{14}	0.15	95	Narrow
Union Carbide	Solution	Fluidized Bed	Gas	3–5	99	Broad
BASF	Gas	Stirred Bed	Gas	4.2	99[b]	Broad
Amoco	Gas	Stirred Bed	Gas	—	99[b]	Broad

[a]Continuously stirred tubular reactors.
[b]estimated value.
Reproduced with permission from *Trends in Polymer Science* [10].

Table 2.5 [10] compares characteristics of the low-pressure conventional liquid slurry and solution processes with the more recent gas-phase production processes.

2.2.2. Environmental Aspects of Production

The potential for fugitive emissions from the manufacture of polyethylenes depend on the particular type of process used. In addition to the possibility of monomer (ethylene) leakage common to all processing plants, possible diluent and co-monomer emissions into the atmosphere need to be closely controlled. However, no major point sources of these emissions are apparent in the manufacture, and in any event the monomer and co-monomer feed materials are relatively nontoxic (acting only as asphyxiants at a high enough concentration). The possible exposure of workers to some of the solvents and diluents, however, is more of a concern as these can act as central nervous system (CNS) depressants at a high enough concentration. The emissions from polyethylene manufacture were estimated at about 13 for HDPE and 1–18 for LDPE (kg of VOC/Mg of plastic resin produced) [11]. Most of the fugitive emissions was monomer, along with small amounts of lower hydrocarbons such as ethane and isobutane. Some particulate emissions also occur during polyethylene production, and a need exists for an audit and an analysis of the health impacts (if any) of these. With improvements in process efficiency, the amount of fugitive emissions (VOC and particulates) from a typical plant have decreased, but the volume of resin produced continues to increase with time, requiring frequent assessment and documentation of emissions. A known toxic material used in the manufacturing process is the chromium(IV) oxide-based catalyst. In high enough doses it can be mutagenic, cause liver damage, affect the CNS, and even be carcinogenic [12]. The recommended 8-h TLV (ACGIH) for the material[8] is 0.05 mg/m^3.

2.3. POLYPROPYLENE

Since the mid-1950s when Ziegler–Natta catalysts that yielded high-molecular-weight resin were discovered, the demand for polypropylene has increased dramatically. The conventional catalyst systems used in its manufacture are quite versatile and can be adapted to produce the homopolymer, block copolymers, and random copolymers with other monomers. With increased production, polypropylene has become cost effective to a point where it could be used in a variety of low-value packaging applications such as containers, closures, biaxially oriented film, and tape. The United States presently accounts for about 25% of the global production capacity of the resin. The recently developed metallocene catalysts have been used for the polymerization of propylene as well. This novel route allows close control of the degree of stereoregularity of the resin and the copolymerization of the propylene with a much wider selection of co-monomers (compared to conventional catalyst systems). Metallocene grades of polypropylene (mPP), including syndiotactic resins, are already available commercially but still constitute only a small fraction of the total production.

Polypropylene is typically manufactured by the direct polymerization of propylene in a low-pressure process employing Ziegler–Natta catalyst systems (typically aluminum alkyls and titanium halides with optional ether, ester, or silane activators). The process can be carried out in liquid or slurry in conventional manufacturing or in the newer gas-phase stirred-bed or fluidized-bed reactors. The polymerization generally yields an isotactic index (generally measured as the percent insolubles in heptane) of 85–99. The isotactic form of the polymer with a high degree of crystallinity (40–60%) is preferred for most practical applications. Isotactic polypropylene (iPP), the principal type used by the polymer processing industry, has a density of about 0.92–0.94 g/cm^3. The weight-average molecular weight of polypropylene from these processes is in the range of 300,000–600,000 with a polydispersity index of about 2–6 [13]. Some atactic polypropylene results as a by-product[9] of the process and has found limited practical use [14]. The atactic form is mostly amorphous and has a density of only about 0.85–0.90 g/cm^3. Small amounts of the syndiotactic form of polypropylene (where the methyl groups on repeat units are located on alternate sides of the chain on adjacent

---

[8] ACGIH TLV: The threshold limit value, averaged over an 8-h workday, determined by a private professional group, the American Conference of Governmental Industrial Hygienists. The Occupational Safety and Health Administration (OSHA) adopted many of ACGIH's recommended limits in the 1970s, but since then, ACGIH has revised some of the standards, and some of these may be different, often less stringent, compared to OSHA requirements.

[9] Atactic (amorphous) polypropylene can be directly synthesized as well. Resins in low- to moderate-molecular weight resins are commercially available (Eastman Chemical Company) for hot-melt adhesive and other applications.

units) are made commercially using the single-site metallocene catalyst and are being evaluated in various applications. The syndiotactic resin has lower crystallinity (30–40%) and are softer, tougher, stronger (higher impact strength and elastic modulus), and relatively more transparent than the isotactic resin.

As with the polyethylenes, about a quarter of the polypropylene is manufactured as copolymers, especially with ethylene. The presence of ethylene units at a level as low as 0.5–3% by weight serves to break up the crystallinity in polypropylene matrix and leads to improved flexibility and clarity. A random copolymer that contains about 1–7% ethylene yields a lower melting, optically clearer material with improved flexibility. Incorporating larger amounts of the ethylene co-monomer gives a heterophasic impact-grade copolymer. Mechanical blending of polyethylene or ethylene–propylene rubber with polypropylene can also yield a comparable high-impact polymer. These can have a higher overall ethylene content of up to 15 wt % as the random copolymer in this case is made with much higher levels of ethylene comonomer. As the name implies, the resin has excellent impact characteristics but a somewhat lower flexural modulus.

As the demand for polypropylene continues to create a large supply of low-cost resin, there is a potential for the resin to be used in applications presently using other types of commodity thermoplastics. Its replacement of polyethylene, however, is unlikely except in applications that demand high-temperature resistance where the use of iPP and its copolymers might be advantageous. With rigid PVC products the chance of such substitution appears to be reasonable except in applications that require routine exposure of the material outdoors. The relatively poor weathering resistance of iPP, despite its comparable mechanical properties (in filled or fiber-reinforced compounds) to that of PVC, limits its use outdoors. However, with rapidly improving polypropylene technology and better compounding techniques, such substitutions and even the replacement of some expensive engineering thermoplastics with polypropylenes might become feasible [15].

2.3.1. Manufacture of Polypropylene

Propylene monomer is produced by catalytic cracking of petroleum fractions or the steam cracking of hydrocarbons during the production of ethylene. Conventional processes in liquid phase and in slurry use stirred reactors and a diluent such as naphtha, hexane, or heptane. The reaction takes place typically at a temperature of about 60–80°C and at 0.5–1.5 MPa, and the final product is obtained as a solid suspension of polypropylene in the liquid phase. Isolation of the resin requires a separation step (such as centrifugation) followed by washing the resin free of residual diluent and drying.

The manufacturing process for polypropylene has undergone many changes since 1957 when the first facility went on stream. In the 1960s the Novolon gas-phase process and the Phillips process for polymerizing liquid propylene were introduced. These processes had the advantage of not using any diluents, but they generally suffer from relatively poorer catalyst performance and some limitations

Table 2.6 Comparison of Low-Pressure Production Processes for Polypropylene

Company	Reaction Phase	Reactor	Solvent	Residence Time (h)	Molecular Weight Distribution	Isotactic Index (%)
Montedison, Mitsui	Liquid slurry	CSTR	Hexane or heptane	20–40	Narrow	98–99
Amoco Shell, Himont	Bulk liquid	CSTR	None	60	Broad	96–98
Union Carbide, BASF	Gas	Stirred bed	None	—	Broad	85–95

Reproduced with permission from *Trends in Polymer Science* [10].

on the stereoregularity of resins. In 1975 with the introduction of improved third-generation catalysts that facilitate the reaction at the same temperature but at the slightly higher pressure of 2.5–3.5 MPa, both optimum yield and stereoregularity could be achieved. These catalysts introduced, by Montedison and Mitsui, could be used with liquid monomer systems in the new gas-phase reactors. The latter technology modeled after the already successful Unipol polyethylene process went on stream in 1985. The flow diagram for the process for polyethylene production, shown in Figure 2.6, applies equally well for polypropylene. Polymerization-grade propylene (usually at a purity of at least 99.8%) is used in place of the ethylene monomer feed. A suitable co-monomer (usually ethylene) is also generally used. Most of the advantages of gas-phase processes cited for polyethylene also apply to the production of polypropylene. Cavanaugh et al. have summarized the characteristics of the different manufacturing processes [10] (see Table 2.6).

Over a third of the polypropylene produced in the United States is ultimately processed into useful products by injection molding. A wide range of resins spanning a melt flow index (MFI) range of 2 to >70 g/10 min is available for the purpose. In North American markets a majority of the polypropylene is injection molded into products or spun into fibers for use in various textile applications. The latter includes sacks made of woven polypropylene strips cut from oriented sheets used for packaging agricultural products. The common molded products include closures, containers, bottles, jars, and crates. A relatively small fraction of the polypropylene (about a tenth) is extruded into film. In applications involving low-temperature use (as with refrigerated packages), the copolymers are preferred over the homopolymer. Both biaxially oriented film (BOPP) and nonoriented packaging films of polypropylene are used in food packaging. The former is used as a barrier film, usually with a surface coating. Nonoriented films are used in general-purpose applications such as apparel bags, bandages, diaper linings, and in sanitary products. Blow molding of polypropylene is also common, and is used in the production of bottles and containers.

2.4. POLY(VINYL CHLORIDE)

Poly(vinyl chloride) (PVC), the second widely used resin in the world (after polyethylene) is made by the polymerization of vinyl chloride monomer (VCM). In theory the chemical structure of the polymer is simple, consisting of the same structure as for polyethylene with one hydrogen in every other —CH$_2$— group being replaced by a chlorine atom

$$-\!\!\sim\!\!CH_2-CHCl-CH_2-CHCl-CH_2-CHCL\sim$$

However, as the repeat unit is asymmetrical because of the presence of only a single chlorine atom, two types of linkages, head to tail and head to head, are possible:

$\sim$CH$_2$—CHCl—CH$_2$CHCl$\sim$ $\sim$CH$_2$—CHCl—CHCl—CH$_2$$\sim$
    Head to tail   Head to head

In general, however, the head-to-tail linkages are predominant (nearly 90%) in the resin. The weight-average molecular weight $\overline{M}_w$ of commercial PVC resins ranges from about 100,000–200,000 and the polydispersity index is about 2.0. The resin has a glass transition temperature of 75–85°C and a crystalline melting point of 120–210°C. The crystallinity in PVC is due to syndiotactic sequences in the polymer and amount to about 7–20% in commercial resins. Resins with higher levels of crystallinity can be obtained by polymerization under specific conditions.

The polymer is susceptible to both photo- and thermal degradation; and, for products intended for outdoor use, the resin has to be compounded with light stabilizers. Such formulations typically contain other additives (such as a thermal stabilizer package to protect the resin during processing), fillers, and lubricants. The compounds not containing any plasticizers or the rigid PVC materials (also referred to as uPVC) are used extensively in building products such as pipes, fittings, siding, window frames, and rainwater products. In unplasticized formulations of PVC intended for outdoor use, an opacifier, usually rutile titania, that effectively absorbs the damaging ultraviolet (UV-B) radiation is incorporated in the formulation to protect the surface from UV-induced degradation. The solar ultraviolet component, particularly the UV-B radiation (290–315 nm) causes light-induced dehydrochlorination and, to a lesser extent, oxidation reactions in inadequately protected formulations. This leads to uneven yellowing discoloration of white vinyl surfaces, "chalking," or the release of titania pigment from the binder at the exposed surface of the material, and loss of impact strength of the product on prolonged exposure. In the United States PVC is used predominantly in the building sector with nearly half the resin sold going into pipe, fitting, and conduit, followed by residential siding.

PVC resin can also be made into a versatile soft pliable rubbery material by incorporating plasticizers such as organic phthalates into the compound. Plasticized PVC (also referred to as pPVC) is used widely as packaging film, roofing membranes, belting, hoses, and cable covering. With pPVC, calendering is employed to produce films and sheets. The resin is also used as a coating on paper or fabric and is made into numerous household products. A small amount of the plasticized film is used in packaging, for instance, in meat wraps where it is approved by the Food and Drug Administration (FDA) for food-contact use. Of the commodity thermoplastics, PVC is the most versatile resin in terms of both the processability and the range of applications.

2.4.1. Manufacture of Poly(vinyl chloride)

The polymerization process was known as far back as the 1930s, but the technology for bulk polymerization of high-quality PVC emerged only in the 1960s. The vinyl chloride monomer (VCM) is made by pyrolysing 1,2-dichloroethylene, which in turn is manufactured by reacting ethylene (from cracking of either petroleum or natural gas) with chlorine from the electrolysis of brine. The hydrogen chloride by-product formed in this dehydrochlorination process is used to produce more of the 1,2-dichloroethylene via the oxychlorination of ethylene in the presence of Cu(II) chloride catalyst. PVC manufacture is the key industrial consumer of chlorine, accounting for about 42% of the production (bleaching of pulp in paper production by comparison accounts for only about 5%).

$$CH_2=CH_2 + Cl_2 \longrightarrow CHCl-CHCl + HCl$$
$$CH_2=CH_2 + 4\ HCl + O_2 \longrightarrow 2\ CHCl-CHCl + H_2O$$

PVC can be made by either bulk polymerization, suspension polymerization, or emulsion polymerization of the VCM. Most of the resin (over 90%) is now[10] made by free-radical-initiated suspension polymerization of VCM. The reactor used for emulsion polymerization of the monomer is also about the same as that for suspension polymerization except that stripping is usually carried out under vacuum. Polymerization of VCM is an exothermic reaction (+410 cal/g) and removal of heat from the system is an important consideration in large-scale manufacture. Controlling the temperature of the reaction is important as it dictates the average molecular weight and the polydispersity of the resin product. This is relatively easier to achieve in suspension polymerization compared to bulk polymerization processes because the former is carried out in a water medium.

In suspension polymerization the vinyl chloride monomer is dispersed in water using a protective colloid or a surfactant to control the final particle size (usually between 130 and 165 μm) and a monomer-soluble initiator (usually an azo compounds or a peroxide) is used. Gelatin, soaps, glycols, and pentaerythritol or their mixtures can be used as dispersing agents in the reaction mixture. The polymerization is usually carried out in a glass-lined reactor with controlled agitation, at a

---

[10] The first process used to polymerize vinyl chloride was a heterogeneous bulk polymerization process where the monomer and initiator were reacted in a rotating cylindrical reactor with steel balls used to remove heat away from the product (the tumbling also served to ground up the resin product).

temperature of 50–75°C and at a pressure of about 0.7 MPa. Oxygen is usually excluded from the reaction vessel to prevent interference with the free-radical polymerization reaction. Vinyl chloride monomer is volatile with a boiling point of −13.4°C and is a hazardous air pollutant. (The reactants are usually maintained under pressure during the process to keep the VCM in a liquid state). Special attention is therefore paid during manufacture to ensure that workers are not exposed to the monomer vapor. For instance, overpressure protection devices to prevent the emission of monomer into the atmosphere control the relief valve emissions from monomer storage tanks. The resin product as it comes off the reactor contains residual monomer at levels that can be as high as 3% by weight and is cleaned by stripping. Steam stripping is typically carried out in a continuous stripping column with steam forced up a column down which the resin slurry flows from the top. The tower is maintained at a temperature above the glass transition temperature T_g, of the resin, to facilitate faster stripping. Resin leaving the column has only parts-per-million levels of the monomer, and most of this is also removed during subsequent drying of the resin in rotary driers. The monomer separated is extracted from the output steam from the column in a monomer recovery operation. After this rigorous cleaning process the residual monomer content of commercial resin is less than 1 ppm. Figure 2.8 shows a schematic representation of the manufacturing process.

2.4.2. Environmental and Safety Issues on PVC

Over the years, two major controversial issues have been associated with large-volume use of PVC plastics. The first has to do with the fate and health impacts of

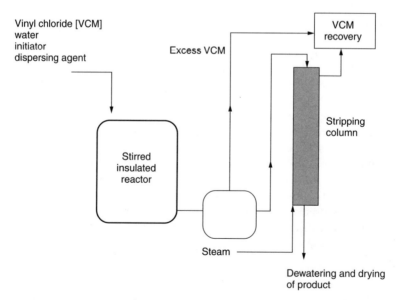

Figure 2.8. Flow diagram for suspension or emulsion polymerization of vinyl chloride.

additives used in vinyl products, particularly the phthalates used as plasticizers and the heavy metal compounds used as thermal stabilizers. Potential toxicity and the environmental consequences of phthalate released from plasticized PVC products have been debated for decades. The second concern is the potential production of dioxins and furans on the incineration of PVC in mixed municipal solid waste streams. These issues have been controversial, with numerous studies that seem to both refute and support these claims being published in the literature.

a) Plasticizers Phthalates, such as the commonly used di-2-ethylhexyl phthalate (DEPH) and diisononyl phthalate (DINP), are used at a 10–50 wt % in soft flexible PVC compositions used in products such as flexible roofing membranes or vinyl sheets. These plasticizers tend to migrate slowly to the surface of the product and can therefore enter the environment or come into human contact. Common plasticizers are indeed found in low levels dispersed in the environment[11] in most parts of the world [16] and generally believed to be even ingested routinely along with food. The low acute and chronic toxicity of these common plasticizers is well established [17, 18]. Based on pronounced biological effects caused by large doses of the plasticizer in animals, various human diseases have been attributed to this low-level intake of plasticizers [19]. The European Commission in 1999 banned the use of DEPH in soft toys used by young (under 3 years) children, encouraging voluntary restrictions on the use of the plasticizer by the manufacturers [20].

The research literature on the topic shows phthalates are not genotoxic. The observation that they cause liver cancer in laboratory rats, however, is well established, but this particular carcinogenic mechanism is not expected to be operative in humans. Authorities such as the World Health Organization's International Agency on Research on Cancer and the expert panel on American Council on Science and Health in the United States [21] both found no evidence of DEHP causing cancer in human populations. PVC is used in medical devices such as blood bags, infusion systems, and in hemodialysis equipment, and it is likely that those using these devices do receive relatively higher doses of the dissolved plasticizer into their bodies. Therefore, the issue of toxicity of common plasticizers is an important one with practical health implications. The prevailing expert opinion in the United States, however, is that the use of plasticized PVC in medical, toy, and other applications are quite safe [19]. Another more recent health concern is endocrine disruption[12] by chemicals, and plasticizers are included in the class of relevant chemical agents. (Also included in the category of endocrine disrupters are compounds that occur in nature, specially the phytoestrogens in plant species.) The contributing effect of the PVC plasticizers to the bioavailable environmental

---

[11] Phthalates do undergo breakdown in the environment via biodegradation (both by aerobic and anaerobic pathways), hydrolysis, and photodegradation.

[12] As the name suggests, these chemicals cause changes in the endocrine system of humans and animals. The specific human effects include low sperm counts, incidence of undescended testicles, and increased susceptibility to testicular and prostrate cancer.

pool of endocrine disruptors is unknown and its relevance to human health has not been established at this time [17, 22].

b) Stabilizer Depending on the region of the world, commonly used photothermal stabilizers in PVC include compounds of lead, tin, barium, zinc and cadmium. Toxicity of all lead compounds causing neurological effects in children, kidney damage, sterility, and even cancer is well known. Cadmium compounds and organotin compounds are also toxic, affecting the nervous system and the kidney.

These metal compounds leaching out of PVC into the environment in landfill situations is a serious potential concern. However, in landfills (as well as in the potable water systems that use PVC pipes), these compounds are for the most part locked in the rigid plastic matrix. Unlike in plasticized systems, no significant rates of migration of these chemicals is unlikely. Consequently, release of heavy metals in any significant quantities into landfill leachate, sewer environments, or water distribution networks is not anticipated [23].

The same is not true, however, for fires that involve PVC materials [24] or for uncontrolled incineration of PVC waste. While the plastic is not particularly flammable in a fire due to its high halogen content (among common polymers, PVC ranks among the best in terms of time to ignition), considerable amounts of the metal compounds can be released [24] into the atmosphere. Research aimed at replacing these metal-based stabilizers with alternatives that are safer in fires is valuable in responding to this real problem associated with plastic fires.

c) Dioxins Dioxins and furans [polychlorinated dibenzo-*p*-dioxins (PCDD) and polychlorinated dibenzofurans (PCDF)] are a class of compounds where some of the members are highly toxic environmental pollutants.[13] They are suspected human carcinogens and may alter endocrine and immune functions. Dioxin was the primary toxic component of Agent Orange (and was also released at Love Canal in Niagara Falls, New York, and at Seveso, Italy.). Some of these compounds occur in nature while others are anthropogenic, most commonly produced during the combustion of materials. However, improvements in incineration processes and reductions in emissions have resulted in a steady decrease of PCDD/PCDF levels in the environment as well as in human milk [25]. Most dioxins enter the human system via food (particularly the meat and dairy products). However, the extent to which these compounds are toxic to human beings (as opposed to animal models for which most data is available) is not entirely clear. Different mammalian test species show widely different susceptibility to dioxins. Studies by Ames et al. [26] concluded that despite the potent toxicity dioxins posed to rodents it was unlikely to be a significant human carcinogen at the levels to which the population is typically exposed.

---

[13] Of the 210 compounds that comprise the class PCDDs and PCDFs, 75 PCDDs and 135 PCDFs are found in the environment as a result of combustion and natural processes. The most toxic of these, based on animal studies, is 2,3,7,8-tetrachlorodibenzo-*p*-dioxin.

The principle source of these toxic emissions is the incineration of mixed waste (including waste containing plastics) [27], but it is not entirely clear if changing the amount of PVC waste in a waste stream would result in a corresponding change in the dioxin emissions [28]. It is generally believed that incineration at a high enough temperature in adequately designed incinerators would minimize dioxin emissions, if any, from the process. Incineration with energy recovery can be an effective means of waste management for mixed waste streams. As waste streams typically include a PVC fraction, it is important to reliably quantify and establish unambiguously the role of PVC in dioxin generation under large-scale incineration.

2.5. POLYSTYRENE

General-purpose polystyrene (also called crystal polystyrene because of the clarity of resin granules) is a clear, hard, glassy material with a bulk density of 1.05 g/cm^{-3}. These desirable physical characteristics, as well as easy moldability, low water absorbancy, and good color range in which the resin was available, made it a popular general-purpose resin. Its brittleness, which limited the range of products in which the resin could be used, was soon overcome when the high-impact toughened grades of polystyrene containing rubber became available. The resin is available as a general-purpose grade, high-impact resin, high-molecular-weight resin (for improved strength), high-heat grades with a higher softening point, and easy flow grades for sophisticated molding applications. High-heat resins are high-molecular-weight resins with melt flow rates of 1.6 g/10 min. The medium and easy flow grades contain 1–4% added mineral oil or other lubricant to obtain higher flowrates of about 7.5 and 16 g/10 min, respectively. Impact-grade resin accounts for about half of the demand for polystyrene and is widely used in injection molding of consumer products. Some copolymers such as (acrylonitrile–butadiene–styrene) copolymers (ABS), styrene–acrylonitrile copolymers (SAN), and styrene–maleic anhydride copolymers (SMA) are also commercially available [29]. Copolymer of styrene with butadiene (SBR) is an important elastomer widely used in passenger tire applications.

2.5.1. Manufacture of Polystyrene

The first commercial production of polystyrene (PS) was carried out in the early 1930s by the Farben Company (Germany) and was soon followed in 1937 by the Dow Chemical Company introducing in the United States a grade called "Styron." Styrene monomer is mainly produced by the dehydrogenation of ethylbenzene made by reacting ethylene and benzene in a Friedel–Crafts reaction using a catalyst system containing aluminum chloride [30]. Yields in excess of 98% are common in this process. The thermal cracking reaction that produces the dehydrogenation is carried out at 630°C in the presence of a catalyst, commonly a mixture of Fe_2O_3, Cr_2O_3, and K_2CO_3. The reaction yields a mixture of products, but the process conditions can be controlled to obtain about 80% conversion.

$$\text{Benzene} \xrightarrow[95°C]{AlCl_3} \text{Ethylbenzene} \xrightarrow[\text{Catalyst}]{630°C} \text{Styrene (CH=CH}_2\text{)}$$

$$\text{Styrene} \longrightarrow \text{Polystyrene } {-}[CH_2{-}CH(C_6H_5)]_n{-}$$

The styrene is separated from the product mix, which also contains unreacted ethylbenzene and other impurities, by vacuum distillation. The monomer can easily autopolymerize into a hard solid and is therefore inhibited from polymerization during storage by mixing in a few parts per million of a free-radical reaction inhibitor (generally t-butyl catechol). A relatively small amount of styrene is also made by the oxidation of ethyl benzene in a process introduced by Union Carbide. The ethylbenzene hydroperoxide formed by oxidation is reacted with propylene to form propylene oxide and 2-phenyl ethanol. The latter compound is dehydrated to obtain styrene.

While bulk or emulsion polymerization can also be used for the purpose, the commercial manufacture of polystyrene is mostly carried out in a solution process using a free-radical initiator. The solvent, typically ethylbenzene, used at a level of 2–30%, controls the viscosity of the solution. High-impact-grade polymer used in injection-molding and extrusion is modified with butadiene rubber incorporated during polymerization. The solvent and residual monomer in the crude resin is removed by flash evaporation or in a devolatilizing extruder (at about 225°C). Figure 2.9 is a schematic of the polymerization process.

Expandable polystyrene is extensively used in the production of foam products and extruded foam sheets. These expandable beads contain a nucleating agent and a blowing agent in addition to the high-molecular-weight polystyrene resin. A hydrocarbon blowing agent, typically pentane at a level of 5–8% by weight or a chlorofluorocarbon (CFC) blowing agent, might be used, The latter, however, is now restricted because of its potential for depleting stratospheric ozone, and the U.S. industry phased out the use of ozone-depleting CFCs in expanded foams in the 1980s. With hydrocarbon blowing agents a bicarbonate/citric acid system might be used as the nucleating agent (with CFCs finely divided talc was commonly used for the purpose). The choice and amount of blowing agent determines the density, and therefore the thermal conductivity, of the foam produced, both important considerations in building applications. Densities down to 1 lb/ft^3 (16 kg/m^3), amounting to a grade of foam that is 97% by volume of air or blowing agent, are possible. The bead is generally expanded with steam or hot air in

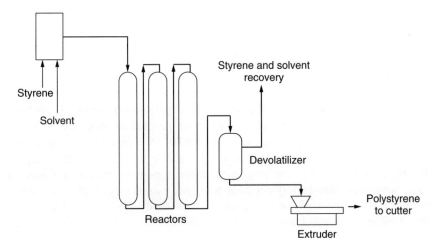

Figure 2.9. Simplified flow diagram for solution polymerization of styrene.

a two-stage process, first to preexpand the beads, and then to further expand and adhere to each other inside a mold. Molded expanded resin is used in polystyrene foam container and cup and packaging material production. Foam sheets in the density range of 2–10 lb at 1 lb/ft^3 (32–160 kg/m^3) can be produced by extrusion. The blowing agent and the resin are combined in a preextruder and formed into sheets by extrusion through a slit die. These sheets may be thermoformed into consumer goods such as packaging trays or plates.

Polystyrene foams are low-density packaging materials that provide exceptional insulation and bioinert containment cost effectively. Its popular use in hot beverage cups and food containers, particularly in the fast-food industry, illustrate these advantages. As such, the product provides very good value for the minute amounts of fossil fuel resources used in its production. (These resources are, in principle, recyclable, as the polystyrene can be separated from waste streams by density separations, solvent extractions, or by pyrolysis into the monomer styrene.) Expanded polystyrene foam products provide a good example of the importance of communicating environmental information relating to plastics to the consuming public. Misconceptions about the use of CFCs in the foams, the difficulty in viable recycling of postconsumer products, and exaggerated concerns over their disposal in landfills have unfortunately limited the use of polystyrene foam in the packaging sector. The development underscores the importance of the plastics industry moving proactively and coherently to educate the consuming public as to environmental attributes of its products.

2.6. POLY(ETHYLENE TEREPHTHALATE)

The U.S. production volume of the aromatic polyester, poly(ethylene terephthalate) (PET), is comparable to that of low-density polyethylene or polystyrene. The resin is used in several key products; a large part of the polyester is converted into fibers for textile applications and into backing materials for audio and video tape recording media. Biaxially oriented polyester film is used in food packaging (e.g., in boil-in-bag foods) and as thermoformed sheets in frozen meal trays that can be heated in a microwave or even a conventional oven. Because of its good insulating properties, PET films are used in electrical devices as well.

The best known product made from aromatic polyester, however, is the blow-molded soda bottle, where the growth has been particularly dramatic during the past few years.[14] This application accounts for 70–80% of the resin consumption and exploits the excellent barrier properties of the resin to carbon dioxide. It also requires the resin to be processed under conditions that would yield an amorphous (transparent) polymer with good gas barrier properties. The number-average molecular weight of the grade of resin used in bottles is about 80,000 (that used in fibers and films is about 45,000).

2.6.1. Manufacture of PET

Poly(ethylene terephthalate) is the condensation polymer made from terephthalic acid and ethylene glycol. The acid or its dimethyl ester is obtained by the oxidation of *p*-xylene, a product from catalytic reforming of naphtha. The glycol is obtained from ethane via the corresponding cyclic oxide. With the availability of purified terephthalic acid since the 1960s direct esterification of the acid in a continuous process is used in commercial production of the polyester [31]:

$$H_3CO-\overset{O}{\underset{\|}{C}}-\underset{}{\bigcirc}-\overset{O}{\underset{\|}{C}}-OCH_3 + HO-(CH_2)_2-OH \longrightarrow$$

$$HO-(CH_2)_2-OOC-\bigcirc-COO(CH_2)_2OH$$

Diester

270–285°C

$$\left[O-(CH_2)_2-O-\overset{O}{\underset{\|}{C}}-\bigcirc-\overset{O}{\underset{\|}{C}}\right]_n$$

In the batch process, a mixture of dimethyl terephthalate and the glycol (slightly over 2 molar proportions) is heated to 150–210°C in the presence of a catalyst, usually acetates of lithium, manganese, and zinc. The product of the ester exchange, the bis(2-hydroxyethyl terephthalate), is melt polycondensed at 270–280°C at pressures of 66–133 psi in the presence of a catalyst. In the

---

[14] In the early 1970s nitrile resins were considered good candidates for packaging carbonated beverages. However, safety issues related to trace amounts of toxic acrylonitrile in the plastic bottles precluded their use in this application.

continuous esterification process a mixture (paste) of terephthalic acid and the glycol (1:1.6 molar ratio) is esterified, usually in two stages at a temperature of 245–250°C. Water is removed as formed and the reaction is carried to 98% completion. The product, mixed with an antimony oxide catalyst, is polymerized, again in a two-stage process at a temperature of 255–298°C.

2.7. FROM RESINS TO THERMOPLASTIC PRODUCTS

Thermoplastic resins are available to the processing industry as pellets of resin. Converting the raw material into useful products can involve separate segments of the plastics industry. As Figure 2.10 suggests, the resin might be compounded by a custom compounder and formed into the final product by a processor or a fabricator. The compounding can also be carried out by the processor in an in-house facility.

The resin raw material needs to be mixed intimately with a variety of chemical additives to impart specific properties to the end product. Additives are used widely in the plastics industry, in nearly all types of plastic products. The use of common plastics in consumer products would not be possible without the use of additives. For instance, vinyl plastics (particularly PVC) undergo easy thermal and photodegradation; no useful products can be made with it if stabilizer additives designed to protect the resin during thermal processing and use were not available. Selecting the appropriate set of additives called for by a given product and mixing these in correct proportion with the resin is referred to as compounding. Table 2.7 lists the common classes of additives used in manufacturing plastic products. To ensure adequate mixing or dispersion of the additive in the mass of plastic, these may be added into the virgin resin as a concentrate (or a masterbatch) already dispersed in the same or a compatible resin. Some additives may also be added directly to the resin mix. In any event, the mixing is accomplished by passing the resin and additive mixture at a temperature

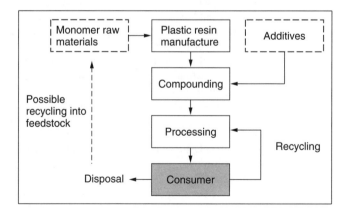

Figure 2.10. Flow diagram illustrating components of plastics industry.

Table 2.7 Common Plastics Additives and Their Functions

Additive	Function	Example
Antiblocking agent	Prevents sticking of thin plastic sheets to each other, or "blocking"	Quartz or silica in polyethylene
Antioxidants	Reduces the rate of autoxidation of the plastic at service temperature	Metal deactivators, peroxide decomposers.
Antistatic agent	Prevents charges on polymer surface leading to static discharge	Quaternary ammonium salts in rigid PVC
Biocide	Prevents growth of microorganisms on plastics	Phenols and chlorinated phenols in coatings
Blowing agent	Used to create polymeric foams.	Inert gases and AIBN* that decompose into N_2 on heating.
Inert filler	Reduces the cost of formulation and changes the color	Chalk used in plastic formulations
Reinforcing filler	Increases the modulus and other properties of a polymer	Carbon black in rubber formulations
Coupling agent	Promotes better adhesion between phases in filled and glass-fiber-reinforced plastics	Organosilanes, titanates, and zirconates
Curing agents	Crosslinks the polymer	Sulfur or organic sulfur compounds in rubber
Flame retardant	Reduces the flammability of plastics products	Borates, and organophosphorous compounds
UV stabilizers	Minimizes the solar UV-B induced degradation of plastics outdoors	Hindered amines and light absorbers
Impact modifier	Increases the impact resistance of plastics	Rubber and thermoplastics in epoxy resin
Lubricant	Minimizes internal and external friction leading to degradation	Ethylene(bis)stearamide used in rigid PVC
Pigments	Colors plastic products	Inorganic pigments, carbon black, and organic pigments
Plasticizer	Softens the plastic and makes it more processable	Phthalates in rigid PVC compounds

*AIBN = 2,2' azobisisobutyronitrile.

high enough to melt the thermoplastic, through a mixing screw in an extruder (a compounding extruder). Care is taken not to overheat or overshear the mix to an extent to cause chemical breakdown of the plastic itself or the additive materials. The now "compounded" resin with the additives evenly distributed within its bulk is repelletized, cooled, dried (where the pelletization is carried out under cooling water), and stored for subsequent processing.

Processing is the final step that converts the compounded material into a useful plastic product. Basically, the compounded resin needs to be melted into a liquid and heated to a temperature that allows easy handling of the fluidized plastic or the "melt." This melt is fed into molds or dies to force the material into required shapes and quickly cooled to obtain the product. Usually, some minor finishing is needed before the product is made available to the consumer. The different processes used and the equipment employed are determined by the type of product being manufactured. A detailed discussion of the various processing techniques available for common thermoplastics is beyond the scope of this discussion. However, the basic principals involved in common processing methods associated with high-volume products will be discussed briefly below.

2.7.1. Extrusion Processing

The most important processing technique for common thermoplastics is extrusion, where the plastic material is melted in a tubular metal chamber and the melt forced through a die. The design of an extruder is not unlike a toothpaste tube (heated, of course, to melt the resin), and tubular products such as plastic rods, plastic tubes, plastic drinking straws, coatings on electrical wire, and fibers for textile applications can be manufactured using an appropriately engineered die. To exert enough pressure to force the viscous melt through the small die orifice, an Archimedean screw is used. Most of the heat needed to melt the resin is derived from the mechanical shearing action of the screw, although external heating is also provided. The screw transports the resin from the inlet (at the hopper) through a long passage with several heating regions into a heated die. The resin passes through a region of the screw (with decreased depth in screw channels) that ensures further mixing and consolidates the melt removing any empty spaces or bubbles in melt prior to reaching the mold. The passage of melt is controlled by a layer of mesh on its way to the entrance of the die; this breaker plate assembly (with screen pack) serves to filter out any particulate debris and to control the melt flow into the die. The design of the die determines the geometric features of the product extruded. Figure 2.11 shows the main features of a simple single-screw extruder, along with three types of common extrusion dies. The simplest die is a precisely drilled hole or a slit yielding a rod or a ribbon product. A slightly more complicated design (a circular orifice with a central solid region) produces pipes and tubing. The first two dies shown in the diagram are for tube (or pipe) products and laminates. The third is a specialized die for coating thermoplastic resins on electrical conductors. As the conductor is drawn through the cylindrical die, it contacts the molten polymer introduced

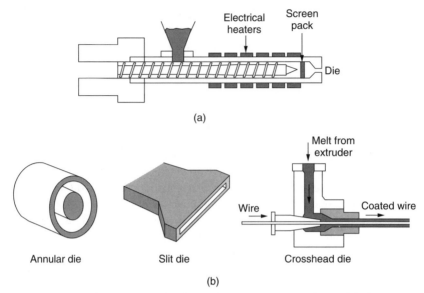

Figure 2.11. Main features of a simple single-screw extruder, along with three types of common extrusion dies.

from the top of the die. Extremely complicated dies are used in the extrusion of complicated profiles, for instance, in plastic window and door frames.

The product emerging from the die is handled by "down stream" equipment that would essentially cool (in case of pipe cut to size) and collect the product for storage. The actual pieces of equipment used for the purpose depend on the type of product manufactured.

2.7.2. Injection Molding

Injection molding is one of the most popular processing operations in the plastics industry. In recent years, more than half the processing machinery manufactured were injection-molding machines. The equipment is basically designed to achieve the melting of the resin, injecting the melt into a cavity mold, packing the material into the mold under high pressure, cooling to obtain solid product, and ejecting the product for subsequent finishing. It is different from extruders in that a mold is used instead of a die, requiring a large force to pack the melt into the mold. A machine is typically classified by the clamping force (which can vary from 1 to 10,000 tons!) and the shot size determined by the size of the article to be manufactured. Other parameters include injection rate, injection pressure, screw design, and the distance between tie bars.

The machine is generally made of (a) a hydraulic system, (b) plasticating and injection system, (c) mold system, and (d) a clamping system. The hydraulic system delivers the power for the operation of the equipment, particularly to open and clamp down the heavy mold halves. The injection system consists of a reciprocating screw in a heated barrel assembly and an injection nozzle. The system is designed to get resin from the hopper, melt and heat to correct temperature, and deliver it into the mold through the nozzle. Electrical heater bands placed at various points about the barrel of the equipment allow close control of the melt temperature. The mold system consists of platens and molding (cavity) plates typically made of tool-grade steel. The mold shapes the plastic melt injected into the cavity (or several cavities). Of the platens, the one attached to the barrel side of the machine is connected to the other platen by the tie bars. A hydraulic knock-out system using ejector pins is built into one of the platens to conveniently remove the molded piece.

The machine operates in an injection-molding cycle. The typical cycle sequence is, first, the empty mold closes, and then the screw movement delivers an amount of melt through the nozzle into it. Once the mold is full, the pressure is held to "pack" the melt well into the mold. The mold is then cooled rapidly by a cooling medium (typically water, steam, or oil) flowing through its walls, and finally the mold opens to eject the product. It is common for this cycle to be closely monitored and to be mostly automated by the use of sophisticated control systems. Figure 2.12 shows a diagram of a simple injection-molding machine indicating the hydraulic, injection, and mold systems. The mold filling (a), compaction (b), cooling (c), and ejection (d) steps are also illustrated in Figure 2.12. Figure 2.13 shows a modern injection-molding machine.

When a multicavity mold designed for several "parts" is used, the ejected product is complex, consisting of runners, a spruce, and flashing that needs to

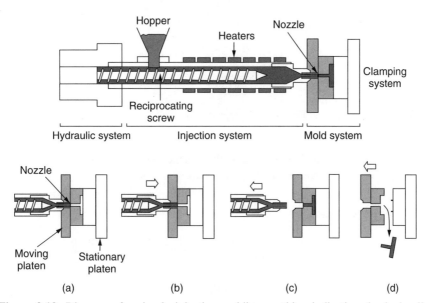

Figure 2.12. Diagram of a simple injection-molding machine indicating the hydraulic, injection, and mold systems.

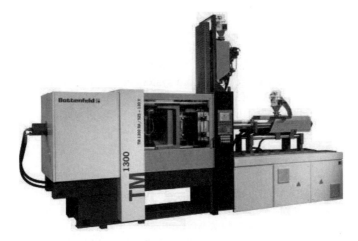

Figure 2.13. Modern injection-molding machine (Courtesy Battenfeld).

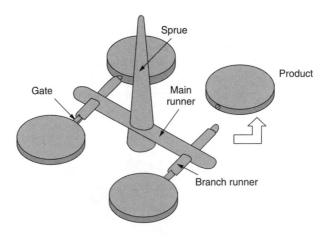

Figure 2.14. Injection-molded piece.

be removed (and recycled) to obtain the plastic product. Figure 2.14 shows a molding with one of the product "parts" removed from it.

2.7.3. Blow Molding

This is the primary processing technique used to fabricate hollow plastic objects, particularly bottles, which do not need a very uniform distribution of wall thickness. It is a secondary shaping technique that inflates the preprocessed plastic (usually extruded) against the inside walls of the mold with a blow pin. In addition to extrusion blow molding, injection blow molding and stretch blow molding are commonly employed. With most polymers, especially when the product size is

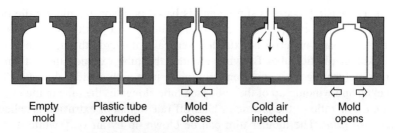

Figure 2.15. Blow molding of plastic bottle.

large, extrusion blow molding is used; while injection blow molding is typically used with smaller products with no handleware. Semicrystalline materials that are difficult to blow are molded by stretch blow molding. Common resins such as PVC, PS, PP, LDPE, HDPE, and PET are blow molded routinely. Figure 2.15 illustrates the steps involved in extrusion blow molding.

In extrusion blow molding, the most common blow-molding process, an extruder is used to produce a thick-walled plastic tube called the *parison*. The parison is extruded directly into a water-cooled cavity mold, which is then closed, and air injected through the top or the neck of the container. The softened polymer in the parison inflates against the wall of the mold, which cools the melt and solidifies it into the mold shape. The mold opens and the part is removed and deflashed to remove any excess plastic. While the wall thickness of the parison itself is uniform, that of the product formed (a bottle) will not be uniform because of its different geometry. This variation in wall thickness needs to be taken into account when designing products intended for blow molding. In this processing cycle most of the time is spent on cooling the mold. Therefore, it is usual to have several molds set up on a rotating table that takes up sections of parison from a single continuous extruder to optimize the process.

In injection blow molding, an injection-molding machine replaces the extruder. In the first stage a parison with the threads of the finished bottle molded in is injection molded onto a core element. The injected parison core is then carried to the next station on the machine, where it is blown up into the finished container as in the extrusion blow-molding process above. In some instances the parison might be stretched inside the mold to obtain a biaxially oriented plastic product. As the parison is injection molded, there is good control of the weight of the final product in this type of blow molding.

In stretch blow molding (for resins such as PET used in soda bottles) an injection-molded preform (usually obtained from a separate specialized vendor) is used. The preform is loaded into a simple machine that heats it to soften the plastic and stretches it inside the mold to shape the plastic into a bottle.

2.7.4. Extrusion Blowing of Film

Extrusion blowing of common plastics such as polyethylenes into film is one of the oldest processing techniques (dating back to the 1930s in the United

States). The basic process is simple and is based on a special annular die that is connected to one or more extruders. In the simple case with a single extruder, the molten plastic material is extruded vertically upwards through the die into a thin-walled plastic tube. Blowing air into the tube expands the soft molten polymer, deforming it circumferentially into a tube with a wider diameter, while the pickup and winding up of the collapsed tube elongates the tube in the machine direction. The ratio of the pickup or haul-off rate to that of extrusion is called the *draw-down ratio*. The tubular film can be blown up by air only while it is soft and soon forms a "freeze line" at a maximum diameter (the ratio of the diameter at the freeze line to that of the annular die is the *blow-up ratio* for the film). To obtain a uniform film, it is crucial to maintain constant extrusion rates and a symmetric stable "bubble" or the inflated cylinder of polymer at all times during processing. Typically the bubble can be 15–30 ft tall and up to several feet in diameter. The processing variables as well as the grade of resin used for film blowing determines the quality and uniformity of the film product. Figure 2.16 shows a photograph of an annular die used in blowing films, and Figure 2.17 shows a diagram of film blowing equipment.

The same process can also be used to produce a multilayered film using several extruders, one for each type of resin used, and a feed block to direct the resin

Figure 2.16. Annular die used in extrusion blowing of plastic film showing the "bubble".

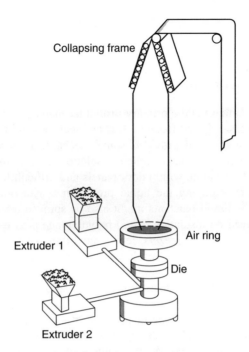

Figure 2.17. Schematic representation of extrusion blowing of plastic film.

into different layers. The layers need to be selected carefully for their processing characteristics as well as their performance in the final product. For instance, in coextrusion of a barrier film for packaging applications, different layers of the film might be selected for different functionality needed in the product.

2.8. POLYURETHANE AND OTHER POLYMERIC FOAMS

Polymeric foams as a class of materials is of particular interest in the present discussion not only because they are an important component of the waste stream but also because ozone depleting substances (ODS) such as Freon were once widely used as auxiliary blowing agents in their manufacture. However, in the United States and in western Europe, Freon is no longer used for the purpose and the practice is being phased out in many other parts of the world. The Montreal Protocol (along with its subsequent associated improvements) has been singularly successful in eliminating the use of major ODSs in polymeric foams and is expected to ensure the phase out of even the low-ODS agents presently allowed for the purpose, in the near future. This section reviews the chemistry of polymeric foams, the role played by Freon, and the substitutes available today for the purpose.

Foam is a polymeric material where gas-filled cells that are either open to or are isolated from each other fill the bulk of the matrix.

2.8.1. Polyurethane Foam

Polyurethane (PUR) foam is the primary insulating material used in refrigerators (residential and industrial) and freezers. It is an ideal material for the application because the rigid foam used provided some mechanical support, and the gas trapped within the cells afforded excellent insulation. It is an insulation material that is convenient to generate within door panels and difficult to access locations in the appliances by pumping the liquid prepolymer with other chemicals for reaction in situ. The liquid reactants might also be sprayed using an appropriate gun to obtain sprayed-on insulation. Rigid polyurethane is also cast as slabstock for use in insulation of large freezers, refrigerated transportation containers, and storage tanks. Rigid polyurethane insulation is used for conservation of heat (as with hot-water heaters) and in a number of consumer products such as coolers. Rigid foams are also finding increasing use as floatation materials in the hull of recreational watercraft.

The appliance market for polyurethane foam was recently estimated[15] at 291 million pounds (North America in 2000) or about 5% of the total annual demand for polyurethanes in this region. As the volume of foam used in appliances as well as in building is tied to the economy, the demand is expected to slowly increase in the long-term as the gross domestic product (GDP) in the region increases. Increasingly stringent energy standards that appliance manufacturers (and possibly builders in the future) will have to comply with may also increase the demand for insulation materials.

In building applications rigid polyurethane foam is used laminated to board stock (or to other substrates such as foil or paper), discontinuous or continuous sandwich panels, slabstock, spray, and as pipe insulation material. Some of these sandwich structures are used in building (and transportation) applications that do not exploit their insulation characteristics. The largest volume building application is in roofing insulation [where both PUR and polyisocyanurate (PIR) foams are used].

Flexible PUR slabstock enjoys a large market in seat cushions, upholstery, and carpet backings. Over 2 billion pounds of the flexible foam is annually produced in the United States alone. Selecting the appropriate density of foam is important to ensure the durability of the product (such as a cushion) as well as its comfort level. The latter is quantified in the industry by measures such as the compression modulus, flex fatigue, resilience, indentation force, and support factor. Polyurethane foam slabstock is manufactured by continuously dispensing accurately metered reactants through a mixing head onto a lined trough on a

---

[15] Data was reported in *Appliance Engineer* (June 2000, p. 68). The survey was sponsored by the Alliance for Polyurethane Industry.

moving belt. The prepolymer reacts exothermically expanding into a foam as it moves forward on the conveyer belt through fittings designed to give it a flat upper surface. The foam can also be made in the batch mode, using a box mold with a floating lid, into which the mixed chemicals are poured and allowed to expand. A variety of organic auxiliary blowing agents (such as pentane, methylene chloride, acetone, and liquid carbon dioxide) are available to replace any ODS blowing agents in use.

Integral skin PUR foams are made by injecting the reactants into a closed (vented) or open mold. Reaction under these conditions produces a foam with a high-density outside "skin" layer and a soft foam interior. Common products such as steering wheels, footwear components, and computer housings generally employ integral skin PUR foams.

Chemistry Polyurethane is produced by the reaction of a polyol with an diisocyanate (or in some instances a polyisocyanate) in the presence of catalysts. The polyols of choice are poly(propylene glycol), block copolymers of ethylene oxide (10–15%) with propylene oxide, or the newer polymer polyols (based on polymers such as polystyrene or styrene–acrylonitrile copolymer). Polyester diols such as polycaprolactone diol can be used in place of the polyether polyol in this reaction. The isocyanate of choice is a mixture of the 2,4 and 2,6 isomers of tolylene di-isocyanate in the ratio of 80:20, generally referred to as 80:20TDI. Other isocyanates such as diphenylmethane di-isocyanate (MDI), hexamethylene di-isocyanate (HMDI), and isophorone di-isocyanate (IPDI) are also used. A tin-based or amine catalyst is used to promote the reaction. Given the wide choice of reactants available, the reaction can yield foams with a range of different mechanical and thermal characteristics.

The basic reactions involved are shown below. The foaming is due to the generation of carbon dioxide from the reaction of water with excess isocyanate present in the system. While the foam density can be controlled to some extent by changing the amount of water, excess water can encourage the formation of relatively stiffer urea bonds, affecting the quality of the foam. The auxiliary blowing agents (ABA) are therefore used to increase foaming to obtain lower density foams that have good insulation properties, thanks to the thermal characteristics of the ABA. ABAs in the foam slowly diffused out during use and were released at the end of their useful life.

$$O=C=N-R_1-N=C-O + HO-(-R_2-O)-H$$
$$\text{Diisocyanate} \qquad\qquad \text{Polyol}$$

$$\longrightarrow -[O-CO-NH-R_1-NH-CO-O-R_2]_n-$$
$$\text{Polyurethane}$$

The polyol is shown above as a small molecule for convenience of illustration. In practice, they are low-molecular-weight polymers with a number-average molecular weight of a few thousand grams per mole.

Abbreviation	Chemical name	Structure
TDI	Tolylene-2,4-diisocyanate	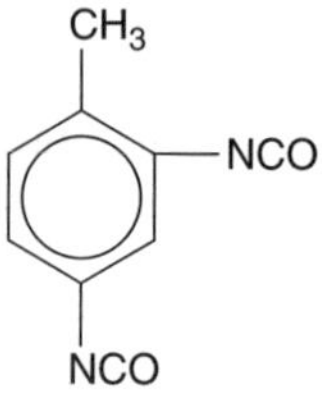 benzene ring with CH_3, NCO (ortho), NCO (para to methyl)
HMDI	Hexamethylene diisocyanate	$OCN-C_6H_{12}-NCO$
IPDI	Isophorone diisocyanate	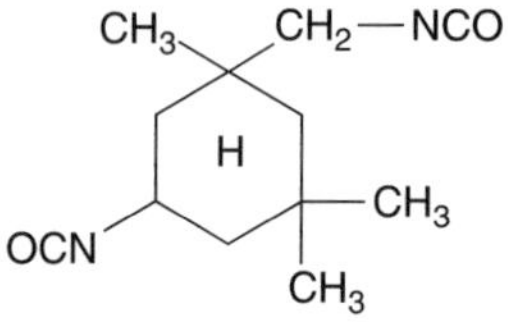 cyclohexane ring with CH_3, CH_2-NCO, H, CH_3, CH_3, OCN
MDI	4,4'-Methylene bisphenylene diisocyanate	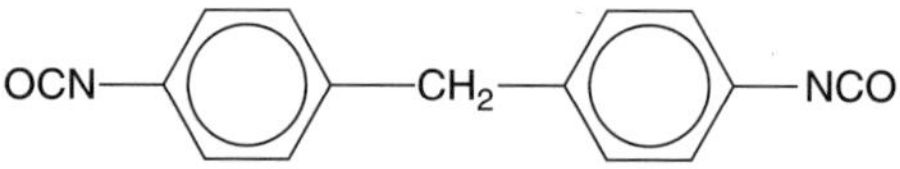 OCN—⟨ ⟩—CH_2—⟨ ⟩—NCO
CHDI	Cyclohexane 1,4-diisocyanate	 OCN—⟨ H ⟩—NCO
PPDI	p-Phenylene, 1,4-diisocyanate	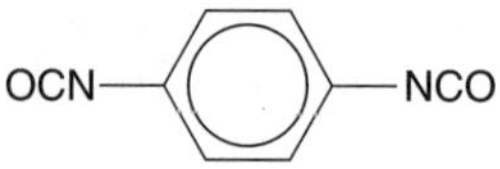 OCN—⟨ ⟩—NCO

The polyols HO—(—R_2—O)—H popularly used are as follows:

    PEG: poly(ethylene glycol)   HO—$(C_2H_4O-)_n$—H
    PPG poly(propylene glycol)   HO—$(C_3H_6O-)_n$—H
    PTG: poly(tetramethylene glycol)   HO—$(C_4H_8O-)_n$—H

The structural formulas of the common polyisocyanates used in the preparation of polyurethanes are shown above. The polyethers prereacted with excess isocyanate to yield an isocyanate end-capped prepolymer are stable viscous liquids under dry conditions. These mixed with water and catalysts produce a soft foam used in upholstery and other applications. The more popular and economic route to producing foam is the "one shot" process where the reactants, catalysts, and additives are all mixed together in a mixing head and spread onto a moving trough to obtain slabstock of the foam. Typically the mixing head is designed to accommodate several streams of feed: the polyols; the isocyanates; catalyst system, water, and additives; and the blowing agents.

Auxiliary Blowing Agents (ABA) Historically, CFCs, particularly CFC-11, CFC-12, CFC-113, and CFC-114, were popularly used in polyurethane foams. In 1992, the U.S. Environmental Protection Agency (EPA) issued its final rule implementing section 604 of the Clean Air Act Amendments of 1990. The rule required the production of class I substances (CFCs, halons, carbon tetrachloride, and methyl chloroform) to be gradually phased out completely by January, 2000 (2002 for methyl chloroform). The phase out now includes hydrochlorofluorocarbon (HCFC) blowing agents presently used in some of the polyurethane foam products. These will be phased out completely in the United States by year 2030 and alternative ABAs that are either low or zero ozone depleting have been identified and put to use. Table 2.8 compares some of the conventional CFC blowing agents with selected alternatives. Those that have even a low ozone-depleting potential are yet likely to be regulated in the future. Roughly a decade after its peak use in 1989, the use of ozone-depleting chemicals in foam-blowing applications have now dropped by more than 75% worldwide.

In flexible slabstock and molded polyurethane foam production, methylene chloride, CO_2 (liquid or gas injection), and water are used as ABAs. In rigid insulation foams CFCs have been replaced with HFC-134a (sometimes blended with HFC 152a) or liquid CO_2. In integral skin polyurethane products, CO_2, water, and hydrocarbons are used for the purpose.

2.8.2. Phenolic Foams

Another thermoset plastic that is used as a foam in building applications is phenol–formaldehyde resins. These foams can be manufactured with about the same range of densities as polyurethanes or polyisocyanurate foams and are good thermal insulators as well. The phenolic foams are self-extinguishing and emit relatively low smoke on burning. But, they are relatively more expensive to produce and are therefore of limited popularity. Phenolic resins are two-part systems consisting of a phenol–formaldehyde condensate [mainly consisting of bis(hydroxyphenyl methane) mixtures] and a hardening agent based on formaldehyde. HCFC-141b is the predominant blowing agent used with phenolic foams [32]. While pentanes can be used as a partial replacement as an ABA, this lowers the insulation efficiency and more importantly reduces the desirable fire performance of the foam. However, the amount of phenolic foams used worldwide is relatively small (less than 5%) and the release of ODS from this source is minimal.

2.8.3. Other Foams

Both polyethylene and polystyrene are used extensively as thermoplastic foam materials in packaging as well as in building applications. Expandable polystyrene

Table 2.8 Characteristics of Common CFCs, CHFCs, and Alternatives for Use with PUR Foams[a]

	CFC-11	CFC-12	HCFC-22	HCFC 141b	HCFC 1142b	CO_2	n-pentane	Cyclopentane
Chemical formula	$CFCl_3$	CCl_2F_2	$CHClF_2$	CCl_2FCH_3	CH_3CClF_2		C_5H_{12}	$(CH_2)_5$
Molecular weight	137	121	86	117	100	44	72	70
Boiling point (°C)	24	−30	−41	32	−10	−139	36	50
Conductivity[b]	7.4	10.5	9.9	8.8	8.4	14.5	14	11
GWP (100 yr)	4000	8500	1700	630	2000	1	11	11
ODP	1	1	0.055	0.11	0.065	0	0	1
Application	Phased out	Phased out	One-component rigid foams	Most rigid foam types	Insulation foam	Appliance, pipe, and spray rigid foam	Insulation foams	Insulation foams
			Insulating rigid foams	Integral skin foams	Integral skin foam	Flexible foams of all types	Integral skin foams	Integral skin foams
				Miscellaneous foams				

[a]Table based on data reported in Table ES-1 and Table 1 in *1998 Report of the Flexible and Rigid Foams Technical Options Committee*, United Nations Environmental Program, Nairobi, Kenya, December 1998.
[b]Conductivity expressed in (mW/m K at 10°C).

beads are made by polymerization of styrene in the presence of pentane (or by incorporation of pentane into preformed beads under pressure). The fraction of about 6% pentane trapped in the bead is the sole blowing agent. These might be steam expanded and molded into products or sheets. The pentane is a VOC and is also flammable (so is polystyrene), limiting the application of these foams in certain situations.

In extruded boardstock manufacture the polystyrene beads and foaming agents are fed into an extruder where the beads expand, incorporate the blowing agent, and are passed through a die. Freon (HCFC-142b and HCFC 22) is used as the primary blowing agent popularly in polystyrene foam boardstock in most parts of the world. While these are low ozone depleting, they also obtain excellent thermal insulation characteristics. Zero ozone-depleting polymer (ODP) alternatives are, however, available for polystyrene foam boardstock manufacture and include HFCs (HFC 134, HFC 134a, and HFC 152a), blends of HFCs and carbon dioxide, and carbon dioxide.

REFERENCES

1. Honk Kong Plastics Technology Center, *Techno Economic and Market Research Study on Hong Kong's Plastics Industry, 1999/2000...2001*, Trade and Industry Department, Hong Kong, 2001, p. 3–1.
2. Franklin Associates, *A Comparison of Energy Consumption by the Plastics Industry to Total Energy Consumption in the United States*, Franklin Associates, Priarie Village, KS, 1990.
3. Franklin Associates, *Comparative Energy Evaluation of Plastics Products and Their Alternatives for the Building and Construction and Transportation Industries*, Franklin Associates, Priarie Village, KS, 1991.
4. Chemistry in Britain, *Good News for Polyolefins, in Chemistry in Britain*, 2001, p. 31.
5. D. E. James, in H. F. Mark et al., eds., *Encyclopedia of Polymer Science and Engineering*, Vol. 6, John Wiley & Sons, New York , 1988, p. 481.
6. D. L. Beach and Y. V. Kissin, in H. F. Mark et al., eds., *Encyclopedia of Polymer Science and Engineering*, Vol. 6, John Wiley & Sons, New York, 1988, p. 486.
7. K. Ziegler, *Angew. Chem.* **67**, 33 (1955).
8. P. D. Ritchie, *Vinyl and Allied Polymers, Vol. 1*, CRC Press, Cleveland OH, 1968, p. 32.
9. W. H. Joyce, in *Energy and Environment in the 21st Century*, Eds. J. W. Tester, D. O. Wood, and N. A. Ferrari, Massachusetts Institute of Technology, Cambridge, MA, 1991.
10. T. J. Cavanaugh and E. B. Nauman, *Trends Polym. Sci.* **3**(2), 48 (1995).
11. Radian Corporation, *Polymer Manufacturing, Technology and Health Effects*, Noyes Data Corporation, Park Ridge, NJ, 1986, p. 716.
12. International Agency for Research on Cancer, *IRC Monograph on the Evaluation of Cancer Risks to Man*, **23**, 205, 325 (1980).
13. D. A. Howe, in J. E. Mark, ed., *Polymer Data Handbook*, Oxford University Press, New York, 1999, p. 781.

14. A. Sustic and B. Pellon, *Adhesive Age* **November**, 17 (1991).
15. M. J. Balow, in H. J. Karian, ed., *Handbook of Polypropylene and Polypropylene Composites*, Marcel Dekker, New York, 1999, p. 555.
16. C. A. Staples, *Chemosphere* **40**, 885 (2000).
17. D. F. Cadogan, *Plast. Rubber Compos.* **28**, 476 (1999).
18. R. M. David, *Toxicol. Sci.* **55**, 433 (2000).
19. M. Sharpe, *J. Environ. Monit.* **2**, 4N–7N (2000).
20. European Commission, *Commission Decision 1999-875-EC*, European Commission, Brussels, Belgium, 1999.
21. C. E. Koop, *A Scientific Evaluation of Health Effects of Two Plasticisers Used in Medical Devices and Toys: A Report from the American Council on Science and Health*, 1999, http://www.medscape.com/pages/public/about/about.
22. N. P. Moore, *Reprod. Toxicol.* **14**, 183 (2000).
23. I. Mersiowsky, *Plast. Rubber Compos.* **28**, 321 (1999).
24. A. A. Meharg, *Toxicol. Ecotoxicol. News* **1**, 117 (1994).
25. F. X. R. E. A. Van Leeuwen, *Chemosphere* **40**, 1095 (2000).
26. B. N. Ames and L. S. Gold, *Environ. Health Perspect. Suppl.* **105**, 865 (1997).
27. UNEP, *Dioxin and Furan Inventories. National and Regional Emissions of PCDD/PCDF*, United Nations Environment Programme Chemicals, Nairobi, Kenya, 1999.
28. H. G. Rigo and A. J. Chandler, *Organohalogen Comp.* **27**, 56 (1996).
29. J. Hahnfeld and B. D. Dalke, in H. F. Mark, C. G. Overberger, G. Menges, eds., *Encyclopedia of Polymer Science and Engineering*, 1989, p. 64.
30. Y. C. Yen and J.-T. Huang, *Styrene and p-Methyl Styrene and Polymers*, SRI International, Menlo Park, CA, 1984.
31. I. Goodman, *Encyclopedia of Polymer Science and Engineering*, Wiley, New York, 1988, p. 43.
32. UNEP, *The 1998 Report of the Flexible and Rigid Foams Technical Options Committee*, Nairobi, Kenya, 1998, p. 23.
33. N. J. Maraschin, in A. L. Brody, ed., *Encyclopedia of Packaging Technology*, Wiley, New York, 1997.

SUGGESTED READING

1. American Institute Of Chemical Engineers, *Safety in Polyethylene Plants* (3 Volumes), January 1978.
2. B. Brewer and G. Epstein, eds., *PVC: Formulation, Compounding and Processing*, T/C Press, October 1991.
3. K. M. Finlayson ed., *Plastic Film Technology: High Barrier Plastic Films for Packaging*, Vol. 1, CRC Press, LLC, April 1990.
4. K. M. Finlayson, ed., *Plastic Film Technology: Extrusion of Plastic Film and Sheeting*, Vol. 2, Technomic Publishing Co., January 1993.
5. R. A. Gsell, H. L. Stein, and J. J. Ploskonka, eds., *Characterization and Properties of Ultra-High Molecular Weight Polyethylene*, American Society for Testing & Materials, February 1998.

6. H. G. Karian, ed., *Handbook of Polypropylene and Polypropylene Composites*, Marcel Dekker, March 1999.
7. E. P. Moore, *Polypropylene Handbook: Polymerization, Characterization, Properties, Processing, Applications*, Hanser-Gardner Publications, July 1996.
8. L. I. Nass and C. A. Heiberger, *Encyclopedia of PVC, Second Edition, Revised and Expanded*, Marcel Dekker, September 1987.
9. L. I. Nass, ed., *Encyclopedia of PVC: Conversion and Fabrication Processes*, 2nd ed. Marcel Dekker, October 1998.
10. K. R. Osborn, *Plastic Films: Technology and Packaging Applications*, CRC Press, LLC, September 1992.
11. A. J. Peacock, *Handbook of Polyethylene: Structures, Properties, and Applications*, Marcel Dekker, January 2000.

CHAPTER 3

POLYMERS AND ENERGY

IAN BOUSTEAD
Boustead Consulting Ltd, Horsham, West Sussex

3.1. INTRODUCTION

Energy has always been an important cost element in polymer production, but the world modeling exercises of the 1960s drew attention to the limited availability of fossil energy and for the first time highlighted potential conservation problems [1, 2]. The oil crises of the mid-1970s sharply focused the dependence of most industries on oil and firmly put oil and gas on the conservation agenda.

The petrochemical industry, and especially the plastics industry, has long been aware of its dependency on fossil hydrocarbons, not only for energy but also for its raw materials. It also recognized that, although engineers had traditionally been concerned with improving the efficiency of individual plants, the true energy dependency could only be properly assessed by looking at all of the production steps, starting with the extraction of raw materials from the earth through to the finished polymer at the factory gate.

One of the first reports of this approach was that of Harold Smith [3] at the World Energy Conference in 1969, and this was later followed by the Imperial Chemical Industries (ICI) analysis of polymers in 1973 and 1975 [4]. The early 1970s also saw the beginning of the subject area that eventually became known as life-cycle assessment, although at the time it was variously known as energy analysis, resource and environmental profile analysis, and ecobalance analysis. Many of the systems examined in the early reports in this area involved the use of plastics in a variety of applications [5, 6]. However, the data gathering was fragmented and relied on the willing collaboration of a few companies that also saw the value of the approach. Many companies guarded their data

---

Plastics and the Environment, Edited by Anthony L. Andrady.
ISBN 0-471-09520-6 © 2003 John Wiley & Sons, Inc.

jealously and regarded the development of energy analysis as a threat to their commercial operations.

This fragmented approach to data collection continued until 1990 when the Association of Plastics Manufacturers in Europe (APME) set up a major program of work aimed at evaluating not only energy use but also the raw materials use and the emission of solid, liquid, and gaseous wastes from the systems used to produce polymers from raw materials in the earth [7]. Similar programs were initiated by the American Plastics Council (APC), although their results have never been published, and by the Japanese Plastics Waste Management Institute (PWMI). [Note that when the APME exercise was initially set up, it was initiated by a division of APME known as the Plastics Waste Management Institute (PWMI), which no longer exists. Occasionally, the APME reports are referenced in the literature as PWMI data and care is needed to avoid confusion with the Japanese PWMI, which still exists.] The primary aim of all these exercises was to provide accurate, plant-based data that were truly representative of the industry as a whole so that the earlier fragmented approach, which could be misleading, was no longer necessary.

3.2. CONTRIBUTIONS TO ENERGY

The total cumulative energy associated with the production of any commodity, when the process traces all steps from the extraction of raw materials in the earth through to finished product at the factory gate was, by international agreement, defined as *gross energy* [8]. This gross energy is made up of four main components:

1. Energy directly consumed by the processing operations
2. Energy used by any transport operations within the processing sequence
3. Energy equivalent of any feedstock taken into the system
4. Energy used by the fuel producing industries in supplying components 1–3, sometimes referred to as precombustion energy

Originally, gross energy was expressed as a single quantity, representing the sum of the different components. However, it was soon realized that this single value could be misleading. For example, a plant in Norway consuming 1 MJ of electricity would exhibit a gross energy of the order of 1.25 MJ, whereas an identical plant in Texas would exhibit a gross energy of the order of 4 MJ. The difference arises from the very high proportion of hydropower in Norway (efficiency 80%) compared with the high proportion of coal-fired power generated in Texas (efficiency 30%). This problem was especially acute for electricity, and so it was generally agreed that the consumption of thermal fuels would be reported separately from electricity consumption.

A further refinement introduced was the separate reporting of the energy directly used by the materials processing operations from that used by the fuel-producing industries responsible for producing and delivering the energy.

3.3. FEEDSTOCK

One problem that directly affects the plastics industry is that the thermal energy component of gross energy included a contribution from hydrocarbons, which are consumed by the process but are not burned, namely, feedstock. Feedstock is the material feed to any operation. For plastics, this feedstock represents materials that could be used as fuels but is, in fact, used as a material. The consumption of hydrocarbon feedstock is qualitatively different to the consumption as hydrocarbon fuels for a number of reasons:

1. Fuels are burned and are therefore gone forever after their use. In contrast, feedstock hydrocarbon is rolled up as a material in the product and, apart from losses and chemical changes during processing, remains available for reclamation and use as a fuel at a later date.
2. When fuels are consumed, they give rise to air emissions such as CO, CO_2, SO_x, and NO_x. In contrast, hydrocarbon feedstock is not burned and therefore gives rise to no such emissions.
3. Fuel energy can be derived from a number of different sources. For example, steam can be generated using oil, gas, coal, hydropower, or nuclear electricity so that there is always considerable scope for fuel substitution. In contrast, feedstock relies on the chemical composition of the material and, for the plastics industry, must be a hydrocarbon. As a consequence the scope for feedstock substitution is severely limited.
4. The minimum feedstock energy is the value equivalent to the mass of hydrocarbon material that is needed to produce the finished polymer. Any unwanted hydrocarbon by-products can be used as a fuel; that is, excess feedstock can be converted to a fuel. As a consequence, most polymer production units show a feedstock consumption that is very close to this expected minimum. Thus any attempts to make polymer production more energy efficient must concentrate on reducing the direct fuel consumption rather than the feedstock consumption.

As a result of these considerations, it is now common practice to report feedstock consumption separately from the consumption of fuels.

When feedstock is expressed as energy, great care is needed when interpreting the results. For example, if a plant takes in 1 kg of hydrocarbon feedstock as crude oil, then the energy equivalent is 45 MJ (the gross calorific value of crude oil). However, if an equivalent plant were to take in 1 kg of hydrocarbon feedstock as natural gas, the feedstock energy would be 54 MJ (the gross calorific value of natural gas). Thus, although both plants consume equal masses of hydrocarbon feedstock, the second plant appears superficially to be the less energy efficient.

3.4. TRANSPORT

The separation of the transport contribution is a relatively recent innovation. In extended systems that involve the use of materials in consumer products, transport can become a significant and, at times, dominant factor in total energy use. For the production of primary materials, such as polymer resins, it is usually relatively insignificant. However, it is usually reported separately for consistency and for use in the evaluation of downstream operations.

3.5. REPRESENTATIONS OF ENERGY

When gross energies are calculated using primary data collected from plant operations, there is sufficient detail within the calculations to satisfy all of the above requirements. Indeed, there is so much information available that if it were all to be reported, the results tables could become so large as to be unmanageable. A considerable amount of work has been done over the last 20 years or so in an attempt to aggregate the results so that they become usable without losing too much of the available useful detail.

One of the first such representations is the energy table shown in Table 3.1 [9]. Within this table, the overall energy requirements are analyzed into a number of groups. First, there is a breakdown by fuel-producing industry. The electricity supply industry is separately identified because, of all fuel supply industries, the electricity industry exhibits the lowest production efficiency. The oil industry is also separately identified, because, although oil fuels are consumed in a variety of different forms, they are all derived from a common source, crude oil, and they all exhibit approximately the same production efficiency.

Finally, all of the remaining fuels are grouped under the heading of other fuels. This group contains coal, coke, gas, and any biological fuels. This group also contains entries for any energy recovered as steam or condensate as well as any energy arising from sulfur burning in the production of sulfuric acid. In this report, the main contribution to the other fuels group is natural gas. Although it is possible to subdivide this group further, the simple representation of Table 3.1 is usually adequate. Each of the fuel-producing industry contributions is further subdivided into energy content of delivered fuel, feedstock energy, energy use in transport, and fuel production and delivery energy.

Table 3.1 Format for Presenting Energy in Terms of the Fuel-Producing Industries[a]

Fuel Type	Fuel Production & Delivery Energy (MJ)	Energy Content of Delivered Fuel (MJ)	Energy Use in Transport (MJ)	Feedstock Energy (MJ)	Total Energy (MJ)
Electricity	0.69	0.38	0.00	0.00	1.08
Oil fuels	0.40	7.99	0.06	25.21	33.66
Other fuels	2.02	7.66	0.06	22.51	32.24
Total	3.11	16.03	0.12	47.71	66.98

[a] The quantitative data refer to the average European energy required to produce 1 kg of ethylene. Totals may not agree because of rounding.

Energy content of delivered fuel represents the energy that is received by the final operator who consumes energy. This is independent of country and is directly related to the technology used in the various processing operations because this governs the demand for energy. This group also contains any entries that arise as a result of the recovery of energy as steam or condensate because this recovered energy will be used by other processes and so appears as a negative entry (credit) in the energy table.

Feedstock energy represents the energy of the fuel bearing materials that are taken into the system but used as materials rather than fuels. The quantities of hydrocarbon feedstocks that are taken into the system are represented in terms of their gross calorific value because, frequently, in the course of processing, some, if not all, of this feedstock may be converted to a fuel. It is a simple matter to convert from feedstock energy to mass if the calorific value is known since the energy content of a feedstock is simply the product (calorific value × mass).

Energy use in transport refers to the energy associated with fuels consumed directly by the transport operations as well as any energy associated with the production of non-fuel-bearing materials, such as steel, that are taken into the transport process.

Fuel production and delivery energy represents the energy used by the fuel-producing industries in extracting the primary fuel from the earth, processing it, and delivering it to the ultimate consumer. This will also include the energy associated with the production of any nonfuel materials (such as steel) that are used in the fuel production process.

The importance of this breakdown is that energy content of delivered fuel and feedstock energy are dependent on the technology used by the process operators. In contrast, the fuel production and delivery energy depends upon the country in which the processes are carried out. Also, the transport column, depends on the geographical location of the plants and has nothing to do with the actual production process. If therefore the aim is to compare technologies or plants that are using the same technology, then the country-dependent data can be stripped out of the results by omitting the fuel production and delivery energy and transport energy columns. Alternatively, by adjusting the values in the fuel production and delivery energy column to suit local conditions, the energy table can be used to illustrate the effect of using the same technology in different geographical locations with different fuel infrastructures.

3.6. GROSS AND NET CALORIFIC VALUES

One aspect of energy calculations that causes many problems in interpretation is the conversion factor used to convert fuel mass to energy. The conversion factor is called the calorific value of the fuel and represents the quantity of energy that can be derived from unit mass of fuel. There are, however, two types of calorific value — gross and net.

Gross calorific value is the energy obtained from a fuel when the combustion products are returned to the thermodynamic standard state and any water produced

128 POLYMERS AND ENERGY

in the combustion reaction is liquid. This definition yields a value equal to the total energy resource available from the fuel. Gross calorific value is sometimes referred to as the high heat value.

Net calorific value is the energy obtained when the combustion products all remain as gases. This definition is widely used in engineering design when the aim is to ensure that any water produced in the combustion reaction leaves the combustion reactor as a gas, thus minimizing corrosion. Since the energy content of gaseous water is not recovered, the net calorific value is lower than the gross calorific value. For natural gas the difference is about 10%, and for oil products the difference is about 5%. Net calorific value is sometimes referred to as the low heat value.

3.7. TYPICAL GROSS ENERGIES FOR POLYMER PRODUCTION

Typical values for the production of a number of different polymers are given in Table 3.2. These values are based on practices in plants across Europe but, given the similarity of technologies used in polymer production worldwide, there is no reason to suppose that different results would be obtained elsewhere when a correction has been made for the fuel infrastructure.

The actual values vary from plant to plant. In practice the direct energy consumption varies by ±10% and, when the fuel infrastructure of the different countries is included, the variations in gross energies are of the order of ±17%. These variations will include effects due to age and size of plant, level of maintenance, and variations in technology as well as accuracy of data reporting.

When considering the variations in production energy, it is important to remember a number of points.

1. Many polymer production units are part of larger chemical complexes. At such locations the goal is often the optimization of the whole site performance rather than individual plant performance.
2. The data of Table 3.2 do not refer to the final production step alone but to all of the processes from the extraction of crude oil and gas through to the production of monomers and finally to polymerization. The final polymerization step commonly accounts for only around 10% of the gross energy.
3. Most polymers are produced as powders or chips. However, some, such as viscose, are produced directly as the finished fiber, and so the spinning process is included in the data of Table 3.2. The data for nylon 66 and PET refer to granules but in practice these may also be produced at the site where they are converted to fiber.
4. Polyurethane is not sold as a finished polymer but as the precursors TDI, MDI, and polyols. These are mixed in the appropriate proportions together with other additives by the converter.
5. The data of Table 3.2 can be used to illustrate how the type of feedstock can markedly affect the gross energy requirement. For example, HDPE

Table 3.2 Gross Cumulative Energy Required to Produce 1 kg of Polymer[a]

	Electricity (MJ)				Oil Fuels (MJ)				Other Fuels (MJ)				
	Fuel Production & Delivery	Energy Content of Delivered Fuel	Transport	Fuel Production & Delivery	Energy Content of Delivered Fuel	Transport	Feedstock Energy	Fuel Production & Delivery	Energy Content of Delivered Fuel	Transport	Feedstock Energy	Total (MJ)
Acrylonitrile–butadiene–styrene copolymer (ABS)												
	4.83	2.16	0.03	0.56	15.82	0.23	17.37	2.86	22.77	0.08	28.41	95.13
Acrylic dispersion												
	6.91	3.09	0.03	0.80	12.49	0.13	22.52	2.74	22.52	0.08	17.37	88.69
Acrylic fiber												
	9.43	4.20	0.07	0.88	21.50	0.16	20.15	5.38	43.62	0.16	28.41	133.97
Ethylene–propylene–diene rubber copolymer (EDPH)												
	5.32	2.57	0.03	1.42	13.04	0.10	30.80	3.26	22.10	0.08	18.16	96.89
Epoxy (liquid resin)												
	17.13	7.93	0.11	0.66	10.25	0.40	15.94	4.94	56.90	0.09	26.24	140.59
MDI[1] (polyurethane precursor)												
	9.97	4.38	0.07	0.47	7.46	0.32	13.17	4.04	34.58	0.09	20.94	95.48
Nylon 6												
	10.43	4.64	0.06	1.12	13.62	0.77	16.41	4.66	45.16	0.14	22.50	119.51
Nylon 6 + 30% glass fiber												
	15.72	6.82	0.09	0.83	10.20	0.66	12.04	4.14	42.38	0.13	16.16	109.16
Nylon 66												
	20.55	8.82	0.12	0.80	11.59	0.31	20.75	5.74	47.38	0.14	28.38	144.58

(*continued overleaf*)

Table 3.2 (continued)

	Electricity (MJ)			Oil Fuels (MJ)				Other Fuels (MJ)				Total (MJ)
Fuel Production & Delivery	Energy Content of Delivered Fuel	Transport	Fuel Production & Delivery	Energy Content of Delivered Fuel	Transport	Feedstock Energy	Fuel Production & Delivery	Energy Content of Delivered Fuel	Transport	Feedstock Energy		
Nylon 66 + 30% glass fiber												
19.42	8.19	0.15	0.66	8.56	0.35	15.35	4.89	42.62	0.13	18.57	118.89
Nylon 66 fiber												
38.55	16.37	0.20	0.93	11.60	0.28	20.86	6.82	62.58	0.15	26.21	184.54
Polybutadiene												
4.36	2.00	0.02	0.93	22.03	0.15	33.18	3.73	26.49	0.18	13.31	106.40
Polycarbonate												
10.38	4.50	0.10	0.48	8.56	0.18	11.01	3.74	50.38	0.07	26.49	115.90
Polyester fiber												
10.96	4.66	0.07	0.94	11.75	0.25	32.97	3.80	19.59	0.14	5.73	90.85
Polyethylene (high density) (HDPE)												
5.69	2.83	0.02	0.49	13.83	0.07	34.81	2.61	5.94	0.24	13.70	80.23
Polyethylene (linear low density)												
2.02	0.89	0.01	0.35	8.31	0.17	11.03	1.41	11.75	0.01	36.42	72.37
Polyethylene (low density) (LDPE)												
7.01	3.82	0.03	0.50	8.48	0.06	21.93	1.83	7.48	0.07	29.62	80.82
Polyethylene terephthalate (PET) (amorphous)												
5.75	2.38	0.03	0.86	10.87	0.26	32.56	3.32	12.77	0.13	5.84	74.78
Polyethylene terephthalate (bottle grade)												
6.02	2.49	0.03	0.88	11.06	0.26	32.56	3.37	13.56	0.14	5.84	76.10

Polymethyl methacrylate (beads)	9.15	4.05	0.06	2.25	21.61	0.42	14.19	4.31	24.70	0.07	26.86	107.65
Polymethyl methacrylate (sheet)	15.84	6.98	0.10	2.40	22.78	0.65	15.25	5.27	34.84	0.08	24.91	129.10
Polyols (polyurethane precursor)	12.54	5.59	0.08	0.50	12.07	0.25	21.22	3.11	23.31	0.11	15.08	93.86
Polypropylene (PP)	3.69	1.86	0.02	0.57	11.56	0.05	29.91	2.78	7.66	0.02	19.12	77.24
Polystyrene (expandable)	2.39	1.07	0.02	0.62	9.94	0.32	22.91	2.60	18.92	0.07	24.89	83.74
Polystyrene (general purpose)	2.58	1.13	0.01	0.71	12.41	0.31	19.51	2.44	19.57	0.06	28.02	86.77
Polystyrene (high impact)	3.31	1.45	0.01	0.74	13.87	0.31	20.98	2.63	20.69	0.07	28.14	92.22
Polyvinyl chloride (emulsion) (PVC)	8.13	3.69	0.03	0.76	9.28	0.16	11.63	2.25	22.05	0.07	9.96	68.01
Polyvinyl chloride (suspension)	7.83	3.66	0.04	0.34	5.18	0.13	11.80	1.76	15.20	0.06	11.12	57.11
Polyvinylidene chloride	16.49	7.38	0.06	0.84	9.56	0.29	12.59	2.94	33.83	0.25	8.18	92.41
Styrene–butadiene dispersion	3.29	1.53	0.02	0.56	14.07	0.19	24.03	2.75	19.02	0.09	22.29	87.84
TDI[2] (polyurethane precursor)	13.65	5.96	0.10	0.79	10.00	0.39	11.29	3.93	42.87	0.10	20.74	109.82
Viscose fiber	19.44	8.84	0.15	1.32	10.30	0.25	0.01	9.16	53.70	0.44	49.64	153.27

[a]Totals may not agree because of rounding. These data are based on average European practices during the 1990s.
[1]MDI = Diphenylmethane diisocyanate.
[2]TDI = Tolylene diisocyanate.

exhibits a consumption of 34.81 MJ of oil feedstock and 13.70 MJ of gas feedstock. If the calorific values of oil and gas are 45 and 54 MJ/kg, respectively, these data correspond to $34.81/45 = 0.774$ kg oil and $13.70/54 = 0.254$ kg gas—a total feedstock input of $0.774 + 0.254 = 1.028$ kg hydrocarbon feedstock per kilogram polymer. Suppose now that all of this feedstock were to be supplied as gas rather than a mixture of oil and gas. The gas feedstock rises to $54 \times 1.028 = 55.15$ MJ, and the gas production energy rises from 2.61 to 8.10 MJ. At the same time, the oil feedstock decreases to zero, and the oil production energy decreases from 0.49 to 0.14 MJ. The overall gross energy therefore increases from 80.23 MJ/kg polymer to 92.37 MJ/kg polymer, an increase of 15%. Conversely, if all of the feedstock had been supplied as oil, then the gross energy would decrease to 76.51 MJ/kg polymer, a fall of 5%. It is generally true that gross energy is always higher when gas feedstock is used, and so any comparison of two polymer production processes must take into account variations in feedstock type.

3.8. CONVERSION PROCESSES

The techniques used to convert polymer granules to final products are well established, and the main inputs are electricity, thermal fuels, and cooling water. Table 3.3 gives values of these parameters for a selection of processes. At the conversion stage various additives and fillers are usually incorporated into the product. In Table 3.3 these are included in the column headed resin.

The data of Table 3.3 cannot be regarded as definitive because of the nature of the conversion industry. In particular, note should be taken of the following points:

1. Unlike polymer production, the conversion industry is fragmented and ranges from large modern plants, continuously producing a single product such as sheet or PET bottles, to small jobbing molders that produce short runs of many different products, usually by injection molding. For these small molders, the frequent stopping, starting, and mold changes on machines that are not always tailored to the job in hand means that they are usually less energy efficient than the large-volume producers who use machines specifically matched to their purpose.
2. Materials losses are typically 1–2% and mainly arise from contaminated waste generated during startup, maintenance, and machine purging. For injection molding there is also a loss arising from the removal of flashing and sprues; such waste can only sometimes be recycled back into the process.
3. Few operators monitor separately the energy consumption of the machines and the energy used in space heating and lighting. Consequently, the data of Table 3.3 refer to plant performance rather than machine performance.

Table 3.3 Data for Production of 1 kg of Selected Products

Polymer	Product	Process	Resin (kg)	Electricity (MJ)	Thermal Fuels (MJ)	Water (kg)
PP	Oriented film	Extrusion	1.01	4.76	4.99	223.0
PET	Bottles	Stretch blow molding	1.00	5.00	—	0.1
PET	Film	Extrusion	1.01	4.87	4.02	0.3
HDPE	Bottles	Blow molding	1.00	6.15	0.03	3.0
HDPE	Pipes	Extrusion	1.00	1.94	0.79	25.0
LDPE	Film	Extrusion	1.00	1.69	0.00	0.3
PP	Miscellaneous	Injection molding	1.01	7.55	13.96	11.1
PVC	Sheet	Calendering	1.00	1.82	1.02	19.1
PVC	Unplasticized sheet	Extrusion	1.04	1.70	0.20	—
PVC	Miscellaneous	Injection molding	1.01	4.95	1.19	22.2
EPS[1]	Trays	Thermoforming	1.02	1.81	0.22	85.0

[1] Expanded polystyrene.

4. Water consumption can vary widely from one plant to another depending upon whether the plant uses a recirculating system or a once-through cooling system. The picture is further complicated in that some plants refrigerate the cooling water before use. This leads to an increase in energy consumption but a decrease in the volume of water needed so that once-through systems can also differ markedly.

3.9. ENERGY AND POLYMER RECYCLING

Increasingly, there is a demand, as with other materials, that plastics should be recycled. For most thermoplastics this is a possible route for postconsumer material, and this is dealt with in more detail in other chapters. However, there is some confusion over the energy savings that can be achieved by plastics recycling, and this can lead to conflicting decisions.

The problem is best illustrated by considering first the recycling of metals. Primary aluminum, for example, is produced by the electrolysis of fused alumina and is an energy-intensive process with a gross energy of 150 MJ/kg aluminum. Suppose that 100% of all aluminum can be recovered, that there are no losses during remelting, and that remelting uses an energy of 20 MJ/kg, then for the average energy required to use 2 kg of aluminum is $150 + 20 = 170$ MJ or the equivalent of 85 MJ/kg. In other words the primary production energy is spread over more than one use. Repeated recovery and remelting would, of course, reduce the average energy per kilogram even further. In practice the average saving will be lower because it is not possible to recover 100% and some melt losses do occur. The example does, however, illustrate the critical point that repeated recycling can lead to energy savings because the initial production energy is spread over more than one use.

For plastics the situation is somewhat different. The total gross cumulative energy required to produce virgin polymer is the sum of the polymer production energy and the feedstock energy. The polymer production energy is the equivalent of the production energy for aluminum, but the feedstock energy has no corresponding component in the aluminum example. Thus when considering polymer recycling, it is essential to keep separate the production energy and the feedstock energy, and different techniques are needed to affect these components.

There are essentially two techniques applied in polymer recycling: mechanical recycling and energy recovery. In mechanical recycling, the postconsumer material is recovered, cleaned, and flaked so that it can be used again to fabricate further products. There may be a deterioration in the properties of the recycled polymer, which means that it is used in products with a less demanding specification; but, nevertheless such recycled material replaces polymer that would otherwise have to be produced from virgin sources. This process is identical to the recovery and reuse of aluminum because the production energy is spread over more than one use. The feedstock energy is unaffected by mechanical recycling. If 1 kg of polyethylene has a feedstock energy of 45 MJ/kg, then this will be present in the virgin polymer and also in the recycled polymer; burning recycled polymer yields the same energy as burning virgin polymer. Feedstock energy can therefore be regarded as a circulating load that remains as long as the polymer exists.

In energy recovery, the postconsumer material is recovered and burned as a fuel. This process essentially recovers that proportion of the feedstock energy that remains in the polymer. In practice, the whole of the feedstock energy is seldom available for recovery. Feedstock energy is the calorific value of the hydrocarbon feed to the polymer production system. However, chemical changes and losses during production mean that the feedstock energy will be different from the calorific value of the final polymer. Despite this, it is clear that energy recovery acts on the feedstock energy and has nothing to do with the production energy.

This distinction between production energy and feedstock energy is very important in recycling for two main reasons:

1. Because the two recovery methods act on different components of the gross energy, they are not mutually exclusive. Mechanical recycling acts on production energy by spreading it over multiple uses. In contrast, energy recovery acts on feedstock energy by attempting to recover it. These two recovery techniques are often presented as if they were mutually exclusive so that a choice must be made between mechanical recycling or energy recovery. In fact this choice need never be made. If the ultimate goal is energy saving, then mechanical recycling should always be followed by energy recovery.

2. To achieve energy savings in mechanical recycling, the recovery and reprocessing of postconsumer waste must use less energy than is needed to produce virgin polymer. To assess whether or not a recycling process is energy saving, it is necessary to identify the window of opportunity for recovery. For HDPE, the gross energy to produce virgin polymer (i.e., production energy plus feedstock energy) is 80 MJ/kg. Of this 48 MJ/kg is feedstock and 32 MJ/kg is production energy. The recovery process therefore must not use more than 32 MJ/kg.

REFERENCES

1. H. M. Meadows, L. M. Meadows, J. Randers, and W. W. Behrens. *The Limits to Growth*, Pan Books, London, 1972.
2. National Academy of Sciences & National Research Council, *Resources and Man—A Study and Recommendations*, W H Freeman, San Francisco, 1969.
3. H. Smith. *Trans. World Energy Conf.* **18**(E), (1969).
4. *The Competitiveness of LDPE, PP and PVC after the 1963 Oil Crisis*, Imperial Chemical Industries, London, 1974, 1975.
5. G. Sundstrom. Investigation of the Energy Requirements from Raw Materials to Garbage Treatment for Four Swedish Beer Packaging Alternatives, Report for Rigello Pak AB, 1973.
6. Franklin Associates, *Family Size Soft Drinks Containers—A Comparative Energy and Environmental Impact Analysis*, Report for Goodyear Tire and Rubber Company, Prairie Village, Kansas, November 1977.
7. I. Boustead. *Ecoprofiles of the European Plastics Industry*, (1) *Methodology* (1992), (2) *Olefin Feedstock Sources* (1993), (3) *Polyethylene & Polypropylene* (1993), (4) *Polystyrene* (1993, 1997), (5) *Co-product Allocation in Chlorine Plants* (1994), (6) *PVC* (1994, 1998), (7) *Polyvinylidene chloride* (1994), (8) *PET* (1995), (9) *Polyurethane Precursors (TDI, MDI, Polyols)* (1996), (10) *Polymer Conversion* (1997), (11) *ABS and SAN* (1997), (12) *Liquid Epoxy Resins* (1997), (13) *Polycarbonate* (1997), (14) *PMMA* (1997), (15) *Nylon 66* (1997), (16) *PET Film Production* (1998), (17) *Polymer Dispersions* (1999), APME/CEFIC, Brussels.
8. International Federation of Institutes of Advanced Studies (IFIAS), Workshop Report No. 6. Stockholm, 1974.
9. I. Boustead and G. F. Hancock. *Handbook of Industrial Energy Analysis*, Wiley, New York, 1979.

PART 2

CHAPTER 4

PLASTICS IN PACKAGING

SUSAN E. M. SELKE
School of Packaging, Michigan State University, East Lansing, Michigan

4.1. INTRODUCTION

In the United States, approximately 25% of all plastics produced are used in packaging. While paper and paperboard is the most commonly used packaging material, plastics are the most rapidly growing. They have extensive use in containers, in flexible packaging, as blisters, trays, and other forms. While they are very often used alone, they also have important applications where they are combined with other materials in coatings or as layers in multilayer structures. In such applications, plastics contribute important properties such as ease of forming, heat sealability, barrier, flexibility, impact strength, light weight, reduced package size, and low cost. However, from time to time, plastics packaging has suffered from negative consumer perceptions ranging from the belief that all plastics are associated with dangerous emissions when they are burned to the fear that components migrating from plastics packaging into products may damage human health. While in many cases there is some factual basis for such concerns for a few plastics, in most cases the fear is both out of proportion to the danger and extended to other plastics materials where there is no reason for concern. The environmental benefits associated with plastics packaging, on the other hand, are rarely appreciated.

Since most packages have a lifetime of less than one year, the series of Environmental Protection Agency (EPA) reports on the U.S. municipal solid waste (MSW) stream, while not precisely targeted at reporting uses of plastics and other packaging materials, provides a useful estimate of overall use of plastics and other materials in packaging, as well as information about the contribution of various types of plastics packaging to waste problems. Figure 4.1 shows

Plastics and the Environment,   Edited by Anthony L. Andrady.
ISBN 0-471-09520-6   © 2003 John Wiley & Sons, Inc.

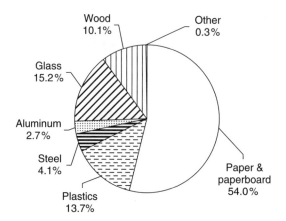

Figure 4.1. Packaging in U.S. municipal solid waste stream, 1998 [1].

the proportions of various types of packaging materials in U.S. MSW generated (prior to recovery) in 1998. As can be seen, plastics accounted for slightly less than 14% of all packaging in MSW by weight [1].

However, the contribution of plastics to landfill is underrepresented by this value. The important measure in a landfill is contribution by volume, not weight. Because of their low density compared to materials such as metals and glass, plastics occupy a greater proportion of landfill space by volume than they do by weight. For a variety of reasons, it is exceedingly difficult to obtain accurate estimates of the contribution to landfill volume of various materials, including plastics packaging. The EPA has published such estimates on occasion, but they must be regarded as only rough approximations. Figures 4.2 and 4.3 show the relative proportions of plastics in discarded packaging (after recycling) in U.S. MSW in 1996 by weight and by volume, respectively. As can

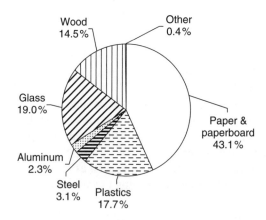

Figure 4.2. Packaging discarded in United States, by weight, 1996 [2].

INTRODUCTION **141**

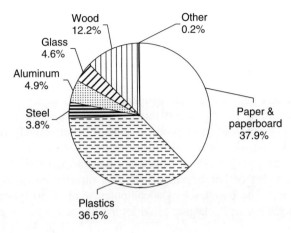

Figure 4.3. Packaging discarded in United States, by volume, 1996 [2].

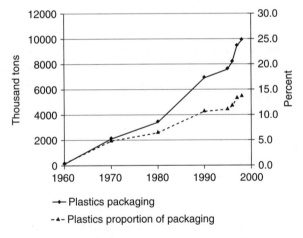

Figure 4.4. Plastics in packaging in U.S. municipal solid waste stream [1].

be seen, plastics represent a much larger proportion by volume (37%) than by weight (18%) [2]. It should be noted that Figures 4.1 and 4.2 are not directly comparable, since Figure 4.1 shows proportions before recovery for recycling, and Figure 4.2 shows proportions after recovery. In 1996, plastics amounted to 11.8% of the packaging in municipal solid waste, before recovery. Figure 4.4 shows the growth in both absolute amounts and proportions of plastics in the packaging component of U.S. municipal solid waste.

In this chapter, we will examine the types of plastics packaging, their fabrication methods, their occurrence in municipal solid waste, environmental impacts (both positive and negative) associated with their use and disposal, and packaging changes that can reduce those environmental impacts. First, we will examine the basic functions of packaging and the role plastic packaging plays in meeting those functions.

4.2. PACKAGING FUNCTIONS

Packaging is a cost item for product manufacturers. Obviously, the reason companies pay for putting their products into packages is because of the functions that the packages provide. Most products cannot be effectively manufactured, distributed, and used without the use of packaging at some point in the process. For products such as soft drinks the necessity for packaging is obvious. No one is likely to want his or her beverage delivered directly into cupped hands at the grocery store! The purchaser of a computer wants it to function correctly when it is brought home. Packaging provides the protection necessary to prevent damage. For products such as automobiles, the need for packaging may be less evident. However, the modern process of automobile manufacturing requires that parts from a number of manufacturing facilities be assembled — and to reach the assembly facility in acceptable condition, the parts must be packaged.

While packaging functions can be grouped in various ways, one common grouping is containment, protection, utility, and communication. While grouping the functions in this way makes discussion easier, it should be emphasized that the boundaries between the groups are not always clear-cut, and in many cases a single package feature serves multiple functions.

Containment is the most basic package function. It is not possible to transport a liquid or granular product without containing it in some fashion. For larger products that could conceivably be transported without a container, containment simplifies the task by making it possible to handle a number of products as a unit — the reason most people bring their groceries home in a bag.

Protection of the product by the package takes a number of forms. Containment itself is likely to provide some degree of protection against product contamination or damage due to exposure to environmental influences such as dust and microorganisms. Depending on the product, protection may be needed from gain or loss of water or other volatile substances; from impact, abrasion, or vibration; from corrosion; from exposure to sunlight; from pilfering, and so forth. In some cases, such as child-resistant packaging or packaging for hazardous materials, packages function to protect humans or the environment from exposure to the product.

Utility refers to functionality or convenience provided by the package for one or more entities involved in product manufacture, distribution, or use. For example, cooking sprays in aerosol cans produce a lubricating effect in a frying pan with much less product than would be required without the aerosol. The carton on a tube of toothpaste permits the retailer to stack the tubes on the shelf without them rolling off. The metal edge on the carton of plastic wrap permits the consumer to tear off the desired amount of the product. Stretch wrap and a pallet permit a load of goods to be handled by a forklift and stored in a rack system in a warehouse.

Communication refers to the messages about a product imparted by a package to those who interact with it. These messages range from basic identification of the product and its manufacturer to subtle (or not so subtle) "buy me" messages communicated to potential purchasers. We commonly think of communication as the printed information about the package, but color and shape often also serve important communication functions. Smell and sound are not often used as communicators at present, but that may change in the near future. We already have scratch-and-sniff features, as well as devices for prescription drug containers that when placed in a special device will read the label information to the user. Bar codes are routinely used to transmit price information at the point of sale, as well as for tracking goods during distribution.

While these functions are common to all types of packages, increasingly product manufacturers are turning to plastics packaging to obtain the same or even improved functionality at lower cost than with alternative materials.

4.3. ADVANTAGES OF PLASTICS IN PACKAGING APPLICATIONS

Plastics of various types offer a number of attributes that contribute to the packaging industry's growing reliance on this material. While the details vary from one application to another, the most significant driving force in increasing plastics use is the corporate bottom line — use of plastics packaging very often increases corporate profitability by lowering costs, increasing sales, or sometimes both. While it is not possible, in the scope of this work, to illustrate all the ways in which use of plastics in packaging can be advantageous, some illustrative examples follow.

When a consumer purchases products from a retailer, very often these products will be packed in a plastic merchandise bag. While paper merchandise bags have not been wiped out of the market, plastic has taken a large portion of the market, and many retailers, especially in the nongrocery segments, no longer offer consumers a choice between paper and plastic bags. The primary reason is that plastic bags are less expensive for the retailer to purchase. This is not their only advantage. Plastic bags take less room in the warehouse and in the distribution vehicle. In many cases the built-in handle offers improved functionality for the consumer. The bags are not susceptible to loss of strength when they get wet, so may provide improved product protection. When they are no longer useful, they take up less space in a landfill than their paper equivalent, where they are stable and do not contribute to land subsidence or production of methane gas. If they are incinerated, they provide a valuable source of energy for conversion into electricity. Of course, on the other hand, unlike paper bags they do not biodegrade in a landfill and do not contribute to production of methane for conversion to heat or electricity. Also, they are manufactured from a nonrenewable resource.

In distribution of automobile parts, reusable plastic totes have, to a considerable extent, displaced corrugated boxes. The plastic containers have a higher initial purchase price but a much longer lifetime. Of course, they must be managed appropriately so that they are not lost or damaged, and so they are at the appropriate location when they are needed. The amount of packaging waste generated by the assembly plant is greatly decreased. The assembly line stays cleaner. Protection of parts from damage may be enhanced. Again, the bottom line is decreased cost.

In packaging of fresh meat, old-fashioned white paper butcher wrap has all but disappeared. Even if the retailer maintains a butcher shop with fresh meat displayed in glass cases, the meat probably arrives at the store in plastic high-barrier vacuum packages that are able to exclude oxygen and microorganisms and prevent loss of water vapor, thereby providing an extended shelf life. For appeal to American consumers, the meat must be oxygenated on display, to change the purple color it has in low-oxygen environments to the bright red color preferred by consumers. Film [usually polyvinyl chloride (PVC)] with high oxygen permeability as an overwrap on a plastic tray [often expanded polystyrene (PS)] makes this possible and allows the consumer to see the meat while providing needed protection from contamination and moisture loss. In contrast, meat packaged in butcher wrap begins to dry out almost immediately. New packaging systems for retail meat use a high-barrier film sealed to a deep-draw tray to maintain a high oxygen atmosphere that extends shelf life while maintaining product visibility and desirable appearance. Instead of the roughly 3 days the meat will stay fresh in the consumer's refrigerator with the PVC film system, this new system allows consumers to keep the meat unfrozen in their refrigerator for about a week — and eliminates the problem of leaky packages, as well.

The market segment of fresh prepared foods depends on the use of plastics to provide the tailored barrier properties and/or modified atmosphere needed to maintain acceptable quality in the product for the required time. Many of these products were not available at all only a few years ago. Now you can purchase salad ready to eat, fresh pasta ready to be cooked that can be kept in your refrigerator for several weeks, cheese already grated, carrots already peeled, and a variety of other products — all dependent on plastics both to maintain quality and to allow the consumers to see the product and be convinced to purchase it.

Shampoo, bath oil, liquid soap, and many other products used primarily in bathrooms are nearly always packaged in plastic containers. Plastics not only are resistant to moisture, they are able to withstand dropping out of slippery hands onto hard surfaces without breaking — and, not inconsequentially, without producing sharp fragments that could impart serious injury and open the way for lawsuits. Plastics can provide transparency if it is desired, or they can be formulated in a variety of colors and shapes to produce the desired image for a product. The plastic closures used on the containers usually include a dispensing feature of some type to facilitate product use and also to decrease spills. Dispensing is also aided by the flexibility of plastic, which allows the container to be squeezed. Further, the plastic containers are considerably lighter in weight and somewhat smaller than the glass or metal packages they replace, which directly translates into savings in both distribution and storage. In addition to savings in fuel use and warehouse space, less distribution packaging (corrugated boxes, pallets, etc.) is needed to deliver the same amount of product.

Computer chips for various types of electronic appliances are very susceptible to damage from static electricity. Special plastic wraps or bags provide a conductive path to dissipate the static charges, thus protecting the chips from damage during parts distribution.

Sterility of medical instruments and other medical devices is a life-and-death issue. Many such devices are intended for only a single use and are sterilized in the package before they are distributed. Very often, the package used to permit in-package sterilization and then maintain this sterility is plastic. Spun-bonded polyethylene (DuPont's Tyvek), for example, permits ethylene oxide sterilization because it is porous to the gas, but the pore sizes are too small to permit microorganisms to enter. Plastics can also be used to permit radiation sterilization or even steam sterilization. For operating room use, plastics can provide ease of opening without production of the airborne fibers that paper may generate.

4.4. TYPES OF PLASTICS PACKAGING

Plastics packaging can be grouped into the general categories of rigid and semi-rigid containers, flexible packaging, and other forms. In the container category, bottles are the most common package type, but plastics are also widely used in tubs, tubes, drums, bins, trays, and other shapes. The flexible packaging category includes wraps, pouches, bags, sacks, envelopes, and similar packages. The "other" category includes, among others, cushioning (loose-fill and molded), blisters, caps, and lids.

As mentioned, the use of plastics in packaging has been growing steadily for a number of years. Rigid plastic containers (bottles, jars, drums, etc.) have replaced materials such as glass and metal. There is also a long-standing trend of replacement of rigid packages with flexible packages. For example, many institutional food service items are now distributed in multilayer pouches rather than in steel cans.

4.4.1. Flexible Packaging

Production of flexible packaging begins with extrusion of plastic resin to yield sheet or film. For film, two main processes are used. In the cast-film process, the resin is extruded through a slit-shaped die onto a water-cooled chrome roller, where it is solidified. After the edges are trimmed off, the film is wound onto a roll. The film may be oriented by stretching it in the machine direction (uniaxially oriented) or in both the machine and cross directions (biaxially oriented) to improve mechanical and barrier properties. This process is also used to produce plastic sheet for thermoforming or other applications.

In the blown-film process, the resin is extruded through an annular die and internal air pressure used to expand the bubble. After the resin solidifies, the bubble is collapsed and may be wound as a tube or slit and wound as rolls of flat film. Since the film is stretched by the take-off rollers as well as by the bubble expansion, it has biaxial orientation.

If a multilayer material is desired, it can be produced by coextrusion or by a subsequent laminating operation. Coextrusion has the advantage of producing a multilayer material in a single step and permits individual layers to be much thinner than is possible in a laminated structure. On the other hand, lamination permits use of nonplastic layers such as paper or foil. Coating, with plastics or nonplastics, can also be used to produce multilayer structures. Metallized film uses an extremely thin layer of aluminum, added in a vacuum deposition process, to greatly enhance barrier or, sometimes, to enhance the appearance of the material. Silicon oxide coatings (SiO_x) are also used to enhance barrier. This technology is not yet widely used in the United States, though it is quite common in Japan. An advantage over metallization is that transparency is preserved, including transparency to microwave radiation so that these materials can be employed for packaging microwave-ready food products. Transparent aluminum oxide and other metal oxide coatings have also been developed.

Films may be used without further conversion as wraps, such as in pallet stretch wrap, or in shrink wrap. Most often, films are converted into bags, sacks, pouches, or envelopes. Usually, the package seams are formed using heat sealing. In many cases, packaging is done in a form–fill–seal operation in which the package is formed and the product filled in a single machine. If the package is printed, as is often the case, the printing is usually done while the plastic material is in the form of roll stock, prior to forming the package. Commonly, some type of surface treatment of the plastic is necessary prior to printing, to enhance ink adhesion. Corona treatment, in which the plastic is exposed to an electrostatic discharge that produces some oxidation of the surface, is used most often.

4.4.2. Blow-Molded Containers

Plastic bottles are generally produced by blow molding. The most common process is extrusion blow molding, in which the plastic resin is extruded through an annular die, forming a hollow tube (parison). The parison is captured in the blow mold, and air pressure is used to expand it into the shape of the mold. This is a very versatile process that can be used to produce a wide variety of shapes and sizes, from tiny containers to 55-gal drums and larger. The containers may have hollow handles, offset necks, and a variety of other features. Some are even produced with two or more compartments in a single container. Extrusion blow-molding results in generation of scrap (flash) from the extra material that must be trimmed off from above the rim of the container and below the bottom (pinch). Handles result in generation of additional scrap from the area between the handle and the body. If a container has an offset neck, a considerable amount of scrap is usually generated from the neck area. In most cases, this flash is automatically collected, ground up, and metered back into the process. The amount of material required in each container can be minimized by controlling the distribution of plastic in the parison, and hence in the finished container, using such techniques as die shaping and parison programming.

Injection blow molding is used primarily for plastic resins that lack sufficient melt strength for extrusion blow molding. Injection blow molding is also used for applications such as pharmaceutical products where very accurate finish (threaded area of the neck) dimensions are required to ensure good sealing of the container. In this two-step molding process, a preform or parison is produced by injection molding, and then it is transferred to a blow-mold for expansion into its finished form. In the injection-molding step, a mold is filled through one or more gates with melted plastic, which is then solidified, and the molded product ejected. The injection-molded parison may be immediately transferred to the blow mold and blown without reheating or may be totally cooled and then reheated and blown at a later time. Injection blow molding produces much less scrap than extrusion blow molding. It also often permits better control over the wall thickness of the finished container.

Stretch blow molding is a process designed to impart biaxial orientation to bottles or jars. In this process, a preform is produced (usually by injection molding–injection stretch blow molding) that is shorter than the final container. During the blowing step, a stretch rod stretches the preform in the vertical direction as it is expanded in the radial direction by blowing. Biaxial orientation can significantly improve the strength and barrier properties of the container.

Coextrusion blow molding is often used to produce multilayer plastic containers. A multilayer parison is produced, and subsequent operations are identical to ordinary blow molding. Reuse of flash is more difficult since it will be multilayer. In many cases, the flash is used in an inner layer in the container, either alone or combined with other recycled plastic. High-density polyethylene (HDPE) laundry product bottles, for example, routinely are made with a three-layer structure in which the inner layer contains flash blended with postconsumer recycled HDPE.

Multilayer injection blow-molded containers have been available for a much shorter time than multilayer extrusion blow-molded containers, as development of the process was more difficult. However, these too are now available. A prominent example is multilayer polyethylene terephthalate (PET)-based ketchup bottles, which contain three layers of PET, and two layers of ethylene vinyl alcohol (EVOH) as an oxygen barrier.

When containers will be exposed to elevated temperatures, such as in packaging of hot-filled products, container distortion may be a problem, especially for biaxially oriented bottles. In this case, a subsequent heat-setting operation may be used to reduce thermal stress and minimize deformation on subsequent exposure to elevated temperatures.

For many products, the labels are placed into the empty mold and sealed to the container during the molding process (in-mold labeling). This eliminates the need for a subsequent labeling step and produces somewhat recessed labels with a better appearance, including improved scuff resistance. In-mold labeling is used most often with extrusion blow molding but can also be used with injection molding.

4.4.3. Injection-Molded Components

Injection molding alone (see description above) is also used to produce some packaging containers and components. Most threaded closures (caps) for bottles are produced in this way since there is no other feasible way to produce accurate threads in plastic caps. Plastic tubs and other shapes are also often produced by injection molding. Injection molding excels at producing very tight tolerances on the finished parts.

In both injection molding and injection blow molding, the plastic flows to the injection mold through a system of runners. In most packaging applications, a variant of injection molding, hot runner molding, is used. In this process, the mold is designed to provide heating of the runners to keep the plastic in them from solidifying. When the mold is emptied and the next cycle begins, the first plastic to enter the mold is the still-molten plastic that was in the runners during the previous molding cycle. This significantly reduces scrap generation and also decreases cycle time.

4.4.4. Thermoformed Shapes

Thermoforming is another common method of producing packages and package components. Most trays and blisters (dome of plastic attached to a backing material to enclose a product) are produced by thermoforming. In thermoforming, a plastic sheet is heated to soften it, and then it is shaped into or over a mold, using vacuum or a combination of vacuum and pressure.

There are a number of thermoforming variations, including the use of male or female (positive or negative) molds, prestretching with air (billow forming), and use of mechanical devices to assist in moving the plastic into the mold (plug-assist). Matched-mold forming uses a combination of male and female molds to produce very good dimensional control. Twin-sheet thermoforming uses two molds with two plastic sheets to produce hollow objects such as pallets. In-line thermoforming builds thermoforming into the extrusion line, so that the plastic need not be totally cooled and rolled up before it is thermoformed, permitting some energy savings. Melt-to-mold thermoforming is a relatively new process that can be regarded as a cross between extrusion and thermoforming. In this process, the melted plastic is delivered to a roller that contains engraved molds, rather than cooled on a smooth roller for subsequent reheating and thermoforming.

Most thermoforming operations produce a relatively large amount of scrap in each cycle. Scrap rates of 30% are not unusual. If the sheet is produced in one location and thermoformed in another, or if the material is multilayer, reuse of scrap is more difficult than in extrusion blow molding.

4.5. COMMON PACKAGING PLASTICS

The major plastic resins used in packaging are high-density polyethylene, low-density and linear low-density polyethylene, polyethylene terephthalate, polypropylene, polystyrene, and polyvinyl chloride. A variety of plastics are used in lesser quantities: nylons or polyamides, polycarbonate, polyethylene naphthalate, polyvinylidene chloride, ethylene vinyl alcohol, polyvinyl alcohol, polyvinyl acetate, polyacrylonitrile, and more. In many applications, copolymers

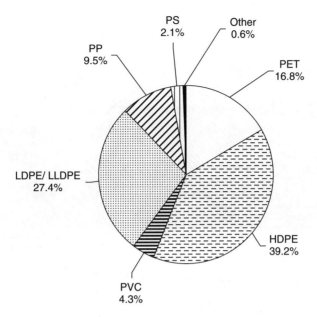

Figure 4.5. Plastics resins in packaging in U.S. municipal solid waste stream, 1998 [1].

and/or multilayer packages containing two or more different polymers are used. Figure 4.5 shows the proportions of major plastics found in packaging in U.S. municipal solid waste in 1998.

4.5.1. High-Density Polyethylene

In 2000, approximately 15,602 million pounds of virgin high-density polyethylene (HDPE) resin were used in the United States and Canada [3]. Packaging film and containers accounted for a major fraction of this use, over 38% if grocery sacks and merchandise bags are included as packaging (Fig. 4.6).

The largest category of HDPE use in packaging, by far, is bottles. For virgin resin, nearly 50% of these bottles were used for food, 6% for motor oil, and the remaining 44% for household and industrial chemicals of various types [3]. Within the food category, the largest fraction is milk and water bottles. In 1998, HDPE milk and water bottles amounted to more than 18% of all plastic containers in the MSW stream. Bottles and containers totaled 31.4% of all HDPE packaging in the MSW stream, and packaging represented over 78% of all HDPE in the MSW stream [1]. HDPE offers reasonable stiffness and strength, excellent moisture barrier, the ability to use pigments to achieve a wide variety of colors, and low cost. Its primary drawback is its lack of transparency and its poor barrier to most gases.

In the United States, recycling of HDPE bottles through curbside collection and dropoff systems is very common. According to the American Plastics Council, more than 20,000 American communities have access to plastics recycling,

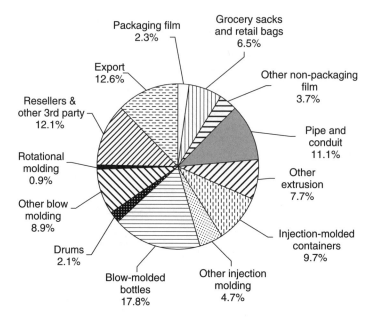

Figure 4.6. Uses of HDPE in United States and Canada, 2000 [3].

and nearly all major communities include HDPE bottles in their collection programs [4]. Recycling of nonbottle HDPE containers, and of other forms of HDPE packaging, is much less common.

Milk and water bottles are the highest value type of recycled postconsumer HDPE because their lack of pigment permits them to have a wide variety of potential uses. They also have the highest recycling rate of HDPE containers, 31.4% in 1998 in the United States. The overall HDPE packaging recycling rate in 1998 was 10.3% [1]. The HDPE bottle recycling rate was 30.7% in 1998 but fell to 29.7% in 1999. The amount of HDPE bottles recycled in 1998 increased to nearly 750 million pounds but did not increase as fast as production [4].

In Alberta, Canada, a voluntary stewardship program for HDPE milk bottles achieved a provincewide recovery rate of 40%, with 16 communities achieving rates of 70% or more. The program provided a guaranteed floor price to collection organizations along with funds for promotional activities [5].

Recycled HDPE is an important material in manufacturing of certain types of HDPE packaging. Most HDPE bottles for laundry products such as detergent and fabric softener are manufactured with a three-layer structure, containing postconsumer recycled HDPE blended with regrind from the bottle production process in the middle layer. Many motor oil bottles contain a blend of postconsumer recycled plastic with virgin plastic in a single-layer structure. Recycled HDPE from milk bottles is often used in manufacture of merchandise bags.

The U.S. Food and Drug Administration (FDA) has issued a letter of nonobjection for at least one process for use of recycled HDPE in a buried inner layer in packaging for certain types of food products.

4.5.2. Low-Density Polyethylene (LDPE) and Linear Low-Density Polyethylene (LLDPE)

About 17,565 million pounds of virgin low-density polyethylene and linear low-density polyethylene were used in the United States. and Canada in 2000 [3], primarily in film applications. Packaging film alone represented over 28% of LDPE and LLDPE use, if carryout bags are included (Fig. 4.7). EPA estimates that LDPE/LLDPE bags, sacks, and wraps accounted for 85.6% of all LDPE and LLDPE packaging in the MSW stream in 1998, and that packaging accounted for 50.7% of all LDPE and LLDPE in the MSW stream [1]. LDPE and LLDPE provide excellent flexibility, good strength, moderate transparency, good moisture barrier, and low cost.

Linear low-density polyethylene came into widespread use in packaging, displacing the older LDPE in many applications, because LLDPE's more uniform linear copolymer structure with a narrower molecular weight distribution resulted in improvement in both tensile and tear properties compared to highly branched LDPE. Despite its higher price per pound, this permitted cost savings because of the ability to use much thinner film (downgauge) while maintaining desired performance. Heat seal performance of LLDPE, however, is inferior to LDPE. In applications such as coatings, LDPE is used much more than LLDPE. In a number of film applications, blends of LDPE and LLDPE are used to obtain the best combination of properties. Often, the term LDPE is used generically to refer to LDPE, LLDPE, and blends of the two materials.

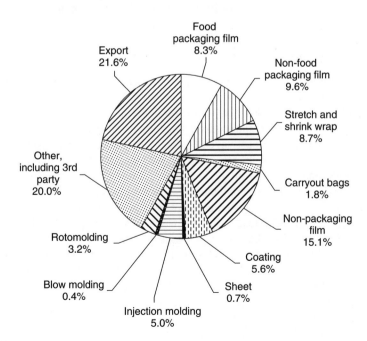

Figure 4.7. Uses of LDPE and LLDPE in the United States and Canada, 2000 [3].

152 PLASTICS IN PACKAGING

The introduction of metallocene catalysts has permitted new variants of LLDPE to be produced, with much improved ability to tailor properties to end-use requirements. LLDPEs that incorporate higher α-olefins as co-monomers have much improved heat seal properties, for instance. These new polymers are likely to further displace LDPE and result in even greater ability to downgauge films in both LDPE and LLDPE applications. These same catalyst systems are likely to also permit source reduction through downgauging in HDPE, PP, and other applications.

Recycling of LDPE and LLDPE is much less common than recycling of HDPE. The EPA reported a recycling rate for LDPE/LLDPE bags, sacks, and wraps of 5.2% in 1998 and an overall LDPE/LLDPE packaging recycling rate of 4.4%. The vast majority of this material was collected from businesses, rather than from individual households, and originated in stretch wrap for pallet loads of goods. Collection of LDPE, LLDPE, and HDPE grocery and merchandise sacks at retail stores used to be fairly common, although it never captured a large portion of the material available. Many retailers have discontinued such programs because of cost and contamination issues. Some drop-off collection is still available for such materials, sometimes through schools or community organizations. Collection of plastic film at curbside is available only in a handful of communities in the United States. In Ontario, Canada, curbside collection of plastic film is common. The Environment and Plastics Industry Council (EPIC) has published a guide to such collection [6].

A major use of recycled LDPE and LLDPE is in the manufacture of trash bags. Some is also used in the manufacture of merchandise bags. A use of growing importance is, mixed with wood fibers, plastic lumber for applications such as window frames.

4.5.3. Polyethylene Terephthalate

Consumption of virgin polyethylene terephthalate (PET) resin in the United States and Canada totaled 5110 million pounds in 2000 [3]. Bottles were by far the largest application (Fig. 4.8). Soft drink bottles continue to be the largest use for PET bottles, but other applications (custom bottles) have been growing at a much faster rate for the past several years. According to EPA estimates, soft drink bottles accounted for 49.4% of all PET packaging in U.S. municipal solid waste in 1998, and 53.2% of all PET containers in municipal solid waste [1].

PET is used in packaging applications where its excellent transparency, stiffness, strength, improved high-temperature performance, and reasonably good barrier properties are sufficient to justify its extra cost compared to HDPE.

According to the EPA, the soft drink bottle recycling rate was 35.4% in 1998, and the overall PET container recycling rate was 23.4% [1]. The American Plastics Council reported a 24.4% recycling rate for PET bottles in 1998, which declined to 22.8% in 1999, continuing a pattern of decline from its high of 50% in 1994. The tonnage of PET bottles recycled has grown during this period but has been outstripped by gains in the amount of PET used.

Like HDPE, PET bottles are collected through most curbside and drop-off recycling programs in the United States. In nine states, PET soft drink bottles

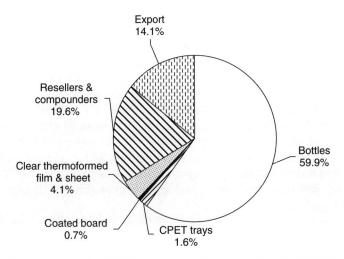

Figure 4.8. Uses of PET in the United States and Canada, 2000 [3].

are subject to a deposit, usually 5 cents. California has a quasi-deposit system that provides a refund value for the containers. In both California and Maine, the deposit/refund value systems have been extended to non-soft-drink beverages. In most cases, the true deposit systems achieve a recovery rate of over 90%.

In Europe, PET recycling rates in 1999 ranged from 70 to 80% in Sweden and Switzerland, down to less than 5% in the United Kingdom.

The U.S. FDA has issued letters of nonobjection for several processes for incorporation of recycled PET in food packaging. Such approval has been granted for systems using the recycled PET blended with virgin PET in direct contact with food, as well as in a buried inner layer. Often, these processes have used PET bottles collected through deposit systems, which have been shown to yield cleaner materials than curbside bottles. Chemically recycled PET that has been broken down to monomer, purified, and repolymerized has also been approved. Recently, Plastic Technologies, Inc. (PTI) of Bowling Green, Ohio, announced the first commercial production in the world of a plastic bottle made of 100% mechanically reprocessed curbside recycled PET.

During the past several years, Coca-Cola, and more recently Pepsi, have come under attack for failure to use recycled PET in the manufacture of soft drink bottles. While incorporation of up to 25% recycled content is common in a few countries, it was never widespread in the United States and all but disappeared by the late 1990s. Coca-Cola has begun to use small amounts of recycled PET in soft drink bottles and in early 2001 announced a plan to use an average of 10% recycled content in all its PET bottles by 2005.

The movement by brewers to introduce beer in enhanced barrier PET bottles has also spawned controversy. PET recyclers were initially quite concerned about the effect of such containers on their processes and profitability, particularly because many of the bottles are amber in color. Environmental and consumer organizations, and also some legislators, voiced concern as well. Commitment by industry to somewhat modify the design of the bottle label and closure, to buy the amber recycled PET, and to include recycled PET in an inner layer in the containers satisfied most processors, as well as some of the other protestors.

In addition to limited use in beverage bottles, recycled PET packaging is used in nonfood bottles, trays, blisters, strapping, and other packaging applications. It also has a wide variety of nonpackaging uses, especially as fiber.

In many of its packaging applications, PET has replaced glass. Usually this has resulted in large reductions in both the volume and weight of packaging used to deliver products. The savings occur not only in the primary package but also in associated distribution packaging, since goods packaged in PET take up significantly less space than those packaged in glass. Energy savings are also usually substantial.

4.5.4. Polypropylene

Polypropylene (PP) is used in packaging in much smaller amounts than HDPE, LDPE/LLDPE, and PET. Its uses are also more evenly divided between various markets. The EPA estimated PP in packaging in municipal solid waste to be 13.8% in containers, 48.9% bags, sacks, and wraps, and 37.2% "other plastics packaging" [1]. A significant fraction of the latter category is caps. The majority of plastic closures (caps) are made of PP. Polypropylene films are often used for packaging applications where a stiff, highly transparent film is desired. Since it was first introduced, PP has captured nearly all the packaging film markets once held by cellophane. In container markets, PP is also used where its improved high-temperature performance, transparency, and stiffness compared to HDPE are an advantage, such as in applications where containers are hot filled. Clarified grades of PP can provide sufficient transparency to enable PP to compete with PET.

Recycling of PP packaging is much less common than recycling of HDPE, LDPE/LLDPE, and PET, at least in part because there is less of it that can be readily recovered. PP is rarely included in either curbside or drop-off collection programs, and there is little recycling of PP from business sources. All-plastic-bottle collection programs, which exist in a few locations, do include PP bottles. The EPA estimated the 1998 recycling rate for PP packaging as 3.2% [1].

Some PP is collected, often inadvertently, when HDPE or PET bottles are collected and have their PP closures in place. On HDPE bottles in particular, this is problematic since HDPE and PP are not separable in common sink-float or hydrocyclone processes designed to separate heavier-than-water from lighter-than-water materials. Even very small amounts of pigmented PP caps cause discoloration in unpigmented HDPE streams. While PP is more readily tolerated in pigmented HDPE streams, large amounts of PP cause unacceptable decreases in performance of the recycled material. PP contamination of PET is not a major issue since the materials can be readily separated.

Just as there is little recycling of PP packaging, there are few uses in packaging for recycled PP.

4.5.5. Polystyrene

Polystyrene (PS) is used in packaging applications in both foamed and highly transparent nonfoamed (crystal) form. High-impact polystyrene (HIPS) is partially a blend and partially a copolymer, designed to have greatly improved impact strength. PS finds applications in containers, film, and, very importantly, as cushioning material. Foamed PS cushioning is produced in molded and die-cut shapes and in loosefill form. Uses of virgin PS in the United States and Canada are shown in Figure 4.9. In packaging materials in MSW in 1998, 28.6% of PS was in containers, 28.6% in bags, sacks, and wraps, and 42.9% in the "other packaging" category that included cushioning materials.

PS is a stiff, brittle material with relatively poor barrier properties. Crystal PS has excellent transparency. In its foamed form, PS has very good cushioning abilities, as well as insulation properties. Cost is relatively low, especially in molded expanded polystyrene (EPS) and extruded foam PS, where a small amount of mass can yield a large volume. PS is readily shaped by injection molding, extrusion, or thermoforming, in either foamed or unfoamed form.

Recycling of PS in the United States is limited almost exclusively to recycling of cushioning materials. The EPA calculated the recycling rate for PS packaging in 1998 at 4.8%. The Alliance of Foam Packaging Recyclers (AFPR) reported a 9.6% recycling rate for EPS packaging in 1999, up from 9.5% in 1998 [7]. The International EPS Alliance Task Force (ITF), launched in 1991, established an international recycling agreement for EPS, which has now been signed by more than 25 countries [8]. A major market for recycled EPS is the manufacture of new EPS cushioning. Loosefill cushioning, especially, can also be reused. Many AFPR members cooperate with mailing service organizations to collect

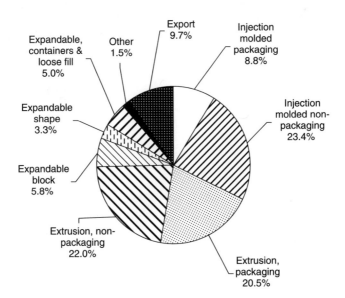

Figure 4.9. Uses of PS in the United States and Canada, 2000 [3].

EPS from consumers for reuse or recycle, and the organization maintains a listing of collection sites [9].

Polystyrene, particularly foamed polystyrene, came under considerable attack from environmental organizations in the 1980s, due to its association with litter and with ozone depletion. While some PS foams, including packaging foams, were manufactured with chlorofluorocarbon (CFC) blowing agents, EPS molded foams never used CFCs. In all industrialized countries, use of CFCs in packaging materials, including PS, was phased out a number of years ago. The blowing agents used today, for both molded and extruded foams, are hydrocarbons or, increasingly, carbon dioxide.

The litter problem remains with us. While it is certainly true that litter is a consequence of unacceptable human behavior, not inherent in the material, when littered, PS does not biodegrade, so it can remain in the environment for a very long period of time. The static cling tendencies of PS also contribute to its propensity to be littered, as can be attested to by anyone who has tried to pour a box of loosefill cushioning into their trash—especially if there was even a gentle breeze at the time!

4.5.6. Polyvinyl Chloride

Packaging is only a small part of the market for polyvinyl chloride (PVC), which has seen declining market share in packaging over the last decade or so. PVC packaging in U.S. municipal solid waste is predominantly in the "other packaging" category (Fig. 4.10), especially in the form of trays and blister packaging. These are used in a variety of applications, including wide use in medical packaging.

The properties of PVC packaging can be manipulated to a very wide degree by plasticization. Therefore, PVC packaging ranges from relatively stiff and brittle containers to very soft flexible film used for wraps. In general, PVC is a highly transparent material, with some tendency to yellow over time, especially if it is

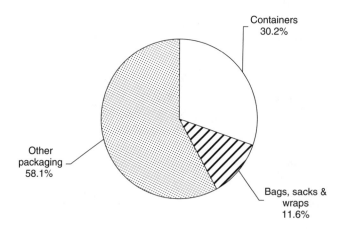

Figure 4.10. PVC Packaging in U.S. municipal solid waste, 1998 [1].

exposed to sunlight. It thermoforms very readily, making it often the material of choice for blister packaging. The barrier performance ranges from reasonably good in bottles to relatively poor in films.

PVC has been under attack from environmental groups on a variety of grounds for a number of years. Because PVC contains chlorine, it may lead to the formation of chlorinated dioxins if it is disposed by incineration. While evidence suggests that the presence or absence of PVC is not very significant in dioxin emissions from well-controlled incineration, less important than combustion conditions, there is reason for concern about its presence in poorly controlled incineration systems. PVC has also been attacked because of concerns related to the carcinogenicity of its vinyl chloride monomer. Increasingly, lead and cadmium stabilizers used in some PVC resins are being restricted or banned, due to concerns about toxic effects of these heavy metals.

Another set of concerns relates to the plasticizers used in PVC, in particular, the endocrine-mimicking effects of phthalate plasticizers. This concern led to widespread abandonment of PVC in soft toys for infants and more recently to calls for discontinuation of the use of phthalate-plasticized bags for intravenous (IV) solutions used in hospitals. In April, 2001, the European Parliament voted for stricter standards on the use of PVC, which is likely to lead to more switching away from PVC packaging [10].

PET (including PET copolymers such as glycol modified PET, PETG) has emerged as a strong competitor to PVC in both bottle and thermoformed sheet markets. PET offers transparency as good or better than PVC, and recent price reductions for PET have made it highly competitive. Additionally, PET does not carry the negative environmental image of PVC.

In the United States, there is very little recycling of PVC packaging materials. As mentioned, most recycling collection programs target only bottles. Since relatively few PVC bottles are available, they are generally not accepted in collection programs, with the exception of those collecting all plastic bottles. Recycling of PVC bottles is also problematic if they are mingled with PET bottles. PET is extremely susceptible to degradation from even very small amounts of PVC. The "look-alike" nature of PVC and PET bottles means that if both are collected, hand sorting is likely to result in some degree of contamination. Very effective automated sorting systems have been developed for PET/PVC separation of intact containers, but it is relatively expensive, and not all separation facilities can afford the capital investment. Contamination of PET from PVC components in packages, such as liners in closures, labels, and the like, can also be a significant problem. Density-based separation systems cannot be used effectively to separate these two resins, as their density ranges overlap.

4.5.7. Other Plastics

As mentioned, a variety of plastics are used in lesser quantities than the major packaging plastics discussed above. These materials are, in general, significantly more expensive but have specific properties that make them important for particular types of applications.

Nylons (polyamides) excel when high-temperature performance is required, such as for boil-in-bag packaging for frozen foods. They provide excellent odor and flavor barrier, moderate oxygen barrier, and excellent strength and toughness.

Ethylene vinyl alcohol (EVOH) and polyvinylidene chloride (PVDC) are the usual choices when excellent oxygen barrier is required. EVOH offers better oxygen barrier when dry than PVDC, but barrier decreases substantially at high humidity. EVOH is also more readily processed than PVDC. Both are commonly used as one component in a multilayer structure, rather than alone.

Polycarbonate (PC) is highly transparent, and very strong and tough. It has been the material of choice for refillable 5-gal water bottles but is beginning to experience competition from PET in this market. It also has a number of applications in medical packaging.

Polyethylene naphthalate (PEN) is a polyester with improved properties compared to PET. It is a better barrier and has improved strength and high-temperature stability, but it is considerably more expensive than PET. So far, its high cost has prevented its widespread use in packaging.

Polyvinyl acetate (PVA) and ethylene vinyl acetate (EVA) are widely used in adhesive formulations and also have some use as packaging films.

Polyacrylonitrile (PAN) in copolymer form is used in high-barrier containers, particularly for household or industrial chemicals.

A number of other specialty polymers are used in packaging as well. Recycling of these materials is essentially nonexistent. Small amounts may be incorporated in mixed-plastic recycling systems that manufacture items such as plastic lumber. The relatively small amounts of these plastics that are available for recovery, and the additional complication that they often form part of multilayer structures, make their recovery and recycling as pure materials impractical in most cases.

4.6. BIODEGRADABLE PLASTICS

When the "solid waste crisis" hit the United States in the 1980s, plastics were often attacked as particular problems because they are nonbiodegradable. There was a perception that biodegradation resulted in recovery of valuable landfill space. Some states even passed laws requiring that certain types of plastics (usually merchandise bags) be degradable.

Degradability therefore became a marketing asset, and a number of manufacturers sold merchandise and grocery bags that claimed to be degradable. Some of these structures were photodegradable, rather than biodegradable, capitalizing on the lack of public understanding of the difference between these characteristics. Others were made from a blend of starch and low-density polyethylene, sometimes with photo-oxidants added. The manufacturers claimed these starch–plastic blends were biodegradable. They claimed that microorganisms, after removing the roughly 6% starch contained in the blend, would be able to attack and consume the LDPE that remained since it would now have greatly increased surface area. However, no convincing evidence in support was produced, and some independent researchers found no evidence of significant biodegradation of the LDPE component.

However, within a few years, evidence mounted that biodegradation is very slow in a well-constructed landfill, designed to protect against groundwater pollution due to generation of contaminated leachate. The combination of lack of oxygen and relatively dry conditions leads to very slow microbiological activity. Even such highly biodegradable substances as food have been found to be still recognizable after 10 years or more in a landfill. Some state attorneys general took legal action against manufacturers of photodegradable trash bags, accusing them of misleading consumers. Several environmental groups called for boycotts of degradable plastics on the same grounds. Degradability was, by many, regarded as a scam.

Since that time, there have been two important developments. First, new plastics have been developed that are truly biodegradable. Second, the use of composting for waste disposal has increased dramatically. Composting is designed to increase microbiological activity and the rate of biodegradation. In most composting operations, the process is managed to assure ample supplies of oxygen and moisture, and temperature may also be maintained at optimum levels. There are also some anaerobic composting operations designed to yield methane as a product. The output from composting is humus, which can be used as mulch or applied to fields. The presence of nonbiodegradable plastics (or other nonbiodegradable materials) is undesirable and can render the compost unusable if contamination levels are too high. Therefore, for waste streams that will be composted, use of biodegradable plastics is an asset. Many municipal composting programs accept yard waste only. However, composting of source-separated residential organics is growing, as is composting of food service waste streams of various types, including restaurant waste. For example, the Brookfield Zoo in the Chicago area recently began composting the wastes generated from its food service operations. The zoo has begun using only biodegradable materials for disposable plates, cups, and utensils in its food service operations to facilitate composting. San Francisco is moving toward citywide collection of source-separated residential organics, which will be composted. In some parts of Europe, there is a long tradition of composting, but it is still quite rare in the United States, except for yard waste.

One of the first high-profile biodegradable plastics to be used in packaging was polyhydroxybutyrate-valerate (PHBV), a natural polyester copolymer produced by certain types of bacteria. Imperial Chemical Industries, Ltd. (ICI) developed a process for controlling the relative amounts of butyrate and valerate units in the polymer by modifying the diet fed to the bacteria and then harvesting the polymer, and sold it under the name Biopol. Properties of the plastic are generally similar to polypropylene. PHBV was used for a relatively short time in Europe by Wella for shampoo bottles. However, PHBV was much more expensive than the synthetic plastics it was competing with; its price ranged around $20/kg. ICI spun off the Biopol business, and it was eventually bought by Monsanto, which withdrew it from the market in 1998. In May, 2001, Metabolix purchased

Biopol and licensed the production technology from Monsanto, with the intent to launch commercial products in late 2001. Other researchers and companies have developed similar bacteria-based polyesters, and there have also been successful efforts to transfer the plastics manufacturing genes to plants. However, so far none of these processes have achieved commercial success.

The starch–polyethylene blends that claimed to be biodegradable have been succeeded by materials containing nearly 100% starch that are truly biodegradable. These have had limited success, primarily in foam form as cushioning materials and in film form for compost bags. Water is used as a plasticizer to make the starch thermoplastic so it can be melt processed. In some cases, relatively small amounts of other biodegradable substances, including plastics, are blended with the starch to improve properties and processability. Many of these materials are water soluble in addition to being biodegradable. The foam cushioning material manufacturers point to the water solubility and lack of static cling as highly desirable in preventing litter problems associated with loosefill polystyrene cushioning.

One of the most promising of the new generation of biodegradable plastics is based on polylactic acid (PLA). Cargill Dow, a joint venture of Dow Chemical and Cargill, has begun selling these materials under the name NatureWorks PLA, with applications targeted at films, thermoformed containers, coating for paper and paperboard, and bottles. Full-scale production is scheduled to start in 2002. In addition to waste disposal benefits, Cargill Dow claims PLA uses 20–50% less fossil fuel resources than conventional plastics and emits less net carbon dioxide. In addition to Cargill Dow, a few other companies manufacture PLA in relatively small amounts, many targeting the high-value medical market rather than packaging.

Several companies are manufacturing synthetic biodegradable polyesters. DuPont's Biomax can be used in film or containers. Showa Highpolymer makes Bionolle for film and containers. Several other companies make similar materials. Prices are generally higher than for competitive nonbiodegradable plastics.

Polyvinyl alcohol (PVOH) is a water-soluble biodegradable plastic that has been available for many years. Its high cost and water solubility limit its suitability for packaging, but it does have important niche markets. For example, it is used to package some agricultural chemicals, enabling the still-packaged product to be placed in a mixing tank, thus avoiding human exposure to hazardous chemicals. Modified grades of PVOH have limited solubility.

In the waste disposal area, the environmental benefits of biodegradable plastics are limited to waste streams that will be composted, items that are associated with litter problems, and items that are apt to get into sewage treatment systems. If wastes will be disposed by landfill or incineration, biodegradability offers no real advantage.

4.7. SOURCE REDUCTION

One powerful way to reduce the environmental impact of a package (or product) is to use less material in its manufacture. That translates immediately into less use of raw materials and less waste at the end of the package life. Usually it also means less use of energy throughout the package life cycle. Since it also almost always translates into lower cost (since less material must be purchased), downgauging of packaging (making the package thinner) has a long history in packaging. In the past, the driving force for this type of source reduction was cost reduction. More recently, environmental benefits have assumed a significant, though usually secondary, role.

A variety of examples can be given of this type of source reduction. Plastic soft drink bottles, for example, are lighter than they used to be. In 1977, when 2-liter PET soft drink bottles were introduced, they weighed 68 g (including the HDPE base cup). In 1999, the redesigned bottles weighed only 49 g, a 27% reduction. With about 5.6 billion of these bottles being sold each year, the total savings are approximately 200 million pounds per year [11].

Procter & Gamble Corp. (P&G) and its bottle supplier, Continental PET Technologies, redesigned the PET bottles P&G used for vegetable oil. The new design used 30% less plastic than the old design, while achieving the same strength, and resulted in reduction of about 2.5 million pounds of plastic per year. In addition, the rectangular design for the new bottle enabled the bottles to fit in the shipper more compactly than the old cylindrical bottles, resulting in the elimination of about 13 million pounds of corrugated board per year [12].

The replacement of LDPE films by thinner LLDPE films for a wide variety of packaging applications has already been mentioned. Metallocene polymers can provide similar benefits.

When source reduction results in less of the same material used, and no other substantive changes, the environmental benefits are clear. Often, however, substantial source reduction is accomplished by changing materials, package style, and a variety of other features. With such changes, a life-cycle analysis approach is the only truly accurate way to evaluate the overall environmental effects of the change. However, such analysis are complex and costly, and there remains no universally accepted way to combine disparate environmental impacts into a usable score or set of scores for evaluating environmental costs and benefits of the two (or more) alternative packaging systems. Nonetheless, a number of studies suggest that, by and large, if substantially less material is used, the overall environmental effect is likely to be positive.

Substantial amounts of source reduction can often be gained by switching from glass or metal to plastics packaging and by switching from rigid or semirigid packaging to flexible packaging (usually containing plastic). While such reduction is usually measured in terms of weight, there is typically a reduction in the volume of packaging material used as well.

When Clorox switched from glass bottles to plastic bottles for barbeque sauce and salad dressing, the new PET bottles weighed 85% less than the old glass bottles. The result was reduction of nearly 30 million pounds of glass per year. Use of corrugated boxes was cut by 2 million pounds per year since the plastic bottles used less space [12]. For example, 18-oz glass jars for peanut butter weigh 10.2 oz each. PET jars of the same capacity weigh only 1.7 oz, an 83% reduction.

Additional savings accrue during distribution, due to the weight difference. It takes three trucks to haul as much peanut butter in glass jars as two trucks can haul in plastic jars [11].

Many medical products have moved from rigid to flexible packages. Kendall Health Care switched from a thermoformed tray to a flexible pouch for a urine collection assembly used in hospitals and achieved a waste volume reduction of about 50%. Medchem changed hemostat packaging from a glass jar inside a foil-lined container to a plastic tray in a pouch and achieved an 80% reduction in packaging by weight [13].

A very rapidly growing type of packaging is stand-up pouches, flexible pouches that are designed to stand erect on the shelf and replace bottles, cans, or cartons. In 1995, Procter & Gamble won the Flexible Packaging Association's Green Glove Award for its stand-up bags for detergent, which used 80% less packaging than the paperboard box they replaced, as well as incorporating 25% postconsumer material [14].

In addition to reduction in package volume or weight, such packages can offer advantages that are less obvious. For example, the aluminum foil/plastic stand-up pouch for Whiskas cat food, which replaced a 10-oz steel can, is reported to require 30% less retorting time than steel cans because the pouch can be heated more quickly and evenly [15]. That translates directly into reduced energy use for the retorting process and probably into a decrease in the amount of cooling water required as well. Kapak Corp., a Minneapolis-based flexible packaging manufacturer, points to flexible packaging advantages in freight savings (one truckload of 1-liter pouches have the same holding capacity as 25 truckloads of rigid packages) and warehouse space (96% savings), as well as energy reduction (75% less manufacturing energy), and a 25:1 ratio for source reduction [16].

Of course, not all packaging reduction innovations meet with consumer success, and what is successful in one country may not be so in another. DuPont Canada estimated in 1990 that switching from HDPE gallon bottles for milk to twin-pack 2-quart LLDPE pouches could reduce the weight of discarded packaging by 58%. They further calculated that replacement of the mix of then-current milk packaging by pouches could reduce the landfill volume of discarded milk packaging by nearly 93% [17]. While such pouches have enjoyed some success in Ontario (where there is a deposit on HDPE milk bottles but not on pouches), they have not met much success in the United States.

4.8. REUSE

Another way to achieve reduction in the amount of packaging requiring disposal is to switch from one-use to reusable packaging. Often, especially for distribution packaging, this involves a switch from paperboard or wood to plastic. Reusable packages often require initial use of more material than disposable packages. However, when the amount of product delivered is factored in, the net result can be very large reductions in the amount of packaging used per unit of product delivered.

As discussed, reusable packaging for automobile parts has become commonplace, and it is growing in other segments as well. Cost reduction is generally the driving factor in making such changes, but the environmental benefits can be considerable. In 1994, INFORM published a study of reusable shipping containers that estimated the average cost per trip for reusable plastic 2-ft^3 containers as 4.4 cents, compared to 53 cents per trip for disposable corrugated. The weight of material used for 1 million shipments in reusable plastic containers was estimated at 11 tons, compared to 750 tons for one-way corrugated boxes [18].

4.9. RECYCLING

When packages and package systems are being designed or modified, relatively small changes can sometimes make the difference in whether the package will be recyclable at the end of its life or whether it will not. (It should be noted that the term *recyclable* means different things to different people — it is used here in its practical sense, of being capable of fitting into existing or developing recycling programs and, hence, likely to be recycled at the end of its useful life.)

First, in the United States, for the most part only plastic bottles are accepted by recycling collection programs accessible to consumers. Therefore, if a consumer package is not a bottle (or jar), there is little likelihood that it will be recycled. Further, the vast majority of recycling programs collect only HDPE and PET bottles. Any other type of bottle is also unlikely to be recycled. Adding new materials to recycling programs generally faces a "which came first — the chicken or the egg" problem. Community organizations understandably do not want to collect materials that have no market. From an environmental perspective, there is no value in using additional resources, including energy, to segregate, collect, and process materials if they are ultimately headed for disposal. On the other side, potential users of the recycled material do not want to invest in developing the knowledge base and process changes necessary if they will not have a reliable source of recycled material that is available at a reasonable price. In the past, legislation has often been utilized to "jump-start" the system, either by mandating collection of materials for recycling, mandating use of recycled material, or creating financial incentives for the establishment of recycling systems and markets for recycled materials.

It is not uncommon for there to be a trade-off between recyclability and source reduction. For example, ground coffee can be effectively packaged in either a steel can, or a multilayer vacuum pouch containing several layers of plastic, including a metallized film. The steel can is recyclable in most locations and has a relatively high recycling rate, 58.4% in 2000 in the United States [19]. The pouch is essentially unrecyclable since virtually all recycling programs will not accept it. However, the pouch represents considerable reduction in the amount of material used, as well as in energy requirements for distribution. Similarly, the LLDPE milk pouches described above are not recyclable in most cases, while the HDPE bottles they replace are recyclable. For this particular case, DuPont claimed that a 90% recycling rate for the bottles would be required to match the reduction in waste volume achieved by use of pouches [17].

When new packages are introduced, proper attention to package design can make the difference in whether the package will be a valued addition to the recycling stream, nonrecyclable and essentially neutral, or a problem for existing recycling systems. In 1995, recommendations for plastic bottle design for recycling were released from the Plastic Redesign Project, funded by the U.S. EPA and the states of Wisconsin and New York (and later California) and involving representatives from industry, government, and plastic bottle recyclers [20]. An updated list of recommendations was released in late 1998 [21]. Among the recommendations are the following:

- Do not pigment natural HDPE bottles such as those for water and milk; do not pigment or tint PET bottles any color other than green.
- Do not use pigmented caps on natural HDPE bottles; make caps and fitments on HDPE bottles compatible with HDPE (except living hinge applications).
- Do not use aluminum caps; do not use aluminum seals unless the consumer can pull the seal completely off.
- Use only water-dispersible adhesives on labels; use labels that have a specific gravity less than 1 on PET bottles; do not use metallized labels on bottles with a specific gravity greater than 1; use PVC and PVDC labels only on PVC containers; ensure that pigments do not bleed from the label during recycling.
- Do not print directly on unpigmented bottles, except date coding.
- Make all layers in multilayer bottles sufficiently compatible so that the material can be sold into high-value end markets.
- Do not use PVC bottles for products that are also packaging in bottles such as PET that look like PVC.

The Association of Postconsumer Plastic Recyclers (APR), a U.S. trade association representing plastics recycling companies, also publishes a list of design guidelines for plastic bottles. Specific sets of guidelines are presented for PET, natural HDPE, pigmented HDPE, PP, and PVC bottles [22]. Guidelines include:

- No paper attachments of any kind; no PVC attachments of any kind except on PVC bottles; no PET attachments on PVC bottles.
- No metal closures; PP or HDPE/EVA closures and closure liners preferred on PET bottles; HDPE, LDPE or PP closures preferred on HDPE, PP, and PVC bottles.

- Sleeves and safety seals should be completely detachable and easily removed in conventional separation systems; shrink sleeves preferred if sleeves are necessary.
- Unpigmented or green preferred for PET; unpigmented preferred for homopolymer HDPE; pigmented preferred for copolymer HDPE; unpigmented preferred for PP.
- Nonbleeding inks and water-soluble or dispersible adhesives preferred.

The APR also operates the Champions for Change program that "invites consumer product companies, technology companies, suppliers, converters and others to test new bottle compositions directly with commercial plastics recyclers" in order to [22]:

- "Help ensure that new materials and designs are compatible with the existing plastics recycling infrastructure;
- Help promote technology transfer so that the recovery of plastic materials keeps pace with new packaging designs; and
- Help sustain the economics of an industry supplying valuable post-consumer recovered material to markets worldwide."

As mentioned, when PET beer bottles were introduced in the United States, recyclers voiced concern, on several grounds, over the effect these bottles would have on PET recycling systems. First, existing processes were set up to handle green and clear bottles, and many of the new bottles were amber. That meant, in many cases, that processes would have to be modified to segregate an additional color, or else the existing product streams would be contaminated and therefore decrease in value. If pure streams of amber PET were produced, the processors were concerned that there would be no market for the material. Second, the bottles contained an aluminum cap. Most soft drink bottles are collected for recycling with the cap in place, and the same would be true for beer bottles. When PET soft drink bottle recycling systems were initiated, all the bottles also had aluminum caps. High-quality recycled PET requires that aluminum contamination be kept extremely low. Separating the aluminum from the PET turned out to be one of the most difficult aspects of recycling system operation. A change in design to PP caps, although not done with the aim of improving recyclability, greatly facilitated PET recycling. Now the processors faced reintroduction of this highly problematic material. Further, the initial bottle design incorporated a metallized label, which added to the threat of unacceptable levels of aluminum contamination. An additional concern, for some container designs being introduced, was the presence of novel barrier materials and their unknown effect on the recycling process. This combination of factors drew criticism, calls for boycott, and even resolutions from local governments to prohibit introduction of the container, before the industry committed to design changes.

In contrast, when Heinz introduced PET ketchup bottles, the company paid explicit attention to ensuring that the bottles were compatible with existing PET soft drink bottle recycling systems. The bottle is manufactured with a five layer structure: PET/EVOH/PET/EVOH/PET. The EVOH is required to obtain sufficient oxygen barrier for the product. The absence of a tie (adhesive) layer between the PET and EVOH causes the material to delaminate when the bottles are ground and the material washed during recycling. The EVOH material is mostly removed during the wash and rinse processes. The small amount that remains is not detrimental to PET properties.

In December, 2000, a new report from the Plastic Redesign Project examined the effect of new plastic packaging innovations on plastic bottle recycling, with a specific focus on whether trends in bottle design could further erode the prices paid to local recycling programs for plastic bottles because of the expense in handling new structures [23]. Mentioned as design changes that have facilitated plastics recycling are elimination of the base cup in 2-liter soft drink bottles, substitution of PET for PVC bottles, substitution of EVOH for PVC liners in closures for carbonated beverages, and development of LDPE shrink labels to substitute for glued-on labels on milk bottles.

The major concerns identified are the increasing use of colors other than green in PET bottles and the growth in use of PET bottles that contain non-PET components to enhance barrier properties. Both of these developments increase the cost of processing recovered plastic bottles and also affect the value of the recovered material. The study estimates that the intermediate processor's cost to manually separate another color from clear and green PET bottles would be about 6 cents per pound, and if mechanical sorting is available, it would be about 1.5 cents per pound. The value of green PET flake currently averages 18.9% lower than clear PET flake. The value of new colors, such as amber from beer bottles and blue from water bottles, would be lower, averaging perhaps 1.1 cents per pound less than clear.

4.10. ENERGY AND ENVIRONMENTAL ASSESSMENTS

As discussed, a thorough evaluation of the environmental impact of alternative types of packaging systems requires a life-cycle assessment (LCA) that takes into account all impacts from "cradle to grave." However, all such analyses must be constrained within a set of boundaries, and the choice of boundaries can make a difference in the conclusions that are drawn. The more comprehensive and detailed is the analysis, the greater is its cost. Further, numerous assumptions are required, especially when moving from the life-cycle inventory that documents inputs and outputs to the impact analysis, which describes the effects of those inputs and outputs. This effort is inextricably tied to value judgments about the relative equivalence of greatly differing types of items listed in the inventory. Even in the life-cycle inventory, assessments will vary in the data used. Some studies opt for use of publicly available data only and are often criticized on the grounds that recent improvements in processes are not recognized. Others use industry sources for recent data to reflect such improvements and are criticized because the data is not available for public scrutiny. In some European countries, government policies require use of government-approved computer-based LCA models, which may not provide all relevant data and assumptions for independent evaluation. Therefore, use of LCA, while theoretically the only fully justifiable method for evaluating the environmental consequences of packaging decisions, in practice still needs considerable development before it can become a useful tool. As a consequence, there has been considerable interest in developing short-cut methods that, while not evaluating all environmental impacts of packaging (or other) alternatives, can more readily be employed for guidance in making environmentally sound choices.

Life-cycle analysis of plastics resins themselves is outside the scope of this chapter (see Chapter 3). Our focus is the additional environmental impact associated with the conversion of plastics resins to packaging materials and the use and disposal of these materials.

One of the most wide-ranging LCA packaging studies is the massive report done by the Tellus Institute on the major packaging materials, including plastics [24]. However, this study examines only polymer production and ultimate disposal. It does not include processing steps such as forming and molding.

A study by Franklin Associates in 1992 examined the energy requirements for plastics and their alternatives in packaging materials [25]. Energy requirements for fabrication were presented for blow molding, injection molding, fabrication of film, and thermoforming and are summarized in Table 4.1. Analyses of energy requirements for resin production and for manufacture, fabrication, and recycling systems for several types of resins and container systems are shown in Table 4.2. Other studies of energy consumption for plastic packaging are also available [26, 27].

A study published in The Netherlands in 1992 [28] evaluated the environmental impact of polystyrene, paper, and porcelain coffee cups. The cups were made from a 50:50 blend of high-impact polystyrene (HIPS) and general-purpose

Table 4.1 Energy Requirements for Plastics Processing [25]

Process	Process Energy (MJ/kg)		Transportation Energy (MJ/kg)	Total Energy (MJ/kg)
	Electricity	Natural Gas		
Blow molding PET	186.5		2.3	188.7
Blow molding non-PET	74.6		2.3	76.8
Injection molding	198.9		7.7	206.6
Film fabrication	78.8		3.6	82.4
Thermoforming of sheet (nonfoam)	64.6		4.5	69.2
Calendering PVC sheet	31.1		3.6	34.7
Molding foam PS	149.2		29.3	178.5
Extrusion and thermoforming foam PS	149.2	1.0	29.3	179.6

Table 4.2 Energy Requirements for HDPE and PET Polymerization, Fabrication, and Recycling [25]

Process	Material Resource Energy (MJ/kg)	Process Energy (MJ/kg)	Transportation Energy (MJ/kg)	Total Energy (MJ/kg)
HDPE resin manufacture	538.5	316.1	19.3	873.9
Manufacture & 100% open-loop recycling of HDPE milk bottles	273.3	321.8	19.0	614.1
PET resin manufacture	447.3	413.3	26.8	887.4
Manufacture & 100% open-loop recycling of PET soft drink bottles	223.7	370.9	23.5	618.1

polystyrene, extruded, and thermoformed. (It should be noted that expanded polystyrene cups were not considered, as they had a very small share of the market.) Transportation of granulated PS to the conversion facility, and of the cups to the point of use, was also evaluated. Disposal impacts were based on 60% landfill and 40% incineration. The calculated environmental impacts for the resin production, conversion, and waste processing stages are shown in Table 4.3. As can be seen, in most cases the inputs and outputs associated with resin production are much larger than those associated with conversion and disposal.

The Association of Plastics Manufacturers in Europe (APME) has produced a series of ecoprofile reports on various plastics, including packaging materials. APME adopted the term *ecoprofile* since the analysis extends only to the point of sale and does not cover disposal. Also, the reports do not include the impact or improvement portions of life-cycle assessment. The reports are publicly available on the association website, *www.apme.org* and are updated periodically.

Table 4.4 presents selected ecoprofile information for high-density polyethylene bottles, produced by extrusion blow molding, and packaged for shipment to the user, and, for comparison, information for HDPE resin. It should be noted that these use somewhat different data sets. Emissions from the conversion process were stated to be insignificant compared to those associated with inputs to the system. Energy use was distributed as shown in Figure 4.11. Of the portion of energy used in bottle production, 77.3% was used in molding the bottles. The remainder was used in water chilling, storage, conveying, resin handling, and deflashing of the container (Fig. 4.12) [29].

Table 4.5 presents selected ecoprofile information for LDPE and LLDPE resin and for LDPE blown film packaged for shipment. The processing operation consumes 1.6926 MJ of electricity per kilogram of film production. It is assumed that 2% of production is waste polymer, which is recycled back into the process. Production of the resin uses nearly 90% of the total energy required to produce film (Fig. 4.13) [29].

Table 4.3 Environmental Impacts Associated with HIPS/PS Coffee Cups in The Netherlands [28]

	Units	Resin Production	Production of Cups	Waste Processing
Raw materials	kg/kg PS	0.941	0.964	
Auxiliary materials	kg/kg PS	0.036	0.044	
Thermal energy	MJ/kg PS	33.459	12.948	−18.400
Air emissions				
CO_2	g/kg PS	—	524.203	298.000
CO	g/kg PS	1.900	0.414	0.356
NO_x	g/kg PS	4.463	1.466	0.580
SO_2	g/kg PS	8.776	2.970	0.384
Hydrocarbon	g/kg PS	25.578	2.506	—
Particulates	g/kg PS	0.902	0.233	0.262
Aldehyde	g/kg PS	0.003	0.00311	—
Organic comp.	g/kg PS	0.005	0.00555	—
NH_3	g/kg PS	0.001	0.000582	—
Chlorophenols	g/kg PS	—	—	0.0000064
PCDD	g/kg PS	—	—	0.00000048
PCDF	g/kg PS	—	—	0.0000010
HCl	g/kg PS	0.032	—	—
Fluoride	g/kg PS	—	0.0000119	—
Water emissions				
Floating material	g/kg PS	—	0.000176	—
Chemical oxygen demand	g/kg PS	0.001	0.000528	—
Oils & fats	g/kg PS	0.542	—	0.00562
Phenols	g/kg PS	0.001	—	
Dissolved substances	g/kg PS	46.170	0.0375	
Metal ions	g/kg PS		0.00000356	
Organic substances	g/kg PS	0.112	—	
NH_3	g/kg PS	—	0.000735	
Sulfate	g/kg PS	—	0.000335	
Nitrate	g/kg PS	—	0.000375	
Chloride	g/kg PS	—	0.0000202	
Na	g/kg PS	—	0.000260	
Landfill volume of waste	cm^3/kg PS	54.775	101.643	600

Selected ecoprofile information for production of PP resin and oriented PP film, packaged for shipment, is shown in Table 4.6. The film data is a weighted average of blown film and cast, tenter-frame oriented film. Production of the resin uses 73.66% of the total energy required (Fig. 4.14). Production of 1.2% contaminated scrap that must be disposed of is incorporated into the calculations [29].

Table 4.7 presents selected ecoprofile information for production of bottle-grade PET resin, PET injection stretch blow-molded bottles, and PET film, packed

Table 4.4 Selected Consumption and Emissions for HDPE Resin and HDPE Bottles [29]

	Units	HDPE Resin	HDPE Bottles
Electricity	MJ/kg	8.25	28.15
Oil fuels	MJ/kg	49.36	49.40
Other fuels	MJ/kg	22.31	22.32
Total energy	MJ/kg	79.92	99.87
Water	kg/kg	55	59
Air emissions			
Dust	mg/kg	2,900	7,100
CO	mg/kg	820	1,500
CO_2	mg/kg	1,700,000	3,000,000
SO_x	mg/kg	14,000	24,000
NO_x	mg/kg	9,900	15,000
Hydrocarbons	mg/kg	5,900	6,200
Methane	mg/kg	5,700	8,300
H_2S	mg/kg	2	2
HCl	mg/kg	48	230
HF	mg/kg	2	13
Metals	mg/kg	8	9
Aromatic hydrocarbons	mg/kg	140	140
Hydrogen	mg/kg	100	100
Water emissions			
COD	mg/kg	200	200
BOD	mg/kg	150	150
Acid (H+)	mg/kg	47	47
Dissolved solids	mg/kg	350	350
Hydrocarbons	mg/kg	51	52
NH_4	mg/kg	11	11
Suspended solids	mg/kg	2,100	2,200
Phenol	mg/kg	4	5
Al^{3+}	mg/kg	<1	<1
Ca^{2+}	mg/kg	21	21
Na^+	mg/kg	370	370
Cl^-	mg/kg	340	340
SO_4^-	mg/kg	49	49
CO_3^-	mg/kg	25	26
Detergent/oil	mg/kg	68	68
Dissolved organics	mg/kg	27	27
Sulfur/sulfide	mg/kg	5	5

for shipment. For bottles, production and delivery of the resin accounts for 78% of the total energy used [29].

Selected ecoprofile information for expandable polystyrene (EPS) and for thermoformed EPS are presented in Table 4.8. In the thermoforming operations studied, trim scrap ranged from 25 to 45%, all of which was recovered and sent

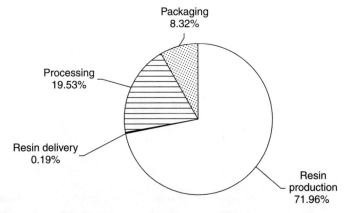

Figure 4.11. Energy distribution in production of HDPE extrusion blow-molded bottles [29].

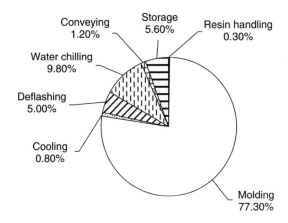

Figure 4.12. Distribution of energy use in HDPE bottle blow molding [29].

for reprocessing. A waste production level of 2% was assumed. Resin production accounted for 80.74% of the total energy requirements (Fig. 4.15) [29].

Recently, a life-cycle assessment for various alternative package systems for yogurt was published by the Center for Sustainable Systems at the University of Michigan. The study concluded that the use of larger size containers and switching from injection molded polypropylene cups to thermoformed PP cups would result in significant decreases in the environmental burden associated with the yogurt packaging [30]. An earlier study evaluated the life-cycle design of several alternative packages for milk. Life-cycle energy comparisons for some of these package systems are shown in Table 4.9 [31].

Several computer programs for life-cycle assessment calculations are now available, primarily in Europe. Some governments require their use, although

Table 4.5 Selected Consumption and Emissions for LDPE Resin, LLDPE Resin, and LDPE Film [29]

	Units	LDPE Resin	LLDPE Resin	LDPE Film
Electricity	MJ/kg	10.42	2.85	16.65
Oil fuels	MJ/kg	31.12	19.96	32.81
Other fuels	MJ/kg	39.04	49.51	42.52
Total energy	MJ/kg	80.57	72.33	91.98
Water	kg/kg	60	124	64
Air emissions				
Dust	mg/kg	2,000	670	3,400
CO	mg/kg	1,100	890	1,600
CO_2	mg/kg	1,900,000	1,400,000	1,900,000
SO_x	mg/kg	8,300	3,000	13,000
NO_x	mg/kg	9,600	5,600	12,000
Hydrocarbons	mg/kg	6,800	3,000	7,200
Methane	mg/kg	5,800	4,500	8,100
H_2S	mg/kg	1	1	3
HCl	mg/kg	56	12	110
HF	mg/kg	3	<1	6
Metals	mg/kg	3	1	4
Aromatic hydrocarbons	mg/kg	29	7	30
Hydrogen	mg/kg	72	13	73
Water emissions				
COD	mg/kg	470	250	790
BOD	mg/kg	130	8	160
Acid (H+)	mg/kg	63	40	65
Dissolved solids	mg/kg	160	60	160
Hydrocarbons	mg/kg	45	47	48
NH_4	mg/kg	8	1	9
Suspended solids	mg/kg	220	240	520
Phenol	mg/kg	3	2	4
Al^{3+}	mg/kg	<1	<1	<1
Ca^{2+}	mg/kg	1	<1	1
Na^+	mg/kg	190	36	190
Cl^-	mg/kg	280	240	300
SO_4^{2-}	mg/kg	87	250	89
CO_3^-	mg/kg	42	220	43
Detergent/oil	mg/kg	180	51	180
Dissolved organics	mg/kg	37	23	38
Sulfur/sulfide	Mg/kg	10	<1	10

use in regulating packaging has been harshly criticized by organizations such as the European Organization for Packaging and the Environment (EUROPEN). EUROPEN contends that LCA is a valuable tool for helping to reduce the overall environmental burdens and contributing to the development of sustainable product life cycles. However, EUROPEN feels LCA is not in itself a decision-making

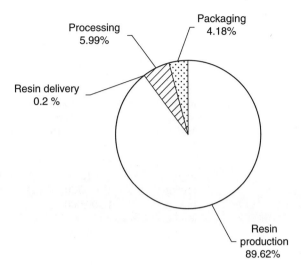

Figure 4.13. Energy distribution in production of LDPE blown film [29].

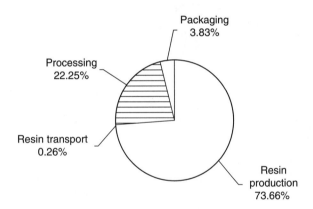

Figure 4.14. Energy distribution in production of OPP film [29].

tool and should not be used to discriminate against certain types of packaging. EUROPEN is especially critical of recent developments in Denmark and Germany where it feels questionable use of life-cycle assessments is being used to justify imposing regulatory sanctions on packaging or packaging materials that rate less favorably [32].

Recently, there has been increasing controversy about the use of plasticizers in plastic packaging materials and the association of these compounds, particularly those in the phthalate family, with cancer and reproductive anomalies. For packaging plastics, these concerns affect primarily PVC and PC. In 2000, a government panel concluded that there is serious concern that di(2-ethylhexyl) phthalate (DEHP) in vinyl medical devices may harm the reproductive organs

Table 4.6 Selected Consumption and Emissions for PP Resin and Biaxially Oriented PP Film (OPP) [29]

	Units	PP Resin	OPP Film
Electricity	MJ/kg	5.31	21.49
Oil fuels	MJ/kg	42.24	44.94
Other fuels	MJ/kg	29.64	36.79
Total energy	MJ/kg	77.19	103.22
Water	kg/kg	61	288
Air emissions			
Dust	mg/kg	1,500	5,300
CO	mg/kg	720	1,600
CO_2	mg/kg	1,900,000	2,700,000
SO_x	mg/kg	13,000	24,000
NO_x	mg/kg	9,600	18,000
Hydrocarbons	mg/kg	2,300	2,900
Methane	mg/kg	6,100	12,000
H_2S	mg/kg	1	3
HCl	mg/kg	33	180
HF	mg/kg	1	10
Metals	mg/kg	7	10
Aromatic hydrocarbons	mg/kg	3	3
Hydrogen	mg/kg	77	150
Water emissions			
COD	mg/kg	180	500
BOD	mg/kg	34	68
Acid (H+)	mg/kg	56	58
Dissolved solids	mg/kg	100	110
Hydrocarbons	mg/kg	51	56
NH_4	mg/kg	10	10
Suspended solids	mg/kg	340	700
Phenol	mg/kg	4	7
Al^{3+}	mg/kg	18	18
Ca^{2+}	mg/kg	1	1
Na^+	mg/kg	250	300
Cl^-	mg/kg	1,300	1,400
SO_4^-	mg/kg	56	58
CO_3^-	mg/kg	31	32
Detergent/oil	mg/kg	69	70
Dissolved organics	mg/kg	65	66
Sulfur/sulfide	Mg/kg	1	1

of critically ill and premature male infants exposed during medical treatment, as well as concern about effects from the pregnant mother's exposure to DEHP and other DEHP exposure during the first several years of the child's life. Possible hazards from six other phthalates used as plasticizers in PVC were rated much lower. PVC bags used for blood, intravenous fluids, enteral food, and parenteral

Table 4.7 Selected Consumption and Emissions for Bottle-Grade PET Resin, Stretch Blow-Molded PET Bottles, and PET Film (OPP) [29]

	Units	PET Resin	PET Bottles	PET Film
Electricity	MJ/kg	8.10	29.21	27.30
Oil fuels	MJ/kg	45.87	46.02	45.27
Other fuels	MJ/kg	23.51	23.53	38.33
Total energy	MJ/kg	77.47	98.75	110.90
Water	kg/kg	26	26	68
Air emissions				
Dust	mg/kg	7,100	12,000	11,000
CO	mg/kg	23,000	24,000	22,000
CO_2	mg/kg	4,300,000	5,600,000	5,500,000
SO_x	mg/kg	43,000	54,000	47,000
NO_x	mg/kg	20,000	26,000	31,000
Hydrocarbons	mg/kg	14,000	15,000	14,000
Methane	mg/kg	10,000	13,000	21,000
H_2S	mg/kg	<1	<1	2
HCl	mg/kg	340	530	460
HF	mg/kg	10	21	18
Metals	mg/kg	220	220	180
Aromatic hydrocarbons	mg/kg	2	2	3
Hydrogen	mg/kg	38	38	350
Water emissions				
COD	mg/kg	2,200	2,200	2,500
BOD	mg/kg	900	900	760
Acid (H+)	mg/kg	26	26	78
Dissolved solids	mg/kg	73	73	250
Hydrocarbons	mg/kg	120	120	110
NH_4	mg/kg	1	1	3
Suspended solids	mg/kg	380	490	820
Phenol	mg/kg	11	12	12
Al^{3+}	mg/kg	1	1	1
Ca^{2+}	mg/kg	<1	<1	130
Na^+	mg/kg	1,600	1,600	2,100
Cl^-	mg/kg	330	340	1,700
SO_4^{2-}	mg/kg	35	35	97
CO_3^-	mg/kg	3	3	21
Detergent/oil	mg/kg	24	24	27
Dissolved organics	mg/kg	10,000	10,000	7,900
Sulfur/sulfide	Mg/kg	<1	<1	<1

nutrition, along with blood and oxygen tubing and respiratory masks, are among the sources of DEHP exposure in an intensive care unit. Alternative materials are available for most of these applications but are often more expensive [33]. Polyolefins, usually PP based, are slowly replacing PVC in medical bag systems,

Table 4.8 Selected Consumption and Emissions for Expandable Polystyrene (EPS) and Thermoformed EPS [29]

	Units	EPS	Thermoformed EPS
Electricity	MJ/kg	3.38	15.70
Oil fuels	MJ/kg	34.04	36.94
Other fuels	MJ/kg	46.23	50.51
Total energy	MJ/kg	83.66	103.14
Water	kg/kg	181	208
Air emissions			
Dust	mg/kg	1,800	3,100
CO	mg/kg	1,600	2,200
CO_2	mg/kg	2,500,000	3,300,000
SO_x	mg/kg	9,700	20,000
NO_x	mg/kg	12,000	18,000
Hydrocarbons	mg/kg	4,500	9,400
Methane	mg/kg	9,500	12,000
H_2S	mg/kg	<1	2
HCl	mg/kg	35	68
HF	mg/kg	2	2
Metals	mg/kg	5	11
Aromatic hydrocarbons	mg/kg	210	220
Hydrogen	mg/kg	83	140
Water emissions			
COD	mg/kg	670	1,200
BOD	mg/kg	140	200
Acid (H+)	mg/kg	39	43
Dissolved solids	mg/kg	110	130
Hydrocarbons	mg/kg	97	110
NH_4	mg/kg	12	13
Suspended solids	mg/kg	740	1,300
Phenol	mg/kg	4	10
Al^{3+}	mg/kg	45	45
Ca^{2+}	mg/kg	1	1
Na^+	mg/kg	700	750
Cl^-	mg/kg	3,400	3,600
SO_4^{2-}	mg/kg	240	240
CO_3^-	mg/kg	160	170
Detergent/oil	mg/kg	55	63
Dissolved organics	mg/kg	51	52
Sulfur/sulfide	Mg/kg	1	1

including tubing, in parts of Europe, although they are still not in common use in the United States [34].

Bisphenol A, a monomer used in the production of polycarbonate, has been suspected of having endocrine-mimicking effects and being associated with premature puberty in girls. At this time, the evidence is inconclusive.

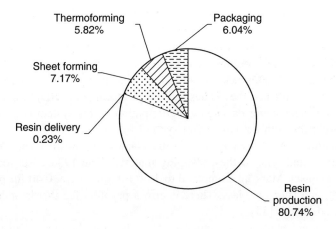

Figure 4.15. Energy distribution of thermoformed expandable polystyrene [29].

Table 4.9 Energy Use in Milk Packaging [31]

Package Type	Energy for Material Production (MJ/1000 gal delivered product)	Life-Cycle Energy (MJ/1000 gal delivered product)
HDPE bottle—single use	6930	7720
LDPE pouch	1550	1700
HDPE bottle—refillable	470	2320
Polycarbonate bottle—refillable	1020	2630

Government agencies and other groups, in the United States and elsewhere, have recommended additional studies on phthalates, bisphenol A, and similar compounds that are known or suspected to have hormone-mimicking or disrupting effects.

4.11. LEGISLATION AND REGULATION

As mentioned, legislation and regulation have been used in efforts to decrease the environmental impact of products or packages, particularly their impact on disposal systems, by increasing recycling and use of recycled content, as well as providing incentives for source reduction. Plastics packaging is often a target of such regulation. The most prevalent approach, globally, is implementation of the philosophy that the entity making the packaging decision should be held responsible for the management of waste packaging and should be required to meet target recycling levels. This idea is variously known as producer pays, producer responsibility or extended product (or producer) responsibility (EPR). Its first major manifestation was in Germany.

Germany's Ordinance on the Avoidance of Packaging Waste requires that manufacturers and distributors take back and recycle or reuse a certain percentage of all packaging materials or participate in an established national waste management program. For distribution packaging, third-party organizations have been set up to handle the take-back and recycling of used packages. For retail packages, industry established the Duales System Deutschland (DSD) to collect, sort, and recycle postconsumer packaging. Participation in this system is identified by labeling packages with the Green Dot symbol.

Participating companies pay fees to the Duales System that vary according to the amount and type of the packaging material. The 1994 costs for plastics were 2.95 Deutsch Mark/kg, compared to 1.50 for aluminum, 0.40 for paper and paperboard, and 0.15 for glass. There is also a per item fee that is added to the weight-based charge [35].

Consumers either place packaging materials bearing the Green Dot in yellow bags or yellow bins for curbside collection or bring them to drop-off locations located near their homes. Companies that do not participate in the Green Dot system are required by law to take back and recycle their own used packaging. Consumers have the right to return such packaging to the retailer where they purchased the goods.

For transport packaging, participation in the Green Dot system is an option, but in most cases manufacturers either arrange their own systems or participate in material-specific multiuser systems that offer better pricing than DSD. In many cases, manufacturers have switched to reusable transport packaging, often using collapsible plastic crates.

The government sets minimum requirements for both collection and recycling of packaging. Currently, a recycling rate of 60% is required for plastics packaging. The Duales System reported that in the year 2000, 589,000 metric tonnes of plastic packaging were recycled, for a savings of about 20 billion MJ of primary energy. This analysis of energy use is the first step in an effort to measure the ecological benefit of packaging recycling. The organization also reports that the amount of sales packaging taken home each year by the average shopper declined by over 13% between 1991 and 1997 [36].

For several years after the German Packaging Ordinance was adopted in 1991, experts predicted the system would fail. However, by the mid-1990s the idea of producer responsibility had instead been adopted by the entire European Union (EU), and continues to expand around the world. A partnership of European Green Dot organizations, PRO EUROPE, was established to grant the right to use the Green Dot in other countries in order to help prevent trade barriers and guarantee free movement of goods. In April, 2001, Hungary became the twelfth European country to adopt the Green Dot as the financing mark for packaging recovery and recycling. Membership also includes, in addition to Germany, European Union member states Belgium, France, Ireland, Luxembourg, Austria, Portugal, and Spain, and non-EU members Norway, Latvia, and the Czech Republic. About 60,000 licensees mark approximately 460 billion articles of packaging with the Green Dot each year. The symbol is protected as a trademark in 170 countries around the world, making it the world's most widely used trademark [36]. On the negative side, the cost of recycling retail plastic packaging is reported to exceed the cost of the plastic raw materials used to make it [37].

Since the cost of participating in the Green Dot system is affected by both the type and amount of packaging material used for a product, there is an incentive for companies to reduce their use of packaging. For example, a German bakery changed from a 160-μm film to a 130-μm film, for a net reduction in packaging film volume of about 20%. The result was a savings in environmental charges of about 25%, totaling more than $100,000 per year [38].

The European Union's Directive on Packaging and Packaging Waste, Council Directive 94/62/EC, adopted in 1994, requires EU member states to take necessary action to recover between 50 and 65% of packaging waste and to recycle between 25 and 45%, with a minimum of 15% for each type of packaging material [39]. As described above, the approach that has been uniformly adopted to reach these goals is some variation of producer responsibility for packaging. For retail packaging, often this involves licensing and use of the Green Dot system. Countries in the EU differ considerably on the amount of packaging recovery, including plastics, that is currently being achieved. In general, countries that only recently implemented the regulations have lower recovery rates. Several countries that are applying for EU membership have also implemented producer responsibility for packaging materials. Norway, which is not an EU member, has also adopted the Green Dot system. The EU is also adopting producer responsibility for certain products, including automobiles and electronics.

The EU in 1999 moved to adopt what are termed "essential requirements" for packaging to establish standards for how packaging must meet environmental demands. The regulation provides that packaged products can be banned if they interfere with recycling programs, are not easily recovered, or could be packaged with less material without impeding performance. There has been considerable difficulty in reaching agreement about what the packaging standards will be. At this writing, while companies have to show proof of compliance with the essential requirements, there is no standard method for doing so, and consequently there is doubt about whether the requirements are legally enforceable.

Several other countries around the world have also adopted mandatory extended product (or producer) responsibility (EPR). Examples include Japan, Taiwan, South Korea, Brazil, and Peru.

Canada adopted a National Packaging Protocol in 1989 that required a 50% reduction in packaging waste going to disposal by 2000. This target was achieved by 1996, four years ahead of schedule, through a variety of mechanisms, differing from province to province. The Canadian Industry Packaging Stewardship Initiative resulted in adoption of a set of Guiding Principles for Packaging Stewardship in 1996. A number of provinces have instituted deposits on selected containers, and there are also a number of voluntary programs run by industry to recover certain types of packaging materials.

In the United States, most action on recovery and recycling of packaging has occurred at the state level. Ten states have programs that impose deposits on certain beverage containers, in most cases on carbonated soft drinks and beer. Maine and California have extended the deposit system to a variety of additional beverages, including bottled water and fruit juice. Maine's system covers all beverages except milk. California's is less inclusive. It should be noted that California's system is actually a refund value system, rather than a deposit system. Distributors pay a per-container fee to a state fund, and consumers then can deliver the empty containers to designated redemption centers for refund. In most deposit states, consumers are charged a per-container deposit, in addition to the item price, and can return the empty container to any retailer handling that brand for return of the deposit. While such systems are vehemently opposed by the beverage industry and retailer groups, they have proven to be quite effective in achieving return of targeted containers for recycling. In most deposit states, redemption rates are 90% or more. California, with its lower refund value and less convenient redemption system, has lower recovery rates. In Michigan, where the deposit is 10 cents per container rather than the more typical 5 cents, return rates are higher.

The extended product responsibility idea has yet to have any substantial impact on packaging for domestic consumption, although it does, of course, affect companies exporting goods to countries with such regulations. Several states have begun to propose or pass EPR for problematic products such as batteries, and Minnesota has an EPR system in place for carpet. Many states have banned certain wastes from disposal facilities. The most common such material is yard waste. Some states prohibit disposing of recyclables. Regulations requiring recycled content or voluntary agreements for minimum recycled content are common for newspaper.

California, Oregon, and Wisconsin have laws that require recycled content in some plastic containers. The Wisconsin law requires at least 10% "recycled or remanufactured" content in plastic containers except those for food, beverages, drugs, and cosmetics. There has evidently been little enforcement of this law, and it has had little effect.

Oregon requires use of 25% postconsumer recycled content in rigid plastic containers (RPCs) unless the recycling rate for plastic containers in the state is at least 25%. There are exemptions for food and medical packaging, source-reduced containers, and some others. Companies can also comply under recycling rate provisions if all containers of a specific type or brand are recycled at a rate of at least 25%. Since the law became effective, the recycling rate for RPCs in the state has been over the 25% minimum, so no packaging modifications have been required. However, the state Environmental Quality Department has warned that recent declines in the plastic recycling rate may cause the rate to drop below the 25% minimum within the next few years, in which case users of RPCs covered by the law would have to show compliance in some other way.

California has a law that is very similar to Oregon's, but its history has been considerably different. For the first year in which compliance was required, the California Integrated Waste Management Board (CIWMB) was unable to reach agreement on an overall RPC recycling rate for the state and ended up adopting a range of rates that spanned the 25% mark. Since that time the rates have been below 25% and falling. Therefore, companies are legally required to comply with the law in some other manner. This effort was complicated considerably by the fact that the rate was determined 2 years after the fact, rather than estimated for the coming year as is done in Oregon. Thus, companies were caught by surprise and found themselves in noncompliance because they had expected to comply under the 25% overall RPC recycling rate provision. The CIWMB has reached compliance agreements with several companies in which the companies admitted noncompliance and put into place plans to come into compliance in subsequent years. In April, 2001, CIWMB announced that five companies operating under such agreements had come into compliance, so no further enforcement action would be taken against them. The board has also revised its procedure for determining rates and is moving toward forecasting rates rather than documenting them after the fact. There is also hope that the expanded beverage container refund system, which went into effect in January, 2000, will boost recycling rates. The most recent recycling rate, for 1999, released in July 2000, showed a 17.9% rate for all rigid plastic containers, and a 24.8% rate for PET. In contrast, the 1995 rates were 24.6% for all rigid plastic containers, and 38.8% for PET. The law provides that PET containers, to be in compliance, must be recycled at a rate of at least 55%. There have also been a series of efforts to toughen the law. One of the most recent efforts, introduced in April, 2001, would require RPC manufacturers to pay a penalty if the recycling rate for the containers was under 50%. So far, none of these efforts have been successful.

REFERENCES

1. U.S. Environmental Protection Agency, *Municipal Solid Waste Generation, Recycling and Disposal in the United States: Facts and Figures for 1998*, EPA530-F-00-024, 2000.
2. U.S. Environmental Protection Agency, *Characterization of Municipal Solid Waste in The United States: 1997 Update*, EPA530-R-98-007, 1998.
3. Resins 2001, *Mod. Plastics* Feb., 27–49 (2001).
4. American Plastics Council, *www.plasticsresource.com*, 2001.
5. Alberta Dairy Council, *www.milkjugrecycling.com*, 2001.
6. EPIC, *Best Practices Guide for the Collection and Handling of Polyethylene Plastic Bags and Film in Municipal Curbside Recycling Programs*, The Environment and Plastics Industry Council, Mississauga, Ontario, 1998.
7. T. Buwalda and C. McLendon, *Alliance of Foam Packaging Recyclers*, Summer 2000, p. 2.

8. EPS Recycling International, *www.epsrecycling.org*.
9. Alliance of Foam Packaging Recyclers, *www.epspackaging.org*.
10. S. Toloken, *Plastics News* April 9, 2001, p. 1.
11. American Plastics Council, *www.plasticsresource.com*.
12. U.S. Environmental Protection Agency, *National Source Reduction Characterization Report for Municipal Solid Waste in the United States*, EPA530-R-99-034, 1999.
13. T. Flaherty, *Pharmaceutical & Medical Packaging News*, July 1996, p. 22.
14. *Packaging Digest*, Flex-packs stand at attention, March 1995, p. 46.
15. *Packaging Strategies*, Retortable SUPs — cost-effective alternative to cans, June 15, 1999, p. 6.
16. Kapak Corporation, *www.kapak.com*.
17. R. Leaversuch, *Modern Plastics*, April 1990, p. 26.
18. D. Saphire, Delivering the Goods: Benefits of Reusable Shipping Containers, *INFORM*, 1994.
19. Steel Recycling Institute, *www.recycle-steel.org*.
20. P. Anderson, S. Kelly, and T. Rattray, *Biocycle*, July 1995, p. 64.
21. The Plastic Redesign Project, Design for Recyclability: Recommendations for the Design of Plastic Bottles, *www.plasticredesign.org*, Nov. 1998.
22. Association of Postconsumer Plastics Recyclers, *www.plasticsrecycling.org*.
23. P. Anderson and D. Stusek, New Plastic Packaging Innovations: Will They Continue their Historic Tradition to Improve Recycling or Will They Make Plastics too Expensive to Recycle? The Plastic Redesign Project, *www.plasticredesign.org*.
24. Tellus Institute, *CSG/Tellus Packaging Study*, Tellus Institute, Boston, MA, 1992.
25. Franklin Associates, *An Energy Study of Plastics and Their Alternatives in Packaging and Disposable Consumer Goods*, Prairie Village, KS, 1992.
26. I. Boustead and G. F. Hancock, *Energy and Packaging*, Ellis Horwood, Chichester, England, 1981.
27. L. L. Gaines, *Energy and Materials Use in the Production and Recycling of Consumer-Goods Packaging*, ANL/CNSV-TM-58, U.S. Dept. of Commerce, Washington, DC, 1981.
28. J. van Eijk, J. W. Nieuwenhuis, C. W. Post, and J. H. de Zeeuw, *Reusable Versus Disposable: A Comparison of the Environmental Impact of Polystyrene, Paper/Cardboard and Porcelain Crockery*, Ministry of Housing, Physical Planning and Environment, The Netherlands, 1992.
29. Association of Plastics Manufacturers in Europe, *www.apme.org*.
30. D. Brachfeld, T. Dritz, S. Kodama. A. Phipps, E. Steiner, and G. Keoleian, *Life Cycle Assessment of the Stonyfield Farm Product Delivery System*, Center for Sustainable Systems, The University of Michigan, Ann Arbor, MI, 2001.
31. D. Spitzley, G. Keoleian, and J. McDaniel, *Life Cycle Design of Milk and Juice Packaging*, EPA/600/SR-97/082, 1997.
32. The European Organization for Packaging and the Environment, *www.europen.be*.
33. B. Hileman, *Chem. Eng. News*, Aug. 7, 2000, p. 52.
34. R. Leaversuch, *Mod. Plastics*, May 2001, p. 88.

35. *Environmental Packaging*, Thompson Publishing Group, Tampa, FL, 2001.
36. Duales System Deutschland AG, *www.gruener-punkt.de/en*.
37. D. Loepp, *Plastics News*, Oct. 26, 1998, p. 5.
38. R. Colvin, *Mod. Plastics*, March 1998, p. 48.
39. European Union, *http://europa.eu.int*.

CHAPTER 5

PLASTICS IN AGRICULTURE

IKRAM HUSSAIN AND HALIM HAMID
Center for Refining and Petrochemicals Research Institute King Fahd University of Petroleum and Minerals Dhahran, Saudi Arabia

5.1. INTRODUCTION

The agricultural and horticultural industries are emerging as major potential consumers of plastics in the form of film, sheet, pipe, and molded products in such areas as agriculture: disease and pest control, water conservation supply and drainage, fertilizer transport, storage and application, crop conservation, livestock rearing, produce collecting and transportation, tools, machinery and equipment, and buildings and structures.

The energy situation also called for a reappraisal of the role of film plastics for reducing heat loss from both plastics and glass greenhouses. The use of plastics in winter crop storage was a neglected field, but now interest is growing for using insulated double-skinned black plastic constructions similar to those employed in mushroom production. Plastics has been effectively used as aids for fruit and vegetable storage in the form of linings for boxes, bags, and bins. Rigid plastics also have an important role in containers for holding and transporting produce from point of harvest to point of consumption. Use of plastics in netting for crop windbreaks, plant pots and containers, soil conditioners (expanded plastics waste), mulch films, and pipe for drip or trickle irrigation will be increasing in the future all over the world.

Single- as well as multiple-span polyethylene tunnels and a rigid polyvinyl chloride (PVC)-rooted structure covering up to a third of an acre are increasingly being used for agricultural cultivation. Greater economy has been achieved using a single structure consisting of posts and tensioned wires with a polyethylene film roof.

Agriculture, horticulture, and related areas of plastics applications consume 2,250,000 tons/year of plastics worldwide.

Plastics and the Environment, Edited by Anthony L. Andrady.
ISBN 0-471-09520-6 © 2003 John Wiley & Sons, Inc.

According to Jean-Claude Garnaud of the International Committee on Plastics in Agriculture [5], in western Europe consumption of plastics in agri-related areas amounts to 700,000 tons/year. Film applications area is dominant in terms of volume usage. Applied Market Information (AMI) estimated that the European market uses about 500,000 tons/year in agricultural films, of which greenhouse films are about 50%, silage and mulch films are 25% each.

The agricultural film market is growing at about 2.5%/year overall, according to Andrew Reynolds of AMI. Greenhouse film applications show the fastest growth because of the demands for early crops of fruits and vegetables. On average 25–37 g of films are consumed per kilogram of produce for crops, melons, strawberries, and tomatoes [44].

Linear low-density polyethylene (LDPE) is the most important polymer, followed by (high-density polyethylene) HDPE. According to Gernaud, western Europe consumes 570,000 tons/year of agricultural LDPE, of which 350,000 tons have been used in films and tubings.

The European Union (EU) countries use 160,000 tons in transparent crop covering, 148,000 tons in black silage film, 60,000 in transparent mulch films, and around 33,000 in stretch films [44].

Low-diameter low-density polyethylene/linear low-density polyethylene (LDPE/LLDPE) based irrigation tubing is another significant agricultural end use. Higher diameter (>32 mm) tubes are usually made with HDPE or PVC. Western Europe uses 100,000 tons of rigid PVC pipes for drainage and 60,000 tons/year of flexible PVC hose piping, though not all can be classified as agriculture end use. Other applications for plastics in agriculture include polypropylene (PP) twine for binding applications and corrugated PP pipes/fittings for irrigation.

High-density polyethylene is currently used in netting applications for tree protection and oyster farming; also HDPE, PP, and polystyrene (PS) foams are used in making packaging crates for vegetable/fruit transportation.

Other polymers find limited agricultural uses. Among these are glass-reinforced plastics (GRP) tanks, silos and vats, polyethylene terephthalate (PET) filament based thermal screens and flat covering sheets. Polycarbonate (PC) and polymethylmethacrylate (PMMA) are used in limited amounts for double- or triple-wall sheets for top of the line greenhouses.

This chapter presents the use of plastics in three major sectors: (i) greenhouses (ii) mulches, and (iii) agricultural irrigation. Since these plastics are meant for long-term usage in harsh outdoor conditions of higher temperatures, ultraviolet (UV) radiations, temperature cycling, and other climatic factors. Therefore, these plastics are stabilized with a number of UV stabilizers including antioxidants, hindered amine light stabilizers (HALS), and Ni quenchers, which provide protection against, heat, UV radiation, temperature cycling during day to night, high wind velocity, high humidity, and other climatic factors that are interacting simultaneously. The most important weather parameter that affects the lifetime of plastics used in agricultural applications are discussed.

Stabilization of plastics, particularly greenhouse films, mulch films, silage, and rigid pipes used in irrigations, are discussed in detail. Some case studies showing the outdoor performance of greenhouse films, mulches, and pipes exposed in near equatorial regions are also discussed in this chapter. The UV stabilizers mentioned in this chapter are state-of-the-art products. Their structures and trade names of these commercial products are shown in the Appendix.

5.2. GREENHOUSE FILMS

Currently, the low-density polyethylene (LDPE) homopolymer and the ethylene–vinyl acetate (EVA) copolymers are the most common plastics materials used for greenhouse cover films applications [4]. However, in Japan, plasticized (PVC) is used to a larger extent. The growing importance of plastics films for greenhouse covers is best illustrated by a few figures with greenhouse films worldwide was approximately 60,000 ha [1]. In 1980 the surface area covered with worldwide greenhouse films was 80,000 ha [1] and reached 220,000 ha in the 1990s [2]. Spain went through a tremendous evolution for the agricultural surface-covered greenhouse: from 15,000 ha in 1983 [3] to an area of 28,000 ha in the 1990s [2]. Although there is some leveling off in the total area covered with greenhouses in southern Europe, the development is likely to continue in other parts of the world, particularly in South America.

The main types of polyethylene films used as greenhouse covers include the following:

1. LDPE is enriched by needed and measured quantities of UV stabilizers that ensure the films a life of some months to 2–3 years.
2. EVA is PE containing different doses of vinyl acetate (VA). The higher the quantity of VA used, the better the absorption for long infrared (IR) (5–35 μm); nevertheless, the mechanical properties may be not quite satisfactory.
3. Thermic IR or modified PE is produced from compounds enriched with special "charges" that absorb other parts of the long IR radiation than VA does.
4. Thermic PE contains low quantities of VA and simultaneously some thermic charges. They possess the advantages of the EVA and of the thermic PE but do not have their disadvantages.
5. Linear PE seems to present some advantages interesting in agriculture [5].

5.2.1. Characteristics and Performance of Coextruded and Blended Films

Coextruded greenhouse films based on two- and three-layer structures are generally made from LDPE, LLDPE, and EVA copolymers.

Each of the layer impart specific properties to a multilayer film. These advantages include absorption of desirable radiation wavelength range, mechanical strength, resistance to agro-chemicals, and the like.

Table 5.1 Characteristics and Performance of Polymers and Their Blends Used for Greenhouse Films

Type of Polymer	Characteristics	Performance E_{50} (kLy)* in Florida
LLDPE	Superior mechanical properties compared to LDPE	260
	Elongation at break is significantly higher than LDPE	
	Stabilization: 0.27 HALS4–0.2% HALS5	
LLDPE/LDPE blends (50/50)	Superior mechanical properties of LLDPE	
	Superior optical properties of LDPE	
	Easier processing compared to LLDPE	
	Stabilization: 0.15% HALS5 + 0.12% UVA1	
	Comparable light stability of blend as compared to its components	145
EVA/LDPE	EVA copolymers are transparent to visible light; allow all wavelength needed for photosynthesis	
	EVA has spectral emission between wavelength of 7–14 μm, which constitute an important part of energy losses from soil and plants	
	EVA has absorption bands in long wavelength IR, in contrast to LDPE; these are infrared barrier film or thermic film	
	EVA up to a 14% VA is used as IR barrier film or thermic film	
	Stabilization: 0.37 HALS5 + 0.3% UVA1	
	Comparable or higher UV stabilization than LDPE	385

*Kilo Langley.

A summary of polymers and their blends used for greenhouse applications are presented in Table 5.1. The EVA/LDPE blends show superior UV resistance as compared to LLDPE or LLDPE/LDPE blends (50/50). Additionally, the EVA also serves as infrared or thermic barrier film.

The partial prevention of the dissipation of the thermal energy (greenhouse effect) from the greenhouse during cool nights from EVA-based blended film is desirable. Thus, the temperature inside the greenhouse does not fall much and the productivity is not affected.

5.2.2. Monolayer and Multilayer Films

Currently, coextruded greenhouse films in the form of multilayer are becoming popular. The reasons for their success is the technical advantages that each film layer can offer. For example, the partial barrier property of EVA to long-range IR

Table 5.2 Characteristics and Properties of Multilayer[a] Greenhouse Films

	Polymer Layer		Months to Retain
First Layer	Second Layer	Third Layer	50% Elongation
LDPE[b]	—	—	18[c]
LDPE[d]	EVA[d]	—	21[c]
LDPE[e]	LDPE[e]	EVA[f]	695[g]

[a] 75 μm thickness of coextruded film.
[b] Stabilization: 0.6% HALS3 + 0.3% UVA1.
[c] Exposure in São Palo (Brazil).
[d] Stabilization: 0.6% HALS3 + 0.3% UVA1 in first layer and 10% in EVA.
[e] Stabilization: 0.2 HALS6 + 0.27% UVA1.
[f] 150 μm thickness of coextruded film with upper layer 40 μm, middle layer 60 μm, lower layer 50 μm.
[g] Exposure in Florida.

Table 5.3 Factors Influencing the Lifetime of Greenhouse Plastic Films

Factors Responsible for Greenhouse Film Performance	Details of Responsible Factors
Polymer type	LDPE, LLDPE, EVA copolymer
Impurities in polymer	Chromophores present, catalyst residues
Weather parameters	UV radiation, temperature, moisture, pollutants, temperature cycling, wind
Film parameters	Thickness
Stabilization	HALS, UV absorbers, antioxidants Ni quenchers
Greenhouse related	Support, frame metal, protection of contact surfaces, design, film tightness
Chemical treatment	Insecticides, pesticides (sulfur containing, chloride based)

wavelength and the mechanical strength and reduced dust accumulation of LDPE or LLDPE can be combined to get two-layer film with certain technical advantages [24, 25]. Table 5.2 presents the characteristics and properties of monolayer versus multiplayer. A study carried out by Hanninger and Roncaglione [26] shows the comparable performance of monolayer and multilayer films having the same thickness. Table 5.2 also shows that the lifetime of coextruded LDPE/EVA films exposed in São Palo, Brazil, and LDPE/LDPE/EVA film exposed in Florida. It has been discussed by Gugumus [22] that the lifetime of the coextruded film LDPE/EVA has intermediate lifetime of each of the constituting single layers.

The lifetime of the greenhouse cover films used outdoor depends mainly on a number of factors. Table 5.3 presents the factors responsible for the outdoor performance of greenhouse films and the details of these factors.

However, the three most important factors that have great influence on the performance of polyethylene greenhouse films are weather parameters, stabilization of plastics, and chemical treatment. The effect of these parameters are discussed in detail in the following sections.

5.3. WEATHER PARAMETERS THAT AFFECT PLASTICS LIFETIME

During natural weathering, plastics are exposed to solar radiation (primarily UV radiation), heat, moisture (dew, rain, humidity), pollutants (aerosols, acid rain, ozone), and wind. These climatic factors vary so widely over Earth's surface that weathering of plastics is not always exactly predictable. Therefore, plastic performance varies with changes in climatic conditions [6].

Lifetime of polymers exposed to natural environment cannot be predicted accurately because of complexity of the natural weathering phenomenon and multitude of weather parameters interacting simultaneously [7]. A brief discussion of the important factors that mainly contribute to degradative processes during natural weathering follows.

5.3.1. UV Radiation

Plastic films exposed to natural weather undergoes photo-oxidative reactions. Sunlight, together with oxygen, initiates the chemical reactions in the polymer matrix. The important aspects of sunlight include sunshine duration, total solar radiation, and UV radiation that may be used in modeling of the degradative processes [6–8].

The solar radiation reaching Earth's surface is only about one-half to two-third of its value at the outer limits of the atmosphere. The most damaging portion of the total solar radiation is the UV section of the radiation, which causes the weather-induced degradation of plastics [8, 9]. The sun sends a continuous spectrum of energy radiation on Earth. The radiation varies in terms of wavelength, and therefore, photon energy. The wavelength has an inverse relationship with quantum energy [10].

The extraterrestrial radiation of the sun is largely constant. The intensity of radiation depends on the angle of incidence. The changes in the intensity as a result of the tilt of Earth's axis and the rotation of Earth in a daily as well as annual cycle are linked to the geographic latitudes where radiations are monitored [11].

The radiation wavelength ranges reaching Earth's surface are from about 290 to 1400 nm [13–15]. The portion of the sun's spectrum between 290 and 400 nm is important to polymers because it includes the highest energy region and is the only portion that can cause direct harm to unmodified polymers [16]. The visible region of the solar spectrum, 400–800 nm, does not usually cause direct harm to polymers but can do so by interaction with sensitizing substances in the polymers. The IR portion, less than 800 nm, is generally considered harmless in a photochemical sense, but it may have a role in the thermal oxidative degradation of some polymers.

The UV portion of radiation largely causes the degradation of polymers; the energy at visible and high wavelengths is too low to damage the chemical structure of a polymer. Because of its chemical structure every polymer is susceptible to photochemical degradation at a particular wavelength. The absorption of UV radiation and its concomitant degradative effects vary for each individual polymer. Stability is strongly dependent on the specific chemical and molecular structure of the polymer; variations in structure lead to differences in the absorption ranges of individual polymers. Theoretically, polyethylene material should not contain functional groups that would be capable of absorbing UV radiation. However, polyethylene absorbs, and gets degraded by, UV radiation. Sources of UV-absorbing chromophores are [39–41]:

1. Polymer structure
2. Catalyst residues
3. Thermal-processing degradation products
4. Antioxidants and transformation products
5. Colorants, fillers, flame retardants, etc.

Most polymers should not absorb at wavelength greater than 300 nm. The above-mentioned sources of UV-absorbing chromophores present in polymers may lead to the initiation of weathering processes [16]. It is a UVB radiation with a shorter wavelength (280–315 nm) that is mainly responsible for photodamaging, discoloration, and loss of mechanical properties of polyethylene. The degree of deterioration of UVB levels on the lifetime of polymer depends on the geographic location of exposure and the tendency of the material to absorb UVB radiation [17]. In Saudi Arabia UV dosage of 170–220 kLy/year is received as compared to 130–140 kLy/year in Florida.

The weathering process depends significantly on temperature along with other factors. Though the maximum temperature attained by the polymers is not sufficient to promote bond cleavage of any structures likely to be found in polyethylenes and other plastics. The role of heat in the weather-induced degradation of plastics is in accelerating processes otherwise induced, such as hydrolysis, secondary photochemical reactions, or the oxidation of trace contaminants [18]. It has been determined that the degradation rates of polymers in the tropical zones far exceed those in temperature zone. Also the simulated weathering experiments have shown that the oxidation rates of polyethylene exposed to 300-nm radiation increase fourfold from 10 to $50°C$ [19].

The annual average incident energy in different parts of the world is reported in the literature [20–22]. Table 5.4 presents the annual average incident energy measured in different parts of the world.

5.3.2. Temperature

The ambient temperature in near-equatorial regions may reach $50–55°C$. However, the plastic samples exposed under such extreme conditions may reach about

Table 5.4 Annual Average Incident Solar Radiation for Different Geographic Locations WorldWide

Places	Incident Solar Radiation, kLy/year
Florida (USA)	140
Arizona (USA)	190
Almeria (Spain)	150
Basel (Switzerland)	100
Dhahran (Saudi Arabia)	180–220 [23]
Melbourne (Australia)	140 [21]
Singapore	140 [21]

77°C. Usually, the plastic surface temperature is significantly higher than the surrounding air temperature. Exposed sample temperature plays an important role in degradation during weathering. The greenhouse films are exposed to heat stress at the point of contact with metallic framework.

5.3.3. Humidity

The moisture in contact with the greenhouse films can have combined effects of unique physical properties with its chemical reactivity [27]. The water in contact with polymers may have three kinds of effects that may contribute to degradative reactions: (i) hydrolysis of labile bonds, such as those of polyesters or polyamides; (ii) physical destruction of the bonds between polymer and a filler, resulting in chalking or fiber bloom; rain may wash away water-soluble degradation products and additives; moisture, can contribute to swelling of certain plastics, and (iii) photochemical effect, in solving the generation of hydroxyl radicals or other reactive species that can lead to promotion of free-radical reactions [27, 28]. In order to assess the role of various weather parameters on the degradation of plastics, it is critical to have accurate climatological data corresponding to the exposure sites.

5.4. STABILIZATION OF PLASTICS

Polymeric films used in greenhouse and mulch film applications have been a subject of great interest to additive manufacturers, film processors, and researchers for over 25 years [29–34].

In the 1960s and 1970s antioxidants and UV absorbers provided basic levels of protection against thermal and UV degradation. During the 1980s, hindered amine light stabilizers (HALS) extended the service lifetime of a range of polyolefin-based applications. Even during the past 10 years development has continued in the area of new structures of HALS, their performance, controlling the negative effects of agro-chemical (particularly acidic pesticides) on the performance of HALS. There are three classes of UV stabilizers generally used by the agri-film processors. These UV-stabilizers include the following [12]:

1. UV absorbers: benzophenones and benzotriazoles
2. Ni quenchers: first compounds used for greenhouse stabilization in combination with UV absorbers
3. Polymeric HALS: have superior UV stability; have disadvantage of possibility of deactivation with acidic pesticides

Ciba has currently produced two new Tinuvin products: Tinuvin 492 and Tinuvin 494. These structures have lower basicity as compared to original hindered amines [31, 35]. Great Lakes Chemicals has added two new HALS, UVASIL 816 and UVASIL 299 (HALS2). These are tertiary HALS that provide higher performance in greenhouse films and greater resistance to pesticide spraying. Tinuvin 123 is particularly designed to give sulfur resistance in rose greenhouses.

5.4.1. Stability of Greenhouse Films Exposed in Florida

In Florida the performance of Ni quenchers may vary depending on their chemical structure and their combinations with UV absorbers, as well as their level of additions in the formulations. Table 5.5 shows some Ni quenchers, such as Ni2, that have superior performance even when the others are combined with benzophenone-type UV absorbers.

5.4.2. Light Stability of Greenhouse Films Exposed in Near-Equatorial Regions

The prediction of lifetime of greenhouse films is not possible without conducting the real-time weathering trials at locations where those films have to be used

Table 5.5 Light Stability of LDPE Blown Films[a] (A)

UV Stabilization	E_{50} (kLy) Florida[b]
Control	32
0.15% Ni-1	23
0.30% Ni-2	48
0.60% Ni-2	54
0.15% Ni-1	125
0.15% UVA1	90
0.15% Ni-2 + 0.15% UVA1	70
0.15% Ni-1 + 0.15% UVA1	150

[a] Blown films 200 μm thick.
[b] E_{50}, energy (kLy) to 50% retained elongation.

[6, 35–40]. The weather is variable from time to time and from place to place; hence the weather trial carried out at a certain location is not fully adequate to ascertain the degradation pattern. Hamid and Prichand [36] conducted an extensive study on weather-induced degradation of LLDPE and hence used a mathematical modeling technique to describe the weathering phenomenon.

Prediction of Lifetime of Greenhouse Films   Amin et al. [63] has developed an empirical model to predict the mechanical properties of greenhouse films exposed for 3 years duration (from 1991–1993) at Dhahran, Saudi Arabia (26.32°N, 50.13°E). Samples were withdrawn periodically and tensile strength and percent elongation at break data were determined along with other properties such as crystalline melting point (T_m) and carbonyl absorbance.

Tensile strength (TS) and percent elongation (PE) data were used to develop empirical equations. Prediction models were developed with exposure time as an independent parameter for both tensile strength and percent elongation at break of exposed greenhouse film, which describes the data accurately as follows:

$$PE = 517.5 - 4.438 T_m - 0.0576 (T_m)^2 \qquad (1)$$

$$TS = 23.88 - 0.149 T_m - 0.000335 (T_m)^2 \qquad (2)$$

Figure 5.1 shows the comparison of experimental and predicted values for percent elongation at break (PE). The coefficient of determination R^2 was 0.89.

Comparison of Greenhouse Film Exposed on Racks and a Greenhouse   Khan and Hamid [62] have conducted a 3-year study on LDPE-based greenhouse films exposed on racks as well as real operating greenhouse at Dhahran, Saudi Arabia. Weathering trials were conducted for 3 years. The LDPE films 180 μm

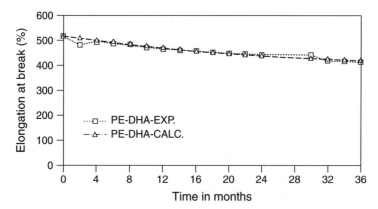

Figure 5.1. Experimental data and predicted values of percent elongation at break as a function of exposure time at Dhahran, Saudi Arabia.

Table 5.6 Typical Meteorological and Radiation Data for Dhahran, Saudi Arabia

Month	Temperature (°C)	Relative Humidity (%)	Solar Radiation (Langley)	UV Radiation (Langley)
January	14.4	81	291.63	12.04
February	16.8	77	361.80	14.71
March	20.9	64	477.21	18.40
April	25.5	65	524.8	21.84
May	31.6	50	602.52	23.65
June	35.8	36	686.52	27.52
July	37.3	42	636.40	25.46
August	36.6	49	609.83	25.80
September	33.2	63	548.16	23.99
October	30.3	73	444.28	20.30
November	23.6	75	365.07	17.03
December	19.0	82	318.89	14.88

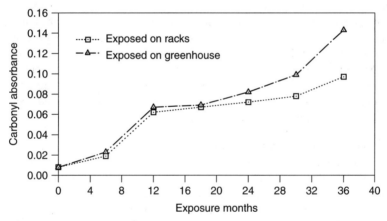

Figure 5.2. Carbonyl absorbance versus exposure time for greenhouse film exposed on model greenhouse and aluminum racks.

thick were stabilized with HALS and Ni quencher. Table 5.6 presents the typical meteorological data for 1993, when exposure trials were carried out.

Chemical changes in terms of carbonyl absorbance was monitored using Fourier transform infrared (FTIR) spectroscopy in the region of 1715–1730 cm^{-1}. Figure 5.2 presents change in carbonyl absorbance of greenhouse films exposed on rack and an operating greenhouse, as a function of exposure time. The films exposed on a greenhouse show a higher growth in carbonyl absorbance than the film exposed on racks. The higher values of carbonyl absorbance after 12 months of exposures indicate that severe photo-oxidative reactions took place. Also the carbonyl absorbance increased for

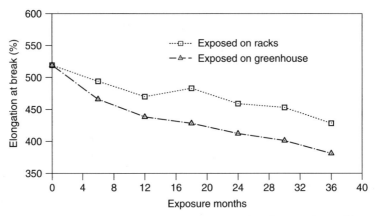

Figure 5.3. Percent elongation at break versus exposure time for greenhouse film exposed on model greenhouse and aluminum racks.

films exposed on both the greenhouse and racks. Figure 5.3 presents change in percent elongation at break versus exposure time for films exposed in racks and greenhouse films. Both the films have shown a downward trend, but a faster drop in measured property is observed for film exposed on operating greenhouse.

The higher rate of degradation on film exposed on greenhouses is attributed to the higher temperatures, humidity, and wind stresses, conditions that prevailed in an enclosed greenhouse tunnel [38, 40]. Furthermore, the higher degradation of greenhouse-exposed films may also be due to sulfur and halogen-based chemicals, which have adverse effects on HALS-stabilized LDPE films.

Effects of Chemical Treatment on Greenhouse Films The crops grown inside a greenhouse tunnel need specialized treatment depending on the type of insects/pests that would attack the plants. Pesticides used for spraying under greenhouse are mainly sulfur- or halogen-based compounds. It has been experienced by researchers that sulfur-containing additives, such as distearyl thiodipropionate (DSTDP) may have an antagonistic effect on the lifetime of HALS-based films [29, 34, 41, 42]. HALS also suffer when used with halogenated flame-retardants [29, 42]. The lifetime of the PE films will depend on the nature of pesticides/insecticides used, frequency of spraying, method of application, and design of the greenhouse.

It has been shown that single spraying of metam sodium during a one-year period is sufficient to degrade it. However, if the soil is covered with a film sheet during application of nematicide and kept down for 4 weeks, the film was protected [43].

A study was conducted by Epacher and Pukawzky [44] on three types of films having three different stabilizer packages exposed to 24 commercial pesticide formulations. These films were soaked for 1 h, 1 day, and 1 week duration. Artificially accelerated weathering on the same films were carried out under dry conditions for 300 h and 600 h inside a Xenotester.

Table 5.7 Effect of Pesticides and Irradiation on the Ultimate Elongation of PE Films

Pesticide	Percent Elongation after Treatment for 1 day and Xenotest Time (h)					
	Package A[a]			Package B[b]		
		300 (h)	600 (h)		300 (h)	600 (h)
Actellic	351.8	394.4	222.3	292.0	247.0	199.2
Chinetrin	314.4	318.3	330.0	282.7	331.4	281.9
Neoron	324.6	356.6	52.9	317.3	215.4	32.5
Sulfur	339.6	309.3	230.5	323.8	181.6	0
Thiovit	356.9	—	301.6	300.1	—	139.6
Topas	311.6	337.3	349.8	316.0	309.9	285.6
Neviken	337.1	278.3	50.3	287.7	130.2	23.7

[a]Package A: Hostavin N30 + Hostavin AR08.
[b]Package B: Tinvin 622 (HALS 5) + Chinassorb 944 (HALS 6).

The effect of soaking time in pesticide use is measured using a UV spectrometer. Table 5.7 shows the effect of pesticide and irradiation on the ultimate elongation of PE films.

This study showed that a large number of pesticides do not interact with the stabilizer package. Some sulfur-containing compounds and organic halogenides initiate oxidation of PE and lead to faster deterioration of mechanical properties. The deterious effect of these was tentatively explained by the generation of alkoxy radicals due to the interaction of sulfur and HALS. Introduction of a UV absorber into a UV stabilizer package improves the light stability and lifetime of a film [44].

5.5. MULCH FILMS

Polyethylene plastic film was first used as mulch in the late 1950s. It helps to accelerate plant growth by increasing soil temperature and stabilizing soil moisture. When scheduled drip (or trickle) irrigation is used, plastic mulch helps maintain optimum soil moisture, aids plant establishment, and promotes excellent crop growth throughout the season [47]. The vegetables on plastic mulch routinely mature 1–2 weeks earlier than those on bare ground. Plastic mulch also helps protect vegetables from decay by preventing contact with contaminated soil. In addition, black plastic and other wavelength-selective having additive to block photosynthetically active light mulches also control most weed growth.

Cantaloupe, tomato, pepper, cucumber, squash, eggplant, watermelon, and okra are high-value vegetable crops that show significant increases in yield and/or fruit quality when grown on plastic mulch. Other crops such as sweet corn, snap bean, and southern pea show similar responses. It promotes yield of maize grown under greenhouses to be 2.4 tons per hectare more than the plants grown conventionally [46, 47]. Table 5.8 shows the major advantages and disadvantages of a number of films.

Table 5.8 Advantages Versus Disadvantages for Mulch Films

Advantage	Disadvantage
Earlier production Soil temperature at the 2-inch depth is increased up to 10°F under black plastic, which leads to production 1–2 weeks earlier	Removal of plastic mulch is time consuming and costly
Reduced leaching of ingredients It caused reduction of fertilizer, nitrogen, potassium, magnesium	Higher annual cost
Reduced weed problems Since black plastic mulch does not transmit light, it prevents the growth of weeds	Higher initial cost
Reduced evaporation Soil water evaporation is reduced under plastic mulch	
Enhanced soil fumigation Improves the effectiveness of soil fumigation by slowing the escape of fumigants from the soil	
Yield increase Results from many factors acting synergistically	

Mulch films constitute the application with the biggest extension of land covered worldwide (about 4 million hectares). Mulch films are designed to have a short life span and normally have less contact with critical agro-chemicals during their service life [49]. Mulch films are made with thickness, between 10 and 80 μm. Usually pigments, mainly carbon black, titanium dioxide, and organic pigments are incorporated into their formulation. However, they require proper light and thermal stabilization additives with intermediate chemical resistance.

5.5.1. Types of Plastic Used for Manufacture of Mulch Films

Black plastic, white plastic, and white-on-black plastic are the types most often used in the United States. Black plastic is usually used in the spring because it warms the soil and prevents the growth of most weeds. For late spring, summer, and fall crops, white or white-on-black plastics are used since these reduce the risk of crop damage from excessive heat. It is advisable, as a general rule, not to use black or clear plastic whenever daytime temperatures average more than 85°F at planting.

Embossed plastic mulch is preferred by many growers because it has superior bed-hugging ability. It does not control weeds and, during late spring and summer growing seasons, excessive heat builds up under clear plastic mulch. Photodegradable plastic mulch is an alternative to conventional plastic mulch film, which poses retrieval and disposal problems. Although photodegradable plastic looks very much like other plastic mulch when it is installed, it can be broken down by ultraviolet sunlight. The actual rate of breakdown depends on several factors including temperature, the proportion of the plastic shaded by the crop, and the amount of sunlight received during the growing season.

Degradable plastic mulches with relatively precise degradation rates have only recently become available. In the past, rapid breakdown was a major problem. Now there are a number of formulations that have a breakdown period suitable for vegetable production in Georgia. New types of plastic mulch are continuously being developed and evaluated by many companies and researchers [56–59]. Low-density polyethylene is generally used in the manufacture of mulch films.

Photodegradable PE films have been made by incorporating varying concentration of certain metal organocomplex, for example, ferric dialkyl dithio carbamate, which acts initially as a UV stabilizer, and then becomes a catalytic activator of photodegradation [48].

5.6. PLASTICS IN SILAGE

Silage films are used primarily to preserve silage and maize. They maintain the nutritional value of the contents and inhibit undesirable fermentation processes. Traditional silage films can be mono- or co-extruded, range from 100 to 200 μm, and are normally pigmented (black or white). In the case of bale wrapping for round or square bales, films are normally very thin (10–25 μm), applied in multiple layers and pigmented (mainly black or white).

For this particular application the films need good resistance to weathering in order to preserve their original mechanical and gas barrier properties, thus ensuring its protective role throughout the duration of outdoor exposure. These films are UV stabilized so that they can retain their original mechanical and other required physical properties. LDPE is a commonly used polymer for silo bunker covers, silage bags, bale wraps, and the like.

5.7. DISPOSAL OF WASTE PLASTIC FILMS

Disposal of plastic films used for greenhouses, mulches, and silage bale wrapping do have some associated problems. Farmers dispose of these plastics through one of the two methods: on-site land filling and burning.

Though recycling is a good way to resolve this problem, the main hurdle in recycling is the cost involved in collection, transportation, and cleaning operation. Moreover, some agri-films contain as high as 50% contamination level by weight. Contamination comes from silage juices, pesticides, product residues, moisture vegetation, dirt, and sand.

200 PLASTICS IN AGRICULTURE

A combination of efforts from different sectors of film producers, farmers, and municipalities will be needed to resolve this problem [60, 61]. The various solutions to curtail the problems of plastic film waste disposal include:

1. Waste reduction of plastic films by farmers.
2. Means to make recycling viable and economical for the farmers need to be determined.
3. Finding cost-effective ways to collect, clean, and store the material and finding end markets for the recycled products.

5.7.1. Disposal of Plastic Films in Near-Equatorial Regions

Disposal of plastic films is one of the main problems faced by municipalities in the cities of Saudi Arabia and other Gulf countries. A study was conducted to evaluate the photodegradability of 50- to 70-μm thin film exposed at the Dhahran, Jeddah, and Baha exposure sites [64].

Untreated, unstabilized films made of blend 75% HDPE and 25% LDPE were exposed at the above three sites for a duration of 4 months. Changes in carbonyl absorbance and mechanical properties were measured on samples withdrawn on a monthly basis. Figure 5.4 presents the change in percent elongation at break versus exposure time. It was conclude that it took about 3 months to lose its original mechanical properties at the Dhahran, Jeddah, and Baha sites.

The rate of degradation is higher at Dhahran as compared to the Jeddah and Baha sites. However, a complete breakdown does not occur within 3 months of natural exposures. It was recommended that if films having thickness closer to 50–70 μm having masterbatch containing starch, pro-oxidant filler is used; then it is possible to have biodegradability simultaneously with photodegradability in this part of the world, where higher UV radiation and temperature leads to earlier

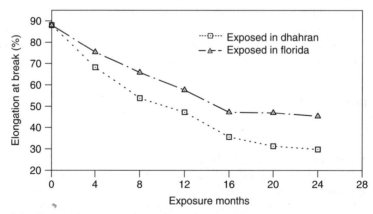

Figure 5.4. Changes in percent elongation at break with respect to exposure time for the exposed PVC samples.

photodegradation of PE films. Thus, a complete breakdown of the polyethylene structure may be achieved within 3–4 months during natural weathering process in Saudi Arabia and other Gulf countries.

5.8. DRIP IRRIGATION SYSTEMS

Plastic pipe constitute the third largest section of plastics used in the agriculture field. Drip or trickle irrigation and capillary watering techniques are being used widely for both greenhouse and field watering of flowers and vegetable crops.

5.8.1. Role of Drip Irrigation

Drip irrigation is thought of as not so much an irrigation system as a total plant support system, meaning that water, fertilizer, and other necessities are delivered over quite different situations, namely a large system suitable for a large grower or corporate farmer, a smaller system that could be used by a reliable operation in a low-capital, low-agronomic-skill environment, which may exist at a village level in a lesser developed country.

The basic system consists of LDPE, HDPE, or PVC hose or tube in which the emitters are installed as per spacing provided during manufacturing. Fertilizers and sanitation agents are kept in specified vessels.

The circulating mains consist of 20- to 25-mm plastic pipe extending from fertilizer vessels and run past every in field valve and then vessel. Also passing through every in-field valve in the system in conjunction with fertilizer mains in a high-pressure clear water main connected to a pressure system, which does not circulate. Injection of fertilizer can be done as per requirement of crop in the field, at a single point as well as at multiple points. The advantages of the system can be summarized as follows:

1. *Saving of Water*   Due to localized application of water to the root of the plant, surface evaporation is reduced, runoff is decreased, and deep percolation loss is avoided resulting in up to 60% saving in water used in conventional irrigation.
2. *Better Yield of Crop*   Increase in root length as well as crop yield has been experienced due to slow and frequent supply of water; for example, a 12% increase in strawberries in California.
3. *Saving in Labor and Energy*   Scientific design of the system using principles of hydraulics would require labor only to start and stop the operation and less energy for pumping less water at lesser pressure than open-field system.
4. *Suitable for Poor Soil*   Both light and heavy soils difficult for ordinary irrigation system can be successfully irrigated by this system.
5. *Weed Growth Minimized*   Growth of weeds is reduced due to partial wetting of soil.

6. *Convenient for Cultural Practices*   The field is always accessible for spraying, weeding, and harvesting.
7. *Less Soil Erosion*   Controlled irrigation of the field leads to less sol erosion problems.
8. *Use of Saline Water*   Due to frequent watering, the soil moisture always remains high and the salt concentration remains below harmful level.
9. *Improve Efficiency of Fertilizers*   Due to reduced loss of nutrient through leaching and runoff water and localized application of fertilizers, the efficiency is greatly increased.

5.8.2. Plastics Used in Drip Irrigation System

Polyethylene pipe, 1/2 or 1/3 inch in diameter with emitter devices inserted in the line in accordance with the plant spacing is used. A simple emitter consists of a heavy walled length of diameter tubing (0.036 inch ID), which is inserted into a punch hole in the PE pipe. The length of tubing that is inserted in the downstream direction determines the rate of flow of water. These lengths may be from 24 to 48 inches.

Another type of perforated wall type of drip irrigation was pioneered by Chapin Watermatics of Watertown, New York. Plastics including HDPE, LDPE, and LLDPE are used in these applications.

About 2 million hectares are under irrigation with low-diameter pipe made of LDPE/LLDPE in Europe. Higher diameter (>32 mm) tubes are usually made in HDPE or PVC. Western Europe used 100,000 tonnes of rigid PVC pipe for drainage and 65,000 tonnes/year of flexible PVC hose piping [45]. Apart from these materials, corrugated PP pipes, fittings, and accessories are also used worldwide.

5.8.3. Stabilization of Polyolefin Pipes

Polyethylene is sensitive to UV radiations, less than polypropylene. Polyolefin-based pipes need stabilization to combat UV light and heat.

Pipes are made by extrusion processing of polyolefin. Fittings are made by injection-molding process. Pipes used in irrigation of greenhouses and field watering are exposed to extreme outdoor weather. Since then pipes are meant to draw lifetime of 2–3 years or more, therefore, they need to have appropriate stabilization.

The HDPE, LDPE, and LLDPE can be stabilized by UV absorbers of type benzophenone or benzotriazole and also combination of HALS and UV-absorbers. Usually pipes used in irrigation are pigmented with white titanium dioxide or carbon black. Titanium dioxide may be used up to an additional level of 0.5 wt%. Each of the HALS and UV absorber may be added up to 0.05 wt% [50–53].

Polypropylene in comparison with polyethylene is more sensitive to UV light. Stabilizers including benzotriazoles together with HALS can be used to provide

UV resistance required for long-term use of pipes. However, large differences in UV stability are found on changing unpigmented to white, yellow, red, and blue pigmented samples. In combination with a UV absorber (benzophenone type), titanium dioxide has a slightly negative effect, whereas chrome yellow and cadmium red show a slightly positive effect [52].

5.8.4. Stabilization of PVC Pipes

Polyvinyl chloride cannot be processed alone without having an appropriate heat stabilizer, lubricants, co-stabilizers, and processing aids. PVC pipes used in outdoor applications are usually stabilized by either of the three types of stabilization systems:

1. Mixed-metal stabilizers
2. Lead (Pb)-based heat stabilizers
3. Organotin stabilizers

The combination of appropriate thermal stabilizers with adequate pigments provide a high level of weathering resistance to PVC products [54]. Titanium dioxide pigment loading up to 5–12 wt% can be used to achieve longer lifetime of the exposed products.

5.8.5. PVC Pipe Weatherability in Near-Equatorial Regions

In this study white pigmented PVC pipes were exposed to outdoor weather in Dhahran, Saudi Arabia, for a duration of 24 months. These pipes were Pb-stabilized together with titanium dioxide pigment. During outdoor exposure, the ambient temperature reaches as high as 48°C together with high UV radiation of

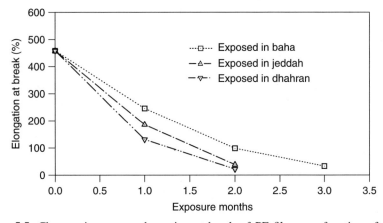

Figure 5.5. Changes in percent elongation at break of PE film as a function of exposure time.

170–190 kLy/year, humidity, dust, and the like [55]. The same pipes were also exposed in Florida for 2 years. Samples were withdrawn periodically. Changes in exposed pipe samples were monitored in terms of FTIR, differential scanning calorimetry (DSC), and mechanical properties testing. Figure 5.5 shows the change in percent elongation at break with respect to exposure time.

REFERENCES

1. J. C. Garnaud, *Plasticuture* **49**, 37–52 (1981).
2. J. C. Garnaud, 13th International Congress of CIPA, Vol. 1, Verona, Italy, March 8–11, 1994.
3. F. R. de Pedro, *Rev. Plas. Mod.* **332**, 161–170 (1984).
4. F. Benoit and N. Ceustermans, Ecological vegetable growing with plastics, *Plasticulture*, **95**, 11–20 (1992/3).
5. P. Dilara and D. Briassovlis, in Proceedings of the International Congress for Plastics in Agriculture (CIPA), Tel Aviv, Israel, 9–15, March, 1997.
6. S. H. Hamid and M. B. Amin, *J. Appl. Polym. Sci.* **55**, 1385–1394 (1995).
7. S. H. Hamid and I. Hussain, in S. H. Hamid, ed., *Handbook of Polymer Degradation*, Marcel Dekker, New York, 2000, pp. 639–727.
8. F. H. Winslow, W. Matreyek, and A. M. Trozzolo, *Polym Preprint, ACS Div. Polym. Chem.* **10**, 1271–1280 (1969).
9. M. R. Kamal, *Polym. Eng. Sci.* **10**, 108–121 (1970).
10. B. Ranby and J. F. Rabek, *Photodegradation, Photooxidation and Photostabilization of Polymers*, Wiley, London, 1975.
11. K. L. Coulson, *Solar Terrestrial Radiation*, Academic, New York, 1975.
12. R. Ghani, Paper presented by Ciba at a seminar at KFUPM/RI, Dhahran, Saudi Arabia, October 4, 1987.
13. A. G. Maadhah, S. H. Hamid, and F. S. Qureshi, in Proceedings of Conference on Utilization of Plastic Materials in Packing and Packaging Industry in the Kingdom, Yanbu, Saudi Arabia, December, 1988.
14. A. L. Andrady, M. B. Amin, A. G. Maadhah, S. H. Hamid, K. Fueki, and A. Torikai, in *United Nations, Environmental Program Environmental Effects Panel Report*, UNEP, Nairobi, Kenya, 1989, Chapter 6, pp. 55–60.
15. P. P. Klemchuk, Antioxidants. Reprint from *Ulman's Encyclopedia of Industrial Chemistry*, Verlag Chemie, Weinheim, Germany A3, 1985, pp. 91–111.
16. A. M. Trozzolo, in W. L. Hawkins, ed., *Photooxidation of Polyolefins, Polymer Stabilization*, Wiley, New York, 1972.
17. A. L. Anrady, M. B. Amin, S. H. Hamid, X. Hu, and A. Torikai, in *Environmental Effects of Ozone Depletion: Assessment*, UNEP, Nairobi, Kenya, UNEP Report, 101–110, 1994.
18. S. H. Hamid, M. B. Amin, A. G. Maadhah, and J. H. Khan, in Proceedings, First Saudi Symposium on Energy, Utilization and Conservation, Jeddah, Saudi Arabia, March 4–7, 1990.
19. F. H. Winslow, W. Matreyek, and A. M. Trozzolo, *Soc. Plast. Eng.* **18**, 766–772 (1972).

20. H. E. Landsberg, H. Lippmann, K. H. Paffen, and C. Troll, *World Maps of Climatology*, 3rd ed., Springer, Berlin, 1966.
21. A. Davis and D. Sims, *Weathering of Polymers*, London, Applied Science, 1983, pp. 1–19.
22. F. L. Gugumus, in S. H. Hamid, ed. *Handbook of Polymer Degradation*, Marcell Dekker, New York, 2000, pp. 39–80.
23. F. S. Qureshi, M. B. Amin, A. G. Maadhah, and S. H. Hamid, *J. Polym. Eng.* **30**, 67–84 (1990).
24. T. Daponte, *Plasticulture* **75**, 5–12 (1987).
25. G. W. Gilby, *Plasticulture* **81**, 19–28 (1989).
26. F. Henninger and L. C. Roncaglione, in XII Congreso International de Plasticos en Agricultura, Granada, Spain, May 3–8, 1992.
27. A. L. Andrady, in Proceedings Symposium Degradable Plastics, Washington, DC, SPI, 22–25, 1987.
28. M. R. Kamal, and R. Saxon, *Appl. Polym Sym.* **4**, 1–28 (1967).
29. F. Gugumus, *Makromol. Chem. Macromol. Symp.* **27**, 25–84 (1989).
30. F. Gugumus, *Angew. Makromol. Chem.* **176/177**, 241–289 (1990).
31. F. Gugumus, *Polym. Degrad. Stabil.* **39**, 117–135 (1993).
32. F. Gugumus, Current trends in mode of action of hindered amine light stabilizers. Polym. *Degrad. Stabil.*, **40**, 167–215, 1993.
33. M. B. Amin, H. S. Hamid, and J. H. Khan, *J. Polym. Eng.* **14**, 253–267 (1995).
34. M. Lee, *European Plastic News* April 24–26, 2000.
35. F. Gugumus, Presented at the 12emes Journees d'Etudes sur le Viellissement des Polymeres, Bandol, France, Sept. 25–26, 1997.
36. S. H. Hamid and W. H. Prichard, *Polym. Plast. Technol. Eng.* **27**, 303–334 (1988).
37. S. H. Hamid, A. G. Maadhah, F. S. Qureshi, and M. B. Amin, *Arab J. Sci. Eng.* **13**, 503–531 (1988).
38. S. H. Hamid, F. S. Qureshi, M. B. Amin, and A. G. Maadhah, *Polym. Plast. Technol. Eng.* **28**, 475–492 (1989).
39. F. S. Qureshi, S. H. Hamid, A. G. Maadhah, and M. B. Amin, *Prog. Rubber Plast Technol.* **5**, 1–14 (1989).
40. S. H. Hamid and W. H. Prichard, *J. Appl. Polym. Sci.* **43**, 561–678, 1981.
41. S. H. Hamid, M. B. Amin, J. H. Khan, and A. G. Maadhah, Paper presented at Industrial Air Pollution Symp., King Saud University, Riyadh, Saudi Arabia, 15–17 Nov. 1993.
42. F. Gugumus, in P. Klemchuk and J. Popsil, eds., *Inhibition of Oxidation Processes in Organic Materials*, Vol. 2, CRC Press, Boca Raton, FL, 1989, pp. 29–162.
43. J. F. Barahona and J. M. G. Vasquez, *Plasticulture* **65**, 3–10 (1985).
44. E. Epacher and B. Pukanszky, *Interactions of Pesticides and Stabilizers in PE Films for Agricultural Use*, Society of Plastic Engineers, Brookfield, USA, ANTEC '99, 1999, pp. 3785–3790.
45. A. Warmington, *European Plastic NEWS* April, 25–26 (2000).
46. L. Boldrin, *Plasticulture* **83**, 51–54 (1989).
47. H. M. G. Vander Werf, *J. Agronomy Crop Sci.* **170**, 261–269 (1993).

48. P. F. Bruins, *Plastics Agricul. Popular Plastics*, April, 15–19 (1978).
49. S. Chakrawarty and L. B. Singh, 20th WEDC Conf., Colombo, Srilanka, 1994, pp. 213–215.
50. J. W. Picket, 5th International Conf. on Advances in Stabilization and Controlled Degradation of Polymers, Zurich, Switzerland, 1983.
51. J. E. Picket, in P. P. Klemchuk, ed., *Polymer stabilization and degradation*, ACS Symp. Ser No. 280, Washington DC, 1985, p. 313.
52. F. Gugumus, in R. Gachter and H. Muller, eds., *Plastic Additives Handbook*, Hanser, New York, 1984, pp. 129–270.
53. G. N. Foster, in A. Patsis, ed., *Advances in the Stabilization and Controlled Degradation of Polymers*, Vol. 1, Basel Technomic, Basel, Switzerland, 1989.
54. E. Barth, *Kunststoffe* **76**, 43 (1986).
55. I. Hussain, S. H. Hamid, and J. H. Khan, *J. Vinyl Additive Tech.* **1**(3), 137–141 (1995).
56. A. Liakatas, *Mausam* **42**(1), 25–28 (1991).
57. F. Mage, *Scientia Horticulture* **16**, 131–136 (1982).
58. B. Loy, *Agricultural Eng.* January 13–16, 1992.
59. G. J. L. Griffin, ed., *Chemistry and Technology of Biodegradable Polymers*, Blackie Academic and Professional, Stockholm, Sweden, 1994.
60. G. Scott, in G. Scott, ed., *Polymer and the Environment*, Royal Society of Chemistry, 1999, pp. 68–92.
61. Local Government Services, Greater Manchester Waste Disposal Authority integrated Waste Management Strategy, Coopers and Lybrand, 1998, (unpublished report).
62. J. H. Khan, and S. H. Hamid, *Polym. Degrad. Stab.* **18**, 137–142 (1995).
63. M. B. Amin, S. H. Hamid, and F. Rahman, *J. Appl. Polym. Sci.* **5**, 279–284 (1995).
64. I. Hussain, J. H. Khan, and S. H. Hamid, The 4th Saudi Eng. Conf., Jeddah, 5, Nov. 1995, pp. 437–442.

APPENDIX: STRUCTURE OF SELECTED UV-STABILIZERS

Abbreviation	Structure	Trade Name
HALS1	R—NH—CH$_2$—CH$_2$—CH$_2$—N(R)—CH$_2$—CH$_2$—N(R)—CH$_2$—CH$_2$—CH$_2$—NH—R R = triazine ring substituted with N(C$_4$H$_9$)-piperidinyl(N-CH$_3$, 2,2,6,6-tetramethyl) groups Chimassorb 119	Tinuvin 492 and Tinuvin 494 (Mixtures of Chimassorb 119: metal oxides/stearates)
HALS2	—(Si–O)$_n$— with CH$_3$ and (CH$_2$)$_3$—O—piperidinyl (2,2,6,6-tetramethyl, N-H) side group	UVASIL 299
HALS3	H$_{17}$C$_8$O—N(piperidinyl)—O—CO—(CH$_2$)$_8$—CO—O—(piperidinyl)N—OC$_8$H$_{17}$	Tinuvin 123
HALS4	HN(piperidinyl)—O—CO—(CH$_2$)$_8$—CO—O—(piperidinyl)NH	Tinuvin 770

(*continued overleaf*)

(continued)

Abbreviation	Structure	Trade Name
HALS5		Tinuvin 622
HALS6		Chimassorb 944
HALS7		Hostavin N 30
UVA2		Chimassorb 81

APPENDIX: STRUCTURE OF SELECTED UV-STABILIZERS

(*continued*)

Abbreviation	Structure	Trade Name
Ni-1	[Ni complex with $H_2N-C_4H_9$ ligand, two phenolate oxygens bridged by S, with neopentyl substituents on aromatic rings]	Cyasorb UV 1084
Ni-2	$\left[(H_3C)_3C-(CH_2)_5\text{-N(-}(CH_2)_5\text{-C(CH}_3)_3)\text{-C(=S)-S}^- \right]_2 Ni^{2+}$ (bis-dialkyldithiocarbamate with two 2,2-dimethyl-heptyl groups on N)	

CHAPTER 6

COATING

LOREN W. HILL
Coating Consultant, Wilbraham, Massachusetts

6.1. ENVIRONMENT COST AND BENEFITS COATINGS

Organic coatings are used primarily for two purposes: to improve appearance and to protect substrates. The protection of substrates contributes tremendous environmental benefits to society by prolonging the useful lifetime of many products and essential components of our infrastructure. Products important to our way of life include houses made of wood and automobiles made of steel, often galvanized. Infrastructure components that are essential to our way of life include steel bridges and pipelines. It would be foolish to try to use these items without coatings. The useful lifetime of uncoated wooden houses or uncoated metal bridges would be very short. Without coatings, the number of trees that would have to be cut down for wooden structures and the amount of iron ore mined and processed for bridges alone would be too great to be practical. Approaches to housing and bridge building would, of course, have to be entirely different and undoubtedly less cost effective than the methods made possible by use of organic coatings.

The value of year 2000 U.S. shipments of paint and allied materials according to Census Bureau [1] figures is $17.84 billion. It has been estimated that the value (sales dollars) of the products protected by these coatings annually is $2 trillion. In many cases, perhaps most, these products would not be marketed at all in uncoated form. The coating material cost is only 0.9% of the value of the products that are protected by the coatings. The cost of paint application varies greatly by end use, but at least in some cases the application cost may be about the same as the cost of materials. Thus, applied coatings may cost about 2.0% of the value of the products they protect. Coatings are obviously extremely cost

---

Plastics and the Environment, Edited by Anthony L. Andrady.
ISBN 0-471-09520-6 © 2003 John Wiley & Sons, Inc.

effective. Coatings are also extremely conservation effective. We tend to think of recycling as a green industry, but application of protective coating is a greener industry because it reduces raw material use without the expenditure of energy of collecting, separating, and reprocessing as required in recycling. Recycling is a good idea, and applying effective coatings is a better idea.

6.2. COATINGS AND THE ENERGY CRISIS

In Chapter 1 the energy crisis is discussed in terms of the big picture of Earth and its environment. One of the main issues is modern people's use of nonrenewable fossil fuels for electric generation (ultimately heating, cooling, lighting, etc.), transportation, and industrial production of goods. Many of the binder materials used in coating today are polymerized forms of monomers that are prepared by industrial synthesis from organic starting materials obtained as petroleum distillates. Historically, paint binders were derived by chemical modifications of plant-derived natural oils, and despite a long-term trend to other binders, natural oils continue to be used at significant levels in coatings. Williams [2] reports that the use of the most prevalent oil-derived binder, namely alkyds, peaked in 1980 at about 650 million pounds/year with a decrease in use to 500 million pounds/year in 2000. Coatings binder materials as a whole increased from 2600 million pounds in 1980 to about 4100 million pounds in 2000. Thus, alkyds represented 25% of total paint binders in 1980 and 12.2% in 2000.

The annual U.S. consumption of fossil fuel including petroleum, natural gas, and coal expressed in oil equivalents was about 7600 million tons in 1999 (Chapter 1) and an estimated 7800 million tons for 2000. In 2000 the 4100 million pounds (2.05 million tons) of fossil fuel used as paint binder represents a nearly negligible 0.026% of total fossil fuel use. One of the 12 principles of green chemistry [3] is to use renewable feedstocks rather than depleting feedstocks such as those from fossil fuel. Binders derived from natural oils qualify as renewable, but the energy cost of agricultural production of the oil seed and energy cost of oil extraction and processing should be included in determining the best binder source for sustainability.

The coatings industry uses energy derived from fossil fuels in coating component synthesis, formulating paints, shipping, applying, and in some cases heat curing. The amount of fossil fuel used in these ways probably exceeds that consumed as a raw material source. Undoubtedly, the coating industry could join nearly all branches of the chemical manufacturing industry in seeking greater efficiency in these uses of fossil fuel.

6.3. COATINGS AND THE MATERIAL CRISIS

Concerns about depletion of key resources for the coatings field include mainly petroleum-based starting materials for binders and mineral sources of titanium dioxide for use as a white pigment. Coatings chemists are likely to join other organic chemists in advocating that petroleum should be conserved to provide a long-term raw material source rather than being used in inefficient ways for energy and transportation. Two types of titanium dioxide that differ in crystal structure are used, rutile and anatase. Rutile is used in larger volume because it has greater hiding power due to its higher refractive index [4(a)]. The rutile form of TiO_2 is sufficiently concentrated in natural deposits to be mined as an ore [5]. Another mineral source is iron titanium oxide, $FeTiO_3$, called ilmenite. Approximately 90% of world reserves of titanium dioxide are in the form of ilmenite. Ilmenite is also the most important ore for production of titanium metal, but by far the largest use of ilmenite is for titanium dioxide white pigment [5]. Occurrence of natural anatase is too widespread for use as an ore. In Chapter 1, it was noted that world reserves of titanium dioxide are estimated at 300 million metric tons, and worldwide annual consumption was 414,000 metric tons in 1997.

6.4. HISTORICAL DEVELOPMENT OF THE COATINGS INDUSTRY

Decorative or artistic paints have a very long history, dating back about 35,000–45,000 years to a time when humans lived in caves. Ancient binders included animal fat, natural waxes, tree exudates such as sap and pitch, and possibly egg white or albumen [6, 7]. In China, artistic painting developed very early. Museum treasures include painted vases dating from 1500 BC to AD 1500. Kumanotani [8] has reported extensively on the drying mechanism of a natural organic fluid called oriental lacquer. It comes from the sap of the lacquer tree (*Rhus venicifera*), which grows in China, Korea, and Japan. Despite extreme skin irritation during collection and processing of the sap, oriental lacquer has been harvested and used on valuable jewelry boxes and other artistic items for centuries.

Protective coatings have a shorter history but still span the Middle Ages to current times. Historically [4(b), 9], binder materials used for coating consisted of natural drying oils or modified forms of drying oils. Early varnishes were prepared by heating oils in the presence of dried tree exudates referred to as gum, rosin, or amber. Edwards [9] gives varnish formulas from the eleventh century and discusses current interest in the varnish finishes on violins made from 1550 to 1750. Some of the varnishes were too solidlike to be applied at room temperature. They had to be heated for application in liquid form. Interestingly, one of the ways to meet solvent regulations today is to reduce viscosity for application by heating the paint instead of adding more solvent. Other varnishes were dissolved in turpentine and thinned with additional turpentine. Turpentine is a low-boiling distillate from the resinous sap of pine trees.

Early binders for pigmented paints usually did not contain rosin, but they were still based on natural drying oils or modified forms of drying oils. "Drying" in the sense used here means conversion from a liquid state to a solid state. Often for coatings, the process of drying includes both solvent loss and chemical reaction. For natural drying oils, the fatty acids attached to glycerol in the triglycerides contain allylic double bonds (—CH=CH—CH$_2$—CH=CH—).

Double bonds of this type react with atmospheric oxygen to form hydroperoxides. The hydroperoxides react further to form oxygen-centered and carbon-centered free radicals. Free-radical chain reactions take place to oxidize the fatty acids and eventually to form a polymerized three-dimensional network structure. Wicks et al. [4(b)] give the structures of eight fatty acids that are currently used or were used previously in coatings. Various fatty acids differ greatly in content of allylic double bonds, with drying oils having high levels and nondrying oils having low levels. Drying oils are obtained by crushing and pressing the seeds of plants such as flax (linseed oil), safflower, soybean, sunflower, castor bean, coconut, and tung oil trees [4(b)].

Drying oils were used directly as the binder in early paints, but later the oils were hydrolyzed to recover the fatty acids from the triglycerides. The largest coatings use of fatty acids is in alkyds, a term intended to indicate a combination of alcohol and acid. Alkyds are polyesters prepared from esterification of mixtures that contain several types of alcohols and several carboxylic acids as monomers. Alcoholic monomers included both diols (two —OH groups per monomer molecule) and polyols (three or more —OH groups per monomer molecule). Acidic monomers included the fatty acids, which are monofunctional, and dicarboxylic acids such as phthalic acid or its anhydride as the acid group monomers. For many years the binder components of most "oil-based paints" were alkyd resin. Alkyds had the advantage of permitting designed variations in the amount of oil and the allylic group content of the oil used. In the early paints based on unmodified drying oils, the dried paint had acceptable mechanical properties for a limited time. Before the optimum time, paint films were too soft and after the optimum time they were too hard and brittle. These changes reflected continued reaction of allylic double bonds throughout a very long period. In contrast, alkyds could be designed to have optimum double-bond content for more uniform properties over an extended period.

Nitrocellulose lacquers were used very early as automobile topcoats. They produced very attractive high-gloss finishes. Since the nitrocellulose molecular weight is high, a very high level of solvent was needed to reach spray viscosity. This resulted in the need for many coats to reach adequate film thickness. Both the high solvent levels and the use of multiple coats contributed to extremely high volatile losses from nitrocellulose lacquers.

The solvent or "thinner" for drying oils and alkyds continued to be turpentine for many years. As use of petroleum increased throughout the industrialized world, the cost of petroleum-derived solvents became very competitive, and turpentine was gradually replaced by petroleum distillates. Solvent mixtures often contained both aliphatic and aromatic hydrocarbons. Commercial aliphatic solvents are mixtures of straight-chain, branched-chain, and alicyclic hydrocarbon [4(c)]. Naphthas used by varnish makers and painters (VM&P) are aliphatic solvents with high volatility (distillation range 120–150°C) [10]. Solvent mixtures of lower volatility are called mineral spirits. Low-boiling (150–210°C) and high-boiling (177–210°C) grades are available [10]. Special grades with low aromatic content were referred to as low-odor mineral spirits. Among the aromatics, benzene is prohibited because of its toxicity whereas toluene and xylene were widely used but are now on the hazardous air pollutants (HAP) list. Although benzene has not been used in coatings for a long time, the replacement of xylene and related aromatics is continuing today as deadlines for compliance to HAP regulations are drawing near at the time of this writing (see Section 6.6).

6.5. BINDERS AND SOLVENTS USED IN MODERN COATINGS

The polymeric part of a paint formula that can be varied and sometimes chemically designed to make the cured paint perform well in a given end use is called the binder or sometimes the vehicle. The binder disperses and encapsulates the pigment in the dried or cured paint film. Adhesion of the binder to the substrate is critical throughout the useful lifetime of the coated article. Drying oil and alkyd-based paints served the coatings industry well for many years, but the need for improvements in outdoor durability and better combinations of hardness and flexibility for many end uses resulted in many new types of binder materials from about 1920 to the present time. Many of the polymers used in other branches of polymer technology have been used with modifications in various types of coatings. Edwards [9] and Walker [11] provide chronological lists of milestones in the coatings industry that include the year of introduction of new binders.

6.5.1. Binders in Thermoset Coatings

As concerns for solvent emissions increased, industrial coatings shifted to thermoset coating systems and architectural coatings shifted to waterborne latex systems. Thermoset systems undergo chemical reaction during cure to increase molecular weight and to improve film properties. In general, high molecular weights are needed in the final film for good properties. Thermoset coatings permit application at lower solvent levels because the binder components can be of lower molecular weight to start. The term *high-solids solvent-borne* (HSSB) coatings is used to designate coatings that require less solvent because binder components of low molecular weight are used. The binder components used in HSSB automotive topcoats are acrylic copolymers that contain crosslinkable sites, often hydroxyl groups. The —OH groups react with another binder component called the crosslinker. Frequently used crosslinkers are etherified melamine formaldehyde (MF) resins or polyfunctional aliphatic isocyanate oligomers. For uses where gloss retention is not so critical, oil-free polyester polyols are often used instead of acrylics along with these same crosslinkers. Binders for coil coatings usually consist of polyester polyols crosslinked with MF resins.

In coil coating a large rolled-up sheet of metal is unrolled and fed continuously into reverse-roll application equipment [4(d)]. Paint is applied by rollers that are slightly wider than the metal strip, which is usually in the range of 24–72 inches wide. The coated metal sheet enters a long oven where heat cure is carried out. As the continuous sheet exits the oven, it is taken up by accumulators to provide time for cooling before the sheet, now coated, is rolled up again. Coil coating is very fast, with line speeds of 100–200 m/min. The oven temperature must be very high because the dwell time is often only about 15–40 s. The combination of roll coating and enclosed oven headspace results in high solvent vapor concentration in the oven effluent. The high concentration is ideal for solvent vapor incineration, and nearly all coil lines are operated with a significant fraction of the heat for curing provided by solvent remediation incinerators.

Several reactive binder types do not require heat for cure. Epoxy and urethane binders are examples of very widely used ambient cure systems. Epoxy coatings are used for industrial maintenance, bridges, and many types of metal primer systems. Epoxy systems are used in combination with phenolic resins, amine functional crosslinkers, and carboxylic acid functional crosslinkers. The rate of reaction between epoxy groups and the crosslinkers in most current uses is fast even at room temperature. These components are packed separately and mixed shortly before use. Reaction begins immediately after mixing and molecular weight and viscosity increase. The time period during which viscosity remains low enough for application is called the potlife.

Urethane binders are used broadly in many different forms for coatings on plastics, metal, and wood. The American Society for Testing and Materials (ASTM) recognizes five types of urethane binders including both two-package and one-package systems. The rate of reaction between isocyanate groups and hydroxyl groups is fast at room temperature and can be controlled to some extent by catalyst type and level. In some industrial operations automated mixing spray equipment is fed from separate containers of the isocyanate and hydroxyl components, which permits use of systems with a short potlife. One-package urethane systems require a strategy to get sufficient storage stability. In moisture cure urethanes the formulation does not contain any co-reactant for the isocyanate groups. Stability results from excluding water. When applied, atmospheric water reacts with isocyanate groups, converting them ultimately to amine groups and carbon dioxide. The amine groups react with other isocyanate groups to cure the binder, and the carbon dioxide is evolved as a "cure volatile." In another type of one-pack system, isocyanate groups are reacted with a blocking agent so that reactivity with the hydroxyl component is greatly reduced. The blocked isocyanate systems require high cure temperatures to cause unblocking so that the —NCO plus —OH reaction can occur.

6.5.2. Binders in Waterborne Coatings

Waterborne coatings require some strategy to reverse binder–water interactions before and after film formation. Water must dissolve or disperse the binder before and during paint application, but after the film is cured it must be resistant to dissolution or swelling by water. In latex paints the binder is dispersed in the form of particles. The interior portion of each particle is hydrophobic, whereas the surface of each particle is made hydrophilic to aid dispersion and to prevent coagulation during storage. The polymer molecules inside the particles can have high molecular weight without contributing to high viscosity because the molecules remain inside the particles. After the film is applied, water is lost by evaporation, the particles move closer together, and eventually come into contact. Then the polymer molecules diffuse across the former boundaries of the particles in a process called coalescence. Many latex paints contain coalescing agents, which are oxygenated organic solvents that are selected to partition in an optimum way to promote interparticle diffusion during coalescence.

A different strategy for reversing binder–water interaction is used in the so-called water-soluble or water-dispersible systems. The same thermoset binder components used in solvent-borne systems are modified by introducing hydrophilic groups such as carboxylic acid groups. The acid groups are neutralized by amine groups from aminoalcohols such as dimethylaminoethanol (DMAE). The ionic sites contribute more to solubilization and/or dispersion than do the —COOH groups alone. During heat cure, part of the amine is driven off and often some of the —COOH groups react with the crosslinker. Both of these changes help convert the system from a slightly hydrophilic dispersion to a hydrophobic cure film. Latex systems are pure white like whole milk, but water-dispersible systems are often completely clear like a true solution. They look like solutions (thus the term "water soluble"), but rheology and other studies indicate that they actually consist of very small particle dispersions [12, 13]. The particles are too small to scatter visible light. Usually there are too few ionic or other solubilizing groups for dispersion in water alone, and coupling solvents are used along with water. The term "coupling solvent" is applied to solvents that tend to prevent phase separation in systems that contain both hydrophilic and hydrophobic components. Most coupling solvents are alcohol ethers such as propylene glycol monomethylether (I). As the structure suggests

(I) $HOCH_2CH(CH_3)OCH_3$

this solvent is made by reacting propylene oxide with methanol. All of the alcohol ethers used in coatings are obtained by reacting various alcohols with either propylene oxide (P) or with ethylene oxide (E). In a commonly used two-letter designation, the oxide is identified first (P or E) and the alcohol next by the first letter in its name: (M) methanol, (E) ethanol, (P) propanol, or (B) butanol. Structure (I) is PM, a considerably shorter name than propylene glycol monomethylether. EB (aka butoxyethanol or ethylene glycol monobutylether or Butyl Cellosolve) was used very extensively in early water-reducible coatings, but toxicity concerns for all members of the ethylene glycol series have resulted in a switch to coupling solvents based on propylene glycol [10]. A remarkable property of many alcohol ethers is their tolerance for mixing with other compounds covering a wide range of polarity. For example, several of these coupling solvents form one phase solution in all proportions with both water (very polar) and heptane (very nonpolar).

Binders used in waterborne latex paints are prepared by emulsion polymerization of mixtures of monomers selected to give the optimum glass transition temperature, T_g [4(e)]. Low T_g contributes to good film formation so that the paint can be applied on cool days. Low-temperature application also depends on the effectiveness of coalescing agents included in the formulation. Low-temperature limits printed on paint can labels vary from about 35 to 50°F. High T_g contributes to increased hardness and toughness of the dry film. Slow volatilization of coalescing agent causes an increase in hardness over an extended period. In the United States most exterior latex paints are prepared from acrylic monomers. The high T_g monomer is usually methylmethacrylate and the low T_g monomers are butyl acrylate and/or ethyl acrylate. Occasionally, some styrene is also included along with the acrylic monomers. Interior latex paints are often prepared from a mixture of vinyl acetate and a low T_g monomer such as butyl acrylate.

Table 6.1 contains data from the U.S. Census Bureau's Current Industrial Report on Paint and Allied Products [1]. Manufacturer's estimated shipments are reported in four categories: architectural coating, product finishes for original equipment manufacturer (OEM), special–purpose coatings, and miscellaneous allied paint products. Shipment volume in gallons and the dollar values for each category for 1995–2000 are given in Table 6.1. Architectural coatings include interior and exterior, primer and topcoat house, barn, and roof paints; floor paints; tinting bases; trim paints; and stains and lacquers. Product finishes for OEM include assembly-plant-applied auto, truck, and SUV finishes (excludes refinish paint), appliance finishes, container and closure coating (beverage and food cans, bottle caps), factory-finished or primed wood or composition board (house siding), aluminum siding and extrusions, machinery and equipment and finishes, and wood furniture and cabinet finishes. The powder coating portions of these OEM finishes were reported separately in pounds and were converted by the Census Bureau to gallons by using a conversion factor of 5 (5 lb = 1 gal) [1]. The powder coating data will be discusses in detail in Section 6.7.3. Special-purpose coatings include traffic marking paints, automotive refinish (body shop use), marine coating for ships, and yachts and pleasure craft, and industrial new construction and maintenance and aerosol paint concentrates for packaging in aerosol containers (spray cans). Allied paint products include paint and varnish removers, pigment dispersions, thinners for lacquers and solvent-borne paints, brush cleaners, some (but probably not all) ink vehicles, and putty and glazing compounds. The two largest categories, architectural coatings and product finishes for OEM, accounted for 43.8 and 30.7% of the shipment volume, respectively.

These two categories were nearly equal in dollar value, representing 36.0 and 34.5% of the total in 2000. The average selling prices calculated simply as dollars/gallon in 2000 for each category were: architectural $9.94, OEM $13.56, special purpose $20.00, and allied paint products $8.41. If the miscellaneous and allied paint products category is subtracted from the total quantity for 2000, the result is 1280.4 million gallons or close to 1.3 billion gallons, which is the number often quoted for 2000.

The changes in quantities shipped from 1995 to 2000 are small, perhaps reflecting a mature industry with slow or no growth. These volumes will reflect changes made in response to regulations. Changes from conventional solvent borne to HSSB will reduce volume shipped, but the dollar value of paint shipped

Table 6.1 Quantity and Value of a Paint and Allied Products [1]

Year	Quantity (millions of gal)	Value (millions of dollars)
Architectural Coatings		
2000	645.6	6,419.9
1999	660.2	6,617.9
1998	631.6	6,115.2
1997	655.6	6,264.9
1996	640.3	6,246.3
1995	621.1	6,041.3
Product Finishes for Original Equipment Manufacturers		
2000	453.2	6,149.2
1999	439.7	6,146.3
1998	428.3	6,098.2
1997	425.4	5,750.7
1996	398.7	5,474.1
1995	376.2	5,263.6
Special-Purpose Coatings		
2000	181.7	3,634.2
1999	174.4	3,532.3
1998	173.3	3,472.0
1997	181.8	2,896.0
1996	208.9	3,263.8
1995	195.1	3,103.0
Miscellaneous Allied Paint Products		
2000	195.0	1,640.9
1999	191.9	1,516.9
1998	210.5	1,612.8
1997	210.0	1,647.9
1996	220.3	1,570.5
1995	215.9	1,5437
Total		
2000	1,475.4	17,844.2
1999	1,466.2	17,813.4
1998	1,443.7	17,298.2
1997	1,472.8	16,559.5
1996	1,468.2	16,544.7
1995	1,408.3	15,951.6

should not be changed significantly because the proportion of higher cost binder components is increased as solvent is removed. The dollar/gallon value should increase, reflecting the fact that the solvent being taken out is often the lowest cost material in the formulation. The volume change in converting from solvent borne to waterborne is not easily predicted in general, perhaps there will be little volume change in most cases. The dollar saving associated with replacing solvent by water may be significant is some cases, but often the binder used in waterborne will be more expensive than that used in conventional solvent-borne paint.

6.6. EFFECT OF GOVERNMENT REGULATIONS ON THE COATINGS INDUSTRY

Development of HSSB and waterborne coatings was well underway by 1950 and predated solvent regulations in the United States [14]. The first volatile organic compound (VOC) regulation, called Rule 66, was passed by the Los Angeles Air Pollution Control Agency in early 1967. The Clean Air Act of 1970 and the Clean Air Act Amendments of 1977 and 1990 have had a huge impact on the coatings industry [15]. Regulations completely changed the path of development of coating technology in some industries (see Section 6.7.3).

6.6.1. Volatile Organic Compound Regulations

One of the main components of smog is ground-level ozone (O_3). Ozone is sometimes categorized as a toxic oxidant. It is an eye irritant and a human health risk at levels sometimes reached in air around U.S. cities. The climate and land formations in and around Los Angeles favor temperature inversions that tend to trap air emissions. Other cities experience high ozone less frequently. Trapped organic compounds are susceptible to photochemical reactions that lead to ozone formation. The atmospheric chemistry of ozone formation is complex involving nitrogen oxides (NO_x), many different organic compounds (represented as RH for simplification), oxygen, and many free-radical intermediates. Wicks et al. [4(f)] have indicated that some of the reactions in ozone generation are

$$RH + \cdot OH \longrightarrow H_2O + R\cdot$$
$$R\cdot + O_2 \longrightarrow ROO\cdot$$
$$ROO\cdot + NO \longrightarrow RO\cdot + NO_2$$
$$NO_2 + h\nu \longrightarrow NO + O$$
$$O_2 + O \longrightarrow O_3$$

The nitrogen oxides shown in this scheme can be the limiting reactant or ozone formation under some conditions, and in such cases reducing VOC has little benefit for reducing ozone formation [4(f)].

Representing all VOCs by a single symbol (RH) in this scheme is a gross oversimplification. Based on chemical kinetics principles, each organic compound is expected to have a characteristic rate constant for the hydrogen abstraction reaction shown as the first step in ozone production above. The rate of ozone production could be quite different for different compounds. Jones [16] has recently described use of a compound-specific approach in regulations proposed to the California Air Resources Board (CARB) for solvent emission from aerosol can coating. Reactivity is expressed on a "maximum incremental reactivity" (MIR) scale. MIR values are given for several solvents in Table 6.2. The MIR values shown here vary widely, and it is reported [16] that the range for a longer list is from 0.01 to 11 g of ozone per gram of solvent. The units of MIR makes it easy to calculate ozone production potential for a formulation from the weight fractions (w_i) of each solvent in the formulation. For example, if a HSSB coating has 55 wt% solids and 45 wt% solvent (mixture) and the solvent mixture is 25 wt% acetone and 20 wt% p-xylene, the "product-weighted maximum incremental reactivity" (PWMIR) for the coating can be calculated as:

$$PWMIR = w_1 \times MIR(1) + w_2 \times MIR(2)$$
$$PWMIR = 0.25 \times 0.43 + 0.20 \times 4.25 = 0.96$$

The units of the 0.96 value are grams of O_3 per gram of coating. This is a very direct way to look at ozone generation. It is evident in Table 6.2 that MIR values for xylene are very high and that at least some oxygenated solvents have very low MIRs. It appears that a solvent replacement strategy would be very effective for ozone prevention. At the time of this writing, adoption of the MIR method had not been finalized by CARB.

In 1970 Congress created the U.S. Environmental Protection Agency (EPA) to combine several government environmental activities [15]. Congress also passed the first national Clean Air Act (CAA) in 1970. EPA smog chamber studies revealed that irradiation of mixtures of organic compounds and oxides of nitrogen produced peroxides and ozone. Nearly all organic compounds were found to produce smog in these chamber studies. Based on these results, the EPA adopted a regulatory objective of reducing emission of all organic compounds from all sources including those from coating production and application. A more scientifically sound course of action would have been to determine the photochemical activity of compounds widely used in coatings. The MIR values obtained by Jones [16] (Table 6.2) suggest that substantial differences exist. In Europe, solvents are classified by their "photochemical ozone creation potential" (POCP) [4(f)]. Various xylene isomers are reported to have very high POCP values as shown for MIR values for xylene isomers in Table 6.2. The large differences in compound-specific data suggest that a solvent substitution approach to ozone prevention would be very effective. Regulations based on compound-specific data would most likely be more effective for meeting the National Ambient Air Quality Standard (NAAQS) for ozone than the current EPA regulations.

Table 6.2 Photochemical Reactivity Based Ozone Production Potential

Solvent	Maximum Incremental Reactivity (grams ozone per gram of solvent)
Acetone	0.43
t-Butyl acetate	0.20
m-Xylene	10.6
p-Xylene	4.25

It also seems likely that the coatings industry could have produced environment friendly coatings at much lower cost.

With the present VOC regulations proof of low photochemical reactivity has been used as a basis for removing certain solvents from VOC consideration. Such solvents are referred to as "exempt." VOC-exempt solvents at this time include acetone, methyl acetate, and parachlorobenzotrifluoride. A proposal to exempt t-butyl acetate is under consideration. Since t-butyl acetate is not on the HAP list, VOC exemption would leave this solvent essentially unregulated. Please note the MIR values of acetone and t-butyl acetate in Table 6.2. These low values support their exemptions. Cooper et al. [17] have studied the use of t-butyl acetate in high-solids formulations of epoxy–amine two-pack coatings. This is a very widely used system for maintenance of industrial equipment and for primers for metal substrates. Conversion to high solids and replacement of xylene have been difficult because the amine crosslinker reacts with most carbonyl containing solvents. Cooper et al. [17] conclude that t-butyl acetate does not react appreciably with amine curatives and that t-butyl acetate can replace substantial amounts of xylene in two-pack epoxy–amine coatings.

Regulation of solvent emissions under the Clean Air Act Amendments (CAAA-70) is based on the National Ambient Air Quality Standards for ozone set by the EPA. The original NAAQS for ozone was 1-h, 0.12 ppm. "To attain this standard, the maximum 1-hour average concentration measured by continuous ambient air monitor must not exceed 0.12 ppm more than once per year, averaged over 3 consecutive years [18]". Areas that did not meet this standard were designated as nonattainment areas. States were required by the CAA to develop state implementation plans (SIPs) to reduce ozone levels in nonattainment areas. The EPA assisted the states in developing their SIPs by issuing guidance documents. In 1977 the EPA issued the first series of Control Technique Guidelines (CTGs). Each CTG applies to a particular segment of the coating industry. Each CTG also has recommended VOC emission limits. The EPA proposed the CTG limits and provided a period for paint industry response and sometimes negotiation before the CTG was adopted. CTGs for nearly all segments are available from EPA. Brezinski [15] provides a table of limits for many segments. For example, the values for auto electrodeposit primer (invariably waterborne) are 1.2 lb VOC/gal minus water or 0.14 kg VOC/liter minus water. The values for an auto topcoat (usually HSSB) are

2.8 lb VOC/gal or 0.34 kg VOC/liter. The goal was to achieve the ozone NAAQS in every state by 1975. Despite very significant ozone reductions in many areas, the goal was not met. The 1977 Clean Air Act Amendments set new dates to achieve NAAQS in every state.

Recently, the EPA changed the NAAQS for ozone. The revised ozone primary Standard [18] is 8 h, 0.08 ppm. "To attain this standard, the 3-year average of the fourth-highest daily maximum 8-hour average of continuous ambient air monitoring data over each year must not exceed 0.08 ppm [18]." Reduction of the ozone limit from 0.12 to 0.08 ppm makes the new NAAQS more difficult to attain. The change in time period for averaging (1-h average for the old, 8-h average for the new) makes the new NAAQS easier to attain. Overall, representatives of the coating industry and other effected industries feel strongly that the change makes attainment more difficult and that the result will be a large increase in the number of nonattainment areas [19]. Industrial group filed suit against the EPA claiming failure to consider the economic impact of its rules and for lack of a scientific basis for these standards. Legal cases eventually reached the Supreme Court and the judges' interpretation of the CAA favored the EPA position regarding authority to revisit and reestablish air quality standards. There is considerable bitterness and outrage surrounding this issue [19]. Effects of the NAAQS changes on the coatings industry were not evident at the time of this writing.

The Clean Air Act Amendments of 1990 (CAAA-90) have had, and in the future will continue to have, a major impact on the coatings industry. Brezinski [15] provides a clear and detailed description of the many new federal and state regulations associated with this comprehensive legislation with emphasis on features that will impact most on the coatings industry. CAAA-90 addresses control of ozone in the atmosphere, avoiding depleting of ozone in the stratosphere, reduction of acid rain and control of HAPs. HAP regulations will be discussed in Section 6.6.2. Title I (Ozone Control in the Atmosphere) directs the EPA to develop CTGs for segments not previously covered and to prepare new (lower limit) CTGs for numerous specified segments. Negotiated rulemaking will be used whereby representatives of EPA, industry, the states, and environmental groups negotiate the content of the proposed rule. CAAA-90, Title I, specifies classification of nonattainment areas according to mid-1991 ozone level: marginal (0.121–0.138 ppm), moderate (0.138–0.160 ppm), serious (0.160–0.180 ppm), severe (0.180–0.280 ppm), and extreme (0.280 ppm and above). Maximum number of years (beginning in 1990) to compliance with ozone NAAQS is mandated for each of the five classes: 3, 6, 9, 15, and 20 years, respectively. The Los Angeles area is the only area known to be in the "extreme" class. Stricter provisions will be imposed depending on the area classification for: VOC monitoring and inventory, revision of SIPs to incorporate RACT (reasonably available control technology) limits and future CTGs for major source, requirement of reviews and permitting for new or modified sources, and several other less general provisions [18].

Provisions for reducing acid rain do not adversely affect the coatings industry. Some cured coatings will benefits from this part of CAAA-90. For example, auto topcoats are damaged by acid rain under certain climatic conditions [20]. Permanent "water" spotting and in more severe cases pitting are caused by drying of acidic rain droplets. Such damage is called "environmental etch."

CAAA-90, Title VI (Stratospheric Ozone Protection), will not affect coatings much because chlorofluorocarbons are not used. One interesting case is 1,1,1-trichloroethane (TCE). Since TCE is VOC exempt, its use in coatings increased significantly in the late 1970s. Later it was found that TCE contributes to stratospheric ozone depletion, and therefore it is on the HAPs list. Title VI provides staged phase out of TCE use as a solvent.

6.6.2. Hazardous Air Pollutants (HAPs) and Right to Know Regulations

Title III (Air Toxics Program) of CAAA-90 directs EPA to evaluate and control the emission of HAPs. Among the 189 compounds identified in CAAA-90 as hazardous (HAPs list), Brezinski [15] identifies 24 that are used in coatings, 17 organic compounds, 6 metal salts, and ammonia. Some of the 17 organics are used as monomers in binder synthesis and therefore are not emitted to the atmosphere. The EPA has authority to delete or add compounds to the list, and industry groups may petition EPA to take compounds off the list (referred to as "delist"). Delisting petitions under consideration at the time of this writing include ethylene glycol monobutylether (EGBE) aka (EB), methyl ethyl ketone (MEK), and methyl isobutyl ketone (MIBK). EGBE has been used in numerous ways in waterborne coatings. MEK and MIBK have been widely used as active solvents in HSSB coatings. A petition to delist methanol was denied on May 2, 2001.

In 1991 the EPA targeted 17 compounds from the HAP list for rapid reduction in emissions by including them in the 33/50 Program. Companies that joined this voluntary program agreed to reduce emissions of targeted HAPs by 33% by 1992 and by 50% by 1995. The target compounds were selected, at least in part, by consideration of use level rather than toxicity level. MEK and MIBK are targeted, and these are among the least toxic of all 189 chemicals on the HAP list as indicated by current petitions to delist these as HAPs. Companies are expected to be reluctant to spend development time and incur expenses in trying to meet 33/50 for MEK an MIBK since these compounds may not remain on the HAP list.

The Emergency Planning and Community Right to Know Act (EPCRA, also known as SARA; 1986, Title III) is also administered by the EPA. The EPCRA does not regulate emissions but rather requires reporting of emissions of compounds that are on yet another list of about 300 compounds, claimed to be toxic. The number of lists maintained by EPA is large. One of the EPA's important documents is the "List of Lists." Under EPCRA, companies are required to keep records and issue annual reports of emissions of compounds on the specified list (SARA 313). Data is compiled and reported annually through the Toxics Release Inventory (TRI).

The Ketones and Oxo Process Panels of the Chemical Manufacturers Association provide a guide [21] to solvent users in 1994 for assistance in understanding

Table 6.3 Regulatory Status of Common Solvents

Solvent	Regulated VOC	HAP List	EPCRA Reportable	33/55 Program
Acetone	Exempt	No	Removed from list	No
Aromatic blends	Yes	Yes	Yes	Yes
n-Butyl alcohol	Yes	No	Yes	No
t-Butyl alcohol	Yes	No	No	No
sec-Butyl alcohol	Yes	No	Yes	No
n-Butyl acetate	Yes	No	No	No
t-Butyl acetate	Petition to exempt	No	No	No
Diacetone alcohol	Yes	No	No	No
Dimethylformamide	Yes	Yes	Added to list	No
Ethyl alcohol	Yes	No	No	No
Ethyl acetate	Yes	No	No	No
Ethylene glycol	Yes	Yes	Yes	No
E-Series glycol ethers	Yes	Yes	Yes	No
EGBE	Petition to exempt	Yes	Yes	No
n-Hexane (in blends)	Yes	Yes	Added to list	No
Isophorone	Yes	Yes	Added to list	No
Methyl ethyl ketone	Petition to exempt	Yes	Yes	Yes
Methanol	Yes(petition denied)	Yes	Yes	No
Methylene chloride	No	Yes	Yes	Yes
Methyl I-butyl ketone	petition to exempt	Yes	Yes	Yes
2-Nitropropane	Yes	Yes	Yes	No
Perchloroethylene	Petition to exempt	Yes	Yes	Yes
n-Propyl acetate	Yes	No	No	No
i-Propyl acetate	Yes	No	No	No
n-Propyl alcohol	Yes	No	No	No
i-Propyl alcohol	Yes	No	No	No
Toluene	Yes	Yes	Yes	Yes
1,1,1, Trichloroethane	No	Yes	Yes	Yes
Xylene (s)	Yes	Yes	Yes	Yes

government regulations. Table 6.3 is an updated version of a table from the guide. At the time of this writing there were 5 petitions to exempt from VOC regulations, and decisions on these 5 could have far-reaching implications for formulation of coatings. In this group, 1 petition had been denied (methanol), and 1 petition had been approved (acetone) at the time of this writing. This group of solvents includes 8 of the 17 compounds that were targeted for rapid emission reduction in the 33/50 Program. Inorganic compounds or compounds not used in coatings made up the remaining 9 compounds in the 33/50 Program. In addition to toxicity considerations, avoiding EPCRA-reportable solvents is desirable because record keeping and report preparation are expensive. Eighteen of the 29 entries in Table 6.3 are EPCRA reportable.

Although the amendment was passed in 1990, the phase in of HAP regulations was spread over an extended period with many scheduled for the 1997–2003 time frame. Control of HAP emissions under CAAA-90, Title III, will be achieved by emission standards set for source categories. The coatings source categories include "surface coating processes" for all processes for which a CTG has been issued or is planned and "manufacture of paints, coatings and adhesives." Emission control regulations are based on MACT (Maximum Achievable Control Technology) on a source by source basis. Title III directs EPA to consider health impact and economic factors in defining MACT limits. While VOC regulations vary from state to state depending on state implementation plans (SIPs), HAP regulations are set by the EPA and apply nationwide. Limits on emission of each HAP as set by EPA are called national emission standards (NESHAP).

CAAA-90, Title V (State Operating Permit Program), consolidates all state and federal rules and regulations under a single document that gives the states authority to monitor and enforce regulations through a permitting process [15]. "Major sources" are defined as sites with the potential to emit 10 tons or more per year of a single pollutant or 25 tons or more per year of all pollutants. To receive a permit, operators of major sources are required to record and report pollutant emissions. These pollutants include all HAPs for which a NESHAP has been established. Proposed state permit programs must be approved by the EPA within 1 year, and operators of major sources must apply for a 5-year permit within 1 year of the approval of the state program.

6.6.3. Regulations of Pigments and Additives

Three of the inorganics on the HAP list are of particular concern in the coatings industry: chromium compounds, mercury compounds, and lead compounds. Hexavalent chromium is very effective at passivating metal substrates. It is used in various forms in chemical pretreatments of metal [4(g)] and as a corrosion inhibiting pigment [4(h)] in primers for metal. The main component in metal pretreatment for auto bodies is zinc phosphate, but the last rinse after phosphate treatment has been chromic acid. Replacements for the chromic acid rinse have not resulted in comparable corrosion resistance. In primers, complex salts involving zinc chromate, potassium chromate, and zinc hydroxide are used as passivating pigments. Pilcher [22] includes replacement of chromium VI compounds as one of the big challenges of the twenty-first century. It is noted that coating suppliers have been working for about 20 years to obtain "chromium-free" systems with good corrosion protection, but success has been limited for steel substrates.

Stoffer et al. [23] recently used cerium-based conversion coatings for corrosion protection of aluminum alloys used extensively for military and commercial aircraft. Results are encouraging for replacement of chromium-based conversion coatings. Work began recently on corrosion protection of steel by cerium ions [23].

White lead pigments have not been used in architectural paints for many years. White lead, $PbCO_3 \cdot Pb(OH_2)$, was used extensively as a white pigment until the

1930s, when it was replaced by zinc pigments (ZnO, Zns, and Zns/BaSO$_4$). Use of all other white pigments dropped off rapidly when TiO$_2$ became available. The extremely high refractive index of TiO$_2$ provided much greater hiding power than any other white pigment [4(a)]. Houses that were first painted before about 1950 may have some early coats that contain lead. Occurrence of lead poisoning is certainly not always traceable to paint, however, because there are many other possible lead sources in our surroundings [24].

Tetra-ethyl lead was used as an antiknock ingredient in gasoline from 1940 to about 1990. It has been estimated that gasoline provided approximately 7 million tons of lead. Lead from this source can be found in the soil especially in high traffic areas. Other sources are glazed pottery, older plumbing, and dental fillings. Despite many possible sources, coatings seem to be the target of choice for litigation [24].

Some lead salts have been used in industrial coating as a corrosion inhibiting additive, but this low-level use has also ended recently so that the term "lead free" can be applied to formulations. Lead replacement in response to HAP listing will undoubtedly continue. Court cases related to very early use of lead may also continue to be a problem.

Mercury compounds were at one time used as a bactericide and a fungicide in waterborne paints. A bactericide is needed to prevent bacteria growth during storage of the sealed paint can. A fungicide is needed to prevent mildew growth on the cured paint film under humid conditions. Wicks et al. [4(i)] give the modern-day replacements for the earlier mercury compounds.

Replacement of mercury compounds in antifouling paints for the hulls of ships is challenging. In previous antifouling paints, the binder served as a carrier with controlled release of a toxic additive, often an organomercury compound, so that a barnacle, for example, was poisoned before it attached itself to the hull. Brady [25] describes a different approach based on low surface energy of the antifouling paint. The top surface of the paint film is designed to have such low surface energy and smoothness (nonadhesive properties) that even the excellent natural adhesive produced by a barnacle will not permanently attach the barnacle to the hull. Good antifouling coatings are effective for reducing fuel consumption in ship propulsion because fouling increases resistance greatly.

There are a lot of additives in most paint formulations each, with a different purpose. In addition to the main components (binders, pigment, and solvent/water), a latex exterior house paint, for example, contains [4(j)]:

Ionic surfactants needed for emulsion polymerization when the latex is prepared
Ionic and nonionic surfactants to stabilized the latex dispersions during storage
Surface-active agents designed for pigment dispersion
Thickeners and rheology control agents for optimizing viscosity for application
Bactericides and fungicides (as discussed above)
Special-purpose pigments for control of gloss
Ethylene glycol or propylene glycol for uniform drying and for antifreeze

Coalescing agents such as the isobutyric ester of 2,2,4-trimethylpentane-1,3-diol

A weak base for pH control (often ammonium hydroxide)

Most of these additives do not have significant regulation issues other than those previously discussed such as the ethylene glycol and ethylene-glycol-based alcohol ether inclusion on the HAP list. The response has been to switch to similar compounds prepared from propylene oxide instead of ethylene oxide. The VOC of a white exterior house paint containing typical examples of the additives noted above is 0.196 kg VOC/liter excluding water [4(h)].

A common additive in HSSB thermoset formulations is a reactive diluent. Reactive diluents are like solvents in that their main purpose is to reduce viscosity for applications. One could classify a solvent as a nonreactive diluent. Reactive diluents are like the binder in that they react during cure, but their molecular weight is lower than that of the binder. A regulatory issue sometimes results from low molecular weight. Molecular weight is one of the considerations in approval of premanufacture notification (PMN) forms. Epoxy groups have some toxicity issues, and the term "oxirane" (another name for epoxy groups) is on the HAP list. Higher molecular weight reduces risk of exposure so that established use of epoxy binders of higher molecular weights does not necessarily ensure approval of the lower molecular weight compounds proposed for use a diluents.

6.7. COATINGS INDUSTRY RESPONSES TO REGULATION

There has been a long-term slow but continuous change to waterborne and high-solids solvent-borne coating. This change started before regulation. Reduction in the use of solvents is an economically as well as environmentally sound policy for the coatings industry. Solvents are lost as the film is formed; therefore, it is obvious that solvents are not essential to performance of the coating. Organic solvents are believed by many in the industry to be needed for getting the paint material evenly applied in a thin, continuous film, but coatings chemists are finding other ways to get the films applied in a useful form. Most coatings chemists, like the majority of sensible members of our society, are concerned about the environmental impact of industrialization. Thus, economic considerations (profitability) and environmental considerations were driving the industry toward solvent reduction before government regulations were adopted. There can be little doubt, however, that regulations significantly accelerated the move to reduce solvent use.

There are widely differing views about the thinking, or lack thereof, that was given to the regulations. Many would claim that the blanket limits on volatile organic compounds represented a nonscientific approach to ozone reduction. The idea that different compounds react at different rates is very basic to the field of chemistry. Regulators probably believed that getting compound-specific data would be too slow, considering the urgent need to reduce ozone levels. Many regulators probably believed that the coatings industry would do very little to reduce solvent emissions without regulation. Representatives of companies that served the paint industry mainly as suppliers of organic solvents had a lot to lose by solvent regulations.

The EPA is, of course, assailed from both sides. Environmental groups attack (usually figuratively) the EPA for not writing stricter regulations and for not enforcing them more vigorously. Coatings industry representatives claim that the regulations are too limiting and that the regulators have not judged the economic impact realistically. The CAAA-90 contains the term "negotiated rulemaking." Representatives of EPA, the coatings industry, and environmental groups are directed to meet to negotiate the content of a proposed rule. There were earlier examples of negotiations also in establishing CTGs for various segments of the coatings industry. Cooperation is also well served by attendance and presentations by EPA personnel at technical conference of the coatings industry such as the Gordon Research Conference on Coatings and Films.

6.7.1. Reduction of Solvent Emission by Selection of Application Method

Transfer efficiency is reported [14] to be an important consideration for controlling solvent emission levels. Estimated transfer efficiency for various methods to be viewed as "only a general guideline" [14] are shown in Table 6.4. Transfer efficiency is defined as the percentage of paint that is actually deposited on

Table 6.4 Estimated Transfer Efficiency for Various Methods of Application

Application Method	Transfer Efficiency (%)	
	Ref. [14]	Ref. [26]
Manual hand spray (air spray)	<20	25
Plural component spray (two-pack)	<20	[a]
High-volume, low pressure air (HVLP)	<40	65
Electrostatic spray—air	(40–90)[b]	60–85
Electrostatic spray—rotary	[a]	65–94
Airless spray	<30	40
Liquid carbon dioxide spray	<30	[a]
Roller coating, direct or reverse	>95	[a]
Dip coating	>80	[a]
Curtain coating	>95	[a]
Flow coating	>80	[a]
Electrodeposition coating (E-coat)	>95	[a]

[a] Not given.
[b] Personal communication, W. J. Blank.

the substrate. Blank [14] and Adams [26] do not give the same estimates, but they are not too far off where comparisons are possible. It is surprising that air spray is still used despite its very low transfer efficiency. The term used for paint that does not wind up on the substrate is "overspray." Solvent is lost from the overspray as well as from the paint that actually gets on the substrate.

A very effective way to reduce solvent emissions is to change from air spray to another type of application. Use of HVLP spray guns or airless spray guns provide some modest improvement over air spray. Electrostatic spray provides a large to very large improvement. The range for electrostatic spray is quite broad in part due to differences in substrate form. For example, higher transfer efficiency is expected for a flat solid panel than for chicken wire. The high end of the range corresponds to equipment supplier claims, and these (85 or 90% efficiencies) can be achieved under ideal conditions. In electrostatic spray, special spray guns or other devices (rotary bell or disk atomizers) impart a static electric charge to atomized paint droplets. The metal substrate is grounded so that the particles are attracted to the substrate surface. Electrostatic spray is the method of choice for auto topcoats. Electrostatic spray was developed for low-conductivity solvent-borne coating, but specialized equipment has been developed to permit electrostatic spray of waterborne coatings.

Liquid carbon dioxide spray [10] may sound a little strange because we are accustomed to dealing with CO_2 as a solid or gas. Most of us have never seen it in liquid form. At easily attainable temperature and pressure, gaseous CO_2 passes through a critical point and becomes liquidlike (called a supercritical fluid). In this state it can be used to replaced organic solvent up to about 30 vol % of the formulation. When the pressure in the application gun is released, liquid CO_2 reverts to its gaseous form, and this transformation is reported to contribute to good atomization of the paint. Atomization is an important step in spray application. The improvement in transfer efficiency over air spray is only modest (Table 6.4). The main advantage is replacement of solvent by a diluent that does not have to be counted as part of the VOC. Carbon dioxide is like water in terms of not being included in VOC [10]. This advantage is balanced against the cost of using special pressurized spray equipment and purchasing the supercritical CO_2 containers.

Application methods that do not involve formation of tiny droplets all have much higher transfer efficiency than air spray. From the list in Table 6.4, applications in continuous liquid form (not droplets) include: direct or reverse roller coating, dip coating, curtain coating, and flow coating. These methods are low cost and environmentally sound, but their use has limits depending on shape and size of the article to be coated. For example, roller coating only works on flat substrates. Application by roller coating is dominant for coil-coated metal (discussed in Section 6.5.1).

Electrodeposition coating in its present form was developed mainly for applying corrosion-resistant primers to automobile body assembly. It is evident in Table 6.4 that it provides very high transfer efficiency. Other advantages are discussed in Section 6.7.2

6.7.2. Conversion to Waterborne and High-Solids Coatings

A switch from solvent-borne to waterborne systems is one of the most frequent responses to government regulation of solvent emissions. One of the reasons that this switch has not gone even farther is that the strategies for making hydrophobic paint films from water systems are only partially successful. Films are often not hydrophobic enough for many demanding end uses. Architectural paints, however, are mainly waterborne, as shown in Table 6.5. Waterborne systems represent 69.0% of the exterior architectural coating shipments and 88.0% of the interior shipments. Census figures [1] indicate that conversion of house paint to a waterborne system is very extensive except for a portion of floor enamels, primer, clear finishes, sealers, and stains for shingles or shakes.

As recently as the 1970s and early 1980s, solvent-based primers were recommended for the "do-it-yourself" house painter for good adhesion over highly weathered (chalky) surfaces. Improvements in formulation have resulted in much improved adhesion of waterborne primers over chalky exterior walls. Now waterborne latex is used by nearly all do-it-yourself house painters for both primer and topcoat.

As noted in Section 6.5.2, the hydrolysis resistance of latex films from emulsion polymers prepared from the vinyl acetate (VA) monomer is marginal for outdoor use. In Europe and recently in the United States, vinyl versatate (II) ($R_1 = $ —$CH_2CH_2CH_2CH_2CH_2CH_2CH_3$ and $R_2 = R_3 = $ —CH_3) has been introduced for use along with vinyl acetate for improved outdoor performance of latex coatings. Vinyl versatate is the vinyl ester of versatic acid, a 10-carbon carboxylic acid of the highly branched structure sometimes called "neo":

(II)
$$R_1-\underset{R_3}{\overset{R_2}{C}}-\overset{O}{\overset{\|}{C}}-O-(H)C=CH_2$$

Table 6.5 Manufacturer's Estimated Shipments of Architectural Coatings in 2000 [1]

	Quantity (million of gallons)	Value (millions of dollars)
Exterior		
Water	167,867	1,769,171
Solvent	75,284	854,645
Interior		
Water	347,281	1,161,796
Solvent	47,521	556,612
Other	6,406	64,825
Total	644,359	6,407,459

Commercial versions contain several isomers in which R_1, R_2, and R_3 are alkyl groups of varying length but with a total of 8 carbon atoms. The mixed carboxylic acid precursors of these highly branched esters are collectively called neodecanoic acid [27], and the ester is then called vinyl neodecanoate (VN). The advantage of VA/VN copolymers over other VA copolymers is greater hydrolysis resistance and increased hydrophobicity [27]. These advantages are realized at relatively low cost. Developments such as this are likely to increase the rate of conversion to waterborne systems.

Use of waterborne coatings for industrial finishes is increasing quite rapidly, but figures are not readily available. Census Bureau tabulations give data for the solvent–water breakdown as in Table 6.5 only for architectural coatings but not for other categories. Coatings consultants provide estimates for the solvent–water breakdown for industrial coatings. For example, for product finishes for OEM, a consultant report indicates that use of waterborne coatings increased from 14% in 1994 to 22% in 1999. Over the same period low-solids solvent-borne (SSB) coatings for OEM decreased from 45% to 23%. Most of the decrease in LSSB coatings was made up by waterborne, but a shift to powder coating (see Section 6.7.3) and to HSSB coatings also contributed.

Changing from solvent-borne binders to waterborne latex binders for product finishes for OEM and for special-purpose coatings has great environmental benefits, and in most cases it has been achieved without sacrifice of performance properties. One of the strategies that made this possible is changing from traditional surfactants to new types of surfactants. Buckman et al. [28] describe the use of self-crosslinking polymeric surfactants in emulsion polymerization. Grade et al. [29] discuss improvements in water resistance obtained by use of nonmigratory surfactants. Another strategy for improving performance of latex binders for use in industrial coatings is to design systems to undergo crosslinking after coalescence. Winnik [30] has described the critical interplay between the rates of interdiffusion and crosslinking in thermoset latex films. Taylor and Klotz [31] recently published data essential for understanding and predicting interdiffusion rates. Their work includes extensive data on the glass transition temperatures and activity coefficients of filming aids used in latex systems. Zhu et al. [32] describe synthesis of a castor acrylated monomer (CAM) that can be copolymerized in a latex binder to provide low glass transition temperature for promotion of interdiffusion followed by oxidative curing to improve final properties.

Several waterborne systems that are neither latex types nor traditional water-reducible types are finding increased use in industrial coatings. Urethane chemistry has been extended to water by use of dispersions and also as two-pack systems having unreacted isocyanate groups. In aqueous polyurethane dispersions (PUDs), urethane oligomers are solubilized with carboxyl groups, as described for other water-dispersible binders, and chain extended with polyfunctional amines [33]. In PUDs the isocyanate groups are fully reacted before dispersion in water, but two-pack waterborne systems with free isocyanate groups have also been introduced. Many of the isocyanate groups react with water to eventually yield amine groups and carbon dioxide, but careful design can promote reaction with a network forming coreactant as well [34]. Blasko et al. [35] recently discussed a new type of acetoacetoxy-functional monomer that can serve as a crosslink site in a copolymer for thermoset waterborne binders. Many papers are presented every year at the International Waterborne, High Solids and Powder Coatings Symposium sponsored by the University of Southern Mississippi, Department of Polymer Science. The 2002 offering is the twenty-ninth annual symposium. The longevity and success of this symposium suggests that coatings technology and environmental considerations have been closely linked for a long time.

Automotive coatings provide several examples of different responses to regulation within a single coating system [20]. The metal substrates in use today are nearly always galvanized. The coating process involves the following:

1. Chemical pretreatment of substrate (also called conversion coating)
2. Electrodeposition of corrosion-resistant primer
3. Primer/surfacer coating
4. Topcoat consisting of basecoat/clearcoat, two-coat one-bake method

The phosphate conversion coating and the need to replace the chromic acid rinse were discussed in Section 6.6.3. Conversion to waterborne cationic electrodeposition primer (step 2) is essentially complete throughout the automotive industry. The electrochemical plating of the primer onto the substrate permits coverage in recessed or partially enclosed areas for greater corrosion protection. The purpose of the primer/surface (step 3) is mainly to give a very smooth underlayer for improved appearance of the topcoat. Solvent emissions in this step have been steadily reduced over an extended period by using HSSB thermoset coatings. A more recent development, still at an early stage of introduction, is the use of powder coating for the primer/surfacer. Of course, powder is already widely used for off-assembly line coating such as under-the-hood parts and wheels. (Powder coating is discussed in Section 6.7.3) Auto, SUV, and light-truck topcoats are almost all of the basecoat/clearcoat type, also called two-coat, one-bake coating. The pigmented basecoat is applied by electrostatic spray, and it is partially dried in a short-time, low-temperature oven (called heated flash). The clearcoat is also applied by electrostatic spray over the still partly wet (called wet-on-wet) basecoat. The two coats that make up the topcoat are then cured at a normal automotive bake cycle, for example about 30 min at 120–125°C. At the present time both solvent-borne and waterborne basecoats are in use, but the change to waterborne basecoats is continuing. Clearcoats are nearly all HSSB thermoset systems. The 'Low-Emission Paint Consortium' (LEPC), involving all users and producers of automotive paint in the United States, has carried out a lot of exploratory work on the use of powder coating for clearcoats [20], but commercial introduction has not occurred in the United States so far. The LEPC is one of 12 consortia under the auspices of the U.S. Council for Automotive Research [20].

Papasavva et al. [36] recently reported a life-cycle analysis (LCA) of paint processes for coating of an SUV of typical size such as a Chevy Blazer. The processes analyzed included application, flashoff, and bake. The new calculations were combined with those from an earlier study [37] that gave the environmental impact associated with the manufacturing of various paint materials. Data included atmospheric emissions, energy consumption, water consumption, water emissions, and solid waste emissions. The baseline system (or scenario) was taken to be solvent-borne primer, waterborne basecoat (white W1), and solvent-borne clearcoat. The baseline system was changed by replacing solvent-borne primers by powder primers, either P1 (powder acrylic) or P2 (powder polyester). In one case the white waterborne basecoat (W1) was replaced by a pewter (silver/gray) waterborne basecoat (W2). The baseline system was further changed by replacing solvent clearcoat by an acrylic powder clearcoat, designated PC. The scenarios analyzed included:

Scenario	Primer/Surfacer	Basecoat	Clearcoat
1. Baseline	S	W1	S
2.	P1	W1	S
3.	P1	W2	S
4.	P2	W1	S
5.	P2	W1	PC

The most obvious expectation is that replacement of solvent-borne coatings by powder will reduce VOC. The painting process contribution to VOC emissions per SUV for scenarios 1 and 5 were 2.0 and 1.2 kg, respectively. Including the VOC emissions contributed from material manufacturing increases these values moderately to 2.14 and 1.27 kg. These VOC values reflect the use of VOC abatement methods as discussed in Section 6.7.5. Differences for the white and pewter colors (scenarios 2 and 3) were small. Energy consumption for synthesis of the acrylic powder primer (P1) is 275 MJ/kg, whereas the corresponding value for the polyester powder primers is much lower, 107 MJ/kg [37]. This difference was still evident in the combined (paint process plus manufacturing) data on energy consumption for scenarios 2 and 4 [36].

6.7.3. Conversion to High-Solids Coatings

Most responses to regulation are clearly defined because the change involves a new chemical system. In high-solids coatings the chemistry is often unchanged except for reduction of the molecular weight of binder components. Molecular weight change can be carried out incrementally so that the shift to higher solids is evolutionary. The change from low-solids solvent-borne (LSSB) coatings to high-solids solvent-borne (HSSB coatings is not easily defined because there has been no agreement about the solvent content cutoff between LSSB and HSSB. Wicks et al. [4(k)] state that there is no single definition of "high solids." Often the cutoff depends on the solvent content that was present in a particular type of coating before attempts to reduce VOC began. Furthermore, as solvent reduction continues, the level corresponding to HSSB status continues to decrease.

For example, if a conventional solvent-borne auto basecoat used in 1970 had a nonvolatile content by volume (NVV) of 30%, then a basecoat used in 1980 at an NVV of 45% may have been called a HSSB basecoat [4(k)]. However, by 1990 it is doubtful that a NVV of 45% would be called HSSB, and a NVV value of 55% might be needed to qualify as HSSB.

The "moving target" aspect of HSSB status can give a misleading picture of the effort to reduce emission by changing to high-solids systems. Consultants report that in 1994, 66% of the coatings volume used for OEM product finishes was solvent borne and the LSSB:HSSB breakdown was 45 to 21%. By 1999, the solvent-borne portion of OEM paint was down to 45% and the LSSB:HSSB split was 23 to 22%. The 1994–1999 comparison for HSSB (21 vs. 22%) appears to indicate very little change to high-solids use, but in fact the change in high-solids definition over time may contribute to the smallness of the apparent change to HSSB. Of course, part of decrease from 66% LSSB coatings in 1994 to 45% in 1999 reflects changes to waterborne and powder coatings as well as to HSSB coatings.

During the early years of regulation in the United States, molecular weight reductions made in binder component to increase the solids content were very substantial. Air quality regulations came later in Europe and Japan, and the regulations were not so sweeping. These differences caused an entirely different course of binder development for OEM automotive coatings in the United States versus the rest of the world. In the 1960s and early 1970s the dominant crosslinkers for automotive coatings worldwide were polymeric (actually oligomeric) butylated melamine formaldehyde (MF) resins [38]. These resins were quite viscous. To facilitate handling, they were diluted by addition of about 20 to 30 wt% solvent, often n-butanol. In the United States, suppliers of MF resins developed methylated and mixed (methyl/butyl) ether crosslinkers of very low molecular weight (approaching monomeric). These resins were so much lower in viscosity that they did not require any solvent for handling [38]. The high-solids MF resins were very reactive in the presence of p-toluene sulfonic acid catalysts [39]. Use of these high-solids MF resins in the United States resulted in large decreases in solvent content of automotive coatings. Reduction in the molecular weights of the acrylic polol co-reactants used with high-solids MF resins resulted in further solvent reductions. When Japanese companies such as Honda and Toyota opened assembly plants in the United States (the so-called transplants), the Japanese paints could not be used here due to high solvent content. Each major Japanese auto painter supplier negotiated an agreement with a U.S. auto paint company to supply their transplants in the United States. Agreement was never reached about which paint had better performance, but Japanese technical personnel in charge of coating at the transplants accepted the high-solids U.S. systems as "a necessary evil."

Use of thermoset crosslinking to build up molecular weight is a much-used strategy to permit use of low-molecular-weight binder components to start. Several studies have explored the possibility of carrying this approach to the point of formulating zero-VOC thermoset liquid coatings. Blank [14] recently summarized and analyzed this quest for the ultimate goal in high solids in a study titled "The Slow and Winding Road to Zero VOC." It is concluded that zero VOC would be possible from liquid coatings if we could use cured coatings that have a very low glass transition temperature, for example a T_g around room temperature or below. Such coatings would be soft and/ or rubbery, of course. For sufficient hardness, automotive coatings of the acrylic melamine type are designed to have values of T_g between 50 and 100°C. It is also pointed out [14] that zero solvent content does not always mean zero VOC. If the binder molecular weight is decreased drastically, then compounds in the low end of the molecular weight distribution are quite volatile and may be lost from the coating during cure. Low-molecular-weight products of the curing reaction, often called cure volatiles, also leave the film during cure and contribute to VOC. Crosslinkers that produce cure volatiles include [14]: MF resins (alcohol and low levels of formaldehyde), blocked isocyanates (released blocking agents such as ketoximes, pyrazol, or alcohols), alkoxy silanes (alcohols), and activated transesterifications (alcohols). Some crosslinking reactions give off water, which does not count toward VOC: hydroxyethylamide + carboxyl, acetoacetate + amine, and a related group including methylol amide, glycoluril, and cyclic urea.

Jones [40] used oligomeric, telechelic (end-functional) polyester polyols to prepare solventless liquid coatings. The absence of any branching on the diol monomers was an important feature to obtain low viscosity. Excessive softness of the cured films was partly overcome by adding low-molecular-weight diluents having ring structures. Use of low levels of water in otherwise solvent-free formulations caused large viscosity decreases such that molecular weight reduction did not have to be carried quite so far. It is not evident that these new leads have been used in commercial systems.

Jones [41] compared reported solvent emissions in 1965 (700 million gallons) and 1997 (600 million gallons), finding a net decrease. If the growth rate in coatings from 1965 to 1997 is taken as 3.5% per year and no effort was made to reduce solvent use, the quantity of solvent would have increased to 2100 million gallons by 1997, 3.5 times higher than the reported 600 million gallons. This difference is believed to reflect the many technical changes directed at reducing solvent emissions. Blank [14] compared the annual growth of solvent-borne paint volumes (3.5%) with the growth in solvent use over the same period (very close to zero growth) from 1985 to 2000. The average solids content was calculated to be 50% in 1985 and 60% in 2000. Both of these calculations indicate very significant progress, but further work is needed (and is underway) in the solvent-borne coating sector.

6.7.4. Powder and Radiation Cure Coatings

Powder coatings use has increased much more rapidly than coatings in general. Growth rates of 10% or more were recorded for powder through most of the 1990s while overall coatings shipments experienced about a 3.0% growth rate early in the 1990s to about zero growth more recently (see Table 6.1). Census

Bureau [1] figures for 2000 indicate shipments of 340 million pounds of powder, which recalculated as gallons at 68 million gallons (5 lb/gal). Powder amounts to 14.9% of product finishes OEM and 5.3% of total coatings shipments. Higher powder use levels are reported by consultants [42], with a suggestion that the numbers reported to the Census Bureau miss some of the growth. Radiation cure coatings use in the United States is reported [42] to be about 85 million pounds, and powder and radiation cure together are stated to represented about 9% of U.S. shipments [42]. Growth for radiation cure is estimated to be 9% annually [42].

Powder coatings can be of the thermoplastic or thermosetting type. Most of the growth is in thermosets. Improvements in powder preparation such as smaller and more uniform particle size have contributed to growth by allowing greater range of film thickness with good control of thickness [4(m)]. A remarkable balance of the effects of heating on thermoset powder has been achieved. Control of melt/flow/cure is a demanding exercise in rheology and chemorheology. Control of curing reaction rates at various stages of powder preparation, storage, and cure is very challenging. Powders are melt mixed in an extruder at elevated temperatures, and during this stage little or no reactions should occur. Later, after application and melt flow at elevated temperatures, we want curing to take place to nearly complete conversion in the crosslinking reaction. The bake temperature is not too much higher than the melt/mix temperature in the extruder. If reaction is too fast during bake, melt/flow will be incomplete and the film will have a rough surface. For a rather extended period of powder coating evolution, cure temperatures were around 200°C [4(m)]. It was incorrectly predicted that reduced cure temperatures would not be possible in part because this temperature was needed for melt/flow. Recent growth has resulted from reduction in cure temperatures. Particle size control and clever use of crystalline materials with sharp melting points have contributed to good melt/flow at lower temperatures.

Most powder coating is carried out on metal substrates, but the reduction in cure temperatures is opening new opportunities for powder coating of heat-sensitive substrates such as wood, fiberboard used in furniture, and plastics [42]. A very recent development is the combination of powder coating with ultraviolet (UV) cure [14, 42, 43]. This hybrid technology will provide further advances for powder coating of heat-sensitive substrates. The requirement of irradiation to set off the cure reaction, of course, means that premature reaction during melt mixing in the extruder and during storage of the powder will be avoided. Independent control of melt/flow and heating and of cure by UV irradiation also provides many advantages in flow and leveling of powder coating.

Use of powder coating in the automotive industry has already reached significant levels as discussed in the life-cycle analysis review in Section 6.7.2. The automotive primer/surfacer powder coatings being introduced in the United States by Chrysler are of the epoxy hybrid type [42]. Thermoset powder coating terminology is a bit obscure. Epoxy hybrid means a combination of bisphenol A (BPA) epoxy resin and a carboxyl-functional polyester resin. Epoxy powder coating means a BPA epoxy resin crosslinked by amine groups, usually by the amine groups in dicyandiamide. The poor weathering properties of BPA make such powders unsuitable for outdoor topcoats. Improved durability can be achieved by replacing BPA epoxies by polyester binders. In Europe a popular combination is a carboxyl-functional polyester crosslinked by triglycidylisocyanurate (TGIC). In the United States a popular combination is hydroxyl-functional polyester crosslinked by a blocked isocyanate. Both of these types are called polyester powder coatings. Concern about toxicity of TGIC has lead to an intense search for a replacement. Hydroxyalkylamide crosslinkers, which like TGIC react with —COOH groups, have had some success, but the evolution of water (versus no cure volatile from TGIC) causes popping problems in some powder coatings. Despite improved durability compared to BPA-containing coatings, polyester powder coatings are not sufficiently durable for automotive topcoats. For auto topcoats, acrylic powder coatings are used. The only commercial automotive use of acrylic powder topcoat is by BMW in Europe [44], but U.S. auto paint suppliers have studied acrylic powders extensively as part of the research of the Low-Emission Paint Consortium [20]. In acrylic powders, an epoxy-functional acrylic monomer, glycydil methacrylate (GMA), is copolymerized with other monomers already used in topcoats. The epoxy groups of the GMA acrylic powder are crosslinked by dibasic acids such as dodecanedioic acid, $HOOC(CH_2)_{10}COOH$ [4(m)]. Commercialization of acrylic powder coating for auto topcoats in the United States represents a large potential market.

For a technology introduced almost 30 years ago, the small current use of 85 million pounds of UV-cured liquid represents a very low growth rate. Perhaps the best chance for UV is to grow along with powder in the hybrid area of UV-cured powder coatings. Current use areas include flexible food packaging, furniture and wood kitchen cabinets, wear layers for vinyl flooring, and abrasion-resistant clear coatings for plastics [42]. A rather different type of radiation cure coating is cationic UV-cured coating [45], in which the UV indirectly generates a super acid. Cationic UV-cured coatings are used to cure coil coatings for highly postformable beverage can ends [43]. As is true for nearly all coatings types, UV-cured coating suppliers discuss opportunities in the automotive coatings market [42, 46].

6.7.5. Solvent Abatement: Incineration or Recovery

If spray application is used and the spray booth is not fully automated so that people must enter the spray booth, large volumes of air are pumped into the spray booth to sweep the solvent vapors away for safety reasons. The booth make-up air causes large reductions in the concentrations of solvent in the air, which in turn makes both incineration of solvent and recovery of solvent very inefficient and expensive. Perhaps the best abatement method is to change to a different type of application method, as described in Section 6.7.1.

Incineration is practical when the solvent concentration in the cure oven is relatively high. High concentrations of solvent can be maintained in situations were there is no exposure of workers. Two examples of this situation are coil coating ovens and wire coating ovens. Nearly all such ovens have incinerators, and the energy value of the emitted solvent is used as part of the fuel for heating the oven. Flammable vapors in the presence of air and heat often explode. Explosion can be avoided by having a mixture too high in the flammable vapor for explosion (upper explosion limit) or too low in the flammable vapor for explosion (lower explosion limit). For incineration a high vapor concentration increases heat recovery efficiency, but the concentration must remain safely below the lower explosion limit [4(n)]. Incineration and change to an alternative coating technology often must be compared as solvent emission reduction strategies. For coil coating with solvent-borne formulations, incineration may be a cost-effective alternative to changing from an LSSB coating to an HSSB coating. For a spray-applied thermoset solvent-borne coating, incineration is not a viable option. Change to HSSB or to another low-emission coating method may be viable choices.

Solvent recovery can be accomplished by adsorption on activated carbon or by solvent vapor condensation [4(n)]. In the adsorption method, solvent-rich air flowing out of the drying oven is directed through a bed of activated charcoal. Some of the solvent adsorbs on the carbon. When the carbon is saturated, the bed is heated to drive off the solvent, which is then often used as fuel for heating the oven. A second recovery method involves condensation of solvent vapors in the oven effluent. In one type of condenser, liquid nitrogen is used to cool the air/vapor flow from the cure oven, and the vapors condense. Efficiency is increased by feeding the gaseous nitrogen from the coolant into the oven, which increases the lower explosion limit in the oven. This improves efficiency by permitting safe operation at higher solvent vapor concentration [4(n)].

Papasavva et al. [36] showed diagrams of auto paint lines that indicated abatement on the solvent-borne primer/surfacer ovens, on the heated flashed zone after waterborne basecoat application, and on the solvent-borne clearcoat oven. The abatement equipment included a "carbon concentrator," which is likely a device for solvent vapor adsorption on activated carbon. A collection efficiency of 90% was reported. The abatement equipment also included a system for oxidizing the vapors to form water and carbon dioxide. These units were called reactive thermal oxidizers (RTOs) [36], and their destruction efficiency was reported to be 95%. For the various scenarios discussed in Section 6.7.2, the abatement efficiencies were reported to range from 52.3 to 62.5%. Emissions from powder coatings ovens are expected to be so low that abatement is not needed, and calculations with and without powder coating oven abatement gave the same results for VOC emission [36].

REFERENCES

1. U.S. Census Bureau, Current Industrial Reports, Paint and Allied Products MA325f(00)-1, July, 2001; *http://www.census.gov/cir/www/ma28f.html*. See also E. W. Bourguignon, *Paint & Coatings Industry*, **XVII**(10), 136 (2001).
2. R. C. Williams, in C. D. Craver and C. E. Carraher, eds., *Applied Polymer Science 21st Century*, Elsevier, New York, 2000, p. 473.
3. S. G. Ritter, *C.&E. News*, **79**(29), 29 (2001).

4. Z. W. Wicks, Jr., F. N. Jones, and S. P. Pappas, *Organic Coating Science and Technology*, 2nd ed., Wiley-Interscience, New York, 1999; (a) p. 371, (b) p. 258 (c) p. 306. (d) p. 541, (e) 154, (f) p. 329, (g) 115, (h) p. 136, (i) p. 565, (j) p. 562, (k) p. 462, (m) p. 486, (n) p. 331.
5. Anon., Mineral Galleries, *http://mineral.galleries.com/minerals/oxides/ilmentie/htrr.*
6. A. E. Rheineck, *J. Paint Technol.* **44**(567), 35 (1972).
7. Oil Colour Chemists Assoc. Australia, *Surface Coating*, Vol. 1, 2nd Ed., Chapman and Hall, New York, 1983, p. 1.
8. J. Kumanotani, *Heterogeneity in the Surface of Oriental Lacquer Films*, XIIth Internat. Conf. Org. Coat. Sci. and Technol., July 7–11, 1986, Athens, Greece, p. 195.
9. K. N. Edwards, in C. D. Craver and C. E. Carraher, eds., *Applied Polymer Science 21st Century*, Elsevier, New York, 2000, p. 440.
10. S. A. Yuhas, in J. V. Koleske, ed., *Paint and Coatings Testing Manual*, 14th ed., ASTM, Philadelphia, PA, 1995, p. 125.
11. F. H. Walker, *Introduction to Polymers and Resins*, 2nd ed., FSCT Series on Coatings Technology, Federation of Societies for Coatings Technology, Philadelphia, PA, 1999.
12. L. W. Hill and B. M. Richards, *J. Coatings Technol.* **51**(659), 59 (1979).
13. L. W. Hill and Z. W. Wicks, Jr., *Pro. Org. Coatings* **8**, 161 (1980).
14. W. J. Blank, *The Slow and Winding Road to "Zero" VOC*, in R. F. Storey and S. F. Thames, eds., Proceedings of the Twenty-Eighth International Waterborne, High-Solids, and Powder Coatings Symposium, New Orleans, LA, Feb. 21–23, 2001, p. 1.
15. J. J. Brezinski, in J. V. Koleske, ed., *Paint and Coatings Testing Manual*. 14th ed., ASTM. Philadelphia, PA 1995, p. 3.
16. D. R. Jones, Polymers and Coatings Program, Calif. Polytechnic State University, personal communication.
17. C. Cooper, P. Galick, S. Harris, D. Porreau, and C. Rodriguez, *J. Coatings Technol.* **73**(922), 21 (2001).
18. EPA website, *http://www.epa.gov/oar/oaqps/pzpmbro/current.htm.*
19. W. Peterson, *Paint & Coatings Industry*, **XVII**(5), 32 (2000).
20. R. A. Ryntz, in R. F. Storey and S. F. Thames, eds., Proceedings of the Twenty-Fifth International Waterborne, High-Solids, and Powder Coatings Symposium, New Orleans, LA, Feb. 18–20, 1998, p. 2.
21. Chemical Manufacturers Association, Ketones and Oxo Process Panel, *A Guide for Solvent Users*, CMA, Washington, DC, 1994.
22. G. R. Pilcher, *J. Coatings Technol.* **73**(921), 135 (2001).
23. J. Stoffer, T. O'Keefe, M. O'keefe, W. Fahrenholtz., T. Schuman, P. Yu, E. Morris, S. Hayes, A. Williams, A. Shahin, and B. Rivera, in R. F. Storey and S. F. Thames, eds., Proceedings of the Twenty-Ninth International Waterborne, High-Solids, and Powder Coatings Symposium, New Orleans, LA, Feb. 6–8, 2002, p. 85.
24. D. Brezinski, editorial, *Paint & Coatings Industry*, **XVII**(9), 12 (2001).
25. R. F. Brady, Jr., *J. Coatings Technol.* **72**(900), 44 (2000).
26. J. Adams, in FSCT Symposium, Louisville, KY, May 1990, (from Ref. 4, p. 420).
27. H. W. Yang, V. Swarup, R. Subramanian, O. W. Smith, and S. F. Thames, in R. F. Storey and S. F. Thames, eds., Proceedings of the Twenty-Seventh International

Waterborne, High-Solids, and Powder Coatings Symposium, New Orleans, LA, March 1–3, 2000, p. 308.

28. A. J. P. Buckman, T. Nabuurs, and G. C. Overbeek, in R. F. Storey and S. F. Thames, eds., Proceedings of the Twenty-Ninth International Waterborne, High-Solids, and Powder Coatings Symposium, New Orleans, LA, Feb. 6–8, 2002, p. 25.

29. J. Grade, T. Blease, E. Armendia, M. J. Barandiaran, and J. M. Asua, in R. F. Storey and S. F. Thames, eds., Proceedings of the Twenty-Ninth International Waterborne, High-Solids, and Powder Coatings Symposium, New Orleans, LA, Feb. 6–8, 2002, p. 145.

30. M. A. Winnnik, *J. Coatings Technol.* **74**(925), 48 (2002).

31. J. W. Taylor and T. D. Klotz, in R. F. Storey and S. F. Thames, eds., Proceeding of the Twenty-Ninth International Waterborne, High-Solids, and Power Coatings Symposium, New Orleans, LA, Feb. 6–8, 2002, p. 181.

32. P. Zhu, O. W. Smith, S. F. Thames, C. Wessels, and R. A. Prior, in R. F. Storey and S. F. Thames, eds., Proceedings of the Twenty-Ninth International Waterborne, High-Solids, and Powder Coatings Symposium, New Orleans, LA, Feb. 6–8, 2002, p. 139.

33. R. Satguru, J. McMahon, J. C. Padget, and R. G. Coogan, *J. Coatings Technol.* **66**(830), 47 (1994).

34. R. Tinsley, T. Nguyen, D. Bolanowska, and V. Adamczyk, in R. F. Storey and S. F. Thames, eds., Proceedings of the Twenty-Seventh International Waterborne, High-Solids, and Powder Coatings Symposium, New Orleans, LA, March 1–3, 2000, p. 201.

35. J. E. Blasko, G. C. Calohoun, R. J. Esser, R. E. Karabetsos, D. T. Krawczak, C. S. Giddings, and D. L. Trumbo, *J. Coatings Technol.* **74**(925), 83 (2002).

36. S. Papasavva, S. Kia, J. Claya, and R. Gunther, *J. Coatings Technol.* **74**(925), 65 (2002).

37. S. Papasavva, S. Kia, J. Claya, and R. Gunther, *Prog. Org. Coatings* **43**, 193 (2001).

38. J. O. Santer, *Prog. Org. Coatings* **12**, 309 (1994).

39. L. W. Hill and Shi-Bin Lee, *J. Coating Technol.* **71**(897), 127 (1999).

40. F. N. Jones, *J. Coatings Technol.* **68**(852), 25 (1996).

41. F. N. Jones, Coatings Research Institute, Eastern Michigan University, personal communication.

42. M. Reisch, *C&E News*, **79**(45), 33 (2001).

43. G. E. Boothand and S. F. Thames, in R. F. Storey and S. F. Thames, eds., Proceedings of the Twenty-Ninth International Waterborne, High-Solids, and Powder Coatings Symposium, New Orleans, LA, Feb. 6–8, 2002, p. 409.

44. *Paint Coatings Journal*, **XVII**(9), 54 (2000) (News Feature).

45. J. V. Koleske, Proceedings Radcure Europe, May 4–7, 1987, Munich, Germany.

46. J. Hess, *Coatings World*, **6**(4), 38 (2002).

CHAPTER 7

WASTES FROM TEXTILE PROCESSING

BRENT SMITH

Cone Mills Professor of Textile Chemistry, NCSU College of Textiles

7.1. INTRODUCTION

This chapter describes wastes from textile processing operations and their sources, based on commercial practice, literature, and experiences related to the United States and Europe. Certainly there are different practices in less developed emerging economies.

In order to convert fibers into consumer products, several processes are required in a production sequence, as indicated in Figure 7.1. This chapter reviews environmental impacts of each of the above textile processing steps after fiber formation. In addition, information is presented on processing auxiliary chemicals, as well as some general ideas concerning process improvements to reduce wastes and environmental impact of processing.

There are fundamentally two approaches to environmental protection: waste treatment and waste prevention. Treatment systems are well known and are beyond the scope of this discussion. Some comments on prevention will be presented. These pollution prevention ideas are similar to the more extensive treatment given in the U.S. Environmental Protection Agency (EPA) *Manual for Best Management Practices for Pollution Prevention in Textile Processing* [1].

One of the most difficult aspects of assessing the overall environmental impact of textile processing is the textile industry's extremely fragmented nature. It is easier to assess environmental effects of individual unit processes than to evaluate entire life cycles of pollution associated with products, [2, 3]. Each step—beginning with design, purchasing, and training, and continuing all the way through the unit manufacturing processes, and finally ending with distribution, merchandising, and cut and sew—has upstream and downstream effects.

Plastics and the Environment, Edited by Anthony L. Andrady.
ISBN 0-471-09520-6 © 2003 John Wiley & Sons, Inc.

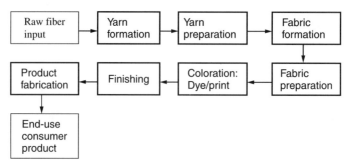

Figure 7.1. Processing steps typically required to convert fibers into end-use textile products.

Examples of global views of waste are given in Figures 7.2 and 7.3 [1]. Participants in the life cycle include dye and chemical suppliers, designers of fabrics, merchandising, maintenance, personnel training, purchasing, unit processing operations (e.g., spinning, weaving/knitting, dyeing/finishing, cut/sew), and regulatory agencies, to name a few.

In some cases, business issues present significant barriers to environmental improvements [4]. Primary disincentives include risk of change, inadequate recognition of benefits due to nonglobal costing procedures, lack of information, varying environmental standards in different locations, and poor communications. To improve environmental performance in the textile supply chain, regulators, suppliers, customers, and manufactures must integrate efforts of vendors, processors, and customers to improve the global life-cycle approach to environmental issues [5].

7.1.1. Contaminants in Fibers

Any additive or contaminant that is part of a fiber is likely to be liberated to the environment during subsequent processing. Therefore, it is worthwhile to examine the nonfiber content of raw textile fibers. These contaminants, even if present in trace concentrations, can contribute significantly due to the massive amount of fibers that are typically used by manufacturers. The environmental aspects of these contaminants are discussed under fabric preparation, where they are typically liberated into the air or wastewater. Contaminants include natural waxes and oils, metals, agricultural residues, added lubricants, tints, unreacted monomer, catalyst residues, colorants, tines, brighteners, delusterants, fiber finishes, and antistatic additives. Ultimately the fibers themselves also become waste when the textile end-use products are discarded.

7.1.2. General Description of Processing Wastes

Textile processing operations produce solid waste, wastewater, and airborne emissions. These are summarized here, and their sources are discussed in the process-by-process review that follows in later sections of this chapter.

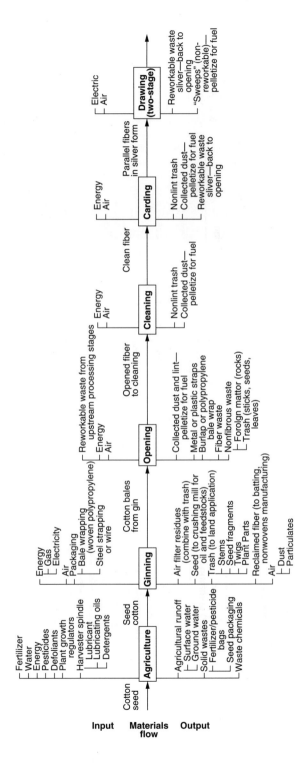

Figure 7.2. Manufacturing waste analysis of mens' knit golf shirt (*continued overleaf*).

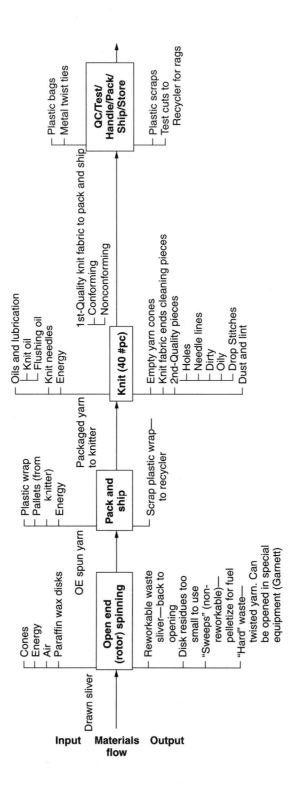

Figure 7.2. (*continued*)

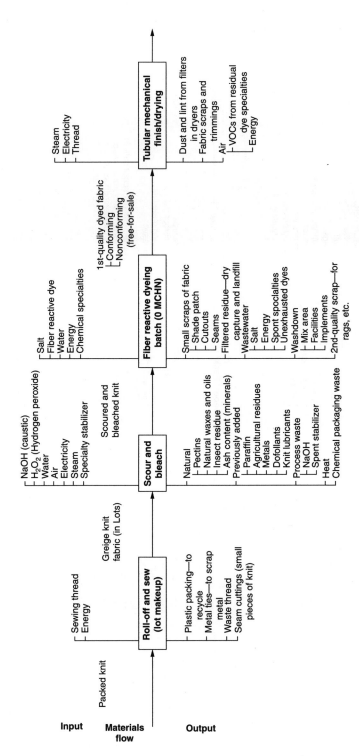

Figure 7.2. (*continued overleaf*)

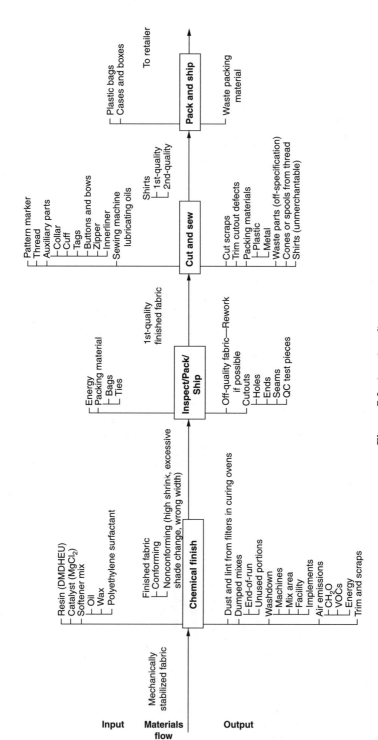

Figure 7.2. (*continued*)

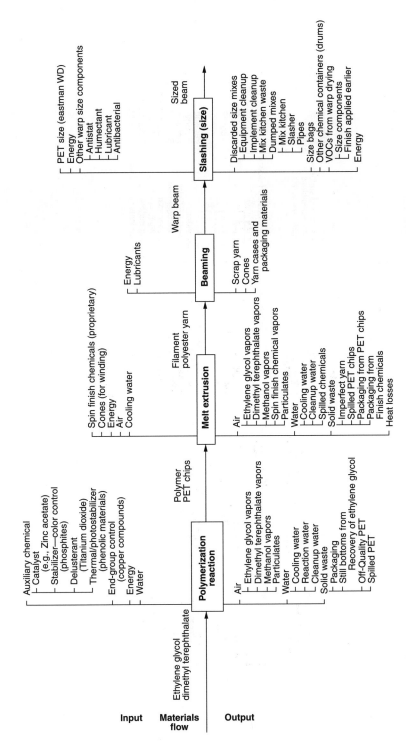

Figure 7.3. Manufacturing waste analysis of women's polyester dress (*continued overleaf*).

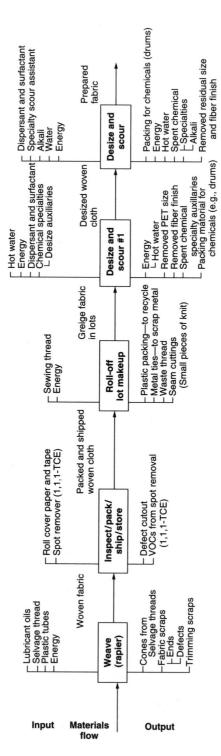

Figure 7.3. (*continued*)

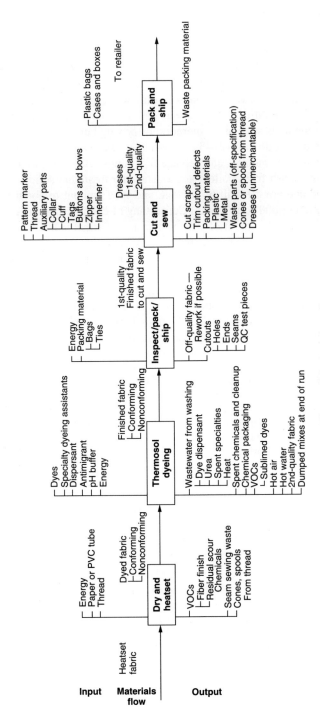

Figure 7.3. (*continued*)

7.1.3. Solid Waste

Solid waste from textile operations comprises mainly drums and other packing materials, ash and sludge, processing wastes (yarn/fiber/fabric), paper cones and tubes, and paper bags. A comprehensive study of solid waste generation in textile operations showed wastes as listed in Table 7.1 [6]. The 290 facilities surveyed produced an average of over 51 tons of solid waste per month each. Of these wastes, 64% went to public landfills, about 23% was recycled, and the rest went to other disposal methods such as incineration or private landfills. Most of the recycled materials are cardboard (2943 tons/month), fiber waste (1881 tons/month), card waste (1646 tons/month), and selvage trimmings (1089 tons/month). Two of the major sources, ash from boilers and wastewater treatment solids, are directly process related and are difficult to prevent, and they amount to about one fourth of the total, or over 6000 tons/month. Conserving of energy can reduce boiler ash. Reduction of chemical content of wastewater and efficient wastewater treatment system operation can result in a decrease of treatment sludge disposal needs.

In another study, a vertically integrated company comprising five plants produced annual waste packaging materials as shown in Table 7.2, which amounted to about 500 tons total [7]. Waste fabric, yarn, and fiber from processing accounts for another major fraction of solid waste. In one multifacility company (spinning, weaving, dyeing, and finishing), annual processing waste amounted to over 1000 tons of fiber, fabric, and yarn. Selvage trimming and seam cut out

Table 7.1 Solid Wastes Produced by Textile Manufacturers [6]

Aluminum cans	Bale wrapping	Boiler ash	Carding waste
Cardboard	Carpet backing	Carpet remnants	Carpet trim
Carpet waste	Compacted trash	Computer paper	Fabric waste
Fiber waste	Garbage	Glass	Hard thread
Hard plastic	Latex foam solids	Metal drums	Office paper
Paper bags	Paperboard drums	Plastic bale wrap	Plastic drums
Plastic film	Plastic containers	Plastic drum liners	Rags
Scrap metal	Scrap wood	Selvage trimmings	Slasher waste
Soft thread	Surface finishing lint	Sweeps	
Wastewater sludges	Wooden pallets	Yarn waste	

Table 7.2 Typical Packaging Wastes [7]

Bale wrap	100 tons
Cardboard boxes	230 tons
Metal drums	3,250 count
Paper bags	41,500 count, or 10 tons
Paper drums	6,850 count
Plastic drums	1,150 count
Wooden pallets	1,550 count

waste in many wet processing operations amounts to as much as 2% of the fabric processed.

7.1.4. Air

Unlike water and solid waste, no comprehensive study has been published on air pollution from textile operations. Textile mills produce atmospheric emissions from all manner of processes, and these have been identified as the second greatest problem for the textile industry [8]. There has been much speculation about air pollutants from textiles but, in general, air emissions data for textile manufacturing operations are not readily available [9–11]. Most published data are mass balance not direct measurements [12, 13]. Direct reading tubes and gas chromatography/mass spectrometry (GC/MS) have been used more recently to get more reliable data [14, 15]. Hopefully, in the future air emissions data will continue to be collected from textile operations, and better definitions of industry norms can be expected. Considerable effort is now underway in that regard [14, 16].

Primarily because of nitrogen and sulfur oxide emissions from boilers, most textile plants are classified as major sources of hazardous and toxic air emissions under EPA regulations. Once classified as a major source, all air emissions (even trace emissions) from unit processes in the facility fall under a high level of regulation. Unit operations of concern for air emissions are printing, coating, finishing, and dyeing. Within these general facility identifications, specific point sources and general area sources generally include boiler stacks, coating operations, ovens, solvent processing units (e.g., dry cleaning), dyeing machines, mix kitchens, drug rooms, and storage tanks.

Fabric coating operations are worth special note in regard to air emissions. Although the number of coating operations is very small, they typically emit a higher level of offensive air pollutants than other types of textile manufacturing. The solvent content of the air in coating ovens can reach levels of several percent by weight. Published data concerning coating solvent losses are not readily available, but it is not unusual to find tens and even hundreds of thousands of pounds of methyl ethyl ketone (MEK), methyl isobutyl ketone (MIK), toluene, xylene, and dimethyl formamide emitted from coating processes in a large facility. These losses come from bulk tank, mix kettles, holding drums, churn mills, mixers, knife coaters, oversprays, and drying and curing equipment. In most cases, capture efficiency is a limiting factor in effectively treating these emissions. Recovery systems are readily available from various suppliers and, as the cost of solvents and supplemental incineration fuels rise, recovery will become more and more attractive economically. A long-term solution is to develop water-based, nonvolatile or less hazardous solvent types.

In addition to coating operations, lower levels of emissions come from widely used processes such as heat setting, thermofixation, and drying and curing ovens in conventional fabric finishing. Typical operating temperatures are 300°F for drying, 360°F for curing of finishes, 400°F for thermofixation of dyes, and 375°F for heat setting. Many materials not normally regarded as volatile will vaporize under these conditions.

Another significant source of air emissions is breathing losses from vented bulk storage tanks. When liquid is drawn out of these tanks, when air cools, or when the ambient atmospheric pressure rises, air enters the tank. When the tank is filled, temperature rises, or the barometric pressure falls, air is expelled. When the material stored in the tank contains volatile components, the expelled air constitutes an emission source.

In addition to point sources, significant air emissions can arise from sources such as solvent-based cleaning activities (cleaning and maintenance), spills, wastewater treatment systems, general production area ventilation, and warehouses (especially formaldehyde-emitting fabric storage). Fugitive emissions and spills can be important, especially for a highly dispersible waste form such as airborne pollution [17].

Aeration of secondary activated sludge biological treatment lagoons strips volatile components of the mixed liquor, and these emit from the waste treatment system as a general area source. Volatile components of spent processing baths (e.g., dye carriers, solvent scouring agents, and machine cleaners) as well as degradation products of these components can reasonably be expected to strip and emit during the treatment process. No data on this have been made public at this time.

Fabric stored in warehouses can produce volatile emissions from process residues, especially printing, dyeing, or finishing chemicals that remain in the fabric. The most important of these is formaldehyde, but others, notably hydrocarbons (e.g., from softeners and wax water repellent finishes), can be present.

Spills can emit volatile air pollutants for months and even years. These are not unusual in off-loading areas and within the berms of bulk storage areas.

Common air pollutant emissions from ovens include mineral oil, knitting oils, fiber finishes, softeners, hydrocarbons, urea from printing or continuous thermofix fiber reactive dyeing, and volatile dyebath additives that are sorbed by substrate and then released during subsequent heat setting, drying, and curing [15, 18, 19].

Fugitive/area emissions include not only formaldehyde and hydrocarbons from warehouses but also processing chemicals from routine handling, wastewater treatment systems, and spills.

7.1.5. Indoor Air Pollution

Indoor air quality can be adversely affected by materials that emit pollutants (primary emitters), as well as by materials that sorb and reemit pollutants [16]. In the United States, the Occupational Safety and Health Administration (USHA) proposed an indoor air quality rule that will affect 21 million workers and that will require the development and implementation of indoor air compliance programs [20]. At this time, primary emissions from several process residues (chemical finishing, dyeing, printing, assembly, and product fabrication) are under study [16, 20]. In addition, sorption/reemission characteristics of textile materials are of interest. Tests of air emissions from various products including draperies have been done. In one study, drapes were found to emit indoor air pollutants

Table 7.3 Emissions Detected from Drapes [21]

From Drapery Material
  Acetone
  2,5-Dimethylfuran
  Benzaldehyde
  1,4-Dioxane
  Benzene
  Ethanol
  Butanol
  Methylene chloride
  p-Xylene
  Dimethyldisulfide
  Decane
  Toluene
  Decenal
  1,1,1-Trichloroethane
  Dichlorobenzene
  Trimethylbenzenes
  1,2-Dichloroethane
  m-Xylene
  Chloroform
  Tetrachloroethene
    plus 100 more VOCs

From the Drapery Linings
  Acetone
  Chloroform
  Decenal
  Ethanol
  Ethyl methyl benzenes
  3-Hexanone
  2-Methylfuran
  Pyrollidine
  Trichloroethene
  Benzene
  Decane
  1,4-Dioxane
  Ethyl Acetate
  Hexanes
  Methylene chloride
  Pyridinone
  Toluene
  m-Xylene
    plus 80 additional VOCs

Table 7.4 Potential Indoor Air Pollutants Found in Textile Finishes [16]

Acetic acid	Acetone
Ethyl acetate	Ethylene glycol
Glyceride-based softener	Formaldehyde
Hydrocarbon wax emulsions	Methanol
Methylhydrogen polysiloxane	Nonylphenol PEG ether
2-Pentanone, 4-methyl-	Polyoxyethylated tridecyl alcohol
Residual monomers (acrylic)	Tetrachloroethylene
Toluene	Triethanolamine

as shown in Table 7.3 [21]. These can result either from process residues or from materials sorbed during fabrication, storage, installation, or consumer use. A survey of finish chemical components listed on material safety data sheets (MSDS) for typical commercial finishes include a variety of somewhat volatile materials that might be emitted, as shown in Table 7.4 [16]. These, plus volatile components of synthetic polymers, fiber finishes, and dyeing and printing process chemical residues may survive the process and later be emitted.

7.1.6. Toxic Air Emissions

The EPA and the states have regulated hazardous air pollutants (HAP) and toxic air pollutants (TAP), as shown in Table 7.5 [1].

Several researchers have speculated about air pollutants potentially of concern in textiles, including volatile organic compounds (VOCs), photochemically reactive materials, TAPs, and HAPs [10, 16, 22, 23]. Most of these have subsequently been detected in air emission testing of textile operations, as listed in Table 7.6 [22, 23]. The list in Table 7.6 gives textile air pollutants that actually have been detected [1, 14]. The amounts of the listed materials are very small in most cases, but because of high-volume nitrogen and sulfur oxide emissions from boilers, many textile plants are classified as major sources of TAPs and HAPs.

Table 7.5 Initial List of 189 Hazardous Air Pollutants Identified in the Clean Air Act Amendments of 1990 [28]

No.	CAS^a No.	Pollutant
1	75070	Acetaldehyde
2	60355	Acetamide
3	75058	Acetonitrile
4	98862	Acetophenone
5	53963	2-acetylaminofluorene
6	107028	Acrolein
7	79061	Acrylamide
8	79107	Acrylic acid
9	107131	Acrylonitrile
10	107051	Allyl chloride
11	92671	4-aminodiphenyl
12	62533	Aniline
13	90040	o-Anisidine
14	1332214	Asbestos
15	71432	Benzene (including benzene from gasoline)
16	92875	Benzidine
17	98077	Benzotrichloride
18	100447	Benzyl chloride
19	92524	Biphenyl
20	117817	Bis(2-ethylhexyl)phthalate (DEHP)
21	542881	Bis(chloromethyl)ether
22	75252	Bromoform
23	106990	1,3-Butadiene
24	156627	Calcium cyanamide
25	105602	Caprolactam
26	133062	Captan
27	63252	Carbaryl
28	75150	Carbon disulfide
29	56235	Carbon tetrachloride

Table 7.5 (*continued*)

No.	CAS[a] No.	Pollutant
30	463581	Carbonyl sulfide
31	120809	Catechol
32	133904	Chloramben
33	57749	Chlordane
34	7782505	Chlorine
35	79118	Chloroacetic acid
36	532274	2-chloroacetophenone
37	108907	Chlorobenzene
38	510156	Chlorobenzilate
39	67663	Chloroform
40	107302	Chloromethyl methyl ether
41	126998	Chloroprene
42	1319773	Cresols/Cresylic acid (isomers and mixture)
43	95487	o-Cresol
44	108394	m-Cresol
45	106445	p-Cresol
46	98828	Cumene
47	94757	2,4-D, salts and esters
48	3547044	DDE
49	334883	Diazomethane
50	132649	Dibenzofurans
51	96128	1,2-Dibromo-3-chloropropane
52	84742	Dibutylphthalate
53	106467	1,4-Dichlorobenze(p)
54	91941	3,3-Dichlorobenzidene
55	111444	Dichloroethyl ether (Bis(2-chloroethyl)ether)
56	542756	1,3-Dichloropropene
57	62737	Dichlorvos
58	111422	Diethanolamine
59	121697	N,N-Diethyl aniline (N,N-Dimethylaniline)
60	64675	Diethyl sulfate
61	119904	3,3-Dimethoxybenzidine
62	60117	Dimethyl aminoazobenzene
63	119937	3,3'-Dimethyl benzidine
64	79447	Dimethyl carbamoyl chloride
65	68122	Dimethyl formamide
66	57147	1,1-Dimethyl hydrazine
67	131113	Dimethyl phthalate
68	77781	Dimethyl sulfate
69	534521	4,6-Dinitro-o-cresol, and salts
70	51285	2,4-Dinitrophenol
71	121142	2,4-Dinitrotoluene
72	123911	1,4-Dioxane (1,4-Diethyleneoxide)
73	122667	1,2-Diphenylhydrazine
74	106898	Epichlorohydrin (1-Chloro-2,3-epoxypropane)
75	106887	1,2-Epoxybutane
76	140885	Ethyl acrylate
77	100414	Ethyl benzene

(*continued overleaf*)

Table 7.5 (*continued*)

No.	CAS[a] No.	Pollutant
78	51796	Ethyl carbamate (urethane)
79	75003	Ethyl chloride (chloroethane)
80	106934	Ethylene dibromide (dibromoethane)
81	107062	Ethylene dichloride (1,2-Dichloroethane)
82	107211	Ethylene glycol
83	151564	Ethylene imine (aziridine)
84	75218	Ethylene oxide
85	96457	Ethylene thiourea
86	75343	Ethylidene chloride (1,1-Dichloroethane)
87	50000	Formaldehyde
88	76448	Heptachlor
89	118741	Hexachlorobenzene
90	87683	Hexachlorobutadiene
91	77474	Hexachlorocyclo-pentadiene
92	67721	Hexachloroethane
93	822060	Hexamethylene-1,6-diisocyanate
94	680319	Hexamethylphosphoramide
95	100543	Hexane
96	302012	Hydrazine
97	7647010	Hydrochloric acid
98	7664393	Hydrogen fluoride (hydrofluoric acid)
99	123319	Hydroquinone
100	78591	Isophorone
101	58899	Lindane (all isomers)
102	108316	Maleic anhydride
103	67561	Methanol
104	72435	Methoxychlor
105	74839	Methyl bromide (bromomethane)
106	74873	Methyl chloride (chloromethane)
107	71556	Methyl chloroform (1,1,1-Trichloroethane)
108	78933	Methyl ethyl ketone (2-butanone)
109	60344	Methyl hydrazine
110	74884	Methyl iodide (iodomethane)
111	108101	Methyl isobutyl ketone (hexone)
112	624839	Methyl isocyanate
113	80626	Methyl methacrylate
114	1634044	Methyl tert butyl ether
115	101144	4,4-Methylene bis (2-chloroaniline)
116	75092	Methyl chloride (dichloromethane)
117	101688	Methylene diphenyl diisocyanate (MDI)
118	107779	4,4'-Methylenedianiline
119	91203	Naphthalene
120	98953	Nitrobenzene
121	92933	4-nitrobiphenyl
122	100027	4-nitrophenol
123	79469	2-nitropropane
124	684935	N-Nitroso-N-methylurea
125	62759	N-Nitrosodimethylamine
126	59892	N-Nitrosomorpholine

Table 7.5 (*continued*)

No.	CAS*a* No.	Pollutant
127	56382	Parathion
128	82688	Pentachloronitrobenzene (quintobenzene)
129	87865	Pentachlorophenol
130	108952	Phenol
131	106503	*p*-Phenylenediamine
132	75445	Phosgene
133	7803512	Phosphine
134	7723140	Phosphorus
135	85449	Phthalic anhydride
136	1336363	Polychlorinated biphenyls (aroclors)
137	1120714	1,3-Propane sultone
138	57578	beta-Propiolactone
139	123386	Propionaldehyde
140	114261	Propoxur (baygon)
141	78875	Propylene dichloride (1,2-Dichloropropane)
142	75569	Propylene oxide
143	75558	1,2-Propylenimine (2-methyl aziridine)
144	91225	Quinoline
145	106514	Quinone
146	100425	Styrene
147	96093	Styrene oxide
148	1746016	2,3,7,8-Tetrachloro-dibenzo-p-dioxin
149	79345	1,1,2,2-Tetrachloroethane
150	127184	Tetrachloroethylene (perchoroethylene)
151	7550450	Titanium tetrachloride
152	108883	Toluene
153	95807	2,4-Toluene diamine
154	584849	2,4-Toluene diisocyanate
155	95534	*o*-Toluidine
156	8001352	Toxaphene (chlorinated camphene)
157	120821	1,2,4-Trichlorobenzene
158	79005	1,1,2-Trichloroethane
159	79016	Trichloroethylene
160	95954	2,4,5-Trichlorophenol
161	88062	2,4,6-Trichlorophenol
162	121448	Triethylamine
163	1582098	Trifluralin
164	540841	2,2,4-Trimethylpentane
165	108054	Vinyl acetate
166	593602	Vinyl bromide
167	75014	Vinyl chloride
168	75354	Vinylidene chloride (1,1-Dichloroethylene)
169	1330207	Xylenes (isomers and mixture)
170	95476	*o*-Xylenes
171	108383	*m*-Xylenes
172	106423	*p*-Xylenes
173		Antimony compounds
174		Arsenic compounds (inorganic including arsine)

(*continued overleaf*)

Table 7.5 (*continued*)

No.	CAS[a] No.	Pollutant
175		Beryllium compounds
176		Cadmium compounds
177		Chromium compounds
178		Cobalt compounds
179		Coke oven emissions
180		Cyanide compounds[b]
181		Glycol ethers[c]
182		
183		Manganese compounds
184		Mercury compounds
185		Fine mineral fibers[d]
186		Nickel compounds
187		Polycyclic organic matter[e]
188		Radionuclides (including radon)[f]
189		Selenium compounds

[a] CAS = Chemical Abstract Service.
[b] X′CN where X = H′ or any other group where a formal dissociation may occur. For example, KCN or Ca(CN)$_2$.
[c] Includes mono- and diethers or ethylene glycol, diethylene glycol, and triethylene glycol R-(OCH$_2$CH$_2$)n-OR′ where
  n = 1, 2, or 3
  R = alkyl or aryl groups
  R′ = R, H, or groups that, when removed, yield glycol ethers with the structure R-(OCH$_2$CH)n-OR. Polymers are excluded from the glycol category.
[d] Includes mineral fiber emissions from facilities manufacturing or processing glass, rock, or slag fibers (or other mineral-derived fibers) of average diameter 1 micrometer or less.
[e] Includes organic compounds with more than one benzene ring that have a boiling point greater than or equal to 100°C.
[f] A type of atom that spontaneously undergoes radioactive decay.

Table 7.6 Air Pollutants Detected from Textile Manufacturing Operations [1, 14]

Acetylaldehyde	Acetic acid	Acrylic monomers
Acrylonitrile	Ammonia	Antimony compounds
Arsenic compounds	Biphenyl	Benzo pyrene
Beryllium compounds	Butadiene	Cadmium compounds
Caprolactam	Carbon monoxide	Chlorine
Chromium compounds	Cobalt compounds	Copper compounds
Dibutyl phthalate	Diethanol amine	Diethyl sulfate
Difluoro methane	Dioxane	Epoxy butane
Ethyl acetate	Ethyl benzene	Ethylene oxide
Ethylene glycol	Formaldehyde	Glycol ethers
Hexane	Hydrocarbons	Hydrogen chloride
Hydroquinone	Lead	Manganese and compounds
Mercury and compounds	Methanol	Methyl ethyl ketone
Methyl isobutyl ketone	Methyl methacrylate	Methylene chloride
Nickel compounds	Nitrogen oxides	Perchloroethylene
Selenium compounds	Sodium chromate	Styrene
Sulfur oxides	Sulfuric acid	Toluene
Trichlorofluoromethane	Trichlorotrifluoroethane	Trichloro ethane
Vinyl acetate	Xylene	Zinc compounds

Disperse dye carriers are occasionally used in polyester processing in older atmospheric dyeing equipment, and they are also used for repair work on defective dyeings. Components of these are given in Table 7.7 [18, 19]. These are typically released from fabrics during subsequent drying and heat-setting operations.

Another important use of solvents is in screen-printing operations. These are used for machine, screen, and squeegee cleaning. In one case study, GC/MS showed the presence of the VOCs shown in Table 7.8 [24].

A great deal of additional in-depth information is presented in the EPA industry preliminary EPA categorization for printing, coating and dyeing [25].

7.1.7. Wastewater

Textile manufacturing is one of the largest industrial producers of wastewater, and many studies have been published on wastewater from textile operations. One excellent and definitive source of accurate and current information is available [26]. Conventional pollutants in textile wastewater are characterized in Tables 7.9–7.11. Table 7.9 characterizes water use amounts based on historical self-monitoring data submitted by mills. Table 7.10 summarizes raw waste concentrations in which the pollution quantity is expressed based on the amount of wastewater discharged (e.g., ppm). Table 7.11 summarizes similar information based on the amount of production (e.g., pounds of pollutant per thousand

Table 7.7 Chemical Constituents Typically Found in Dye Carriers [18, 19]

Biphenyl	Butyl benzoate
4-Chlorotoluene	1,2-Dichloro benzene
1,4-Dimethyl naphthalene	Ethyl benzene
Isopropanol	Methyl benzoate
2-Methyl 3-hydroxymethyl benzoate	Methyl cresotinate (methyl p-toluate)
2-Methyl naphthalene	Methyl salicylate
Naphthalene	Ortho phenyl phenol
Perchloro ethylene	Phenyl ether
Phthalate esters	1,2,4-Trichlorobenzene
Xylene (all isomers)	

Table 7.8 VOC Emissions from Solvents Used in Screen Printing [24]

Trimethyl cyclohexane	Propyl cyclohexane
Hexyl cyclohexane	Ethyl cyclohexane
Methyl ethyl cyclohexane	Butyl cyclohexane
Decahydro naphthalene	Butyl ethyl cyclopentane
Methyl decane (3 isomers)	Xylene (3 isomers)
Ethyl benzene	Trimethyl benzene (3 isomers)
Methyl ethyl benzene	and other hydrocarbons

Table 7.9 Water Consumption and Wastewater Discharge Volumes by Subcategory

Subcategory	Water Usage (L/kg)			Water Usage (gal/lb Production)			Discharge (Median Mill)		No. of Mills
	Min.	Med.	Max.	Min.	Med.	Max.	m³/day	MGD	
1. Wool scouring	4.2	11.7	77.6	0.5	1.4	9.3	103	0.051	12
2. Wool finishing	110.9	283.6	657.2	13.3	34.1	78.9	1,892	0.500	15
3. Low water use processing	0.8	9.2	140.1	0.1	1.1	16.8	231	0.061	13
4. Woven fabric finishing									
a. Simple processing	12.5	78.4	275.2	1.5	9.4	33.1	636	0.168	48
b. Complex processing	10.8	86.7	276.9	1.3	10.4	33.2	1,533	0.405	39
c. Complex processing plus desizing	5.0	113.4	507.9	0.6	13.6	60.9	636	0.168	50
5. Knit fabric finishing									
a. Simple processing	8.3	135.9	392.8	0.9	16.3	47.2	1,514	0.400	71
b. Complex processing	20.0	83.4	377.8	2.4	10.0	45.2	1,998	0.528	35
c. Hosiery processing	5.8	69.2	289.4	0.7	8.3	34.8	178	0.047	57
6. Carpet finishing	8.3	46.7	162.6	1.0	5.6	19.5	1,590	0.420	37
7. Stock and yarn finishing	3.3	100.1	557.1	0.4	12.0	66.9	961	0.254	116
8. Nonwoven finishing	2.5	40.0	82.6	0.3	4.8	9.9	389	0.100	11
9. Felted fabric finishing	33.4	212.7	930.7	4.0	25.5	111.8	564	0.149	11

Table 7.10 Median Raw Waste Concentrations by Subcategory (Based on Effluent Volumes)

Subcategory	BOD (mg/L)	COD (mg/L)	COD/BOD	TSS (mg/L)	O&G[a] (mg/L)	Phenol (µg/L)	Chromium (µg/L)	Sulfide (µg/L)	Color APHA Units
1. Wool scouring	2,270	7,030	3.1	3,310	580	ID	ID	ID	ID
2. Wool finishing	170	590	3.5	60	ID[b]	ID	ID	ID	ID
3. Low water use processing	293	692	2.4	185	ID	ID	ID	ID	ID
4. Woven fabric finishing									
a. Simple processing	270	900	3.3	60	70	50	40	70	800
b. Complex processing	350	1,060	3.0	110	45	55	110	100	ID
c. Complex processing plus desizing	420	1,240	3.0	155	70	145	1,100	ID	ID
5. Knit fabric finishing									
a. Simple processing	210	870	4.1	55	85	110	80	55	400
b. Complex processing	270	790	2.9	60	50	100	80	150	750
c. Hosiery processing	320	1,370	4.5	80	100	60	80	560	450
6. Carpet finishing	440	1,190	2.7	65	20	130	30	180	490
7. Stock and yarn finishing	180	680	3.77	40	20	170	100	200	570
8. Nonwoven finishing	180	2,360	13.1	80	ID	ID	ID	ID	ID
9. Felted fabric finishing	200	550	2.75	120	30	580	ID	ID	ID

[a] O&G = Oil and grease.
[b] ID = Insufficient data to report values.

Table 7.11 Median Raw Waste Loads by Subcategory (Based on Production Volume)

Subcategory	BOD (kg/kkg)	COD (kg/kkg)	TSS (kg/kkg)	O&G[a] (kg/kkg)	Phenol (g/kkg)	Chromium (g/kkg)	Sulfide (g/kkg)
1. Wool scouring	41.8	1,289	43.1	10.3	ID[b]	ID	ID
2. Wool finishing	59.8	204.8	17.2	ID	ID	ID	ID
3. Low water use processing	2.3	14.5	1.6	ID	ID	ID	ID
4. Woven fabric finishing							
a. Simple processing	22.6	92.4	8.0	9.1	8.2	4.3	7.6
b. Complex processing	32.7	110.6	9.6	3.8	7.7	2.6	12.5
c. Complex processing plus desizing	45.1	122.6	14.8	4.1	13.1	20.9	ID
5. Knit fabric finishing							
a. Simple processing	27.7	81.1	6.3	4.0	8.7	7.8	13.0
b. Complex processing	22.1	115.4	6.9	3.5	12.0	4.7	14.0
c. Hosiery processing	26.4	89.4	6.7	6.6	4.2	6.4	23.8
6. Carpet finishing	25.6	82.3	4.7	1.1	11.3	3.4	9.4
7. Stock and yarn finishing	20.7	62.7	4.6	1.6	15.0	12.0	27.8
8. Nonwoven finishing	6.7	38.4	2.2	ID	ID	0.5	ID
9. Felted fabric finishing	70.2	186.0	64.1	11.2	247.4	ID	ID

[a] O&G = Oil and grease.
[b] ID = Insufficient data to report values.

pounds of production). Specific sources of pollutants found in textile wastewater are addressed later in the process-by-process review.

In general, textile wastewater treatment methodologies are mature and well developed. Improving treatment processes dominated the environmental attention of the industry and the regulators in the past. At this time, several new issues are emerging for the textile industry to address [27, 28]. These include hard-to-treat wastes, for example, color in dyeing and printing wastewater, salt in dyeing wastewater, toxic and hazardous air emissions, trace levels of metals in dyeing wastewater, and aquatic toxicity of wastewater.

7.1.8. Hazardous Waste

Most textile operations typically produce no hazardous waste as a routine matter, with the exception of a few unit operations (e.g., coating, dry cleaning, solvent scouring, solvent-based coating operations, and laboratory and maintenance activities). The economic incentive to eliminate hazardous wastes from industrial processes is great, and most textile mills have screened out potential hazardous waste sources. Perhaps the most important general source of hazardous waste in textile processing is spills or process excursions that produce unexpected hazardous waste. The greatest potential spill hazard for hazardous waste in textile operations is bulk chemical off-loading and storage areas.

Some operations have used perchloroethylene for batch dry cleaning as well as 1,1,1-trichloroethylene as a spot removal in the final inspection department. These materials are very efficient in removing oils; however, they have the potential to produce a routine hazardous waste, even if recycled, due to the production of distillation residues.

Coating operations based on latex materials and solvents (e.g., MEK, MIK, acetone, toluene, xylene) are the largest hazardous waste generators in the textile industry. However, there are very few such operations. These fabrics are used in various products including fire hoses, tarpaulins, offset printing blankets, geotextiles, barrier protective clothing, tennis shoe uppers, rain gear, and landfill liners. These are produced in many cases by dissolving or softening natural or synthetic rubber or latex materials in mixed solvents as noted above. These solutions or plasticized materials are then applied (e.g., by spreading or spraying) to fabrics, then dried or cured. These mix residues are hazardous wastes. Water-based emulsions for coating have met with some success but are not applicable in all situations.

Many cleaning, maintenance, and construction chemicals are hazardous and become hazardous waste when discarded or spilled. These include insecticides, boiler chemicals, cooling tower treatment chemicals, weed killers, biocides, machine cleaners, paint strippers, and floor finishes.

The most common way the textile industry avoids producing hazardous wastes is by prescreening chemicals prior to use and avoiding the use of materials that will potentially produce hazardous waste.

7.2. YARN FORMATION

Typically, fibers are formed into either spun or continuous filament yarns. In the case of the latter, the yarn formation process comprises drawing, texturizing, and winding, during which the continuous filament stock is heated, stabilized, and given specific characteristics appropriate to its particular end use. The main pollution resulting from this process is air emissions from the volatile components of fiber finishes previously applied. Since these are very high temperature processes, many compounds not normally regarded as volatile will, in fact, be vaporized.

In the case of spun yarns, there are basically two types of waste to be considered. The first is solid waste, primarily reworkable and nonreworkable soft fiber waste, hard waste, and packaging materials (e.g., bale wrap and straps, cones, yarn cases). The second is spinning additives, which are later removed and become waste from subsequent processes. These include tints, antistatic agents, and lubricants.

Yarn formation processes produce surprisingly little waste, as is often the case in commodity operations because raw material cost and utilization is very closely related to profitable performance. The main types of waste from spinning operations are packing materials (polypropylene bale wrap), bale straps (plastic or metal), process waste (reworkable, nonreworkable, hard waste), and packing materials (cones, yarn cases) [29]. Reworkable waste is clean fiber waste that is directly reused in the opening line through a designated waste hopper. It is standard practice in nearly every spinning mill to have at least one waste hopper in the opening line. Reworkable waste, which originates from nearly every process in the spinning mill, is collected, delivered to waste containers, and then returned to the opening line. Nonreworkable waste is generally produced in various cleaning processes (e.g., opening, cleaning, and carding) and consists of soiled or damaged fiber waste. Hard waste is yarn that is highly twisted and sized; thus, the fibers are very difficult to open and reuse in the normal spinning process.

In one company with six spinning operations, the waste was equivalent to 70 bales of fiber a day [30]. A 1986 U.S. Department of Energy study estimated the total amount of fiber (polyester and cotton) waste in the United States to be 45 million pounds annually from the spinning mills. In addition, another 60 million pounds originated from cutting room waste in cut-and-sew operations, as well as 350 million pounds of discarded consumer goods [31]. The amount of waste depends on input fiber quality and the methods of processing. Over the years, the nonlint content of cotton, for example, has decreased significantly due to improved ginning and agricultural procedures such as improved mechanical harvesting machines. In addition, most cotton gins now have several stages of fine cleaners, which remove more trash, and thus give reduced nonlint content for the fiber bales going to the spinning mill.

The net result is that nonreworkable waste generation from spinning processes has steadily declined from 15% in the 1930s, to 10% in the 1950s, to 3 to 5% in the 1990s [29]. This nonreworkable waste level depends on the grade of cotton and the product being made. Modern lint reclamation equipment can recover part of this waste to give ultimately 1.5% or less waste to discard. This 1.5% or so of trash is generally sold for reuse applications (e.g., filler for padded mailing envelopes) or compressed into high-density cylindrical pellets for boiler fuel.

In addition to the waste discussed above, which is produced directly in the spinning operation, two additional waste issues are important. One is packaging waste (which is typically reused) and the other is chemical additives that must be removed later. Additives such as tints and lubricants are often applied during spinning operation and must be dealt with later during preparation for printing, dyeing, and finishing. Bale wrap, cones, cases, and strapping are generally reused or recycled (e.g., plastic reusable cones and reusable yarn pallets with shrink wrap).

7.3. YARN PREPARATION

After formation, yarns must be prepared for fabric formation. In the case of knitting yarns, this involves clearing the major defects, lubricating the yarns with winding emulsion or mineral oil, and winding the yarns onto cones for the knitting creel. Any lubricants, tints, or other chemical processing assistants added at this stage eventually become wastewater pollution during subsequent knit fabric preparation processes.

Weaving yarns, on the other hand, must be sized using a chemical mixture added to warp yarns in the slashing process prior to weaving. This size material is almost always removed after weaving in a downstream wet process called desizing. Size improves the toughness of the yarn as well as its abrasion and bending behavior. As a result, abrasion, fuzzing, static buildup, shedding of lint, stretch, breaking, creep, and entanglement are suppressed in the weaving process. Modern weaving machines run at very high speeds and therefore place extreme demands on the yarns in the loom. Proper warp sizing is therefore a critical factor in fabric quality [32].

The main component of warp size is usually starch, polyvinyl alcohol (PVOH) or a blend of the two. Other size materials are sometimes used commercially to a lesser extent. These other, less often used types, include derivatized natural sizes and synthetic sizes listed in Table 7.12. In addition to the main size components

Table 7.12 Textile Warp Size Materials

Derivatized Natural Sizes
    Carboxymethyl cellulose (CMC),
    Hydroxyethyl cellulose (HEC)
    Starch ether derivatives (SE)
    Starch ester derivatives

Synthetic Sizes
    Polyvinyl acetate (PVAc)
    Polyacrylic acid (PAA)
    Polyester (WD)

Table 7.13 Additives Found in Warp Size Recipes

Adhesives and binders
   Natural gums (locust bean gum, tragasol)
   Gelatin
   Soya protein
   Casein
   Acrylates
   CMC

Antistatic agents

Antisticking agents, to reduce fouling of equipment
   Waxes
   Mineral oils
   Sulfated tallow
   Pine oil
   Kerosene
   Stoddard solvent

Biocides and mildew inhibitors, to improve shelf life of woven goods
   Orthophenyl phenol (OPP)

Defoamers, to suppress foam in sizing operations
   Zinc and calcium chloride
   Light mineral oil
   Isooctyl alcohol

Deliquescents and humectants, to protect against overdrying
   Zinc and calcium chloride
   Polyalcohols (polyethylene glycol)
   Glycerine
   Propylene glycol and ethylene glycol
   Diethylene glycol
   Urea

Emulsifiers, dispersants, and surfactants, to stabilize size mixes
   Nonionic ethylene oxide compounds

Lubricants and softeners improve bending and frictional characteristics of the yarns
   Fats, waxes, oils
   Tallow and sulfated tallow
   Butyl stearate
   Glycerine
   Mineral oil

Thinning agents, to increase penetration
   Enzymes
   Oxidizers
   Perborates, persulfates, peroxides, and chloramides

Tints, for identification of warps
   Fugitive coloring agents

Weighters
   Clay

listed in Table 7.12, size recipes also contain other additives depending on the specific type of yarn, fabric, loom, and other commercial factors, as shown in Table 7.13. These are typically removed during wet processing; thus all of these materials appear in waste streams from fabric preparation operations. In fact, the main source of biological oxygen demand (BOD), chemical oxygen demand (COD), and total organic carbon (TOC) in textile wet processing effluent is from size materials added to the yarns during slashing. Undesirable materials (e.g., zinc salts, OPP) are generally not used in the United States and Europe but regularly appear in greige fabrics (unfinished fabric just off the loom or knitting machine) imported from emerging economies. Many of the above (e.g., waxes, oils, tallow, ortho phenyl phenol, isooctyl alcohol, glycerine, and urea) are volatile at the high temperatures used in certain subsequent processes; therefore, their can potentially contribute to air pollution.

Warp size itself is a high-volume waste stream that deserves attention because it essentially all becomes a waste from the desizing operation after weaving. About 200 million pounds of warp size are used each year in the United States alone. Of this, about two thirds is starch and about one third is other types, mostly PVOH. PVOH can be recovered, but for economic and technical reasons, only about one third of all recoverable sizes are actually recovered. Thus over 90% of warp size, or 180 million pounds a year in the United States, is preplanned waste [1].

The main pollutants from the slashing operation itself are packaging materials, unused portions of size mixes, machine cleaning agents, and fiber lint and yarn waste. Typically 6% or more of the weight of the goods is added as size [33]. Efforts in pollution control from warp sizing operations include waste treatment, work practice audits, proper use of implements, spill avoidance and control, repairing leaks, proper clean up, and correct disposal of chemical residues.

7.3.1. Chemical and Process Alternatives

There are several warp size alternatives as noted above (i.e., starch, PVOH, CMC, PVAc, PAA, and polyester). These materials are quite different in BOD, TOC, and COD, as well as degradability. Also PVOH and CMC are recoverable, whereas starch is not. In general, it is not feasible to recover mixed sizes. Typical BOD values and add-on levels are given in the Table 7.14 [33].

7.4. FABRIC FORMATION

Textile fibers are typically converted into fabrics by knitting or weaving processes, although other methods are sometimes used (e.g., hydroentanglement, needle punching, and braiding). One additional major specialized textile area is carpets, which may be formed in several ways such as weaving and tufting.

In general, little waste is produced in weaving and knitting, compared to other processing steps (e.g., spinning or finishing). Most is solid waste from trimmings or from packaging materials. Carpet manufacture, on the other hand, produces

Table 7.14 BOD and Typical Add-on Levels of Warp Size Materials [33]

Size Material	BOD (ppm)	Typical Add On (%)
PVAc	10,000	
PVOH	10,000–16,000	5.2
PVOH/CMC (3:1)	17,500	5.2
CMC	30,000	5.2
HEC	30,000	
Starch/PAA (5:4)	295,000	7.8
Alginate	360,000	
Starch ether	360,000	
Starch/PVOH (5:1)	405,000	7.8
Starch	470,000–650,000	11.7

Additive	BOD (ppm)	Aquatic Toxicity LC_{50} (ppm)
Surfactants	10,000–1,000,000	<1–28
Urea	90,000	>1,000
Glycerin	640,000	>1,000
Oils and waxes	100,000–1,500,000	
Biocide	Test fails	
DEG	60,000	>1,000

considerable waste. Also the manufacture of certain industrial fabrics (e.g., paper making felts and filter materials) can produce large amounts of trimming waste. Most knitting and weaving operations recycle packaging materials (e.g., cardboard cases, cones, and tubes). In addition to the waste directly produced, the fabric formation process is critical to downstream quality and efficiency. Thus it has a strong impact on pollution arising from those subsequent processes.

One particular type of weaving machine, the water jet loom, produces wastewater as well as air pollution from drying operations. Generally the water from water jet weaving and from vacuum extraction prior to drying is filtered and recycled.

Knitting operations are not regarded as significant pollution sources, but, like sizing and weaving, they have the potential to impact downstream processes. Downstream waste is reduced by attention to fabric design, fabric quality, lubricant, or size amount and type.

Fabric quality is important because off quality (e.g., holes, dirt, filling bands, barre) becomes waste at the end of the sequence of manufacturing processes. Also, processing efficiency at the finishing plant demands greige fabrics of high quality. Many types of wet processing equipment (e.g., bleaching ranges, jet dyeing machines) are sensitive to physical defects in fabric. Fabric contaminated with dirt, oils, tints, stains, knitting oils, and flushing oils require more chemical- and energy-intensive scouring and preparation procedures. Most fabric guidance

Table 7.15 Carpet Wastes and Dollar Value for One Typical Manufacturer [34]

Type of Waste	Annual Value
Backing selvage	$96,000
Mitter selvage	$92,000
Backing seams	$71,000
Seam trim	$53,000
Other trim, selvage and samples	$190,000
Miscellaneous	$10,000
Annual total	$512,000

and straightening systems do not function properly if holes or major fabric defects are present in the goods. This contributes to waste and reworks.

Carpet waste amounts to about 2% of the total annual production of 900 million square yards, or 18 million square yards of waste. The dollar value of this waste is about $100 million. One carpet manufacturing operation that produced about 8 million square yards of carpet annually was found to have wastes as shown in Table 7.15 [34].

7.5. TEXTILE WET PROCESSING CHEMICALS

Sources for chemical pollution from textile wet processing includes dyes and pigments, specialty chemical, commodities (including water), and incoming substrates. Typically, the largest pollution source is the incoming substrate. However, there are a considerable number of specialty processing assistants, as well as commodity chemicals that are used to prepare, color, and finish greige fabrics [35]. Specialty chemicals typically are proprietary mixtures of commodity chemicals and surfactants designed to accomplish specific purposes in processing.

7.5.1. Chemical Specialties

Chemical specialties are used to control processes or in some cases compensate for inadequacies in equipment or substrate design. For example, dyebath lubricants are used to prevent abrasion, creasing, and cracking during dyeing when the cloth is incompatible with the dyeing machine's fabric transport system. Another common use of chemical specialties is to alleviate problems or deficiencies caused by other chemicals. For example, defoamers are used to suppress foam resulting from other chemicals, and bath stabilizers are added to prevent bath precipitation when incompatible chemical specialty materials are used together. Specialties and their components are too numerous to describe in detail here, but further information is given on a process-by-process basis. Surfactants are very important components of nearly every chemical processing assistant, therefore the environmental impact of surfactants in textile wet processing can be discussed in some general terms.

Surfactants are widely used in textiles for a multitude of purposes. Most surfactants eventually find their way into the wet processing wastewater at some point. They are an important contributor to aquatic toxicity, BOD, COD, and TOC in textile wastewater. Finally, there are many types to choose from, and selecting the correct one requires a good understanding of several subtle issues with respect to performance as well as environmental protection.

There are four basic classifications of textile surfactants: cationic, anionic, nonionic, and amphoteric [10]. Overall, usage is about 59% anionic (25% natural soaps, 34% synthetic), 33% nonionic, 7% cationic, and 1% amphoteric [36]. Alkylphenol ethoxylate (APE) surfactants are one of the largest groups of nonionic surfactants, accounting for over 400 million pounds per year in the United States [37]. These four categories include major product types shown in Table 7.16 [10, 31, 37, 38].

In many cases, surfactant is applied to the fiber for rewetting, lubricity, antistatic, or other purposes, while in other cases the surfactant affects the processing solution in some way, such as stabilization of an emulsion or dispersion. Surfactant residues are present in fiber, yarn, or fabric from applications in upstream operations. Oils and waxes on natural fibers, fiber finishes on synthetic fibers, winding emulsion on yarn, coning oil, yarn finish, knitting oil, and warp size are examples of surfactant-containing residues found on textile substrates [10]. In addition to uses in which the surfactant is applied to the substrate, many textile processes use water-insoluble textile processing assistants that are applied from aqueous emulsions. Essentially all chemical specialties contain surfactants to improve solubility and dispersability and to suspend water-insoluble materials

Table 7.16 Surfactants in Textile Processing Chemical Specialties [10, 31, 37, 38]

Nonionic	
Tertiary thiol ethoxylate (TTE)	Emulsifier, wetter, scour
Diethanol cocoamide (DEC)	Scouring, dyeing, lubricant, softener
Linear alcohol ethoxylate (LAE)	Emulsifier, wetter, scouring
Alkylphenol ethoxylate (APE)	Emulsifier, wetter, scouring
Anionic	
Dodecyl benzene sulfonic acid (DDBSA)	Scouring, foaming agent
Naphthalene sulfonic acid (NSA)	Dispersant for disperse dyes
Sodium lauryl sulfate (SLS)	Scouring, afterwash
Sulfated ethoxylated alcohol (SEA)	Foam, scour, wet, retarder
Cationic	
Alkyl dimethyl benzyl ammonium chloride (ADBAC)	Antistat
Tallow amine ethoxylate (TAE)	Dyeing, antistat, decoupler
Amphoteric	
Cocoamphocarboxy propionate (CCP)	Scouring wool and silk

in processing baths. Surfactants are used in wet processing to ensure complete wetting and penetration of processing solutions, or in some cases as foaming agents [10]. In addition, surfactants are used for wet processing applications such as dyeing, dispersing, emulsifying, bath lubricants, desizing, detergents, scouring, mercerizing, bleaching, dye retarders and levelers, and finishing [37–40]. A wide variety of types are available, which allows for selection of less polluting alternatives for these purposes [38]. Cationic surfactants are used in textiles and account for as much as 12% of all fabric softener use [41]. They lower the surface tension of water and assist in emulsion, dispersion, and foam stabilization [39]. Of 400 million pounds per year of APE use, 82% is ethoxylated nonyl phenol of which about 80–90 million pounds of APE are treated and discharged to rivers each year [37].

In major textile-producing countries, about 10% of all surfactants are consumed in textile uses. The ultimate fate of 70% of these is in wastewater [36]. Kravitz reported surfactants in raw textile waste at concentrations of about 50–200 ppm [38]. Other studies of raw textile wastewater have found concentrations of specific surfactants of 1.8–2.4 ppm [37]. Concentration in treated effluent depends on the percent degradation that occurs during treatment [38, 40].

Surfactants are a major source of aquatic toxicity in textile wastewater and, in order to avoid this problem, toxic surfactants must be either eliminated from use or degraded via biological treatment. Surfactants vary widely in their characteristics with respect to biodegradation; therefore, surfactant substitutions in textile processes have a great effect on wastewater treatability, as well as residual toxicity after treatment.

The treatability and degradability of surfactants has been a major topic of study in the textile industry over the last few years. Kravitz, Moore, Huber, Achwal, and Naylor have all published data related to this, as have the Chemical Manufacturers Association and the EPA [1, 37–39, 41, 42].

Kravitz studied three types of nonionic ethoxylate surfactant for degradability: linear alcohol ethoxylate (LAE), APE, and tertiary thiol ethoxylate (TTE). LAE was 70% degraded in bench-scale tests and was reportedly harmless to the environment after treatment. In these bench-scale treatment and testing studies, APE passed through the treatment systems as did TTE, being degraded to only 25–30% [43]. Later work by the same author attributed degradability characteristics of surfactants to the hydrophobe part of the molecule. Failure to completely degrade resulted in problems such as foaming in the receiving waters and aquatic toxicity. APE was found to degrade much slower at low temperature (e.g., 8°C) than at room temperature, whereas LAE degradability was much less temperature dependent [43]. In general, laboratory studies have confirmed that APE degrades slowly and that intermediate forms of incompletely degraded APE in some cases reportedly are more toxic than undegraded APE [37].

More recently, an industry group comprising of major surfactant suppliers studied the fate of APE in the real-world environment by measuring APE concentration in raw textile wastewater and in 30 rivers using special chromatographic analytical techniques for low-level analysis in the parts per billion (ppb) range. APE from textile operations was 94–97% degraded in real-world waste treatment systems [37]. That was far greater than reported in bench-scale tests previously reported [38, 39, 43]. In wastewater with APE concentrations of about 2 ppm, the effluent after treatment in real-world systems was immeasurably low on about 70% of samples. Based on worst-case models, in terms of discharge locations and low water flows, almost all in-stream concentrations were below the no-observable-effect concentration.

To a large extent, surfactant choices control the toxicity of raw textile wastewater [39]. Surfactants vary in their aquatic toxicity as well as treatability depending on specific features of the molecular structure. Moore reported that aquatic toxicity correlates with hydrophyllic/lipophyllic balance of the surfactant. He also determined that water hardness is a contributing factor in aquatic toxicity of

Table 7.17 Aquatic Toxicity of Surfactants [42, 65, 66, 67][a]

Class	Type	LC_{50} (ppm)
Nonionic		
	TTE	17.32
		28.4
	DEC	2.4
	LAE	5.4
		0.8
		0.5 to 150
	APE	12.5
		2.9, 1.6
		1.3 to 1000+
Anionic		
	DDBSA	19.9
	NSA	not reported
	SLS	27.8
	SEA	20.2
		0.4 to 400
Cationic		
	ADBAC	11.9
	TAE-15 mol EO	4.1
	TAE-150 mol EO	66.1
Amphoteric		
	CCP	159.1

[a] Variation data are due to differences in the type of test (static vs. flow through), species tested, and other constituents of waste and specific surfactant selection. Values given are LC_{50} in ppm; therefore lower numbers represent higher toxicity.

surfactant solutions [39]. Other work suggests that salt, a persistent common component of textile wastewater, has a synergistic effect in promoting aquatic toxicity of raw textile process wastewater [44]. Toxicity data for a number of surfactants are shown in Table 7.17 [42, 45–48].

7.5.2. Chemical Commodities

In view of the large-scale use of commodity chemicals in textile manufacturing, surprisingly little has been written about controlling the resulting pollution. Commodities widely used in textile wet processing includes acids, alkalis, electrolytes, oxidizers, organic solvents, and reducing agents. Daily commodity use for processing 100,000 pounds of fiber reactive dyed and finished cotton woven fabric would typically be as in shown Table 7.18.

Because of very high consumption of commodities, impurities at the ppm level can be significant. Unlike specialties, commodities are chemicals of known composition and are sold based on market competition for that material. In many cases, there are specialty versions of commodities, such as specialty sequesterants, used in place of ethylene diamine tetra acetic acid (EDTA); specialty buffers, used in place of caustic or soda ash; and specialty bleaches, used in place of hydrogen peroxide. Some of the most common commodities in use in textile operations are shown in Table 7.19.

Table 7.18 Commodity Chemicals Typically Used to Process 1000 lb of Cotton

Raw Material	Quantity (lb)
Salt	80,000
Acid and alkali	15,000
Peroxide	2,000
Silicate	3,000

Table 7.19 Commodity Chemicals Typically Used in Textile Processing

Acids—mineral	Hydrochloric, sulfuric, phosphoric acid
Acids—organic	Formic, acetic, oxalic
Alkali	Caustic, soda ash, trisodium phosphate, sodium bicarbonate
	Ammonia, sodium silicate, trisodium polyphosphate
Buffering salts	Monosodium phosphate
Electrolytes	Common salt, Glauber's salt, Epsom salt
Oxidizers	Peroxide, sodium chlorite, sodium hypochlorite
	Percarbonate, perborate, periodate, air
Reducing agents	Sodium hydrosulfite, bisulfite, thiosulfate

7.6. SUBSTRATE PREPARATION (PRETREATMENT)

As the first step in the wet processing sequence, substrates are typically cleansed of contaminants that inhibit sorption or interfere with reaction of the processing solutions used for dyeing, printing, and finishing. A wide variety of contaminants are present in raw fibers [48]. As the fiber is heated or scoured, these contaminants are liberated into water and air as pollutants. Due to the massive amounts of fibers used, impurities present even in trace quantities can be important pollutants. The total polyester/cotton fiber wastes (from all sources in the United States) is estimated to be 4.5×10^8 lb annually [31].

Many natural fiber impurities present potential pollution problems, including natural waxes and oils (BOD, COD, TOC, FOG (fats, oils and grease)), metals (aquatic toxicity, treatment system inhibition), agricultural residues (aquatic toxicity) and, lubricant residues (BOD, COD, TOC, FOG).

7.6.1. Natural Fibers: Cotton and Wool

Raw cotton contains pesticides, fertilizer, and defoliants as shown in Table 7.20, and metals as shown in Table 7.21 [1]. These are removed during preparation and become part of the wastewater stream.

There are many reports of environmental impacts arising from wool processing. One of the more significant is pesticide residues that are released into the wool processing wastewater during wool processing [49–51]. Wimbush reports that pentachlorophenol was found at a level as high as 100 ppm in consumer products on wool carpets, as shown in Table 7.22 [52].

7.6.2. Synthetic and Regenerated Fibers

As supplied to the primary textile industry, synthetic fibers contain several types of impurities, including finishes (e.g., antistatic, lubricants), synthesis by products (e.g., monomer, oligomer, catalyst), additives (e.g., antistatic, lubricant, humectant), delusterants, and fluorescent brighteners. These can vaporize or wash off during wet processing operations, producing aquatic and air pollution, toxicity, metals contamination of wastewater and treatment sludges, BOD, COD, and TOC. Tests of wastewater effluent from mills that exhibited aquatic toxicity identified many specific fiber contaminants [40, 45]. Fiber extracts also exhibited high BOD and COD, as shown in Table 7.23. Note that the COD:BOD ratio of the acrylic extract was very high, indicating the potential for pass-through during treatment and subsequent discharge to the environment. Specific compounds shown in Table 7.24 have been identified in synthetic fiber extracts. Many of these are undesirable in wastewater or are VOCs. Metals extracted from synthetic fibers are shown in Table 7.25 [45].

7.6.3. Preparation Processes

The process of cotton preparation comprises singeing, desizing, scouring, and bleaching and perhaps mercerizing. Synthetic substrate preparation typically

Table 7.20 Results of Pesticide Residue Sampling in Cotton From Growing Regions Worldwide

Source	Total DDTs (1.0000)	Lindan (0.5000)	HCB (0.1000)	Qunitozen (1.0000)	Dicofol (2.0000)	Methoxychlor (10.0000)	Endosulfan (30.0000)	Tetradifon (1.5000)	Total Limit MST (1.0000)
Argentina	—[a]	0.004	—	—	—	—	—	—	0.0040
Argentina	—	0.002	—	—	—	—	—	—	0.0020
Australia	0.031	0.002	—	—	—	—	—	—	0.0330
Columbia Acala	—	0.002	—	—	—	—	—	0.024	0.0260
Côte d'Ivoire	—	—	—	—	—	—	—	—	0.0000
Greece	0.037	0.005	—	—	—	—	—	—	0.0420
Israel	—	0.006	—	—	—	0.168	—	—	0.1960
Mali	0.014	0.002	—	—	0.022	—	—	—	0.1600
Mexico	—	—	—	—	—	—	—	—	0.0000
Mexico-Mante	—	—	—	—	—	—	—	—	0.0000
Paraguay	0.041	0.002	—	—	—	—	—	—	0.4300
Peru del Cerro	0.009	0.003	—	—	—	—	—	—	0.1200
Peru-Pima	0.075	0.002	—	0.002	—	—	—	—	0.0790
Sudan-Rahad Acala sg	—	0.001	—	—	—	—	—	—	0.0010
Sudan-Shambat	—	0.002	—	—	—	—	0.107	—	0.1090
Syria	—	0.003	—	—	—	—	—	—	0.0030
Tadzhikistan ELS rg	0.046	—	—	—	—	—	—	—	0.0460
Chad	—	0.002	—	—	—	—	—	—	0.0020
Turkey	—	0.004	—	—	—	—	—	—	0.0040
Turkey	0.027	0.004	—	0.006	—	—	—	—	0.0370
Turkey Hatay	—	—	—	—	—	—	0.02	—	0.0230
Turkmenistan	—	0.003	—	—	—	—	—	—	0.0000

(continued overleaf)

Table 7.20 (continued)

Source	Pesticide (TLV for vegetable foodstuffs in mg/kg)								
	Total DDTs (1.0000)	Lindan (0.5000)	HCB (0.1000)	Qunitozen (1.0000)	Dicofol (2.0000)	Methoxychlor (10.0000)	Endosulfan (30.0000)	Tetradifon (1.5000)	Total Limit MST (1.0000)
Turkmenistan ELS rg	0.059	—	—	—	—	—	—	—	0.0590
United States: Arizona	—	0.0008	0.0002	—	—	—	—	—	0.0010
United States: California	—	0.006	—	—	—	—	—	—	0.0006
United States: MOT	—	0.004	—	—	—	—	—	—	0.0040
United States: Pima	—	0.005	—	—	—	—	—	—	0.0050
United States: Pima	—	0.0007	—	—	—	—	—	—	0.0007
United States: Texas	0.008	0.0007	0.0003	—	—	—	—	—	0.0018
Uzbekistan	—	0.004	—	—	—	—	—	—	0.0040
Uzbekistan ELS rg	—	—	—	—	—	—	—	—	0.0000
Zimbabwe Albar	0.019	—	—	—	—	—	—	—	0.0190
Zimbabwe Delmac rg	0.008	0.003	—	0.006	—	—	—	—	0.0170

[a]Not detected.

Table 7.21 BOD, COD, and Metals in Cotton Extracts [1]

Pollutant in Extract	Two Samples Were Analyzed (ppm)	
BOD	514	848
COD	956	1693
Cu	0.048	0.050
Sb	0.003	Not detected
Ti	2.11	<0.5
Mn	0.140	0.104
Mg	1.632	1.741
Cu	2.29	3.16
Vd	0.070	0.032
Al	33.91	29.45
Cr	0.095	0.142
Cs	0.0076	0.0083
Zn	7.38	7.35
Co	0.017	0.0024

Table 7.22 Pentachlorophenol Content of Raw Wool [52]

Origin	Level Detected (ppm)
Australia	0.091
Europe	4.46
New Zealand	0.010
South Africa	3.19
South America	162.5

Table 7.23 BOD and COD for Extracts from Synthetic Fibers [45]

Fiber Extract	COD (ppm)	BOD (ppm)	COD/BOD Ratio
Acrylic	1139	155	7.3
Dacron (R) polyester	271	88	3.1
Nylon	6417	1803	3.6
Trivera (R) polyester	319	92	3.5
Wool	13470	3322	4.1

Table 7.24 Compounds Identified in Synthetic Fiber Extracts [45]

Polyester
   3-Methyl cyclo pentanone
   Alcohols ($C_{12}-C_{18}$)
   Carboxylic acids ($C_{12}-C_{24}$)
   Diethyl ketone
   Dodecanol
   Esters of $C_{14}-C_{24}$ carboxylic acids
   Hexanone
   Hydrocarbons
   Hydrocarbons ($C_{14}-C_{19}$)
   Methyl isobutyl ketone
   Several phthalate esters
   Tetra-hydro-2,5-dimethyl-cis-furan
   Several esters of carboxylic acids ($C_{12}-C_{18}$)

Acrylic
   Hydrocarbons (several $C_{15}-C_{19}$)
   Esters of carboxylic acids ($C_{17}-C_{22}$)
   Alcohols
   Phthalate esters
   N,N-dimethyl acetamide
   Other nitrogenous compounds

Nylon 6
   Diphenyl ether
   Hydrocarbons ($C_{16}-C_{20}$)
   Carboxylic acids ($C_{14}-C_{18}$) and dicarboxylic acids
   Various esters of carboxylic acids ($C_{10}-C_{18}$)
   Alcohols ($C_{20}-C_{22}$)
   Other nitrogenous compounds

Table 7.25 Metals Extracted from Synthetic Fibers [45][a]

Fiber	As	Cr	Cu	Pb	Zn
Polyester	0.06	0.01	0.03	bdl	0.117
Acrylic	bdl	bdl	0.025	bdl	0.087
Nylon 6	bdl	bdl	0.034	0.010	0.148

[a] All are reported in ppm, bdl means below detection limit of ICP analysis.

includes desizing, scouring, and heat setting. During these processes, many of the above-described pollutants are liberated as water or air pollutants. Preparation processes, especially desizing and scouring, typically produce over half of the pollution resulting from textile wet processing, and most of that comes from contaminants on the incoming substrate. Since preparation is in essence a cleansing operation, large amounts of water are used, and in fact preparation is the major contributor of water volume in textile wet processing. Most preparation processes (except for wool) are alkaline, thus high pH or alkalinity is another important water pollutant from preparation processes. Scoured-off impurities also produce BOD, COD, metals, aquatic toxicity, and the like. The surfactants used for scouring also produce BOD, COD, and aquatic toxicity.

Preparation is usually the highest volume single process in a mill; thus economics and pollution control greatly favor continuous rather than batch preparation processes [53]. Preparation is also critical to everything that happens later in the mill; so the quality of preparation influences the overall quality of work. Latent defects introduced during preparation cause all manner of reworks, off-quality; and other problems [54].

To reduce the amount of wastewater generated, several equipment and process options are available including countercurrent washing, low bath ratio batch bleaching, cold-batch bleaching, bleach bath reuse, continuous horizontal washers, continuous knit bleaching ranges, combining bleaching and scouring for cotton, and combining scouring and dyeing for synthetics [1].

7.6.4. Singeing

Singeing is a dry process for woven goods (only) to improve the surface appearance and reduce pilling. It has essentially no pollutants associated with it, other than a very small amount of air emissions from the burner flames.

7.6.5. Desizing

Size materials are applied to the warp yarns in woven goods in order to facilitate weaving. In the desizing operation, insoluble starch is converted by an enzyme into soluble sugars that are very easily attacked by bacteria in the waste treatment systems; therefore they have a very high BOD. Desizing of starch-sized fabrics often accounts for more BOD than all other processes in the finishing mill combined [40].

Alternative sizes include polyvinyl alcohol, carboxymethyl cellulose, polyacrylic acid, and others. These are simply washed off with alkaline solutions. They become part of the wastewater stream and are very resistant to biodegradation. They contribute to the COD, TOC, and total dissolved or suspended solids in the wastewater (see Tables 7.12–7.24).

7.6.6. Scouring

Cotton scouring removes waxes, oils, and the like from cotton greige goods using strong alkali to saponify natural oils and surfactants to emulsify and suspend nonsaponifiable impurities in the scouring bath. Wool scouring removes naturally occurring oils from the wool using surfactants. These oils are often recovered for other uses. Scouring of synthetic fabrics is generally done with weakly alkaline surfactant solutions to remove water-soluble or dispersible materials. The main pollutants from these operations are alkali, surfactants, and removed impurities, as described above.

7.6.7. Bleaching

In bleaching, colored impurities that did not scour out are decolorized by an oxidation process typically involving peroxide at high pH. The pollutants from this process are minimal. Impurities in the goods have been previously removed in desizing and scouring, and the reaction product of the peroxide reaction is water. The main pollutant is a large volume of wastewater. There is a minor amount of pollution associated with the use of bleaching chemicals (e.g., silicate, sequesterant, and surfactant).

7.6.8. Mercerization

Mercerization is a process of treating cotton and cotton/polyester woven goods with 20% caustic solution under tension to enhance dyeability, appearance, and luster. The pollutant is wastewater of very high alkalinity or neutralization salts. This waste stream can be recycled through evaporative caustic reclamation systems, but that is rare in commerce. Another potential reuse strategy for this waste stream is scouring.

7.6.9. Heat Setting

Heat setting is typically used to stabilize fabrics with high polyester content. During heat setting, volatile components of previously applied fiber finishes produce air pollution when vaporized by heat setting [28]. These sometimes are condensed and reclaimed for use as fuel as an appurtenant to heat recovery.

7.6.10. Summary

Pollutants from preparation comprise mainly high volume water use, alkalinity, BOD (COD, TOC), and small amounts of offensive pollutants (e.g., metals, aquatic toxicity, air toxics) generated from fiber contamination. Table 7.26 shows

Table 7.26 BOD from Preparation Processes [40]

Process	BOD per 1000 Production
Singe	0
Desize starch	67
Desize starch mixed size	20
Desize PVA or CMC	0
Scouring	40–50
Bleaching with peroxide	3–4
Bleaching with hypochlorite	8
Mercerizing	15
Heatsetting	0

the BOD from preparation processes [40]. Most of the BOD comes from sizes, knitting oils, and natural impurities that are removed from the greige fabrics. Recovery and reuse of synthetic sizes can reduce BOD and dissolved or suspended solids in effluent.

7.7. COLORATION: DYEING AND PRINTING

Color may be imparted to various forms of textile substrates, including direct incorporation of color into the polymer, or dyeing of stock, yarn, fabric, garment, hosiery, carpet, and the like. Typically the application of dyes comprises several steps, that is, surface sorption, diffusive penetration into fibers, fixation, and washing. This is true of various classes of dyes shown in Table 7.27.

Pigment coloration may be done during fiber formation or later by surface application in which there is little or no penetration into the fiber, and the attachment of color to the substrate is primarily through the use of binders, such as latex.

Each fiber type has affinity for specific classes through ionic bonding, hydrogen bonding, and van der Waals/London forces or by simple solubility effects. Specific dyeing conditions and auxiliary chemicals are required for the application of the various classes of dyes to particular substrates. These are summarized in Table 7.27 [1]. Of course, the main waste of concern from wet processing operations is large amounts of water. It typically requires over 250 pounds of water to produce a pound of dyed and finished textile fabric. The main source of dyeing wastewater is from after-washing operations. In addition to the fibers listed in Table 7.27, there are many others, some of which are not practical to dye, such as asbestos, polypropylene, aramid, and glass. They must be colored in the polymer melt before extrusion or colored later with pigments and binders.

Fixation levels vary between dye classes, substrates, dyeing machines, and procedures. Table 7.28 shows typical dye fixation for some common situations. The unfixed dye is washed out and causes colored wastewater [1].

7.7.1. Product Design

In considering the environmental impact of textile wet processing, it is important to recognize the nature of the textile supply chain. It is not unusual to see fibers handled by 5–10 different companies during the manufacturing process, between the fibrous state and the end-use product. This places severe constraints on the abilities of manufacturers to address environmental issues.

Essentially all textile products are custom-made to order. The type and amount of pollution produced in a dyeing or printing process is determined by many factors, the most important of which is product design. In terms of color and dyeing, most textile end-use product design is based on fashion and aesthetics. Textile color designers are experts in fashion, marketing, and the like, but few have expertise to understand the environmental implications of their design decisions. Designers' end-use requirements (e.g., cost, fastness, shade range, quality) are

284 WASTES FROM TEXTILE PROCESSING

Table 7.27 Dye Classes and Fibers

Dye Class	Fibers Applied to	Chemicals Used	Mechanism	Pollution Concern
Acid	Wool (hair) Nylon Silk	Acetic acid	Affinity by ionic bonds	Color pH
Azoic (Naphthol)	Cellulose	Diazotization	Coupling within the fiber	
Basic	Acrylic Cationic polyester	Acetic acid	Affinity by ionic bonds	Aquatic toxicity pH
Direct	Cellulose	Salt, alkali	Affinity by weak interaction	Color, salt
Disperse	Polyester Acetate, triacetate Other synthetics	Dispersant	Simple solubility	Color Organics Antimony
Fiber reactive	Cellulose Wool (hair)	Salt, alkali	Reaction with fiber	Salt pH
Mordant[a]	Natural fiber	Heavy metals	Binding to sorbed metal	Metals
Pigment	All	Latex	Binding in surface film	Suspended solids Color
Solvent[a]	Synthetic	Organic solvents	Simple solubility	Organics
Sulfur	Cellulose	Redox reagents	Insolubilizing by oxidation	Sulfide Redox chemicals Color
Vat	Cellulose	Redox reagents	Insolubilizing by oxidation	Redox chemicals Color

[a]Rare in commerce.

Table 7.28 Typical Dye Fixation [1]

Class	Typical Fixation (%)	Fibers Typically Applied to
Acid	80–93	Wool, nylon
Azoic	90–95	Cellulose
Basic	97–98	Acrylic
Direct	70–95	Cellulose
Disperse	80–92	Synthetic
Premets	95–98	Wool
Reactive	50–80	Cellulose
Sulfur	60–70	Cellulose
Vat	80–95	Cellulose

being met by the dyer or printer by selecting specific machines, processes, and dye classes, as well as specific dyes within a dye class. All of these decisions are important factors in the type and amount of pollution produced from the manufacture of a textile product. The fragmented nature of the textile supply chain is a major factor in pollution.

7.7.2. Metals in Dyes

The largest chemical class of dyes used in exhaust dyeing processes is azo, followed by anthraquinone, triphenyl methane, and other types (e.g., stilbene and phthalocyanine). Very few (e.g., only 2% commercial direct dyes) have metals as an integral part of the dye chromophore. However, some dyes have low-level metal impurities that are present incidentally, rather than by necessity in terms of functionality and color. Mercury-based compounds are used as catalysts in dye manufacture, which is sometimes present as a trace residue. The EPA in its regulation of metals makes no distinction between metal that is bound up as an integral part of the dye structure and free metal. Metals found as integral parts of dye chromophores comprise mainly cobalt, copper, and chrome. Up to half of dye is not fixed in some applications. In addition quite a bit of dye is released to the environment as discards of pad solutions, unused portions of mixes, cleanup of machines and implements, small-scale spills during routine handling, drum washing, and after-washing.

Dye manufacturers are now considering the environmental impact of the dyes they are producing, in addition to the traditional considerations of economy, higher wet fastness, and high tinctorial value [55]. Most dyes that might potentially cause damage to the environment have been eliminated from commerce. Surprisingly, there is no environmental classification system for dyes, and well over half are of undisclosed structure. Dyestuffs can be synthesized based on safer intermediates.

Many anthraquinone dyes are derived by sulfonation in the presence of mercury catalysts. This may cause problems with pollution during manufacture. Some dye manufacturers use mercury-free manufacturing practices of these dyes [55].

7.7.3. Dyeing Processes — Continuous and Batch

Dyeing processes can be classified in many ways, one of which is continuous versus batch dyeing. A continuous dyeing machine, called a "continuous dye range," typically comprises many chambers or tanks through which the substrate passes in sequence. A different step of the dyeing process is accomplished in each chamber, for example, application of dye solution, application of reactive chemical baths (redox, alkali), steaming (for penetration into, or reaction with, fibers), dry heat to accelerate reaction, washing-off or soaping-off unfixed dye, and chemical after-treatments to improve fixation. The range typically runs continuously at high speed (e.g., over 100 yards/min) until the entire supply of substrate is processed. During the running time, chemical solutions are fed into the various chambers to replenish treatment solutions consumed during that process step. These continuous dyeing machines are usually very large and therefore require large amounts of substrate (e.g., hundreds of yards of fabric) for thread-up. Therefore, continuous dyeing is typically used for large quantities of substrate, due to the fact that there are large startup losses of substrate during the time it takes to reach a consistent steady state.

Batch dyeing, on the other hand, processes fixed amounts of substrate in a machine, usually with a single treatment chamber. A batch (e.g., 100–1000 lb) of substrate is placed in the chamber and various solutions are imported for the different steps of the dyeing process. After each step, the processing solution is discarded. This method lends itself well to small amounts of substrate. However, batch dyeing is slower and produces more waste and batch-to-batch color variation, which is not generally seen in continuously dyed goods.

Compared to batch dyeing operations, continuous dyeing has better consistency, less pollution, and lower cost for long runs (e.g., typically over 10,000 lb of substrate, depending on the range and substrate). But for short runs the reverse is true [56].

In continuous dyeing (and in printing), pollution results primarily from failure to achieve 100% dye fixation; discarding excess chemical solution mixes; machine cleaning waste; startup, stop-off, and color/substrate change losses; and implement cleanup and handling losses.

In batch dyeing, the main sources of pollution are spent processing baths, which are discarded after each step of the process. Reuse of batch processing baths is possible but not widely practiced. Continuous baths are not discarded as often (usually only at the beginning and end of a run), and water reuse, especially countercurrent washing, is commonplace. Fixation efficiencies are generally higher in continuous dyeing than in batch dyeing. The continuous method is becoming ever more popular commercially due to improvements such as reduction of startup losses, easier machine cleaning (e.g., Teflon-coated parts), and better substrate handling to facilitate dyeing of delicate substrates (e.g., knits) [57].

7.7.4. Role of Quality

In addition to the above, another very important factor in the pollution potential from coloration processes is the percent of "right first-time" dyeings. Corrective measures such as reworks (e.g., redye, stripping) and shade adjustments (e.g., top-ups, adds) are very chemically intensive and have a much lower expectation of success than the first-time dyeing. Every situation is different, but typically batch dyeing operations with 5% or less reworks are considered good. Reworks over 10% are in most cases clearly excessive [3].

7.7.5. Batch Dyeing Machines

The following summarizes some of the main environmental implications of various processes, machines, dye classes, and the like. Table 7.29 gives some typical characteristics of batch dyeing machines [58].

Table 7.29 Typical Characteristics of Batch Dyeing Machines [58]

Machine	Bath Ratio	Substrate	Washing
Beam	10:1	Open-width fabric or yarn	Fair
Beck	17:1	Rope fabric	Good
Garment	50:1	Garments	Good
Hosiery	40:1	Hosiery	Good
Jet	8:1 to 12:1	Rope fabric	Very good
Jig	5:1	Open-width fabric	Poor
Laboratory	40:1	Fabric or yarn	Fair
Package	10:1	Yarns wound on packages	Good
Paddle	40:1	Garments, rugs, etc.	Good
Skein	17:1	Yarn skeins	Good

7.7.6. Bath Ratio in Batch Dyeing

The bath ratio (or liquor ratio) is the mass (weight) of dye solution divided by the mass (weight) of substrate. The amount of pollution produced in coloration of the manufacture of a textile depends strongly on this factor. Boilers are the main source of air pollution from textile wet processing operations. The amount of air pollution depends on the amount of steam required to heat the bath. Low-bath-ratio dyeing machines thus consume less energy and produce less air pollution.

Generally, textile coloration requires the use of dyes and chemicals, some of which are based on the amount of the bath and others on the amount of fabric. Amounts of chemicals that act on the bath (e.g., salt, buffers, acid, alkali, lubricants, dispersing agents, and surfactants) generally are based on the weight of the bath. Certain other chemicals (e.g., dyes and exhaustible finishes) act on the cloth and are based on the weight of goods. Lower bath ratio machines have less bath per unit of substrate processed; therefore; they produce less chemical pollution from residual processing chemicals in the baths. Also, fixation efficiencies of dyes are inversely related to bath ratio; therefore, low-bath-ratio batch dyeing produces less color pollution. Low-bath-ratio dyeing machines reduce the amount of energy consumption, air pollution, water pollution, and also reduce the wastewater treatment system size requirements, as there is less water to treat [59–61].

On the other hand, most of the water in textile coloration processes is used for washing, not for the dyebath itself, therefore the correlation between bath ratio and water use is not perfect. A machine with mechanical configuration that facilitates high washing efficiency, due to high bath–substrate contact and high agitation might accomplish washing with less water, bath ratio notwithstanding [40].

Certain modern jet and package dyeing machines called "ultra low liquor ratio," or ULLR [62], have been designed to operate 5:1 and in some cases even as low as 3:1 by utilizing short and compact piping, low-volume pumps, and better space utilization of substrate in the kier [61]. This reduces energy and water consumption, as well as all of the chemicals based on the bath, plus improved dye fixaton. But washing efficiency may suffer, as there is little free water in the system to promote agitation and efficient washing of contaminants from the substrate after dyeing [40, 62].

7.7.7. Chemical Handling Systems

The manner of measuring and dosing the chemicals into the process can greatly influence the amount of pollution that results. Computerized dosing systems meter chemicals directly to a dyeing (or other) bath according to some sort of control strategy, as opposed to the old practice of manually adding chemical into the bath. Research systems now under development sense the dye exhaustion as it occurs and adjust the dosing profile of salt and alkali accordingly. Automated color kitchens reduce working losses from implement cleanup and disorderly work practices, but in addition, the exact correct amount of mix is made, reducing discards and the probability of off-quality dye work. This applies particularly to continuous dyeing operations and printing. Well-designed dosing systems do not make up any excessive amounts of mixes, thus reducing startup and stop-off waste. Shipping, storage, handling, and delivery systems can also contribute to spill potential and associated pollution [63].

Due to the importance of right first-time dyeings, improved controls with programmable capabilities to improve existing control protocols as well as to implement entirely new control strategies are being used to good advantage to reduce pollution.

7.7.8. Pollutants Associated with Dye Classes

The data given in Table 7.30 are typical for each dye class. There are, within each class, subclasses that have variances in application procedures (e.g., different pH, temperature, salt requirements). Also the dyeing method and substrate have influence. Therefore, there is a large variation in these reported values. The following summarizes typical wastewater characteristics observed from various dye classes, as reported in [1]. One particularly noteworthy feature of these data is the extremely large variations that occur. This is typical of textile wastewater, which is highly variable from mill to mill, season to season, and so forth.

There is significant interest in the use of natural dyes to reduce the environmental impact of dyeing processes. These are not in widespread use and suffer from several disadvantages. The shade range that can be produced with natural dyes is very limited. They are useful primarily for natural, not synthetic, fibers. The fastness, especially to light, of many natural dyes is inferior. The cost in many cases is high. The pollutant characteristics of natural dyes are not well known. Many traditional natural dyes required mordants (copper or iron) for application to natural fibers. In many cases, the dye affinity is low. Nevertheless, these represent a significant possibility for improvement and are under study at this time.

Table 7.30 Typical Pollutants for Various Dye Classes [1]

Class	Substrate (typical)	Method (typical)	ADMI Color Units	TOC (mg/L)	BOD (mg/L)	pH	Cl⁻ (mg/L)	TSS (mg/L)	TDS (mg/L)
Acid	Wool, nylon	Batch	370	210	135	4	14	14	1086
			3200	315	240	5	33	9	2028
			4000	400	570	7	Nil	5	1750
Azoic (Naphthol)	Cotton	Continuous	2415	170	300	9	7630	387	10900
Basic	Acrylic	Batch	5600	255	210	5	27	13	1469
Direct	Cotton	Batch	2730	140	15	7	61	26	2669
Disperse	Polyester, synthetic	Batch or continuous	100	130	78	7	27	14	395
			215	240	159	8	28	39	771
			315	300	234		33	101	914
			1245						
Fiber Reactive	Cotton	Batch or continuous	1390	150	102	9	57	9	691
			3890	230		11	9800	32	12500
Sulfur	Cotton	Batch or continuous	450	400	990	4	42	34	2000
Vat	Cotton	Batch or continuous	1910	265	294	12	190	41	3945

7.7.9. Acid Dyes

Acid dyes are used for coloring nylon, wool, and certain other natural polymers. They are fairly low molecular weight (typically 300–500 g/mol) anionic materials, usually the sodium salts of sulfonic acids.

Acid dyes are typically applied in batch processes from acidic baths (pH 3–7), with high bath exhaustion. These dyes are very water soluble, making machine cleaning fairly easy and not requiring much in the way of chemicals. One subclass of acid dyes that is used for navy and black shades on wool (premetallized) contain cobalt and chromium. These dyes show aquatic toxicity due to their high metal content and are therefore more damaging to the environment than nonmetallized acid dyes [64].

Acid dyes are applied in the following manner:

Fill batch dyeing machine.
Add acid, lubricants, and surfactants to dyeing machine.
Add substrate to batch dyeing machine.
Paste-up dye in cool water under acidic conditions.
Dissolve dye paste in hot water.
Add dissolved dye to dyeing machine.
Heat to dyeing temperature (typically 190°F).
Run 30 min, maintain pH if necessary with additional acid
Cool and drain.
Apply fixative (syntan or chrome for certain wool shades).
Wash (multiple times).

The acid converts the —NH_2 end groups of the polymer to —NH_3^+, thus providing a site for strongly binding the anionic dye. The dye exhaustion and fixation is usually high, typically 80–95%.

Dyeing wastewater from acid dyeing operations typically has characteristics as shown in Table 7.30. Aquatic toxicity of such dye waste is typically low. There are important subclasses of acid dyes that contain metals as part of the dye structure, and these can show up in the waste stream due to handling losses, washing off, and incomplete exhaustion onto fibers. In addition, there is a subclass of acid dyes called "chrome" dyes that only achieve their full shade development (usually navy or black shades) and their desired wet and light fastness when chrome III salts are added to the bath. These can cause high chromium levels in wastewater. There are several steps that can be taken to control chrome discharges from such dyeings.

The probability of right first-time dyeing with acid dyes in very high due to the generally good leveling behavior of these dyes, if properly applied.

Aside from the premetallized dyes, the main pollution concern from acid dyeing are BOD (from organic acids) as well as low pH (3–7, depending on dye and application method). In addition, the pH in acid dyeing wastewater can vary rapidly, shocking wastewater treatment systems and reducing efficiency, which may produce waste treatment system upsets or aquatic toxicity [65].

Research is underway to develop premetallized acid dye replacements based on iron instead of other more harmful metals, for example, substituting iron for cobalt in Acid Red 182 and Acid Blue 171, and also iron for chromium in Acid Black 172. These iron-based dyes are nonmutagenic and, of course, do not introduce undesirable metals in dyeing wastewater [1].

The use of these metallized dyes is not the dyer's choice, as they are required in order to match certain shades, as specified by the designer.

Another important group of acid dyes is chrome dyes, a special subclass of dyes for wool that produce a beautiful range of black and navy shades. According to Shaw [49] and to Duffield [66], about 70% of wool dyeing today uses heavy metals, mainly chrome. Conventional chrome dyeing of wool produces about 155 ppm chrome in spent baths, but special methods can be used to bring chrome levels well under 35 ppm [49–51, 66]. These baths, when mixed with other wastewaters from the operation give chrome levels of 15 and 0.7 ppm, respectively [5].

These chrome discharges are toxic in wastewater and also can accumulate in waste treatment sludges, thus rendering the sludges hazardous and complicating the issue of sludge wasting.

7.7.10. Basic Dyes for Acrylic and Certain Modified Polyesters

Basic dyes are cationic materials that have high affinity (typically over 6 kcal/mol) for binding to anionic sites (typically SO_3^- in acrylic and certain other fibers (e.g., copolymer versions of polyester and nylon). They are usually applied by batch dyeing procedures from acidic baths at temperatures of 200–220°F, using a procedure similar to the following:

Fill batch dyeing machine with water.
Add chemical auxiliaries.
Add substrate.
Paste basic dye in cool acetic acid.
Dissolve basic dye paste in hot water.
Add dissolved basic dye to machine.
Adjust dyebath pH to 4.5.
Add 1 g/L salt if dyeing temperature is to be above 210°F.
Heat to 205°F (atmospheric machine) or 220°F (pressure).
Run for exhaust (typical 45–60 min).
Cool to 140°F.
Drop bath, refill machine with water, and wash as needed.

Basic dyebaths have very high exhaustion, essentially 100%. Cationic materials, including basic dyes, typically have high aquatic toxicity, but the high dyebath exhaustion prevents discharge of significant amounts of basic dye from spent batch dyebaths. They can enter the waste stream, however, through spillage, discards, and disorderly work practices. Basic dyes do not contain metals.

7.7.11. Direct Dyes for Cotton, Rayon, and Other Cellulosic

These are anionic materials very similar to acid dyes but with higher molecular weight. They are applied to pure cellulosic substrates from neutral to weakly alkaline baths (pH $\sim$ 8) or to blends (e.g., polyester–cotton) from weakly acidic baths (pH $\sim$ 6) by procedures typically similar to the following:

Fill machine with water.
Add dyeing auxiliary chemicals to batch dyeing machine.
Add prepared substrate.
Paste direct dye in cool water.
Dissolve direct dye paste in hot water.
Add predissolved direct dyes to machine.
Adjust pH to 8 with soda ash.
Heat to dyeing temperature (typically 1854°F).
Add one fourth of the required salt; continue heating.
Add one fourth salt; continue heating to 205°F.
Add one half salt.
Run at 205°F for 15 min.
Cool to dyeing temperature (typically 185°F).
Run 20 min for exhaust.
Cool to 150°F, drop bath, and wash with cool water.
Refill machine with water; add fixative.
Wash as needed.

These dyes require only about 2–20 g/L of salt, typically common salt or Glauber's salt. Dyebath exhaustion is typically 85% or better, if proper temperature and salt concentrations are used.

Direct dyes bind to cellulose at nonspecific dye sites on the surface of naturally occurring crystalline areas within the fiber. The binding mechanism is hydrogen bonding or van der Waals forces. This provides for fairly weak binding (typically 2 kcal/mol). As a result, these dyes wash off during end use unless fixatives are applied. These fixatives may be of several types, typically resinous dicyandiamide crosslinking materials.

Direct dyes typically have low aquatic toxicity and good solubility. A few direct dyes, notably turquoise and green dyes, contain metals (e.g., copper as a phthalocyanine complex).

7.7.12. Disperse Dyes for Polyester, Acetate, and Other Synthetics

Disperse dyes are materials that have very low water solubility (typically 10^{-7} mole fraction). Disperse dyes are supplied as finely divided powders containing dispersing agents (e.g., naphthalene sulfonic acid and lignin sulfonate), which keep the particles suspended in the dyebath. They are applied to polyester and other synthetic materials from neutral to slightly acid baths by procedures typically similar to the following:

- Fill batch dyeing machine with water.
- Add chemical auxiliaries.
- Load substrate into machine.
- Paste disperse dye in cool water with dispersing agent.
- Dilute disperse dye paste in warm (not hot) water.
- Add predispersed dyes to batch dyeing machine.
- Set pH to 6 with acetic acid.
- Heat to 260°F (pressure).
- Run for exhaust (typically 45–60 min).
- Cool to 200°F, release pressure (if pressure machine).
- Cool to 140°F, drop bath, and wash.

Older equipment not capable of heating above 210°F may require the use of dye carriers to accelerate the rate of disperse dyeing. These carriers are emulsions of organic materials, (e.g., biphenyl, ortho phenyl phenol, trichlorobenzene, methyl naphthalene, methyl cresotinate, etc.). Many of the ester-type carrier materials are by-products from fiber manufacturing. These carrier materials are rarely used in modern dyeing operations, having been eliminated due to their water and air pollution potential, safety hazard, and cost.

Reductive after-scouring, typically with alkali (soda ash or caustic) and hydrosulfite, is a common practice on polyester and polyester–cotton blends. This is done to remove fugitive disperse dye from the surface of the cotton component of the blend. This discharge of hydrosulfite can cause high immediate oxygen demand, as well as shock loading of waste treatment systems [55]. New diester disperse dyes require only alkaline (not reductive) after-clearing.

7.7.13. Fiber-Reactive Dyes on Cotton and Other Cellulosic

Of all dye classes, fiber reactives have been the fastest growing and the most popular over the last 30 years. They typically give an exceptionally large shade range with brilliant colors possible in all hues, and the fastness is outstanding. But, as currently constituted and applied, the pollution potential of these is perhaps the greatest of all dye classes. They require huge amounts of salt, and the fixation efficiency is poor, with most fiber-reactive batch dyeings utilizing only one half to two thirds of the available color. The rest is discharged as colored wastewater. Current efforts to improve the environmental performance of fiber-reactive dyes are to reduce the need for salt, to improve washing-off performance, to increase fixation, and to develop machines for continuous application of fiber-reactive dyes for small lots and for delicate substrates (e.g., knits). The low degree of fiber-reactive dye fixation is a major contributor to the color pollution in wastewater [5].

Bifunctional reactives have been introduced to increase fixation, which raises fixation levels from 50 to 60% presently to the improved levels of 70–80%. Increasing reactivity (tri and polyfunctionality) does not increase fixation correspondingly [62]. Further improvements in affinity will result in higher fixation.

When dyeing fiber-reactive dyes on cotton, some unfixed, hydrolyzed reactive dye remains in the fiber to be washed off, requiring quite a lot of water. An alternative is to limit the washing, then fix the small remaining residue of hydrolyzed fiber-reactive dye with fixing agents [58].

Fiber-reactive dyes are water-soluble anionic colorants similar to acid dyes except that they contain a moiety that can react with nucleophylic fiber sites, including specifically the primary $C_{(6)}$—OH of cotton or —NH_2 of wool. The two types of reactive groups that are commercially important today are chloro triazine and vinyl sulfone. Some fiber-reactive dyes contain both types. Also other variants exist, for example, chloro quinoxaline or pyrimidine. These are applied in batch dyeing machines from neutral baths with large amount of salt to cause exhaustion of the dye onto the substrate. Once exhausted, alkali is added to generate the cellulosate anion for the nucleophylic substitution (triazine) or Michael addition (vinyl sulfone) reaction. Prior to the addition of alkali, the binding is through weak hydrogen bonding or van der Waals forces. Once reacted the dye is covalently bound in the fiber. Because of the low physical affinity (typically less that 2 kcal/mol) of these dyes for the substrates, and the competing hydrolysis reaction, fixation efficiency is low, resulting in large amounts of color in the effluent. Fixation of color typically is about 60–90%. The major pollution problems for this class are salt and color. Both of these are especially difficult to treat and tend to persist and pass through wastewater treatment systems. Toxicity of the dyes is typically low, and few contain metals. The degree of difficulty for applying fiber-reactive dyes is high, and in some operations the rework rate can become excessive (over 10%).

There are many variants of fiber-reactive procedures. The following is a typical "conventional" version:

Fill batch dyeing machine with water.
Add dye auxiliary chemicals.
Add substrate.
Paste fiber-reactive dyes in cool water.
Dissolve dye paste in hot water.
Add predissolved fiber-reactive dyes to batch dyeing machine.
Heat, while adding one fourth of salt.
Continue heating, while adding one fourth salt, to dyeing temperature.

Heat to dyeing temperature (120–200°F depending on type).
Add one half salt while holding at dyeing temperature.
Add alkali in portions (1/4, 1/4, 1/2).
Run 45 min for reaction.
Cool to 140°F, drop bath, and wash thoroughly.
Refill machine, add fixative if required, and run 10 min.
Wash as needed by repeating above drop/fill.

Salt requirements are 30–100 g/L depending on shade, which is much greater than that required for direct dyeing. The dyeing temperatures may be anywhere from 120 to 200°F depending on dye reactivity. The alkali type and amount depend on the specific reactive group. Caustic, soda ash, trisodium phosphate, and combinations are used. The pH in the reactive stage of the dyeing is typically well above 11.

Important process alternatives include continuous dyeing or pad-batch dyeing of fiber-reactives, which requires no salt and results in higher fixation efficiencies. But the limitations of the continuous process eliminate its use on knits and small lots [40, 63, 67].

7.7.14. Naphthol (Azoic) Dyes for Cotton and Other Cellulosic

Naphthol dyes are supplied in two parts: a coupling component (substituted beta naphthol) and a diazo component. Both parts are water soluble, but when reacted, they form a larger insoluble colorant molecule. This is entrapped in the fiber, giving the desired color as well as outstanding fastness to washing and bleaching. These dyes are typically applied in a continuous process by treating coupler-impregnated fabric with the diazo component. A typical continuous application process comprises the following steps:

Prepare first pad bath: aqueous solution of diazo salt.
Prepare second pad bath: naphthol coupling component.
Pad prepared fabric through first (diazo salt) bath.
Dry (thermal, steam-heated cans, or infrared).
Pad fabric through second (coupler) bath.
Wash, in several stages.
Dry on steam cans.
Cool and wind up on roll.

Because of the reaction of diazo salt with coupler, these dyes have almost total resistance to washing off. The coupler solution is prepared with alkali, but aside from that, little is required in the way of chemical auxiliaries.

An important environmental and safety concern is the presence of toxic or carcinogenic impurities in the coupler components.

7.7.15. Pigments for All Fibers

Pigments are insoluble coloring matters typically supplied as predispersed pastes. The pigment colorant is added to a recipe of latex binder and alginate or other antimigrant. These are popular for dyeing light shades continuously and also for printing. In the printing recipes, polymeric thickeners are used to ensure proper rheology. A typical application procedure is

Prepare pad bath: dispersion of pigment and binder.
Pad fabric through bath.
Dry on steam cans.
Cool and wind up on roll.

These are not dyed by exhaust methods. Primary environmental concerns are suspended solids and color. The suspended solids result from the foaming and coagulation of the binder, antimigrant, and thickeners. Due to their large hydrodynamic radius and neutral buoyancy, these solids are difficult to treat as they will not settle nor float for efficient skimming.

7.7.16. Sulfur Dyes for Cotton and Other Cellulosic

Sulfur dyes are complex mixtures of insoluble chromophore structures that contain polysulfide linkages. They are typically supplied as dispersions in water or as prereduced solutions in water. During the sulfur dyeing process, these are reduced to —S^-, which render the dye temporarily soluble. These are then reoxidized to the polysulfide state, which is insoluble. The oxidized form of the dye is thus trapped within the fiber structure, giving color that is very fast to washing and bleaching. Sulfur dyes are commercially applied by either batch or continuous methods. The following procedure is typical:

Fill batch dyeing machine with water.
Add fabric.
Add 5 g/L sulfide reducing agent.
Add prepasted dye.
Heat to 190°F, and add 30 g/L salt.
Run 30 min.
Overflow wash until clear.
Fill machine with water.
Add 1 g/L of 35% peroxide.
Heat to 190°F.
Run 10 min.
Cool to 140°F, drop bath, and wash with nonionic surfactant.
Add 0.1 g/L acetic acid (56%).
Heat to 185°F, run 10 min, and cool to 140°F.
Water wash.

Traditionally, sulfur dyes were reduced in the dyehouse by boiling the dye with soda ash and sodium sulfide to render them soluble [68]. Then, in the reduced form, they were applied to cotton; then reoxidized later in the process to produce fast dyeings. A by-product is foul smelling sulfur dioxide, as well as sulfides in the wastewater. In the 1990s, new types of sulfur dyes were introduced that feature lower sulfide content, thus less sulfides in wastewater and less hydrogen sulfide air pollution. The chemical nature of these proprietary reducers is not disclosed, but it appears from the MSDS information to be an organo-sulfur reducing agent [68].

Modak reported that corn sugar waste (reducing sugars) from a starch manufacturing operation is useful for sulfur dye reduction [53]. In one reported case, textile wastewater sulfide concentration was reduced from 30 to 2 ppm, with associated reduction in aquatic toxicity. An increase in BOD resulted but was easily handled by the textile wastewater treatment system, whereas the sulfide waste was not amenable to waste treatment. The zone settling velocity in the secondary clarifiers improved as a result of the decrease in sulfide, thus increasing waste treatment efficiency. Odors were reduced. The corn starch manufacturer saved $12,000 on waste treatment system expansion and $2400 in operating expenses, had the waste stream not been reused [53].

7.7.17. Vat Dyes for Cotton and Other Cellulosic

Vat dyes, like sulfur dyes, are insoluble anthraquinoid or indigoid colorants, which are rendered temporarily soluble by reduction to a leuco form. Vats are supplied as dispersions in water or as prereduced solutions in water. During the vat dyeing process, they are reduced from aromatic $>C=O$ to $>C-O^-$, which renders the dye temporarily soluble. These are then reoxidized to an insoluble state trapped within the fiber structure, giving color that is very fast to washing and bleaching. Vat dyes, like sulfur dyes, are commercially applied by either batch or continuous methods. The following procedure is typical:

Fill batch dyeing machine with water.
Add substrate.
Add prepasted dye.
Heat to 175°F and run 5 min.
Add 2 g/L aqueous ammonia (concentrated ammonium hydroxide).
Add 12 g/L sodium hydrosulfite reducing agent.
Run 25 min.
Wash until clear.
Add 2 g/L hydrogen peroxide (35%).
Heat to 120°F, run 10 min, and wash.
Fill machine.
Add 0.2 g/L nonionic surfactant and 0.1 g/L acetic acid (56%).
Heat to 185°F, run 10 min, cool to 140°F, and wash.

Vat dyes are of two types—indigoid (e.g., indigo) and anthraquinoid—and are mainly available in the blue and green shade range. These dyes are useful as replacements for metal-containing fiber-reactive and direct dyes for turquoise and green shades.

7.7.18. Dyebath Reuse

In the 1960s, about 10–16% of textile wastewater was reclaimed or recycled [40]. Recent improvements, including dyebath reuse, make it possible to increase reuse dramatically, with corresponding cost savings and waste reduction. Some textile facilities now recycle well over half of their water from processes such as batch processing bath reuse, noncontact cooling water, washing (countercurrent washing), steam condensate recovery, and the like. No current study of water reuse in textile mills has been published. Dyebath reuse has been shown to reduce water use, BOD, COD, and TOC loadings up to 33% [40]. Also, the dyebath reuse concept has a return on investment in the form of dye, chemical, and energy savings that pretreatment does not. Savings, installation costs, and operating expenses are site specific, but a typical payback period is 13–20 months [40].

One extremely simple technique that saves not only water but also, in some cases, BOD is to reuse the final bath from one batch dyeing cycle to load the next lot. This works well in situations where the same shade is being repeated or where the nature of the dyeing system is such that the dyeing machine is fairly clean. This is particularly useful in acid dyeing of nylon hosiery.

7.7.19. Waterless Coloration Technologies

In the area of preparation and dyeing, there are waterless processes based on supercritical carbon dioxide fluid (SCF) technology. These use no water at all, and drying is simply a matter of allowing the carbon dioxide to evaporate, which happens immediately upon releasing the supercritical pressure. Since there is no water or pollution associated with the process and the CO_2 evaporates without any applied heat, energy is saved [5]. Intensive research is underway to commercialize this process.

Waterless mercerization can be accomplished by the use of liquid ammonia, which is all recovered and reused. Mercerization is effective and is practiced on a fairly wide commercial scale.

Other laboratory developments include ink jet printing and the use of powder colorants and xerographic printing techniques for waterless textile coloration processes. Each of these reduces not only water pollution but also air pollution and energy consumption due to the elimination of drying processes.

7.7.20. Bulk Systems/Auto Dispensing

Many textile operations use bulk storage tanks and automatic dispensing for commodity chemicals, due to the high volume and the price advantage of buying in bulk. The use of bulk systems prevents pollution by reducing small-scale spills, handling losses, implement cleanup, container washing and disposal, and the like. These pollution and cost savings are balanced against the small probability of a large-scale spill from a bulk storage area.

7.8. FINISHING

Textiles are made from a very limited list of a dozen or so structural fibers. From these, products must be made with millions of different specific end-use properties. In many cases the inherent properties of the fiber itself and substrate structure do not inherently impart the desired properties of comfort and functionality, for example, water repellent, stain block, soil release, crease resistant, flame retardant, antifungal, rot resistant, stiff, soft, antistatic, stretch, recovery, antiseam slippage, and enhanced tearing strength. Thus, fabrics must be finished either to overcome the inadequacies of design or to fine tune the attributes of fabrics, not only for the end uses noted above but also for facilitating cut-and-sew operations (e.g., needle lubrication, anticurl).

There are hundreds of finishing chemicals, as well as mechanical finishes, including calendaring, compacting, decatizing, heat setting, napping, pressing, sanding, Sanforizing, shearing, and sueding.

Waste from finishing operations are comprised of the substances shown in Table 7.31. The volume of waste from finishing is quite low compared to preparation, dyeing, and printing. Solid waste in the form of rags, scraps, and selvage trimmings are generally collected and sold to fiber recyclers. Selvage trimmings are salable as raw materials for recyclers and also for braided rugs or other craft activities. Water pollution from finishing can be reduced by careful attention to accurate calculation and make up of mixes to minimize discards [64] and proper work practices in handling chemicals, in particular, drum washing and bulk systems including delivery tanker trucks.

Table 7.31 Finishing Wastes

Solid
    Fabric scraps and trimmings from selvages, seams
    Fiber dust and fragments from napping, shearing, sanding, sueding
    Packaging materials such as paper tubes, empty chemical drums
Liquid
    Discarded mixes
    Obsolete or outdated chemical disposal
    Implement, equipment, and facility cleanup water
Vapors
    Exhaust gases from gas-fired drying and curing ovens
    Boiler emissions for electrical and steam-heated ovens

Table 7.32 Observed Air Pollutants and Typical Sources in Finishing [1]

Air Pollutant	Source (volatilization from ...)
Acetic acid	Residue from dyeing or printing
Acrylic monomers	Residue from hand builder
Biphenyl	Residue from dye carrier
Carbon monoxide	Incomplete oxidation of fuel
Dibutyl phthalate	Residue from dye carrier
Ethylene oxide	Breakdown of wetting agent
Formaldehyde	Breakdown of crosslinking resin
Glycol ethers	Softeners
Hexane	Softeners, wax water repellent
Hydrocarbons	Softeners, wax water repellent, fiber finish, yarn lubricants
Hydrogen chloride	Catalyst, machine cleaner, reduction of chlorinated organics
Methanol	Crosslinking reaction product, wetting agent
Methyl ethyl ketone	Machine cleaning solvent
Methyl methacrylate	Hand builder impurity
Methylene chloride	Machine cleaning solvent
Perchloroethylene	Machine cleaning solvent, dye residue
Toluene	Machine cleaning solvent
Trichloro ethane	Machine cleaning solvent, spot remover
Vinyl acetate	Hand builder impurity
Xylene	Machine cleaning solvent, dye residue

A more significant environmental issue is air emissions from finishing operations. Table 7.32 gives some observed air pollutants and typical sources in finishing [1]. In addition to these, there are visible emissions arising from vaporization of oil softeners and volatilization of fiber finish residues and knitting/winding lubricants.

Cotton and other cellulosic fabrics, and blends, usually require reactive crosslinking finishes to provide improved bending properties (i.e., crease behavior) and shrinkage. The products of choice for crosslinking cellulose are N-methylol compounds, which are made by reacting urea with formaldehyde and other additives. In application, storage, and use, these reactive N-methylol crosslinkers release formaldehyde to the air.

There are many choices in terms of softeners, including natural and synthetic. The main types are listed in Table 7.33 [69, 70]. Several alternatives are used to avoid the pollution associated with resins and softeners as noted above. There are mechanical finishing alternatives for shrinkage control, as well as improved chemical substitutions. For example, fatty acid softeners are very biodegradable [69, 70]. Quaternary types have high aquatic toxicity [69]. Mineral oil and paraffin wax softeners may still be in use even though polyethylene glycol (PEG) or polyethylene oxide (PO) products are available that do not smoke when heated, thereby reducing air emissions from dryers [59]. Paraffin and polyethylene types are not biodegradable. Reactive silicones are very well fixed and will not wash

Table 7.33 Types of Finishing Softeners [70]

Type	Subgroup	Comment
Fatty materials	Anionic	Biodegradable. Some methods of production use metal catalyst, thus may contain residual metal impurities.
Fatty materials	Cationic	High aquatic toxicity.
Petrochemical	Hydrocarbon	Produce hydrocarbon emissions from ovens.
Petrochemical	Alkene oxides	Can contain volatile ethylene oxide impurity.
Petrochemical	Polyethylene	Can produce hydrocarbon emissions from drying and curing ovens. Nonbiodegradable.
Silicone	Reactive	Durable
Silicone	Nonreactive	Nondurable

Table 7.34 Examples of Mechanical Finishing Options

Fabric	Property	Mechanical Methods
Knit	Shrinkage control	Compacting
	Smoothness	Pressing, decatizing
	Width control	Knit goods calendaring
	Shrinkage control	Heatsetting (synthetics only)
Woven	Shrinkage control	Sanforizing, decatizing
	Smoothness, luster	Woven goods calendar
	Softness	Breaking, sanding, napping, sueding

off of the fiber, whereas other types will wash off in home laundering of the textile products. In some cases, cellulase enzymes can remove surface roughness in yarns and fabrics, thus creating smoothness and lubricity without the use of any chemical additives at all. Certain yarn types (i.e., ring spun yarns) are inherently soft, while rotor spun yarns are "scratchy" and require more softeners to provide comfort for apparel end uses.

A wide variety of end-use properties can be imparted by mechanical means. Table 7.34 gives an abbreviated list of mechanical finishing options. Any complete treatment of this would be far beyond the scope of this document. There are dozens of mechanical finishing processes. These mechanical finishing processes have several advantages, not the least of which is that no chemicals are involved, although sometimes lubricants may be required (e.g., for napping). Another advantage is that when there is no reactive chemistry involved, the dyed shades are not affected during finishing.

Of course, finishing without chemicals results in less pollution because there are no mix discards, no pads to dump or clean, no chemically contaminated machines to clean, no chemical residues to release in storage and use, no obsolete chemicals to dispose of, no drums to wash and recycle, and no finish components vaporizing in dryers.

7.8.1. Novel Finishing Practices

Low add-on finishes conserve energy and pollution (less chemicals needed for a given level of performance) [71]. The method also conserves energy and speeds up production. Because there is less water used, the chemicals (if applied properly) are more evenly distributed throughout the fabric. Thus they more efficiently stabilize the fabric.

Foam technology is used to apply stain release chemicals to carpet. Liquid solutions normally used for treatment are replaced with foams. Wet pickup is lowered to as little as 8% in special applications, compared to 100% wet pickup in conventional methods [72]. The minimum pickup level for good finish penetration on cotton and cotton blends is about 45% [71]. The reduction in water use at one plant was 20 million liters per year, and energy was cut 10% overall [72]. However, foam and the surfactants necessary to produce it have their own set of problems in wastewater.

Softeners and other such surface finishes can be added by oversprays and other techniques, which have very low add on and which require no dumping of residual pad liquors at the end of a run.

7.8.2. Wool Finishing

Four high-priority areas were identified by a study group of the International Wool Secretariat (IWS) (49–51). They were pesticide residues in wastewater from wool from pesticides applied to sheep, discharge of mothproofing agents from wool carpet manufacture, halo-organics from wool shrink proofing, and chromium from chrome dyeing. These are being regulated and the aquatic toxicity from wool operations due to these types of materials must be reduced [49, 73, 74]. Also, as previously mentioned, a significant concern is agricultural use of pentachlorophenol pesticide.

Insect resistance (mothproofing) for wool has traditionally used chemicals that are being phased out due to regulations [73, 74]. The usual mothproofing procedures require permethrin, which is controlled in many parts of the world. It is reportedly biodegradable, has low mammal toxicity (humans), and high aquatic toxicity [73]. Mothproofing of wool is most often done by adding permethrin to the dyebath and causing it to exhaust onto the wool [75]. The spent dyebaths, when discarded, cause aquatic toxicity.

Several chemical alternatives have been reported [1]. Chlorphenylid was previously widely used but now has been withdrawn due to environmental problems in its manufacture. Flucofuron is not fully evaluated commercially at this time and is known to be ineffective against certain pests, but evaluations are still underway. Cycloprothrin has good performance and has aquatic toxicity three orders of magnitude less than permethrin [75]. Diphenylurea, on the other hand, is lower in aquatic toxicity but less biodegradable [73]. Cyfluthrin is effective but has been withdrawn due to textile mill workers' reactions to the chemical. Sulcofuron has low affinity in some application methods and is not effective against certain pests.

In any case, discharge of processing bath residues is an environmental problem. Applications typically are made commercially by exhaust processes and continuous (e.g., spray) processes. The discharges of mothproofing agents from finishing operations can be reduced, but never completely eliminated, so long as these mothproofing agents are used. Treatment of the spent (or left over) residual processing baths with alkali hydrolyzes the toxic chemical contents. The amounts of agents used should be carefully controlled to give the desired result with the minimum applications. Another new alternative now under investigation is nitromethylene. Studies on mixtures of nitromethylene and permethrin show promise and are continuing [73].

Due to regulations prohibiting certain mothproofing agents and limitations cited above, the remaining type of choice is pyrethroid based. Lower polluting mothproofing possibilities are being pursued on three avenues. First, new application methods involving microemulsion spray/centrifuge techniques with recycle of pesticide solution are being studied [73]. Second, new pesticides with low aquatic toxicity are being evaluated and introduced into commerce. Third, nonpesticide mothproofing methods are also under development.

The use of microemulsion spray/centrifuge techniques can reduce the discharges of permethrin to as low as 1.7 g of permethrin discharged per ton of wool treated for the microemulsion spray/centrifuge technique, compared to 8.0 g of permethrin discharged per ton of wool treated for the conventional hank treatments in dyebaths [49–51].

New, enar zero-pollution discharge application methods would include the use of low liquor ratios, microemulsions, accurately metering minimum necessary amounts, better process controls, and the like. Research continues on these alternatives and promising developments are underway.

Wastewater from shrink proofing of wool contains halo-organics, which later appear in drinking water supplies as absorbable organic halogens. One measured aggregate discharge was 39 ppm [50]. Research is underway to develop nonchlorine shrink proofing methods, but so far, there are none.

7.9. PRODUCT FABRICATION

Once a raw fiber has been converted into a finished fabric, it then must be cut and joined into an end-use product or into a combination with other materials into a textile-containing product (e.g., furniture, automotive seat, briefcase). This typically produces large amounts of fabric waste. In addition, any inefficiencies at the point of product fabrication result in production of upstream waste without realization of the equivalent amount of product. Pattern marker cutting efficiency is a critical factor in the amount of cutting room waste generated. Cutting room waste levels are variable depending on product style, panel adjacency requirements, and other product specific factors. Table 7.35 gives some typical values.

Table 7.35 Typical Cutting Room Waste

Type of Goods	% Waste Typical	Best (upper limit of utilization)
Denim	16–24	6
Knits open width	13–16	11
Knits tubular	25–27	22
Woven—other than denim	6–28	No consensus

As an example of the magnitude of this waste, about 800 million yards of denim are produced in the United States each year, or a total of well over 0.5 billion pounds of denim. Fabric utilization efficiency in cut-and-sew operations ranges from the about 72 to 94%. For denim, the utilization is about 84%, resulting in about 16% of denim production or roughly 100 million pounds annually. Cutting pattern efficiency depends strongly on garment design factors such as shape and seam location, size assortment as required by retail sales, and fabric width and other technical considerations. The textile industry has generally done a good job of material utilization and optimizing cutting efficiency. Some cutting room waste is recycled into end uses like papermaking.

7.10. CONCLUSION

In developed economies, the textile industry has placed great emphasis on compliance with environmental regulations, which are strictly enforced. The record of the industry in general has been exemplary. Many manufacturers have gone far beyond required levels of environmental protection. Excellent waste treatment systems and fairly sophisticated pollution prevention programs are the norm. Further improvements in these settings will develop along one of two lines of attack.

First is the transfer of known technology from more sophisticated manufacturers to smaller companies with less resources to spend on environmental protection. This includes information on environmentally friendly machinery, process design and optimization, training programs, auditing procedures, and the like. This technology transfer process is well underway. In addition, successful methods of regulation and treatment and prevention of pollution are being transferred to emerging economies.

Second is expansion of the concepts of pollution prevention to a level transcending unit process boundaries, and even transcending facility or corporate boundaries. This is sometimes very difficult because of business issues but is beginning to see some progress.

There is good reason for optimism that the textile industry's environmental performance will continue to improve even more in the future [76]. One important aspect of this is the global nature of the textile supply chain that results in complex influences, both positive and negative, on environmental performance. One aspect of this has previously been mentioned in relationship to design of textile products, which is typically outside of the manufacturers' control. Other aspects include communications, joint ventures, technology transfer programs, financial considerations, and promulgation of consuming economy standards to developing manufacturing areas. Availability of chemical safety data rapidly through computer network access is an important factor. Environmental standards placed on imported goods either through national customs regulations or by marketing companies in consuming economies can encourage manufacturers in developing regions to adopt environmental standards of the consuming economies. There are many examples of this, including European EcoLabel, ISO 14000, and the American Business for Social Responsibility initiatives. Also, financial organizations that invest in developing economies realize that their investments are best served by encouraging environmentally sound practices. This all points to improved environmental performance by the textile manufacturing industry in the future.

REFERENCES

1. U.S. EPA. 1996. Best Management Practices for Pollution Prevention in the Textile Industry. EPA/625/R-96/004. Washington, DC (September).
2. Fava, J. A., and A. Page. 1992. Application of product life-cycle assessment to product stewardship and pollution prevention programs. Water Sci. Tech. 26(1–2):275.
3. Glover, B., and L. Hill, 1993. Waste minimization in the dyehouse. Textile Chem. And Colorist (June). p. 15.
4. Chambers, D. 1993. Waste minimization: A corollary. Textile Chem. And Colorist 25(9):14.
5. Horstmann, G. 1993. The green dyer—fiction or reality. Australasian Textiles (January/February).
6. Lejeune, T. H. 1993. Future issues in solid waste management. In: Proceedings of the Conference for Executives and Managers on Environmental Issues Affecting the Textile Industry, Charlotte, NC (June 14–15). North Carolina Department of Environment, Health and Natural Resources, Raleigh, NC.
7. Smith, B. 1989, unpublished case study.
8. Mohr, U. 1993. Ecology must be dealt with. Australasian Textiles (January/February). p. 45.
9. Zeller, M. V. 1975. Instrumental techniques for analyzing air pollutants generated in textile processing. Textile Horizons (January).
10. Smith, B. 1989. A workbook for pollution prevention by source reduction in textile wet processing. Office of Waste Reduction, North Carolina Department of Environment, Health, and Natural Resources, Raleigh, NC.
11. Goodman, G. A., J. J. Porter, and C. H. Davis, Jr. 1980. Volatile organic compound source testing and emission control. Clemson University Review of Industrial Management and Textile Science (January).
12. U.S. EPA. 1988. Estimating chemical releases from textile dyeing. EPA/560/4-88/004h. Washington, DC (February).
13. Smith, B. 1986. Identification and reduction of pollution sources in textile wet processing. North Carolina Department of Natural Resources and Community Development, Pollution Prevention Pays Program, Raleigh, NC.

14. McCune, E. G. 1994. Facility total emissions summary: Annual air pollutant emissions inventory for 1993. State of North Carolina, Department of Environment, Health, and Natural Resources, Division of Environmental Management, Raleigh, NC.
15. Castle, M. 1992. A novel approach to practical problems. J. Soc. Of Dyers and Colourists (July/August). p. 306.
16. Smith, B., and V. Bristow. 1994. Indoor air quality and textiles: An emerging issue. Amer. Dyestuff Reporter (January). p. 37.
17. Berglund, R. L., and G. E. Snyder. 1990. Waste minimization: The sooner the better. Chemtech (June). p. 740.
18. Mock, G. N. 1984. Fundamentals of dyeing and printing. North Carolina State University, Raleigh, NC.
19. Kulube, H. M. 1987. Residual carrier components in exhausted textile dyebaths. Master's thesis, Department of Textile Chemistry, North Carolina State University, Raleigh, NC.
20. Leovic, K. W., J. B. White, and C. Sarsony. 1993. EPA's indoor air pollution prevention workshop. Presented at the 86th Annual Air & Waste Management Association Meeting, Denver, CO (June).
21. Bayer, C. 1992. Indoor environment testing using dynamic environmental chambers. ITEA Journal (December).
22. Zeller, M. V. 1975. Instrumental Techniques for Analyzing Air Pollutants in Textile Processing. Textile Horizons (January).
23. Goodman, G. A. et al. 1980. Volatile organic compound source testing and emission control. Clemson University Review of Industrial Management and Textile Science (January).
24. Smith, B. 1987. Identification of pollutant sources at Superba Printing, Mooresville, North Carolina. Office of Waste Reduction, North Carolina Division of Environmental Management, Raleigh, NC.
25. U.S. EPA. 1998. Preliminary Industry Characterization: Fabric Printing, Coating and Dyeing. EPA Draft. Washington DC (July).
26. U.S. EPA. 1979. Development document for effluent limitations guidelines and standards for the textile mills: Point source category (proposed). EPA/440/1-79/0226. Washington, DC (October).
27. Wagner, S. 1993. Improvements in products and processing to diminish environmental impact. COTTECH Conference, Raleigh, NC (November 11–12).
28. Smith, B. 1994. Future pollution prevention opportunities and needs in the textile industry. In: Pojasek, B., ed. Pollution prevention needs and opportunities. Center for Hazardous Materials Research (May).
29. Miraldi, R. V. 1958. Cotton processing waste: A mill survey. National Cotton Council of America, Utilization Research Division (August).
30. Nityanand, A. R. 1984. Waste investigation and control for a spinning mill. Textile Horizons (September). p. 79.
31. Cates, D. 1986. Conversion of polyester/cotton industrial wastes to higher value products. DOE/ID/12521-1 (DE87002509). Prepared for U.S. Department of Energy, Office of Industrial Programs by North Carolina State University, Raleigh, NC.
32. Milner, A. J. 1992. The importance of size selection. Australasian Textiles (March/April). p. 33.

33. Smith, B. 1992. Reducing pollution in warp sizing and desizing. Textile chem. And Colorist (June). p. 30.
34. Palmer, J. 1983. Waste management: How to turn carpet waste into bottom line profits. Carpet and Rug Indus. (October). p. 8.
35. American Association of Textile Colorists and Chemists (AATCC). 2001 Buyer's guide. Textile Chem. And Colorist (July).
36. Achwal, W. B. 1990. Environmental aspects of textile chemical processing, parts 1 and 2. Colourage (September). p. 40.
37. Naylor, C. G. 1992. Environmental fate of alkylphenol ethoxylates. Soap/Cosmetics/Chem. Specialties (August).
38. Kravetz, L. 1989. Selection of surfactants which have minimal impact on the environment. In: Proceedings of the Pollution Prevention by Source Reduction in Textile Wet Processing Conference (May 23). Jointly sponsored by the Textiles Chemistry Department, North Carolina State University, and the North Carolina Division of Environmental Management, Raleigh, NC.
39. Moore, S. B., R. A. Diehl, J. M. Barnhardt, and G. B. Avery, 1987. Aquatic toxicities of textile surfactants. Textile Chem. And Colorist (May). p. 29.
40. Smith, B. 1986. Identification and reduction of pollution sources in textile wet processing. North Carolina Department of Natural Resources and Community Development, Pollution Prevention Pays Program, Raleigh, NC.
41. Huber, L. H. 1984. Ecological behavior of cationic surfactants from fabric softeners in the aquatic environment. JAOCS (February). p. 377.
42. Chemical Manufacturers Association (CMA). 1991. Alkylphenol ethoxylates: Human health and environmental effects. Interim report of the Alkylphenol and Ethoxylates Panel of the Chemical Manufacturers Association (December). Washington, DC: CMA.
43. Kravetz, L., J. P. Salanitro, P. B. Dorn, K. F. Guin, and K. A. Gerchario, 1986. Environmental aspects of nonionic surfactants (draft report). Presented at the American Association of Textile Chemists and Colorists International Conference Exhibition, Atlanta, GA (October 31).
44. Wernsman, D. unplished manuscript (Managing the Environmental Impact of Textile Chemicals).
45. Lee, J. 1995. Masters Thesis at N.C. State University. Pollution Reduction in Textile Manufacturing.
46. Moore, S. B., R. A. Diehl, J. M. Barnhardt, and G. B. Avery. 1987. Aquatic toxicities of textile surfactants. Textile Chem. And Colorist (May). p. 29.
47. Kravetz, L., J. P. Salanitro, P. B. Dorn, K. F. Guin, and K. A. Gerchario. 1986. Environmental aspects of nonionic surfactants. Draft report. Presented at the American Association of Textile Chemists and Colorists International Conference Exhibition, Atlanta, GA (October 31).
48. Kravetz, L. 1989. Selection of surfactants which have minimal impact on the environment. In: Proceedings of the Pollution Prevention by Source Reduction in Textile Wet Processing Conference (May 23). Jointly sponsored by the Textiles Chemistry Department, North Carolina State University, and the North Carolina Division of Environmental Management, Raleigh, NC.
49. Shaw, T. 1989. Environmental issues in the wool textile industry. International Wool Secretariat (IWS) Development Center Monograph. IWS, West Yorkshire, England.

50. Shaw, T. 1989. Environmental issues in the wool textile industry. International Wool Secretariat (IWS) Development Center Monograph, IWS, West Yorkshire, England.
51. Shaw, T. and D. Allanach, 1989. Mothproofing and the environment. International Wool Secretariat (IWS) Development Center Monograph, IWS, West Yorkshire, England.
52. Wimbush, J. M. 1989. Pentachlorophenol in wool carpets — investigating the source of contamination. International Wool Secretariat (IWS) Development Center Monograph, IWS, West Yorkshire, England.
53. Modak, P. 1991. Environmental aspects of the textile industry: A technical guide. Draft report. Prepared for the United Nations Environment Programme (March).
54. Smith, B., and J. Rucker, 1987. Water and textile wet processing — Part I. American Dyestuff Report (July). p. 15
55. Leaver, A. T., G. Brian, and P. W. Leadbetter. 1992. Recent advances in disperse dye development and applications. Textile Chem. And Colorist 24(1):18.
56. Sommerville, W. N. 1988. Economic analysis of short lot dyeing techniques. Master's thesis, Department of Textile Chemistry, North Carolina State University, Raleigh, NC.
57. Smith, B. 1989. Source reduction by new equipment. Presented at Pollution Resource Reduction in Textile Wet Processing. Raleigh, NC (May 23–24).
58. Smith, B. 1989. ATI's dyeing and printing guide. Amer. Textiles Int. (ATI) (February).
59. Norman, P. I., and R. Seddon, 1991. Pollution control in the textile industry — The chemical auxiliary manufacturer's role, part 1. J.Soc. Dyers and Colourists (April). p. 150.
60. Fulmer, T. D. 1992. Cutting costs and pollution: Save energy and fight pollution at the same time. Amer. Textiles Int. (March). p. 38.
61. Brenner, E., T. Brenner, and M. Scholl. 1993. Saving water and energy in bleaching tubular knits. Amer Dyestuff Reporter (March). p. 76.
62. Cunningham, A. 1993. Ultra-low liquor ratios key to reducing water and chemicals in exhaust dyeing. Australasian Textiles (January/February). p. 39.
63. Houser, N. et al. 1994. Pollution Prevention and U.S. Textiles. American Textile Institute (ATI) (March). p. 28.
64. Horning, R. H. 1981. Carcinogenicity and azo dyes. Presented at the Textile Industry and the Environment Symposium, Washington, DC (March 30–31)
65. Richardson, S. 1991. Multimedia environmental concerns in warp sizing: Low tech approaches to waste reduction, North Carolina Pollution Prevention Program, Raleigh, NC (February).
66. Duffield, P. A., J. M. Wimbush, and P. F. A. Demot. 1990. Wool dyeing with environmentally acceptable levels of chromium in effluent. International Wool Secretariat (IWA) Development Center Monograph. IWS, West Yorkshire, England.
67. Stone, R. L. 1979. A conservative approach to dyeing cotton. Cotton, Inc., Monograph. Cotton, Inc., Raleigh, NC.
68. Cook, F. L. 1991. Environmentally friendly: More than a slogan for dyes. Textile World (May). p. 84.
69. Patel, H. 1993. Synthetic softener developments and the environment. Australasian Textiles (January/February). p. 48.

70. Tomasino, C. 1992. Chemistry and technology of fabric preparation and finishing. Professor's course packet for TC-310. Textiles Chemistry Department, North Carolina State University, Raleigh, NC (September).
71. Smith, B. 1985. Determining optimum wet pickup in low add-on finishing. American Dyestaff Reporter (May). p. 13.
72. Powell, D. 1992. New foam technology makes pollution prevention pay. Australasian Textiles (November/December).
73. Haas, J. 1993. Mothproofing still possible within environmental laws. Australasian Textiles (January/February). p. 43.
74. Allanach, D. 1989. The insect resist treatment of carpet yarns using low volume zero pollution technology. International Wool Secretariat (IWS) Development Center Monograph. IWS, West Yorkshire, England.
75. Russell, I. M. 1989. Cycloprothrin, an environmentally safer pyrethroid for industrial insect resist treatment of wool? International Wool Secretariat (IWS) Development Center Monograph, IWS, West Yorkshire, England.
76. Smith, B., Global Environmental Trends: Greening of the Textile Supply Chain, American Dyestuff Reporter, Volume 87 number 9, September 1998.

PART 3

CHAPTER 8

ENVIRONMENTAL EFFECTS ON POLYMERIC MATERIALS

NORMA D. SEARLE
Deerfield Beach, Florida

8.1. INTRODUCTION

Most polymeric materials exposed to the environment are subject to deterioration caused by the combination of all weather factors, including solar radiation, heat/cold, moisture (solid, liquid, and vapor), oxygen, and atmospheric contaminants. However, the actinic radiation of the sun is the critical factor since it generally initiates the reactions that are subsequently promoted by the other weather factors. The ultraviolet (UV) radiation absorbed by colorless as well as colored polymeric materials and the visible radiation absorbed by colored polymers have sufficient energy to break many types of chemical bonds. Bond breakage is an essential step in the photodegradation process. The environmental effect on polymeric materials, referred to as *weathering*, depends on (1) the specific wavelengths and amount of solar radiation the materials are capable of absorbing, (2) the strength of their chemical bonds in relation to the photon energies of the solar radiation absorbed, and (3) the reaction processes promoted by heat, moisture, and other weather factors following bond breakage.

Exposure to the outdoor environment affects not only the polymeric material itself but also other components within the matrix, such as dyes, pigments, processing additives, and stabilizers. Complex interactions of the combination of weather factors with the polymer and its components result in irreversible changes in the chemical structures and physical properties in a direction that generally changes both the appearance and mechanical properties and reduces the useful life of the material. Most organic polymeric materials require the use of one or more types of stabilizers for protection against the effects of the environment in order

---

Plastics and the Environment, Edited by Anthony L. Andrady.
ISBN 0-471-09520-6 © 2003 John Wiley & Sons, Inc.

to maintain serviceability for a reasonable length of time. Tests of weatherability are an essential aspect of the development of new and improved products as well as for quality control of production lots and certification of stability specifications. Because of the length of time required for tests under natural weather conditions of the weatherability of materials designed to have relatively good weather resistance, outdoor accelerated tests as well as laboratory accelerated tests that simulate the effects of natural weather have been developed. Adequate testing necessitates information on the critical weather factors and appropriate measurements of the changes produced in properties that are important to the usefulness of the products.

8.2. WEATHER FACTORS AND THEIR EFFECTS ON POLYMERIC MATERIALS

8.2.1. Terrestrial Solar Radiation

The spectral energy of solar radiation on the Earth's surface ranges from about 298 nm in the UV region to about 2500 nm in the near-infrared (NIR) region. The total energy consists of direct energy from the sun and scattered radiation from the sky. The proportion of the latter is greatest for the shortest wavelengths. The spectral energy distribution and the intensity of the radiation vary with geographic location, altitude, season, time of day, and atmospheric conditions as well as with angle of incidence. The ultraviolet portion is most sensitive to these variables and the shorter the wavelength, the more it is affected. The fraction of UV (298–400 nm) in total solar radiation varies between about 1 and 5%, depending on conditions. At all latitudes, the altitude of the sun, and thus total solar irradiance, is highest at solar noon. Figure 8.1 [1] shows the effect of season on the UV spectral power distribution of solar radiation at 41°N latitude. Due to change in angle of the sun with season, the short wavelength cut-on of direct normal solar radiation shifts from below 300 nm in the summer to about 310 nm in the winter.

It is a generally accepted fact that because shorter wavelengths have higher photon energies associated with them, they have a greater actinic effect on materials. Thus, shorter wavelengths are capable of breaking stronger bonds and more types of bonds. However, only wavelengths absorbed by a material can cause damage. Regardless of the intensity of the incident radiation on the surface of a material, it cannot do any harm if the material is not capable of absorbing it. The amount of incident radiation of each wavelength absorbed depends on the spectral absorption coefficients of the material and on the spectral emission properties of the radiation source, that is, the wavelengths emitted and their intensities. Although the absorption coefficient of polymeric materials generally increases with decrease in wavelength, the shortest wavelengths of terrestrial solar radiation have very little effect on most polymeric materials because of their very weak intensities.

The relative destructive effects of the incident wavelengths of a specific radiation source on a material is graphically represented by the "activation

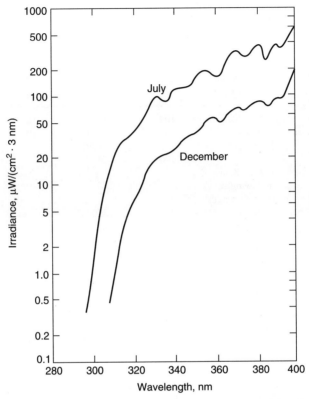

Figure 8.1. Seasonal variation of direct normal solar UV spectral irradiance at 41°N latitude at noon under clear-sky conditions. (Reprinted with permission of the Optical Society of America [1].)

spectrum" of the material. It depends on both the relative amounts of radiation of each wavelength absorbed and the quantum efficiencies of degradation by the absorbed wavelengths. Since quantum efficiencies cannot be predicted from fundamental principles, the activation spectrum must be determined experimentally. Table 8.1 [2] lists data based on activation spectra of a number of types of polymeric materials exposed to solar simulated radiation, that is, to borosilicate-glass-filtered xenon arc radiation. The columns labeled "Max. Change" show the spectral regions of the source mainly responsible for causing the optical and mechanical changes measured. Data obtained by the spectrographic and sharp cut-on filter techniques are listed separately. The techniques have been described previously [3–7]. Activation spectra based on exposure to solar radiation were also obtained for some of the polymers. Generally, the activation spectra were similar to those based on the filtered xenon arc source. One of the exceptions was polycarbonate. It showed a very strong effect by radiation near 300 nm based on the artificial source, which

Table 8.1 Activation Spectra Maxima Based on Borosilicate-Glass-Filtered Xenon Arc Radiation

Polymer	Measured Change	mils	Filter Technique Max. Change (nm)	mils	Spectrographic Technique Max. Change (nm)
Acrylonitrile– butadiene–Styrene	Yellow bleaching	100	340–360	10	330
	impact str.	100	>380	10	380–400
	Impact str.	100	350–380[a]	—	—
		100	>380[b]	—	—
Nylon—6	UV, modulus	—	—	?	390, 450
Polyamides (aromatic)	Yellowing & tensile strength	—	—	?	360, 370, 414[c]
Polyarylate	Yellowing	3	350	—	—
		60	385	—	—
Polycarbonate	UV at 340 nm	28	<300; 310–340	4.5	295; 310–340
	Yellowing	28	<300; 310–340	4.5	295; 310–340
Polyethylene	Yellowing	—	—	4	310
	C=O	—	—	4	340
Polyethylene– naphthalate	Yellowing FTIR, modulus	—	—	5	370–380
Polypropylene	UV	10	—	15	295, 330, 370
	C=O	60	320	15	340–380
	Tensile str.	60	320–350[a]	—	—
	Tensile str.	—	360–380[b]	—	—
Polystyrene	Yellowing	125	300–330	125	319
Polysulfone	UV at 330 nm	1	310	1	305
	Yellowing	1	320	1	310–320
	C=O; OH	—	—	1	330
Polyurethane (arom.)	Yellowing	—	—	?	350–415[d]
Polyvinyl chloride	Yellowing	40	300–320	2	308–325[d]

[a] Short exposure.
[b] Extended exposure.
[c] Films and fibers.
[d] Various samples.

was missing in the activation spectrum based on solar radiation. Because of the very much lower intensity of these wavelengths in solar radiation, they have a negligible effect in spite of the fact that they are strongly absorbed by the polycarbonate chromophore and are capable of causing significant degradation. Longer wavelengths are weakly absorbed by polycarbonate and its impurities [3].

The wavelength sensitivity will often vary with the type of degradation measured. For example, for some polymers, wavelengths shorter than 350 nm are responsible for optical changes, but mechanical changes in the same polymers are caused by wavelengths longer than 350 nm, often extending into the visible region. In the case of polypropylene and acrylonitrile–butadiene–styrene, the spectral region mainly responsible for changes in mechanical properties shifts to longer wavelengths with increase in exposure. In these polymers, as in all aliphatic-type polymers, ultraviolet absorbing impurities are the only species responsible for initiating degradation by solar radiation. As exposure progresses, UV-absorbing impurities with longer-wavelength-absorbing capabilities are formed.

The maximum sensitivity of aromatic-type polymers exposed to solar radiation or to a light source with similar spectral power distribution shifts to longer wavelengths with increase in thickness. Thus, the data in Table 8.1 shows that wavelengths primarily responsible for yellowing polyarylate shift from the 350- to the 385-nm spectral region with increase in thickness from 3 to 60 mils. The shift is the result of a combination of factors: (1) the main absorption band of the polymer broadens with thickness, (2) with broadening of the band, the long wavelength edge extends into the spectral region in which the intensity of the source increases with wavelength, and (3) the percent increase of light absorbed with thickness is greater in the spectral region of weak polymer absorption, that is, in the longer wavelength region, because of the logarithmic nature of the absorption process.

The shift in wavelength sensitivity to longer wavelengths with increase in thickness is a function of the type of polymer as well as the type of light source. The effect is greatest for an aromatic polymer exposed to solar or solar simulated radiation. It would be much smaller for an aromatic-type polymer exposed to a light source such as the fluorescent UVA-340 lamp in which the intensity decreases with increase in wavelength longer than 350 nm. The shift would be negligible for aliphatic-type polymers exposed to solar radiation since their main absorption bands are below the cut-on of terrestrial solar radiation. The latter is absorbed only by the impurities in aliphatic-type polymers.

The shift in activation spectrum with thickness of an aromatic polymer exposed to solar radiation demonstrates the importance of testing these materials in the form in which they will be used in practice. The type of ultraviolet absorber required to screen the harmful wavelengths and its effectiveness will differ with the thickness of the aromatic-type polymeric tested. Due to differences among different types of polymeric materials in the effect of thickness on their wavelength sensitivities, the form in which they are tested can change their stability ranking.

8.2.2. Temperature

Temperatures are generally highest in the climatic zones of Earth in which solar irradiance is highest. Both irradiance and temperature are usually highest near the equator and decrease with increasing latitude. The temperature of an exposed material depends on the amount of radiation absorbed, the emissivity of the material, thermal conduction within it, and exchange of heat with the surroundings through conduction and convection. The surface temperatures of exposed materials are higher than that of the surrounding atmosphere. A large portion of the absorbed radiation is converted to heat, and the amount absorbed is closely linked to color, with white materials absorbing only about 20% of the incident energy and black about 90%. Therefore, the darker the color, the higher the temperature of the material. Surface temperatures have been reported to reach 77°C in exposed plastic specimens [8] and 120°C inside a closed automobile exposed to sunlight [9].

Temperature can affect the weathering of polymeric materials in a variety of ways. Elevated temperatures can significantly influence the destructive effects of solar radiation on polymeric materials by accelerating the rate of the secondary reactions and altering the reaction processes that follow the primary photochemical step of bond breakage. For example, higher temperatures can alter the mechanism of degradation by increasing the rate of diffusion of oxygen and water into the material and can alter the rates of secondary reactions by increasing the mobility of radical fragments and other intermediates. Higher temperatures accelerate hydrolysis reactions, while low temperatures cause condensation to form on the material as dew. The temperature of exposed materials can also influence their stability ranking because of differences among materials in the effect of temperature on the secondary reactions. Daily and seasonal temperature cycling can cause mechanical stress in composite systems, such as between a coating and substrate or between coating layers, due to mismatch in the thermal expansion coefficients. Temperature cycling often results in cracking and loss of adhesion of the coating. Freeze/thaw cycling or thermal shocks due to cool rain hitting hot, dry surfaces will induce mechanical stress that can cause structural failures in some systems or accelerate degradation already initiated.

8.2.3. Moisture

All materials used outdoors are exposed to the influence of moisture, which, in combination with solar radiation, contributes significantly to the degradation of many polymeric materials. The form and amount of moisture vary widely, depending on the geographic area and ambient temperature. Moisture can take the form of humidity, dew, rain, frost, snow, or hail. Moisture contributes to the weathering of polymeric materials both by reacting chemically in hydrolytic processes and by imposing mechanical stresses when it is absorbed or desorbed [10]. It can also act as a solvent or carrier, for example, in leaching away plasticizers or in transporting dissolved oxygen. Examples of chemical reactions of water are hydrolysis reactions of labile bonds such as those in polyesters, polyamides, polycarbonates, and other heteroatom-containing polymers and its promotion of chalking of titanium dioxide (TiO_2) pigmented coatings and building products exposed to solar radiation. Chalking results from the release of TiO_2 particles at the surface when the organic binder is degraded by the hydroxyl and perhydroxyl free radicals formed in the reaction between water, oxygen, and the titanium and hydroxyl ions produced on exposure of the pigment to UV radiation. Experience shows that chalking is most prevalent when moisture at the surface is highest; little to no chalking occurs in dry environments. Some polymeric materials exposed to high humidity degrade at an accelerated rate due to the plasticizing action of water, which enhances the accessibility of atmospheric oxygen into the material.

Periodic cycling of moisture in the form of humidity or liquid water creates mechanical stresses in many materials due to the water concentration gradient. In general, moisture absorption causes reduction in mechanical properties, such as strength and stiffness of composite materials. Absorbed water, by expanding the volume of the surface layers, results in compressive stresses on the outside and tensile stresses in the bulk. Reduction in volume of surface layers during drying is resisted by the hydrated inner layers, creating bulk compressive and surface tensile stress gradients. Cracking and loss of adhesion of coatings results. The span of time over which precipitation occurs and the time the sample surface is exposed to wetness are more important in the effect of rain on materials than the total amount of precipitation. The penetration depth of moisture into the material depends on its diffusion coefficient and the period of cycling. Thus the effect of the moisture is substantially greater when the total amount of precipitation is distributed over a longer time period. Since water absorption is a diffusion-controlled process, it may take weeks or months to reach equilibrium in thick forms of polymeric materials because of the slow diffusion rates. However, moisture uptake is temperature dependent, the rate increasing with increase in temperature.

Solar radiation has a pronounced effect on the humidity-induced stresses in that it causes formation of polar or hydrophilic groups, which increase the tendency of the material to absorb moisture. Many photochemical aging processes initiated by solar radiation cause embrittlement of the surface, thus enhancing the tendency to crack under tensile stresses during the drying period. High relative humidity levels in conjunction with heat, as, for example, in tropical and subtropical climates, often promote microbial growth. Mold, mildew, and other microbiological and botanical agents can play a significant role in material degradation. All are examples of the combined action of moisture with other weather factors causing degradation of polymeric materials.

8.2.4. Oxygen

The importance of oxygen in the weathering process is attested to by the fact that most polymer failures that occur during outdoor exposure are due to photooxidation reactions. Photooxidation is most prevalent on the surface. It is significantly reduced toward the center of thick samples because of the limited supply of oxygen in that region. The penetration of oxygen into the polymer is related to its rate of diffusion, which depends on temperature, polymer type, and morphology. It has been shown that oxygen diffusion, not radiation, is the rate-controlling process in photooxidation of polyolefin plaques at depths at which the rate of oxygen consumption is greater than the rate at which it can be replenished from the environment [11–14].

Oxygen enhances the effect of solar radiation in a number of ways: (1) It forms a complex with conjugated unsaturated hydrocarbons that strongly absorbs UV radiation, thus increasing the amount of solar radiation absorbed. (2) It reacts with the carbon-centered free radicals produced by solar radiation to form peroxy radicals. The destructive effect of the radiation is multiplied manyfold by the chain reactions propagated by the peroxy radicals. (3) The reaction of oxygen in its ground state with the triplet states of sensitizers such as ketones and certain dyes produced by absorption of solar radiation forms singlet oxygen. The latter, a very reactive form of the molecule, is responsible for the rapid deterioration of many materials, particularly those with conjugated unsaturation such as natural rubber and synthetic elastomers.

8.2.5. Atmospheric Pollutants

Ozone is present in Earth's atmosphere both as a result of UV photolysis of oxygen in the upper atmosphere and as a result of reaction between terrestrial solar radiation and atmospheric pollutants such as nitrogen oxides and hydrocarbons from automobile exhausts. It is a powerful oxidant that can react rapidly with elastomers and other unsaturated polymeric materials to cause stiffening and cracking, particularly under mechanical stress. Other common air pollutants include sulfur oxides, hydrocarbons, nitrogen oxides, and particulate matter such as sand, dust, dirt, and soot. Some of these may react directly with organic materials but have a much more severe effect in combination with other weather factors. For example, dilute sulfuric acid is formed only when sulfur dioxide (SO_2) and water on the surface of materials is exposed to solar radiation. It causes rapid discoloration of pigments as well as crosslinking [10] and embrittlement [15] of polymers.

Acid rain, an important consequence of pollutants generated by modern industrial societies, has been shown to damage both organic and inorganic materials exposed to the environment. Acid rain enhances hydrolytic degradation and thus is an important factor in weathering of polymeric materials in which the mechanism includes hydrolysis. Acids have also been shown to interfere with the use of hindered amine light stabilizers (HALS) used to improve the light stability of acrylic urethane clearcoats. It is believed that acids may reduce the effectiveness of HALS by reacting with it to form salts, which are then washed out of the coating. The results of the action of acidic pollutants and radiation on automotive coatings is described by Schulz and Trubiroha [16]. While acids generally act synergistically with radiation to accelerate the effects of weathering, acid precipitations can also slow the aging processes in polymers [17].

8.3. ENVIRONMENTAL STABILITY AND DEGRADATION MECHANISMS OF POLYMERIC MATERIALS

Prolonged outdoor exposure of all organic polymeric materials leads to appearance changes, such as loss of gloss, formation of surface crazes and cracks, chalking, yellowing and fading, as well as to breakdown of mechanical properties, including impact strength, tensile strength, and elongation, properties that are particularly important in some applications. However, environmental stabilities of commercial polymeric materials differ widely because of (1) differences in their chemical structures and thus in both their absorption properties and resistance to the absorbed radiation and (2) differences in the types and amounts of impurities introduced during manufacturer and as a result of thermal treatment in processing and fabrication. The UV characteristics of polymers, the susceptibility of unstabilized and commercial polymers to degradation, and reported outdoor lifetimes are summarized by Kamal and Huang [18] and Searle [19].

Since aliphatic polymers, for example, polyolefins and polyvinyl chloride (PVC), do not have structural chromophores capable of absorbing solar UV radiation, they should be stable when exposed to the environment as pure materials. The UV absorption curves of 2 mil films of polyethylene and poly(vinyl chloride) are shown in Figure 8.2 [3] along with the spectral emission of terrestrial solar radiation. The absorption band of the structural components of aliphatic-type polymers generally peaks below 220 nm, and the tail of the band barely extends to 298 nm, the shortest wavelength reported for terrestrial solar radiation. However, the presence of solar UV absorbing impurities such as catalyst residues, organic contaminants, and thermal oxidation products, primarily hydroperoxides and carbonyl groups attached to the polymer chain, sensitize the aliphatic-type polymers to photooxidation. The impurities are not detectable in the spectral curves because of their low concentration and weak absorption, but the small amount of radiation absorbed initiates rapidly propagating free-radical chain reactions that cause extensive photooxidation. Many aliphatic types are more sensitive to outdoor

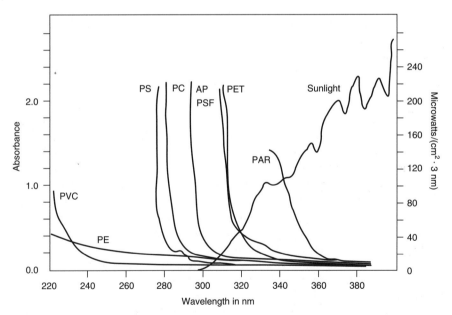

Figure 8.2. Ultraviolet absorption spectra of 0.05-mm polymer films and spectral irradiance of July noon sunlight (direct beam) at 41°N latitude: AP = aromatic polyester; PAR = polyarylate; PC = polycarbonate; PE = polyethylene; PET = poly(ethylene terephthalate); PS = polystyrene; PSF = polysulfone; PVC = poly(vinyl chloride). (Reprinted with permission of Technomic Publishing Co. [4].)

exposure than most aromatic-type polymers in spite of the fact that the latter are capable of absorbing much more solar UV radiation. For example, polyolefins and PVC are less stable than polycarbonate and aromatic polyesters.

The chromophoric structures in most aromatic-type polymers are capable of strongly absorbing wavelengths longer than 290 nm but vary in the long-wavelength limit of the radiation they absorb. Figure 8.2 compares the UV absorption curves of 2-mil films of various aromatic-type polymers. Because of the strong absorption in this region, the full absorption bands of 2-mil films are not measurable. The spectra show only the long-wavelength edge of the bands but demonstrate the differences among the aromatic polymers in the wavelengths of solar UV radiation they can absorb. For example, the long-wavelength edge of the absorption bands of polystyrene and polycarbonate barely extend into the solar region, but polyarylate can absorb wavelengths as long as 380 nm. The long-wavelength limit of absorption also depends on the thickness, shifting to longer wavelength with increase in thickness. Thus, the total amount of solar UV and the energies of the photons absorbed differ among the aromatic-type polymers. The shorter wavelengths represent higher energy photons, which are capable of breaking stronger and more types of bonds, thus causing more degradation. However, differences in stability are not directly related to the amounts of radiation and photon energies absorbed. Polymers that absorb more solar radiation are not necessarily less photochemically stable. Degradation depends on the ability of the radiation to break chemical bonds, and thus on the relation between the photon energies and the bond strengths, as well as on the reactions that follow bond breakage.

Since the structural components of many of the aromatic types absorb solar radiation directly, degradation results from two major reaction mechanisms: one pathway is initiated by photolysis of the structural chromophores, which occurs in the absence of oxygen; the other pathway involves photooxidative chain reactions initiated by the energy absorbed by the impurity chromophores. Following direct photolysis of the aromatic structure by wavelengths shorter than the longest wavelength absorbed by the aromatic species, a photo-Fries rearrangement to a new chromophore often results in addition to oxidation of the radical species generated. Longer wavelengths can only be absorbed by impurities and defects, similar to those in the aliphatic-type polymers. The photooxidation reactions involve hydroperoxidation of the aliphatic carbon atoms of the polymer and often oxidation of the phenyl ring. Some of the polymers for which dual mechanisms of degradation have been reported are polycarbonate, aromatic polyesters, polyurethanes, and aromatic as well as aliphatic polyamides. The relative importance of the reactions initiated by the short versus long wavelengths depend on the spectral power distribution of the source of radiation.

The rubber-modified aromatic polymeric materials, for example, high-impact polystyrene (HIPS) and acrylonitrile–butadiene–styrene (ABS) are the least stable because of the rapid photooxidation of the rubber component, which sensitizes the polymer to further oxidation. Diene-based rubbers are also susceptible to degradation by ozone, which causes chain scission of the main-chain unsaturation in the polymer surface, resulting in reduction of surface strength. The photostability of commercial polymeric materials is also affected by the presence of pigments and fillers. Some pigments, such as the organic yellow and red types as well as some types of inorganic pigments, act as sensitizers and thus decrease the stability of the polymer. Other pigments, for example, phthalocyanine blue, phthalocyanine green, rutile titanium dioxide, and others, generally act as stabilizers, particularly of thick sections, by absorbing UV radiation and thus preventing it from reaching the deeper layers of the polymer. The visible radiation absorbed by colored pigments can also accelerate the rate of degradation by increasing the temperature of the polymeric material. The darker the pigment, the higher the temperature. Thus, the screening protection provided by dark pigments can be offset by the increased rate of degradation at the higher temperatures.

Environmental stabilities and types of changes that occur on weathering depend to a large extent on the chemical reactions that follow bond breakage, that is, on the mechanism of degradation. It varies with the type of polymer, the incident wavelengths, particularly in the case of aromatic-type polymers, and the form in which the polymeric material is used, that is, fiber, film, or plaque, which determines predominance of surface versus bulk reactions. Photooxidation is primarily a surface phenomenon because of the importance of oxygen and its slow diffusion through the polymer as well as the fact that thermal oxidation products formed during processing and fabrication are concentrated on the surface. In aliphatic polymers that do not contain strongly UV-absorbing additives so that solar UV can penetrate to the back surface, photooxidation proceeds from both the front and back surfaces where the availability of oxygen is highest. In the case of aromatic-type polymers, degradation caused by the shorter wavelengths does not penetrate much beyond the surface layers because of the strong absorption of these wavelengths. The longer wavelengths are more weakly absorbed and can affect bulk properties if the article is sufficiently thick to absorb a significant fraction of these wavelengths. Photooxidation of a polymer in the solid state is dependent on the rate of oxygen diffusion into the polymer and the formation of charge-transfer complexes between the polymer and oxygen molecules. In thin forms of polymers, oxygen diffusion is not rate determining. In thick polymer sections, and particularly at higher irradiation intensities and/or temperatures, oxygen may become deficient in the inner layers, increasing the role of alkyl radicals and decreasing the role of alkoxyl radicals in polymer oxidation. Therefore, different strategies for inhibition of degradation are necessitated in these areas of the polymer. When oxygen is present, the rate of oxidation depends on the reactivity of oxygen molecules with the macroradicals. Degradation mechanisms and stabilities of various types of commercial polymers are summarized in the next sections.

8.3.1. Polyolefins

The photostability of different types of polyolefins varies depending on their chemical structures, crystallinity, and morphology. Polypropylene (PP) is especially sensitive to UV radiation and must be stabilized against the effects of solar radiation in order to be serviceable outdoors. Degradation on outdoor exposure is manifested by loss of gloss, formation of surface crazes, chalking, and breakdown of its mechanical properties. The light stability is influenced by degree of orientation and crystallinity, sample thickness, pigmentation, and the presence of fillers.

Polyethylene (PE) is inherently less sensitive to oxidative attack than PP, but stabilization of PE is also mandatory for outdoor use. The stability varies with the type of polyethylene and manufacturing process. Linear low-density polyethylene (LLDPE) (1-octene comonomer) is significantly less sensitive to photooxidation than low-density polyethylene (LDPE) with comparable density and molecular weight [20, 21]. Generally, LDPE is less susceptible to photooxidation than high-density polyethylene (HDPE). The most fundamental difference between polyethylene homopolymers and polypropylene is the behavior of hydroperoxides toward photolysis. On photooxidation, hydroperoxides accumulate in PP, but decrease rapidly on UV exposure of PE. In copolymers of polyethylene with vinyl acetate, the stability depends on the content of vinyl acetate. The higher the content, the more the copolymers act like polyvinyl acetate, which is more susceptible to photooxidative degradation than polyethylene.

Among the impurities and defect structures in polyolefins that can absorb solar UV radiation to initiate photodegradation, hydroperoxides are the most important during the early stages of photooxidation. Although they are present in very small amounts and therefore absorb very little solar radiation, the autocatalytic oxidative chain reactions initiated rapidly cause significant photooxidation of the polymers and ultimately lead to catastrophic failure. The rate-determining step is generally the abstraction of hydrogen atoms from other polymer molecules by the macroalkylperoxyl radicals that form. The resulting macrohydroperoxides are readily dissociated by heat, light, or metal ions to produce alkoxyl or hydroxyl macroradicals. In addition to abstraction of hydrogen atoms by these radicals to continue the chain reaction by forming new macroalkyl radicals, the alkoxyl radicals can undergo unimolecular scission reactions causing cleavage of the polymer backbone, thus generating further radicals and reducing the size of the polymer molecule. The bimolecular termination reactions of the radicals can lead to crosslinking resulting in an increase in molecular mass. Polypropylene undergoes mainly chain scission, whereas polyethylene undergoes predominantly crosslinking [20, 22, 23].

8.3.2. Polystyrene (PS)

The polystyrene chromophore absorbs very little terrestrial solar radiation. The cutoff varies with thickness but is considered to be at 280–300 nm. The formation of a charge-transfer (CT) complex between oxygen and polystyrene can extend the absorption to 340 nm or longer wavelengths [24], and absorption of commercial grades has been reported to extend into the visible region to wavelengths as long as 500–550 nm [25]. On exposure to the environment, the UV/visible radiation absorbed by the impurities introduced during polymerization and processing initiate photooxidation reactions leading to appreciable discoloration, brittleness, and loss in mechanical strength. Thus, the thermal history of PS has a significant effect on its sensitivity to the environment.

An acetophenone-type aromatic ketone attached to the polymer chain has been identified as one of the main thermal oxidation products responsible for the effect of solar radiation on the polymer. Also, peroxide groups in the polymer chain, formed by copolymerization of oxygen with styrene during synthesis (by the radical process only [26]), absorb solar UV and readily photolyze, leading to chain scission and photosensitization of the degradation of polystyrene. Other photosensitizing groups include hydroperoxides, conjugated double bonds, and traces of monomer. The nature of the free radicals generated depends on the wavelengths and conditions of irradiation [27] as well as on the polymerization process [26].

Abstraction of hydrogen from polystyrene by the aromatic ketones raised to the high-energy triplet state by absorption of UV is considered to be the most important initiation mechanism for polystyrene photooxidation [28]. Following formation of tertiary carbon free radicals, the oxidation of polystyrene proceeds according to the general scheme with formation of hydroperoxides, a primary photooxidation product. Decomposition of the tertiary hydroperoxides is considered the most probable cause for the observed chain scissions. The acetophenone-type structures are also formed as a product of the photooxidation reactions. Energy transfer from the excited aromatic ketones to hydroperoxides, followed by their decomposition, is another important initiation mechanism. The sensitized hydroperoxide decomposition increases in importance with increase in polymer oxidation levels and becomes the dominant initiation reaction [20]. The sensitization of polystyrene photooxidation by singlet oxygen formed by UV excitation of the $PS-O_2$ charge-transfer complex has been proposed as an alternate initiation mechanism [24]. The yellow color in photodegraded polystyrene is generally attributed to conjugated polyenes in the polymer backbone. It has also been attributed to alpha-diketone groups [29] and to isomerization or ring opening reactions of the benzene moiety [24].

In styrene–acrylonitrile (SAN) copolymers photooxidation is also initiated by acetophenone groups formed during processing [30]. However, in styrene copolymers containing polybutadiene, for example, impact polystyrene and ABS, the thermally produced hydroperoxides are considered responsible for photooxidative degradation [31, 32]. These copolymers are particularly sensitive to oxidation because the rubber phase, polybutadiene, is very sensitive to hydroperoxidation and is believed to sensitize the photooxidation of polystyrene. Carbonyl compounds do not seem to play an appreciable role in initiation of the photooxidation of these polymers [31]. The loss of impact strength of ABS resins on photooxidation has been correlated with the oxidation of the polybutadiene phase, that is, with the decrease of the double-bond content of the resins [33, 34]. It has been suggested that destruction of the elastomeric properties of the rubber phase is due to crosslinking of the polybutadiene phase [35]. Due to the slow diffusion of oxygen into the material, photoproducts are localized essentially in the surface layers where most of the double bonds disappear.

8.3.3. Aliphatic Polyamides (Nylons)

Exposure of aliphatic polyamides (nylons) to the environment causes discoloration and appreciable reduction in tensile strength and average molecular weight. Since the polyamide itself can absorb short-wavelength solar UV radiation, dual mechanisms of degradation initiated by sunlight have been identified in the aliphatic polyamides [36, 37]. Direct photolysis of the polyamide structure, followed by reactions of the resulting free radicals, is initiated by wavelengths as long as 340 nm absorbed by the structural components. Longer wavelengths, absorbed by impurities and defects, can only initiate photooxidation reactions. This is in contrast to other aliphatic-type polymers, such as the polyolefins, which degrade only via a photooxidation mechanism. Although the photolytic reaction is the more important weathering process [38], the thermal history of nylon polymers has a significant effect on the impurities present, and thus on the light stability.

Direct excitation of the amide chromophore and scission of the carbon–nitrogen bond adjacent to the methylene group in the amide linkage, —N(H)—CO—, is an important weathering process. In addition to reduction in the chain length, the methylene and amino radicals formed can abstract hydrogen atoms from adjacent polymer chains resulting in crosslinking, and they can also react with oxygen to undergo oxidation reactions, perhaps through a peroxide intermediate. The yellow coloration of photooxidized nylon polymers is attributed to the formation of pyrrole-type compounds [37, 39].

Wavelengths extending into the visible region [38] can excite defect and impurity chromophores in the polymer chain to initiate photooxidation reactions. In contrast to these reactions in other polymers, the hydroperoxides do not propagate new oxidation chains. Instead, following photochemical homolysis of the peroxidic bond, the free radicals formed preferentially recombine to imides and water. N-1-hydroxylated groups, which form to a minor extent, are thermally unstable above 60°C [40]. The main route of disappearance of the imide groups involves hydrolysis by the water that is formed in situ in close proximity to the imide groups. Thus, hydrolysis of the imide groups, which contributes to scission of the polymer chain, can occur even under "dry" conditions. Chain scission and reduction in molecular weight predominate over crosslinking in nylons. However, the susceptibility of the imide group to hydrolysis varies with the chain length of the nylon and its ability to absorb water. For example, nylon 6 absorbs more moisture than nylon 11 and shows greater changes in mechanical properties with increased humidity levels in the environment. The loss in impact strength of nylon samples was shown to clearly reflect the influence of absorbed water and seasonal humidity conditions of the exposure site [38]. When TiO_2 is incorporated, in the presence of moisture it tends to accelerate photooxidation [40]. Some dyes also act as photosensitizers.

8.3.4. Aromatic Polyamides (Polyaramids)

Aromatic polyamides are more susceptible to the effects of sunlight than the aliphatic types. Significant yellowing and rapid decrease in molecular weight and in mechanical properties result from exposure of unprotected aromatic polyamides to the environment. Aromatic polyamides are affected by wavelengths well into the visible portion of the spectrum. It has been shown that wavelengths in the 400- to 500-nm region cause an appreciable amount of deterioration [6, 38]. Similar to the aliphatic types, the aromatic types are also subject to two mechanisms of degradation: (a) direct photolysis of the polyamide structure by the radiation absorbed by the structural features and (b) photooxidation initiated by the radiation absorbed by the impurities. Because the structural chromophores absorb longer wavelengths than the aliphatic types, their direct photolysis can be initiated by longer wavelengths, and this degradation pathway is even more important than in the aliphatic types. Following cleavage of the amide bond, the main reactions in the absence of oxygen are a photo-Fries rearrangement to a 2-aminobenzophenone structure in the polymer and crosslinking [41, 42]. When oxygen is present, it reacts with the radical pair forming oxygenated products, thus competing with the photo-Fries rearrangement and reducing crosslinking and yellowing. Due to the high degree of orientation in polyamide fibers and thus the slow rate of diffusion of oxygen, photo-Fries products are formed to a large extent when the fibers are exposed to UV even in the presence of oxygen.

8.3.5. Poly(Vinyl Chloride)

Although the structural units of poly(vinyl chloride) (PVC) do not absorb solar UV radiation, the environment has a deleterious effect on the polymer. Its instability is due to the UV-absorbing impurities formed during polymerization and processing and thus to its thermal history. The primary UV-absorbing chromophores formed thermally are conjugated polyene sequences of various lengths produced by dehydrochlorination. Their spectral absorption properties include the full solar UV plus some short-wavelength visible radiation, giving the polymer its yellow (or darker) coloration. The thermal oxidation products, carbonyl groups, peroxides, and particularly hydroperoxides, common to most polymers, are all important photosensitizers for PVC degradation. Absorption of UV and some visible radiation by the thermally produced defects initiates further dehydrochlorination that proceeds ziplike in the polymer chain to form additional polyene structures. These structures have very large extinction coefficients [43] and therefore rapidly become the predominant absorbing species.

The excitation energy can be transferred from one polyene to another, causing chemical reactions that cleave the most labile bond, that is, the allylic C−Cl bond. Multistep photochemical excitations result in the formation of polyenic sequences with growing conjugation lengths and deepening of the color. The relative numbers of polyenes of different lengths is very dependent on the spectral power distribution of the light source. Irradiation by a light source "rich" in short wavelengths, such as the fluorescent UVB source, will favor the development of shorter polyenes. Irradiation by natural sunlight or a source that simulates its spectral irradiance will provoke the formation of a higher concentration of longer polyenes and thus the "blackening" of PVC [44].

The polyenic sequences are readily photooxidized in the presence of molecular oxygen. Reactions proceed primarily by a free-radical chain mechanism leading to chain scission (C−C bond cleavage) and crosslinking and result in deterioration of the mechanical and electrical properties. Photooxidation of the polyene sequences is accompanied by photobleaching, the extent depending on the emission spectrum of the light source [45]. Sources that emit only a narrow range of UV wavelengths and no visible radiation cannot excite all the polyene sequences and effect less photobleaching. Photooxidation also takes place via reaction of oxygen with the chlorine free radicals formed when the allylic C−Cl bonds cleave.

Due to the slow diffusion of oxygen into PVC, weathering is primarily restricted to the surface layers. Oxidation products develop only to a depth of about 200 μm from the surface, whereas polyenes are formed to a depth of about 400 μm [46]. The latter act as UV absorbers, shielding the bulk of the polymer from UV radiation. Progressive destruction of surface layers results in loss of gloss, embrittlement, and loss of impact strength. In impact PVC (ABS, MBS, or MABS modified), rapid photooxidation of the elastomer component is observed, leading to an acceleration of PVC degradation [47]. In flexible PVC, the loss of plasticizer during weathering results in property deterioration. In contrast to some other polymeric materials, moisture is not an important factor in the degradation of PVC itself. However, the effect of moisture on additives can be a significant factor in the weathering of PVC. For example, the photocatalytical activity of titanium dioxide in TiO_2-pigmented PVC exposed outdoors depends on the presence and amount of rain.

8.3.6. Polycarbonates

Polycarbonate (PC) is more stable to UV radiation than polyolefins and PVC. However, prolonged outdoor exposure of PC leads to changes in the molar mass, reduction in mechanical properties, particularly impact strength, and yellowing, properties that can be very important in some applications. The photodegradation mechanism of bisphenol-A (4,4′-isopropylidenebisphenol) polycarbonate depends on the initiating wavelengths. Degradation occurs via a photolytic pathway when PC is directly excited by wavelengths absorbed by the aromatic carbonate chromophoric groups. This mechanism is generally believed to be initiated by wavelengths shorter than about 310 nm, but the long wavelength limit increases with thickness. Direct scission of one of the C−O bonds of the carbonate group leads to the formation of two free radicals followed by a photo-Fries rearrangement to a polymeric phenyl salicylate type of structure. Absorption of somewhat longer UV wavelengths by the latter results in scission of the second C−O bond of the carbonate group, a second photo-Fries rearrangement and formation of an orthodihydroxybenzophenone type of structure in the polymer. In the presence of oxygen, some of the radicals react with it, leading to typical photooxidation chain reactions involving peroxyl radicals and abstraction of hydrogen atoms from the macromolecular chain to form hydroperoxides. Yellowing of polycarbonate by this mechanism is attributed, in part, to the rearranged product, but especially to the photooxidation products, orthoquinone and orthodiphenoquinone, which are highly yellow colored [48–51].

Initiation of photooxidation of PC by longer wavelengths is generally attributed to UV absorption by the defects and impurities, which are primarily thermal oxidation products. It has been suggested [50] that radical chain photooxidation processes caused by solar radiation may also result from the photo-Fries reactions initiated by the short wavelengths. The hydroperoxide intermediate that forms absorbs longer wavelength UV than the original impurities and photodegrades to a variety of oxidation products. Photoinduced oxidation on the side chains and phenyl rings [52] result in a mixture of highly colored ring oxidation species that are mainly responsible for the yellow color in polycarbonate by this mechanism [49, 50, 53–59]. Charge-transfer complexes of O_2/PC are believed to contribute significantly to photooxidation in the early stages [20]. Increase in the rate of photooxidation by hydrolysis becomes important on prolonged exposure [60]. The relative amount of degradation by the photo-Fries mechanism versus the autocatalytic photooxidation process when PC is exposed outdoors will vary with location, season, angle of exposure, and atmospheric conditions since these factors influence the short-wavelength cutoff of solar radiation. In blends of PC with PP, PC stabilizes PP when it is present at low concentrations (~10%), but when PC is the major component, PP sensitizes the photooxidation of PC.

8.3.7. Polyesters

The photooxidative stability of aliphatic polyesters is generally higher than that of polyolefins, although similar degradation mechanisms are involved. The thermoplastic aromatic polyesters, polyethylene terephthalate (PET) and polybutylene terephthalate (PBT), are intrinsically more light stable than polyurethanes or polyamides. However, prolonged weathering alters the chemical and mechanical properties because of chain scission at the carbon–oxygen single bond of the ester group in the polymer backbone. It leads to yellowing, surface crazing, brittleness, and reduction in tensile strength. Due to the strong absorption of solar UV, it does not penetrate into the inner layers and surface deterioration predominates [61]. The limitation of oxidation to the first 50 μm in PBT is due to absorption of UV and not to the diffusion of oxygen [62]. PET fibers lose their elasticity, elongation, and tensile strength and eventually brittleness results. In thicker forms the deeper polymer layers are more or less unattacked.

Similar to the photodegradation of polycarbonate and polyamides, dual mechanisms occur in the aromatic polyesters, with the relative contributions dependent on the incident wavelengths. Direct photolysis of the ester groups results from exposure to wavelengths absorbed by the structural components. Depending on the thickness, photolytic reactions can be initiated in PET by wavelengths as long as 360 nm and in PBT by wavelengths as long as 350 nm [20, 52]. Chain scissions result in reduction in average molecular weight. However, recombination of the free radicals that form lead to crosslinking as well as to benzophenone-type and conjugated structures responsible for yellowing. A photo-Fries rearrangement of the chemical structure has been identified in PET but not in PBT.

When oxygen is present, hydroperoxides are the main products formed during the initial stages of photooxidation. Their decomposition leads mainly to chain scission and products of photooxidation, some of which contribute to the yellowing of PET and PBT. In PET, crosslinking reactions under photooxidation conditions were found to be insignificant in comparison with chain scission reactions [63, 64]. Although in the presence of oxygen the yellowing due to the photolytic reactions is reduced by photobleaching, photooxidation products are formed three times as fast [52]. In PBT photo-induced oxidative processes are of much less importance than photolysis of the structural component. Hydroperoxides are present in fairly low concentration because of the low reactivity of oxidative sites [62]. Photooxidation reactions are also initiated by the longer wavelengths absorbed by the impurities. Thermoplastic polyester elastomers are very sensitive to photooxidation, mainly due to the presence of polyether segments. The rate of photooxidation increases with the content of polyether segments [64]. The main photooxidation products are hydroperoxides and their decomposition products.

Since the ester group in polyesters is subject to hydrolysis, moisture can have a very significant influence on photodegradation reactions. Water reacts chemically to form lower molecular weight polymers and some monomers, thus reducing the molecular weight and affecting tensile strength, elongation, and ductility [65]. In some cases, the hydrolysis reaction that takes place after water has been absorbed and diffused into the polymer becomes autocatalytic [66]. Loss in strength is related primarily to the tendency to absorb surface moisture. Exposure of PET yarns and fabrics to UV radiation under high-humidity conditions accelerates crazing, loss in ductility, and strength. It has been shown that with increase in relative humidity, weight loss of thin films of polyester resins increase, rather than decrease [67]. The effect of humidity varies with the type of resin. The resistance to water and oxidation is conferred by neopentyl glycol, methacrylate esters, and especially by isophthalic acid compositions.

8.4. STABILIZATION OF POLYMERIC MATERIALS AGAINST ENVIRONMENTAL EFFECTS

8.4.1. Types of Stabilizers

The commercialization of many types of organic polymers for use in outdoor applications has been possible due to the successful development of stabilizers that function in different ways. Some control the amount of actinic radiation that reaches the bulk of the polymer and others inhibit the chemical reactions that follow absorption of the radiation. The former include inorganic pigments and carbon black, usually referred to as light screeners, and ultraviolet absorbers. The degradation processes initiated by the radiation absorbed are inhibited in a number of ways. The excited molecular species are deactivated by additives, called *excited-state quenchers*, that prevent initiation of chemical reactions by transferring and dissipating the absorbed energy harmlessly. Photoantioxidants, such as decomposers of hydroperoxides, are used to transform labile intermediates into more stable compounds, thus preventing their photodecomposition and the formation of free radicals. An important type of inhibition is the scavenging of free radicals that have been allowed to form.

Photostabilizers are used in very low concentrations but can significantly lengthen the lifetime of most organic polymeric materials. They are generally classified according to their principle mode of action. However, some stabilizers contribute to the protection of polymers by more than one mechanism. For example, hydroxyphenylbenzotriazoles deactivate the excited states of carbonyl groups in addition to screening UV radiation. Initially, only screening agents and UV absorbers were available as light stabilizers. Later, nickel chelates offered additional protection, but the introduction of sterically hindered amine light stabilizers (HALS), developed more than 20 years ago, significantly advanced the ability to protect polymers against the environment.

For effective stabilization, the additives should have optimum physical properties, such as solubility in the polymer, resistance against volatility and leaching, compatibility with the polymer and other additives, and good photostability. Physical incompatibility can result in significant loss of the stabilizer from the polymer, particularly from thin specimens [68]. In the past two decades improvements have focused on developing low volatility, higher molecular weight stabilizers, as well as some with functional groups that could be grafted to the polymer to reduce stabilizer loss. A recently introduced approach to reduce volatility and extractability, which has shown some promise, is the grafting of a variety of classes of stabilizers to polysilanes with functional groups [69].

Thermal stabilizers used during processing and fabrication play an important role in protecting polymers against environmental effects by reducing the formation of ultraviolet absorbing impurities that sensitize and accelerate degradation under service conditions. Thermal stabilizers that are photochemically stable also protect polymers against thermooxidation by the environment. However, this section is primarily concerned with the stabilizers that either screen or interfere with the photochemical effects of solar radiation.

Light Screeners and Ultraviolet Absorbers All light-screening agents protect by restricting UV penetration to very short distances from the surface and thus confining the damage to thin surface layers. To be effective in stabilizing the polymer, the screening agent must have a larger absorption coefficient than the polymer for the actinic wavelengths. Stable forms of carbon black are very effective in stabilizing many outdoor grades of polymers. It strongly absorbs all wavelengths in both the UV and visible regions. In addition to the screening function, the highly condensed aromatic ring structure and impurities on the carbon black particle surfaces give it radical scavenging and antioxidant abilities [70, 71]. Recent work has demonstrated that it can also be an effective quencher of excited states [72]. However, it can have prooxidant effects under certain conditions. The carbon black properties are determined by its mode of manufacture and are related to the type and particle size as well as its concentration and dispersion in the matrix. It can provide effective protection for high-density polyethylene (HDPE) at a concentration as low as 0.05% [73]. Stable forms of titanium dioxide are also effective in reducing the degradation of many polymers caused by solar radiation. It strongly absorbs most of the actinic UV. In general, the efficiency of protection increases with decreasing particle size of the pigment [74], but the effect is limited by the agglomeration of very small particles into clusters that are not easily dispersed. Photoactive forms of these pigments as well as other types of pigments, for example, cadmium sulfide, promote photodegradation by photosensitizing the polymer [47].

Ultraviolet absorbers were among the first organic stabilizers used. They are colorless compounds that strongly, but selectively absorb ultraviolet radiation and harmlessly dissipate it as heat so that it does not lead to photosensitization. They are also characterized by their very good stability to the absorbed radiation. However, based on the UV absorption mechanism alone, they can only provide limited protection to surface layers and thin samples, for example fibers and films. In accordance with the Beer–Lambert absorption law, the amount of radiation reaching any particular layer diminishes exponentially with the distance from the exposed surface. Thus, the effectiveness of protection via screening of the actinic radiation from the polymer by the UV absorber increases with sample thickness. Protection by UV absorbers is most effective when the additive is concentrated on the surface, such as when it is incorporated in a thin film coextruded over the polymer [75].

Although the main mechanism of protection by UV absorbers is competitive absorption of the radiation harmful to the polymer, many also protect by interaction with products formed on UV exposure, such as by quenching (deactivation) of excited states and by scavenging free radicals, for example, alkoxyl radicals. These mechanisms are independent of sample thickness and are applicable to surface layers. They are effective in thin films (about 50 μm) and retard changes in surface-related properties, such as loss of gloss, chalking, surface roughening, and the like. In contrast, UV absorbers only become effective for film thicknesses greater than 100 μm. Protection by the screening mechanism has a greater effect on mechanical properties than on surface changes [20].

Derivatives of the 2-hydroxybenzophenones and 2-hydroxybenzotriazoles have been the most commonly used UV absorbers. Hydrogen bonding of the ortho-hydroxy groups is essential for their spectroscopic properties and photostability [23]. The absorbed energy is dissipated through internal conversion mechanisms involving reversible formation of the hydrogen-bonded ring. The oxanilide-type UV absorbers provide only minor protection through the UV absorption mechanism. The main mechanism has been reported [76] to be quenching of excited states in photooxidation reactions by chemically combining with carbonyl groups or other chromophores in the ground state. They are used in combination with other UV absorbers to effectively protect polyolefins. The hydroxyphenyltriazines, which recently became available, are used mainly for protection of engineering plastics. They have higher extinction coefficients than the benzophenones and benzotriazoles and are assumed to protect efficiently by the UV absorption mechanism [77]. The structural features of various types of UV absorbers and their UV absorption characteristics are described by Gugumus and Zweidel [77] and Pickett [78].

Excited-State Quenchers and Hydroperoxide Decomposers Deactivation (quenching) of excited electronic states of potential photosensitizers by energy transfer to an acceptor provides a method for dissipation of some of the actinic energy absorbed by the chromophore. The quenchers dissipate the energy either in the form of heat or as fluorescent or phosphorescent radiation. For example, deactivation of excited carbonyl species by transition-metal chelates reduces backbone cleavage caused by carbonyl photolysis. Metal chelates can also deactivate singlet oxygen, a highly active form of oxygen. The latter forms when energy is transferred to the ground state of oxygen from a chromophore that has absorbed UV or visible radiation. Excited-state quenchers are particularly useful in stabilizing thin sections, such as films or fibers, since their action is independent of the thickness of the materials to be protected. Some nickel-containing stabilizers that quench excited states may also act as radical scavengers and/or as hydroperoxide decomposers in polypropylene and polyethylene [79, 80].

Hydroperoxides formed during processing play a crucial role in the photooxidative degradation of polymers. They are more potent initiators of photodegradation than carbonyl groups. An important method of stabilization is decomposition of the hydroperoxides into inactive products to prevent generation of radicals. It can be achieved by reaction with phosphite esters or nickel chelates or by catalytic action of compounds such as dithiocarbamates and mercaptobenzothiazoles and others [81]. Metal complexes of sulfur-containing compounds are very efficient hydroperoxide decomposers in catalytic quantities and also act as efficient quenchers of excited states.

Free-Radical Scavengers Free-radical scavengers, additives that trap oxidation propagating free radicals, limit the damage caused by actinic radiation by breaking the oxidation chain. The importance of UV stabilization via reduction of the free-radical concentration has been known for many years [70, 82]. A number of compounds have been used for this purpose [23]. The HALS, the latest type to be introduced, are very effective for a number of types of polymers and are being used extensively as photostabilizers in commercial polymers, either alone or in combination with a UV absorber [83, 84]. They are a unique class of stabilizers in that the parent compound has to be oxidized to the corresponding stable nitroxyl radical before becoming effective. The nitroxyl radical inhibits photooxidation by scavenging alkyl radicals in competition with oxygen to form alkoxy hindered amine derivatives. The latter then further inhibit oxidation by reacting with alkyl peroxy radicals, thus reducing their ability to abstract a hydrogen atom from the polymer and propagate free-radical chain reactions. In the process the nitroxyl radicals are regenerated from the corresponding hydroxylamine and alkylhydroxylamine [23, 85]. Thus, efficient stabilization is realized through trapping both alkyl and alkyl peroxyl radicals and regeneration of some of the active species, which prolongs the lifetime of the stabilizer.

A number of other stabilizing reactions have also been proposed for HALS [20, 81]. Some investigators believe that they catalyze hydroperoxide decomposition [85]. Gugumus reports [20, 86–88] that the HALS-induced decomposition of hydroperoxides has been demonstrated unequivocally with preoxidized polypropylene in the solid state at room temperature and in the absence of light. However, others believe that this mechanism is too slow to significantly contribute to stabilization at moderate temperatures. Another role attributed to HALS is inhibition of the catalytic action of metal impurities in the polymer by forming complexes with them, thus preventing formation of metal ion–hydroperoxide complexes that initiate autooxidation. HALS are generally of little use as stabilizers if acids are present because they react to form salts, which offer no weather resistance.

The HALS stabilizers neither act as UV screens nor excited state (singlet oxygen or triplet carbonyl) quenchers. Their UV-stabilizing action is antagonized by the presence of antioxidants that remove alkyl peroxyl radicals and hydroperoxides. However, because they do not stabilize polymers during processing, antioxidants and thermal stabilizers have to be incorporated along with the HALS. Effective synergism has been achieved during processing when some secondary and tertiary hindered amines are used in combination with both aliphatic and aromatic phosphites [89].

Combinations of Stabilizers Combinations of different types of stabilizers are often used to reduce degradation through protection by more than one mechanism. In many cases, the combination gives synergistic protection, that is, it provides better protection than the sum of the effects of each when used alone. The combination of HALS with a UV absorber, such as a benzophenone or benzotriazole type, was an important new development in stabilization of polyolefins and has since been used for other polymers. However, the effect of combining a UV absorber with HALS cannot be predicted. It depends on the mechanism of degradation of the polymer, which can vary with the manufacturing process, and the mechanism(s) of protection by the stabilizer. In some cases, antagonism results so that the effectiveness of one of the stabilizing additives is diminished by the presence of the other. For example, the combination of a benzophenone-type UV absorber with a low-molecular-weight HALS was shown to be antagonistic in one type of HDPE but synergistic in another type of HDPE [76]. The effectiveness of the combination also depends on thickness. Synergism is almost never manifested in very thin films. For example, synergism was exhibited in 0.1-mm films of LDPE and PP but not in tapes of PP, which are only 0.05 mm thick [76].

The combination of UV absorbers, each acting only by a screening mechanism, can be expected to have just an additive effect. However, synergism has been exhibited by combinations of UV absorbers from different classes due to the simultaneous protection by other mechanisms. For example, combinations of oxanilides with either the benzophenones, benzotriazoles, or hydroxyphenyltriazines are strongly synergistic in polyolefins. Combinations of benzophenones and benzotriazoles can be synergistic or antagonistic, or just additive, depending not only on the type of polymer but on the manufacturing process. Thus, the combination is synergistic in Phillips-type high-density polyethylene but antagonistic in the Ziegler type. Similarly, it is antagonistic in the conventional-type linear low-density polyethylene but synergistic in the metallocene-type product [76–90].

Specific combinations of low- and high-molecular-mass HALS as well as the combination of two kinds of polymeric HALS have given pronounced synergistic protection to polyolefins [91]. However, the combination of two low-molecular-mass HALS usually does not have more than an additive effect and some specific compounds may have a strong antagonistic effect. Some combinations of low- and high-molecular-mass HALS may also be antagonistic to one another. The performance of the combinations of two HALS stabilizers cannot be predicted but must be determined experimentally.

Synergistic and antagonistic reactions can occur between stabilizers or antioxidants and pigments in both thermal and photochemical degradation. For example, in photooxidative stabilization of polypropylene, rutile titanium dioxide was shown to be synergistic with hindered phenolic antioxidants and HALS but antagonistic with benzotriazole and benzophenone absorbers [92]. In the presence of pigments, such as titanium dioxide and phthalocyanine blue, the contribution of the UV absorbers is usually too small to be of any use. The effect can even become antagonistic. However, with some organic pigments, especially the yellow and red ones, a UV absorber is added to protect not only the polymer but, more importantly, the pigment. White and Turnbull [81] have reviewed studies on synergistic and antagonistic actions of different types of stabilizers, primarily in polyolefins.

8.4.2. Stabilization of Polymers

Polyolefins In addition to stabilization of polyolefins against thermal oxidation to reduce the sensitivity to light, stabilization against exposure to light is required for articles to be used outdoors as well as those intended for indoor use [93]. Light stabilizers include UV absorbers of the benzotriazole and the benzophenone types (except for thin sections), HALS, and nickel-containing stabilizers. The latter are used for thin sections such as tapes and films and for surface protection. The type of stabilizer is dictated by the type of polyolefin, its thickness, application, and desired lifetime of the article [20].

The performance of low-molecular-weight HALS for stabilization of polypropylene fibers, films, and tapes is superior to that of other types of light stabilizers even at one tenth the concentration. The performance increases significantly in the concentration range 0.05–1.2%. However, when a pigment is present that acts as UV screener, the contribution of HALS to the light stability of PP multifilaments is considerably reduced. Polymeric HALS is superior to low-molecular-weight HALS in heat-treated fiber as well as in fiber treated with acrylic latex. The inferior performance of low-molecular-weight HALS is due to its tendency to migrate with heat treatment in the former case and to its extraction by the acrylic latex in the latter case.

The combination of HALS with a UV absorber is used in films of polypropylene and polyethylene as well as in thick sections. In films of LDPE, nickel quenchers were commonly used with a UV absorber, except in a very thin film, in which a higher concentration of nickel stabilizer is superior to the combination. The low-molecular-weight HALS are not sufficiently compatible with LDPE at the concentrations necessary, possibly as high as 2%, for the required protection. Incompatibility of HALS with LDPE has been overcome with the development of polymeric HALS. It is considerably better than either the UV absorber or nickel quencher or combinations of the two. For thicker films (100–200 μm), the combination of a benzophenone-type UV absorber with polymeric HALS is significantly superior to an equivalent amount of polymeric HALS. The type of stabilizers used for linear low-density polyethylene (LLDPE) and ethyl vinyl acetate (EVA) copolymer are similar to those for LDPE. Since LLDPE has superior mechanical properties (elongation at break and tensile strength), thinner films can be used for most applications, and the loss of UV stability with reduction in thickness has to be compensated for by improving the stabilization system.

For UV stabilization of thick sections of all polyolefins, the combination of HALS and a UV absorber are used. The poor efficiency of HALS in unpigmented plaques is considerably improved with the addition of a UV absorber to protect the bulk of the material. However, in titanium-dioxide-pigmented polyethylene, the UV absorber is surpassed by HALS at much lower concentration. Both the stability of the polymer and the lightfastness of pigment are improved. The effect of adding a UV absorber is only marginal because the role of the UV absorber is taken over by the pigment. The effect of 0.5% of the pigment is much greater than the effect of 0.05% UV absorber. For LDPE and LLDPE, only polymeric or high-molecular-mass HALS and some UV absorbers can be used, but for PP, low-molecular-weight HALS may be better than the polymeric, especially in the unpigmented polymer.

Elastomers Stabilization with additives other than carbon black is limited to unpigmented or light-colored elastomers. Substituted salicylanilides protect natural rubber (NR) against UV by screening, and various phenols and aromatic amines inhibit hydroperoxide formation on photooxidation of polybutadiene. Combinations of phenolic antioxidants with benzotriazole UV absorbers and low-molecular-weight HALS are used for stabilization of thermoplastic rubbers (styrene block copolymers) intended for hot melt and solvent-based sealants. High loadings of a UV absorber in combinations with an antioxidant and HALS improves the light stability of a polyurethane sealant.

Styrenic Polymers Stabilization of polystyrene and its copolymers is necessary for articles expected to be exposed to solar radiation or indoor fluorescent lighting. Because of the significant role played by thermal oxidation products in the effect of these sources on the polymers, thermal stabilization at the processing stage is required to reduce their sensitivity to light. The use of a phenolic antioxidant was shown to increase the retention of mechanical properties and, when used in combination with a HALS, further improvement was obtained [94]. It was also shown that certain phenolic antioxidants used in combination with a UV absorber afforded greater light stability than the UV absorber alone [95]. However, the combination of a low-molecular-weight HALS with a UV absorber, such as the benzotriazole-type, provided synergistic stabilization and superior performance over the earlier formulations [20], even for ABS and impact-resistant polystyrenes that are very sensitive to oxidation because of the butadiene component. HALS effectively reduces the fast loss of impact and tensile strength due to surface degradation, and the UV absorber protects the deeper layers that contribute more significantly to the yellowing. Protection by the benzophenone- and benzotriazole-type UV absorbers has been attributed to both the screening mechanism and to transfer of energy from the excited carbonyl groups, that is, the quenching mechanism [96].

Polyamides: Aliphatic and Aromatic Polyamides require stabilization to reduce the appearance and mechanical changes caused by exposure to the environment. Various types of stabilizers are used, often in combination, in both aliphatic and aromatic types to effectively protect them. Carbon black is a very effective stabilizer. It has been reported that nylons containing carbon black retained a reasonable proportion of their tensile properties after 12 years in the tropics [38]. Reduction in yellowing, an important practical problem, is achieved by the addition of ultraviolet stabilizers or suitable pigments or dyes [20, 97]. Good light stability was also achieved with HALS, both as free molecules and bound to the chain ends [98]. However, the best performance of nylons is obtained when it is stabilized with the combination of a UV absorber, a hindered amine stabilizer, and a phenolic antioxidant [20], analogous with results found with other thermoplastics such as polyolefins and styrenic polymers. In aromatic polyamides, benzotriazole-type UV absorbers act by quenching the excited states as well as by absorption of UV radiation [20]. The addition of a metal deactivator was shown [98] to effectively inhibit the photosensitized oxidation of nylon 6,6 by iron oxide, a known trace-metal salt introduced during the manufacturing process. The addition of cupric ion to aromatic polyamide fibers at concentrations of about 2.5% was shown [99] to significantly retard photodegradation, although it was insufficient to stabilize the polymer for certain long-term end-use applications.

Polyvinyl Chloride (PVC) The light stability of PVC without added light stabilizers is a function of the thermal stabilizer system. Because of the high temperatures required for its compounding and fabrication, good thermal stabilizers must be used during processing in order to provide a product with adequate weatherability [100, 101]. Scavenging of HCl and inhibition of formation of hydroperoxides or promotion of their decomposition to nonradical species are the most important stabilizing mechanisms. Ba/Cd salts and organic tin carboxylates confer some UV stability to rigid PVC on outdoor exposure by inhibiting the formation of UV-absorbing groups during processing. However, for adequate performance, the use of stabilizers that protect against the damaging effects of sunlight is mandatory, especially for transparent and translucent articles, even with the use of a good thermal stabilizer system.

Outdoor performance can be greatly improved by the addition of an ultraviolet absorber, such as the orthohydroxybenzotriazole or orthohydroxybenzophenone type, particularly when used in combination with an HCl scavenger [101]. In pigmented PVC, appropriate thermal stabilizers in combination with adequate pigments can confer sufficient UV stability for numerous outdoor applications. For example, rutile TiO_2 at high loadings, up to 12% and more, imparts superior light stability. In some cases, addition of a UV absorber is required for best performance. Some pigmented formulations, such as those containing photosensitive forms of TiO_2 and pigments having extractable iron as a contaminant, catalyze degradation of PVC. The metal can be introduced as a component of the darker color pigments or as an impurity in one of the additives [102–104].

Low-molecular-weight HALS has almost no effect, except in impact-modified PVC in which HALS in combination with a benzotriazole-type UV absorber outperforms UV absorbers used alone in the same concentration. In TiO_2-pigmented impact-modified PVC, HALS is much more effective than UV absorbers. In transparent plasticized PVC films used in greenhouses, stabilization by HALS is comparable to that of the benzophenone- and benzotriazole-type UV absorbers. However, the best performance is obtained with the combination of HALS and a UV absorber.

Polycarbonate (PC) Protection of polycarbonate materials against the effects of solar radiation is mandatory to reduce yellowing and loss of tensile strength. Incorporating a UV-absorbing stabilizer, typically a benzotriazole type, has been shown to effectively protect a 1-mil film of bisphenol A polycarbonate (BPA-PC) against both forms of degradation caused by exposure to UV radiation [105]. A non-UV-absorbing stabilizer, such as one of the nonreactive types of hindered amines, used either alone or in combination with the UV absorber, offers some improvement by reducing surface degradation. However, for adequate protection against the environmental effect on impact strength, the surface requires either a coating containing a high concentration of an ultraviolet absorber or impregnation with a solution containing the UV absorber. Thermal stabilization during processing to reduce the initial color of a PC article is usually accomplished with a phosphite additive. The yellow color formed during processing is reduced on initial exposure, particularly if the light source emits long-wavelength UV and short-wavelength visible radiation.

Polyesters UV absorbers have been used as efficient stabilizers of the linear polyesters, polyethylene terephthalate (PET) and polybutylene terephthalate (PBT) as well as for improving the lightfastness of dyes in polyester fibers during outdoor exposure. Effective stabilization of high-molecular-weight PET fibers by a UV absorber was shown [106] to depend on the level of uptake and its uniform distribution throughout the fiber cross section. A benzophenone-type stabilizer provided considerably better protection than the benzotriazole- and benzotriazine-type stabilizers because it diffused throughout the fiber whereas the other two stabilizers only penetrated peripherally. Since the UV absorbers act as carriers for the dye, the diffusion pattern of the UV stabilizer also influences the penetration of the dye and its protection by the UV absorber.

Low-molecular-weight HALS, a hydroperoxide inhibitor, has been shown to considerably reduce photooxidation of PBT [107]. It can be expected that the combination of a UV absorber with HALS will be particularly effective. There is little data on stabilization of thermoplastic polyester elastomers, but the stabilizers used to protect polymers that degrade by free-radical chain oxidations should also be efficient for these types. Thus, UV absorbers of the benzotriazole and benzophenone types may be used for their screening capabilities, and synergistic effects can be expected with the addition of HALS to reduce the rate of chain oxidations.

8.5. ENVIRONMENTAL WEATHERING TESTS

It is well known that the type of environment to which a material is exposed varies with geographic location because of differences in intensity and spectral power distribution of solar radiation, temperature, moisture, and atmospheric pollutants. Tests are commonly carried out in environments that have the most severe conditions, such as subtropical or desert climates, in order to determine resistance to the worst conditions. Most natural and accelerated outdoor weathering tests in the United States are carried out in either (or both) south Florida or central Arizona. These have the two most important "bench mark" environments where materials typically fail fastest due to intensification of the weather elements responsible for degradation. South Florida has a subtropical climate that is particularly destructive to materials that are sensitive to moisture. The central Arizona desert has become a worldwide recognized standard exposure environment for testing in a climate typical of the hot and dry conditions of the desert. In the summer, solar irradiance and temperatures are higher than those in south Florida. The large daily and summer-to-winter temperature swings also differentiate Arizona from south Florida environments. Other test sites are in tropical and temperate climates, industrial areas, and salt air locations. Often, the same materials are tested in several different climates.

Because of seasonal and year-to-year climatological variability of outdoor conditions in any one location, repeated testing during different seasons and over a period of at least two years is usually recommended. Results conducted for less than 12 months will often depend on the particular season of the year in which they begin. In order to take into account the variability of weather, the performance of test materials should be evaluated in comparison with that of control materials having similar composition and construction and known performance, rather than in terms of absolute property change. The use of two control materials, one having good durability and one having poor durability, and at least two, preferably three, replicate specimens of each test and control material is recommended.

Outdoor weathering tests are commonly characterized as natural or accelerated. The term *natural* is generally used for outdoor exposure on fixed-angle racks in locations and orientations that maximize the effects of weathering components, particularly solar radiation. Temperature and wetting are also intensified by various techniques of natural outdoor testing exposures to produce higher degradation rates than materials experience under normal end-use conditions. The variables and their effects are described in American Society for Testing and Materials (ASTM) standards and other references [2, 108–111].

Techniques to further optimize annual radiant exposure to solar radiation and temperature of specimens have included changes in exposure angle with seasonal changes of the sun's path and the use of a motor-driven, follow-the-sun rack, the Track Rack. The latter is an equatorial mount, which is designed to maintain noontime conditions from sunrise to sunset by keeping the sample surface at a constant near-normal angle to the direct solar beam. It is used primarily in arid environments such as central Arizona where the direct beam component of solar radiation is high. Its efficiency is low in subtropical environments where the diffuse portion of solar radiation is a significant portion of the total irradiance. The option of adding water spray to Track Rack exposures provides intensification of the three critical weather factors, radiant energy, temperature, and moisture, which can significantly accelerate degradation rates over static exposures for many materials.

In spite of optimization of weather factors in natural weather tests, long periods of outdoor exposure are required to evaluate the durabilities of many of the new polymeric materials and formulations that have been developed with vastly improved stability over previous types. Therefore, *accelerated* outdoor exposures that further increase temperature and/or solar irradiance and moisture are often used. Accelerated outdoor weathering of polymeric materials are described in ASTM D4364 [112], ASTM D4141 [113], and SAE J1976 [114]. Options include: (a) black-box exposure and its modifications for higher panel temperatures and longer wetness times than exposures using open or backed sample racks and (b) exposures using Fresnel reflectors for increased irradiance and temperature.

In black-box exposures, the test panels form the top surface of an open aluminum box painted with flat black paint on the outside. The box is typically positioned at 5° from the horizontal, facing the equator, but can be positioned at any angle. It provides the high temperatures encountered by surface coatings, vinyl rooftops, and decorative trim on automobiles exposed to direct sunlight. The insulating effect of the warmed air trapped in the box dampens out temperature variations due to passing clouds and intermittent breezes. The black box also maintains surface condensation longer after daybreak, thus providing greater probability of interactions between irradiance and moisture than in other types of exposures.

The heated black box is a modification that produces significantly higher panel temperatures than those produced by the black box. It is equipped with heaters and blowers to heat the air inside to a specified temperature during daytime hours.

Heating of the backside of the panels with warm air reduces the daily variability of specimen temperatures. The heater is turned off at night to allow specimens to cool down and thus allows for formation of condensation to simulate in-service conditions. It is most useful for exposures conducted in the late fall, winter, and early spring.

The combination of a heated black box covered with 3-mm-thick clear tempered safety glass (automotive side-window glass) and the Track Rack system is referred to as the CTH Glas-Trac (Controlled Temperature and Humidity, Under Glass, Sun Tracking). The equipment was designed primarily to determine the durability of automotive interior materials and to use various windshield types for the glass cover. The exposure cabinet is controlled at $70 \pm 5°C$ air temperature during daylight hours and at $38 \pm 5°C$ air temperature and $75 \pm 10\%$ relative humidity during nighttime hours. Exposures are timed in megajoules per square meter of total UV (295–385 nm). Details of the test are given in SAE J2230 [115].

The maximum acceleration of aging processes in outdoor weathering is obtained by exposure on a Fresnel-reflector panel rack that provides high-intensity solar radiation by following the sun and reflecting the sun's rays from an array of 10 flat mirrors onto a single target area of the test material. A blower forces air across the target to cool the test specimens and generally limits sample surface temperatures to about 15°C above the maximum temperature of equatorially mounted samples exposed to unconcentrated normal incidence radiation. Testing of samples exceeding 13 mm (0.5 in.) in thickness is not recommended because the cooling may not be sufficient for such samples. Periodic spraying with deionized water is designed to simulate and accelerate the results of conventional testing in semihumid subtropical and temperate regions. Nighttime spray cycles simulate rain and dew. Tests carried out in the absence of a programmed moisture cycle are intended to simulate conventional exposure testing in desert, arid, and semiarid regions. The test can be used to intensify either direct exposures or those behind glass. Descriptions of the device and guidelines for its use are given in ASTM D4364 [112] and ASTM G90 [116].

Since the device concentrates only the direct rays of the sun and not the diffuse radiation, it requires clear atmospheric conditions with little moisture, such as is prevalent in Arizona. The solar flux is approximately eight times as high as on an equatorial mount without mirrors, but acceleration factors over conventional outdoor tests based on elapsed time to a predetermined change in property have been reported to vary from 2 to 11 for various polymer types and compositions [8]. The effectiveness of the Fresnel-reflector accelerated weathering machines depends on the quantity and quality of the UV in the direct beam component of solar radiation. The acceleration factor will be small for materials that are degraded primarily by short solar UV wavelengths because of the deficiency of the machine in this region compared with global solar radiation. This is due to exclusion by the solar concentrator of diffuse solar radiation, which contains much of the short-wavelength solar UV, and to lower reflectance of the mirrors at short wavelengths. The acceleration factor will also vary with the material tested because all materials are not affected equally by increased irradiance and temperature. Thus, the stability rankings may differ from that of natural exposure. Good correlations as well as discrepancies have been reported between Fresnel-reflector exposures and standard outdoor tests based on stability ranking of polymeric materials [117–122]. Any large intensification of weather stress factors over those present under end-use conditions must be used with caution to avoid changing the types and mechanisms of degradation or altering stability rankings due to differences among materials in their response to intensified stress factors.

8.6. LABORATORY-ACCELERATED WEATHERING TESTS

Laboratory-accelerated test devices that utilize artificial light sources have been used for more than 80 years to support the development of progressively more weatherable formulations as well as for quality control and specification tests. The devices allow control of the three main test parameters, that is, radiation, heat, and moisture, for more consistent exposure conditions compared with their variability outdoors. The consistency of exposure conditions is often monitored with a weathering reference material, such as the AATCC blue wool standards [123] or the polystyrene chips [124] to improve reproducibility of laboratory tests. Often, problems related to reproducibility can be traced to differences in replicate specimens because of nonuniformity in materials or differences among polymer batches. Specimens cut from the same sheet can vary in sensitivity because of differences in content of UV-absorbing impurities or in nonuniform distribution of stabilizers. Aliphatic-type polymers are particularly prone to effects of nonuniform distribution of impurities since only the impurities are capable of absorbing terrestrial solar radiation. Acceleration over real-time weathering is realized by continuous exposure to defined conditions, uninterrupted by diurnal and seasonal cycles or variations in local weather conditions. Further acceleration can be achieved by intensification of irradiance, temperature, and moisture. However, the tests are only useful if they simulate the effects of outdoor exposure along with accelerating the rate of degradation. The validity of the tests requires that the chemical and physical changes induced in materials and the stability rankings of materials are representative of the effects of the weather. For this reason, the relative intensities of the stress factors that influence the degradation process must be similar to those that exist in the natural environment so that the complex interactions of their effects are reproduced.

The single most important consideration when conducting laboratory-accelerated weathering tests is the spectral power distribution (SPD) of the radiation source. Both the absorption of light, which is a prerequisite to degradation, and bond breakage, the primary photochemical step following absorption of light, are wavelength dependent. The wavelengths absorbed and the amount of radiation absorbed at each wavelength are dependent on the relation between the absorption properties of the material and the SPD of the light source. Therefore, matching the SPD of terrestrial solar radiation, that is, the incident wavelengths and their relative intensities, is critical to reproducing the degradation caused by exposure to natural weather conditions. If the spectral emission properties of the accelerated test source differ from that of solar radiation in the UV and visible regions, the mechanism and type of degradation as well as the stability ranking of materials can be distorted compared with the effects of the natural environment. A close match to the short wavelength cut-on of solar radiation is essential, particularly when testing polymers such as polycarbonate or isophthalate-based esters and others that are very sensitive to small changes in radiation in this region. It is now an established fact that differences in the ratio of long- to short-wavelength UV irradiance can alter the mechanism of degradation of many aromatic-type polymers, as well as the stability ranking of these materials. It can also distort the antagonistic reactions produced in some polymers by these spectral regions (see Section 8.6.2). Misleading information is often obtained when the full range of actinic solar UV and visible radiation is not adequately reproduced in laboratory-accelerated tests [125]. Nevertheless, the SPD of some of the light sources commonly used in artificial weathering devices has very little resemblance to the SPD of hemispherical solar radiation on Earth's surface, referred to as "daylight."

Due to differences in mechanism of degradation with initiating wavelengths, the effectiveness of light stabilizers that work by interfering with the mechanism of degradation will depend on the light source used for testing it. For example, hindered amine stabilizers (HALS) have been shown [48] to protect polycarbonate and aromatic polyurethanes only against photooxidation by long-wavelength UV. These stabilizers are ineffective against the photo-Fries reactions promoted by short wavelengths. Therefore, their effectiveness in protecting these polymers against solar radiation cannot be reliably determined using a light source that has shorter wavelength radiation than sunlight.

8.6.1. Light Sources in Laboratory-Accelerated Test Devices

The light sources commonly used for laboratory-accelerated weathering tests include filtered xenon arcs, two types of filtered carbon arcs, fluorescent UV lamps, and metal halide lamps. The spectral power distributions are shown in Figures 8.3 through 8.6 in comparison with solar radiation. The spectral emission characteristics of the sources differ significantly in both the UV and visible regions.

Filtered Xenon Arcs Xenon arc radiation appropriately filtered for the UV region closely matches terrestrial solar radiation in the short-wavelength solar cut-on region and provides good simulation to it in both the UV and visible regions. The xenon arc has become established worldwide as the radiation source used for optimum simulation of the effects of daylight, both direct and through window glass, depending on the type of filters used. Figure 8.3 shows the SPD of a water-cooled xenon arc filtered with the combination of a near-infrared absorbing glass cylinder (CIRA) and a soda lime glass cylinder (C) compared with the SPD of

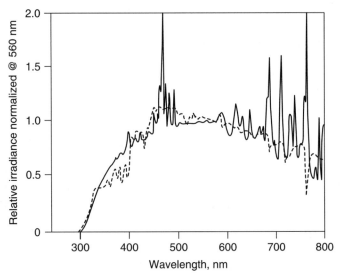

Figure 8.3. Spectral power distributions of the water-cooled xenon arc with coated infrared absorbing (CIRA) quartz inner and soda lime glass outer filters (—) and noon daylight in Miami, FL at 26°S exposure during the spring equinox (---). (Courtesy of Atlas Material Testing Technology LLC.)

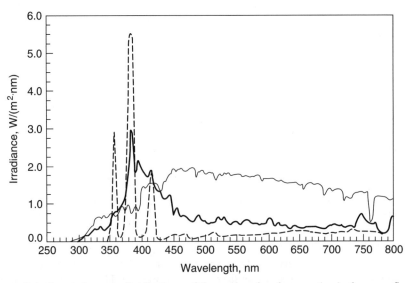

Figure 8.4. Spectral power distributions of the enclosed carbon arc (---), the open flame carbon arc with Corex D® Filters (—), and noon daylight in Miami, FL at 266°S exposure during the spring equinox (—). (Courtesy of Atlas Material Testing Technology LLC.)

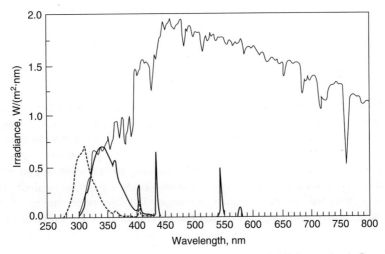

Figure 8.5. Spectral power distributions of fluorescent UVB-313 lamps (---), fluorescent UVA-340 lamps (—), and noon daylight in Miami, FL at 26°S exposure during the spring equinox (—). (Courtesy of Atlas Material Testing Technology LLC.)

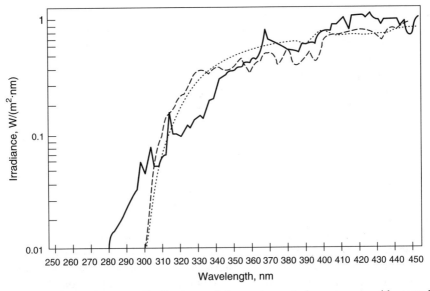

Figure 8.6. Spectral power distributions of the water-cooled xenon arc with coated infrared absorbing (CIRA) quartz inner and soda lime glass outer filters (·····), the filtered HMI metal halide lamp (—), and daily average of daylight in Miami, FL measured at normal incidence during the vernal equinox under clear sky conditions (---). (Courtesy of Atlas Material Testing Technology LLC.)

noon daylight in Miami, Florida, measured at a 26°-tilt angle during the spring equinox. The two spectra are compared by normalization at 560 nm since the absolute intensity of the xenon arc varies with the voltage setting.

Most xenon arc weathering devices control irradiance through electrical power management and have capabilities for periodic water spray or immersion and temperature and humidity control during both light and dark periods. The irradiance in watts per square meter (W/m^2) in a specified spectral range is photometrically monitored, and the lamp power is automatically adjusted to maintain constant irradiance. In the past, the irradiance level specified in most ASTM standards has been 0.35 $W/(m^2 \cdot nm)$ measured at 340 nm. This represents the daily average of the irradiance measured from early morning to late afternoon at normal incidence in Miami under clear-sky conditions during the vernal equinox. However, it is about half the peak irradiance measured at solar noon and less than the irradiance incident on materials during the periods of 3 or 4 h before and after noon in Miami and Arizona. Therefore, more recently, the irradiance specified in many ASTM standards is at least 0.5 $W/(m^2 \cdot nm)$ at 340 nm. Test duration is often specified in terms of radiant exposure (irradiance multiplied by time) expressed in joules per square meter (J/m^2). Procedures for exposure of polymeric materials in xenon arc weathering devices are given in ASTM standards D2565 [126], D4459 [127], D6695 [128], and G155 [129].

Carbon Arcs Figure 8.4 compares representative UV/visible spectral emissions of two types of carbon arc sources with the same daylight spectrum used in Figure 8.3. The SPD of the enclosed carbon arc differs significantly from that of terrestrial solar radiation in both the UV and visible regions. In the UV, at wavelengths longer than 350 nm, it has much greater intensity than daylight, and at shorter wavelengths it has much weaker intensity than daylight. Thus, it would have a much weaker effect than solar radiation on materials that absorb only short-wavelength UV radiation, but a stronger effect on materials that also absorb long-wavelength UV radiation. Therefore, the stability ranking of materials that differ in ultraviolet absorption characteristics in these regions can be expected to be distorted by this source compared with exposure to daylight. Procedures for using enclosed carbon arc devices for exposure of polymeric materials are given in ASTM standards D5031 [130], D6360 [131], and G153 [132].

The spectral emission of the open-flame carbon arc (also referred to as the Sunshine carbon arc) through Corex D type of borosilicate glass filters is shown in Figure 8.4. These filters are commonly used to reduce the excessive short-wavelength UV radiation of the unfiltered arc but transmit some shorter wavelengths than are present in solar radiation. Other types of sharp cut-on glass filters can be used to more closely match the solar cut-on or to simulate daylight through window glass. The open-flame carbon arc has much higher long-wavelength UV intensity and very much weaker intensity of visible radiation than daylight. However, it gives a much better match to daylight in the 300- to 360-nm region and deviates less at wavelengths longer than 360 nm than the enclosed carbon arc. Exposure to the unfiltered open flame carbon arc for faster testing has often produced reversals in stability ranking compared with outdoor exposure because of the unnatural short wavelength UV radiation present. The latter can also alter the mechanism and type of degradation. Procedures for using filtered open-flame carbon arc devices for exposures of polymeric materials are given in ASTM standards D822 [133], D1499 [134], and G152 [135].

Both types of carbon arc devices have capabilities for periodic water spray on the samples and condensation during a dark period as well as for humidity and temperature control. Irradiance is not monitored or electrically controlled as in xenon arc devices, but exposure is maintained constant by the requirement of daily replacement of the carbon rods. Carbon arc exposures, particularly using the enclosed carbon arc, are largely being replaced with xenon arc tests.

Fluorescent UV Lamps The emission properties of two types of fluorescent UV lamps, UVB-313 and UVA-340, are shown in Figure 8.5 in comparison with the daylight spectrum shown in the two previous figures. The emission of the UVB lamp has no relation to the spectral distribution of solar radiation on Earth's surface. A large portion of its energy is at shorter wavelengths than the solar cut-on, and its intensity decreases sharply at wavelengths longer than 313 nm. Most of the energy is below 340 nm. This type of lamp had been used because it could rapidly test the sensitivity of many polymeric materials to UV radiation due to the high flux of short wavelengths. However, because of the excessive short wavelength radiation, exposures to these lamps often caused reversals in stability rankings of polymers and errors in the performance of stabilizers compared with outdoor tests. These highly energetic wavelengths can initiate different mechanisms and types of degradation than the wavelengths present in solar radiation.

The UVA-340 lamps are gradually replacing UVB lamps because of their good match to the solar cut-on. However, they only give good simulation of solar radiation in the region from 300 to about 340 nm and, in common with all fluorescent UV lamps, lack the long-wavelength UV, visible, and near-infrared radiation present in daylight and xenon arc radiation. Compared with stability ranking by daylight, reversals have been reported for colored materials exposed to these lamps [125]. Also, in addition to the lack of actinic effects on materials sensitive to long-wavelength UV and visible radiation, heating of colored materials by absorbed visible and near-infrared radiation is absent. Thus, in contrast to the temperature variations of differently colored materials exposed to a source that emits radiation in the latter regions, all materials exposed to fluorescent UV radiation attain the same (ambient) temperature. For materials that vary in their sensitivity to heat, stability rankings can differ from those based on exposure to solar or solar-simulated radiation.

The fluorescent UV sources are typically incorporated into fluorescent UV/condensation devices in which dark periods with condensation are alternated with periods of UV radiation. The temperature is generally controlled at different levels during each of these periods. The devices may also incorporate periodic water spray on the front surface of the samples during exposure to radiation but do not control relative humidity. Some of the modern types of these devices have capabilities for irradiance control. Procedures for exposing polymeric materials in devices with fluorescent UV lamps are given in ASTM standards D4329 [136], D4587 [137], and G154 [138].

Metal Halide Lamp The UV emission curve of the borosilicate filtered Osram HMI® metal halide lamp is shown in Figure 8.6 [111] compared with the emission curves of daylight and filtered xenon arc radiation. It has a multiline spectrum that can be considered to be a continuum for purposes of material testing. It gives a relatively good simulation of terrestrial solar radiation in the UV region above 300 nm but requires additional filtering of the shorter wavelength radiation for better simulation of daylight. The high efficiency and low infrared output of this lamp eliminate the need for water cooling, thus making these lamps ideally suited for use in large-scale multiple-source arrays and effective in thermal loading studies. However, metal halide lamps have technical problems, which make them difficult to use in weathering devices. One is the dependence of the SPD of the radiation on the temperature of the lamp, thus requiring as constant a temperature as possible in the vicinity of the lamp. Due to the effect of temperature on the SPD, the latter changes with change in power. Therefore, the ability to alter the level of irradiance by changing the power is limited to about 5–10%. Reduction in irradiance is accomplished either by the use of close-meshed wire filters or by increase in the distance between the lamp and sample. The other problem is the variation in SPD from one lamp to another of the same type. It is remedied by measuring the SPD of each lamp and selecting one that is applicable.

8.6.2. Effect of SPD of Radiation Source on Weathering

Since the wavelengths responsible for degradation of a specific polymeric material vary with the SPD of the radiation source, inadequate simulation of the emission properties of the natural source by the laboratory-accelerated test source can have significant consequences on stability testing [139]. Due to different screening requirements, the artificial test source would not provide valid data on the protective effectiveness of a specific UV absorber against solar radiation. Further, because of differences among polymeric materials in their absorption properties and response to the absorbed radiation, differences in the emission properties of light sources can alter the stability ranking of materials.

Differences among light sources in the relative intensities of short to long wavelengths responsible for degradation can alter the predominant mechanism of degradation of many types of polymeric materials. Therefore, test results would differ in the effectiveness of stabilizers that act by interfering with the mechanism of degradation. For example, HALS can only protect polycarbonate and aromatic polyurethane against photooxidation, the mechanism initiated by long-wavelength UV. They are ineffective against the photo-Fries rearrangement caused by short wavelengths. Thus, evaluation of the performance of HALS in these polymers by using a radiation source that primarily emits short-wavelength UV will give misleading information on their ability to protect the polymers against solar radiation. Differences in the relative intensities of short versus long actinic wavelengths can also alter the type of degradation such as yellowing versus bleaching, bond scission versus crosslinking, and chalking versus cracking and crazing in TiO_2-pigmented systems.

8.6.3. Irradiance in Laboratory Weathering Tests

Exposures in laboratory weathering tests to irradiance levels much higher than those encountered under natural exposure conditions can significantly shorten test time. However, simulation of the effects of natural weathering on the materials would be compromised if the high irradiance levels alter the mechanism of degradation. Also, since all materials are not affected equally by increased irradiance, it can change the stability ranking of materials. The effect of irradiance level on the rate of degradation is complex. It varies with type of material, type of stabilizers present, and wavelengths responsible for the degradation. For most materials, the rate of degradation is rarely a linear function of the level of irradiance. For example, the rates of photooxidation of both polypropylene and polyethylene have been shown to be proportional to various fractional powers of the light intensity, ranging from the square root to the first power [140]. At high intensities of light, the quantum yield of degradation, that is, the amount of degradation per photon absorbed, will often be less than at low intensities. In a free-radical process this is explained in part by the "cage effect." As a result of the high concentrations of free radicals formed at the high irradiance levels, recombination occurs rapidly so that reaction with oxygen or other molecules is reduced. Thus, doubling the intensity does not necessarily double the rate of degradation. In photooxidation reactions, oxygen diffusion may become the rate-limiting step at high irradiance levels.

8.6.4. Temperature in Laboratory Weathering Tests

While degradation can be accelerated by testing at temperatures higher than materials are exposed to under normal use conditions, caution must be exercised to avoid producing unrealistic test data. Often, a different degradation mechanism is triggered at high temperatures. For example, certain secondary reactions may occur only at high temperatures due to increase in rate of decomposition of hydroperoxides or in rate of diffusion of oxygen or free radicals formed in primary processes. Some reactions that take place at high temperatures will generally occur at only a very low rate, or not at all, at lower temperatures. Any change from the mechanism of degradation produced by natural weathering conditions precludes simulation of the effects of natural exposures.

Since the effect of temperature varies with type of polymer and its formulations, temperatures different than those encountered in end-use environments can distort the stability rankings of materials in addition to causing unrealistic aging behavior. For example, it was shown [141] that change in air temperature from 30 to 60°C in an artificial weathering test changed the rank order of the stabilities of polyamide, polypropylene, and polyester yarn based on reduction in tensile strength. Exposure at 30°C showed that polyamide is most stable and polyester least stable. At the higher temperature, polyester was more stable than polypropylene.

The ability to control temperature along with irradiance is generally included in the basic design of most laboratory-accelerated weathering devices. The approximate temperature of dark-colored specimens is measured and regulated with the use of a black panel sensor of either the uninsulated or insulated type described in ASTM G151 [142]. It serves to control the air temperature that, in combination with the surface heat due to absorbed radiation, provides the black panel temperature specified in the test method. The heat caused by absorbed radiation depends on the visible and infrared absorption properties of the materials as well as on the spectral power distribution of the source and the irradiance level. Generally, the black panel temperature specified is the maximum temperature that dark samples attain under use conditions. Currently, the highest black panel temperature specified is 89°C during the light-only period in a xenon arc device used for tests on automotive interior materials such as seat covers, dashboards, and the like [143].

8.6.5. Moisture in Laboratory Weathering Tests

Moisture, in combination with radiation, is a very important factor in accelerating degradation of many types of polymeric materials, both by chemical reactions and by imposing mechanical stresses, particularly on composite materials. The processes of photoinitiated oxidation and hydrolysis are interrelated, with high humidity enhancing photooxidative degradation. For example, in a study on the effect of relative humidity (RH) on the photodegradation of an acrylic melamine coating, both the rate and magnitude of chain scission and oxidation were shown to increase with increase in RH [144]. While some polymers are themselves sensitive to moisture, often the moisture sensitivity of formulation ingredients can have a significant effect on weathering in the presence of moisture. For example, some plasticizers in PVC can be hydrolyzed under hot humid conditions by traces of the hydrogen chloride released on exposure of PVC to UV radiation, resulting in significant loss in elongation.

Moisture can be provided in laboratory-accelerated weathering devices in the form of controlled humidity, condensation, or wetting by water spray or immersion. The type of moisture present can influence the type and rate of degradation. Many test procedures include more than one type of moisture. Commonly, the front surface of samples is periodically sprayed with cool, deionized water simultaneous with exposure to light. This imparts a thermal shock to the samples but does not simulate natural weather conditions in which rain is unaccompanied by exposure to radiation. The formation of condensation on samples during the dark period is an effective means of simulating nighttime conditions in an environment with high moisture content, such as in Florida. It is accomplished by providing high-humidity conditions while cooling the back of the samples with ambient air or a water spray. High humidity can be as effective as liquid water. In devices in which wetting is by immersion, the sample is immersed in water while the surface is exposed to light. Since water absorption by polymers from a humid atmosphere or by direct wetness is a diffusion-controlled process, the frequency and duration of the exposure to moisture is often a critical parameter. Weathering processes can be accelerated by more frequent swelling and contraction than occurs in practice. The length of the required wet period depends on the water diffusion coefficient and water absorption capacity of the material as well as on its thickness. Water spray, immersion, or humidity periods ranging up to hours may be required for some polymeric types.

8.7. LABORATORY-ACCELERATED VERSUS ENVIRONMENTAL WEATHERING TESTS

Any accelerated weathering test, either in the laboratory or under environmental conditions, only approximates field exposure. However, the closer the simulation of the natural critical weather factors and the relation between them, the better the correlation in test results between laboratory and outdoor tests and with end-use performance. Acceleration of one factor alone can distort failure modes and stability ranking of materials. Correlation between laboratory accelerated and environmental weathering tests is generally based on similarity in performance ranking of a series of materials by the two types of tests. It has also been based on the similarity in type(s) of degradation or on the profiles of the graphed data for property change versus time or radiant exposure. Ideally, the accelerated test should satisfy all correlation criteria in order to provide a reliable evaluation of the performance of a polymeric material under natural weathering conditions. Good correlation with one exposure site is not necessarily applicable to other sites because of differences in exposure conditions among sites. The latter is also often responsible for poor correlation between test results at different exposure sites. Correlation also varies with the type of material and its formulations.

Correlations between laboratory and environmental weathering tests depend on simulation of the spectral power distribution of the natural source as well as simulation of the type and frequency of wetting, surface temperature of materials, humidity during exposure to the radiation source, and the light/dark cycle. Just as outdoor weathering test results can depend on the season of the year in which the test is started, in laboratory tests the type of degradation and stability rankings can depend on whether the test is started during a wet or dry period of exposure to radiation or during a dark period with moisture. Simulation of the atmospheric acid precipitation in outdoor exposures has been shown to improve correlation between laboratory-accelerated and outdoor tests based on stability ranking and type of degradation for various types of polymeric materials including sealants and automotive clearcoats [145–148]. If intensification of any stress factor in the accelerated test changes the mechanism of degradation, correlation with performance under environmental conditions cannot be expected. For adequate correlation with outdoor exposure, accelerated laboratory conditions should not provide an acceleration factor greater than about 10. Unfortunately, the greater the acceleration, the poorer the correlation. Since performance rankings can vary with the amount of radiant exposure, correlation should be determined at an optimum exposure period [149]. The extent of correlation may also depend on the property measured as the criterion of degradation.

Laboratory-accelerated tests have been used primarily for research and development and for qualifications of materials. Appropriate laboratory-accelerated tests have the potential of providing information to predict lifetimes under natural weathering conditions, but it is largely an unsolved weathering problem because it is not a simple task. The most direct approach is to determine an *acceleration factor* relating exposure times required for the accelerated and natural weathering tests to produce the same property change. Due to the complexity of the effect of intensification of weather factors on a material, the acceleration factor cannot be estimated but must be determined experimentally. Since acceleration factors are material dependent and must be determined for each material type and formulation, a single acceleration factor cannot be established for a laboratory test to predict lifetimes for a variety of materials [2].

A practical proposal for using this approach is to determine the acceleration factor during the early stages of weathering using a sensitive analytical technique for measuring chemical changes [150, 151]. Assuming the relation between the chemical changes and macroscopic property changes is the same for the accelerated and natural exposure, the acceleration factor can be used to extrapolate the slower macroscopic changes produced by the accelerated test to determine time to failure outdoors. Simms [152] developed a curve-fitting method to determine the *acceleration shift factor* (ASF) to account for changes in the relation between accelerated and natural exposure with progression of weathering. The technique requires that the profiles of the graphed property change versus exposure are similar for the two types of exposures.

Regression analysis is another technique for deducing lifetimes by natural weathering from lifetimes determined by laboratory-accelerated tests. It is based on measurements of property changes as a function of exposure by both types of tests. It depends on good linear relations over the period measured or finding the mathematical expression between property change and exposure that gives a linear relationship [153]. However, the same limitations exist as those inherent in using acceleration factors. Other empirical approaches as well as mathematical modeling techniques have been summarized [2]. Although laboratory-accelerated weathering tests have played an important role in development of weatherable polymeric materials, it is a complimentary technique to outdoor weathering tests. Its usefulness depends on how closely it reproduces the chemical and macroscopic changes produced by the slower outdoor exposures. Therefore, before drawing any final conclusions concerning the ability of a polymer to withstand the outdoor environment based on artificial weathering tests, it is necessary to validate the results by conducting outdoor exposure tests for a reasonable length of time.

REFERENCES

1. R. C. Hirt, R. G. Schmitt, N. D. Searle, and A. P. Sullivan, *J. Opt. Soc. Am.* **50**, 704–713 (1960).
2. N. D. Searle, in J. Kroschwitz, ed. *Encyclopedia Polymer Science and Technology*, 3rd ed., Wiley, New York, 2002, "Weathering."
3. N. D. Searle, in S. H. Hamid, ed., *Handbook of Polymer Degradation*, 2nd ed., Marcel Dekker, New York, 2000, Chapter 16.
4. N. D. Searle, in A. V. Patsis, ed., *International Conference on Advances in the Stability and Controlled Degradation of Polymers*, Vol. 1, Technomic, Lancaster, PA, 1989, pp. 62–74.
5. N. D. Searle, in D. Kockott, ed., *International Symposium on Natural and Accelerated Weathering of Organic Materials*, Essen, Germany, Sept. 28–29, 1987, Part B, Atlas SFTS, Lochem, The Netherlands, 1988, pp. 1–15.
6. L. D. Johnson, W. C. Tincher, and H. C. Bach, *J. Appl. Polym. Sci.* **13**, 1825–1832 (1969).
7. R. M. Fischer, W. D. Ketola, R. M. Dittmar, and R. M. King, *Preprint Polymeric Mat. Sci. Eng.* **83**, 136–139 (2000); ACS Nat'l Mtg., Aug. 20–24, 2000.
8. B. L. Garner and P. J. Papillo, *Ind. Eng. Chem. Prod. Res. Dev.* **1**, 249–253 (1962).
9. D. Clauson, in 75th Annual IFAI Convention, Nov. 1987, Industrial Fabrics Assoc. Int'l, St. Paul, Minn., 1988, pp. 96–110 (Preprint).
10. M. E. Nichols and C. A. Darr, *ACS Symp. Ser.* **722**, 333–353 (1999).
11. G. C. Furneaux, K. J. Ledbury, and A. Davis, *Polym. Degrad. Stab.* **3**, 441–432 (1980/1981).
12. A. V. Cunliffe and A. Davis, *Polym. Degrad. Stab.* **4**, 17–37 (1982).
13. G. E. Schoolenberg, J. C. M. deBruijn, and H. D. F. Meijer, in Proc. 14th Int. Conf. on Advances in Stabilization and Degradation of Polymers, Lucerne, 1992, p. 25.
14. G. Yanai, A. Ram, and J. Miltz, *J. Appl. Polym. Sci.* **59**, 1145–1149 (1996).
15. D. Patil, R. D. Gilbert, and R. E. Fornes, *J. Appl. Polym. Sci.* **41**, 1641–1650 (1990).
16. U. Schulz and P. Trubiroha, in R. J. Herling, ed., *Durability Testing of Nonmetallic Materials ASTM STP 1294*, American Society for Testing and Materials, West Conshohocken, PA, 1996, pp. 106–120.
17. P. Trubiroha and U. Schulz, *Polym. Composites* **5**, 359–367 (1997).
18. M. R. Kamal and B. Huang, in S. H. Hamid, M. B. Amin, and A. G. Maadhah, eds., *Handbook of Polymer Degradation*, Marcel Dekker, New York, 1992, Chapter 5.
19. N. D. Searle, in H. Mark, Y. Bikales, C. Overberger, and G. Menges, eds., *Encyclopedia Polymer Science and Engineering*, 2nd ed., Vol. 17, Wiley, New York, 1999, "Weathering."
20. F. Gugumus, in J. Pospisil and P. Klemchuk, eds., *Oxidation Inhibition in Organic Materials*, Vol. II, CRC Press, Boca Raton, FL, 1989, Chapter 2.
21. C. Andrei, V. Tomescu, I. Hogea, P. Ioana, and V. Dobrescu, Preprints IUPAC Macromolecular Sympos. 285 (1983).
22. P. P. Klemchuk, in S. H. Hamid, ed., *Handbook of Polymer Degradation*, 2nd ed., Marcel Dekker, New York, 2000, Chapter 12.
23. S. Al-Malaika, in *Encyclopedia Polym. Sci. and Technol.*, John, New York, 2002, "Stabilization."

24. J. F. Rabek and B. Ranby, *J. Polym. Sci., Polym. Chem. Ed.* **12**, 273–294 (1974).
25. S. I. Kuzina and A. I. Mikhailov, *Eur. Polym. J.* **34**(8), 1157–1162 (1998).
26. N. A. Weir, in N. S. Allen and J. F. Rabek, eds., *New Trends in the Photochemistry of Polymers*, Elsevier Applied Science, London, 1985, Chapter 11.
27. S. I. Kuzina and A. I. Mikhailov, *Eur. Polym. J.* **29**(12), 1589–1594 (1993).
28. G. Gueskens and C. David, *Pure Appl. Chem.* **51**, 2385 (1979).
29. O. B. Zapolskii, *Vysokomol. Soyed.* **7**(4), 615–620 (1965).
30. G. Gueskens and P. Bastin, *Polym. Degrad. Stab.* **4**, 111 (1982).
31. A. Ghaffar, A. Scott, and G. Scott, *Eur. Polym. J.* **12**, 615–620 (1976).
32. G. Scott and M. Tahan, *Eur. Polym. J.* **13**, 981–988 (1977).
33. H. E. Bair, D. J. Boyle, and P. G. Kelleher, *Polym. Eng. Sci.* **20**(15), 995–1001 (1980).
34. M. Ghaemy and G. Scott, *Polym. Degrad. Stab.* **3**, 233–242 (1981).
35. J.-L. Gardette, B. Mailhot, and J. Lemaire, *Polym. Degrad. Stab.* **48**, 457–470 (1998).
36. J. Lemaire, J.-L. Gardette, A. Rivaton, and A. Roger, *Polym. Degrad. Stab.* **15**, 1–13 (1986).
37. J. F. McKellar and N. S. Allen, *Photochemistry of Man-Made Polymers*, Elsevier Science Publishing Co., Inc., New York, NY, 1979, Chapter 2.
38. A. Davis and D. Sims, *Weathering of Polymers*, Applied Science, New York, 1983, Chapter 7.
39. D. Braun and S. Kull, *Angew. Macromol. Chem.* **86**, 171–180 (1980).
40. A. Roger, D. Sallet, and J. Lemaire, *Macromolecules* **19**(3), 579–584 (1986).
41. D. J. Carlsson, L. H. Gans, and D. M. Wiles, *J. Polym. Sci. Polym. Chem. Ed.* **16**(9), 2353–2363, 2365–2376 (1978).
42. D. J. Carlsson, R. D. Parnell, and D. M. Wiles, *J. Polym. Sci. Polym. Lett. Ed.* **11**, 149–155 (1973).
43. C. Decker, *Eur. Polym. J.* **20**(2), 149–155 (1984).
44. J.-L. Gardette and J. Lemaire, *Polym. Degrad. Stab.* **16**, 147–158 (1986).
45. J.-L. Gardette and J. Lemaire, *J. Vinyl Tech.* **15**(2), 113–117 (1993).
46. U. Gesenhues, *Polym. Degrad. Stab.* **68**(2), 185–196 (2000).
47. G. Scott and M. Tahan, *Eur. Polym. J.* **13**(12), 989–996 (1977).
48. A. Factor and M. L. Chu, *Polym. Degrad. Stab.* **2**, 203–223 (1980).
49. E. Ong and H. E. Bair, *Polym. Preprints* **20**(1), 945–948 (1979).
50. A. Factor, W. V. Ligon, and R. J. May, *Macromolecules* **20**, 2461–2468 (1987).
51. A. Factor, in *Advances in Chemistry Series* No. 249, Am. Chem. Soc., Washington, DC, 1996, Chapter 5, pp. 59–76.
52. A. Rivaton and J.-L. Gardette, *Die Angewandte Makromolekulare Chemie* **261/262**(4627), 173–188 (1998).
53. A. Rivaton, D. Sallet, and J. Lemaire, *Polym. Degrad. Stab.* **14**, 1–22 (1986).
54. D. T. Clark and H. S. Munro, *Polym. Degrad. Stab.* **4**, 441–457 (1982).
55. D. T. Clark and H. S. Munro, *Polym. Degrad. Stab.* **8**, 195–211 (1984).
56. D. R. Bauer, J. L. Gerlock, and D. F. Mielewski, *Polym. Degrad. Stab.* **36**, 9–15 (1992).

57. H. S. Munro and R. S. Allaker, *Polym. Degrad. Stab.* **11**, 349–358 (1985).
58. A. Torikai, T. Mitsuoka, and K. Fueki, *J. Polym. Sci. Part A; Polym. Chem.* **31**, 2785–2788 (1993).
59. K. B. Abbas, *J. Appl. Polym. Sym.* **35**, 345–360 (1979).
60. A. Rivaton, D. Sallet, and J. Lemaire, *Polym. Degrad. Stab.* **14**, 23 (1986).
61. P. Blais, M. Day, and D. M. Wiles, *J. Appl. Polym. Sci.* **17**, 1895–1907 (1973).
62. A. Casu and J.-L. Gardette, *Polymer* **36**(21), 4005–4009 (1995).
63. M. Day and D. M. Wiles, *J. Appl. Polym. Sci.* **16**, 175–189 (1972).
64. M. H. Tabankia, J.-L. Philippart, and J.-L. Gardette, *Polym. Degrad. Stab.* **12**, 349–362 (1985).
65. N. Grassie and G. Scott, *Polymer Degradation and Stabilization*, Cambridge University Press, London, 1985, pp. 213–216.
66. D. E. Duvall, *Polym.-Plastics Technol. Eng.* **34**(2), 227–242 (1995).
67. K. G. Martin, Div. Bldg. Research, CSIRO, Victoria, Australia, Report 28, 1974.
68. R. Iyengar and B. Schellenberg, *Polym. Degrad. Stab.* **61**, 151–159 (1998).
69. R. L. Gray and C. Neri, *Die Angewandte Makromolekulare Chemie* **252**(4513), 55–68 (1997).
70. G. Scott, *Atmospheric Oxidation and Antioxidants*, Elsevier, Amsterdam, 1965, p. 287.
71. M. Liu and A. R. Horrocks, *Polym. Degrad. Stab.* **75**(3), 485–499 (2002).
72. J. M. Pena, N. S. Allen, M. Edge, C. M. Liauw, I. Roberts, and B. Valange, *Polym. Degrad. Stab.* **70**, 437–454 (2000).
73. S. W. Bigger and O. Delatycki, *J. Mater. Sci.* **24**, 1946 (1989).
74. A. L. Andrady and A. R. Schultz, *J. Appl. Polym. Sci.* **33**, 1389–1395 (1987).
75. J. E. Pickett, *J. Appl. Polym. Sci.* **33**, 525 (1987).
76. F. Gugumus, *Polym. Degrad. Stab.* **75**, 309–320 (2002).
77. F. Gugumus and H. Zweifel, eds., *Plastics Additives Handbook*, 5th ed., Hanser, Munich, 2001, Chapter 2.
78. J. E. Pickett, in S. H. Hamid, ed., *Handbook of Polymer Degradation*, 2nd ed., Marcel Dekker, New York, 2000, Chapter 5.
79. D. J. Carlsson and D. M. Wiles, *Macromolecules* **7**(3), 259–262 (1974).
80. R. P. R. Ranaweera and G. Scott, *Eur. Polym J.* **12**(8), 591–597 (1976).
81. J. R. White and A. Turnbull, *J. Mater. Sci.* **29**(3), 584–613 (1994).
82. J. H. Chaudet and J. Tamblyn, *J. Appl. Polym. Sci.* **8**, 1949 (1964).
83. F. Gugumus, in G. Scott, *Developments in Polymer Stabilization*, Vol. I, Applied Science, London, 1979, p. 261.
84. H. K. Muller, in S. Al-Malaika, ed., *Reactive Modifiers in Polymers*, Blackie Academic Professional, London, 1997, p. 55.
85. G. Gueskens and D. M. McFarlane, *J. Vinyl Additive Technol.* **5**(4), 186–194 (1999).
86. D. J. Carlsson, K. H. Chan, D. M. Wiles, and J. Dunmis, *Preprints, Org. Coat. Appl. Polym. Sci. Proc.* **46**, 457–462 (1982).
87. D. M. Wiles, in Int. Conf. on Advances in the Stabilization and Controlled Degradation of Polymers, Lucerne, Switzerland, 1982.

88. D. J. Carlsson, K. H. Chan, J. Dunmis, and D. M. Wiles, *J. Polym. Sci. Polym. Chem. Ed.* **20**, 575–582 (1982).
89. I. Bauer, W. D. Habicher, C. Rautenberg, and S. Al-Malaika, *Polym. Degrad. Stab.* **48**(3), 427–440 (1995).
90. A. J. Chirinos-Padron, P. H. Hernandez, N. S. Allen, C. Vasilion, G. P. Marshall, and M. Poortere, *Polym. Degrad. Stab.* **19**(2), 177–189 (1987).
91. F. Gugumus, *Polym. Degrad. Stab.* **75**(1), 295–308 (2002).
92. N. S. Allen, M. Edge, T. Corrales, and F. Catalina, *Polym. Degrad. Stab.* **61**(1), 139–149 (1998).
93. F. Gugumus, in S. H. Hamid, ed., *Handbook of Polymer Degradation*, 2nd ed., Marcel Dekker, New York, 2000, Chapter 1.
94. G. Gueskens, in N. Grassie, *Developments in Polymer Degradation-3*, Elsevier Science Publishing Co., Inc., New York, NY, 1981, Chapter 7.
95. C. Savides, J. A. Stretanski, and L. R. Castello, *Adv. Chem. Series* **85**, 287–306 (1968).
96. G. A. George, *J. Appl. Polym. Sci.* **18**(1), 117–124 (1971).
97. N. S. Allen and J. F. McKellar, *J. Polym. Sci. Macromolec. Rev.* **13**, 241 (1978).
98. N. S. Allen, M. J. Harrison, and M. Ledward, *Polym. Degrad. Stab.* **23**, 165–174 (1989).
99. H. C. Bach and J. R. Sechrist, *Appl. Polym. Sym.* **9**, 177–185 (1969).
100. E. D. Owen, in N. S. Allen, ed., *Developments in Polymer Photochemistry*, Vol. 3, Elsevier Science Publishing Co., Inc., New York, NY, 1982, Chapter 5.
101. J. B. Adeniyi and G. Scott, *Polym. Degrad. Stab.* **17**, 117–129 (1987).
102. S. Girois, *J. Vinyl Additive Technol.* **5**(4), 218–230 (1999).
103. G. T. Peake, Antec '95, Conference Proceedings, Vol. III, 1995, pp. 3265–3268.
104. G. T. Peake, *J. Vinyl Technol.* **2**, 184–186 (1996).
105. T. Thompson and P. P. Klemchuk, *Polym. Preprints* **34**(2), 176–177 (1993).
106. S. B. Ruetsch, X. X. Huang, D. R. Salem, and H. D. Weigmann, *Textile Res. J.* **66**(4), 185–195 (1996).
107. H. Tabankia and J.-L. Gardette, *Polym. Degrad. Stab.* **14**, 351–365 (1986).
108. ASTM G7, *Annual Book of ASTM Standards*, Vol. 14.04, American Society for Testing and Materials, West Conshohocken, PA.
109. ASTM G24, *Annual Book of ASTM Standards*, Vol. 14.04, American Society for Testing and Materials, West Conshohocken, PA.
110. ASTM D1435, *Annual Book of ASTM Standards*, Vol. 8.01, American Society for Testing and Materials, West Conshohocken, PA.
111. K. Hardcastle and N. D. Searle, in R. A. Ryntz, ed., *Plastics and Coatings, Durability, Stabilization, Testing*, Hanser Gardner, Cincinnati, Oh, 2001, Chapter 9.
112. ASTM D4364, *Annual Book of ASTM Standards*, Vol. 8.03, American Society for Testing and Materials, West Conshohocken, PA.
113. ASTM D4141, *Annual Book of ASTM Standards*, Vol. 6.01, American Society for Testing and Materials, West Conshohocken, PA.
114. SAE J1976, *1998 SAE Handbook*, Vol. I, Society of Automotive Engineers, Warrendale, PA, 1998.

115. SAE J2230, *1998 SAE Handbook*, Vol. I, Society of Automotive Engineers, Warrendale, PA, 1998.
116. ASTM G90, *Annual Book of ASTM Standards*, Vol. 14.04, American Society for Testing and Materials, West Conshohocken, PA.
117. M. P. Morse, in G. G. Schurr, ed., *Permanence of Organic Materials, ASTM STP 781*, American Society for Testing and Materials, West Conshohocken, PA, 1982, pp. 43–66.
118. G. A. Zerlaut and M. L. Ellinger, *J. Oil Colour Chem. Assoc.* **64**(10), 387–397 (1981).
119. G. A. Zerlaut, in W. E. Brown, ed., *Testing of Polymers*, Vol. 4, Wiley-Interscience, New York, 1969, pp. 10–34.
120. R. J. Martinovich and G. R. Hill, in M. R. Kamal, ed., *Weatherability of Plastic Materials, Appl. Polym. Symp. No. 4*, Wiley-Interscience, New York, 1967, pp. 141–154.
121. J. B. Howard and H. M. Gilroy, *Polym. Eng. Sci.* **9**(4), 286–294 (1969).
122. C. R. Caryl, in W. E. Brown, ed., Testing of Polymers, Vol. 4, Wiley-Interscience, New York, 1969, pp. 379–397.
123. *AATCC Technical Manual*, Vol. 63, American Association of Textile Chemists and Colorists, Research Triangle Park, NC, 1988, pp. 33–42.
124. SAE J1960, *1998 SAE Handbook*, Vol. I, Society of Automotive Engineers, Warrendale, PA, 1998.
125. N. D. Searle, *Atlas SunSpots* **24**(48), 1–4 (1994); *Surface Coatings (Australia)*, 28–30 (1997).
126. ASTM D2565, *Annual Book of ASTM Standards*, Vol. 8.02, American Society for Testing and Materials, West Conshohocken, PA.
127. ASTM D4459, *Annual Book of ASTM Standards*, Vol. 8.03, American Society for Testing and Materials, West Conshohocken, PA.
128. ASTM D6695, *Annual Book of ASTM Standards*, Vol. 6.01, American Society for Testing and Materials, West Conshohocken, PA.
129. ASTM G155, *Annual Book of ASTM Standards*, Vol. 14.04, American Society for Testing and Materials, West Conshohocken, PA.
130. ASTM D5031, *Annual Book of ASTM Standards*, Vol. 6.01, American Society for Testing and Materials, West Conshohocken, PA.
131. ASTM D6360, *Annual Book of ASTM Standards*, Vol. 8.03, American Society for Testing and Materials, West Conshohocken, PA.
132. ASTM G153, *Annual Book of ASTM Standards*, Vol. 14.04, American Society for Testing and Materials, West Conshohocken, PA.
133. ASTM D822, *Annual Book of ASTM Standards*, Vol. 6.01, American Society for Testing and Materials, West Conshohocken, PA.
134. ASTM D1499, *Annual Book of ASTM Standards*, Vol. 8.01, American Society for Testing and Materials, West Conshohocken, PA.
135. ASTM G152, *Annual Book of ASTM Standards*, Vol. 14.04, American Society for Testing and Materials, West Conshohocken, PA.
136. ASTM D4329, *Annual Book of ASTM Standards*, Vol. 8.03, American Society for Testing and Materials, West Conshohocken, PA.

137. ASTM D4587, *Annual Book of ASTM Standards*, Vol. 6.01, American Society for Testing and Materials, West Conshohocken, PA.
138. ASTM G154, *Annual Book of ASTM Standards*, Vol. 14.04, American Society for Testing and Materials, West Conshohocken, PA.
139. N. D. Searle, in W. D. Ketola and D. Grossman, eds., *Accelerated and Outdoor Durability Testing of Organic Materials, ASTM STP 1202*, American Society for Testing and Materials, West Conshohocken, PA, 1994, pp. 52–67.
140. T. M. Kollmann and D. G. M. Wood, *Polym. Eng. Sci.* **20**, 684–687 (1980).
141. B. J. Tabor and J. C. Wagenmakers, *Melliand Textilberichte* **1**, 9–14 (1992).
142. ASTM G151, *Annual Book of ASTM Standards*, Vol. 14.04, American Society for Testing and Materials, West Conshohocken, PA.
143. SAE J1885, 1998 *SAE Handbook*, Vol. I, Society of Automotive Engineers, Warrendale, PA, 1998.
144. T. Nguyen, J. W. Martin, E. Byrd, and N. Embree, *Polymeric Mat. Sci. Eng.* **83**, 118–119 (2000); *ACS Fall Mtg.*, August 2000.
145. U. Schulz and P. Trubiroha, in R. J. Herling, ed., *Durability Testing of Nonmetallic Materials, ASTM STP 1294*, American Society for Testing and Materials, West Conshohocken, PA, 1996, pp. 106–120.
146. P. Trubiroha and U. Schulz, *Polym. Polym. Composites* **5**(5), 359–367 (1997).
147. U. Schulz, P. Trubiroha, T. Boettger, and H. Bolte, 8th Int'l Conf. on Durability of Bldg. Mat'ls and Components, May 30–June 3, 1999.
148. K. M. Wernstahl and B. Carlsson, *J. Coatings Tech.* **69**(865), 69–75 (1997).
149. R. M. Fischer, SAE Tech. Paper Ser. No. 841022, West Coast Int'l Mtg. & Expos., San Diego, CA, 1984, Soc. Auto. Eng., Warrendale, PA, 1984, pp. 1–9.
150. D. R. Bauer, J. L. Gerlock, and R. A. Dickie, *Prog. Org. Coat.* **15**, 209–221 (1987).
151. J. L. Gerlock, D. F. Mielewski, and D. R. Bauer, *Polym. Degrad. Stab.* **20**, 123–134 (1988).
152. J. A. Simms, *J. Coat. Technol.* **59**(748), 45–53 (1987).
153. F. Gugumus, Preprint, Sympos. on "Polymer Stabilization and Degradation: Problems, Techniques and Applications," Manchester, England, Sept. 1985.

CHAPTER 9

BIODEGRADABLE POLYMERS

STEPHEN P. MCCARTHY

Department of Plastics Engineering, University of Massachusetts at Lowell

9.1. INTRODUCTION

The definition of *biodegradable* polymer varies greatly among scientists, manufacturers, and consumers. An American Society for Testing and Materials (ASTM) definition of a biodegradable plastic involves a "plastic designed to undergo a significant change in its chemical structure under specific environmental conditions resulting in a loss of some properties "..." in which the degradation results from the action of naturally-occurring micro-organisms such as bacteria, fungi, and algae" [1]. Concerns with environmental fate will impose the additional requirement for properly designed biodegradable materials of the complete "mineralization" or disappearance of the degradation products into CO_2, H_2O, CH_4, or biomass without the production of harmful intermediates. The time frame required for biodegradation will be mandated by the disposal method and conditions. The increase in composting and anaerobic bioreactor technology will produce specific environmental conditions and lead to specific requirements for biodegradable plastics packaging. This chapter attempts to review suitable biodegradable materials that may currently be used. This list is only a partial listing based on commercially produced products that are accepted at this point in time as being biodegradable and mineralized in time frames compatible with existing waste disposal methods. It is accepted that the discussion in this chapter is short lived since the future will reveal evidence for biodegradation of additional materials that are currently produced, commercialization of new

---

Plastics and the Environment, Edited by Anthony L. Andrady.
ISBN 0-471-09520-6 © 2003 John Wiley & Sons, Inc.

biodegradable materials, as well as the development of novel disposal technologies for solid waste.

9.2. JUSTIFICATION

Plastics have come under attack in recent years due to their high visibility in the solid waste crisis. Nondegradable plastics packaging is blamed for shortening the life expectancy of commercial landfills, increasing the operational cost, contaminating the environment, and posing a threat to animal and marine life. Plastics account for approximately 15–17% of the $19 billion food-packaging market, and it is predicted to increase to 50% by the year 2000 [2]. Nearly 10% of the plastics used in packaging are used as coatings on other materials, including paper [2]. Increasing legislative and economic imperatives and public perception of paper being. natural and plastics being "foreign" have led to the explosive drive to recycle plastics packaging or to make it "biodegradable."

The number of landfills, used for the disposal of solid waste, in the United States has decreased from 18,500 in 1979 to 6000 in 1988 [2]. A recent study of marine pollution commissioned by the Environmental Protection Agency (EPA) found that 56% of the 170,000 items collected were made from plastics [2]. State and federal environmental legislation has been initiated to address the problems created by the disposal of plastic packaging. These measures range from a requirement of recyclable or degradable packaging to a ban on all plastics food packaging [2]. The major route for disposal of municipal solid waste is burial in sanitary landfills. Under landfill conditions, oxygen becomes depleted due to diffusion limitations, and biodegradation proceeds via a syntrophic association of hydrolytic and fermentative bacteria, obligate proton reducing acetogenic bacteria, and methanogens [3, 4]. Although the intentional minimization of water in traditional landfills will hinder biodegradation, new technologies are currently being adopted to enhance biological activities and create controlled anaerobic bioreactors [5, 6].

Aerobic composting of municipal solid waste is experiencing tremendous growth in the United States. The number of municipal solid waste composting facilities has increased from 4 in 1987, to 8 in 1990, to 16 in 1991, and over 100 are currently in the construction or planning stages [7]. It should be recognized that an increase in composting will never be driven by biodegradable plastics packaging but rather by the organic waste fraction of the municipal solid waste stream (approximately 50%), which through aerobic composting can result in a highly stabilized organic matter for agricultural or landscaping purposes [8]. Incineration remains a viable technology with a lack of public support. Recycling of plastics is a developing technology with tremendous public support. The future increase in plastic items to be recycled will be driven by economics or mandated by legislation. The question of whether plastics packaging should be biodegradable or recyclable needs to be addressed. It is clear that neither solution will solely solve the solid waste problem in an economical manner. The solution is to use both, each in those areas that make the most economical and environmental sense. A broad view of the total "life cycle" of the material needs to be addressed from the acquisition of raw materials to disposal. Within this broader life-cycle assessment, the opportunity to utilize renewable resources particularly for single-use biodegradable packaging becomes extremely attractive.

9.3. REQUIREMENTS FOR DESIGN AND MANUFACTURE

Performance, processability, and price comprise the typical requirements for biodegradable polymers. Performance is based on the ability to perform the necessary functions during the service life as well as the disposal performance, biodegradability. Mechanical strength, processability, moisture and oxygen barrier properties, interactions with the product, printability, sterilizability, transparency, and inert to environmental exposures are a few of the properties necessary to evaluate for the service performance. Due to the variety of applications, it is expected that numerous polymers or formulations will be required to best serve each application.

The biodegradation requirements will be determined by the disposal method and involve organisms, substrate exposure, and environmental factors. The availability of appropriate organisms, co-metabolism, and generation of appropriate enzymes for the biodegradation of the specific polymer need to be considered. The accessibility of the polymer, physical form, molecular weight, branching, degree of crystallinity, and interactions with other components also play an important role. The environmental factors include temperature, oxygen availability, moisture level, pH, and nutrients. These requirements need to be balanced to achieve the optimum packaging. For example, water solubility may enhance the degradability but hinder its application. The design should take into account any additives that may be harmful to the environment. These include cadmium-based pigments, heavy-metal stabilizers, and flame retardants.

The processability requirements of the material will be dictated by the manufacturing process, which is related to the thickness desired and the required properties. Melt processing techniques such as blown film extrusion, calendering, injection molding, thermoforming, and the like will each require variations in melt viscosity and melt strength. The processing of the material can also affect the mechanical properties as well as the biodegradation of the material.

9.4. BIODEGRADABLE POLYMERS FROM RENEWABLE RESOURCES

The materials that make up the renewable resources are agricultural based and based on biosynthesis.

9.4.1. Agricultural

Cellulose

Cellulose is the most abundant of naturally occurring polymers and comprises at least one-third of the vegetable matter in the world [9]. Before World War II, cellulosics were the most important class of thermoplastics [10]. The cellulose content of plants varies from plant to plant, with cotton containing approximately 90% cellulose whereas average wood has about 50% [9]. Cellulose is a linear condensation polymer consisting of anhydroglucose units joined together by β-1,4-glycosidic bonds [11]. The average degree of polymerization (DP) reported for cellulose ranges from a value of 153,300 for California cotton [12] to 2000 for *Valonia sp.* [13]. The biodegradation of cellulose has been more extensively studied more than any other polymer. In general, the lower molecular weight polymer is more readily biodegraded. Cellulose is readily biodegradable and is mineralized by many microorganisms due to the activity of the cellulase enzyme complex, which catalyzes the hydrolysis and/or oxidation of cellulose resulting in the formation of cellobiose, glucose, and finally mineralization [11]. Several enzymes have been shown to act synergistically in the breakdown of cellulose with endocellulases attacking the amorphous regions while the cellulases cause random chain scission [14, 15]. The lignin component, which is often associated with cellulose, can biodegrade via oxidative pathways catalyzed by lignases, laccase, and alcohol oxidase [16, 17]. Due to the high crystallinity and interchain hydrogen bonding, cellulose is insoluble and thermally degrades before melting. Therefore cellulosic packaging materials are subdivided into two categories, regenerated, and modified.

Regenerated cellulose film, cellophane, can be cast from the solution of cellulose xanthate formed through the viscose process after which the film is subsequently hydrolyzed back to cellulose. Since the end product is essentially cellulose, it is readily biodegradable [11]. Cellophane is typically plasticized with ethylene glycol, propylene glycol, or glycerol. To improve the barrier properties, cellophane is often coated with polyvinylidene chloride, which would hinder the biodegradability.

Modified cellulose of commercial significance involves primarily the cellulose esters and ethers. The cellulose esters include cellulose acetate, cellulose propionate, cellulose acetate-butyrate, cellulose nitrate. Cellulose ethers include ethyl cellulose as a melt, processable-grade and water-soluble derivatives. Historically, it was accepted that cellulose derivatives that had a degree of substitution above 1.0 were not biodegradable [18–20]. However, cellulose acetate has been found to be biodegradable in both aerobic compost and anaerobic bioreactor environments [21–23]. The potential for additional, biodegradable, cellulose derivatives is currently being explored.

Cellulose acetate films can be cast from a solution of water, acetone, or chloroform depending on the degree of substitution (DS). Highly substituted cellulose acetate (DS > 2.5) can be melt processed with the addition of a plasticizer. The properties of cellulose acetate vary depending on molecular weight, degree of substitution, and amount and type of plasticizer. The properties of films made from cellulose acetate have an average tensile strength of 24–76 MPa, elongation of 5–55%, and water absorption of 1–3%. Cellulose acetate films, with a degree of substitution of 1.7 and 2.5, were found to biodegrade in an anaerobic bioreactor and a aerobic compost environment [21, 23]. In addition, the bacterium *Pseudomonas paucimobilis* has been isolated and was found to be capable of growth on cellulose acetate with DS values of 1.7 and 2.5 as the sole carbon source [22]. The complete mineralization of cellulose acetate has also been confirmed [24].

Starch

[Chemical structure of Amylopectin]

Starch is a carbohydrate that is synthesized in the organs of plants as a reserve food supply for periods of dormancy, germination, and growth [25]. Starch is the second most abundant "renewable" substance, after cellulose. Starch can be considered a condensation polymer of glucose consisting of two types, amylose, a linear-chain molecule of α-1,4-linked D-glucose, and amylopectin a branched polymer of α-1,4-linked D-glucose with a 1,6-linked D-glucose branched. Due to its function as a food storage, starch is readily biodegraded through enzyme-catalyzed hydrolysis by a number of enzymes [26]:

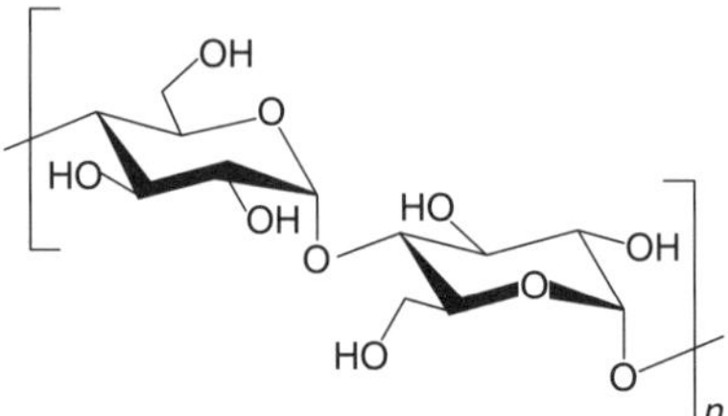

Amylose

Typically, in nature, the amylopectin is semicrystalline and the amylose amorphous. Various processing techniques have been developed resulting in reducing or eliminating the amylopectin crystallinity and resulting in amylose complexation [27]. The most notable of these technologies are the production of destructurized starch and thermoplastic starch. Destructurized starch, originally developed by Warner Lambert and currently owned by Novamont, is claimed to involve the use of water, typically greater than 5%, and mechanical shearing to form a more easily processable structure [28]. Thermoplastic starch is claimed to utilize a plasticizer such as glycerol in a "substantially water-free" starch to produce a thermoplastically processable starch and is currently produced by Biotech in Germany [29]. Earthshell has developed a composite material consisting of starch, calcium carbonate, and cellulose fibers for use in disposable food service items. Novamont in Italy produces starch and vinyl alcohol copolymer blends with starch contents greater than 60% under the trademark Mater-Bi [30, 31]. These materials are available in commercial grades suitable for blown film, injection molding, blow molding, thermoforming, and extrusion and are reported to be totally biodegradable and insoluble in water [31].

Konjac Konjac is a natural polysaccharide found in plant tubers from *Amorphophallus konjac* and produced commercially by FMC (Philadephia, PA). Konjac is a copolymer of glucose and mannose (1:1.6) linked β-1.4 with random acetylation of approximately $\frac{1}{12}$ monomer units. The polymer is water soluble and as such the applications are limited. Chemical modifications of the konjac as well as liquid crystalline properties have been reported [32].

Konjac

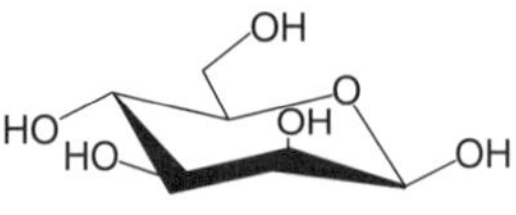

Mannose 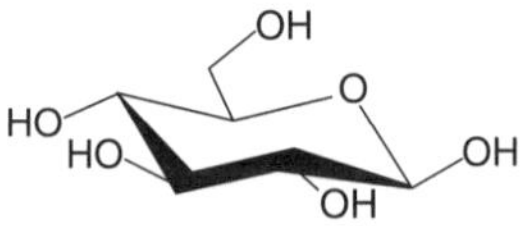   Glucose

9.4.2. Biosynthesis

Polyhydroxyalkonoates (Bacterial Polyesters) The family of polyhydroxyalkanoates are polymers produced as intracellular storage materials in a variety of bacteria grown under physiologically stressed conditions. Poly-β-hydroxybutyrate-*co*-β-hydroxyvalerate (PHBV) is a semicrystalline aliphatic polyester that was initially produced commercially by Imperial Chemical Industries, PLC (London, England, ICI) from the bacteria *Alcaligenes eutrophus*. The homopolymer of poly-β-hydroxybutyrate is melt processable with a melting point of 185°C and a glass transition temperature of -4°C. The melting point and glass transition temperatures decrease with valerate content up to 22% valerate, which is currently the highest valerate content produced commercially by Metabolix. The higher valerate-content copolymers show increases in the melt processability as well as the flexibility and elongation with a decrease in the tensile strength. The effect of crystallinity on the enzymatic degradation of PHBV has shown a dramatic decrease in the degradation rate with increasing crystallinity [33].

$$\left[-O-\underset{\underset{CH_3}{|}}{CH}-CH_2-\overset{O}{\overset{\|}{C}}- \right]_n$$

Poly(β-hydroxybutyrate)

$$\left[-O-\underset{\underset{\underset{CH_3}{|}}{\underset{CH_2}{|}}}{CH}-CH_2-\overset{O}{\overset{\|}{C}}- \right]_n$$

Poly(β-hydroxyvalerate)

$$\left[\left[-O-\underset{\underset{CH_3}{|}}{CH}-CH_2-\overset{O}{\overset{\|}{C}}- \right]_n \left[-O-\underset{\underset{\underset{CH_3}{|}}{\underset{CH_2}{|}}}{CH}-CH_2-\overset{O}{\overset{\|}{C}}- \right]_m \right]_o$$

Poly(β-hydroxybutyrate-*co*-β-hydroxyvalerate)

Pullulan Pullulan, a biodegradable polysaccharide first described in 1959 [34], is a water-soluble extracellular neutral glucan synthesized by the fungus *Aureobasidium pullulans*, more commonly referred to as *Pullularia pullulans*. The structure of pullulan is proposed to consist predominantly of maltotriose units linked via α-1,6-glycosidic bonds [35]. Pullulan containing 5–7% maltotetraosyl units has been reported for several strains of *A. pullulans*. Pullulan can be solution cast from aqueous media, although it has been difficult to process by conventional melt-processing techniques [36]. According to the digestion test of internal enzymes and the feeding test of rats, pullulan is an indigestible polysaccharide such as cellulose or pectin. Pullulan, like starch, begins to degrade thermally at approximately 250°C, then carbonizes, and generates neither extreme heat nor toxic gas [36, 37]:

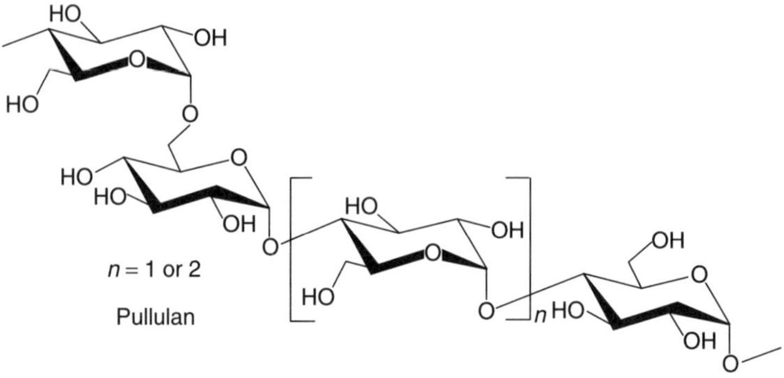

n = 1 or 2
Pullulan

Pullulan can be melt processed by conventional melt-processing methods with the addition of a plasticizer. Plasticizers such as glycerol and ethylene glycol have been shown to reduce the glass transition temperature to below 160°C with the addition of more than 10% plasticizer. Modification of pullulan has also been accomplished by acetylation [35].

Chitin and Chitosan Chitin is a naturally occurring polysaccharide derived primarily from the exoskeleton of shellfish and is described as β-1,4-linked 2-acetamido-2-deoxy-D-glucose. Chitin is also found in insects and filamentous fungi [38, 39]. Chitin has been shown to be readily biodegraded by microorganisms producing chitinases and lysozyme and is subsequently mineralized [40, 41]. Due to the intermolecular hydrogen bonding, chitin is insoluble in water and is not melt processable:

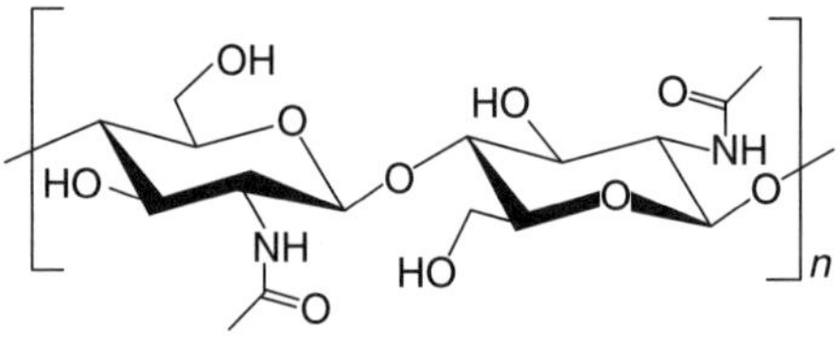

Chitin

Chitosan is deacetylated chitin that is swollen with water and dissolves in a water acetic acid mixture. This polymer is produced commercially from the base-catalyzed deacetylatlon of shellfish waste by Protan (Drammen, Norway). Chitin and chitosan exhibit good mechanical properties as well as low permeabilities:

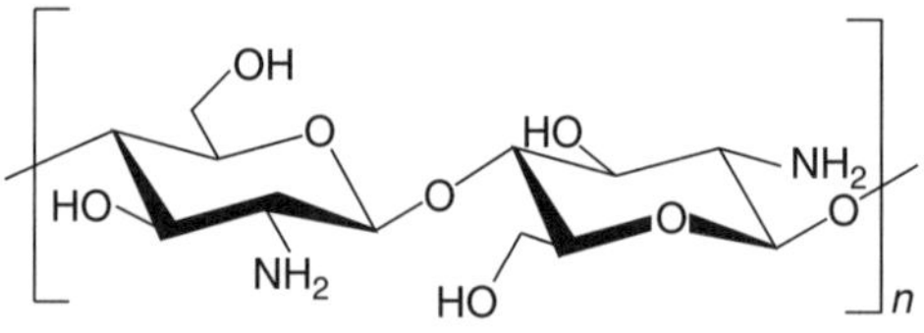

Chitosan

Chitosan can be crosslinked using epochlorohydrin [42]. Chitosan has also been found to exist naturally, being synthesized by zygomycete fungi as part of their cell wall. Chitosan has been shown to be biodegraded by chitosanases [40, 41].

Polylactic Acid   Polylactic acid (PLA) is a thermoplastic, aliphatic polyester that can be synthesized from biologically produced lactic acid. Currently, the major production of polylactic acid is from the ring-opening polymerization of the lactide [43, 44]. This material has been used extensively in the medical field for sutures, staples, and the like and as such is very expensive. Recently, April 2002, Cargill-Dow has opened a large-scale production facility whereby PLA is being produced at a low cost for nonmedical applications:

$$\left[-O-CH(CH_3)-\underset{\underset{O}{\|}}{C}- \right]_n$$
Polylactic acid

The poly-L-lactic acid shows high melting points and good mechanical properties. Polylactic acid degrades by hydrolysis, which has been shown to be accelerated by many enzymes. Recently, it has been found to be biodegradable in a compost environment [43, 45]. The thermoplastic material can be made stereo specific or racemic to yield different properties.

9.5. BIODEGRADABLE POLYMERS FROM PETROLEUM-DERIVED PRODUCTS

Polycaprolactone   Poly-ε-caprolactone (PCL) is a semicrystalline, thermoplastic, linear aliphatic polyester synthesized by the ring-opening polymerization of ε-caprolactone and is produced commercially by Union Carbide–Dow (Midland, MI) and Rhone-Poulenc (Collegeville, PA). This polymer has a melting point of approximately 62°C and a glass transition temperature of approximately −60°C. PCL is readily degraded and mineralized by a variety of microorganisms [45]:

Polycaprolactone

$$\left[-O-(CH_2)_5-\underset{\underset{O}{\|}}{C}- \right]_n$$

The degradation mechanism that has been proposed is hydrolysis of the polymer to 6-hydroxycaproic acid, an intermediate of ω-oxidation, and then β-oxidation to acetyl-SCoA, which can then undergo further degradation in the citric acid cycle [45].

Polyvinyl Alcohol   Polyvinyl alcohol (PVA) is a semicrystalline, carbon backbone polymer synthesized from the alcoholysis of polyvinyl acetate:

$$\left[\begin{array}{c} CH_2 - CH \\ | \\ OH \end{array} \right]_n$$

Polyvinyl alcohol

The properties depend on the molecular weight and the degree of hydrolysis. The crystallinity increases with the increase in hydrolysis due to the hydroxyl groups being small enough to fit into the polyethylene lattice, whereas the atactic polyvinyl acetate is amorphous. The common commercial grades of PVA include water-soluble PVA, which contains a degree of hydrolysis of 88%, and water-insoluble PVA with a degree of hydrolysis >98%. Polyvinyl alcohol is a strong polymer with 125 MPa tensile strength reported for high-molecular weight, high-crystallinity grades at low moisture content. Water is a plasticizer for PVA so the properties change significantly at various humidities. PVA has been found to be biodegradable and mineralized in various environments, although most of the work has been conducted on water-soluble PVA [46].

Poly(butylene/ethylene succinate/adipate/terepthalate)   Recently, there has been development and production of polyesters consisting of various ratios of butane diol and/or ethylene glycol reacted with succinic acid and/or adipic acid and/or terephthalic acid to form a family of high-strength tough biodegradable materials. The amount of terephthalate must be kept below approximately 50% and randomly distributed in order obtain significant biodegradation rates [47]. The glass transition temperature as well as the melting point, crystallinity, and biodegradation rate can be varied systematically depending on the monomer ratios allowing a great degree of versatility. Various versions of these polymers are produced commercially by Showa High Polymer in Japan, BASF in Europe, and Eastman Chemical and DuPont in the United State.

9.6. FUTURE DEVELOPMENTS

This chapter is limited in scope in that there are other materials currently being developed and new materials that will be developed in the future. Various biodegradation testing has been performed on these materials [48–154]. In addition to the creation of new materials, the tailoring of properties to meet specific requirements will be accomplished through blending of two or more biodegradable polymers together and/or compounding with biodegradable plasticizers and/or biodegradable reinforcements or fillers [155–199]. As this field grows, the development of new products will naturally increase. Many

Table 9.1 Commercially Available Biodegradable Polymers

Category	Company (Trade Name)
Use of Natural Polymer	
Starch	National Starch (ECO-FOAM)
	Nihon Corn Starch (Cornpol)
	Oji Seitai (Eco Foam)
	Nissei (Eco Ware)
	Nihon Shokuhin-Kako (Placorn)
Starch-based blends	Novamont (Materi-Bi)
	Biotech
	Chisso (Novon)
	Earthshell
Cellulose acetate	Eastman Chemical (Eastman CA)
	Daicel Kagaku (CelGreen PCA)
	Nihon Shokubai (Lunare ZT)
Chitosan-cellulose	Aicello Kagaku (Doron CC)
Bacterial Polymer	
Poly-3-hydroxybutyrate	Metabolix (Biopol)
	Mitsubishi Gas Chemicals (Biogreen)
Polysaccharide	Takeda (Cardran)
	Hayashihara (Pullulan)
Synthetic Polymer—Renewable	
Polylactic acid	Cargill Dow (NatureWorks)
	Mitsui Chemicals (LACEA)
	Shimadzu Seisakusho (Lacty)
Synthetic Polymer—Petrochemical	
Poly(ethylene/butylene/succinate/adipate)	Showa Highpolymer (Bionolle)
	SK Chemicals (Sky Green)
Polybutylenesuccinate/terephthalate	BASF (Ecoflex)
	Eastman Chemicals (EasterBio)
	Dupont (Biomax)
Polybutylenesuccinate/carbonate	Mitsubishi Gas Chemicals (Iupec)
Polycaprolactone	UCC-Dow (Tone)
	Solvay (CAPA)
	Daicel Kagaku (CelGreen PH)
Polyethylenesuccinate	Nihon Shokubai (Lunare SE)
Polyvinyl alcohol	Air Products
	Kureha (Poval)
	Nihon Gosei Kagaku (Gosenol)

of these blends have been studied as to the miscibility and/or compatibility of various biodegradable polymers with one another. In general, miscibility will tend to decrease the biodegradation rate for two hydrophobic polymers but increase the degradation rate where at least one of the components is water soluble. Plasticization tends to increase the biodegradation rate especially where the plasticizer is readily biodegradable. Table 9.1 lists some of the current manufacturers of biodegradable polymers.

REFERENCES

1. ASTM, *Annual Book of ASTM Standards*, Vol. 8.01, D883, American Society for Testing and Materials, West Conshohocken, PA, 2000, p. 176.
2. M. N. Helmus, *Spectrum, Environmental Issues and Opportunities*, Arthur D. Little Decision Resources, Feb. 1988, 3–1.
3. M. T. Wolin and T. L. Miller, Methanogens, in A. L. Demain, and N. A. Solomon, eds., *Biology of Industrial Microorganisms*; Buttterworths, Boston, 1985, pp. 189–221.
4. W. Gujer and A. J. B. Zehnder, *Wat.Sci. Technol.* **15**, 127–167 (1983).
5. M. A. Barlaz, D. M. Schaefer, and R. K. Ham, *Appl. Environ. Microbiol.* **56**, 56–65 (1989).
6. G. P. Smith, B. Press, D. Eberiel, R. A. Gross, S. P. McCarthy, and D. L. Kaplan, *Polym. Mat. Sci. Eng.* **63**, 867–870 (1990).
7. J. Glenn and R. Spencer, *Biocycle* November, 34–84 (1991).
8. J. H. Crawford, in P. N. Cheremisinoff and R. P. Ouellette, eds., *Biotechnology: applications and Research*, Technomic, Lancaster, PA, 1985, pp. 68–77.
9. J. A. Brydson, *Plastic Materials*, Butterworth, Oxford, UK, 1987, p. 546.
10. J. A. Brydson, *Plastic Materials*, Butterworth, Oxford, UK, 1987, p. 9.
11. P. Finch and J. C. Robert, in T. P. Nevell and S. H. Zeronian, eds., *Cellulose Chemistry and Its Applications*, Ellis Horwood, West Sussex, England, 1985, pp. 312–343.
12. D. A. I. Goring and T. E. Timell, *Tappi* **45**, 454 (1969).
13. M. Marx-Figini, *Biochim. Biophys. Acta* **177**, 27 (1969.
14. G. Halliwell and M. Griffin, *Biochem J.* **128**, 1183 (1973).
15. L. E. R. Berghem and L. G. Peterson *Eur. J. Biochem.* **37**, 21 (1973).
16. M. Tein and T. K. Kirk, *Science* **221**, 661 (1983).
17. M. Shimada and T. Higuchi, in D. N. Hon and N. Shiraish, eds., *Wood and Cellulosic Chemistry*, Marcel Dekker, New York, 1992, p. 557.
18. E. T. Reese, *Ind. Eng. Chern.* **49**, 89–92 (1957).
19. E. B. Cowling, in E. W. Reese, ed., *Enzymatic Hydrolysis of Cellulose and Related Materials*, Macmillan, New York, 1963, pp. 1032.
20. R. G. H. Siu, *Microbial Decomposition of Cellulose*, Reinhold, New York, 1951.
21. J. Gu, M. Gada, S. McCarthy, R. Gross, and D. Eberiel, *Polym. Mat. Sci. Eng.* **67**, 351–352 (1992).
22. M. Nelson, S. McCarthy, and R. Gross, *Polym. Mat. Sci. Eng.* **67**, 139–140 (1992).

23. J. Gu, S. McCarthy, G. Smith, D. Eberiel, and R. Gross, *Polym. Mat. Sci. Eng.* **67**, 230–231 (1992).
24. C. M. Buchanan, R. M. Gardner, and R. J. Komarek, *J. Appl. Polym. Sci.* **47**, 1709–1719 (1993).
25. J. J. M. Swinkels, Sources of Starch, Its Chemistry and Physics, Starch Conversion Technology, Vol. 15, G. M. A. Van Beyum and J. A. Roels, eds., 1985.
26. J. J. Marshall, ed., *Mechanism of Saccharide Polymerization and Depolymerization*, Academic, New York, 1980, p. 55.
27. R. Shogren and B. Jasberg, *J. Environ. Polym. Degrad.* **2**(2), 99–110 (1994).
28. J. P. Sachetto and R. F. T. Stepto, U.S. Patent 4,900,361 (Eur. Pat. Appl. 0 118240) (1990).
29. I. Tomka, U.S. Patent No. 5,362,777 (1994).
30. C. Bastioli, V. Bellotti, L. Del Giudice, and R. Lombi, Pc[Int. Pat. Appl. WO 91/02024 (1991).
31. C. Bastioli, V. Bellotti, L. Del Giudice, and G. Gilli, *J. Environ. Polym. Degrad.* **1**(3), 181–191 (1993).
32. V. Davè and S. McCarthy, *J. Environ. Polym. Degrad.* **5**(4), 237–243 (1997).
33. M. Parikh, R. A. Gross, and S. P. McCarthy, *Polym. Mat. Sci. Eng.* **66**, 408–410 (1992).
34. H. Bender, J. Lehmann, and K. Wallenfels, *Biochem. Biophys. Acta* **36**, 309 (1959).
35. S. Yuen, *Process. Biochem.* **7**, 22 (Nov. 1974).
36. H. Matsunaga, K. Tsuji, and T. Saito, U.S. Patent. 4,045,388 (Aug. 30, 1977) (to Hayashibara Biochem. Labs., Inc.).
37. J. N. Boyer and R. S. Wolfe, *Biological Bull.* **165**(2), 505 (1993).
38. S. Arcidiacono and D. L. Kaplan, *Biotechnol. Bioeng.* **39**, 281–286 (1992).
39. S. Arcidiacono, S. J. Lombardi, and D. Kaplan, in G. Skjak-Brack, T. Anthosen, and O. Sandford, eds., *Chitin and Chitosan*, Elsevier, London, 1989, pp. 319–332.
40. D. L. Kaplan, J. M. Mayer, S. J. Lombardi, B. Wiley, and S. Acidiacono, *Polym. Preprints Am. Chem. Soc. Div. Polym. Chem.* **63**, 732–735 (1990).
41. J. M. Mayer, M. Greenberger, D. H. Ball, and D. L. Kaplan, *Polym. Mat. Sci. Eng.* **63**, 732–735 (1992).
42. J. M. Mayer and D. L. Kaplan, U.S. Patent 5,015,293 (1991).
43. E. S. Lipinsky, and R. G. Sinclair, *Chem. Eng. Prog.* **82**, 26–32 (1986).
44. D. H. Lewis, in M. Chasin and R. Langer, eds., *Biodegradable Polymers as Drug Delivery Systems*, Drugs and the Pharmaceutical sciences, Marcel Dekker, New York, Vol. 45, 1990, pp. 1–41.
45. L. J. E Potts, in *Encyclopedia of Chemical Technology*, 2nd ed., Suppl. Volume Wiley, Interscience, New York, 1984, p. 626.
46. U. Witt, R.-J. Mueller, and W.-D. Deckwer, *J. Environ. Polym. Degrad.*, 1997, p. 81–9.
47. Y. Yakabe and M. Kitano, in Y. Doi and K. Fukuda, eds., *Biodegradable Plastics and Polymers*, Elsevier Science, New York, 1994, p. 331.
48. OECD, *Expert Group Determination of the Biodegradability of Anionic Synthetic Surface Active Agents*, Organization for Economic Cooperation and Development, Paris, 1971.

49. R. T. Wright, A. W. Bourquin, and P. H. Pritchards, eds., *Microbial Degradation of Pollutants in the Marine Environment*, USEPA, Gulf Breeze, FL, 1979, p. 119.
50. L. H. Stevenson, *Microbiol. Ecol.* **4**, 127 (1978).
51. R. G. Austin, in S. A. Barenberg, J. L. Brash, R. Narayan, and A. E. Redpath, eds., *Degradable Materials: Perspectives, Issues and Opportunities*, CRC Press, Boca Ration, FL, 1990, p. 237.
52. G. Iannotti, N. Fair, M. Tempesta, H. Neibling, F. H. Hsieh, and R. Mueller, in S. A. Barenberg, J. L. Brash, R. Narayan, and A. E. Redpath, ed., *Degradable Materials; Perspectives, Issues and Opportunities*, CPC Press, Boca Raton, FL, 1990, p. 425.
53. S. M. Goheen and R. P. Wool, *J. Appl. Polym. Sci.* **42**, 2691–2701 (1991).
54. W. J. Bailey, V. Kuruganti, and J. S. Angle, in J. E. Glass and G. Swift, eds., Agricultural and Synthetic Polymers, Biodegradability and Utilization, vol. 433, ACS Symposium Series, Washington, DC, 1990, p. 149.
55. K. S. Lee and R. D. Gilbert, *Carbohydr. Res.* **89**, 162 (1981).
56. M. M. Lynn, V. T. Stannett, and R. D. Gilbert, *J. Polym. Sci., Polym. Chem. Ed.* **19**, 1967 (1980).
57. S. L. Kim, V. T. Staratett, and R. D. Gilbert, *J. Macromol. Sci.* **7**, 101 (1979).
58. K. W. King and M. I. Vessal, *Adv. Chem. Ser.* **95**, 7 (1969).
59. T. K. Ng, A. Ben-Bessat, and J. G. Zeikus, *Appl. Environ. Microbiol.* **41**, 1337 (1981).
60. D. Groleau and C. W. Forsberg, *Can. J. Microbiol.* **27**, 517 (1981).
61. K. Omiya, K. Nokura, and S. J. Shimizu., *Ferment. Technol.* **61**, 25 (1983).
62. K. Yamane, H. Suzuki, and K. Nisizawa, *J. Biochem.* **61**, 19 (1970).
63. K. Osmundsvag and J. Goksor, *Eur. J. Biochem.* **57**, 405 (1975).
64. D. S. Chalal and W. D. Gray, in edited by A. H. Walters and J. J. Elphick, eds., *Biodeterioration of Materials*, Elsevier, New York, 1968, p. 584.
65. M. P. Levi and E. B. Cowling, in A. H. Walters and J. J. Elphick, eds., *Biodeterioration of Materials*, Elsevier, New York, 1968, p. 575.
66. M. Streamer., K. E. Eriksson, and B. Pettersson, *Eur. J. Biochem.* **51**, 607 (1975).
67. T. Kanda, K. Wakabayashi, and K. Nisizawa, *J. Biochem.* **79**, 977 (1976).
68. M. Paice, M. Desrochers, D. Rho, L. Jurasek, C. Roy, C. F. Rollin, E. DeMiguel, and M. Yaguchi, *Biotechnology* **2**, 535 (1984).
69. A. Hutterman and A. Noelle, *Holzforsch.* **36**, 283 (1982).
70. B. Bucht and K. F. Eriksson, *Arch. Biochem. Biophys.* **129**, 416 (1969).
71. T. Hiroi, *Mokuzai Gakkaishi* **27**, 684 (1981).
72. G. Keilich, P. J. Bailey, E. G. Afting, et al., *Biochem. Biophys. Acta.* **185**, 392 (1970).
73. E. B. Cowling and W. Brown, *Adv. Chem. Ser.* **95**, 152 (1969).
74. T. L. Highley, *Wood and Fiber* **5**, 50 (1973).
75. B. C. Sison, W. J. Schubert, and F. F. Nord, *Arch. Biochem, Biophys.* **69**, 502 (1957).
76. N. J. King, in A. H. Walters and J. J. Elphick, eds., *Biodeterioration of Materials*, Elsevier, New York, 1968, p. 558.

77. M. Ishihara and K. Shimizu, *Mokuzai Gakkaishi* **30**, 79 (1984).
78. S. Doi, *Mokuzai Gakkaishi* **31**, 843 (1995).
79. H. Shimazono, *Arch. Biochem. Biophys.* **83**, 206 (1959).
80. G. Keilich, P. Bailey, and W. Liese, *Wood Sci. Technol.* **4**, 273 (1973).
81. J. Eriksen and J. Goksor, *Eur. J. Biochem.* **77**, 445 (1977).
82. T. M. Wood and S. I. McCrae, *Adv. Chem. Ser.* **181**, 181 (1979).
83. R. D. Brown, Jr. and L. Jurasek, *Adv. Chem. Ser.* **181**, 399 (1979).
84. T. M. Wood, S. I. McCrae, and C. C. MacFarlane, *Biochem. J.* **198**, 51 (1980).
85. T. M. Wood and S. I. McCrae, *Carbohydrate Res.* **57**, 117 (1977).
86. S. Murao and R. Sakamoto, *Agric. Biol. Chem.* **43**, 1791 (1979).
87. A. Ikeda, T. Yamamoto, and M. Funatsu, *Agric. Biol. Chem.* **37**, 1169 (1973).
88. M. R. Coudray, G. Canevascini, and H. Meier, *Biochem. J.* **203**, 277 (1982).
89. D. R. Whitaker, *Arch. Biochem. Biophys.* **43**, 253 (1953).
90. C. E. Warnes and C. I. Randles, *Ohio J. Sci.* **77**, 224 (1983).
91. N. G. Antrim, *Microb. Bcol.* **6**, 317 (1981).
92. F. Young, R. L. Bell, and P. A. Carroad, *Biotechnol. Bioeng.* **27**, 769 (1985).
93. M. Srikantiah and K. C. Mobankurnar, *Indian J. Microbiol.* **20**, 216 (1981).
94. C. Jeuniaux, J. C. Bussers, M. E. Voss-Foucart, and M. Poulicek, in R. A. Muzarelli, C. Jeuniaux, and G. W. Gooday, eds., *Chitin in Nature and Technology*, Plenum, New York, 1996, p. 516.
95. R. A. Smucker, in R. A. Muzarelli, C. Jeuniaux, and G. W. Gooday, eds., *Chitin in Nature and Technology*, Plenum, New York, 1986, p. 254.
96. J. Rodriguez, M. I. Perex-Leblic and F. Laborda, in R. A. Muzarelli, C. Jeuniaux, and G. W. Gooday, eds., *Chitin in Nature and Technology*, Plenum, New York, 1986, p. 102.
97. S. Hara, Y. Yamamura, Y. Fuji, et al., in Proc. 2nd Int. Conf on Chitin and Chitosan, S. Hirano and S. Takura, eds., July 12–14, 1982, Sapporo, Japan, The Japanese Soc. of Chitin and Chitosan, p. 125.
98. I. Chet, E. Cohen, and I. Elster, in R. A. Muzarelli, C. Jeuniaux, and G. W. Gooday, eds., *Chitin in Nature and Technology*, Plenum, New York, 1986, p. 231.
99. A. Ohtakara, H. Ogata, Y. Taketomi, and M. Mitsutomi, in J. P. Zikakis, eds., *Chitin, Chitosan, and Related Enzymes*, Academic, Orlando, FL, 1984, p. 147.
100. A. Hedges and R. S. Wolf, *J. Bacteriol.* **120**(2), 944 (1974).
101. A. Ohtakara, H. Ogata, Y. Taketomi, et al. in J. P. Zikakis, ed., *Chitin, Chitosan and Related Enzymes*, Academic, Orlando, FL, 1984, p. 147.
102. J. S. Price and R. Storck, *J. Bacteriol.* **124**(3), 1574 (1975).
103. M. Shimada and M. Takahashi, in D. N.-S. Hon and N. Shiraishi, eds., *Wood and Cellulosic Chemistry*, Marcel Dekker, New York, 1992, p. 625.
104. J. M. Gould, S. H. Gordon, L. B. Dexter, and C. L. Swanson, in J. E. Glass and G. Swift, eds., *Agricultural and Synthetic Polymers Biodegradability and Utilization*, Vol. 433, ACS Symposium Series, Washington, DC, 1990, p. 65.
105. M. A. Cole, in J. E. Glass and G. Swift, eds., *Agricultural and Synthetic Polymers—Biodegradability and Utilization*, ACS Symposium Series 433, American Chemical Society, Washington, DC, 1990, p. 76.

106. A. Corti, G. Vallini, A. Pera, F. Cioni, R. Solaro, and E. Chiellini, in M. Vert, J. Feijen, A. Albertsson, G. Scott and E. Chiellini, eds., *Biodegradable Polymers and Plastics*, Royal Soc. Chemistry, Cambridge, England, 1992, p. 245–248.
107. A. C. Albertsson, C. Barnsted, and S. Karlsson, *J. Environ. Polym. Deg.* **1**, 241 (1993).
108. F. Kawai, in J. E. Glass and G. Swift, eds., *Agricultural and Synthetic Polymers—Biodegradability and Utilization*, Vol. 433, ACS Symposium Series, American Chemical Society, Washington, DC, 1989, p. 110.
109. F. Kawai and H. Yamanaka, *Arch. Microbiol.* **146**, 125 (1986).
110. F. Kawai, Japanese Patent 208289 (1987).
111. S. Matsumara, S. Maeda, J. Takahashi, and S. Yoshikawa, *Kobunshi Ronbunshu* **45**, 317 (1988).
112. Y. Toikawa, T. Ando, T. Suzki, and K. Takeda, in J. E. Glass and G. Swift, eds., *Agricultural and Synthetic Polymers—Biodegradability and Utilization*, Vol. 433, ACS Symposium Series, American Chemical Society, Washington, DC, 1989, p. 136.
113. V. Benedict, W. J. Cook, P. Jarrett et al., *J. Appl. Polym. Sci.* **28**, 327 (1983).
114. M. Kimura, K. Toyota, M. Iwatsuki, et al., in Y. Doi and K. Fukuda, eds., *Biodegradable Plastics and Polymers*, Elsevier Science, New York 1994, pp. 92, 237.
115. L. Tilstra and D. Johnsonbaugh, *J. Environ. Polym. Deg.* **1**, 257 (1993).
116. P. Lefebvre, A. Daro, and C. David, *Macromol. Sci., Pure Appl. Chem.* **A32**, 867 (1995).
117. H. Sawada in Y. Doi and K. Fukuda, eds., *Biodegradable Plastics and Polymers*, Elsevier Science, New York, 1994, p. 299.
118. M. Tsuji and Y. Omoda, in Y. Doi and K. Fukuda, eds., *Biodegradable Plastics and Polymers*, Elsevier Science, New York, 1994, p. 345; England, **31**, 1200 (1992).
119. Y. Toikawa, T. Ando, T. Suzuki, and K. Takeda, in J. E. Glass and G. Swift, eds., *Agricultural and Synthetic Polymers. Biodegradability and Utilization*, Vol. 433, ACS Symposium Series, American Chemical Society, Washington, DC, 1989, p. 136.
120. D. F. Gilmore, R. C. Fuller, B. Schneider, R. W. Lenz, N. Lotti, and M. Scandola, *J. Environ. Polym. Deg.* **2**, 49 (1994).
121. J. Mergaert, A. Wouters, J. Swings, et al., in M. Vert, J. Feijen, A. Albertsson, G. Scott and E. Chiellini, eds., *Biodegradable Polymers and Plastics*, Royal Soc. Chemistry, Cambridge, England, 1992, p. 267.
122. M. Gada, R. A. Gross, and S. P. McCarthy, Y. Doi and K. Fukuda, eds., *Biodegradable Plastics and Polymers*, Elsevier Science, New York, 1994, p. 177; C. Bastioli, V. Bellotti, M. Camia, Del Gludice, and A. Rallis, *Ibid.*, p. 204.
123. R. J. Muller, J. Augusta, T. Walter, et al. in M. Vert, J. Feijen, A. C. Albertsson, G. Scott and E. Chiellini, eds., *Biodegradable Polymers and Plastics*, Royal Soc. Chemistry, Cambridge, England, 1992, p. 1491.
124. H. Eya, N. Iwaki, and Y. Otsuji, in Y. Doi and K. Fukuda, eds., *Biodegradable Polymers and Plastics*, Elsevier Science, New York, 1994, p. 337.
125. D. Jendrossek, I. Knoke, R. B. Habibian, A. Steinbüchel, and H. G. Schlegel, *J. Environ. Polym. Deg.* **1**, 53 (1993).
126. A. Nishida and Y. Tokiwa, *J. Environ. Polym. Deg.* **1**, 65 (1993).
127. G. Tomsi and M. Scandola, *J. Macromol. Sci., Pure Appl. Chem.* **A32**, 671 (1995).

128. D. F. Gilmore, S. Antoun, R. W. Lenz, and R. C. Fuller, *J. Environ. Polym. Deg.* **1**, 269 (1993).
129. Z. Fillip, *Europ. J. Appl. Microbiol. Biotechnol.* **7**, 277 (1979).
130. Z. Fillip, *Europ. J. Appl. Microbiol. Biotechnol.* **5**, 225 (1978).
131. S. Owen, R. Kawamura, S. Owen, M. Masaoka, R. Kawamura, and N. Sakota, *J. Macromol. Sci., Pure Appl. Chem.* **A32**, 843 (1995).
132. S. J. Huang, M. S. Roby, C. A. Macri, and J. A. Cameron, in M. Vert, J. Feijen, A. Albertsson, G. Scott and E. Chiellini, *Biodegradable Polymers and Plastics*, Royal Soc. Chemistry, Cambridge, England, 1992, p. 149.
133. S. J. Huang, C. A. Macri, M. Roby, C. Benedict, and J. A. Cameron, *ACS Symp. Ser.* **172**, 471 (1981).
134. J. E. Guillet, *Adv. Chem. Ser.* **169**, 1 (1978).
135. N. B. Nykvist, in Proc. of Degradability of Polymers and Plastics Conference, Inst. Electrical Engineering, London, England, 1973, p. 18.
136. A. C. Albertsson, Z. G. Banhidi, and L. L. Beyer-Ericcsson, *J. Appl. Polym. Sci.* **22**, 3434 (1979).
137. A. C. Albertsson and Z. G. Banhidi, *J. Appl. Polym. Ser.* **25**, 1655 (1980).
138. A. C. Albertsson and S. Karlsson, *J. Appl. Polym. Sci.* **35**, 1289 (1988).
139. I. H. Cornell, A. M. Kaplan, and M. R. Rogers, *J. Appl. Polym. Sci.* **29**, 2591 (1984).
140. L. Kravetz, in J. E. Glass and G. Swift, eds., *Agricultural and Synthetic Polymers — Biodegradability and Utilization*, Vol. 433, ACS Symposium Series, American Chemical Society, Washington DC, 1989, p. 96.
141. S. Matsumara and T. Tanaka, *J. Environ. Polym. Deg.* **2**, 89 (1994).
142. R. Engler and S. H. Carr, *J. Polym. Sci., Polym. Phys. Ed.* **11**, 313 (1973).
143. S. A. Bradley, S. H. Engler, and S. H. Carr, in M. A. Golub and J. A. Parker, eds., *Polymeric Materials for Unusual Service Conditions*, Wiley, New York, 1973, p. 269.
144. R. Benner, A. F. Macubbin, and R. E. Hodson, *Appl. Envir. Microbiol.* **47**, 999 (1994).
145. J. Gu, D. Eberiel, S. P. McCarthy, and R. A. Gross, *J. Environ. Polym. Deg.* **1**, 281 (1993).
146. R. A. Gross, J. Gu, D. Eberiel, and S. McCarthy, *J. Macromol. Sci., Pure Appl. Chem. A* **32**, 613 (1995).
147. A. M. Buchanan, D. Dorschet, R. M. Gardner, R. J. Komareck and A. W. White, *J. Macromol. Sci. Pure Appl. Chem. A* **32**, 683 (1995).
148. A. M. Buchanan, R. M. Gardner, and R. J. Komarek, *J. Appl. Polym. Sci.* **47**, 1709 (1993).
149. S. Mauuniam, K. Amay, and S. Yoshikawa, *J. Environ. Polym. Deg.* **1**, 23 (1993).
150. A. Tsuchii, T. Suzuki, and Y. Takahara, *Agri. Biol. Chem.* **42**, 1217 (1978).
151. A. Tsuchii, T. Suzuki, and Y. Takahara, *Agri. Biol. Chem.* **41**, 2417 (1977).
152. A. Tsuchii, T. Suzuki, and Y. Takahara, *Agri. Biol. Chem.* **43**, 2441 (1979).
153. R. T. Darby and A. M. Kaplan, *Appl. Microbiol.* **6**, 900 (1968).
154. A. Bastioli, V. Bellotti, M. L. Camia, L. Del'Giudice, and A. Rallis, in Y. Doi and K. Fukuda, eds., *Biodegradable Plastics and Polymers*, Elsevier Science, New York, 1994, p. 200–13.

155. *Plastics News*, "Food for Thought: Tomorrow's Feedstocks," March, 2000.
156. W. Ma, A. Ranganathan, and S. McCarthy, in A.-C. Albertson, E. Chiellini, J. Feijen, G. Scott, and M. Vert, eds., *Degradability, Renewability and Recycling—Key Functions for Future Materials*, Wiley-VCH, 1999, pp. 63–72.
157. S.-M. Li and S. McCarthy, *Biomaterials* **30**, 35–44 (1999).
158. D. Donabedian and S. P. McCarthy, *Macromolecules*, **31**(4), 1032–1039 (1998).
159. T. J. Adamczyk, D. C. Broe, R. E. Farrell, J. S. Lee, B. L. Daniels, and S. P. McCarthy, 58th Society of Plastics Engineers Annual Technical Conference Proceedings, Vol. 45, Number 1, May 1999, pp. 3255–3259.
160. C. L. Yue, R. Kumar and S. P. McCarthy, *Am. Chem. Soc. Polym. Preprints* **39**(2), 132–133 (August 1998).
161. V. Dave, M. Sheth, S. P. McCarthy, J. Ratto, and D. Kaplan, *Polym. Polym. Preprints* **39**(5) 1139–1148 (1998).
162. M. Parikh and R. Gross, *J. Injection Molding Tech.* **2**(1), 30–36 (1998).
163. G. V. Laverde and S. McCarthy, *Soc. Plastic Engineers Ann. Techn. Conf.* **44**, 2515–2519 (May 1998).
164. W. Ma and S. McCarthy, *Soc. Plastics Engineers Ann. Techn. Conf.* **56**, 2542–2545 (1998).
165. S. P. McCarthy, 5th International Workshop on Biodegradable Plastics and Polymers, (ISBP98) Stockholm, Sweden, (June 1998).
166. M. Sheth, R. A. Kumar, V. Dave, R. A. Gross, and S. P. McCarthy *J. Appl. Polym. Sci.* **66**, 1495–1505 (1997).
167. V. Dave, S. P. McCarthy, R. A. Gross, L. V. Labrecque, and R. A. Kumar, *J. Appl. Polym. Sci.* **66**, 1507–1513 (1997).
168. H. Cai, V. Dave, R. A. Gross, and S. P. McCarthy, *J. Polym. Sci., Polym. Phys.* **40**, 2701–2708 (1996).
169. V. Ghiya, V. Dave, R. A. Gross, and S. McCarthy, *J. Mat. Sci., Pure Appl. Chem.* **A33**(5), 627–638 (1996).
170. A. Gajria, V. Dave, R. A. Gross, and S. McCarthy, *Polymer* **37**(3), 437–444 (1996).
171. C. L. Yue, R. A. Gross, and S. McCarthy, *Polym. Degrad. Stabil.* **51**, 205–210 (1996).
172. G. Rocha and S. McCarthy, *Med. Plastics Biomat.* May/June, 44–48 (1996).
173. S. P. McCarthy, *AIChE Preprints*, **5**, 205–209 (1996).
174. C. L. Yue, V. Dave, R. A. Gross, and S. McCarthy, *Poly. Preprints, Am. Chem. Soc., Div. Polym. Sci.*, **36**, 418–419 (1995).
175. H. Cai, V. Dave, R. A. Gross, and S. McCarthy, *Polym. Preprints, Am. Chem. Soc., Div. Polym. Sci.* **36**, 422–423 (1995).
176. V. Ghiya, V. Dave, R. A. Gross, and S. McCarthy, *Polym. Preprints, Am. Chem. Soc., Div. Polym. Sci.* **36**, 420–421 (1995).
177. M. Sheth, V. Dave, R. A. Gross, and S. McCarthy, *Soc. Plastics Eng., Tech. Papers* **40**, 1829–1833 (1995).
178. A. M. Gajria, V. Dave, R. A. Gross, and S. McCarthy, *Soc. Plastics Eng., Tech. Papers* **40**, 2042–2045 (1995).
179. L. V. Labrecque, V. Dave, R. A. Gross, and S. McCarthy, *Soc. Plastics Eng., Tech. Papers* **40**, 1819–1823 (1995).

180. H. Cai, V. Dave, R. A. Gross and S. McCarthy, *Soc. Plastics Eng., Tech. Papers* **40**, 2046–2050 (1995).
181. C. L. Yue, S. McCarthy, and R. A. Gross, *Soc. Plastics Eng., Tech. Papers* **53**, pp. 2033–2036 (1995).
182. P. Dave, T. Jahedi, D. Eberiel, R. A. Gross, and S. McCarthy *Polym Degrad. Stab.* **45**, 197–203 (1994).
183. V. Dave, J. A. Ratto, D. Rout, R. A. Gross, D. L. Kaplan, and S. McCarthy, *Polym. Preprints, Am. Chem. Soc., Div. Polym. Sci.* **35**(2), 448–449 (1994).
184. G. Rocha, R. A. Gross, and S. P. McCarthy, *Polym. Preprints, Am. Chem. Soc., Div. Polym. Sci.* **33**(2), 454–455 (1992).
185. A. Lisuardi, A. Schoenberg, M. Gada, and S. McCarthy, *Polym. Mat. Sci. Eng.* **67**, 298–300 (1992).
186. S. T. Tanna, R. A. Gross, and S. McCarthy, *Polym. Mat.: Sci. and Eng.* **67**, 294–295 (1992).
187. D. H. Donabedian, R. A. Gross, and S. P. McCarthy, *Polym. Mat.: Sci. Eng.* **67**, 301–302 (1992).
188. P. Dave, M. Reeve, C. Brucato, R. Gross, and S. McCarthy, *Soc. of Plastics Eng.* **37**, 982 (1991).
189. P. Dave, R. A. Gross, C. Brucato, S. Wong, and S. McCarthy, in C. G. Gebelein, ed., *Biotechnology and Polymers*, Plenum, New York, 1991, pp. 53–62.
190. P. Dave, R. Gross, and S. McCarthy, *Soc. of Plastics Eng., Tech. Papers* **36**, 1439 (1990).
191. P. B. Dave, N. J. Ashar, R. A. Gross and S. P. McCarthy, *Polym. Preprints, Am. Chem. Soc., Div. Polym. Sci.* **31**(1), 442–443 (1990).
192. P. B. Dave, R. A. Gross, C. Brucato, S. Wong, and S. McCarthy, *Polym., Mat. Sci. Eng.* **62**, 231 (1990).
193. D. B. Parasar, M. Parikh, M. S. Reeve, R. A. Gross and S. McCarthy, *Polym. Mat. Sci. Eng. (Am. Chem. Soc.)* **63**, 726 (1990).
194. S. M. Mccartin, B. Press, R. Gross, D. Eberiel, and S. P. McCarthy, *Polym. Preprints* **31**(1), 439 (1990).
195. X. Song and S. McCarthy, *Annu. Tech. Conf.—Soc. Plast. Eng.*, 59th, **3**, 2546–2549 (2001).
196. R. E. Farrell, T. J. Adamczyk, D. C. Broe, J. S. Lee, B. L. Briggs, R. A. Gross, S. P. McCarthy, and S. Goodwin, in *Biopolymers from Polysaccharides and Agroproteins*, ACS Symp. Ser., American Chemical Society, Washington, DC, 2001, p. 786.
197. P. Canale, S. Mehta, and S. McCarthy, *J. Appl. Med. Polym.* **5**(2), 65–71 (Autumn 2001).
198. B. Koroskenyi and S. McCarthy, *Biomacromolecules* **2**(3), 824–826 (2001).
199. S. M. Li, X. H. Chen, and R. A. Gross, *J. Mat. Sci: Mat. Med.* **11**, 227–233 (2000).

CHAPTER 10

PLASTICS IN THE MARINE ENVIRONMENT

MURRAY R. GREGORY
Department of Geology, The University of Auckland

ANTHONY L. ANDRADY
Research Triangle Institute

10.1. INTRODUCTION

Marine debris is generally defined as "any manufactured or processed solid waste material (typically inert) that enters the marine environment from any source." *Marine litter* and *floatables* are considered equivalent terms [1]. Marine pollution is defined as "the introduction by man, directly or indirectly, of substances into the marine environment (including estuaries) resulting in such deleterious effects as harm to living resources, hindrance to marine activities including fishing, impairment of quality for use of sea water and reduction of amenities" [2]. Marine debris is dominated by persistent synthetic materials, most of which are plastics.

For over two millennia humankind has indiscriminately discarded its wastes across and around the margins of oceans and inland seas, as well as lakes, rivers, and other waterways. While populations were small and dispersed, and the quantities for disposal were mostly biodegradable, the environmental and other consequences were of minimal significance. Indeed, until recent times there was general acceptance that because of their geographic expanse the oceans had an infinite capacity to assimilate wastes of all kinds. Furthermore, many authorities considered oceanic waters were self-cleansing. Awareness of environmental and management problems to be faced in marine environments developed slowly through the 1950s; but, by the 1960s it was widely accepted that all was not well with the oceans and their health was being questioned [2–6]. Plastic products are

---

Plastics and the Environment, Edited by Anthony L. Andrady.
ISBN 0-471-09520-6 © 2003 John Wiley & Sons, Inc.

Table 10.1 Summary of Desirable Properties in Modern Plastic Materials That Have Also Made Them an Environmental Nuisance

Property	Plastic Debris Items
Strength, durability, and relative inertness; virtually indestructibility in the marine environment	Fishing gear, rope, six-pack ring-strapping bands, and boat construction
General resistance to microbial and other degradation processes	Fishing gear including traps (crab pots); food packaging material; tampon applicators
Transparency, biological inertness, low density, and cost effectiveness	Packaging material for food and for bait; bags.
Ability to be constructed as a low-density foam that is generally negatively buoyant	Insulated beverage cups, bait boxes, and floatation devices
Low cost and disposability	Numerous consumer applications leading common litter (e.g., candy wrappers, cigarette filters)

an indispensable and inescapable part of contemporary life, and ones for which there is an ever-expanding demand. Unfortunately, the very properties of plastics that modern societies find so desirable (Table 10.1) has ensured their proliferation in the marine milieu, and they have become the dominant component of recently recorded marine debris. Numerous surveys at widely separated localities around the world have consistently demonstrated that on an item-by-item basis, plastic materials typically comprise 60–80% of marine debris litter accumulating in the wrack of sandy shores [7–9]. The pollution, environmental, sociologic, and other problems of plastic and other persistent materials (including glass, metallic cloth, and certain paper items) discarded in or reaching marine waters have been addressed and analyzed in detail at a series of international workshops and conferences between 1985 and 1995 [1, 10–13]. Issue papers are also available for a more recent gathering that focused on derelict fishing gear [14]. Aluminum cans are also a common element in marine debris [15].

Goldberg [16, 17] has suggested plastics to be one of several marine pollution problems likely to become more serious as the twenty-first century progresses. Others that he places in this category include nutrients and eutrophication, algal blooms and biotoxins, alien organism introductions and pathogens, and environmental estrogens and synthetic hormones. These apparently disparate factors share the following common attributes: persistence and extended residence times (or half-lives) leading to continued slow accumulation, gradually increasing fluxes, wide dissemination reflecting nonpoint sources, and easy dispersal. The problems created are chronic and potentially global, rather than acute and local or regional.

10.2. PLASTIC LITTER AND OTHER MARINE DEBRIS

10.2.1. Categories and Sources

The chemical structures and the mechanical characteristics of plastics generally found in marine debris, including polystyrene (PS), low- and high-density polyethylene (LDPE and HDPE), polypropylene (PP), polyethylene terephthate (PET), and polyvinyl chloride (PVC) were included in Chapter 2. Investigations of anthropic marine debris commonly recognize and record several types of material, in each of which separate categories may be identified. Typical survey sheet examples include solid or hard plastics (26 categories, e.g., beverage bottles, sheeting, fishing line, cigarette butts, etc.), foamed plastic (9 categories), glass (5 categories), rubber (5 categories), metal (8 categories), paper (8 categories), wood (5 categories) and cloth [18, 21]. From these categories "dirty dozen" listings of marine debris items have been identified. While there are local and regional variations in these, and also minor changes over time, there is also a degree of global uniformity [18]. Since first being recorded in the 1990 survey [19] cigarette butts (i.e., filters typically made of the synthetic polymer cellulose acetate fiber) have consistently topped the dirty dozen list. However, their conspicuous presence on popular tourist or recreational sandy beaches had been noted much earlier, for example, the South Pacific island Kingdom of Tonga [20] and Malta (Gregory, pers. observation). It is often convenient to recognize four broad categories of plastic debris based on size and function and location [21, 22].

1. Microlitter is inconspicuous, fine plastic detritus with a size range of very fine sand to coarse silt usually found in the marine sediment. On washing it passes through a 500-μm sieve but is retained on one at 63 μm.

    Zitko and Hanlon [23] noted that finely granulated polyethylene, polypropylene, and polystyrene particles were found in some Canadian hand cleaners and were a likely marine pollutant. Later, Gregory [24] identified similar products from New Zealand and also noted several quality cosmetic preparations, such as facial scrubs and exfoliants that carry varying amounts of comparable fine plastic detritus. Gregory [24] also drew attention to the plastic media used in the air blasting technique for stripping paint from aircraft and cleaning delicate engine parts and that utilizes particles in the same size range. Photo- or oxidative degradation and embrittlement of discarded plastics, whether afloat on the high seas or lying on the shore, may also lead to further release of material in this size category.

2. Mesolitter covers plastic debris material in the size range <5–10 mm. It is dominated by small translucent and transparent, ovoidal-to-rounded and rod-shaped granules of virgin resins that are the raw material or feedstock of the plastics manufacturing industry. Mesolitter also includes irregular partly degraded fragments of plastics of the same size range, shredded

during recycling operations. The term *nurdles* has been coined for some of this material (C. Moore, pers. communication).
3. Macrolitter covers fragments of plastic material up to 10 or 15 cm across that is readily visible to the naked eye during shoreline surveys and that may also be recognized from shipboard sighting transects in calm weather. This category includes small items and fragments generated from the physical breakdown of larger fabricated items. Confectionery wrappings, drinking straws, bottle caps, and pieces of Styrofoam and convenience food packaging also fall in this category.
4. Megalitter is material measuring decimeters or more across and includes fragments and fabricated items visually identifiable during shipboard sighting surveys. Typical examples are floats from fishing operations, crates and boxes, netting ropes and hawsers, strapping loops, and a variety of plastic bottles and containers.

These may either have a land-based or an ocean-based origin. In both cases, disposal or loss may have been deliberate (whether legal or illegal) or accidental. Land-based debris, commonly associated with metropolitan and urban areas, includes industrial waste and domestic refuse, as well as a diversity of packaging materials, plastic bottles, and other containers of all kinds. Confectionary and convenience food wrapping casually discarded by recreational visitors are also characteristic. These materials reach the sea by way of natural watercourses, storm water drainage outlets, and sewage outfalls. Riverine processes may bring plastic and other litter from distant hinterland and upland sources and make a significant contribution to marine debris on some shores [25], as well as in coastal and oceanic waters. Harbors and inshore waters adjacent to large industrial and urban centers can be important temporary repositories for significant quantities of marine debris [7, 26, 27].

Much of the ocean-based debris is readily identified for it includes waste materials generated by fishers (trawl netting and webbing, labeled fish crates and boxes, etc.) as well operational wastes from commercial shipping activities and ocean-going vessels. Some galley wastes may be difficult to distinguish from land-based domestic refuse. Nevertheless, although composition and types of plastic and other persistent synthetic items in marine debris accumulations may help in identifying possible sources, judgment is seldom unequivocal. The results of some early censuses of beached marine debris reflected collector bias, and the claimed importance of fisheries-related sources were overly generous [21].

The first marine debris workshop [10] focused on derelict and/or discarded fishing gear. Some time previously attention had been drawn to the role merchant vessels played in discharging plastics on the high seas [28]. By the time of the third debris conference in 1994 [1], it was acknowledged that land-based sources of marine debris were of more importance than marine-based ones [28–32]. Following GESAMP [2] there is common acceptance that 70–80% of all marine pollution has land-based sources.

10.2.2. Sampling and Surveys

Shoreline surveys of amounts and types of marine debris, generally with emphasis on plastics and other persistent synthetic materials, and often undertaken in tandem with local community beach cleanup activities have been commonplace since the mid-1980s. A national U.S. beach cleanup day initiated by the Center for Marine Conservation (now the Ocean Conservancy) in 1986 has become a global project involving over 500,000 volunteers from more than 90 countries, with compilations of their results presented annually [33]. While standard record sheets have generally been adopted, the scientific rigor of many volunteer surveys is not great, for without standardized classification procedures, interregional comparisons are of questionable validity [8, 9, 34]. Both *standing crop* and *accumulation rate* approaches have their advocates, although the former is more common. Quantities are most commonly recorded in true or proportional (%) numbers of items or weight of material per unit length or area and based on continuous, systematic, and representative or random transects. Solving the input/output equation and establishing a litter budget model has proven difficult and is seldom attempted. A successful example is the litter flux model described by Bowman et al. [35].

Volume measures are generally estimates. In some instances types and amounts are simply ranked in order of abundance. The approach adopted may reflect time availability, number of individuals involved, and survey objectives. It will take into consideration factors such as site topography, meteoric and oceanographic conditions, relative amounts of marine debris and seaweed or other natural litter present, and also where to set boundaries [36].

Establishing quantities of marine debris afloat on the high seas is particularly difficult. One approach has been adapted from the ornithologists studying seabird distributions where shipboard observers conduct sighting surveys, but these only record debris in the macro- and megalitter [37]. Such surveys are generally nonsystematic, being made from ships of passage [38]. Data on finer material in the micro- and mesolitter categories comes from surface-towed neuston and pleuston nets of varying design [39–42].

10.2.3. Quantities and Distribution

Other than rather speculative comments on its possible environmental impacts [24] plastic microlitter has not been subject to any detailed investigations. Numerous small flakes of plastic resulting from environmental degradation that fall into in this category have been identified after careful searching of the material floated off bulk beach sand samples from South Africa, New Zealand, and Bermuda (Gregory, unpublished). The mesh size used in most surface-towed neuston nets [40, 43] is generally too coarse to retain this material, or there has been a failure to identify it. It is, however, a highly visible component in neuston samples taken from the North Pacific gyre by Moore et al. [42]. Despite pelagic plastic microlitter being so seldom recorded, there can be little doubt that its distribution in surface waters has become global like that of virgin plastic pellets described below.

Virgin plastic granules or pellets have been found in varying quantities on all shores, adjacent to populated and industrialized areas of both the Northern and Southern Hemispheres. Examples include the United States [7], the Mediterranean [44, 45], South Africa [46], and New Zealand [47]. In some instances the numbers recorded are so great that they have been recorded as TNTC ("too numerous to count" [48]). These granules are also present, often in significant numbers on the shores of isolated oceanic islands lying distant from possible manufacturing and processing sources [8, 49] as well as afloat in remote surface waters (Table 10.2). Plastic and other persistent synthetic macro- and megalitter materials are similarly widely distributed from Alaska to Antarctica [1, 10–13]. It is evident these materials have long since effected a global distribution and are now broadcast across all oceanic waters.

Through mechanical attrition, often assisted by photodegradation, plastics in the marine milieu may undergo very slow embrittlement and breakdown into inconspicuous, very fine detritus, which ultimately disappears from view [47]—but not from the environment. Plastic litter accumulating along the strand may be buried by moving beach sand or spread across the backshore to be interred by drifting sand. Should fore-dunes be scarfed in later episodes of erosion, and/or adjacent dune fields suffer deflation, once buried (and out of sight plastic) may be exhumed and recycled to the surface. Strong on-shore winds may blow lighter plastics such as sheeting, bags, and foamed materials from the coast to considerable distances inland. During its passage unsightly shredded plastic may come to hang on and blanket vegetation from shoreline mangals, through and across coastal hinterlands [50]. In a riverine setting this has become known as the "christmas tree" effect [51].

It has been known for some time that positively to neutrally bouyant plastic materials settle to the seafloor [52, 53] and Goldberg's [17] comments on "plasticising the sea floor" have drawn wider attention to the environmental significance

Table 10.2 Summarized Examples of Densities (maximum numbers or range) of Virgin Plastic Granules (and/or other plastic mesolitter) in Surface Waters of the Atlantic and Pacific Oceans

Locality	Density (number/km^2)	Reference
1. Western North Atlantic	160,000	40
2. North Atlantic (north of Sargasso Sea)	80–700	101
3. Sargasso Sea	>10,000	101
4. Cape Basin (South Atlantic)	1,000–2,000	100
5. South Africa (coastal waters and Agulhas current)	3,640	65
6. New Zealand (in-shore waters)	>10,000–>4,000	41
7. South Pacific	20–>2,000	41
8. Western North Pacific	316,000	102
9. North Pacific (central) Gyre	334,000	42

of this (somewhat unlikely) sink. Mechanisms by which this floating plastic can settle to the seafloor are poorly understood. Oshima's [54] observation of a fleet of flimsy white plastic supermarket shopping bags, upended and hovering at depths of 2000 m and drifting in suspended animation like an assemblage of ghosts is an example of this. It has been suggested that rapid and heavy fouling of freely floating plastics may be accompanied by density increases of sufficient magnitude to see them sink [55, 56]. Cleaning of fouled surfaces by grazers may lead to cyclic, yo-yo-like episodes of submergence and resurfacing until permanent settlement to the seafloor [55]. Passive collection of nonliving particulate detritus on plastic surfaces may also lead to the density increases necessary to take some marine debris to the sea bed [57] with or without the need to invoke down-welling and entrainment processes. In general, plastic litter and other marine debris settling to oceanic depths in areas other than those swept by strong currents is doomed for permanent entombment in slowly accumulating muddy sediment.

10.3. BIOLOGICAL AND ENVIRONMENTAL IMPACTS

10.3.1. Entanglement and Ingestion

The consequences of larger marine vertebrates being entangled in and/or ingesting discarded or lost plastic and other persistent synthetic debris items were first brought before an international workshop audience in 1984 [10]. Since that time the magnitude of the problem has become widely acknowledged. In particular, where entangled animals, particularly whales, seals, dolphins, turtles, and seabirds, have been left stranded at the shoreline or observed to be in difficulty in coastal waters, harrowing media images often receive wide publicity and draw strong emotive responses. The biologically harmful effects of entanglement that have been identified include death, suppurating skin lesions, ulcerating body wounds and other injuries, as well as general debilitation from interruption to feeding activities or through failed predator avoidance. Those resulting from ingestion include blockage of the intestinal tract followed by satiation and perhaps ultimately to starvation and death and damage to delicate internal tissues. In either circumstance there may be general deterioration in quality of life and reduced reproductive performance [21, 58, 59].

Harper and Fowler's [60] record of storm-killed prions on New Zealand shores reveals that ingestion of plastic mesolitter (mostly colorless small pellets of virgin polyethylene resin) had become locally significant as early as 1958. Other and later observations have indicated a global pattern to plastic consumption to oceanic foraging seabirds as well as those of the shore (e.g., North Atlantic [61–63], South Atlantic [64, 65], North Pacific [66, 67], equatorial Pacific [68], South Pacific [60, 69], and Southern Ocean [41, 70]. This widespread behavior suggests that plastic is being mistakenly identified as food, or it is possibly an unwitting substitute for pumice, which finds use as crop stones by many seabirds. Feeding strategies of some seabirds suggest selective preference for some types or color of plastics [67]. For instance, red and pink plastic artifacts such as bottle tops and small fishing floats, are common in albatross regurgitations on several Southern Ocean islands. It has been implied that these items have been mistaken for their similarly colored krill prey [41, 71]. Because plastics are typically inert and for the reason that many seabirds are seldom worried by a diet largely of sharp fish bones, Laist's [59] suggestion that entanglement is "a far more likely cause of mortality than ingestion-related interactions" is not surprising. Robertson and Bell [72] and Moser and Lee [63] had previously found no evidence that ingestion of plastic artifacts and resin granules, even when in significant quantities, had other than minimal impact on seabird health. These remarks, however, are in conflict with later comments made herein on adsorbed polychlorinated biphenyls (PCBs).

Discarded plastic bags and sheeting appear to have a fatal attraction for many turtles and some whales. These materials, probably mistaken for medusoid jellyfish, can block the alimentary canal and cause death by starvation. Among the more unusual interactions encountered are death of an endangered laysan finch through drowning in a beach-cast, rainwater-filled, plastic container [73], and entanglement of a horse in stranded trawl netting [74].

Compilations of worldwide data by Laist [59] despite being biased toward land-based observations, demonstrate that large numbers (and a high proportion) of marine vertebrate and some larger invertebrate taxa are subject to entanglement and ingestion (see Table 10.3). Laist's (1997) catalog [59] of examples, comprehensive and exhaustive as it may seem, is incomplete given that of 31 Southern Hemisphere seabirds known to ingest plastic [41] at least 6 were not included.

10.3.2. Epibionts and Associated Biota

Discarded plastic materials, like all floating debris, whether natural or artificial, provide important substrates for a variety of sessile encrusting and fouling organisms, and also attract a diverse and varied motile biota [75]. For a number of opportunistic colonizers, the hard surface of passively drifting plastic artifacts provide a suitable surrogate for naturally floating objects such as logs, pumice, and some surface-dwelling and free-swimming marine vertebrates. Studies from the western North Atlantic, South Pacific, and higher latitudes have generally demonstrated that bryzoans, with >60 species identified, are numerically and spatially the dominant encrusting organisms of beach-cast pelagic plastics [47, 56, 75–78]. Other commonly recorded taxa include serpulid tubeworms, barnacles, bivalve molluscs, coralline algae, and occasional foraminiferans. However, these reports are biased toward those organisms with hard shells and that are resistant to desiccation. Studies of freely drifting and moored plastic artifacts reveal a community typified by soft-bodied and fleshy fouling organisms that include algae, hydroids with minor sponges, ascidians, and sea anemones [55, 56, 103]. Pelagic plastic substrates can also be colonized by a variety of motile invertebrate grazers and predators including crabs, amphipods, gastropods, polychaete worms and chitons, and occasional echinoderms [103]. A

Table 10.3 Summary of Entanglement and Ingestion Numbers Recorded for Major Groups of Marine Animals[a]

Group	Number of Species	Entanglement	Ingestion
Sphenisciformes (penquins)	6	6	1
Podicipediformes (grebes)	2	2	—
Procellariiformes (albatrosses, petrels, shearwaters)	63	10	62
Pelecaniformes (pelicans, boobies, gannets, cormorants, frigatebirds, tropicbirds)	17	11	7
Charadriiformes (shorebirds, skuas, gulls, terns, auks)	51	23	40
Other coastal birds	5	5	—
Mysticete (baleen whales)	6	6	2
Odontocete (toothed whales)	22	5	21
Phocidae (earless or true seals)	8	8	1
Otariidae (sea lions, fur seals)	11	11	1
Carnivora	1	1	—
Fish	61	35	34
Crustaceans (crabs, lobster)	9	9	

[a]Based on the compilations of Laist [59], Appendices 1–4.

number of fish species and several marine mammals are drawn to plastics, either for shelter, protection and feeding, or curiosity.

The biological communities recorded from pelagic plastic megalitter items, or aggregations, have adapted to a pseudo-planktic life-style similar to those commonly associated with drifting seaweeds such as *Sargassum*. The diversity and richness of epibiont species on marine debris varies latitudinally, being greatest in equatorial regions and least in polar waters [75, 78]. There is evidence that plastic substrates may provide a multispecies habitat for a variety of organisms that in nature would occupy different ecological niches and would rarely be encountered together. It may include a disjunct assemblage of algal dwellers, rocky shore taxa, and pelagic species as well as taxa otherwise restricted to sheltered harbor shores, or high-energy, open-coast environments [75]. Furthermore, the presence of sexually mature, brooding individuals or colonial species on drifting marine debris substrates (e.g., gastropods, bryozoans, coralline algae) suggests a capacity to colonize, reproduce, and pass through several life cycles or generations while passively drifting on the high seas at the whim of wind, wave, and tide.

Thus taxa with short-lived and/or nonfeeding larval stages could survive at great distances from their usual ecologic niches. Through this mechanism, pelagic plastic litter may be a significant vector in the dispersal of sessile and motile marine organisms, and in a manner similar to that of macrofauna rafting on floating clumps of seaweed in surface waters off Iceland [75]. Winston et al. [79] described two typical examples. Beach debris on a Florida shore carrying a bryozoan (*Thalamoporella evelinae*), previously unrecorded at that location, was already well known from Brazilian waters, and numerous specimens of a well-known tropical Indo-Pacific oyster (*Lopha cristagalli*) attached to a mass of nylon rope that had stranded on a southern New Zealand shore. These, and similar observations, suggest that drift plastics could play a direct role in the introduction of exotic or aggressive alien taxa, with potentially damaging environmental consequences for sensitive or at-risk shallow littoral and coastal ecosystems [21, 75, 78]. It has also been argued that terrestrial ecosystems, particularly those of remote or isolated small islands, could be endangered by vertebrate species (e.g., rodents, mustellids, cats, etc.) by rafting on matted plastic megalitter [21, 75].

Plastic sheeting reaching the seafloor can inhibit gas exchange between substrate pore waters and overlying seawater. This may induce anoxia and hypoxia in blanketed sediments [16, 17]. Interference with meiofaunal and floral community compositions and densities has also been reported [80].

10.3.3. Ecotoxicology

In several early studies it was noted that pelagic virgin polystyrene and polyethylene resin pellets in surface waters of the northwestern Atlantic and South Pacific Oceans contained polychlorinated biphenyls (PCBs) [47, 81, 82]. At that time PCBs were commonly used as plasticizing additives in manufacturing processes, and this was generally considered the source. However, it was also suggested that they could have been adsorbed and concentrated from ambient levels in seawater [47, 81]. Only recently has the latter possibility been experimentally demonstrated and confirmed [83]. PCBs and other organochlorine contaminants together with their toxic breakdown products may also have been originally incorporated in plastic manufacturing for several reasons, including colorants, stabilizers, softeners, and antioxidants. Ingestion of particulate plastic materials in the micro- and smaller sizes of the macrolitter categories by oceanic seabirds as well as some small prey fish is widely documented. Herein may lie a source for the bioaccumulation of these toxicants that is globally widespread in many seabirds [84] and in some albatrosses may lead to egg shell thinning and reduced reproduction capacity [85].

Air blasting with fine sand-sized, granulated plastic media is a recently developed technology for stripping paint and other coatings from metallic surfaces, and cleaning aircraft and delicate machine or engine parts [24]. This material can be recycled up to 10 times before its cutting capacity is lost. In so doing, it may become significantly contaminated with environmentally hazardous heavy metals such as cadmium, chromium, lead, and mercury. Unless disposal of the used waste is in a secure landfill site, it may make its way to the ocean by way of storm-water and sewage drainage systems, natural water courses, or aerosol dispersal, and as aeolian dust. Heavy metals, including those already identified and other anthropogenic contaminants are known to accumulate in the sea surface microlayer of estuarine, coastal, and open ocean waters. Ecosystems of these environments are complex and sensitive to pollution [86, 87]. They are considered important nurseries for the juveniles and pelagic larvae of many commercially important species and whose productivity is under global threat [86, 88]. It has been suggested that pelagic plastic micro- and mesolitter could be a not insignificant vector by which accumulative contaminants such as heavy metals, PCBs, other organochlorine compounds are transferred to filter feeding and other invertebrates (and to ultimately pass on to higher trophic levels) [24, 89].

10.4. DEGRADATION OF PLASTICS AT SEA

Depending on their chemical structure, plastic materials undergo a combination of thermal oxidation, photo-oxidative degradation, biodegradation, and in some instances even hydrolysis on exposure to the marine environment. Common types of plastics encountered in the marine environment, however, break down primarily through photothermal oxidation processes. The chemical pathways associated with the weathering of plastics at sea are essentially the same as those encountered in outdoor exposure. Some of these chemical changes result in a decrease of mechanical properties of the material to an extent that is dependent upon exposure duration. For items of functional fishing gear such as trawl webbing or crab pots, the rate at which this degradation takes place determines its useful service life.

Premature gear failure, especially during use, can have financial costs that far exceed value of damage to the gear itself. It is therefore crucial that the gear in active use retains its mechanical strength while in water and during storage or intermittent exposure to sunlight on vessel decks. Ideally, faster rates of environmental degradation are desirable only when the active gear turns into plastic marine debris and leads to minimization of damage to marine life and aesthetic impacts on coastal and beach environments. Designing plastic materials that satisfy both these conflicting requirements is technically challenging. No plastic available today provides a cost-effective and reliable means of addressing this problem. An understanding of breakdown mechanisms and technologies that allow a faster degradation of plastics waste at sea are therefore of particular interest.

The composition of plastics debris encountered at sea is similar to that of litter on land except for waste material derived from fishing operations. This is not surprising in that most marine plastics debris originate as beach litter. Fishing-related plastics debris consists mainly of netting and damaged or lost crab pots, monofilament sections, Styrofoam bait boxes, and "loops" of plastic (usually polyester or vinyl) strapping bands. Since the mid-1940s the entire fishing industry has switched from gear made of natural fibers that readily biodegraded in seawater to gear made of synthetic fiber. A primary concern is "ghost fishing" by persistent discarded gear, especially in the case of negatively buoyant nylon netting. Entanglement of marine mammals and birds in floating polyolefin netting such as trawl webbing is also a major concern. Long sections of nylon drift nets can be lost during fishing, intentionally abandoned (e.g., when challenged by authorities in restricted waters) or discarded because of damage. Key properties of the various synthetic fibers used in fishing gear are summarized in Table 10.4. This demonstrates the desirable properties such as high wet strength and low water absorption that makes plastics gear particularly suited for application in the fishing industry. Other attributes not quantified in the table include low cost over gears' life cycle and the low visibility in water resulting in relatively higher catching efficiency of plastics gear. It is estimated that the catching efficiency of transparent nylon monofilament net is 2–12 times greater than that of a conventional cotton or flax net [90]. From an environmental perspective the major drawback of plastic gear is its lack of biodegradability under water and slow photodegradation when exposed floating in water.

10.4.1. Defining Degradation

Degradation implies loss of useful properties following chemical changes in plastic materials. In the marine environment only two major influences are likely to bring about these change over practical time scales: solar radiation and slow

Table 10.4 Characteristics of Synthetic Yarns Used in Commercial Fishing Gear[a]

Fiber	Density (g/cm^3)	Wet Breaking Strength (% of dry strength)	Weight in Water (% of dry weight)	Percent Moisture Gain at 65% Humidity	Shrinkage in Water at 100°C (%)
Nylon 6	1.14	85–95 (continuous filament)	12	4	10
Nylon 66	1.14	85–95	12	4	12
Polyester	1.38	100	28	0.4	8
Polyethylene	0.96	110	Buoyant	0	5–10
Polypropylene	0.91	100	Buoyant	0	3
PVC	1.35–1.38	100	26–28	0.3	40 or more
PVDC	1.7	100	41	0.4	3
PVA	1.3	77	23	5	2

[a]Based on G. Klust, *Netting Materials for Fishing Gear*, FAO Fishing Manuals, Food and Agricultural Organization of the United Naions, Fishing News Books Ltd, Surray, England, 1982. p. 21.

thermal oxidation. In the case of negatively buoyant plastic materials, any degradation is solely a response to the latter mechanism, as ultraviolet wavelengths in sunlight are readily absorbed by water. Light-induced degradation (photodegradation or photothermal oxidation) of floating plastic debris can significantly reduce the mechanical strength of these materials. Environmental biodegradation is a slow reaction for most plastics generally encountered in the marine environment. Except for a few special types not generally used in marine applications (such as cellophane or aliphatic polyesters), most common plastics are exceptionally bioinert. Definition and classification of the terms *degradation* and *biodegradation* was recently discussed by Andrady [91]. The extent of degradation in plastics can be quantified directly in terms of the chemical changes or indirectly in terms of losses in useful physical or mechanical properties. Plastic material is technically said to be "fully degraded" once the polymer structure no longer exists, with the material being transformed invariably into small (nonpolymer) molecules. Various techniques, including spectroscopy, are available to quantify the extent of degradation to plastics material in terms of chemical changes. The full conversion of all breakdown products into carbon dioxide, water, and small inorganic molecules, through continued photo- and biodegradation, is termed *mineralization*. Standard tests are also available to measure the mineralization behavior of plastics; these rely on quantification of carbon dioxide evolved during the process.

From a practical standpoint the term *degradation* in the context of the present discussion can have different shades of meaning. If the interest is limited to preventing damage to species of larger animals or birds due to entanglement in net fragments, strapping loops, or six-pack rings, embrittlement[1] of the plastic constitutes a valid end-point. A plastic material, embrittled through photooxidative degradation, may be so weak that entangled animals can break free and escape with little harm done. This degree of deterioration[2] allows the material to be broken down by attrition into particle sizes too small to pose serious aesthetic problems.

The embrittled plastics, however, are not always benign in marine and coastal ecosystems. Depending on fragment size, such material can be ingested by organisms ranging from whales to birds and even small filter feeding invertebrates. The physiologic and other biologic effects of this have been discussed previously.

10.4.2. Deterioration of Plastics in Seawater

The degradation of plastics in air and in seawater shows some key similarities as well as differences:

---

[1] A plastic material is embrittled when its tensile extensibility falls below about 5%. A plastic degraded to this point falls apart on handling.
[2] The term *disintegration* is used to describe the process of physical breakdown of plastics regardless of the mechanism involved. When disintegration is due to a degradation process, the molecular weight of the plastic material is usually reduced.

1. The chemical pathways involved in photooxidative or other photodegradation of polymers in air and in the marine aqueous environment are essentially similar.
2. Unlike plastics exposed on land, those exposed floating on the sea's surface do not suffer from "heat buildup" due to absorption of infrared radiation in sunlight, as the ocean acts as an efficient heat sink. As a result bulk temperatures of these materials remain much lower than those concurrently expected with on land exposure at the same location. (With on-land exposures, for instance, it is possible for dark-colored plastics to have a surface temperature that is higher by as much as 30°C compared to the ambient air.) This will significantly slow rates of degradation at sea.
3. Foulant coverage on the surface of floating plastics shield the material from exposure to sunlight yielding a slower rate of degradation relative to exposure on land. Also, the sample may not be evenly degraded because of the nonuniformity of surface fouling.

The net effect of these factors is illustrated in Figure 10.1, which compares the loss in extensibility of polyethylene samples exposed floating in seawater to that exposed in air at the same location in Biscayne Bay, Florida. Exposure over an approximately 2-month period was negligible for samples exposed floating on water, while those exposed in air nearly embrittled over the same period. Observations were not continued until embrittlement was obtained for the sample

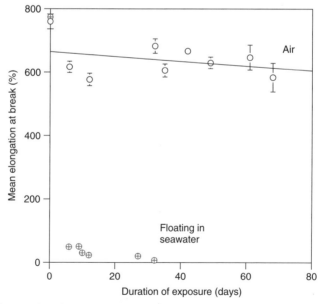

Figure 10.1. Deterioration of polyethylene film samples exposed outdoors in air and floating in seawater in Biscayne Bay, Florida.

Table 10.5 Change in Average Molecular Weight with Extensibility on Outdoor Exposure of Photodegradable LDPE Six-Pack Ring Material [92]

Exposure Time (days)	Extensibility (%)	Mol. Weight M_n 10^{-3} (g/mol)	Mol. Weight M_n 10^{-3} (g/mol)
0	775 (34)	52.8	223.0
3	54 (6)	23.3	67.3
6	46 (6)	15.1	45.1
9	24 (9)	11.6	34.7
13	25 (4)	10.7	38.3
16	18 (3)	10.6	35.0

exposed in seawater because of interference from extensive fouling. However, even when embrittlement of the plastic does occur, the molecular weight of residual fragments remains quite high [92]. This is illustrated for a degradable six-pack ring material made of an ethylene–carbon monoxide copolymer and a metal-catalyzed LDPE film exposed at Cedar Knolls, New Jersey (Table 10.5). While the data shown is for degradation in air, the relationship between molecular weight and extensibility of the material is expected to be substantially similar for degradation in sea water where the same photooxidative mechanisms operate.

10.4.3. Degradation of Floating Plastics Debris

Pegram and Andrady [93] confirmed the slower environmental degradation of plastics in seawater to be true of plastic and rubber materials typically found in marine debris (trawl web fragments, polyester strapping tape, six-pack rings, and latex-rubber balloons). An important plastic debris item reported in nearly all beach surveys is expanded polystyrene foam fragments. These are mostly fragments of foam packaging materials discarded by beach goers, and sections of bait or fish boxes left behind by fisherman. On exposure to sunlight, Styrofoam yellows and forms a brittle surface layer that easily brushes off as a fine powder. This brittle layer turns deep yellow in color over time and to some extent shields underlying layers of polystyrene from light and therefore from further photodegradation. The thickness of the underlying undegraded (i.e., not brittle) polystyrene layer is an approximate measure of the extent of degradation. Using this measure, Styrofoam was found to degrade faster when exposed in seawater than when exposed at the same location in air [94]. The same was true when degradation was monitored using tensile strength as a measure of the extent of deterioration.

This is in contrast to the behavior of other types of plastics discussed above. It is likely to result from progressive removal of the embrittled layer of polystyrene foam by seawater continuously exposing fresh material to sunlight. Such a process would indeed accelerate the photodegradation but would not be operative for stationary samples exposed in air. This interpretation is consistent with changes in molecular weight of the polymer during weathering as determined by GPC (gel permeation chromatography). Samples exposed in seawater showed about the same (or marginally slower) decrease in average molecular weights relative to those exposed on land. Removal of low-molecular-weight species in the degraded layer by seawater leaves behind mostly the higher molecular weight undegraded plastic in the exposed sample. Therefore, the measurement overestimates the average molecular weight for samples exposed at sea.

10.5. PHOTODEGRADABLE PLASTICS AS A MITIGATION STRATEGY

The use of enhanced-photodegradable[3] plastics in packaging products likely to be discarded in the marine environment has been suggested as a means of reducing the impact of marine plastic debris [89]. With products such as plastic bags or six-pack material likely to be discarded at sea or on beaches, this would allow the material to disintegrate in a relatively short time scale, minimizing their encounter with marine species. If these technologies, primarily designed to work in land environments, performed adequately in seawater as well, this approach could significantly reduce the entanglement-related hazard posed by some types of plastics waste.

Accelerated photodegradation of polyolefins (as well as polystyrene) can be obtained economically by either using copolymers containing ketone moieties or by using transition-metal catalysts in the material. In the former approach [95, 96] a copolymer of ethylene with carbon monoxide ($\sim 1\%$) is typically used in place of the ethylene homopolymer. The ketone group introduced into the main chain is able to absorb solar ultraviolet (UV-B) radiation and undergo chain scission (primarily via the Norrish II intramolecular elimination reaction). The chemistry is applicable even where the ketone group is on a side chain as with the ethylene-*co*-vinyl ketone polymers [97]. An alternate means of accelerating photooxidative degradation of polyolefins is to use a trace amount of a transition-metal compound [98], usually iron or manganese carboxylate, as a prooxidant. The metal compounds catalytically dissociate hydroperoxides accelerating the oxidative reactions and hence the chain scission reactions associated with them.

A recent study compared the performance of two types of enhanced photodegradable polyethylene film samples under marine exposure conditions [99]. One of them was a copolymer of ethylene and carbon monoxide ($\sim 1\%$) (ECO) typically used in six-pack rings, and the other was a polyethylene film containing low levels of a metal compound catalyst (MX). The degradable samples as well as control samples of polyethylene film were exposed at two locations, Biscayne Bay, Florida, and Puget Sound, Washington, both in air and floating in

---

[3] All plastics materials are photodegradable on exposure to solar radiation, and therefore the term *photodegradable plastics* is inaccurate as it implies the existence of a nonphotodegradable variety. The term *enhanced-photodegradable plastic* is more appropriate as it refers to a plastic chemically treated to obtain faster rates of photodegradation.

seawater. The average global solar radiation values during the period of exposure were 14.2 and 18.2 MJ/m², respectively, for the two locations. The average seawater temperature in Biscayne Bay and Puget Sound sites were 30°C and 17°C, respectively. The average elongation at break, a property particularly sensitive to degradation, was measured periodically to follow the progress of deterioration of the plastic samples. The study allowed the following conclusions to be drawn (Table 10.6) [92]:

1. As might be expected, the control samples deteriorated at a much slower rate in seawater compared to in air when exposed at the same location.
2. The enhanced-photodegradable polyethylene breaks down faster than the control plastic sample in seawater. Figure 10.2 shows the loss in elongation at break for a enhanced photodegradable polyethylene film to that of a regular polyethylene control film, both exposed floating in seawater in Biscayne Bay. In these experiments, the ECO copolymer material embrittled in about 35 and 90 days, in Biscayne Bay and Puget Sound, respectively. The control sample decreased in extensibility by only about 10–23% during the same duration of exposure. The metal catalyzed polyethylene (MX) also showed enhancement in deterioration, embrittling in about 100 days at Puget Sound, during which period the control sample lost only about 40% of its extensibility.
3. The degree of enhancement in breakdown obtained with these films (relative to regular polyethylene films of the same thickness) in marine exposures was clearly significant. But the degree of enhancement was considerably less than for similar sample exposures in air. This is generally attributed

Table 10.6 Empirical Rate Constant for the Loss in Extensibility of Polyethylene Sheet Samples Exposed in Air (B_A) and Floating in Seawater (B_N)

System	Location	Exposure	B (days^{-1})	(B_A/B_N)
ECO copolymer[a]	Biscayne Bay	Air	69	1.5
		Floating	45	
	Puget Sound	Air	40	3.6
		Floating	11	
	Beaufort, NC[b]	Air	44	4.0
		Floating	11	
Polyethylene/MX[a]	Biscayne Bay	Air	72	2.7
		Floating	27	
	Puget Sound	Air	54	3.0
		Floating	18	

[a]ECO copolymer refers to sheets (0.42 mm) of enhanced-photodegradable polyethylene used in commercial photodegradable six-pack rings and was supplied by ITW HiCone Ltd (Chicago, IL). Polyethylene/MX refers to films (0.03 mm) of polyethylene containing iron compounds as prooxidant additive and was supplied by Plastigone Inc. of Miami, FL.
[b]This is a freshwater exposure location.

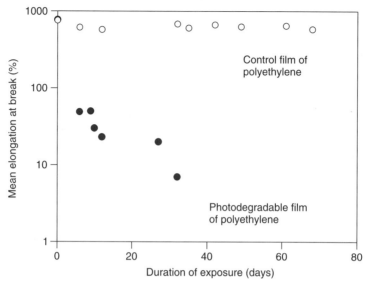

Figure 10.2. Deterioration of enhanced photodegradable polyethylene film and control polyethylene film exposed floating in seawater in Biscayne Bay, Florida.

to the heat buildup resulting in higher sample temperatures for samples exposed in air. Several exposures carried out at a freshwater location also showed the enhanced-photodegradable plastics to function adequately in freshwater as well.[4]

Gradient B of a semilogarithmic plot of percent extensibility versus the duration of outdoor exposure of the sample provides a good measure of its rate of deterioration (see Table 10.6). Of particular interest is the ratio of B_A, the gradient for the plastic samples exposed in air, to B_N, that for the same sample exposed in the marine environment. This ratio of $(B_A/B_N) > 1$ obtained for these locations indicated the disintegration of the enhanced-photodegradable polyethylene samples exposed under marine conditions to be very much slower than for exposure in air; the samples exposed in air degraded at a rate of 50–400% faster. This is consistent with the earlier observation that the rate of photodegradation of plastics is relatively slower when the samples were exposed floating in seawater. While their performance was relatively slower in seawater, enhanced-photodegradable polyethylenes did function adequately under marine exposure conditions. The improvement in the rate of degradation is significant in that the photodegradable six-pack ring material embrittled within 5–10 weeks of marine exposure as opposed to years of exposure needed by the same product made out of regular polyethylene resin. Regular polyethylene samples of comparable dimensions

---

[4] In most parts of the United States near the coast, the six-pack carriers are now required to be made with enhanced-photodegradable plastics materials.

exposed as control samples at the same locations and suffered minimal losses in extensibility within the period of observation. These latter samples are more likely to foul and sink to the benthic regions continuing to pose a threat to marine life, long before it is disintegrated. In terms of preventing entanglement-related damage and reducing the aesthetic problems associated with floating plastics debris, the enhanced-degradable technology provides a practical and valuable mitigation option.

Despite the encouraging results, it is important to point out that photodegradable plastics, while helping to ease entanglement-related impacts, do not offer a complete solution to the problem. Rapid disintegration of the plastic into smaller particles does little to alleviate (or can even increase) the risk of ingestion of plastics by marine birds and other species. While it can perhaps help prevent ingestion-related distress in the larger visible species (such as birds and turtles), the impact on the smaller animals (such as invertebrates, specially the filter feeders) must correspondingly increase.

10.6. CONCLUSIONS

The problem of plastics accumulation in the marine environment is a serious ecological concern. While fishing activity and land-based plastics debris continue to add plastics into the oceans, there are no reliable mechanisms to remove this persistent debris from the marine environment in any practical time scale. Although the ecological impacts of floating debris such as net fragments is appreciated because of high visibility, those due to negatively buoyant plastics accumulating in benthic environments are yet to be well understood. There are no technologies that are effective and economical to address the issue in the short term. Control of disposal of the plastics at sea, either by public education or via regulatory approaches (such as MARPOL Annex V) is particularly important to preserve the ocean environment until adequate and economical technical responses to the problem are developed.

REFERENCES

1. J. M. Coe and D. B. Rogers, *Marine Debris: Sources, Impacts, and Solutions*, Springer, New York, 1997.
2. GESAMP (Group of Experts on the Scientific Aspects of Marine Pollution), *The State of the Marine Environment*, Blackwell Scientific, London, 1991.
3. R. Carson, *The Sea Around Us*, 2nd ed., Oxford University Press, New York, 1961.
4. T. Heyerdahl, *The 'Ra' Expeditions*, George Allen and Unwin, London, 1971.
5. E. D. Goldberg, *The Health of the Oceans*, UNESCO Press, Paris, 1976.
6. M. E. Huber, *Marine Pollution Bull.* **38**, 435–438 (1999).
7. W. R. Trulli, H. K. Trulli, and D. P. Redford, in R. S. Shomura and M. L. Godfrey, eds., Proceedings of the second International Conference on Marine Debris, 2–7 April 1989, Honolulu, Hawaii, pp. 309–324.

8. M. R. Gregory and P. G. Ryan, in J. M. Coe and D. B. Rogers, eds., *Marine Debris: Sources, Impacts, and Solutions*, Springer, New York, 1997, pp. 49–66.
9. C. A. Ribic and L. M. Ganio, *Marine Pollution Bull.* **32**, 554–557 (1997).
10. R. S. Shomura and H. O. Yoshida, eds., Proceedings of the Workshop on the Fate and Impact of Marine Debris, Honolulu, Hawaii U.S. Department of Commerce, NOAA-TM-NMFS-SWFSC-54, 1985.
11. D. A. Wolfe, *Marine Pollution Bull.* **18**, 303–305 (1987).
12. D. L. Alverson and J. A. June, eds., *Proceedings of the North Pacific Rim Fisherman's Conference on Marine Debris*, Natural Resources Consultants, Seattle, WA, 1988.
13. R. S. Shomura and M. L. Godfrey, eds., Proceedings of the Second Conference on Marine Debris, Honolulu, Hawaii. U.S. Department of Commerce. NOAA-TM-NMFS-SWSFC-154, 1990.
14. Anon., International Marine Debris Conference on Derelict Fishing Gear and the Ocean Environment: Issue Papers. Honolulu, Hawaii, August 6–11, 2000, Hawaiian Islands Humpback Whale Sanctuary, Kihei, Hawaii, 2000.
15. L. B. Cahoon, *J. Coastal Res.* **6**, 479–483 (1990).
16. E. D. Goldberg, *Chem. Ecol.* **10**, 3–8 (1995).
17. E. D. Goldberg, *Environ. Tech.* **18**, 195–202 (1997).
18. CMC (Center for Marine Conservation), *1994 International Coastal Cleanup Results*, Center for Marine Conservation, Washington, DC, 1995.
19. CMC (Center for Marine Conservation), *1991 International Coastal Cleanup Results*, Center for Marine Conservation, Washington, DC, 1992.
20. R. Cheshire, Pollution sources survey of the Kingdom of Tonga. South Pacific Regional Environment Programme, Topic Review 19, South Pacific Commission, Nouma, New Caledonia, 1984.
21. M. R. Gregory, *J. Roy. Soc. New Zealand*, **21**, 83–100 (1991).
22. L. A. Thornton and N. L. Jackson, *Marine Pollution Bull.* **36**, 705–701 (1998).
23. V. Zitko and M. Hanlon, *Marine Pollution Bull.* **22**, 41–42 (1991).
24. M. R. Gregory, *Marine Pollution Bull.* **32**, 867–871 (1996).
25. A. T. Williams and S. L. Simmons, *J. Coastal Res.* **13**, 1159–1165 (1997).
26. D. P. Redford, W. R. Trulli, and H. K. Trulli, *Chem. Ecol.* **7**, 75–92 (1992).
27. ICNZT (Island Care New Zealand Trust), Reducing the Incidence of Stormwater Debris and Street Litter in the Marine Environment: A Co-operative Community Approach, Island Care New Zealand Trust, Auckland, New Zealand, 1996.
28. P. V. Horsman, *Marine Pollution Bull.* **13**, 167–169 (1982).
29. A. Nollkaemper, *Marine Pollution Bull.* **24**, 8–12 (1992).
30. M. Schumacher, P. Hoagland, and A. Gaines, *Marine Policy* **20**, 99–121 (1996).
31. M. Liffman and L. Boogaerts, in J. M. Coe, and D. B. Rogers, eds., *Marine Debris: Sources, Impacts and Solutions*, Springer, New York, 1997, pp. 359–366.
32. A. Siung-Chang, *Environ. Geochem. Health* **19**, 45–55 (1997).
33. A. D. Wilson, *Sea Tech.* **42**(8), 7 (2001).
34. G. Rees and K. Pond, *Marine Pollution Bull.* **30**, 103–108 (1995).
35. D. Bowman, N. Manor-Samsonov, and A. Golik, *J. Coastal Res.* **14**, 418–432 (1998).

36. K. Velander and M. Mocogni, *Marine Pollution Bull.* **38**, 1134–1140 (1999).
37. S. Matsumura and K. Nasu, in J. M. Coe and D. B. Rogers, eds., *Marine Debris: Sources, Impacts and Solutions*, Springer, New York, 1997, pp. 15–24.
38. R. V. Grace, Oceanic Debris Observations in the Indian Ocean Whale Sanctuary and Eastern Mediterranean Sea, Report to International Whaling Commission Scientific Committee SC/46/0, Puerto Vallarta, Mexico, 1994.
39. E. J. Carpenter and K. L. Smith, *Science* **175**, 1240–1241 (1972).
40. J. B. Colton, F. D. Knapp, and B. R. Burns, *Science* **185**, 491–497 (1974).
41. M. R. Gregory, in G. P. Glasby, ed., *Antarctic Sector of the Pacific*, Elsevier Oceanography Series, Vol. 51, Elsevier, Amsterdam, 1990, pp. 291–324.
42. C. J. Moore, S. L. Moore, M. K. Leecaster, and S. B. Weisberg, Southern California Coastal Water Research Project, Annual Report, 1999–2000, 2001, pp. 120–124.
43. L. M. Stevens, Marine Plastic Debris: Fouling and Degradation, unpublished M.Sc Thesis, The University of Auckland, Auckland, New Zealand, 1992.
44. J. G. Shiber, *Marine Pollution Bull.* **10**, 28–30 (1979).
45. J. G. Shiber, *Marine Pollution Bull.* **13**, 409–412 (1982).
46. P. G. Ryan and C. L. Maloney, *South African J. Sci.* **86**, 450–452 (1990).
47. M. R. Gregory, *New Zealand J. Marine Freshwater Res.* **12**, 399–414 (1978).
48. D. P. Redford, W. R. Trulli, and H. K. Trulli, *Chem. Ecol.* **7**, 75–92 (1992).
49. M. R. Gregory, *Tane* **37**, 201–210 (1999).
50. N. Wace, in L. P. Zann and D. C. Sutton, ed., *The State of the Marine Environment Report for Australia. Technical Annex: 2 Pollution*, Great Barrier Reef Marine Park Authority for Department of the Environment, Sport and Territories, Ocean Rescue 2000 Programme, Canberra, ACT, Australia, 1995, pp. 73–87.
51. A. T. Williams and S. L. Simmons, *J. Coastal Conservation* **2**, 63–72 (1996).
52. A. Hollstrom, *Nature* **255**, 622–623 (1975).
53. A. T. Williams, S. L. Simmons, and A. Fricker, *Marine Pollution Bull.* **26**, 404–405 (1993).
54. S. Oshima, *Hydro Intl.* **4**(5), 73 (2000).
55. S. Ye and A. L. Andrady, *Marine Pollution Bull.* **22**, 608–613 (1991).
56. L. M. Stevens, M. R. Gregory, and B. A. Foster, in D. P. Gordon, A. M. Smith, and J. A. Grant-Mackie, ed., *Bryozoans in Space and Time*, Proceedings of the 10th International Bryozoology Conference, Wellington, New Zealand, National Institute of Water and Atmospheric Research, 1996, pp. 321–340.
57. J. J. Powlik, *Sarsia* **80**, 229–236 (1995).
58. D. W. Laist, *Marine Pollution Bull.* **18**, 319–326 (1987).
59. D. W. Laist, in J. M. Coe and D. B. Rogers, ed., *Marine Debris: Sources, Impacts, and Solutions*, Springer, New York, 1997, pp. 99–139.
60. P. C. Harper and J. A. Fowler, *Notornis* **34**, 65–70 (1987).
61. S. I. Rothstein, *Condor* **75**, 344–355 (1973).
62. H. Hays and G. Cormons, *Marine Pollution Bull.* **5**, 44–46 (1974).
63. M. L. Moser and D. S. Lee, *Colonial Waterbirds* **15**, 83–94 (1992).
64. R. W. Furness, *Environ. Pollution Ser. A Ecolog. Biolog.* **16**, 261–272 (1985).

65. P. G. Ryan, A. D. Connell, and B. D. Gardner, *Marine Pollution Bull.* **19**, 174–176 (1988).
66. K. W. Kenyon and E. Kridler, *Auk* **86**, 339–343 (1969).
67. R. H. Day, D. H. S. Wehle, and F. C. Coleman, in R. S. Shomura and H. O. Yoshida, eds., Proceedings of the Workshop on the Fate and Impact of Marine Debris, Honolulu, Hawaii, 1984. U.S. Department of Commerce. NOAA-TM-NMFS-SWSFC-54, 1985, pp. 344–386.
68. D. G. Ainley, L. B. Spear, and C. A. Ribic, in R. S. Shomura and M. L. Godfrey, eds., Proceedings of the Second International Conference on Marine Debris, Honolulu, Hawaii, 1989, U.S. Department of Commerce. NOAA-TM-NMFS-SWSFC-154, 1990, pp. 653–664.
69. S. Reed, *Notornis* **28**, 239–240 (1981).
70. D. G. Ainley, W. R. Fraser, and L. B. Spear, in R. S. Shomura and M. L. Godfrey, eds., Proceedings of the Second International Conference on Marine Debris. Honolulu, Hawaii, 1989, U.S. Department of Commerce, NOAA-TM-NMFS-SWSFC-154, 1990, pp. 682–691.
71. P. A. Prince, *Ibis*, **122**, 476–488 (1980).
72. C. J. R. Robertson and B. D. Bell, in J. P. Croxall, P. G. H. Evans, and R. W. Schreiber, eds., *Status and Conservation of the World's Seabirds*, International Council for Bird Protection, Technical Publication, No. 2, Cambridge, England, 1984, pp. 573–586.
73. M. P. Morin, *'Elepaio,'* **47**, 107–108 (1987).
74. Z. Lucas, *Marine Pollution Bull.* **24**, 192–199 (1992).
75. J. E. Winston, M. R. Gregory, and L. M. Stevens, in J. M. Coe and D. B. Rogers, eds., *Marine Debris: Sources, Impacts and Solutions*, Springer, New York, 1997, pp. 81–97.
76. M. R. Gregory, *Marine Environ. Res.* **10**, 399–414 (1983).
77. J. E. Winston, *Marine Pollution Bull.* **13**, 348–351 (1982).
78. D. K. A. Barnes and W. G. Sanderson, Proceedings of the 11h International Bryozoology Association Conference, Panama City, Panama, 2000, pp. 154–160.
79. A. Ingólfsson, *Marine Biol.* **122**, 13–21 (1995).
80. P. Uneputty and S. M. Evans, *Marine Environ. Res.* **44**, 233–242 (1997).
81. E. J. Carpenter and K. L. Smith, *Science* **175**, 1240–1241 (1972).
82. E. J. Carpenter, S. J. Anderson, G. R. Harvey, H. P. Miklas, and B. B. Peck, *Science* **178**, 749–750 (1972).
83. Y. Mato, T. Isobe, H. Takada, H. Kanehiro, C. Ohtake, and T. Kaminuma, *Environ. Sci. Tech.* **35**, 318–324 (2001).
84. P. G. Ryan, *Marine Environ. Res.* **25**, 249–273 (1988).
85. J. P. Ludwig, C. L. Summer, H. J. Auman, V. Gauger, D. Bromley, J. P. Giesy, R. Rolland, and T. Colborn, in G. Robertson and R. Gales, eds., *Albatross Biology and Conservation*, Surrey Beatty, Norton, Australia, 1998, pp. 225–238.
86. J. T. Hardy, *Prog. Oceanogra.* **11**, 307–328 (1982).
87. J. T. Hardy, *Marine Environ. Res.* **23**, 223–225 (1987).
88. J. N. Cross, J. T. Hardy, J. E. Hose, G. P. Hershelman, L. D. Antrim, R. W. Gossett, and E. A. Crecelius, *Marine Environ. Res.* **23**, 307–323 (1987).

89. A. L. Andrady, in Proceedings of the SPI Symposium on Degradable Plastics, Washington, DC, June10th, 1987, p. 22.
90. G. Klust, *Netting Material for Fishing Gear*, 2nd ed., Adlard and Sons. Surrey, England, 1982, p. 175.
91. A. L. Andrady, *J. Macromolec. Sci. Rev. Macromol. Chem. Phys.* **C34**(1), 25–75 (1994).
92. A. L. Andrady, J. E. Pegram, and Y. Tropsha, *J. Environ. Polym. Degrad.* **1**(3), 171 (1993).
93. J. E. Pegram and A. L. Andrady, *Polym. Degrad. Stab.* **26**, 333 (1989).
94. A. L. Andrady and J. E. Pegram, *J. Appl. Polym. Sci.* **42**, 1589 (1991).
95. J. E. Guillett, *Pure Appl. Chem.* **30**, 135 (1972).
96. M. Heskins, and J. E. Guillett, *Proc. Royal Soc. (London) A*, **2333**, 153 (1968).
97. S. K. Li and J. E. Guillett, *J. Polym. Sci. Polym. Chem. Ed.* **18**, 2221 (1980).
98. G. E. Scott, *Polymers and the Environment*, Royal Society of Chemistry, Herts, United Kingdom, 1999, p. 100.
99. A. L. Andrady, J. E. Pegram, and Y. Song, *J. Environ. Polym. Degrad.* **1**(2), 117 (1993).
100. R. J. Morris, *Marine Pollution Bull.* **11**, 164–166 (1980).
101. R. J. Wilber, *Oceanus* **30**(3), 61–68 (1987).
102. R. H. Day, D. G. Shaw, and S. E. Ignell, in R. S. Shomura and M. L. Godfrey, eds., Proceedings of the Second International Conference on Marine Debris, 2–7 April 1989, Honolulu, Hawaii, U.S. Department of Commerce, NOAA-TM-NMFS-SWFSC-154, 1990, pp. 247–266.

CHAPTER 11

FLAMMABILITY OF POLYMERS

ARCHIBALD TEWARSON
FM Global Research

11.1. INTRODUCTION

Polymers are used in the manufacturing of products for applications such as construction (buildings and furnishings), transportation (ground-based and outer space vehicles), and consumer and industrial goods (packaging, electrical and electronic equipment, medical devices, sports and leisure articles, and others). Thermal and various environmental exposures of polymers generally lead to the degradation and deterioration of the polymers [1–9]. Degradation of the polymers may be considered as any type of modification of a polymer chain involving the main-chain backbone or side groups or both. These changes are often of a chemical nature (requiring the breaking of primary valence bonds), leading to lower molecular weight, crosslinking, and cyclization. The following types of degradation can occur in a polymer due to the effects of the environmental factors [1–9]:

- *Thermal*   Degradation initiated by high-temperature exposure
- *Oxidative*   Degradation initiated by exposure to oxidizing agents including air
- *Radiative*   Degradation initiated by exposure to radiative energy
- *Mechanical and Ultrasonic*   Degradation initiated by machine-type operations (e.g., mastication, grinding, ball milling, roll milling, and ultrasonic vibrations)
- *Chemical*   Degradation initiated by exposure to chemicals
- *Biological*   Degradation initiated by exposure to living organisms, usually microorganisms, such as fungi and bacteria; fungi, including bacteria, widely distributed throughout the world in soil, water, and air

---

Plastics and the Environment,   Edited by Anthony L. Andrady.
ISBN 0-471-09520-6   © 2003 John Wiley & Sons, Inc.

Polymers are formulated to satisfy specific end-use requirements for the lifetime of the products:

- For the satisfactory performance of the products (determined by measuring mechanical, thermal, electrical, and other properties and comparing them with the specified limits)
- For eliminating or minimizing the deleterious effects of the environment on the products (determined by measuring the properties associated with the effects of temperature, light, oxygen, ozone, humidity, acids, smoke, aerosols, and others and comparing their values with the specified limits)
- For eliminating or minimizing the fire hazards from the burning products (determined by measuring the flammability of the polymers and comparing the values with the specified limits)
- For safe disposal of the products without any adverse effects to the environment, such as recycling and subsequent repolymerization, recycling to olefinic feedstock by pyrolysis, continued burial in landfill sites, incineration, and use of environmentally degradable polymers
- For eliminating or minimizing contamination of air, water, and soil by the fires of the polymers and their suppression, for example, retention of water used for fire fighting [10, 11]

This chapter deals with the subject of flammability of polymers where thermal, oxidative, and radiative degradations all play very significant roles. Flammability of polymers is a complex subject and has been discussed in other books and review articles by various authors [12–22].

According to *Webster's New World Dictionary & Thesaurus*: "the word flammable means easily set on fire; that will burn readily or quickly: term now preferred to INFLAMMABLE in commerce and industry." "Easily set on fire" refers to the ignition behavior, "burn readily" refers to the combustion behavior, and "burn quickly" refers to the fire propagation behavior of a material. Thus, flammability and fire behavior of materials are used synonymously when describing the ignition, combustion, and fire propagation behaviors of materials separately or in combination.

The ignition, combustion, and fire propagation behaviors of polymer are examined in various flammability test standards, where polymers are intentionally degraded, and the degradation products are ignited and burned under controlled exposure and environmental conditions. Various countries and agencies promulgate these types of flammability test standards[1]:

- Australia (Standards Australia, SA)
- Canada (Canadian General Standards Board, CGSB)

---

[1] Based on the Internet search for flammability using the software from HIS, Englewood, Colorado, 2001.

- China People's Republic (China Standards Information Center, CSIC)
- China Republic of China–Taiwan (Bureau of Standards, Metrology and Inspection, BSMI)
- Europe International Electrotechnical Commission (IEC); European Committee for Electrotechnical Standardization (CENELEC); European Committee for Standardization (CEN); International Standards Organization (ISO)
- Finland (Finnish Standards Association, SFS)
- France (Association Europeene Des Constructeurs De Materiel Aerospatial, AECMA; Association Francaise De Normalisation, AFNOR)
- Germany (Deutsches Institut Fur Normung, DIN)
- Italy (Ente Nazionale Italiano Di Unifacazione, UNI)
- Japan (Japanese Standards Association, JSA)
- Korea (Korean Standards Association, KSA)
- New Zealand (Standards New Zealand, SNZ)
- Nordic Countries (Nordtest: Denmark, Finland, Greenland, Iceland, Norway, and Sweden)
- Russia (Gosudarstvennye Standarty State Standard, GOST)
- South Africa (South African Bureau of Standards, SABS)
- United Kingdom (British Standards Institution, BSI; Civil Aviation Authority, CAA)
- United States (American Society for Testing and Materials, ASTM; Building Officials & Code Administrators International Inc, BOCA; Department of Transportation, DOT; Electronic Industries Alliance, EIA; FM Global Approval; Institute of Electrical and Electronics Engineers, IEEE; Military-MIL; National Aeronautical and Space Administration, NASA; National Fire Protection Association, NFPA; Underwriters Laboratories, UL)

Many of the above standards from various countries or agencies are either very similar, slightly modified versions of each other, or completely different tests. The most common specifications for testing and measurements for small-scale flammability tests consist of:

- *Sample*   About 50–300 mm in length, 10–100 mm in width, and 3–25 mm in thickness
- *Sample Orientation*   Horizontal or vertical
- *Sample Exposure*   Small flame or radiant heat
- *Sample Environment*   Normal air and/or normal air with nitrogen or oxygen (to decrease or increase the oxygen concentration) flowing in the same or opposite direction to flame spread
- *Test Measurements*
  - Ignition time and flame extinguishment

- Dripping of burning or nonburning polymer melt and ignition of combustibles in close proximity to the test sample
- Light obscuration by smoke (specific optical density[2])
- Rate and extent of flame spread and surface charring, minimum heat flux or oxygen concentration required for flame spread
- Heat release rate
- Release rate of smoke and other products

All fires are initiated by exposure of the polymer surface to heat in the presence or absence of a pilot flame. The exposure leads to flaming or nonflaming fires with release of heat and products of complete and incomplete combustion.

11.2. HEAT EXPOSURE OF A POLYMER

In a fire, as thermoplastics[3] and elastomers[4] are exposed to external or internal heat flux, they generally undergo softening and melting followed by the release of vapors to the environment without significant surface charring. Exposure of thermosets[5] to external or internal heat flux in a fire generally results in surface charring and release of vapors to the environment. The polymer vapors mix with air and form a combustible or a noncombustible mixture, depending on the chemical composition of the polymer vapors. As the polymer vapor–air mixture comes in contact with a hot surface or a small flame, the combustible mixture ignites and a flame is established at the surface, while the noncombustible mixture does not ignite. The establishment of a sustained flame at the surface is defined as *ignition*. After ignition, vapors generated from the polymer continue to burn, whereas the noncombustible mixture continues to be released without ignition. These two processes are defined as *flaming* and *nonflaming combustion*, respectively.

The polymer surface location for ignition and establishment of a sustained flame is defined as the *ignition zone*. The sustained combustion extends beyond the ignition zone if the heat flux from the flame of the burning polymer is of sufficient magnitude to ignite the polymer surface ahead of the flame front. The

---

[2] The specific optical density, D_s is expressed as $(V/AL)\log_{10}(100/T)$, where V is the volume of the closed chamber (ft^3 or m^3), A is the exposed area of the specimen, (ft^2 or m^2), L is the optical path length through the smoke (ft or m), and T is the percent light transmittance as read from the light-sensing instrument.

[3] *Thermoplastics* are linear or branched polymers that can be melted upon the application of heat. They can be molded and remolded into virtually any shape [13].

[4] *Elastomers* are chemically or physically crosslinked rubbery polymeric polymers that can easily be stretched to high extensions and rapidly recover their original dimensions. Thermoplastic elastomers can be molded and remolded [13].

[5] *Thermosets* are rigid polymers having short network polymers in which chain motion is greatly restricted by a high degree of crosslinking. They are intractable once formed and degrade rather than melt upon the application of heat [13].

continued extension of the sustained combustion on the polymer surface is defined as *flame spread*. Release of heat and smoke and consumption and charring of the polymer associated with flame spread is defined as *fire propagation*.

Under all three fire stages of ignition, combustion, and fire propagation, heat and products of complete and incomplete combustion are released. Release of heat in a fire is responsible for thermal hazards and release of products of complete and incomplete combustion and polymer vapors themselves are responsible for non-thermal hazards [20]. It is generally believed that both thermal and nonthermal hazards decrease with increase in the resistance to softening, melting, vaporization or decomposition, and release of polymer vapors, ignition, combustion, fire propagation, and release of products of complete and incomplete combustion.

11.3. RELEASE OF POLYMER VAPORS

The release rate of polymer vapors is expressed as [21, 22]:

$$\dot{m}'' = \dot{q}''_n / \Delta H_g \qquad (11.1)$$

where $\dot{m}''$ is the release rate of the polymer vapors per unit surface area of the polymer (g/m²s) and $\dot{q}''_n$ is the net heat flux to the surface (kW/m²) defined as:

$$\dot{q}''_n = \dot{q}''_e + \dot{q}''_f - \dot{q}''_{rr} \qquad (11.2)$$

where $\dot{q}''_e$ is the external heat flux per unit surface area of the polymer (kW/m²), $\dot{q}''_f$ is the flame heat flux per unit surface area of the polymer (kW/m²), $\dot{q}''_{rr}$ is the surface re-radiation loss per unit surface area of the polymer (kW/m²), and ΔH_g is the heat of gasification of the polymer (kJ/g), expressed as [21–23]:

For thermoplastics and elastomers:

$$\Delta H_g = \int_{T_a}^{T_m} c_s \, dT + \Delta H_m + \int_{T_m}^{T_v} c_l \, dT + \Delta H_v \qquad (11.3)$$

For thermosets:

$$\Delta H_g = \int_{T_a}^{T_d} c_s \, dT + \Delta H_d \qquad (11.4)$$

where ΔH_m is the heat of melting of the polymer (kJ/g), ΔH_v is the heat of vaporization of the polymer (kJ/g), ΔH_d is the heat of decomposition of the polymer (kJ/g), T_a, T_m, T_v, and T_d are initial, melting, vaporization, and decomposition temperatures of the polymer, respectively (°C), c_s and c_l are the heat capacity of the original solid and molten polymer, respectively (kJ/g·K).

Relationships in Eqs. (11.1)–(11.4) show that the release rate of polymer vapors, which is very critical in governing the fire hazard, depends on the heat flux from the external heat source $\dot{q}''_e$, and from its own flame, $\dot{q}''_f$, and heat losses

408 FLAMMABILITY OF POLYMERS

from the surface, $\dot{q}''_{rr}$, and on the heat of gasification, ΔH_g. All four parameters depend on the generic nature and thermophysical properties of the polymers and the environmental conditions.

11.4. POLYMER MELTING

As a polymer is heated by an external or internal heat source, it undergoes softening and melting leading to the formation of polymer melt. The polymer melt either flows away from the heat source, drips as burning molten drops, igniting other materials in close proximity or collects and burns as a liquid pool fire (one of the most hazardous conditions in a fire). Standard flammability tests have been promulgated by various agencies to characterize softening, melting, and dripping of burning polymer melt drops and ignition of the materials in close proximity as polymer is exposed to heat (such as the exposure of small samples of polymer to Bunsen burner in the UL 94 [24]).

The softening, melting, and flow of polymer melt depend on the polymer morphology (amorphous and crystalline nature of the polymer) [25]. Amorphous polymers lack sufficient regularity in packing of the chains compared to the crystalline polymers. Amorphous polymers generally exist as hard, rigid, and glassy below their glass transition temperature[6] and as soft, flexible, rubbery materials above the glass transition [25]. The density of a polymer increases with its degree of crystallinity. Density is often the single parameter that is most clearly related to the physical and mechanical properties of the polymers [25]. Properties dependent on crystallinity (e.g., stiffness, tear strength, hardness, chemical resistance, softening temperature, yield point) tend to increase with increasing density for many polymers [25].

The softening, melting, and flow of polymer melt are characterized by the glass transition temperature (T_{gl}), melting temperature (T_m),[7] and melt flow index (MFI),[8] respectively [13, 26–29]. T_{gl} is the property of the amorphous region, whereas T_m is the property of the crystalline region. Tables 11.1 and 11.2 list these values for selected generic polymers and for polymers from parts of a minivan (1996 Dodge Caravan), respectively. In Table 11.1 T_{gl}, T_m, and MFI values are listed, where data are taken from Refs. [4, 13, 26–29] and in Table 11.2, T_{gl} and T_m values are listed where data are taken from Ref. [30].

---

[6] Glass transition temperature is the lowest temperature at which a polymer can be considered softened and possibly flowable [25].

[7] The melting temperature is a temperature at which the thermal energy in a solid material is just sufficient to overcome the intermolecular forces of attraction in the crystalline lattice so that the lattice breaks down and the material becomes a liquid, i.e., it melts.

[8] The melt flow index (also called melt index or melt flow rate) is the number of grams of a polymer that can be pushed out of a capillary die of standard dimensions (diameter 2.095 mm; length 8.0 mm) under the action of standard weight (such as 2.16 kg for polyethylene at 190°C) in 10 min (ASTM standard 1238).

Table 11.1 Glass Transition Temperature, Melting Temperature, and Melt Flow Index of Polymers[a]

Polymer	T_{gl} (°C)	T_m (°C)	MFI (g/10 min)
Ordinary Polymers			
Polyethylene low density (PE-LD)	−125	105–110	1.4
PE high density (PE-HD)		130–135	2.2
Polypropylene (PP) (atactic)	−20	160–165	21.5
Polyvinyl acetate (PVAC)	28	103–106	
Polyethylene terephthalate (PET)	69	250	
Polyvinyl chloride (PVC)	81		
Polyvinyl alcohol (PVAL)	85		
PP (isotactic)	100		
Polystyrene (PS)	100		9.0
Polymethylmethacrylate (PMMA)	100–120	130; 160	2.1, 6.2
High-Temperature Polymers			
Polyetherketone (PEK)	119–225		
Polyetheretherketone (PEEK)		340	
Polyethersulfone (PES)	190		
Halogenated Polymers			
Perfluoro-alkoxyalkane (PFA)	75	300–310	
TFE, HFP, VDF fluoropolymer	200	115–125	20
TFE, HFP, VDF fluoropolymer	400	150–160	10
TFE, HFP, VDF fluoropolymer	500	165–180	10
Polyvinylidene fluoride (PVDF)		160–170	
Ethylenechlorotrifluoroethylene (ECTFE)		240	
Ethylene tetrafluoroethylene (ETFE)		245–267	
Perfluoroethylene–propylene (FEP)		260–270	
MFA		280–290	
Tetrafluoroethylene (TFE)	130	327	

[a]Data are taken from Refs. [4, 13, 26–29]. HFP, hexafluoropropylene; VDF, vinylidene fluoride; MFA, copolymer of TFE and perfluoromethyl vinyl ether (PMVE: Hyflon).

Table 11.2 Glass Transition, Melting, and Valorization/Decomposition Temperatures and Heat of Melting of Polymers from Parts of a Minivan[a]

Part Description	Polymer	T_{gl} (°C)	T_m (°C)	T_v/T_d (°C)			ΔH_m (J/g)
				Initial	Major	Secondary	
Headliner							
Backing top layer	Polyethylene terephthalate (PET)	−6	76, 99, 254		305	337	5
High-density foam	Polyetherurethane (PEU)		A*		279	505	
Low-density foam	Polyester polyurethane (PESPU)	−12	A	264	363	442	
Fabric, exposed	Nylon 6	42	221	276	380	358	57
Instrument Panel							
Foam	PEU	−2	A		248	510	
Cover	Polyvinyl chloride (PVC)		A		269	430	
Structure	Polycarbonate (PC)	−9	A		397	531	
Shelf, main panel	PC	140	A	404	454	548	
Shelf, foam—small seals	PEU		A		252	291	
Resonator							
Structure	Polypropylene (PP)		164		296	325	66
Intake tube	Ethylene–propylene–diene monomer (EPDM), elastomer		A	317	447	543	
Effluent tube	EPDM		A	349	572	452	

Kick Panel Insulation							
Foam	PEU			A		262	
Backing	PVC			A		257	363
Air Ducts							
Small ducts	Polyethylene (PE)			128	293	404	375	149
Large ducts	PP			A	303	341	375	63
Steering Column Boot							
Inner boot	Natural rubber (NR)			A	267	418	495	
Cotton shoddy	Mixture of cotton, polyester, and other fibers			155		305	397	19
Outer interior boot	Polyether–polyester elastomer	−5		A	276	339	406	
Brake Fluid Reservoir							
Reservoir	PP			164		308		70
Cap	PP	−14		165		293		76
Wire Harness							
Tube	PE			128	267	356	406	166
Door Lock Contact							
Structure	Poly(acrylonitrile–butadiene–styrene) (ABS)	−11, 112		A		373		

(*continued overleaf*)

Table 11.2 (continued)

Part Description	Polymer	T_{gl} (°C)	T_m (°C)	T_v/T_d (°C) Initial	T_v/T_d (°C) Major	T_v/T_d (°C) Secondary	ΔHm (J/g)
Windshield Wiper Tray							
Structure	Sheet molding compounds (SMC)		A		341	723	
Sound Reduction Fender Insulation							
Low-density foam	Polystyrene (PS)		A	233	339	266	
High-density foam	PS		A	257	432	288	
Hood Liner Insulation							
Back	PET, cellulose, epoxy		250		300	397	5
Face	PET		245	310	394	370	7
Wheel Well Cover							
Fuel tank shield	PP		166		310	368	60
HVAC Unit Door							
Thermostat	PVC		A		267	440	
Structure	Nylon 66		259		421	514	39
Seal	Thermoplastic polyolefin (TPO)		157	284	396	391	31
Structure	Nylon 66	82	257		428	526	31

Part	Material							
Seal	TPO				154	284	406	15
Structure	Nylon 66	82, 42		259		425	500	35
Seal	TPO				154	284	406	18
HVAC Unit								
Cover	PP	53		124, 162	288	325		49
Seal, foam	ABS–PVC			A		240	435	
Seal-backing	Ethylene vinyl acetate (EVA)	39		A	361	466	433	
Top main housing	PP			A		NM		
Bottom main housing	PP			A		NM		
Fan top cover	PP			A		NM		
Fan bottom cover	PP			165		298	334	57
Cover—direction control	PP			A		NM		
Deflector for airflow	PP			A		NM		
Actuator casing	PP	2		164		274	298	59
Housing	PP			163	288	339		50
Seals—large and small	ABS–PVC			A	228	238	430	
Defogger tube	TPO			A	286	334	372	
Fuel Tank								
Tank	PE			128	291	387	438	161
Hoses	Nylon 12			171	224	418	382	49
Threads/seal for fuel pump	PE			A	291	382	418	
Headlight								
Lens	PC			A	—	411	514	

(continued overleaf)

Table 11.2 (continued)

Part Description	Polymer	T_{gl} (°C)	T_m (°C)	T_v/T_d (°C) Initial	T_v/T_d (°C) Major	T_v/T_d (°C) Secondary	ΔH_m (J/g)
Backing	PC	144	A		450	517	
Retainer	Polyoxymethylene (POM)		174		252		161
Bulb support structure	Polyimide (PI)	207	A	517	558		
Leveling mechanism	PC	144	A		413	522	
Battery Casing							
Top	PE–PP	−15	128, 167		284		19, 64
Sides and bottom	PE–PP	−16	129, 166		305	433	19, 64
Cover	PP	−7	164		346	384	54
Hood to Cowl Weather Stripping							
Foam	EPDM		A		276	416	
Rubber base	EPDM		A		334	416	
Bulkhead Insulation Engine Side							
Exterior/face	Mixed fibers (cotton, nylon 66, glass)		232	48	288	339	5
Insides	PVC–Glass		A		261	391	
Support structure	PVC–Hydrocarbon elastomer		174		267	387	
Grommet, wire cap	PE–HD		A	334	565	435	

[a]From Ref. [30].
*Amorphous.

The differences in the T_{gl} and T_m values for polymers in Tables 11.1 and 11.2 indicate significant variations in the fire behaviors of the polymers, especially between the ordinary and advanced engineered polymers. Ordinary polymers, such as polypropylene, burn with much higher intensity fires, releasing large amounts of products of complete and incomplete combustion compared to the fires of the advanced engineered polymers. An example is shown in Figure 11.1 for polypropylene. In the figure, heat release rate data for a 100-mm diameter and 25-mm thick horizontal solid slab of polypropylene (PP), measured in our laboratory, are presented. The slab was exposed to 50 kW/m² of external heat flux in the ASTM E2058 Fire Propagation Apparatus, shown in Figure 11.2 [31]. PP was burned in normal air under well-ventilated condition.

The steady-state combustion was achieved in about 400 s after ignition and remained until 900 s, where PP was burning as a solid with formation and burning away of very small liquid bubbles at the surface. Between about 1000 and 1100 s after ignition, the PP slab melted rapidly, forming a deep molten liquid pool on top of the solid polymer. Beyond 1150 s after ignition, the deep molten liquid pool of PP was burning as a boiling liquid pool fire with heat release rate increasing exponentially. At about 1300 s the heat release rate decreased rapidly as all the sample had been consumed. The heat release rate for the steady-state combustion of solid polymer is about one third the rate for the combustion of boiling liquid pool over the solid polymer surface.

In a large-scale study of fires of selected parts from a minivan (1996 Dodge Caravan [30]), it was found that a significant aspect of the burning behavior of the

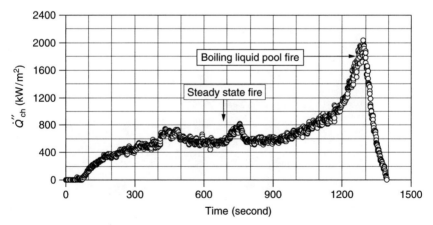

Figure 11.1. Chemical heat release rate for a 100-mm diameter and 25-mm thick horizontal slab of polypropylene exposed to 50 kW/m² of external heat flux in normal air, under well-ventilated condition, in the ASTM E2058 Apparatus [31]. Airflow rate: 2.9×10^{-3} m³/s. Data up to about 900 s are for the combustion with very small bubbles formed at the surface of the solid slab of PP. Beyond about 1150 s, the data are for the combustion with deep liquid pool over the solid slab of PP. Data were measured in our laboratory.

Figure 11.2. The ASTM E 2058 Fire Propagation Apparatus [31].

polymers was the development of a polymer melt pool fire [32]. The observation was consistent with the polymer melt flow characteristics of the polymers (T_{gl} and T_m values of PP, PC, PET, and sheet molding compound (SMC) in Table 11.2). See the nomenclature list at the end of this chapter for abbreviations used in the discussion.

11.5. POLYMER VAPORIZATION/DECOMPOSITION

On heat exposure, a thermoplastic or an elastomer undergoes softening, melting, flow of polymer melt, and vaporization (or decomposition), whereas a thermoset vaporizes or decomposes. As the polymer vaporizes/decomposes, a combustible vapor–air mixture is generated that ignites and the polymer starts burning. The vaporization/decomposition characteristic of a polymer is governed by its thermal stability characterized by its vaporization/decomposition temperature, $T_v(T_d)$.

The factors that govern T_{gl} and T_m, namely the chain rigidity and strong interchain forces also govern $T_v(T_d)$ [4]. The $T_v(T_d)$ values for the polymers from various minivan parts are included in Table 11.2, where data are taken from Ref. [30] and in Table 11.3 for generic polymers, where data are taken from Ref. [33]. The $T_v(T_d)$ values are in the range of 240–572°C in Table 11.2 for polymers from the minivan parts and in the range of 270–789°C in Table 11.3 for generic polymers, suggesting significant differences in the thermal stability of the polymers. Thermal stability is one of the major parameters affecting the fire behavior of polymers.

For sustaining combustion in polymer fires, $T_v(T_d)$ values as well as ΔH_g values need to be satisfied by the external/internal heat source. The ΔH_g value combines all the energy requirements associated with the softening, melting, and vaporizing/decomposition of a polymer. Its value can be obtained [23]: (1) by quantifying each of its components in Eqs. (11.3) and (11.4), such as by the differential scanning calorimetry (DSC), and (2) by the direct quantification technique using some of the standard flammability test apparatuses, such as the ASTM E2058 [31].

For the direct quantification of the ΔH_g values in ASTM E2058 [31], polymer samples (100 mm^2 or 100 mm in diameter and 25 mm thick) are exposed to external heat flux in a co-flowing air with oxygen concentration reduced to below 10% to maintain nonflaming combustion condition [21–23]. By measuring the release rate of polymer vapors under the nonflaming steady-state combustion condition, the ΔH_g value of a polymer is calculated from Eq. (11.1). In the calculations, $\dot{q}''_e$ is the applied external heat flux, $\dot{q}''_f$ is zero, and $\dot{q}''_{rr}$ is the external heat flux value where release rate of polymer vapors is zero. Table 11.4 lists ΔH_g and $\dot{q}''_{rr}$ values for selected polymers determined in this fashion in the ASTM E2058 apparatus [21–23].

11.6. IGNITION OF POLYMER VAPORS

When a polymer is exposed to external or internal heat sources of sufficient strength, energy requirements to vaporize/decompose the polymer are satisfied and a combustible or noncombustible vapor–air mixture is created near the surface of the polymer. The combustible vapor–air mixture autoignites or is ignited by some heat source such as a flame near the surface, and a self-sustained flame is established near the surface. This is defined as the ignition of the polymer.

Table 11.3 Char Yield, Vaporization/Decomposition Temperatures, Limiting Oxygen Index, and UL 94 Ratings for Polymers[a]

Polymer	$T_v(T_d)$ (°C)	Char Yield (%)	LOI (%)	UL 94 Ranking[b]
Ordinary Polymers				
Poly(α-methylstyrene)	341	0	18	HB
Polyoxymethylene (POM)	361	0	15	HB
Polystyrene (PS)	364	0	18	HB
Polymethylmethacrylate (PMMA)	398	2	17	HB
Polyurethane elastomer (PU)	422	3	17	HB
Polydimethylsiloxane (PDMS)	444	0	30	HB
Polyacrylonitrile–butadiene–styrene (ABS)	444	0	18	HB
Polyethylene terephthalate (PET)	474	13	21	HB
Polyphthalamide	488	3	(22)	HB
Polyamide 6 (PA6)—Nylon	497	1	21	HB
Polyethylene (PE)	505	0	18	HB
High-Temperature Polymers				
Cyanate ester of Bisphenol-A (BCE)	470	33	24	V-1
Phenolic triazine cyanate ester (PT)	480	62	30	V-0
Polyethylene naphthalate (PEN)	495	24	32	V-2
Polysulfone (PSF)	537	30	30	V-1
Polycarbonate (PC)	546	25	26	V-2
Liquid crystal polyester	564	38	40	V-0
Polypromellitimide (PI)	567	70	37	V-0
Polyetherimide (PEI)	575	52	47	V-0
Polyphenylenesulfide (PPS)	578	45	44	V-0
Polypara(benzoyl)phenylene	602	66	41	V-0
Polyetheretherketone (PEEK)	606	50	35	V-0
Polyphenylsulfone (PPSF)	606	44	38	V-0
Polyetherketone (PEK)	614	56	40	V-0
Polyetherketoneketone (PEKK)	619	62	40	V-0
Polyamideimide (PAI)	628	55	45	V-0
Polyaramide (Kevlar)	628	43	38	V-0
Polybenzimidazole (PBI)	630	70	42	V-0
Polyparaphenylene	652	75	55	V-0
Polybenzobisoxazole (PBO)	789	75	56	V-0
Halogenated Polymers				
Polyvinyl chloride (PVC)	270	11	50	V-0
Polyvinyl lidenefluoride (PVDF)	320–375	37	43–65	V-0
Polychlorotrifluoroethylene (PCTFE)	380	0	95	V-0
Fluorinated cyanate ester	583	44	40	V-0
Polytetrafluoroethylene (PTFE)	612	0	95	V-0

[a]Data are taken from Ref. [33].
[b]V, vertical burning; HB, horizontal burning.

Table 11.4 Surface Re-radiation Loss and Heat of Gasification of Polymers[a]

Polymer/Sample	$\dot{q}_{rr}''$ (kW/m^2)	ΔH_g (kJ/g) DSC	ΔH_g (kJ/g) ASTM E2058
Ordinary Polymers			
Filter paper	10		3.6
Corrugated paper	10		2.2
Douglas fir wood	10		1.8
Plywood/fire retarded (FR)	10		1.0
Polypropylene (PP)	15	2.0	2.0
Polyethylene (PE), low density	15	1.9	1.8
PE—high density	15	2.2	2.3
Polyoxymethylene (POM)	13	2.4	2.4
Polymethylmethacrylate (PMMA)	11	1.6	1.6
Nylon 6,6	15		2.4
Polyisoprene	10		2.0
Acrylonitrile–butadiene–styrene (ABS)	10		3.2
Styrene–butadiene	10		2.7
Polystyrene (PS) foams	10–13		1.3–1.9
PS granular	13	1.8	1.7
Polyurethane (PU), foams—flexible	16–19	1.4	1.2–2.7
PU foams—rigid	14–22		1.2–5.3
Polyisocyanurate (PIU) foams	14–37		1.2–6.4
Polyesters/glass fiber	10–15		1.4–6.4
PE foams	12		1.4–1.7
High-Temperature Polymers			
Polycarbonate (PC)	11		2.1
Phenolic foam	20		1.6
Phenolic foam (FR)	20		3.7
Phenolic/glass fibers	20		7.3
Phenolic-aromatic polyamide	15		7.8
Halogenated Polymers			
PE/25% chlorine (Cl)	12		2.1
PE/36% Cl	12		3.0
PE/48% Cl	10		3.1
Polyvinyl chloride (PVC), rigid	15		2.5
PVC plasticized	10		1.7
Ethylene–tetrafluoroethylene (ETFE)	27		0.9
Perfluoroethylene–propylene (FEP)	38		2.4
Ethylene–tetrafluoroethylene (ETFE)	48		0.8–1.8
Perfluoro–alkoxyalkane (PFA)	37		1.0

[a] Taken from Refs. [21–23].

The ignition resistance of polymers has been investigated in detail both experimentally and theoretically [14, 15, 21, 22, 34, 35]. It is well recognized that the ignition behavior of a polymer is governed by its actual thickness (d) relative to the thermal penetration depth (δ) [14, 15, 34, 35]. The thermal penetration depth is expressed as:

$$\delta = \sqrt{\alpha t} = \sqrt{(k/\rho c)t} \qquad (11.5)$$

where α is the thermal diffusivity of the polymer (mm/s), t is the heat exposure time of the polymer surface (s), k is the thermal conductivity of the polymer (W/m·K), ρ is the density of the polymer (g/cm^3), and c is the heat capacity of the polymer (J/g K). For thermally thick conditions, $\delta < d$ and for thermally thin conditions, $\delta > d$. Polymers are generally physically thick in their end-use applications, and thus the ignition under thermally thick condition is more common than under thermally thin condition.

In some of the standard flammability tests, ignition time or time to flame extinguishment is specified as the ignition resistance criterion. In some of the tests a single measurement is made, while in others multiple measurements are made for the ignition time or time to flame extinguishment, using either a burner flame or an external heat source with a pilot. The data from the standard tests where ignition time is measured at several external heat flux values [31, 36–38] can be used to derive the ignition properties of the polymers under thermally thick condition.

The following is the most commonly used expression for the relationship between the ignition time and external heat flux based purely on the thermal arguments [14, 15, 21, 22, 34, 35]:

$$1/t_{ig}^{1/2} = a(\dot{q}_e'' - \dot{q}_{cr}'')/(T_{ig} - T_a)\sqrt{k\rho c} \qquad (11.6)$$

where t_{ig} is the ignition time (s), a is the polymer surface absorptivity,[9] $\dot{q}_{cr}''$ is the critical heat flux (CHF)[10] per unit polymer surface area (kW/m^2), T_{ig} is the ignition temperature (°C), and T_a is the ambient temperature (°C). The term $(T_{ig} - T_a)\sqrt{k\rho c}$ is defined as the thermal response parameter (TRP) of the polymer (kW·s$^{1/2}$/m^2) [21, 22]. CHF and TRP are defined as the ignition properties of the polymers.

The ignition properties of polymers can be obtained by using the standard test procedure of exposing the polymer surface in a horizontal or vertical orientation to external heat flux, in the presence of a pilot flame and measuring the ignition time [31, 36–38]. Figure 11.3 shows an example of the ignition test for a circular sample, about 100 mm in diameter and 25 mm in thickness in the ASTM E2058 Apparatus. Figure 11.4 show an example of the ignition data.

---

[9] In Eq. (11.6), the polymer surface absorptivity, a, is taken as unity when the ignition experiments are performed with sample surfaces coated black, such as in the ASTM E2058 Apparatus [31].
[10] CHF value is taken as the external heat flux value at which there is no ignition under quiescent airflow conditions.

Figure 11.3. About 100-mm diameter and 25-mm thick polymer sample with black-coated surface exposed to external heat flux in the presence of a pilot flame in the ASTM E2058 Apparatus [31]. Release of polymer vapors prior to ignition can be noted.

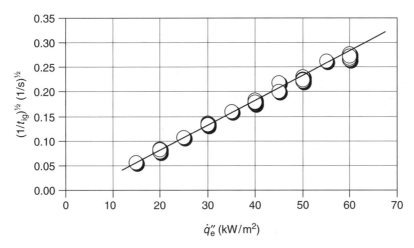

Figure 11.4. Relationship between the time to ignition measured in the ASTM E2058 Apparatus and external heat flux for 100-mm square and 25-mm thick slab of PMMA. The surface was coated black in the experiments. Data are taken from Ref. [39].

These data were measured in the ASTM E2058 Fire Propagation Apparatus under quiescent airflow conditions for about 100 mm² and 25-mm thick slab of polymethylmethacrylate (PMMA), with surface coated black [39]. The experimental TRP value, TRP_{Exp}, is taken as the inverse of the slope of the linear relationship between $(1/t_{ig})^{1/2}$ and $\dot{q}_e''$, for black coated surface ($a = 1.0$).

The ignition properties from ignition time versus external heat flux data for a large number of polymers have been determined [19, 21, 22, 39–44]. Tables 11.5 and 11.6 list the values of the ignition properties obtained from the ignition data measured in ASTM E2058 fire propagation apparatus for selected polymers from parts of minivan and ordinary and advanced engineered polymers [21, 22, 39–41]. Table 11.7 lists the ignition property values obtained from the ignition time with sustained combustion for at least 10 s at various heat flux data measured in ASTM E1354 cone calorimeter and reported in Refs. [19, 42–44]. The data in Tables 11.5 to 11.7 show that the CHF values vary in the range of 10–50 kW/m² and the TRP_{Exp} values in the range of 73–1111 kW·s$^{1/2}$/m², showing very wide variations in the ignition resistance of polymers.

In addition to determining the experimental TRP_{Exp} values, they can also be calculated for polymers with known k, ρ, c, $T_v(T_d)$ and T_{ig} values. These values are available in the literature for many polymers [25–27, 29, 30, 39, 45–48], some of which are listed in Tables 11.5 and 11.6 along with the TRP_{Exp} and its calculated value, TRP_{Cal}. In these tables, TRP_{Cal} values calculated from $T_v(T_d)$ as well as T_{ig} values with known k, ρ, and c have been listed.

Table 11.6 also includes the $\sqrt{k\rho c}$ values calculated from k, ρ, and c values from the literature. The average $\sqrt{k\rho c}$ values for each class of solid polymers in the table vary within a narrow range, suggesting that they are independent of the generic natures of the solid polymers.[11] The average value of $\sqrt{k\rho c}$ for ordinary solid polymers is $0.778 \pm 18\%$, for high-temperature polymers; it is $0.640 \pm 15\%$, and for highly halogenated solid polymers it is $0.624 \pm 18\%$. The overall average value of $\sqrt{k\rho c}$ for solid polymers is $0.781 \pm 18\%$. Thus, the $\sqrt{k\rho c}$ component of TRP can be considered as approximately constant and independent of the generic nature of the polymer. The variations in the TRP values thus would mainly be due to the variations in the T_{ig} values.

The T_{ig} values of polymers can be measured directly using standardized tests or estimated from the CHF values.[12] The T_{ig} values estimated in this fashion from the CHF values have been reported in the literature [21, 22, 39–41], some of which are included in Tables 11.5 and 11.6 along with the T_v/T_d values taken from previous tables. It can be noted that for ordinary and high-temperature polymers, $T_v(T_d) \approx T_{ig}$ and the TRP_{Cal} values calculated from $T_v(T_d)$ values and from T_{ig} values are similar. For highly halogenated polymers, $T_{ig} \gg T_v(T_d)$ and

---

[11] The $\sqrt{k\rho c}$ values of polymers modified for use as foams, expanded elastomers, and fabrics are significantly lower and variable than the values for the solid polymers.

[12] $T_{ig}(°C) \approx [(\dot{q}_{cr}'')^{0.25} \times 364] - 273$, assuming heat losses to be mainly due to re-radiation and polymer surface acting as a black body and an ambient temperature of 20°C.

Table 11.5 Thermo-Physical and Ignition Properties of Polymers from Parts of a Minivan[a]

Part Description	Polymer	$T_v(T_d)$ (°C)	CHF (kW/m²)	T_{ig}^b (°C)	$\rho \times 10^{-3}$ (kg/m³)	c (kJ/kg·K)	$k \times 10^3$ (kW/m·K)	TRP$_{Exp}$	TRP$_{Cal}$ $T_v(T_d)^c$	TRP$_{Cal}$ T_{ig}^b
Head Liner										
Backing top layer	PET	305	—	—	0.69	1.56	0.04	—	57	—
Fabric, exposed	Nylon 6	380	20	497	0.12	2.19	0.24	154	90	120
Instrument Panel										
Foam	PEU	248	—	—	0.11	1.77	0.04	—	19	19
Cover	PVC	269	10	357	1.20	1.37	0.14	263	120	162
Structure	PC	397	—	—	1.12	1.68	0.18	—	219	—
Shelf, main panel	PC	454	20	497	1.18	1.51	0.27	357	301	331
Shelf, foam—small seals	PEU	252	—	—	0.09	1.56	0.06	—	22	—
Resonator										
Structure	PP	296	10	374	1.06	2.08	0.23	277	197	252
Intake tube	EPDM	447	—	—	1.15	1.75	0.30	—	331	—
Effluent tube	EPDM	572	—	—	1.16	1.39	0.36	—	421	—
Kick Panel Insulation										
Foam	PEU	262	—	—	0.02	1.65	0.02	—	7	—
Backing	PVC	257	10	374	1.95	1.14	0.25	215	177	264

(*continued overleaf*)

Table 11.5 (continued)

Part Description	Polymer	$T_v(T_d)$ (°C)	CHF (kW/m²)	T_{ig}^b (°C)	$\rho \times 10^{-3}$ (kg/m³)	c (kJ/kg·K)	$k \times 10^3$ (kW/m·K)	TRP$_{Exp}$	TRP$_{Cal}$ $T_v(T_d)^c$	TRP$_{Cal}$ T_{ig}^b
Air Ducts										
Small ducts	PE	404	—	—	0.95	2.03	0.31	—	297	—
Large ducts	PP	341	15	443	1.04	1.93	0.31	333	253	334
Brake Fluid Reservoir										
Reservoir	PP	308	—	—	0.90	2.25	0.19	—	179	—
Cap	PP	293	—	—	0.90	2.48	0.21	—	187	—
Wire Harness										
Tube	PE	356	—	—	0.95	2.12	0.37	—	290	—
Windshield Wiper Tray										
Structure	SMC	341	20	497	1.64	1.14	0.37	483	267	397
Sound Reduction Fender Insulation										
Low-density foam	PS	339	—	—	0.90	1.70	0.17	—	163	—
High-density foam	PS	432	20	497	0.13	1.62	0.10	146	60	69

				Hood Liner Insulation				
Face	PET	394	10	374	0.66	1.32	0.09	174	103	97
				Wheel Well Cover				
Fuel tank shield	PP	310	15	443	0.93	2.20	0.20	288	186	271
				HVAC Unit Door				
Seal	PP/EPDM	396	—	—	0.93	1.96	0.20	—	227	—
Structure	Nylon 66	428	—	—	1.50	1.69	0.58	—	495	—
Seal	TPO	284	—	—	0.97	1.87	0.13	—	128	—
				HVAC Unit				
Cover	PP	325	—	—	1.19	1.90	0.39	—	286	—
Seal, foam	ABS-PVC	240	19	487	0.10	1.35	0.02	73	10	22
Top main housing	PP	NM	—	—	NM	NM	—	310	—	—
Fan bottom cover	PP	298	—	—	1.21	1.76	0.39	—	253	—
Actuator casing	PP	274	—	—	1.11	1.95	0.34	—	218	—
Seals	ABS-PVC	238	—	—	0.07	2.02	0.15	—	32	—
Defogger tube	PP/EPDM	334	—	—	0.97	1.87	0.33	—	243	—
				Fuel Tank				
Tank	PE	387	15	443	0.94	2.15	0.30	454	286	329
Hoses	Nylon 12	418	—	—	1.04	1.79	0.18	—	230	—

(*continued overleaf*)

Table 11.5 (continued)

Part Description	Polymer	$T_v(T_d)$ (°C)	CHF (kW/m²)	T_{ig}^b (°C)	$\rho \times 10^{-3}$ (kg/m³)	c (kJ/kg·K)	$k \times 10^3$ (kW/m·K)	TRP_{Exp}	TRP_{Cal} $T_v(T_d)^c$	TRP_{Cal} T_{ig}^b
Headlight										
Lens	PC	411	—	—	1.19	2.06	0.20	—	274	—
Backing	PC	450	20	497	1.20	2.18	0.22	434	326	362
Retainer	POM	252	—	—	1.41	1.92	0.27	—	198	—
Leveling mechanism	PC	413	—	—	1.18	1.10	0.19	—	195	—
Battery Casing										
Top	PE-PP	284	—	—	0.91	1.98	0.17	—	146	—
Sides and bottom	PE-PP	305	—	—	0.88	2.15	0.21	—	180	—
cover	PP	346	15	443	0.90	2.22	0.23	323	221	286
Hood to Cowl Weather Stripping										
Foam	EPDM	276	—	—	0.44	2.30	0.07	—	70	—
Rubber base	EPDM	334	—	—	0.41	1.51	0.21	—	113	—
Bulkhead Insulation Engine Side										
Insides	PVC—Glass	261	—	—	1.00	1.05	0.23	—	119	—
Support structure	PVC—HE	267	—	—	1.60	1.24	0.10	—	110	—
Grommet, wire harness cap	HDPE	565	—	—	1.21	1.48	0.45	—	490	—

[a] Ignition data are taken from Refs. [30, 39, 41].
[b] T_{ig} values from CHF were used in the calculation of TRP values.
[c] T_v/T_d values were used in the calculation of TRP values, which are included in the parentheses.

Table 11.6 Thermo-Physical and Ignition Properties of Polymers[a]

Polymer	$T_v(T_d)$ (°C)	CHF (kW/m²)	T_{ig} (°C)	ρ (g/cm³)	c (J/g·K)	k (W/m·K)	$(k\rho c)^{0.5}$ (kW·s$^{1/2}$/m²·K)	TRP$_{Exp}$	TRP$_{Cal}$ From $T_v(T_d)$	TRP$_{Cal}$ From T_{ig}
					Ordinary Polymers					
PE-1	387	15	443	0.94	2.15	0.30	0.779	454	286	327
PE-2	404	—	—	0.95	2.03	0.31	0.773	—	297	—
PE-3	356	—	—	0.95	2.12	0.37	0.863	—	290	—
PP-1	308	—	—	0.90	2.25	0.19	0.620	—	179	—
PP-2	293	—	—	0.90	2.48	0.21	0.685	—	187	—
PP-3	325	—	—	1.19	1.90	0.39	0.939	—	286	—
PP-4	298	—	—	1.21	1.76	0.39	0.911	—	253	—
PP-5	274	—	—	1.11	1.95	0.34	0.858	—	218	—
PP-6	310	15	443	0.93	2.20	0.20	0.640	288	186	271
PP-7	346	15	443	0.90	2.22	0.23	0.678	323	221	287
PP-8	296	10	374	1.06	2.08	0.23	0.712	277	197	252
PP-9	341	15	443	1.04	1.93	0.31	0.789	333	253	334
PE–PP-1	284	—	—	0.91	1.98	0.17	0.553	—	146	—
PE–PP-2	305	—	—	0.88	2.15	0.21	0.630	—	180	—
PVC-1	257	10	374	1.95	1.14	0.25	0.745	215	177	264
Nylon 66	428	—	—	1.50	1.69	0.58	1.213	—	495	—
Nylon 12	418	—	—	1.04	1.79	0.18	0.579	—	230	—
PMMA	330	10	374	1.19	2.09	0.27	0.819	274	254	290
POM	252	10	374	1.41	1.92	0.27	0.855	250	198	303
EPDM-1	447	—	—	1.15	1.75	0.30	0.777	—	332	—
EPDM-2	572	—	—	1.16	1.39	0.36	0.762	—	421	—

(*continued overleaf*)

Table 11.6 (continued)

Polymer	$T_v(T_d)$ (°C)	CHF (kW/m²)	T_{ig} (°C)	ρ (g/cm³)	c (J/g·K)	k (W/m·K)	$(k\rho c)^{0.5}$ (kW·s$^{1/2}$/m²·K)	TRP$_{Exp}$	TRP$_{Cal}$ From $T_v(T_d)$	TRP$_{Cal}$ From T_{ig}
EPDM-3	565	—	—	1.21	1.48	0.45	0.898	—	489	—
SMC-ester	341	20	497	1.64	1.14	0.37	0.832	483	267	397
TPO-1	334	—	—	0.97	1.87	0.33	0.774	—	243	—
						Average	0.778		—	—
						Standard Deviation	0.140		—	—
High-Temperature Advanced Engineered Polymers										
PSF	537	30	580	1.24	1.30	0.28	0.672	469	347	376
PEEK	606	30	580	1.32	1.80	0.25	0.771	550	452	432
PC-1	397	—	—	1.12	1.68	0.18	0.582	—	219	—
PC-2	454	20	497	1.18	1.51	0.27	0.694	357	301	331
PC-3	411	—	—	1.19	2.06	0.20	0.700	—	274	—
PC-4	450	20	497	1.20	2.18	0.22	0.759	434	326	362
PC-5	413	—	—	1.18	1.10	0.19	0.497	—	195	—
PC-6	546	30	580	1.20	1.20	0.21	0.550	455	289	308
PC-7	546	30	580	1.20	1.20	0.21	0.550	455	289	308
PEI	575	25	540	1.27	1.40	0.22	0.625	435	347	325
						Average	0.640		—	—
						Standard Deviation	0.094		—	—
Highly Halogenated Polymers Advanced Engineered Polymers										
PTFE	401	50	700	2.18	1.00	0.25	0.738	654	281	502
FEP	401	50	700	2.15	1.20	0.25	0.803	680	306	546
ETFE	337	25	540	1.70	0.90	0.23	0.593	481	188	308

PCTFE	369	30	580	2.11	0.90	0.22	0.646	460	226	362
ECTFE	401	38	613	1.69	1.00	0.15	0.503	450	192	299
PVDF	348	40	643	1.70	1.30	0.13	0.536	506	176	334
CPVC	348	40	643	1.50	0.90	0.22	0.545	435	179	340
Average							0.624			
Standard Deviation							0.112			
Foams and Expanded Elastomers										
PEU-1	248	—	—	0.11	1.77	0.04	0.088	—	20	—
PEU-2	252	—	—	0.09	1.56	0.06	0.092	—	21	—
PEU-3	262	—	—	0.02	1.65	0.02	0.026	—	6	—
ABS–PVC-1	240	19	487	0.10	1.35	0.02	0.052	73	11	24
ABS–PVC-2	238	—	—	0.07	2.02	0.15	0.146	—	32	—
PS-1	339	—	—	0.09	1.70	0.17	0.161	—	51	—
PS-2	432	20	497	0.13	1.62	0.10	0.145	146	60	69
EPDM-1	276	—	—	0.44	2.30	0.07	0.266	—	68	—
EPDM-2	334	—	—	0.41	1.51	0.21	0.361	—	113	—
TPO-2	284	—	—	0.97	1.87	0.13	0.486	—	128	—
TPO-3	396	—	—	0.93	1.96	0.20	0.604	—	227	—
PVC-2	269	10	374	1.20	1.37	0.14	0.480	263	119	162
PVC—glass	261	—	—	1.00	1.05	0.23	0.491	—	118	—
PVC—elastomers	267	—	—	1.60	1.24	0.10	0.445	—	110	—
Fabrics										
PET-1	394	10	374	0.66	1.32	0.09	0.280	174	105	99
PET-2	305	—	—	0.69	1.56	0.04	0.207	—	59	—
Nylon 6	380	20	497	0.12	2.19	0.24	0.251	154	90	120

[a] Data are taken from Refs. [21, 22, 30, 39–41, 46].

Table 11.7 Ignition Time Measured in the ASTM E1354 Cone Calorimeter and Thermal Response Parameter Values Derived from the Data[a]

Polymers	Ref.	Ignition Time (s) $\dot{q}_e''$ (kW/m^2)									TRP$_{\text{Exp}}$ (kW·s$^{1/2}$/m^2)
		20	25	30	40	50	70	75	100	
Ordinary Polymers										
High-density polyethylene (HDPE)	44	403	—	171	91	58	—	—	—	364
Polyethylene (PE)	42	403	—	—	159	—	47	—	—	526
Polypropylene (PP)	44	120	—	63	35	27	—	—	—	291
Polypropylene (PP)	42	218	—	—	86	—	41	—	—	556
PP/glass fibers (1082)	43	—	168	—	—	47	—	23	13	377
Polystyrene (PS)	42	417	—	—	97	—	50	—	—	556
PS foam	44	NI[b]	—	73	28	18	—	—	—	168
PS— fire retarded (FR)	42	244	—	—	90	—	51	—	—	667
PS foam—FR	44	NI	—	77	40	24	—	—	—	221
Nylon	42	1923	—	—	65	—	31	—	—	333
Nylon 6	44	700	—	193	115	74	—	—	—	379
Nylon/glass fibers (1077)	43	—	193	—	—	53	—	21	13	359
Polyoxymethylene (POM)	42	259	—	—	74	—	24	—	—	357
Polymethylmethacrylate (PMMA)	42	176	—	—	36	—	11	—	—	222
Polybutylene terephthalate (PBT)	42	609	—	—	113	—	59	—	—	588
Polyethylene terephthalate (PET)	42	718	—	—	116	—	42	—	—	435
Acrylonitrile–butadiene–styrene (ABS)	44	299	—	130	68	43	—	—	—	317
ABS—FR	42	212	—	—	66	—	39	—	—	556
ABS—PVC	42	5198	—	—	61	—	39	—	—	357
Vinyl thermoplastic elastomer	42	NI	—	—	1271	—	60	—	—	294

Material										
Polyurethane (PU) foam	42	12	—	—	—	—	—	—	—	76
Thermoplastic PU—FR	42	302	—	—	60	—	38	—	—	500
EPDM/Styrene acrylonitrile (SAN)	42	486	—	—	68	109	36	1	—	417
Polyester/glass fibers (30%)	44	NI	—	—	309	38	—	—	—	256
Isophthalic polyester	44	256	—	115	59	77	—	—	—	296
Isophthalic polyester/glass fiber (77%)	44	480	—	172	91	38	—	—	—	426
Polyvinyl ester	44	332	—	120	78	38	—	—	—	263
Polyvinyl ester/glass fiber (69%)	44	646	—	235	104	78	—	—	—	444
Polyvinyl ester/glass fiber (1031)	43	—	278	—	—	74	—	34	18	429
Polyvinyl ester/glass fiber (1087)	43	—	281	—	—	—	—	22	11	312
Epoxy	44	337	—	172	100	62	—	—	—	457
Epoxy/glass fiber (69%)	44	320	—	120	75	57	—	—	—	388
Epoxy/glass fiber (1003)	43	—	198	—	—	50	—	73	19	555
Epoxy/glass fiber (1006)	43	—	159	—	—	49	—	23	14	397
Epoxy/glass fiber (1040)	43	—	—	—	—	18	—	13	9	512
Epoxy/glass fiber (1066)	43	—	140	—	—	48	—	14	9	288
Epoxy/glass fiber (1067)	43	—	209	—	—	63	—	24	18	433
Epoxy/glass fiber (1070)	43	—	229	—	—	63	—	30	23	517
Epoxy/glass fiber (1071)	43	—	128	—	—	34	—	18	10	334
Epoxy/glass fiber (1089)	43	—	535	—	—	105	—	60	40	665
Epoxy/glass fiber (1090)	43	—	479	—	—	120	—	54	34	592
Epoxy/graphite fiber (1091)	43	—	NI	—	—	—	—	53	28	484
Epoxy/graphite fiber (1092)	43	—	275	—	—	76	—	32	23	493
Epoxy/graphite fiber (1093)	43	—	338	—	—	94	—	44	28	554

(*continued overleaf*)

Table 11.7 (continued)

Polymers	Ref.	Ignition Time (s) $\dot{q}_e''$ (kW/m²)									TRP_{Exp} (kW·s$^{1/2}$/m²)
		20	25	30	40	50	70	75	100	
Cyanate ester/glass fiber (1046)	43	—	199	—	—	—	—	20	10	302
Acrylic/glass fiber	44	553	—	252	148	58	—	—	—	180
Kydex Acrylic paneling—FR	42	200	—	—	38	101	12	—	—	233

High Temperature Polymers

Polycarbonate (PC-1)	42	NI	—	—	182	—	75	—	—	455
PC-2	42	6400	—	—	144	—	45	—	—	370
Crosslinked polyethylene (XLPE)	42	750	—	—	105	—	35	—	—	385
Polyphenylene oxide (PPO), polystyrene (PS)	42	479	—	—	87	—	39	—	—	455
PPO/glass fibers	42	465	—	—	45	—	35	—	—	435
Polyphenylenesulfide (PPS)/glass fibers (1069)	43	—	NI	—	—	105	—	57	30	588
PPS/graphite fibers (1083)	43	—	NI	—	—	—	—	69	26	330
PPS/glass fibers (1084)	43	—	NI	—	—	244	—	70	48	623
PPS/graphite fibers (1085)	43	—	NI	—	—	173	—	59	33	510
Polyarylsulfone/graphite fibers (1081)	43	—	NI	—	—	122	—	40	19	360
Polyethersulfone/graphite fibers (1078)	43	—	NI	—	—	172	—	47	21	352
Polyetheretherketone (PEEK)/glass fibers (30%)	44	NI	—	—	390	142	—	—	—	301
PEEK/graphite fibers (1086)	43	—	—	—	—	307	—	80	42	514
Polyetherketoneketone (PEKK)/glass fibers (1079)	43	—	NI	—	—	223	—	92	53	710
Bismaleimide (BMI)/graphite fibers (1095)	43	—	237	—	—	—	—	42	22	513

Material										
Bismaleimide (BMI)/graphite fibers (1096)	43	—	503	—	—	141	—	60	36	608
Bismaleimide (BMI)/graphite fibers (1097)	43	—	NI	—	—	—	—	66	37	605
Bismaleimide (BMI)/graphite fibers (1098)	43	—	NI	—	—	110	—	32	27	515
Phenolic/glass fibers (45%)	44	423	—	214	—	165	—	—	—	683
Phenolic/glass fibers (1099)	43	—	NI	—	—	121	—	33	22	409
Phenolic/glass fibers (1100)	43	—	NI	—	—	125	—	—	40	728
Phenolic/glass fibers (1101)	43	—	NI	—	—	210	—	55	25	382
Phenolic/glass fibers (1014)	43	—	NI	—	—	214	—	73	54	738
Phenolic/glass fibers (1015)	43	—	NI	—	—	238	—	113	59	765
Phenolic/glass fibers (1017)	43	—	NI	—	—	180	—	83	43	641
Phenolic/glass fibers (1018)	43	—	NI	—	—	313	—	140	88	998
Phenolic/graphite fibers (1102)	43	—	NI	—	—	—	—	79	45	684
Phenolic/graphite fibers (1103)	43	—	NI	—	—	104	—	34	20	398
Phenolic/graphite fibers (1104)	43	—	NI	—	—	187	—	88	65	982
Phenolic/PE fibers (1073)	43	—	714	—	—	129	—	28	10	267
Phenolic/aramid fibers (1074)	43	—	1110	—	—	163	—	33	15	278
Polyimide/glass fibers (1105)	43	—	NI	—	—	175	—	75	55	844
Wood										
Douglas fir	42	254	—	—	34	—	12	—	—	222
Hemlock	44	307	—	73	35	19	—	—	—	175

(*continued overleaf*)

Table 11.7 (continued)

Polymers	Ref.	Ignition Time (s) $\dot{q}_e''$ (kW/m²)									TRP$_{Exp}$ (kW·s$^{1/2}$/m²)
		20	25	30	40	50	70	75	100		

Textiles

Polymers	Ref.	20	25	30	40	50	70	75	100	TRP$_{Exp}$ (kW·s$^{1/2}$/m²)
Wool	44	24	—	15	11	9	—	—	—	232
Acrylic fiber	44	52	—	28	16	19	—	—	—	175

Halogenated Polymers

Polymers	Ref.	20	25	30	40	50	70	75	100	TRP$_{Exp}$ (kW·s$^{1/2}$/m²)
Polyvinyl chloride (PVC) flexible-1	42	117	—	—	27	—	11	—	—	244
PVC flexible-2	42	102	—	—	21	—	15	—	—	333
PVC flexible-3 (LOI 25%)	44	119	—	61	41	25	—	—	—	285
PVC—FR flexible-1	42	236	—	—	47	—	12	—	—	222
PVC—FR flexible-2	42	176	—	—	36	—	14	—	—	263
PVC—FR (Sb$_2$O$_3$) flexible-4 (LOI 30%)	44	136	—	84	64	37	—	—	—	397
PVC—FR (triaryl phosphate) flexible-5 (LOI 34%)	44	278	—	103	69	45	—	—	—	345
PVC—FR (alkyl aryl phosphate) flexible-6 (LOI 28%)	44	114	—	72	49	35	—	—	—	401
PVC rigid-1	42	5159	—	—	73	—	45	—	—	385
PVC rigid-2	42	3591	—	—	85	—	48	—	—	417
PVC rigid-3	42	5171	—	—	187	—	43	—	—	357
PVC rigid-1 (LOI 50%)	44	NI	—	487	276	82	—	—	—	388
PVC rigid-2	44	NI	—	320	153	87	—	—	—	390
Chlorinated PVC (CPVC)	42	NI	—	—	621	—	372	—	—	1111

[a] Data taken from Refs. [19, 42–44].
[b] NI, no ignition.

434

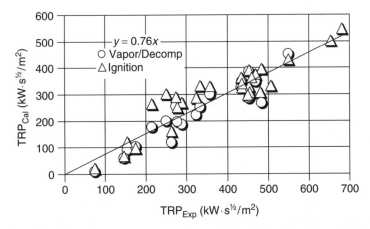

Figure 11.5. Thermal response parameter values calculated from the thermophysical properties and vaporization (decomposition) and ignition temperatures of polymers versus the values obtained from measured time to ignition in ASTM E2058 apparatus [31]. TRP_{cal} values from $T_v(T_d)$ for highly halogenated polymers are not included.

TRP_{Cal} values calculated from $T_v(T_d)$ values and from T_{ig} values are significantly different.

The TRP_{Cal} values are in general lower than the TRP_{Exp} values, as shown in Figure 11.5 (TRP_{Cal} values from $T_v(T_d)$ values for highly halogenated polymers are not included as they are significantly lower than the T_{ig} values). The data in Figure 11.5 show that TRP_{Cal} values are about 24% lower than the TRP_{Exp} values. It thus appears that the purely thermal arguments based on the ignition relationship in Eq. (11.6) is not sufficient to account for the ignition behavior of polymers. There is a need to incorporate the chemical arguments for the ignition behavior of polymers.

Currently there is lack of understanding of the chemical effects on the ignition behavior of polymers. The experimental conditions, therefore, cannot be controlled precisely, and the effects are apparent in the lack of repeatability in the ignition time measurements, especially for polymers where chemical effects are stronger, such as the high-temperature and highly halogenated advanced engineered polymers. Also a better understanding of the chemical effects on the ignition behavior of polymers would lead to the development of more effective fire-retardant/fire-hardening agents for the polymers.

11.7. COMBUSTION OF POLYMER VAPORS

Combustion is a process where polymer vapors react chemically with oxygen from air in the reaction zone of the flame, generating heat and products of complete and incomplete combustion. The fire intensity and thermal and nonthermal hazards in fires depend on the release rates of polymer vapors,

Table 11.8 ASTM Small-Scale Test Apparatuses for the Measurements of Release Rates of Heat and Fire Products and Light Obscuration by Smoke

Design/Test Conditions	ASTM E906 [37]	ASTM E2058 [31]	ASTM E1354 [36]	ASTM E662 [49][a]
Apparatus Design and Capacity				
Airflow	Co-flow	Co-flow/natural	Natural	None
Oxygen concentration (%)	21	0–60	21	21
Co-flow airflow velocity (m/s)	0.49	0–0.146	NA	NA
External heaters	Silicone carbide	Tungsten–quartz	Electrical coils	Electrical coils
External heat flux (kW/m^2)	0–100	0–65	0–100	25
Sampling duct flow (m^3/s)	0.04	0.035–0.364	0.012–0.035	NA
Sample (mm)—horizontal	110 × 150	100 × 100	100 × 100	NA
Sample (mm)—vertical	150 × 150	100 × 600	100 × 100	76 × 76
Ignition source	Pilot flame	Pilot flame	Spark plug	Pilot flame
Ventilation controlled	No	Yes	No	No
Flame radiation simulation by O$_2$	No	Yes	No	No
Heat release rate capacity (kW)	8	50	8	NA
Measurements				
Ignition time	Yes	Yes	Yes	No
Release rate of vapors	No	Yes	Yes	Yes
Release rate of fire products	Yes	Yes	Yes	No
Light obscuration by smoke	Yes	Yes	Yes	Yes
Gas-phase corrosion	No	Yes	No	No
Fire propagation	No	Yes	No	No
Chemical heat release rate	Yes	Yes	Yes	No
Convective heat release rate	Yes	Yes	No	No
Radiative heat release rate	No	Yes	No	No
Flame extinction—water, Halon and alternates	No	Yes	No	No

[a] Closed test chamber: 914 mm (36 inches) wide, 610 mm (24 inches) deep, and 914 mm (36 inches) high.

heat, and products and the chemical nature of the polymer vapors and products. Standard test apparatuses and procedures have been developed to measure these fire characteristics [19, 21, 22, 31, 36, 37, 52, 49–52]. The capabilities of the four popular standard test apparatuses used to quantify these characteristics are listed in Table 11.8.

Three factors that have the most dominant effects on the fire intensity and thermal and nonthermal hazards are (1) availability of air (oxygen), (2) fire properties, generic nature, and mode of vaporization (decomposition) of the polymers, and (3) flame and external heat flux and surface re-radiation loss.

11.7.1. Availability of Oxygen

The availability of oxygen is expressed in terms of an equivalence ratio, Φ [21, 22, 53]. Equivalence ratio is the total mass of oxygen available for combustion per unit total mass of the polymer vapors burned, normalized by the mass stoichiometric oxygen-to-fuel[13] ratio. For $\Phi \leq 1$, fires are fuel lean (well ventilated), and there is little effect of the amount of oxygen available for combustion. For $\Phi > 1$, fires are fuel rich (ventilation controlled), and the amount of oxygen available for combustion governs the combustion process.

Release rates of heat and products of complete of combustion (CO_2) increase with decrease in the Φ value, whereas release of products of incomplete combustion (smoke, CO, and other toxic compounds) increase with increase in the Φ value.

11.7.2. Fire Properties of Polymers

The release rates of heat and products from the combustion of solid polymers are expressed as [21, 22]:

$$\dot{G}''_j = (y_j/\Delta H_g)(\dot{q}''_e + \dot{q}''_f - \dot{q}''_{rr}) = f_j(\Psi_j/\Delta H_g)(\dot{q}''_e + \dot{q}''_f - \dot{q}''_{rr}) \qquad (11.7)$$

$$\dot{Q}''_{ch} = (\Delta H_{ch}/\Delta H_g)(\dot{q}''_e + \dot{q}''_f - \dot{q}''_{rr}) = \chi(\Delta H_T/\Delta H_g)(\dot{q}''_e + \dot{q}''_f - \dot{q}''_{rr}) \qquad (11.8)$$

where $\dot{G}''_j$ is the mass generation rate of product j per unit surface area of the polymer (g/m^2·s), y_j is the yield of product j (g/g), f_j[14] is the generation efficiency of product j, Ψ_j is the maximum possible mass stoichiometric yield of product j (g/g), $\dot{Q}''_{ch}$ is the chemical heat release rate[15] per unit surface area of the polymer (kW/m^2), ΔH_{ch} is the chemical heat of combustion (kJ/g), ΔH_T is the net heat of complete combustion of the polymer (kJ/g), χ is the combustion efficiency, which is the ratio of ΔH_{ch} to ΔH_T. The term $y_j/\Delta H_g$ is defined as the product generation parameter PGP (g/kJ), and the term $\Delta H_{ch}/\Delta H_g$ is defined as the heat release parameter (HRP, kJ/kJ) [21, 22].

The release rates of polymer vapors, heat, and products are measured at various external heat flux values in normal air as well in air with variable oxygen concentration and flow rate in the standard test apparatuses [31, 36, 37]. An illustration of the combustion test is shown in Figure 11.6. This test was performed in

---

[13] Combustible polymers, solids, liquids, and gases in general are defined as fuels.
[14] f_j is the fuel carbon atom conversion efficiency. It is the ratio of the experimental yield of a product j to the maximum possible stoichiometric yield of the product.
[15] Chemical heat release rate is the actual heat that is released in the combustion of a polymer [21, 22]. It has a convective and a radiative component. Chemical heat release rate is always less than the heat release rate for complete combustion as the polymers do not burn completely. Chemical heat release rate is determined from the mass generation rate of CO_2 corrected for the mass generation rate of CO (defined as the carbon dioxide generation calorimetry) and from the mass depletion rate of O_2 (defined as the oxygen consumption calorimetry).

Figure 11.6. Combustion of a liquid fuel soaked in jute cloth wick inside a 100-mm diameter and 25-mm deep Pyrex glass dish in the ASTM E2058 Apparatus. Combustion was performed in normal air at a flow rate of 2.9×10^{-3} m³/s inside a quartz glass tube with no external heat flux.

our laboratory in the ASTM E2058 apparatus in normal air without the external heat flux for a liquid-fuel-soaked wick inside a 100-mm diameter and 25-mm deep Pyrex glass dish.

Examples of the release rates of heat and products measured in the ASTM E2058 Fire Propagation Apparatus and the ASTM E1354 Cone Calorimeter, taken from Refs. [19, 41–44] are listed in Tables 11.9 and 11.10. Table 11.9 lists release rates of products and heat for polymers from parts of a minivan measured in ASTM E2058 Apparatus in normal air and 50 kW/m² of external heat flux, where data are taken from Ref. [41]. Table 11.10 lists heat release rates measured in ASTM E1354 Cone Calorimeter for various polymers at three external heat flux values, where data are taken from Refs. [42–44]. The HRP ($\Delta H_{ch}/\Delta H_g$) values are also listed in Table 11.10, which are obtained from the slopes of the lines representing the relationship between $\dot{Q}''_{ch}$ and $\dot{q}''_e$ [Eq. (11.8)].

Table 11.9 Peak Release Rates of Heat and Products from the Combustion of Polymers from Parts of a Minivan Measured in the ASTM E2058 Fire Propagation Apparatus[a]

Part Description	Polymers	$\dot{G}''_j$ (g/m²·s)				$\dot{Q}''_{ch}$ (kW/m²)
		CO	CO₂	HC[b]	Smoke	
Head Liner						
Fabric, exposed	Nylon 6	0.40	22.5	<0.01	0.66	301
Instrument Panel						
Cover	PVC	1.42	37.3	0.21	3.08	527
Shelf, main panel	PC	1.26	44.7	0.08	3.29	486
Resonator						
Structure	PP	1.00	66.2	0.08	2.09	926
Intake tube	EPDM	0.60	18	0.01	0.91	242
Kick Panel Insulation						
Backing	PVC	1.05	15.6	0.12	1.64	219
Air Ducts						
Large ducts	PP	1.95	78.8	0.36	2.86	1110
Steering Column Boot						
Inner boot	NR	0.79	29.2	0.05	3.34	396
Cotton shoddy	Cott/polyester[c]	0.53	36.3	0.03	1.51	488
Brake Fluid Reservoir						
Reservoir	PP	3.52	88.3	1.17	3.35	1254
Windshield Wiper Tray						
Structure	SMC	0.61	25.5	0.03	2.26	345
Sound Reduction Fender Insulation						
High-density foam	PS	0.64	28.1	0.07	2.34	381
Hood Liner Insulation						
Face	PET	0.27	9.42	0.02	0.44	125

(*continued overleaf*)

Table 11.9 (continued)

Part Description	Polymers	$\dot{G}''_j$ (g/m²·s)				$\dot{Q}''_{ch}$ (kW/m²)
		CO	CO_2	HC[b]	Smoke	
Wheel Well Cover						
Fuel tank shield	PP	1.43	77	0.16	2.35	1078
HVAC Unit						
Cover	PP	0.78	54	0.08	1.73	755
Seal, foam	ABS–PVC	0.6	11.5	0.03	1.08	158
Fuel Tank						
Tank	PE	2.34	91.4	0.57	2.04	1296
Headlight						
Lens	PC	1.74	51.2	0.21	4.75	559
Battery Casing						
Cover	PP	1.64	71.4	0.23	2.4	1004
Bulkhead Insulation Engine Side						
Grommet, wire harness cap	HDPE	3.95	93.1	1.4	2.52	1341

[a] Combustion in normal air at 50 kW/m² of external heat flux in the ASTM E2058 apparatus. Data are taken from Ref. [41].
[b] HC, total hydrocarbons.
[c] Cott/polyester: mixture of cotton, polyester, and other fibers.

The data for the release rates of heat and products, such as listed in Tables 11.9 and 11.10, can be used to derive the fire properties of the polymers [21, 22]. The average values of y_j and ΔH_{ch} can be obtained as follows:

$$y_j = \frac{A \int_{t_0}^{t_n} \dot{G}''_j}{A \int_{t_0}^{t_n} \dot{m}''} \tag{11.9}$$

$$\Delta H_{ch} = \frac{A \int_{t_0}^{t_n} \dot{Q}''_{ch}}{A \int_{t_0}^{t_n} \dot{m}''} \tag{11.10}$$

where t_0 is vaporization or ignition time (s) and t_n is the time for the end of vaporization/combustion (s).

The yields of products and chemical heat of combustion derived from the release rates of products and heat from the fire propagation apparatus and the cone calorimeter are listed in Tables 11.11, 11.12, and 11.13, where data are

Table 11.10 Peak Heat Release Rate Measured in the ASTM E1354 Cone Calorimeter[a]

Polymers	Ref.	$\dot{Q}_{ch}''$ (peak) (kW/m²) $\dot{q}_e''$ (kW/m²)								$\Delta H_{ch}/\Delta H_g$ (kJ/kJ)
		20	25	30	40	50	70	75	100	
Ordinary Polymers										
High-density polyethylene (HDPE)	44	453	—	866	944	1133	—	—	—	21
Polyethylene (PE)	42	913	—	—	1408	—	2735	—	—	37
Polypropylene (PP)	44	377	—	693	1095	1304	—	—	—	32
Polypropylene (PP)	42	1170	—	—	1509	—	2421	—	—	25
PP/glass fibers (1082)	43	—	187	—	—	361	—	484	432	6
Polystyrene (PS)	42	723	—	—	1101	—	1555	—	—	17
Nylon	42	517	—	—	1313	—	2019	—	—	30
Nylon 6	44	593	—	802	863	1272	—	—	—	21
Nylon/glass fibers (1077)	43	—	67	—	—	96	—	116	135	1
Polyoxymethylene (POM)	42	290	—	—	360	—	566	—	—	6
Polymethylmethacrylate (PMMA)	42	409	—	—	665	—	988	—	—	12
Polybutylene terephthalate (PBT)	42	850	—	—	1313	—	1984	—	—	23
Acrylonitrile–butadiene–styrene (ABS)	44	683	—	947	994	1147	—	—	—	14
ABS	42	614	—	—	944	—	1311	—	—	12
ABS—FR	42	224	—	—	402	—	419	—	—	4
ABS–PVC	42	224	—	—	291	—	409	—	—	4
Vinyl thermoplastic elastomer	42	19	—	—	77	—	120	—	—	2

(continued overleaf)

Table 11.10 (continued)

Polymers	Ref.	$\dot{Q}''_{ch}$ (peak) (kW/m²) $\dot{q}''_e$ (kW/m²)									$\Delta H_{ch}/\Delta H_g$ (kJ/kJ)
		20	25	30	40	50	70	75	100	
Polyurethane (PU) foam	42	290	—	—	710	—	1221	—	—	19
EPDM/styrene acrylonitrile (SAN)	42	737	—	—	956	—	1215	—	—	10
Polyester/glass fibers (30%)	44	NI[b]	—	—	167	231	—	—	—	6
Isophthalic polyester	44	582	—	861	985	985	—	—	—	20
Isophthalic polyester/glass fibers (77%)	44	173	—	170	205	198	—	—	—	2
Polyvinyl ester	44	341	—	471	534	755	—	—	—	13
Polyvinyl ester/glass fibers (69%)	44	251	—	230	253	222	—	—	—	2
Polyvinyl ester/glass fibers (1031)	43	—	75	—	—	119	—	139	166	1
Polyvinyl ester/glass fibers (1087)	43	—	377	—	—	—	—	499	557	2
Epoxy	44	392	—	453	560	706	—	—	—	11
Epoxy/glass fibers (69%)	44	164	—	161	172	202	—	—	—	2
Epoxy/glass fibers(1003)	43	—	159	—	—	294	—	191	335	2
Epoxy/glass fibers(1006)	43	—	81	—	—	181	—	182	229	2
Epoxy/glass fibers (1040)	43	—	—	—	—	40	—	246	232	2
Epoxy/glass fibers (1066)	43	—	231	—	—	266	—	271	489	3
Epoxy/glass fibers (1067)	43	—	230	—	—	213	—	300	279	1
Epoxy/glass fibers (1070)	43	—	175	—	—	196	—	262	284	2
Epoxy/glass fibers (1071)	43	—	20	—	—	93	—	141	202	2
Epoxy/glass fibers (1089)	43	—	39	—	—	178	—	217	232	2
Epoxy/glass fibers (1090)	43	—	118	—	—	114	—	144	173	1
Epoxy/graphite fibers (1091)	43	—	NI	—	—	—	—	197	241	2
Epoxy/graphite fibers (1092)	43	—	164	—	—	189	—	242	242	2

Epoxy/graphite fibers (1093)	43	—	105	—	—	171	—	244	202	3
Cyanate ester/glass fibers (1046)	43	—	121	—	—	130	—	196	226	2
Kydex acrylic paneling, FR	42	117	—	—	176	—	242	—	—	3
High-Temperature Polymers										
Polycarbonate (PC-1)	42	16	—	—	429	—	342	—	—	21
PC-2	42	144	—	—	420	—	535	—	—	14
Crosslinked polyethylene (XLPE)	42	88	—	—	192	—	268	—	—	5
Polyphenylene oxide (PPO), polystyrene (PS)	42	219	—	—	265	—	301	—	—	2
PPO/glass fibers	42	154	—	—	276	—	386	—	—	6
Polyphenylenesulfide (PPS)/glass fibers (1069)	43	—	NI	—	—	52	—	71	183	3
PPS/graphite fibers (1083)	43	—	NI	—	—	—	—	60	80	2
PPS/glass fibers (1084)	43	—	NI	—	—	48	—	88	150	2
PPS/graphite fibers (1085)	43	—	NI	—	—	94	—	66	126	1
Polyarylsulfone/graphite fibers (1081)	43	—	NI	—	—	24	—	47	60	1
Polyethersulfone/graphite fibers (1078)	43	—	NI	—	—	11	—	41	65	0.3
Polyetheretherketone (PEEK)/glass fibers (30%)	44	NI	—	—	35	109	—	—	—	7
PEEK/graphite fibers (1086)	43	—	—	—	—	14	—	54	85	1
Polyetherketoneketone (PEKK)/glass fibers (1079)	43	—	NI	—	—	21	—	45	74	1
Bismaleimide (BMI)/graphite fibers (1095)	43	—	160	—	—	—	—	213	270	1
Bismaleimide (BMI)/graphite fibers (1096)	43	—	128	—	—	176	—	245	285	2

(continued overleaf)

Table 11.10 (continued)

Polymers	Ref.	$\dot{Q}''_{ch}$ (peak) (kW/m²) $\dot{q}''_e$ (kW/m²)									$\Delta H_{ch}/\Delta H_g$ (kJ/kJ)
		20	25	30	40	50	70	75	100	
Bismaleimide (BMI)/graphite fibers (1097)	43	—	NI	—	—	—	—	172	168	(1)
Bismaleimide (BMI)/graphite fibers (1098)	43	—	NI	—	—	74	—	91	146	1
Phenolic/glass fibers (45%)	44	—	—	423	214	165	—	—	—	1
Phenolic/glass fibers (1099)	43	—	NI	—	—	66	—	102	122	1
Phenolic/glass fibers (1100)	43	—	NI	—	—	66	—	120	163	2
Phenolic/glass fibers (1101)	43	—	NI	—	—	47	—	57	96	1
Phenolic/glass fibers (1014)	43	—	NI	—	—	81	—	97	133	1
Phenolic/glass fibers (1015)	43	—	NI	—	—	82	—	76	80	(1)
Phenolic/glass fibers (1017)	43	—	NI	—	—	190	—	115	141	1
Phenolic/glass fibers (1018)	43	—	NI	—	—	132	—	56	68	1
Phenolic/graphite fibers (1102)	43	—	NI	—	—	—	—	159	196	2
Phenolic/graphite fibers (1103)	43	—	NI	—	—	177	—	183	189	0.2
Phenolic/graphite fibers (1104)	43	—	NI	—	—	71	—	87	101	1
Phenolic/PE fibers (1073)	43	—	NI	—	—	98	—	141	234	3
Phenolic/aramid fibers (1074)	43	—	NI	—	—	51	—	93	104	1
Phenolic insulating foam	44	—	—	—	17	19	—	29	—	1
Polyimide/glass fibers (1105)	43	—	NI	—	—	40	—	78	85	1

Wood									
Douglas fir	42	237	—	221	—	196	—	—	(—)
Hemlock	44	233	—	218	236	243	—	—	(—)
Textiles									
Wool	44	212	—	261	307	286	—	—	5
Acrylic fibers	44	300	—	358	346	343	—	—	6
Halogenated Polymers									
PVC flexible-3 (LOI 25%)	44	126	—	148	240	250	—	—	5
PVC-FR (Sb$_2$O$_3$) flexible—4 (LOI 30%)	44	89	—	137	189	185	—	—	5
PVC-FR (triaryl phosphate) flexible—5 (LOI 34%)	44	96	—	150	185	176	—	—	5
PVC rigid-1	42	40	—	—	175	—	191	—	3
PVC rigid-2	42	75	—	—	111	—	126	—	2
PVC rigid-3	42	102	—	—	183	—	190	—	2
PVC rigid-1 (LOI 50%)	44	NI	—	90	107	155	—	—	3
PVC rigid-2	44	NI	—	101	137	157	—	—	3
Chlorinated PVC (CPVC)	42	25	—	—	84	—	93	—	1

[a]Data taken from Refs. [19, 42–44].
[b]No ignition.

Table 11.11 Average Yields of Products and Heat of Combustion for Polymers from Parts of a Minivan from the Data Measured in the ASTM E2058 Fire Propagation Apparatus[a]

Part Description	Polymers	y_j (g/g)				ΔH_{ch} (kJ/g)
		CO	CO_2	HC[b]	Smoke	
Head Liner						
Fabric, exposed	Nylon 6	0.086	2.09	0.001	0.045	28.8
Instrument Panel						
Cover	PVC	0.057	1.72	0.005	0.109	24.4
Shelf, main panel	PC	0.051	1.86	0.002	0.105	20.2
Resonator						
Structure	PP	0.041	2.46	0.002	0.072	34.6
Intake tube	EPDM	0.045	2.51	0.001	0.100	33.8
Kick Panel Insulation						
Backing	PVC	0.061	1.26	0.006	0.070	17.4
Air Ducts						
Large ducts	PP	0.056	2.52	0.004	0.080	35.5
Steering Column Boot						
Inner boot	NR	0.061	1.87	0.003	0.130	25.6
Cotton shoddy	Cott/Polyester[c]	0.039	2.17	0.002	0.087	29.4
Brake Fluid Reservoir						
Reservoir	PP	0.058	2.41	0.011	0.072	33.9
Windshield Wiper Tray						
Structure	SMC	0.061	1.86	0.003	0.100	25.5
Sound Reduction Fender Insulation						
High-density foam	PS	0.064	1.8	0.002	0.098	24.6
Hood Liner Insulation						
Face	PET	0.041	1.47	0.003	0.022	20

Table 11.11 (continued)

Part Description	Polymers	y_j (g/g) CO	CO_2	HC^b	Smoke	ΔH_{ch} (kJ/g)
Wheel Well Cover						
Fuel tank shield	PP	0.054	2.45	0.002	0.065	34.5
HVAC Unit						
Cover	PP	0.057	2.49	0.002	0.060	35
Seal, foam	ABS–PVC	0.089	1.62	0.001	0.060	22.6
Fuel Tank						
Tank	PE	0.032	2.33	0.005	0.042	32.7
Headlight						
Lens	PC	0.049	1.67	0.004	0.113	18.2
Battery Casing						
cover	PP	0.045	2.59	0.004	0.071	36.2
Bulkhead Insulation Engine Side						
Grommet, wire harness cap	HDPF	0.064	2.67	0.012	0.058	38.2

[a] Combustion in normal air at 50 kW/m² of external heat flux in the ASTM E2058 apparatus. Data are taken from Ref. [41].
[b] HC, total hydrocarbons.
[c] Cott/polyester: mixture of cotton, polyester, and other fibers.

taken from Refs. [19, 21, 22, 40–44]. Table 11.11 contains the data from the fire propagation apparatus for the polymers from parts of a minivan taken from Ref. [41]. Table 11.12 contains the data from the fire propagation apparatus for the ordinary, high-temperature, and halogenated polymers taken from Refs. [21, 22, 40]. Table 11.13 lists the data from the cone calorimeter for the ordinary, high-temperature, and halogenated polymers taken from Refs. [19, 42–44]. The yield of smoke in Table 11.13 was calculated from the average extinction area at 5 min into the test using the following relationship [21]:

$$y_s(g/g) = 0.0994 \times (\text{average extinction area}) \times 10^{-3} \quad (11.11)$$

In fires, polymers can burn as a solid with formation and burning of very small bubbles of polymer melt at the surface or with char formation at the

Table 11.12 Average Heat of Combustion and Yields of Products for Polymers from the Data Measured in the ASTM E2058 Fire Propagation Apparatus[a]

Polymer	Composition	y_j (g/g) CO	CO_2	HC[b]	Smoke	ΔH_{ch} (kJ/g)
Ordinary Polymers						
Polyethylene (PE)	CH_2	0.024	2.76	0.007	0.060	38.4
Polypropylene (PP)	CH_2	0.024	2.79	0.006	0.059	38.6
Polystyrene (PS)	CH	0.060	2.33	0.014	0.164	27.0
Polystyrene foam	$CH_{1.1}$	0.061	2.32	0.015	0.194	25.5
Wood	$CH_{1.7}O_{0.73}$	0.004	1.30	0.001	0.015	12.6
Polyoxymethylene (POM)	$CH_{2.0}O$	0.001	1.40	0.001	0.001	14.4
Polymethylmethacrylate (PMMA)	$CH_{1.6}O_{0.40}$	0.010	2.12	0.001	0.022	24.2
Polyester	$CH_{1.4}O_{0.22}$	0.075	1.61	0.025	0.188	20.1
Nylon	$CH_{1.8}O_{0.17}N_{0.17}$	0.038	2.06	0.016	0.075	27.1
Flexible polyurethane foams	$CH_{1.8}O_{0.32}N_{0.06}$	0.028	1.53	0.004	0.070	17.6
Rigid polyurethane foams	$CH_{1.1}O_{0.21}N_{0.10}$	0.036	1.43	0.003	0.118	16.4
High-Temperature Polymers						
Polyetheretherketone (PEEK)	$CH_{0.63}O_{0.16}$	0.029	1.60	0.001	0.008	17.0
Polysulfone (PSO)	$CH_{0.81}O_{0.15}S_{0.04}$	0.034	1.80	0.001	0.020	20.0
Polyethersulfone (PES)		0.040	1.50	0.001	0.021	16.7
Polyetherimide (PEI)	$CH_{0.65}O_{0.16}N_{0.05}$	0.026	2.00	0.001	0.014	20.7
Polycarbonate (PC)	$CH_{0.88}O_{0.19}$	0.054	1.50	0.001	0.112	16.7
Halogenated Polymers						
PE +25% Cl	$CH_{1.9}Cl_{0.13}$	0.042	1.71	0.016	0.115	22.6
PE +36% Cl	$CH_{1.8}Cl_{0.22}$	0.051	0.83	0.017	0.139	10.6
PE +48% Cl	$CH_{1.7}Cl_{0.36}$	0.049	0.59	0.015	0.134	5.7
Polyvinyl chloride (PVC)	$CH_{1.5}Cl_{0.50}$	0.063	0.46	0.023	0.172	7.7
Chlorinated PVC	$CH_{1.3}Cl_{0.70}$	0.052	0.48	0.001	0.043	6.0
Polyvinylidene fluoride (PVDF)	CHF	0.055	0.53	0.001	0.037	5.4
Polyethylene–tetrifluoroethylene (ETFE)	CHF	0.035	0.78	0.001	0.028	7.3
Polyethylene–chloro-trifluoroethylene (ECTFE)	$CHCl_{0.25}F_{0.75}$	0.095	0.41	0.001	0.038	4.5
Polytetrafluoroethylene (TFE)	CF_2	0.092	0.38	0.001	0.003	2.8
Polyfluoroalkoxy (PFA)	$CF_{1.6}$	0.099	0.42	0.001	0.002	1.8
Polyfluorinated ethylene propylene (FEP)	$CF_{1.8}$	0.116	0.25	0.001	0.003	1.0

[a] Data taken from Refs. [21, 22, 40].
[b] HC, total gaseous hydrocarbon.

Table 11.13 Average Effective (Chemical) Heat of Combustion and Smoke Yield Calculated from the Data Measured in the ASTM E1354 Cone Calorimeter[a]

Polymers	Ref.	ΔH_{ch} (MJ/kg)	y_{sm} (g/g)
Ordinary Polymers			
High-density polyethylene (HDPE)	44	40.0	0.035
Polyethylene (PE)	42	43.4	0.027
Polypropylene (PP)	44	44.0	0.046
Polypropylene (PP)	42	42.6	0.043
PP/glass fibers (1082)	43	NR[b]	0.105
Polystyrene (PS)	42	35.8	0.085
PS—FR	42	13.8	0.144
PS foam	44	27.7	0.128
PS foam—FR	44	26.7	0.136
Nylon	42	27.9	0.025
Nylon 6	44	28.8	0.011
Nylon/glass fibers (1077)	43	NR	0.089
Polyoxymethylene (POM)	42	13.4	0.002
Polymethylmethacrylate (PMMA)	42	24.2	0.010
Polybutylene terephthalate (PBT)	42	20.9	0.066
Polyethylene terephthalate (PET)	42	14.3	0.050
Acrylonitrile–butadiene–styrene (ABS)	44	30.0	0.105
ABS	42	29.4	0.066
ABSa–FR	42	11.7	0.132
ABS–PVC	42	17.6	0.124
Vinyl thermoplastic elastomer	42	6.4	0.056
Polyurethane (PU) foam	42	18.4	0.054
Thermoplastic PU-FR	42	19.6	0.068
EPDM/Styrene acrylonitrile (SAN)	42	29.0	0.116
Polyester/glass fibers (30%)	44	16.0	0.049
Isophthalic polyester	44	23.3	0.080
Isophthalic polyester/glass fibers (77%)	44	27.0	0.032
Polyvinyl ester	44	22.0	0.076
Polyvinyl ester/glass fibers (69%)	44	26.0	0.079
Polyvinyl ester/glass fibers (1031)	43	NR	0.164
Polyvinyl ester/glass fibers (1087)	43	NR	0.128
Epoxy	44	25.0	0.106
Epoxy/glass fibers (69%)	44	27.5	0.056
Epoxy/glass fibers (1003)	43	NR	0.142
Epoxy/glass fibers (1006)	43	NR	0.207
Epoxy/glass fibers (1040)	43	NR	0.058
Epoxy/glass fibers (1066)	43	NR	0.113
Epoxy/glass fibers (1067)	43	NR	0.115

(*continued overleaf*)

Table 11.13 (*continued*)

Polymers	Ref.	ΔH_{ch} (MJ/kg)	y_{sm} (g/g)
Epoxy/glass fibers (1070)	43	NR	0.143
Epoxy/glass fibers (1071)	43	NR	0.149
Epoxy/glass fibers (1089)	43	NR	0.058
Epoxy/glass fibers (1090)	43	NR	0.086
Epoxy/graphite fibers (1091)	43	NR	0.082
Epoxy/graphite fibers (1092)	43	NR	0.049
Cyanate ester/glass fibers (1046)	43	NR	0.103
Acrylic/glass fibers	44	17.5	0.016
Kydex acrylic paneling, FR	42	10.2	0.095
High-Temperature Polymers and Composites			
Polycarbonate (PC)-1	42	21.9	0.098
PC-2	42	22.6	0.087
Crosslinked polyethylene (XLPE)	42	23.8	0.026
Polyphenylene oxide (PPO) polystyrene (PS)	42	23.1	0.162
PPO/glass fibers	42	25.4	0.133
Polyphenylenesulfide (PPS)/glass fibers (1069)	43	NR	0.063
PPS/graphite fibers (1083)	43	NR	0.075
PPS/glass fibers (1084)	43	NR	0.075
PPS/graphite fibers (1085)	43	NR	0.058
Polyarylsulfone/graphite fibers (1081)	43	NR	0.019
Polyethersulfone/graphite fibers (1078)	43	NR	0.014
Polyetheretherketone (PEEK)/glass fibers (30%)	44	20.5	0.042
PEEK/graphite fibers (1086)	43	NR	0.025
Polyetherketoneketone (PEKK)/glass fibers (1079)	43	NR	0.058
Bismaleimide (BMI)/graphite fibers (1095)	43	NR	0.077
Bismaleimide (BMI)/graphite fibers (1096)	43	NR	0.096
Bismaleimide (BMI)/graphite fibers (1097)	43	NR	0.095
Bismaleimide (BMI)/graphite fibers (1098)	43	NR	0.033
Phenolic/glass fibers (45%)	44	22.0	0.026
Phenolic/glass fibers (1099)	43	NR	0.008
Phenolic/glass fibers (1100)	43	NR	0.037
Phenolic/glass fibers (1101)	43	NR	0.032
Phenolic/glass fibers (1014)	43	NR	0.031
Phenolic/glass fibers (1015)	43	NR	0.031
Phenolic/glass fibers (1017)	43	NR	0.015
Phenolic/glass fibers (1018)	43	NR	0.009
Phenolic/graphite fibers (1102)	43	NR	0.039
Phenolic/graphite fibers (1103)	43	NR	0.041
Phenolic/graphite fibers (1104)	43	NR	0.021
Phenolic/PE fibers (1073)	43	NR	0.054

Table 11.13 (*continued*)

Polymers	Ref.	ΔH_{ch} (MJ/kg)	y_{sm} (g/g)
Phenolic/aramid fibers (1074)	43	NR	0.024
Phenolic insulating foam	44	10.0	0.026
Polyimide/glass fibers (1105)	43	NR	0.014
Wood			
Douglas fir	42	14.7	0.010
Hemlock	44	13.3	0.015
Textiles			
Wool	44	19.5	0.017
Acrylic fiber	44	27.5	0.038
Halogenated Polymers			
PVC flexible-3 (LOI 25%)	44	11.3	0.099
PVC-FR (Sb$_2$O$_3$) flexible-4 (LOI 30%)	44	10.3	0.078
PVC-FR (triaryl phosphate) flexible-5 (LOI 34%)	44	10.8	0.098
PVC rigid-1	42	8.9	0.103
PVC rigid-2	42	10.8	0.112
PVC rigid-3	42	12.7	0.103
PVC rigid-1 (LOI 50%)	44	7.7	0.098
PVC rigid-2	44	8.3	0.076
Chlorinated PVC (CPVC)	42	5.8	0.003

^{a}Data taken from Refs. [19, 42–44].
^{b}Not reported.

surface or as boiling liquid pools. Some of these behaviors are illustrated in Figure 11.1 for the combustion of PP[16] slab (100 mm in diameter and 25 mm in thickness) in normal air and 50 kW/m^2 of external heat flux (presented in a previous section).

Between about 400 and 900 s in Figure 11.1, PP was burning at the steady state as a solid polymer with formation and burning away of very small polymer melt bubbles at the surface with no accumulation of the polymer melt at the surface. The flames were completely lifted off the surface due to the high release rate of polymer vapors resulting in $\dot{q}_f'' \approx 0$. From Eq. (11.8) and properties of PP,[16] the predicted steady-state $\dot{Q}_{ch}''$ value for the combustion of solid PP is 780 kW/m^2, which agrees well with the experimental value.

Beyond 1150 s in Figure 11.1, a deep PP melt layer had accumulated at the surface and was burning as a boiling liquid pool fire with the heat release rate

---

[16] For PP, $\Delta H_{ch} = 38.6$ kJ/g; $\Delta H_g = 2.0$ kJ/g, $\dot{q}_{rr}'' = 15$ kW/m^2 [21].

Table 11.14 Composition, Molecular Weight, and Combustion Properties of Hydrocarbons and Alcohols[a]

Fluid	Composition	M (g/mole)	T_{ig} (°C)	T_b (°C)	ΔH_{ch} (kJ/g)	ΔH_g (kJ/g)	HRP (kJ/kJ)	y_{co} (g/g)	y_{sm} (g/g)
Gasoline	NA[b]	NA	371	33	41.0	0.482	85	0.010	0.038
Hexane	C_6H_{14}	86	225	69	41.5	0.500	83	0.009	0.035
Heptane	C_7H_{16}	100	204	98	41.2	0.549	75	0.010	0.037
Octane	C_8H_{18}	114	206	125	41.0	0.603	68	0.010	0.038
Nonane	C_9H_{20}	128	205	151	40.8	0.638	64	0.011	0.039
Decane	$C_{10}H_{22}$	142	201	174	40.7	0.690	59	0.011	0.040
Undecane	$C_{11}H_{24}$	156	NA	196	40.5	0.736	55	0.011	0.040
Dodecane	$C_{12}H_{26}$	170	203	216	40.4	0.777	52	0.012	0.041
Tridecane	$C_{13}H_{28}$	184	NA	—	40.3	0.806	50	0.012	0.041
Kerosine	$C_{14}H_{30}$	198	260	232	40.3	0.857	47	0.012	0.042
Hexadecane	$C_{16}H_{34}$	226	202	287	40.1	0.911	44	0.012	0.042
Mineral oil	NA	466	NA	360	NA	—	72	NA	NA
Motor oil	NA	NA	NA	NA	29.3	0.473	62	NA	NA
Corn Oil	NA	NA	NA	393	NA	22.3	0.413	54	NA	NA
Benzene	C_6H_6	78	498	80	27.6	0.368	75	0.067	0.181
Toluene	C_7H_8	92	480	110	27.7	0.338	82	0.066	0.178
Xylene	C_8H_{10}	106	528	139	27.8	0.415	67	0.065	0.177
Methanol	CH_4O	32	385	64	19.1	1.005	19	0.001	0.001
Ethanol	C_2H_6O	46	363	78	25.6	0.776	33	0.001	0.008
Propanol	C_3H_8O	60	432	97	29.0	0.630	46	0.003	0.015
Butanol	$C_4H_{10}O$	74	343	117	31.2	0.538	58	0.004	0.019

[a]Data are taken from Ref. [54].
[b]NA, not available.

increasing rapidly. For the burning of polymer melt, the ΔH_g value would be different than the value for the solid PP, probably closer to the value for the higher molecular weight liquid hydrocarbons listed in Table 11.14 taken from Ref. [54] (about 0.690–0.911 kJ/g). The ΔH_{ch} value of the PP melt, however, is not expected to change, as it is independent of the physical state of the fuel as can be noted in Table 11.14. Thus the predicted $\dot{Q}''_{ch}$ value for steady-state combustion of PP melt with $\dot{q}''_f \approx 0$ from Eq. (11.8) is in the range of 1500–2000 kW/m², compared to the experimental $\dot{Q}''_{ch}$ value of 2000 kW/m² just before all PP is consumed. This is consistent with the observations from large-scale fire tests of selected parts from a minivan, discussed earlier where it was found that a significant aspect of the burning behavior of the polymers was the development of a polymer melt pool fire below the part [32].

In Table 11.9, the heat release rates from minivan parts made of PP and PE, which burned as liquid pool fires in the tests, are in the range of 1004–1341 kW/m². This range is in the lower end of the 1500–2000-kW/m² range predicted before for the liquid pool fires of PP and similar to the behavior noted in the large-scale fires of the parts reported earlier [32]. Other polymer parts in Table 11.9 made of charring type of polymers burned as a solid with much reduced heat release rates within the predicted range for the combustion of solid polymers.

Data in Table 11.10 from the Cone Calorimeter indicate that the burning behaviors of polymers are similar to those indicated by the data in Table 11.9 from the Fire Propagation Apparatus. Ordinary polymers (which are thermoplastics and melt easily) have very high heat release rates in the range predicted for the liquid pool fires. For example, for the boiling liquid pool fires of PE, PP, nylon 6, and ABS, $\dot{Q}''_{ch}$ values at 50 kW/m^2 are in the range of 1133–1304 kW/m^2 from the Cone Calorimeter (Table 11.10) and 1004–1341 kW/m^2 from the Fire Propagation Apparatus (Table 11.9).

The heat release rates for thermoplastics with glass fibers and charring-type thermoplastics, high-temperature polymers, and halogenated polymers from the Cone Calorimeter (Table 11.10) are in the range predicted for the burning of solid polymers and similar to those from the Fire Propagation Apparatus (Table 11.9).

The release rates of heat and products are interrelated and can be expressed as [from Eqs. 11.7 and 11.8)]:

$$\dot{Q}''_{ch}/\dot{G}''_j = \Delta H_{ch}/y_j = (\chi/f_j)(\Delta H_T/\Psi_j) \tag{11.12}$$

$$\dot{G}''_{j,1}/\dot{G}''_{j,2} = y_{j,1}/y_{j,2} = (f_{j,1}/f_{j,2})(\Psi_{j,1}/\Psi_{j,2}) \tag{11.13}$$

The f_j and χ values depend on the generic nature and molecular weight (M) of the polymers, air-to-fuel mass stoichiometric ratio (s), equivalence ratio (Φ), and experimental conditions. The values of Ψ_j and ΔH_T on the other hand depend on the generic nature of atoms attached to the carbon atom polymer backbone and are independent of the experimental conditions.

The relative release rates of products and heat can be calculated from Eqs. (11.12) and (11.13) using data for M, s, ΔH_T and Ψ_j, such as listed in Tables 11.15, 11.16, and 11.17 for ordinary, high-temperature, and halogenated polymers, respectively, and generalized relationships for f_j and χ at different Φ values [21, 53]. Alternately, χ and f_j values can be derived from Ψ_j, ΔH_g, and ΔH_T values determined independently from the chemical composition measurement techniques, differential scanning calorimetry and oxygen bomb calorimetry, respectively, and experimental values of $y_j/\Delta H_g$ and $\Delta H_{ch}/\Delta H_g$. The experimental values are determined from the slopes of the relationships between the measured release rates of products and heat versus external heat flux [Eqs. (11.7) and (11.8)].

The fire properties are useful for fire hazard analysis as well as for understanding the combustion behavior of polymers. For example, yield of CO_2 (y_{CO_2}) and ΔH_{ch} are associated with the completeness of combustion, whereas yields of CO and smoke (y_{CO_2} and y_{sm}) are associated with the incompleteness of combustion. Relationships between these parameters are shown in Figures 11.7–11.9, using the data measured under well-ventilated conditions in our laboratory in ASTM E2058 apparatus [21, 22, 40, 41].

Table 11.15 Composition, Molecular Weight, Stoichiometric Air-to-Fuel Ratio, Net Heat of Complete Combustion, and Maximum Possible Stoichiometric Yields of Major Products for Ordinary Polymers

Polymers	Composition	M (g/mole)	s (g/g)	ΔH_T (kJ/g)	Ψ_j (g/g) CO_2	CO	HC[a]	Smoke
Polyethylene (PE)	CH_2	14.0	14.7	43.6	3.14	2.00	1.14	0.86
Polypropylene (PP)	CH_2	14.0	14.7	43.4	3.14	2.00	1.14	0.86
Polystyrene (PS)	CH	13.0	13.2	39.2	3.38	2.15	1.23	0.92
Polystyrene foam	$CH_{1.1}$	13.1	13.4	39.2	3.36	2.14	1.22	0.92
Wood	$CH_{1.7}O_{0.73}$	25.4	5.8	16.9	1.73	1.11	0.63	0.48
Polyoxymethylene (POM)	$CH_{2.0}O$	30.0	4.6	15.4	1.47	0.93	0.53	0.40
Polymethylmethacrylate (PMMA)	$CH_{1.6}O_{0.40}$	20.0	8.2	25.2	2.20	1.40	0.80	0.60
Polyester	$CH_{1.4}O_{0.22}$	16.9	10.1	32.5	2.60	1.65	0.95	0.71
Polyvinyl alcohol	$CH_{2.0}O_{0.50}$	22.0	7.8	21.3	2.00	1.27	0.73	0.55
Polyethylene terephthalate (PET)	$CH_{0.80}O_{0.40}$	19.2	7.2	23.2	2.29	1.46	0.83	0.63
Polyacrylonitrile–butadiene–styrene (ABS)	$CH_{1.1}N_{0.07}$	14.1	13.2	38.1	3.13	1.99	1.14	0.85
Nylon	$CH_{1.8}O_{0.17}N_{0.17}$	18.9	11.2	30.8	2.33	1.48	0.85	0.63
Flexible polyurethane foam	$CH_{1.8}O_{0.32}N_{0.06}$	19.7	9.4	25.3	2.24	1.42	0.81	0.61
Rigid polyurethane foam	$CH_{1.1}O_{0.21}N_{0.10}$	17.8	9.8	25.9	2.47	1.57	0.90	0.67

[a] HC, hydrocarbons.

Figure 11.7 shows that y_{CO_2} values decrease and y_{CO} values increase with changes in the nature of the polymer from ordinary to high temperature to highly halogenated. A similar effect is observed for each generic polymer as combustion conditions change from well ventilated to ventilation controlled (increase in the Φ value).

Figure 11.8 shows a linear relationship between ΔH_{ch} and y_{CO_2}. The slope of the line in Figure 11.8 is the chemical heat of combustion per unit mass of CO_2 released, which is approximately constant and is independent of the generic nature of the polymers. Figure 11.9 shows that there is a linear relationship between y_{CO} and y_{sm} for ordinary polymers and chlorinated ordinary polymers (PE with 25, 36, and 48% chlorine and PVC containing 56% chlorine). For most of the high-temperature polymers and halogenated polymers (halogen atom >56%), there is no relationship between y_{CO_2} and y_{sm}. The reduced y_{sm} values are probably due to char formation and reduced ratio of H atoms relative and C and halogen atoms.

All the properties in this section, individually or in combination with other properties, provide the understanding of the flammability behaviors of polymers and assessment of various hazards in fires.

Table 11.16 Composition, Molecular Weight, Stoichiometric Air-to-Fuel Ratio, Net Heat of Complete Combustion, and Maximum Possible Stoichiometric Yields of Major Products for High-Temperature Polymers

Polymers	Composition	M (g/mole)	s (g/g)	ΔH_T (kJ/g)	Ψ_j (g/g)						
					CO_2	CO	HC	Smoke	HCN	NO_2	SO_2
Polyetherketoneketone (PEKK)	$CH_{0.60}O_{0.15}$	15.0	9.8	30.3	2.93	1.87	1.07	0.80	0.00	0.00	0.00
Polyetherketone (PEK)	$CH_{0.62}O_{0.15}$	15.0	9.9	30.2	2.93	1.86	1.07	0.80	0.00	0.00	0.00
Polyetheretherketone (PEEK)	$CH_{0.63}O_{0.16}$	15.2	9.7	30.4	2.90	1.84	1.05	0.79	0.00	0.00	0.00
Polyamideimide (PAI)	$CH_{0.53}O_{0.20}$	15.7	9.0	24.2	2.80	1.78	1.02	0.76	0.00	0.00	0.00
Polycarbonate (PC)	$CH_{0.88}O_{0.19}$	15.9	9.7	30.1	2.76	1.76	1.01	0.75	0.00	0.00	0.00
Polyethylenenaphthalate (PEN)	$CH_{0.71}O_{0.29}$	17.4	8.2	—	2.54	1.61	0.92	0.69	0.00	0.00	0.00
Polyphenylene oxide (PPO)	$CHO_{0.13}$	15.0	10.9	—	2.93	1.87	1.07	0.80	0.00	0.00	0.00
Polybutanediol terephthalate (PBT)	$CHO_{0.33}$	18.3	8.2	—	2.41	1.53	0.88	0.66	0.00	0.00	0.00
Polybenzoylphenylene	$CH_{0.62}O_{0.08}$	13.9	11.1	—	3.18	2.02	1.16	0.87	0.00	0.00	0.00
Phenol-formaldehyde	$CHO_{0.14}$	15.2	10.6	—	2.89	1.84	1.05	0.79	0.00	0.00	0.00
Polybenzimidazole (PBI)	$CH_{0.60}N_{0.20}$	15.4	12.0	30.8	2.86	1.82	1.04	0.78	0.35	0.60	0.00
Polyphenylenesulfide (PPS)	$CH_{0.67}S_{0.17}$	18.0	10.2	28.2	2.44	1.55	0.89	0.67	0.00	0.00	0.59
Polyphenylenebenzobisoxazole (PBO)	$CH_{0.43}O_{0.14}N_{0.14}$	16.6	9.7	—	2.65	1.68	0.96	0.72	0.23	0.39	0.00
Polyimide (PI)	$CH_{0.45}O_{0.23}N_{0.09}$	17.4	8.6	25.5	2.53	1.61	0.92	0.69	0.14	0.24	0.00
Polyetherimide (PEI)	$CH_{0.65}O_{0.16}N_{0.05}$	16.0	9.8	28.4	2.76	1.75	1.00	0.75	0.09	0.16	0.00
Polyaramide	$CH_{0.71}O_{0.14}N_{0.14}$	16.9	10.1	—	2.60	1.66	0.95	0.71	0.22	0.38	0.00
Polyphenylenebenzobisoxazole	$CH_{0.71}O_{0.14}N_{0.14}$	16.9	10.1	—	2.60	1.66	0.95	0.71	0.22	0.38	0.00
Aramid-arylester copolymer	$CH_{0.71}O_{0.14}N_{0.14}$	16.9	10.1	24.4	2.60	1.66	0.95	0.71	0.22	0.38	0.00
Polysulfone (PSF)	$CH_{0.81}O_{0.15}S_{0.04}$	16.4	9.8	29.4	2.68	1.71	0.98	0.73	0.00	0.00	0.14
Polyphenyleneethersulfone (PES)	$CH_{0.67}O_{0.25}S_{0.08}$	19.3	8.0	24.7	2.28	1.45	0.83	0.62	0.00	0.00	0.27

Table 11.17 Composition, Molecular Weight, Stoichiometric Air-to-Fuel Ratio, Net Heat of Complete Combustion, and Maximum Possible Stoichiometric Yields of Major Products for Halogenated Polymers

Polymers	Composition	M (g/mole)	s (g/g)	ΔH_T (kJ/g)	Ψ_j (g/g)					
					CO_2	CO	HC	Smoke	HCl	HF
Carbon–Hydrogen–Chlorine Atoms										
PE + 25% Cl	$CH_{1.9}Cl_{0.13}$	18.5	10.7	31.6	2.38	1.52	0.87	0.65	0.25	0.00
PE + 36% Cl	$CH_{1.8}Cl_{0.22}$	21.5	8.9	26.3	2.05	1.30	0.74	0.56	0.37	0.00
Polychloropropene	$CH_{1.3}Cl_{0.30}$	23.8	7.2	25.3	1.85	1.18	0.56	0.50	0.45	0.00
PE + 48% Cl	$CH_{1.7}Cl_{0.36}$	26.3	7.0	20.6	1.67	1.06	0.61	0.46	0.49	0.00
Polyvinyl chloride (PVC)	$CH_{1.5}Cl_{0.50}$	31.0	5.5	16.4	1.42	0.90	0.52	0.39	0.58	0.00
Chlorinated PVC	$CH_{1.3}Cl_{0.70}$	37.8	4.2	13.3	1.16	0.74	0.42	0.32	0.67	0.00
Polyvinylidenechloride ($PVCl_2$)	$CHCl$	48.0	2.9	9.6	0.92	0.58	0.33	0.25	0.75	0.00
Carbon–Hydrogen–Fluorine Atoms										
Polyvinyl fluoride	$CH_{1.5}F_{0.50}$	23.0	7.5	13.5	1.91	1.22	0.70	0.52	0.00	0.43
Polyvinylidene fluoride (PVDF)	CHF	32.0	4.3	13.3	1.38	0.88	0.50	0.38	0.00	0.63
Polyethylene–trifluoroethylene (ETFE)	CHF	32.0	4.3	12.6	1.38	0.88	0.50	0.38	0.00	0.63
Carbon–Fluorine Atoms										
Polytetrafluoroethylene (TFE)	CF_2	50.0	2.7	6.2	0.88	0.56	0.32	0.24	0.00	0.00
Polyperfluoroalkoxy (PFA)	$CF_{1.6}$	42.6	3.2	5.0	1.03	0.66	0.38	0.28	0.00	0.00
Polyfluorinatedethylenepropylene (FEP)	$CF_{1.8}$	46.2	3.0	4.8	0.95	0.61	0.35	0.26	0.00	0.00
Carbon–Hydrogen–Chlorine–Fluorine Atoms										
Polyethylenechloro–trifluoroethylene (ECTFE)	$CHCl_{0.25}F_{0.75}$	36.0	3.8	12.0	1.22	0.78	0.44	0.33	0.25	0.42
Polychlorotrifluoroethylene (CTFE)	$CCl_{0.50}F_{1.5}$	58.0	2.4	6.5	0.76	0.48	0.28	0.21	0	0

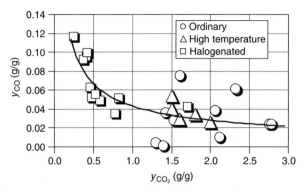

Figure 11.7. Relationship between the yields of CO and CO_2 for the combustion of polymers under well-ventilated conditions. Relationship for polyethylene with 25, 36, and 48% chlorine and PVC is similar to that for the ordinary nonhalogenated polymers.

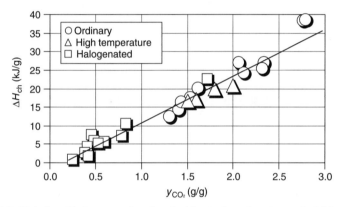

Figure 11.8. Relationship between the chemical heat of combustion and yield of CO_2 for the combustion of polymers under well-ventilated conditions.

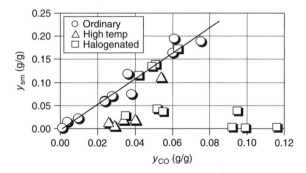

Figure 11.9. Relationship between the yields of smoke and CO for combustion of polymers under well-ventilated conditions. Relationship for polyethylene with 25, 36, and 48% chlorine and PVC is similar to that for the ordinary nonhalogenated polymers.

11.7.3. Heat Flux

The magnitude of $\dot{q}_f''$ and $\dot{q}_{rr}''$ depends on the chemical nature of the polymer, fire size, surface area of the polymer burning, and the environmental conditions [14, 15, 21, 22, 55–58]. The influence of $\dot{q}_f''$ and $\dot{q}_{rr}''$ on the magnitude of the release rates of products and heat is enhanced by hot walls, ceiling, and nearby burning objects [i.e., $\dot{q}_e''$ value in Eqs. (11.7) and (11.8)].

The $\dot{q}_f''$ values have been measured in small and large-scale fires [55–57]. The $\dot{q}_f''$ values can also be determined indirectly by using the measured steady-state release rate of polymer vapors in the combustion of solid polymers along with the known values of ΔH_g and $\dot{q}_{rr}''$ for the polymers in Eq. (11.1) [21, 22, 55–61]. Examples are shown in Figures 11.10 and 11.11. In Figure 11.10, $\dot{q}_f''$ values for

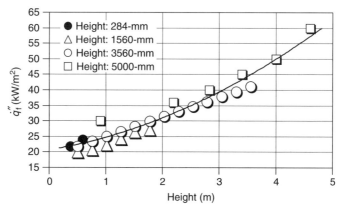

Figure 11.10. Flame heat flux transferred back to the polymer surface in combustion of vertical slabs of polymethylmethacrylate. Data are taken from Refs. [58–61].

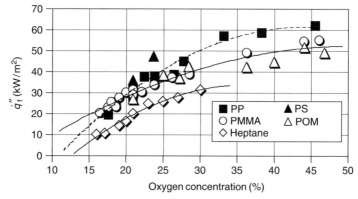

Figure 11.11. Flame heat flux transferred back to the surface versus oxygen concentration for the steady-state combustion of 100-mm diameter and 25-mm thick slabs of polymers and pool of heptane in the ASTM E2058 Apparatus. Data are taken from Ref. [55].

PMMA are plotted against the height (fire size) for upward burning wall fire, where data are taken from Refs. [58–61]. The $\dot{q}_f''$ values increase with height due to increase in fire size. In Figure 11.11, $\dot{q}_f''$ values for polymers burning as horizontal pool fires, in the ASTM E2058 apparatus, are plotted against the O_2 concentration in the co-flowing air, where data are taken from Ref. [55]. The $\dot{q}_f''$ values increase with O_2 concentration of the co-flowing air. The increase in the $\dot{q}_f''$ values is due to increase in the flame temperature and soot concentration and decrease in the diffusion flame height (or reduction in the soot residence time in the flame), creating an ideal path for enhanced flame radiation [21, 22, 55].

The increase in the $\dot{q}_f''$ values with increasing O_2 concentration in air co-flowing around the polymer burning as a diffusion flame has been used to simulate the $\dot{q}_f''$ values, typical of large-scale fires in the ASTM E2058 apparatus [21, 22, 55]. Data obtained in this fashion [21–23, 55] are listed in Table 11.18 along with some of the $\dot{q}_f''$ values measured in normal air in large-scale fires. The data show that in addition to the fire size, the $\dot{q}_f''$ values also depend on the chemical nature of the polymer and its products and molecular weight of the polymer [21–23, 55, 62]:

Table 11.18 Asymptotic Flame Heat Flux Values for the Combustion of Liquids and Polymers[a]

Fuels		$\dot{q}_f''$ (kW/m^2)	
Physical State	Name	$O_2 > 30\%$ (Small Scale)	Normal Air (Large Scale)
Aliphatic Carbon–Hydrogen Atom Containing Fuels			
Liquids	Heptane	32	41
	Hexane	—	40
	Octane	—	38
	Dodecane	—	28
	Kerosine	—	29
	Gasoline	—	35
	JP-4	—	34
	JP-5	—	38
	Transformer fluids	23–25	22–25
Polymers	Polyethylene (PE)	61	—
	Polypropylene (PP)	67	—
Aromatic Carbon–Hydrogen Atom Containing Fuels			
Liquids	Benzene	—	44

(*continued overleaf*)

Table 11.18 (*continued*)

Fuels		$\dot{q}_f''$ (kW/m^2)	
Physical State	Name	$O_2 > 30\%$ (Small Scale)	Normal Air (Large Scale)
	Toluene	—	34
	Xylene	—	37
Polymers	Polystyrene (PS)	75	71
Aliphatic Carbon–Hydrogen–Oxygen Atom Containing Fuels			
Liquids	Methanol	22	27
	Ethanol	—	30
	Acetone	—	24
Polymers	Polyoxymethylene (POM)	50	—
	Polymethylmethacrylate (PMMA)	57	60
Aliphatic Carbon–Hydrogen–Nitrogen Atom Containing Fuels			
Liquids	Adiponitrile	—	34
	Acetonitrile	—	35
Aliphatic Carbon–Hydrogen–Oxygen–Nitrogen Atom Containing Fuels			
Liquid	Toluene diisocyanate	—	28
Polymers	Polyurethane foams (flexible)	64–76	—
	Polyurethane foams (rigid)	49–53	—
Aliphatic Carbon–Hydrogen–Halogen Atom Containing Fuels			
Polymers	Polyvinyl chloride (PVC)	50	—
	Ethylene tetrafluoroethylene (ETFE)	50	—
	Perfluoroethylene-propylene (FEP)	52	—

[a]Data are taken from Refs. [21–23, 55].

- The $\dot{q}_f''$ values are high for fuels with unsaturated aliphatic and aromatic chemical structures such as for butylene, propylene, ethylene, polypropylene, and polystyrene. These fuels burn with highly sooty and luminous flames.
- The $\dot{q}_f''$ values are moderate for fuels with saturated higher molecular weight aliphatic and oxygenated chemical structures such as propane and polymethylmethacrylate. These fuels burn with moderately sooty and luminous flames.

- The $\dot{q}_f''$ values are low for fuels with saturated lower molecular weight and oxygenated chemical structures such as methane, methanol, and polyoxymethylene. These fuels burn with weakly sooty and close to nonluminous flames.
- The $\dot{q}_f''$ values are low for fuels vaporizing as low-molecular-weight monomers, such as the liquids compared to the fuels vaporizing as higher molecular weight monomers and oligomers, as can be noted by the differences in the $\dot{q}_f''$ values of liquids and polymers in Table 11.18.

11.8. FIRE PROPAGATION

Fire propagation is the most critical stage in a fire in terms of hazard assessment, and thus most of the standard test methods use it for the flammability assessment criterion for polymers. Fire propagation represents growth of the combustion process including surface flame spread, nonflaming fire growth, and the fireball in premixed flame propagation [14, 15, 34, 35]. It can be considered as an advancing ignition front in which the leading edge of the flame acts as the source of heat and the source of ignition. It can occur on a horizontal surface, an inclined surface, and on a vertical surface parallel to or opposite to the airflow direction (upward/concurrent or downward or lateral fire propagation). One of the following fire propagation behaviors may be observed for the polymers:

- *Nonpropagating* There is no fire propagation beyond the ignition zone.
- *Decelerating* Fire propagation rate[17] beyond the ignition zone decreases and propagation stops before covering the entire polymer surface.
- *Propagating* Fire propagates beyond the ignition zone until the entire polymer surface is involved on fire.
- *Accelerating* Fire propagation rate beyond the ignition zone increases rapidly covering the entire polymer surface and far beyond with flames in a relatively short time.

For propagating fires, the leading edge of the flame transfers heat ahead of the zone, raises the surface temperature to the ignition temperature of the polymer (satisfying the CHF value), and maintains the temperature until polymer vapors ignite (satisfying the TRP value) [14, 15, 21, 22, 34, 35]. The heat flux provided by the leading edge of the flame depends on the chemical heat release rate. Thus, fire propagates at different rates depending on the heat release rate.

Figures 11.12 and 11.13 illustrate nonpropagating and rapidly propagating fires between two vertical parallel panels of polymers [63]. The panels are about 0.61 m (2 ft) wide, 2.44 m (8 ft) high, and 25 mm (1 inch) thick separated by

---

[17] Fire propagation rate is the velocity at which the ignition front travels over the surface.

462 FLAMMABILITY OF POLYMERS

Figure 11.12. Nonpropagating fire between two vertical parallel panels of a polymer for test duration of 15 min [63]. The panels are about 0.61 m (2 ft) wide, 2.44 m (8 ft) high, and 25 mm (1 inch) thick separated by 0.30 m (1 ft). Ignition source is a 60-kW, about 0.30-m wide, 0.61-m long, and 0.30-m high propane sand burner. The tip of the flame from the burner reaches a height of about 0.91 m (3 ft). Marks on the scale are in feet.

0.30 m (1 ft). The ignition source is a 60-kW, about 0.30-m wide, 0.61-m long, and 0.30-m high propane sand burner. The tip of the flame from the burner reaches a height of about 0.91 m (3 ft). Marks on the scale are in feet.

The common measurements that are made in the standard flammability tests for fire propagation are:

Figure 11.13. Rapidly propagating fire between two vertical parallel panels of a polymer [63]. Panels are about 0.61 m (2 ft) wide, 2.44 m (8 ft) high, and 25 mm (1 inch) thick separated by 0.30 m (1 ft). Ignition source is a 60-kW, about 0.30-m wide, 0.61-m long, and 0.30-m high propane sand burner. Tip of the flame from the burner reaches a height of about 0.91 m (3 ft). Marks on the scale are in feet. The photograph was taken a few seconds before flames extended far beyond top of the panels.

- Extent of fire propagation on a horizontal or vertical surface
- Melting, dripping, and ignition of materials in close proximity to the sample by the hot burning molten droplets
- Minimum heat flux or surface temperature for flame spread
- Minimum oxygen concentration for flame spread

- Flame spread or fire propagation rate
- Heat release rate
- Smoke release rate

The above measurements made in the fire propagation tests are used in different ways to set up the standard test criteria, such as discussed in the following sections for the selected standard tests as illustrations.

11.8.1. The FMVSS 302 Test for Flammability of Vehicle Interior Materials [64]

The test is a horizontal fire propagation test performed in a metal cabinet using 102-mm-wide and 356-mm-long polymer sample. Fire propagation rate and burning time are used as the selection criteria. A polymer that does not burn nor transmit a flame front across its surface at a rate of more than 102 mm/min (1.7 mm/s) is acceptable as a vehicle interior material. If a polymer stops burning after 60 s of heat exposure and has not burned more than 51 mm from the point where the timing was started, it is considered to meet the burn-rate requirement.

As intended, the test criteria are applicable to small polymer parts in the interior of vehicles exposed to small ignition sources. The test is not intended to be applied to large exposed polymer surface areas or polymers in the engine compartment exposed to high-intensity ignition sources. The test method has not been developed to predict the fire behavior of polymers expected in actual fires but rather to screen out polymers with higher resistance to fire propagation from those with low resistance.

11.8.2. UL 94 Standard Test Methodology for Flammability of Plastic Materials for Parts in Devices and Appliances [24]

In this test methodology, both horizontal burning (HB) and vertical burning (V) behaviors of polymers are examined. For horizontal burning test for classifying materials, 127-mm (5-inch) long and 12.7-mm (0.5-inch) wide samples with maximum thickness of 12.7 mm (0.5 inch) are placed on top of a wire gauge and ignited by a 30-s exposure to a Bunsen burner at one end.

For 94HB materials: (1) the burning rate does not exceed 38.1 mm/min (1.5 inches/min) over a 76-mm (3.0-inches) specimen having a thickness of 3–13 mm (0.12–0.50 inch), (2) the burning rate does not exceed 76 mm/min (3.0 in/min) over a 76-mm (3.0-inch) span for specimens having thickness less than 3 mm (0.12 inch), or (3) cease to burn before reaching 102 mm (4.0 inches).

For classifying materials as 94V-0, 94V-1, or 94V-2 in the vertical burning test, 127-mm (5-inches) long and 13-mm (0.5-inch) wide specimens with thickness limited to 13 mm are used. The bottom edge of the specimen is ignited by a 5-s exposure to a Bunsen burner with a 5-s delay and repeated five times until the sample ignites. The 94V-0, 94V-1, and 94V-2 polymer classification criteria are:

Criterion	94V-0	94V-1	94V-2
(A) Flaming combustion time after removal of the test flame (s)	≤10	≤30	≤30
(B) Total flaming combustion time after 10 test flame applications for each set of 5 specimens (s)	≤50	≤250	≤250
(C) Burning with flaming or glowing combustion up to the holding clamp	No	No	No
(D) Dripping flaming particles that ignite the dry absorbent surgical cotton located 12 inches (305 mm) below the test specimen	None	None	Yes
(E) Glowing combustion persisting for more than 30 s after the second removal of the test flame (s)	None	≤60	≤60

The relative resistance of materials to burning according to UL 94 is HB < V-2 < V-1 < V-0. Examples of the UL 94 classification of polymers are listed in Table 11.3 along with their peak vaporization (decomposition) temperature, char yield, and limiting oxygen index (LOI) values, where data are taken from the Federal Aviation Administration (FAA) research study on polymers for aircraft [33]. All the ordinary polymers listed in the table, which generally have low fire resistance, are classified as HB. Most of the high-temperature and halogenated polymers listed in the table that generally have high fire resistance are classified as V-0.

As intended, the test criteria are applicable to small polymer parts in metallic devices and appliances exposed to small ignition sources and not to devices and appliances made entirely of polymers and exposed to high-intensity ignition sources. The test method has not been developed to predict the fire behavior of polymers expected in actual fires but rather to screen out polymers with higher resistance to fire propagation from those with low resistance.

11.8.3. ASTM D 2863-70 Test Methodology for Limited Oxygen Index of Materials [65]

The limited oxygen index (LOI) is the minimum O_2 concentration at or below which there is no downward fire propagation for a vertical polymer sheet inside a glass cylinder, with gas flowing in an upward direction. In the test, 70- to 150-mm (2.8- to 5.9-inch) long, 6.5-mm (0.26-inch) wide, and 3-mm (0.12-inch) thick vertical polymer samples are ignited at the top and fire propagates in a downward direction.

The test is performed at several O_2 concentrations to determine the minimum O_2 concentration at and below which there is no downward fire propagation. LOI values are listed in Table 11.19 taken from Table 11.3 and from Refs. [22] and [66]. The LOI values are arranged on the basis of the generic nature of polymers.

Table 11.19 Limited Oxygen Index (LOI) Values at 20°C for Polymers[a]

Polymers	LOI	Polymers	LOI
Ordinary Polymers		Polyetherketoneketone (PEKK)	40
		Polypara(benzoyl)phenylene	41
Polyoxymethylene	15	Polybenzimidazole (PBI)	42
Cotton	16	Polyphenylene sulfide (PPS)	44
Cellulose acetate	17	Polyamideimide (PAI)	45
Natural rubber foam	17	Polyetherimide (PEI)	47
Polypropylene	17	Polyparaphenylene	55
Polymethylmethacrylate	17	Polybenzobisoxazole (PBO)	56
Polyurethane foam	17	*Composites*	
Polyethylene	18		
Polystyrene	18	Polyethylene/Al_2O_3(50%)	20
Polyacrylonitrile	18	ABS/glass fibers (20%)	22
ABS	18	Epoxy/glass fibers (65 %)	38
Poly(α-methylstyrene)	18	Epoxy/glass fibers (65%)–300 °C	16
Filter paper	18	Epoxy/graphite fibers (1092)	33
Rayon	19	Polyester/glass fibers (70%)	20
Polyisoprene	19	Polyester/glass fibers(70%)–300°C	28
Epoxy	20	Phenolic/glass fibers (80%)	53
Polyethylene terephthalate (PET)	21	Phenolic/glass fibers(80%)–100°C	98
Nylon 6	21	Phenolic/Kevlar (80%)	28
Polyester fabric	21	Phenolic/Kevlar (80%)–300°C	26
Plywood	23	PPS/glass fibers (1069)	64
Silicone rubber (RTV, etc.)	23	PEEK/glass fibers (1086)	58
Wool	24	PAS/graphite (1081)	66
Nylon 6,6	24–29	BMI/graphite fibers (1097)	55
Neoprene rubber	26	BMI/graphite fibers (1098)	60
Silicone grease	26	BMI/glass fibers (1097)	65
Polyethylenephthalate (PEN)	32	*Halogenated Polymers*	
High-Temperature Polymers			
		Fluorinated cyanate ester	40
Polycarbonate	26	Neoprene	40
Nomex	29	Fluorosilicone grease	31–68
Polydimethylsiloxane (PDMS)	30	Fluorocarbon rubber	41–61
Polysulfone	31	Polyvinylidene fluoride	43–65
Polyvinyl ester/glass fibers (1031)	34	PVC (rigid)	50
Polyetheretherketone (PEEK)	35	PVC (chlorinated)	45–60
Polyimide (Kapton)	37	Polyvinylidene chloride (Saran)	60
Polypromellitimide (PI)	37	Chlorotrifluoroethylene lubricants	67–75
Polyaramide (Kevlar)	38	Fluorocarbon (FEP/PFA) tubing	77–100
Polyphenylsulfone (PPSF)	38	Polytetrafluoroethylene	95
Polyetherketone (PEK)	40	Polytrichlorofluorethylene	95

[a] Data are taken from Table 11.3 and Refs. [22, 66].

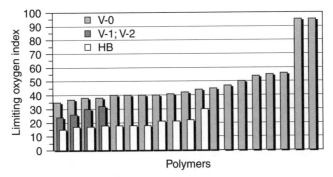

Figure 11.14. Relationship between limiting oxygen index (LOI) and UL 94 classification of polymers. HB, horizontal burning, V, vertical burning.

As discussed previously, $\dot{q}_f''$ values increase with increase in the O_2 concentration of air flowing around a diffusion flame [21, 22, 55]. The increase is due to an increase in the flame temperature and soot concentration and decrease in the flame height (or reduction in the soot residence time in the flame), creating and ideal path for enhanced flame radiation [21, 22, 55]. Thus, polymers with higher LOI values are expected to have higher resistance to fire propagation, as they require higher flame heat flux for flame spread.

The LOI values and UL 94 classification of polymers are interrelated as shown in Figure 11.14. The LOI values for V-0 polymers are ≥ 35, whereas the LOI values are < 30 for polymers classified as V-1, V-2, and HB. As intended, the LOI test is applicable to polymers with small surface areas such as in small parts of metallic devices and appliances exposed to small ignition sources, similar to UL 94 test.

The LOI test method has been not been developed to predict the fire behavior of polymers expected in actual fires but rather to screen polymers for low and high resistance to fire propagation. For the majority of high-temperature and highly halogenated advanced engineered polymers the LOI values are ≥ 40. These polymers have high resistance to ignition, combustion, as well as fire propagation, independent of fire size and ignition source strength [21, 22, 39, 40].

11.8.4. ASTM E162-98 (Also D3675-98): Standard Test Method for Surface Flammability of Materials Using a Radiant Energy Source [67]

In the test, 460-mm long (1.5-ft) long, 150-m (0.5-ft) wide and up to 25-mm (1-inch) thick vertical polymer sample is used. The vertical polymer sample is placed at an angle in front of a 460-mm (1.5-ft) × 300-mm (1-ft) radiant panel. The top of the polymer sample is closer to the radiant panel, forcing the ignition near its upper edge and the flame front to progress in the downward direction. The maximum output of the radiant panel is equivalent to a black body temperature of $670 \pm 4°C$ (45 kW/m^2). The measurements made in the test consist of:

468 FLAMMABILITY OF POLYMERS

- Time of the arrival of the flame at each of the 75-mm (3-inch) marks on the sample holder, which is used to calculate the flame spread factor F_s.
- Maximum temperature rise of the stack thermocouples. It is used to calculate the heat evolution factor Q.

The test data are used to calculate a flame spread index (I_s) expressed as $Q \times F_s$. The I_s values represent downward fire propagation and heat release rate characteristics of polymers and decrease with increase in the resistance to fire propagation and heat release rate. Table 11.20 lists I_s values of some selected polymers as illustrations, where data are taken from Refs. [43] and [68]. The I_s values vary from a high of 2220 to 0, suggesting large variations in the flame spread behavior of polymers.

Many standard tests specify the I_s value as an acceptance criterion of polymers. For example, for structural composites inside naval submarines [43] and for passenger cars and locomotive cabs [69, 70], the following I_s values are specified:

- $I_s < 20$ for structural composites inside naval submarines
- $I_s \leq 25$ for cushions, mattresses, and vehicle components made of flexible cellular foams for passenger cars and locomotive cabs and thermal and acoustic insulation for buses and vans
- $I_s \leq 35$ for all vehicle components in passenger cars and locomotive cabs and for seating frame, seating shroud, panel walls, ceiling, partition, windscreen, HVAC ducting, light diffuser, and exterior shells in buses and vans
- $I_s \leq 100$ for vehicle light transmitting polymers in passenger cars and locomotive cabs

The above-listed criteria for the I_s values <20 suggest that structural composites for inside naval submarines are expected to have high resistance to fire propagation and heat release. Also polymers used in passenger cars, locomotive cabs, buses, and vans with I_s values ≤25 as well as ≤35 are expected to have relatively higher resistance to fire propagation and heat release rate compared to the ordinary polymers with I_s values ≤100 under low heat exposure conditions.

This test method also has not been developed to predict the fire behavior of polymers expected in actual fires but rather to screen polymers for low and high resistance to fire propagation under conditions of lower intensity heat exposure.

11.8.5. ASTM E84-00a (NFPA 255) Standard Test Method for Surface Burning Characteristics of Building Materials [71]

In this 10-min test, a 7.3-m (24-ft) long and 0.51-m (20-inch) wide horizontal polymer sample (total area of 3.25 m²/36 ft²) is used inside a 7.6-m (25-ft) long, 0.45-m (17-inch) wide and 0.31-m (12-inch) deep tunnel. Two gas burners, located 0.19 m (7 inch) below the polymer surface and 0.31 m (12 inch) from one end of the tunnel are used as ignition sources. The two burners release 88 kW of heat creating a gas temperature of 900°C near the specimen surface. The flames from the burners cover 1.37 m (4.5 ft) of the length and the entire

Table 11.20 Flame Spread Index Values for Materials from ASTM E162 Test[a]

Material	Thickness (mm)	Flame Spread Index, I_s
Polyurethane polyether rigid foam	—	2220
Polyurethane polyether flexible foam	—	1490
Polyurethane polyester rigid foam, FR	—	1440
Polyurethane polyester flexible foam, FR	—	1000
Polyurethane polyester rigid foam, FR	—	880
Acrylic, FR	3.2	376
Polystyrene	1.7	355
Polyester/glass fibers (21%)	1.6	239
1087 Vinyl ester/glass fibers	—	156
Plywood, FR, exterior	6.4	143
Polystyrene, rigid foam	—	114
Phenolic, laminate	1.6	107
Red oak	19.1	99
Polyester-FR/glass fibers (27%)	2.4	66
Phenolic/polyethylene fibers (1073)	—	48
Epoxy/glass fibers (1066)	—	43
Phenolic/aramid fibers (1077)	—	30
Vinyl ester/glass fibers (1031)	—	27
Epoxy/glass fibers (1092)	—	23
Phenolic/graphite fibers (1103)	—	20
Bismaleimide/graphite fibers (1096)	—	17
Polystyrene, rigid foam, FR	—	13
Bismaleimide/graphite fibers (1095)	—	13
Nylon/glass fibers (1077)	—	13
Epoxy/glass fibers (1067)	—	12
Bismaleimide/graphite fibers (1097)	—	12
Epoxy/glass fibers (1089)	—	11
Epoxy/glass fibers (1091)	—	11
Polyurethane polyether flexible foam, FR	—	10
Polyvinyl chloride (PVC)	3.7	10
Polyarylsulfone/graphite fibers (1081)	—	9
Polyphenylenesulfide/glass fibers (1084)	—	8
Polyphenylenesulfide/glass fibers (1069)	—	7
Phenolic/glass fibers (1017)	—	6
Phenolic/graphite fibers (1102)	—	6
Phenolic/glass fibers (1100)	—	5
Phenolic/glass fibers (1101)	—	4
Phenolic/glass fibers (1014)	—	4
Phenolic/glass fibers (1015)	—	4

(*continued overleaf*)

Table 11.20 (*continued*)

Material	Thickness (mm)	Flame Spread Index, I_s
Phenolic/glass fibers (1018)	—	4
PVC, FR	3.7	3
Bismaleimide/graphite fibers (1098)	—	3
Phenolic/graphite fibers (1104)	—	3
Polyphenylenesulfide/glass fibers (1083)	—	3
Polyphenylenesulfide/glass fibers (1085)	—	3
Polyetheretherketone/graphite fibers (1086)	—	3
Polyetheretherketone/glass fibers (1079)	—	3
1105 Polyimide/glass fibers (1105)	—	2
Phenolic/glass fibers (1099)	—	1
Asbestos cement board	4.8	0

[a] Data taken from Refs. [43, 68].

Table 11.21 Flame Spread Index for Materials from ASTM E84 Test[a]

Material	Flame Spread Index
Plywood, fir, exterior	143
Douglas fir plywood	91
Rigid polyurethane foam	24
Douglas fir plywood/FR	17
Composite panel	17
Type X gypsum board	9
Rigid polystyrene foam	7

[a] Data taken from Ref. [71].

width or a 0.63-m² (7-ft²) area of the sample. Air enters the tunnel 1.4 m (54 in) upstream of the burner at a velocity of 73 m (240 ft)/min. The test conditions are set such that for red oak flooring control material, flame spreads to the end of the 7.3-m (24-ft) long sample in 5.5 min or a flame spread rate of 22 mm/s.

In the test, measurements are made for the percent light absorption by smoke flowing through the exhaust duct, gas temperature [7.0 m (23 ft) from the burner] and location of the leading edge of the flame as functions of times. The measured data are used to calculate the flame spread and smoke developed indices from the flame spread distance–time and percent light absorption–time areas, respectively. Examples of the data obtained from the ASTM E84 test are listed in Table 11.21, where data are taken from Ref. [71].

The flame spread index represents the resistance to fire propagation and decreases with increase in the resistance. The intent of the test method is to separate materials with higher fire propagation resistance from those with lower resistance and is not intended to predict the fire propagation behaviors of materials in actual fires. The test method does not provide any information on the melting of polymers leading to pool fires and thus is limited in its application.

This standard test method is one of the most widely specified tests, for example, classification of interior finish in buildings in the National Fire Protection Association (NFPA) 101 Life Safety Code [72]:

- Class A Interior Wall and Ceiling Finish: *flame spread index (FSI), 0 to 25*; *smoke developed index (SDI), 0 to 450*
- Class B Interior Wall and Ceiling Finish: *FSI, 26 to 75*; *SDI, 0 to 450*
- Class C Interior Wall and Ceiling Finish: *FSI, 76 to 200*; *SDI, 0 to 450*

A compilation of the interior finish requirements for various occupancies in buildings in the NFPA 101 Life Safety Code [72] is included in Table 11.22.

Table 11.22 Interior Finish Classification Limitations Compiled in NFPA 101 Life Safety Code [72][a]

Occupancy	Exits	Access to Exits	Other Spaces
Assembly, new >300 occupant load	A	A or B	A or B
Assembly, new ≤300 occupant load	A	A or B	A, B, or C
Assembly, existing >300 occupant load	A	A or B	A or B
Assembly, existing ≤300 occupant load	A	A or B	A, B, or C
Educational, new	A	A or B	A or B; C on partitions
Education, existing	A	A or B	A, B, or C
Day-care centers, new	A I or II	A I or II	A or B NR
Day-care centers, existing	A or B	A or B	A or B
Group day-care homes, new	A or B	A or B	A, B or C
Group day-care homes, existing	A or B	A, B, or C	A, B or C
Family day-care homes	A or B	A, B, or C	A, B or C
Health care, new	A or B	A or B; C on lower portion of corridor wall	A or B; C in small individual rooms
Health care, existing	A or B	A or B	A or B
Detention and correctional, new	A; I	A; I	A, B or C
Detention and correctional, existing	A or B; I or II	A or B; I or II	A, B or C

(*continued overleaf*)

Table 11.22 (*continued*)

Occupancy	Exits	Access to Exits	Other Spaces
1- and 2-Family dwellings, lodging or rooming houses	A, B or C	A, B or C	A, B or C
Hotels and dormitories, new	A; I or II	A or B; I or II	A, B, or C
Hotels and dormitories, existing	A or B; I or II	A or B; I or II	A, B, or C
Apartment buildings, new	A; I or II	A or B; I or II	A, B, or C
Apartment buildings, existing	A or B; I or II	A or B; I or II	A, B, or C
Mercantile, new	A or B	A or B	A or B
Mercantile, existing, class A or B	A or B	A or B	Ceilings-A or B; Walls-A, B, or C
Mercantile, existing, class C	A, B or C	A, B or C	A, B or C
Business and ambulatory health care, new	A or B; I or II	A or B; I or II	A, B, or C
Business and ambulatory health care, existing	A or B	A or B	A, B or C
Industrial	A or B	A, B or C	A, B or C
Storage	A or B	A, B or C	A, B or C

[a]A, FSI, 0–25; SDI, 0–450; B, FSI, 26–75; SDI, 0–450; C, FSI, 76–200; SDI, 0–450; I, CRF > 4.5 kW/m^2; II, 2.2 <CRF < 4.5 kW/m^2.

11.8.6. ASTM E648-99 (NFPA 253) Standard Test Method for Critical Radiant Flux of Floor- Covering Systems Using a Radiant Heat Energy Source [73]

In the test, 1.0-m (39.4-inch) long and 0.20-m (7.9-inch) wide horizontal sample is exposed to radiant heat flux in the range of 1–11 kW/m^2 from a 30° inclined radiant panel all contained inside a chamber. The heat flux is at 11 kW/m^2 at the sample surface that is closer to the radiant heater. The radiant flux decreases as the distance between the sample surface and the radiant heater increases to the lowest value of 1 kW/m^2. The sample surface exposed to 11 kW/m^2 is ignited by a pilot flame, and flame spread is observed until the flame is extinguished at some downstream distance due to decrease in the radiant flux. The radiant flux at this distance is defined as the critical radiant flux (CRF) of the sample.

In ASTM E648, fire propagates on the surface of the sample due to heat flux contributed by the leading edge of the flame and the radiant heater. With increasing distance, the radiant heat flux decreases, until the combined flame and radiant heat flux cannot satisfy the flame spread requirements and the flame is extinguished. This test method was developed as a result of need for flammability

Table 11.23 Critical Radiant Flux for Carpets from ASTM E 648[a]

Fiber Weight (oz)	Style	Fiber Type	Yarn	Adhesive	CRF (kW/m²)
28	Cut pile	Nylon 6,6	BCF[b]	Nu Broadlok II[c]	18.5
26	Loop pile	Polypropylene	BCF[b]	Supra STIX 90[c]	2.8
50	Cut pile	Nylon 6,6	Staple	Supra STIX 90[c]	4.6
				Supra STIX 90[d]	4.8
				Supra STIX 90[e]	4.7
28	Cut pile	Nylon 6,6	Staple	Supra STIX 90[c]	3.0
28	Loop pile	Nylon 6,6	BCF[b]	Nu Broadlok II[c]	17.5
50	Cut pile	Wool	Staple	Supra STIX 90[c]	6.4
24	Loop pile	Nylon 6	BCF[b]	Supra STIX 90[c]	3.2
24	Loop pile	Nylon 6	BCF[b]	Supra STIX 90[c]	3.1

[a] Preheat time, 2 min. Data are taken from Ref. [75].
[b] BCF, bulk continuous filament.
[c] Substrate, Sterling board (high-density inorganic fiber-reinforced cement board).
[d] Substrate, Ultra Board.
[e] Eter board.

standard for carpets and rugs to protect the public against fire hazards [74]. As a result several carpet systems were tested by this standard [74–76]. Table 11.23 lists the CRF values for selected materials taken from Ref. [75].

This standard test method is specified for the classification of the interior floor finish in buildings in the NFPA 101 Life Safety Code (Table 11.22) [72]:

- Class I Interior Floor Finish: *critical radiant flux (CRF)*, >4.5 kW/m^2
- Class II Interior Floor Finish: *CRF, greater than 2.2 but less than 4.5 kW/m^2*

The CRF represents the resistance of the polymer to fire propagation; it increases with increase in the fire resistance. It characterizes the ability of the leading edge of the flame front in providing sufficient heat flux ahead of the flame front to satisfy the fire propagation requirements of the materials. The intent of the test method is to separate materials with higher fire propagation resistance from those with lower resistance and is not intended to predict the fire propagation behaviors of materials in actual fires.

11.8.7. ASTM E1321 Standard Test Method for Determining Material Ignition and Flame Spread Properties (Lateral Ignition and Flame Spread Test, LIFT) [38]

This test method determines the material properties related to piloted ignition of a vertically oriented sample under a constant and uniform heat flux and to lateral flame spread on a vertical surface due to an externally applied radiant-heat flux. For the ignition test, a 155-mm (6-inch) square sample is exposed to a nearly uniform heat flux, and the time to flame attachment is measured. For the flame spread test, a 800-mm (31-inch) long and 155-mm (6-inch) wide horizontal sample turned vertically on its side is used.

The sample is placed in front of a 280-mm (11-inch) × 483-mm (19-inch) radiant heater with a 15° orientation to the heater such that the sample surface is exposed to decreasing heat flux. The heat flux is highest at the sample end closer to the heater, where a pilot flame is provided for igniting the vapors. The flame spread data is correlated by the following relationship [15, 35, 38]:

$$V = \Omega/k\rho c(T_{ig} - T_a)^2 \qquad (11.14)$$

where V is the flame (pyrolysis front) velocity (m/s) and Ω is the flame-heating parameter (kW^2/m^3). The test provides data that is used to derive T_{ig}, $k\rho c$, Ω, and minimum temperature for flame spread, $T_{s,min}$. Table 11.24 lists the values reported in Ref. [35]. Results from the lateral flame spread tests in the ASTM E1321 apparatus are similar to downward flame spread, except for materials with excessive melting and dripping [35]. The data in Table 11.24 are generally representative of common construction or interior finish materials [15].

The properties derived from ASTM E1321 provide information about the flame spread characteristics of materials and can serve as an indication of their hazardous characteristics [38]. The test results provide material fire parameters that correspond to property data required by theories of surface flame spread [38]. The analysis may be used to rank materials performance by some set of criteria applied to the correlation; or the analysis may be employed in fire risk growth models to develop a more rational and complete risk assessment for wall materials [38].

11.8.8. Clean Room Flammability Standard for The Semiconductor Industry (NFPA 318 [77] FMR 4910 Test Standard [78] and UL 2360 Test Standard [79])

The NFPA 318 deals with the protection of clean rooms whereas the FMR 4910 and UL 2360 deal with the flammability of polymers for the clean rooms. In the FMR 4910 test standard, ASTM E2058 Fire Propagation Apparatus (FPA) [31] is used, whereas ASTM E1354 Cone Calorimeter [36] is used in the UL 2360 test standard. Both test standards evaluate the fire propagation and smoke release behaviors of the polymers. For polymers for which fire propagation behavior cannot be defined clearly, both test standards use a large-scale parallel panel test [40, 63, 78]. Figures 11.12 and 11.13, discussed in an earlier section, illustrate the parallel panel test.

In both the test standards, fire propagation and smoke release propensity of polymers are quantified. Three types of tests are performed: ignition, combustion, and fire propagation. In the ignition and combustion tests, horizontal samples as squares [100 mm (4 inch)] or circles [100 mm (4 inches) diameter], 3 mm (0.1 inch) to 25 mm (1 inch) in thickness are used. For fire propagation tests,

Table 11.24 Effective Flame Spread Properties Derived from the ASTM E1321 Test[a]

Material	T_{ig} (°C)	$k\rho c$ (kW²·s/m⁴·K²)	Ω (kW²/m³)	$T_{s,min}$ (°C)	$\Omega/k\rho c$ (m·K²/s)	$(\Omega/k\rho c)/(T_{ig}-T_a)^2$ (mm/s)
Synthetic Polymers						
Polyisocyanurate foam (5.1 cm)	445	0.02	4.9	275	245	36.4
Foam, rigid (2.5 cm)	435	0.03	4.0	215	133	20.3
Polyurethane foam, flexible (2.5 cm)	390	0.32	11.7	120	37	6.2
PMMA type G (1.3 cm)	378	1.02	14.4	90	14	2.5
PMMA polycast (1.6 mm)	278	0.73	5.4	120	7	1.8
Polycarbonate (1.5 mm)	528	1.16	14.7	455	13	1.6
Carpets						
Carpet (acrylic)	300	0.42	9.9	165	24	5.3
Carpet #2 (wool, untreated)	435	0.25	7.3	335	29	4.4
Carpet (nylon/wool blend)	412	0.68	11.1	265	16	2.6
Carpet #1 (wool, stock)	465	0.11	1.8	450	16	2.3
Carpet #2 (wool, treated)	455	0.24	0.8	365	3	0.5
Natural Polymers						
Plywood, plain (1.3 cm)	390	0.54	12.9	120	24	4.1
Gypsum board, (common)(1.3 mm)	565	0.45	14.4	425	32	3.7
Gypsum board, FR (1.3 cm)	510	0.40	9.2	300	23	3.0
Plywood, plain (6.4 mm)	390	0.46	7.4	170	16	2.7
Fiberglass shingle	445	0.50	9.0	415	18	2.7
Douglas fir particle board (1.3 cm)	382	0.94	12.7	210	14	2.4
Hardboard (3.2 mm)	365	0.88	10.9	40	12	2.3
Hardboard (nitrocellulose paint)	400	0.79	9.8	180	12	2.1
Asphalt shingle	378	0.70	5.3	140	8	1.3
Fiber insulation board	355	0.46	2.2	210	5	0.9
Particle board (1.3 cm stock)	412	0.93	4.2	275	5	0.7
Hardboard (6.4 mm)	298	1.87	4.5	170	2	0.5
Hardboard (gloss paint)(3.4 mm)	400	1.22	3.5	320	3	0.5
Gypsum board, wallpaper (S142M)	412	0.57	0.79	240	1	0.2

[a] Data are taken from Ref. [35].

100-mm (4-inch) wide, 300-mm (12-inch) high and 3-mm (0.1-inch) to 25-mm (1-inch) thick vertical samples are used.

The ignition and combustion tests are performed in normal air, whereas the fire propagation test is performed in 40% oxygen concentration in the FMR 4910 test standard and in normal air in the UL 2360 test standard. Heat flux values used in the tests are 0–60 kW/min² for the ignition tests in the FMR 4910 test standard and 0–75 kW/m² in the UL 2360 test standard. In the combustion tests, sample surface is exposed to 50 kW/m² in both the test standards. In the fire propagation tests in the FMR 4910 test standard, only the bottom 20% of the vertical sample is exposed to 50 kW/m² in the presence of a pilot flame.

Ignition time and release rates of heat and products are measured in the tests. The fire propagation behavior of polymers is expressed in terms of a fire propagation index (FPI) [21, 22, 40, 78, 80]:

$$\text{FPI} = 1000 \frac{(0.42 \dot{Q}_{\text{ch}}/w)^{1/3}}{[(T_{\text{ig}} - T_{\text{a}})\sqrt{(k\rho c)}]} \quad (11.15)$$

where $\dot{Q}_{\text{ch}}$ is the chemical heat release rate (kW), w is the sample width (m), and $(T_{\text{ig}} - T_{\text{a}})\sqrt{k\rho c}$ is the thermal response parameter (TRP) as discussed previously. FPI is in $(\text{m/s}^{1/2})/(\text{kW/m})^{2/3}$ and is proportional to $V^{1/2}$ and to $\dot{m}''$ (as shown in Figure 11.15 [21, 22, 40, 80–86]). The FPI concept was developed based on the upward flame spread theories [14, 15, 34, 35].

The FPI values have been determined in the ASTM E2058 in the co-flowing air having a 40% oxygen concentration for polymers, electrical cables, and conveyor belts [21, 22, 40, 41, 80–86]. The FPI values for selected polymers, taken from Refs. [21, 22, 40, 41, 81–83], are listed in Tables 11.25 and 11.26. Correlation

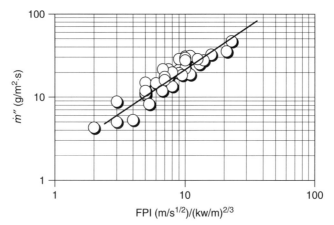

Figure 11.15. Relationship between the generation rate of polymer vapors in combustion in normal air at 50 kW/m² and FPI determined for fire propagation in 40% oxygen concentration. Data were measured in our laboratory in the ASTM E2058 Fire Propagation Apparatus [21, 22, 40, 80–86].

Table 11.25 Fire Propagation and Smoke Release Characteristics of Polymers[a]

Fire Propagation		Smoke Release		
Polymer	FPI	Polymer	SDI	Smoke Quality
Ordinary Polymers				
Polystyrene	34	Polystyrene	5.6	Copious, black
PVC–PVC cable	36	PVC–PVC cable	4.1	Copious, black
Polypropylene (PP)	32	PE–PVC cable	3.8	Copious, black
PE–PVC cable	28	Polybutylene terephthalate	2.2	Copious, black
Polybutylene terephthalate	32	Fire-retarded polypropylene	2.1	Copious, black
Polymethylmethacrylate	31	Silicone–PVC cable	2.0	Copious, black
Fire-retarded polypropylene	30	Polypropylene (PP)	1.8	Very large, black
Silicone–PVC cable	17	Acrylonitrile–butadiene–styrene	0.80	Small, black
Polyoxymethylene	15	Polyester/glass fibers (70%)	0.68–0.91	
Wood slab	14	Polymethylmethacrylate	0.62	Small, light grayish
Polyester/glass fibers (70%)	10–13	Polyoxymethylene	0.03	Very small, grayish-white
Acrylonitrile–butadiene–styrene	8	Wood slab	0.20	Very small, grayish-white
High-Temperature Polymers				
Polyetherimide (PEI)	8	Polycarbonate	2.1	Copious, black
Phenolic/kevlar fibers (84%)	8	Polyphenyleneoxide	1.6	Very large, black
Epoxy/glass fibers (65–76%)	5–11	Epoxy/glass fibers (65–76%)	0.61–2.1	
Epoxy/graphite (71%)	5	Epoxy/graphite (71%)	0.54	
Highly modified PP	4–5	Cyanate/graphite (73%)	0.41	
Highly modified PVC	1–4	Polyetheretherketone (PEEK-1)	0.40	Very small, grayish-white
Phenolic/glass fibers (80%)	3	Phenolic/kevlar fibers (84%)	0.33	
Cyanate/graphite (73%)	4	PPS/glass fiber (84 %)	0.29	
PPS/glass fiber (84 %)	3	Highly modified PP	0.19–0.40	
Epoxy/phenolic/glass fibers (82%)	2	Epoxy/phenolic/glass fibers (82%)	0.18	

(continued overleaf)

Table 11.25 (continued)

Fire Propagation		Smoke Release		
Polymer	FPI	Polymer	SDI	Smoke Quality
Polycarbonate	14	Polysulfone (PSO)	0.18	
Polyphenyleneoxide	9	Polyetherimide (PEI)	0.15	Very small, grayish-white
Polysulfone (PSO)	9	Polyethersulfone (PES)	0.15	
Polyetherimide (PEI)	8	Phenolic/glass fibers (80%)	0.07	
Polyethersulfone (PES)	7	Phenol-formaldehyde	0.06	
Polyetheretherketone (PEEK-1)	6	Highly modified PVC	0.03–0.29	
Phenol-formaldehyde	5	PEEK-2	0.03	Very small, grayish-white
PEEK-2	4			
		Halogenated Polymers		
PE-25% chlorine	15	PE-25% chlorine	1.7	Very large, black
PVC (flexible)	16	PVC (flexible)	1.6	Very large, black
PE-36% Cl	11	PE-36% Cl	1.5	Large, black
PE-48% Cl	8	PE-48% Cl	1.4	Large, black
ETFE (Tefzel)	7	ETFE (Tefzel)	0.18	
PVC-D (rigid)	7	PVC-D (rigid)	0.70	Small, grayish
PVC-E rigid	6	PVC-E rigid	0.30	Very small, grayish-white
PVC-F (rigid)	4	PVC-F (rigid)	0.30	Very small, grayish-white
Polyvinylidenefluoride	4	ECTFE (Halar)	0.15	
PTFE, Teflon	4	Polyvinylidenefluoride	0.12	Very small, grayish-white
PVDF (Kynar)	4	PVDF (Kynar)	0.12	
ECTFE (Halar)	4	TFE (Teflon)	0.01	
TFE (Teflon)	4	PFA (Teflon)	0.01	
PFA (Teflon)	2	FEP (Teflon)	0.01	
FEP (Teflon)	3	CPVC (Corzan)	0.01	
CPVC (Corzan)	1	TFE, Teflon	0.01	Very small, grayish-white

[a]Data are taken from Refs. [21, 22, 40, 41, 81–83].

Table 11.26 Estimated Fire Propagation Index and Smoke Development Index for Polymers from Parts of a Minivan[a]

Fire Propagation			Smoke Release		
Part Description	Polymer	FPI	Part Description	Polymer	SDI
HVAC unit	ABS-PVC	>30	HVAC unit	ABS-PVC	3.4
Fender foam	PS	27	Fender foam	PS	2.6
Head liner fabric	Nylon 6	26	Instrument panel	PVC	1.6
Hood liner face	PET	23	Kick panel insulation	PVC	1.3
Kick panel insulation	PVC	18	Headliner fabric	Nylon 6	1.2
Instrument panel	PVC	15	Instrument panel	PC	1.2
Resonator structure	PP	14	Resonator structure	PP	1.1
Wheel well cover	PP	13	Headlight lens	PC	1.0
HVAC unit	PP	12	Air ducts	PP	0.88
Battery cover	PP	12	Battery cover	PP	0.85
Instrument panel	PC	11	Wheel well cover	PP	0.85
Air ducts	PP	11	Windshield wiper	SMC	0.80
Headlight lens	PC	9	HVAC unit	PP	0.72
Fuel tank	PE	8	Hood liner face	PET	0.51
Windshield wiper	SMC	8	Fuel tank	PE	0.34

[a] For estimations, data are taken from Ref. [41]; FPI is in $(m/s^{1/2})/(kW/m)^{2/3}$; SDI is in $(g/g)(m/s^{1/2})/(kW/m)^{2/3}$.

between the FPI values and vertical fire propagation behaviors are illustrated by the data in Table 11.27. In the table, FPI values are from the small-scale tests (ASTM E2058 Apparatus), and visual observations and extent of fire propagation are from the large-scale parallel panel tests for polymers and electrical cables, where data are taken from Refs. [21, 22, 40, 80–84]. The correlation between small- and large-scale vertical fire propagation data suggests the following:

- For FPI ≤ 6: flames are close to extinction conditions and fire propagation is limited to the ignition zone.
- For 6 < FPI ≤ 10: fire propagation is decelerating and stops short of the sample length.
- For 10 < FPI ≤ 20: there is fire propagation beyond the ignition zone, the rate increasing with the FPI value.
- For FPI > 20: fire propagation beyond the ignition zone is rapid.

The FPI values in Tables 11.25–11.27 indicate that the fire propagation behavior for ordinary polymers is a propagating type, whereas for most of the high-temperature and highly halogenated polymers (≥48% chlorine) it is either the nonpropagating or decelerating type.

The smoke generation characteristic of polymers during fire propagation is expressed by a smoke development index (SDI) [21, 22, 40, 83]. The SDI concept

Table 11.27 Fire Propagation Index from Small-Scale Tests (ASTM E2058) and Mode and Extent of Propagation in Large-Scale Parallel Panel Tests for Various Polymers[a]

Large-Scale Test	Polymer	FPI[b]	Propagation Mode[c]	Propagation Beyond the Ignition Zone (% of total height)[c]
4.9-m long × 0.61-m wide parallel Marinite sheets covered with electrical cables	PVC–PVDF	7	Decelerating	2
	XLPE–EVA	7	Decelerating	45
	XLPE–Neoprene	9	Decelerating	14
	XLPO–XLPO	9	Decelerating	45
	PE–PVC	20	Accelerating	100
2.4-m high × 0.61-m wide parallel slabs of polymeric materials	PVDF	4	None	0
	PVC-B	4	None	0
	PVC-A	6	None	0
	ETFE	7	Decelerating	14
	PVC-C	8	Decelerating	12
	FR-PP	30	Accelerating	100
	PMMA	31	Accelerating	100
	PP	32	Accelerating	100

[a] Data taken from Refs. [21, 22, 40, 80–84].
[b] FPI is in $(m/s^{1/2})/(kW/m)^{2/3}$.
[c] Determined from the measurements for the ignition zone length (in depth burning of the sample), fire propagation length (thin pyrolyzed surface layer), and total height of the sample.

was developed on the basis of the relationship between FPI and $\dot{m}''$ (Fig. 11.15) and the relationship between $\dot{m}''$ and the release rate of smoke, $\dot{G}''_{sm}$:

$$\dot{G}''_{sm} = \dot{m}'' y_{sm} \quad (11.16)$$

where $\dot{G}''_{sm}$ is in kg/m²·s.

$$\dot{m}'' \propto FPI \quad (11.17)$$

From Eqs. 11.16 and 11.17:

$$\dot{G}''_{sm} \propto FPI \times y_{sm} \propto SDI \quad (11.18)$$

The relationship between $\dot{G}''_{sm}$ and SDI is shown in Figure 11.16 where data are from Refs. [21, 22, 40, 80–86].

In SDI, one component (i.e., FPI) is related to the fire propagation characteristic of the polymer (as shown by the data in Tables 11.25–11.27), whereas the other component (y_{sm}) relates the relationship between the generic nature of the polymer and its smoke generation characteristic, as shown by the data in Table 11.28.

The SDI values of selected polymers taken from Refs. [21, 22, 40, 41, 81–83], are listed in Tables 11.25 and 11.26 along with their FPI values. Visually observed

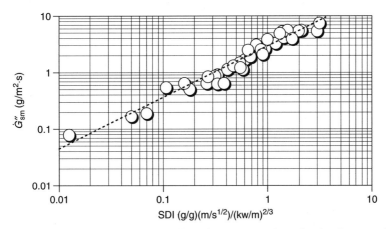

Figure 11.16. Relationship between smoke release rate and smoke development index for polymers. Data were measured in our laboratory in ASTM E2058 Fire Propagation Apparatus [21, 22, 40, 80–86].

Table 11.28 Smoke Yield for Various Gases, Liquids, and Solid Fuels[a]

State	Generic Name	Generic Nature	Smoke Yield (g/g)
Gas	Methane to butane	Aliphatic—saturated	0.013–0.029
	Ethylene, propylene	Aliphatic—unsaturated	0.043–0.070
	Acetylene, butadiene	Aliphatic—highly unsaturated	0.096–0.125
Liquid	Alcohols, ketones	Aliphatic	0.008–0.018
	Hydrocarbons	Aliphatic	0.037–0.078
		Aromatic	0.177–0.181
Solid	Cellulosics	Aliphatic (mostly)	0.008–0.015
	Synthetic Polymers	Aliphatic—oxygenated	0.001–0.022
		Aliphatic—highly fluorinated	0.002–0.042
		Aliphatic—unsaturated	0.060–0.075
		Aliphatic unsaturated—chlorinated	0.078–0.099
		Aromatic	0.131–0.191

[a]Data taken from Refs. [21, 22].

smoke quality is also listed in Table 11.26. The development of smoke increases with the SDI values.

Based on the research on fire propagation and smoke development [20–22, 40, 41, 80–84, 86], the following criteria are used in the FMR 4910 test standard for the selection of polymers for clean rooms of the semiconductor industry [78]:

- $FPI \leq 6$ $(m/s^{1/2})/(kW/m)^{2/3}$
- $SDI \leq 0.4$ $(g/g)(m/s^{1/2})/(kW/m)^{2/3}$

REFERENCES

1. L. Reich and S. S. Stivala *Elements of Polymer Degradation*, McGraw-Hill, New York, 1971.
2. A. L. Andrady, in J. E. Mark, ed., *Physical Properties of Polymers Handbook*, American Institute of Physics, Woodbury, New York, 1996, Chapter 40, pp. 547–555.
3. K. Dawes and L. C. Glover, in J. E. Mark, ed., *Physical Properties of Polymers Handbook*, American Institute of Physics, Woodbury, New York, 1996, Chapter 41, pp. 557–576.
4. W. J. Welsh in J. E. Mark, ed., *Physical Properties of Polymers Handbook*, American Institute of Physics, Woodbury, New York, 1996, Chapter 43, pp. 605–614.
5. L. J. Suggs and A. G. Mikos, in J. E. Mark, ed., *Physical Properties of Polymers Handbook*, American Institute of Physics, Woodbury, New York, 1996, Chapter 44, pp. 615–624.
6. A. L. Andrady, in J. E. Mark, ed., *Physical Properties of Polymers Handbook*, American Institute of Physics, Woodbury, New York, 1996, Chapter 45, pp. 625–635.
7. R. R. Kunz, in J. E. Mark, ed., *Physical Properties of Polymers Handbook*, American Institute of Physics, Woodbury, New York, 1996, Chapter 46, pp. 637–642.
8. Y. H. Mariam and K. Feng, in J. E. Mark, ed., *Physical Properties of Polymers Handbook*, American Institute of Physics, Woodbury, New York, 1996, Chapter 47, pp. 643–665.
9. G. Swift, "Polymers, Environmentally Degradable," *Kirk-Othmer Encyclopedia of Chemical Technology*, Wiley, New York, Online Posting Data, December 4, 2000.
10. H. Holemann, in *Fire Safety Science, Proceedings of the Fourth International Symposium*, International Association for Fire Safety Science, Society of Fire Protection Engineers, Quincy, MA, 1994, pp. 61–77.
11. S. D. Miles, G. Cox, M. N. Christolis, C. A. Christidou, A. G. Boudouvis, and N. C. Markatos, in *Fire Safety Science, Proceedings of the Fourth International Symposium*, International Association for Fire Safety Science, Society of Fire Protection Engineers, Quincy, MA, 1994, pp. 1221–1232.
12. *SFPE Handbook of Fire Protection Engineering*, 2nd ed., The National Fire Protection Association Press, Quincy, MA, 1995.
13. J. E. Mark, ed., *Physical Properties of Polymers Handbook*, American Institute of Physics, Woodbury, New York, 1996.
14. D. Drysdale, *An Introduction to Fire Dynamics*, Wiley, New York, 1985.
15. J. G. Quintiere, *Principles of Fire Behavior*, Delmar, New York, 1998.
16. C. F. Cullis and M. M. Hirschler, *The Combustion of Organic Polymers*, Clarendon, Oxford, UK, 1981.
17. R. Friedman, *Principles of Fire Protection Chemistry*, 2nd ed., National Fire Protection Association, Quincy, MA, 1989.
18. J. W. Lyons, *The Chemistry and Uses of Fire Retardants*, Wiley-Interscience, New York, 1970.
19. V. Babrauskas and S. J. Grayson, ed., *Heat Release in Fires*, E & FN Spon, Chapman Hall, New York, 1992.
20. A. Tewarson, *J. Fire Sci.* **10**, 188–241 (1992).

21. A. Tewarson, in *SFPEH and book of Fire Protection Engineering*, 2nd ed., National Fire Protection Association Press, Quincy, MA, 1995, Section 3, Chapter 4, pp. 3-53–3-124.
22. A. Tewarson, in J. E. Mark, ed., *Physical Properties of Polymers Handbook*, American Institute of Physics, Woodbury, New York, 1996, Chapter 42, pp. 577–604.
23. A. Tewarson and R. F. Pion, *Combustion and Flame*, **26**, 85–103 (1976).
24. UL 94, *Standard for Tests for Flammability of Plastics Materials for Parts in Devices and Appliances*, 3rd ed., Underwriters Laboratories, Northbrook, IL, June 12, 1989.
25. W. J. Welsh, in J. E. Mark, ed., *Physical Properties of Polymers Handbook*, American Institute of Physics, Woodbury, New York, 1996, Chapter 29, pp. 401–407.
26. C. A. Harper, ed., *Handbook of Plastics and Elastomers*, McGraw-Hill, New York, 1975.
27. H. Domininghaus, *Plastics for Engineers—Materials, Properties, Applications*, Hanser, New York, 1988.
28. L. Mandelkern and R. G. Alamo, in J. E. Mark, ed., *Physical Properties of Polymers Handbook*, American Institute of Physics, Woodbury, New York, 1996, Chapter 11, pp. 119–137.
29. J. Scheirs, ed., *Modern Flouropolymers*, Wiley Series in Polymer Science High Performance Polymers for Diverse Applications, Wiley, New York, 2000.
30. I. A. Abu-Isa, D. R. Cummings, and D. LaDue, Thermal Properties of Automotive Polymers I. Thermal Gravimetric Analysis and Differential Scanning Calorimetry of Selected Parts from a Dodge Caravan, Report R&D 8775, General Motors Research and Development Center, Warren, MI, June, 1998.
31. ASTM E 2058-00, *Annual Book of ASTM Standards 2000*, Section Four Construction, Vol. 4.07, American Society for Testing and Materials, West Conshohocken, PA, February 10, 2000, pp. 1084–1108. Also NFPA 287, *National Fire Codes*, Vol. 6, National Fire Protection Association, Quincy, MA, 2000, pp. 287-1–287-28.
32. T. J. Ohlemiller and J. R. Shields, Technical Report NISTIR 6143, National Institute of Standards and Technology, Gaithersburg, MD, August 1998, US National Highway Traffic Safety Administration Docket Number NHTSA-1998-3588-26, December 02, 1998.
33. R. E. Lyon, Solid-state Thermochemistry of Flaming Combustion, Technical Report DOT/FAA/AR-99/56, Federal Aviation Administration, Airport and Aircraft Safety, Research and Development Division, William J. Hughes Technical Center, Atlantic City, NJ, July 1999.
34. A. C. Fernandez-Pello and T. Hirano, *Combustion Sci. Tech.* **32**, 1–31 (1983).
35. J. G. Quintiere, in *The SFPE Handbook of Fire Protection Engineering*, 2nd ed., National Fire Protection Association Press, Quincy, MA, 1995, Section 2, Chapter 14, pp. 2-205–2-216.
36. ASTM E 1354-99, *Annual Book of ASTM Standards 2000*, Section Four Construction, Vol. 4.07, American Society for Testing and Materials, West Conshohocken, PA. January 10, 1999, pp. 826–843. Also NFPA 271, *National Fire Codes*, Vol. 6, National Fire Protection Association, Quincy, MA, 2000, pp. 271-1–271-22.
37. ASTM E 906-99, *Annual Book of ASTM Standards 2000*, Section Four Construction, Vol. 4.07, American Society for Testing and Materials, West Conshohocken, PA. February 10, 1999, pp. 733–758. Also NFPA 263, *National Fire Codes*, Vol. 6, National Fire Protection Association, Quincy, MA, 2000, pp. 263-1–263-22.

38. ASTM E 1321-97a, *Annual Book of ASTM Standards 2000*, Section Four Construction, Vol. 4.07, American Society for Testing and Materials, West Conshohocken, PA, June 10, 1997, pp. 789–804.
39. A. Tewarson, I. A. Abu-Isa, D. R. Cummings, and D. E. LaDue, *Fire Safety Science, Sixth International Symposium*, 2000, pp. 991–1002. U.S. National Highway Traffic Safety Administration Docket Number NHTSA-1998-3588-71, December 13, 1999.
40. A. Tewarson, M. M. Khan, P. K. S. Wu, and R. G. Bill, *J. Fire Mat.* **25**, 31–42 (2001).
41. A. Tewarson Technical Report J.I. OB1R7.RC, Factory Mutual Research, Norwood, MA. October 1997, US National Highway Traffic Safety Administration Docket Number NHTSA-1998-3588-1, July 17, 1998.
42. M. M. Hirschler, in V. Babrauskas and S. J. Grayson, ed., *Heat Release in Fires*, E & FN Spon, Chapman Hall, New York,, 1992, pp. 375–422.
43. U. Sorathia and C. Beck, *Improved Fire- and Smoke-Resistant Materials for Commercial Aircraft Interiors A Proceedings*, Committee on Fire- and Smoke-Resistant Materials, National Materials Advisory Board, Commission on Engineering and Technical Systems, National Research Council, Publication NMAB-477-2, National Academy Press, Washington, DC, 1995, pp. 93–114.
44. M. J. Scudamore, P. J. Briggs, and F. H. Prager, *Fire Mat.* **15**, 65–84 (1991).
45. R. A. Orwoll, in J. E. Mark, ed., *Physical Properties of Polymers Handbook*, American Institute of Physics, Woodbury, New York, 1996, Chapter 7, pp. 81–89.
46. Y. Wen, in J. E. Mark, ed., *Physical Properties of Polymers Handbook*, American Institute of Physics, Woodbury, New York, 1996, Chapter 9, pp. 101–109.
47. Y. Yang, in J. E. Mark, ed., *Physical Properties of Polymers Handbook*, American Institute of Physics, Woodbury, New York, 1996, Chapter 10, pp. 111–117.
48. I. A. Abu-Isa, Report R&D 8869, General Motors Research and Development Center, Warren, MI. US National Highway Traffic Safety Administration Docket Number NHTSA-1998-3588-39, March 29, 1999.
49. M. Janssens, *SFPE Handbook of Fire Protection Engineering*, 2nd ed., National Fire Protection Association Press, Quincy, MA, 1995, Section 3, Chapter 2, pp. 3-16–3-36.
50. V. Babrauskus, in *SFPE Handbook of Fire Protection Engineering*, 2nd ed., National Fire Protection Association Press, Quincy, MA, 1995, Section 3, Chapter 2, pp. 3-37–3-52.
51. *Aircraft Materials Fire Test Handbook*, Final Report DOT/FAA/AR-00/12, Federal Aviation Administration, April 2000.
52. ASTM E 662-97, *Annual Book of ASTM Standards 2000*, Section Four Construction, Vol. 4.07, American Society for Testing and Materials, West Conshohocken, PA. February 10, 1999, pp. 684–706. Also NFPA 258, *National Fire Codes*, Vol. 6, National Fire Protection Association, Quincy, MA, 2000, pp. 258-1–258-19.
53. A. Tewarson, F. H. Jiang, and T. Morikawa, *Combustion and Flame* **95**, 151–169 (1993).
54. A. Tewarson, Technical Report OB1R7.RC (1998), Factory Mutual Research, Norwood, MA, August 1998.
55. A. Tewarson, J. L. Lee, and R. F. Pion, *Eighteenth Symposium (International) on Combustion*, Combustion Institute, Pittsburgh, PA, 1981, pp. 563–570.

56. J. de Ris, *Seventeenth Symposium (International) on Combustion*, Combustion Institute, Pittsburgh, PA, 1979, pp. 1003–1016.
57. P. Joulain, *Twenty-Seventh Symposium (International) on Combustion*, Combustion Institute, Pittsburgh, PA, 1998, pp. 2691–2706.
58. J. G. Quintiere, M. Harkleroad, and Y. Hasemi, *Combustion Sci. Tech.* **48**, 191–222 (1986).
59. L. Orloff, J. L. de Ris, and G. H. Markstein, *Fifteenth Symposium (International) on Combustion*, Combustion Institute, Pittsburgh, PA, 1974, pp. 183–192.
60. L. Orloff, A. T. Modak, and R. L. Alpert, *Sixteenth Symposium (International) on Combustion*, Combustion Institute, Pittsburgh, PA, 1976, pp. 1345–1354.
61. P. K. S. Wu and A. Tewarson, *Interflame96—Seventh International Fire Science and Engineering Conference*, March 26–28, 1996, St. John's College, Cambridge, UK, 1996, pp. 159–168.
62. I. Glassman, *Twenty-Second Symposium (International) on Combustion*, Combustion Institute, Pittsburgh, PA. 1988, pp. 295–311.
63. P. K. S. Wu and J. L. Chaffee, Technical Report J.I.0003000780, Factory Mutual Research, Norwood, MA, Feb. 2000.
64. Number 571.302, Standard No. 302: *Flammability of Interior Materials*, 49CFR Ch.V (10-1-98 Edition), 1998.
65. ASTM D 2863-70, *Annual Book of ASTM Standards 2000*, Section Eight Plastics, Vol. 8.02 Plastic (II), American Society for Testing and Materials, West Conshohocken, PA, 1999, pp. 154–167.
66. NFPA 53, 1999 Edition, *Recommended Practice on Materials, Equipment, and Systems Used in Oxygen-Enriched Atmospheres*, National Fire Protection Association, Quincy, MA, 1999.
67. ASTM E 162-98, *Annual Book of ASTM Standards 2000*, Section Four Construction, Vol. 4.07, American Society for Testing and Materials, West Conshohocken, PA, March 10, 1999, pp. 619–627.
68. C. J. Hilado, *Flammability Handbook for Plastics*, Technomic, Stanford, CT, 1969.
69. Test Procedures and Performance Criteria for the Flammability and Smoke-Emission Characteristics of Materials Used in Passenger Cars and Locomotive Cabs, *Fed. Reg.*, Rules and Regulations, Vol. 64, No. 91, Wednesday, May 12, 1999.
70. Department of Transportation, Federal Transit Administration, Docket 90-A. Recommended Fire Safety Practices for Transit Bus and Van Materials Selection, *Fed. Reg.*, Vol. 58, No. 201, Wednesday, October 20, 1993.
71. ASTM E 84-00a, *Annual Book of ASTM Standards 2000*, Section Four Construction, Vol. 4.07, American Society for Testing and Materials, West Conshohocken, PA, April 10, 2000, pp. 557–574. Also NFPA 255, *National Fire Codes*, Vol. 6, National Fire Protection Association, Quincy, MA, 2000, pp. 255-4–255-17.
72. NFPA 101 Life Safety Code, Chapter 10 Annex, *National Fire Codes—A Compilation of NFPA Codes, Standards, Recommended Practices and Guides*, Vol. 5, National Fire Protection Association, Quincy, MA, 2000, pp. 101-306–101-307.
73. ASTM E 648-99, *Annual Book of ASTM Standards 2000*, Section Four Construction, Vol. 4.07, American Society for Testing and Materials, West Conshohocken, PA, March 10, 1999, pp. 670–683. Also NFPA 253, *National Fire Codes*, Vol. 6, National Fire Protection Association, Quincy, MA, 2000, pp. 253-4–253-15.

74. I. A. Benjamin and C. H. Adams, *Fire J.* **70**(2), 63–70, March (1976).
75. S. Davis, J. R. Lawson, and W. J. Parker, Technical Report NISTIR 89–4191, National Institute of Standards and Technology, Gaithersburg, MD, October 1989.
76. K. Tu, and S. Davis, Technical Report NBSIR 76–1013, National Bureau of Standards, Gaithersburg, MD, 1976.
77. NFPA 318, *National Fire Codes*, Vol. 6, National Fire Protection Association, Quincy, MA, 2000, pp. 318-1–318-22.
78. FMRC Test Standard 4910, *Clean Room Materials Flammability Test Protocol*, Factory Mutual Research, Norwood, MA, September 1997.
79. UL 2360, *Standard Test Method for Determining the Combustibility Characteristics of Plastics Used in Semiconductor Tool Construction*, Underwriters Laboratory, Northbrook, IL, 2000.
80. A. Tewarson, and M. M. Khan, *Twenty-Second Symposium (International) on Combustion*, Combustion Institute, Pittsburgh, PA, 1988, pp. 1231–1240.
81. A. Tewarson, and D. Macaione, *J. Fire Sci.* **11**, 421–441 (1993).
82. A. Tewarson, Technical Report ARL-CR-178, Army Research Laboratory, Aberdeen Proving Ground, MD, 1994.
83. A. Tewarson, R. G. Bill, R. L. Alpert, A. Braga, V. DeGiorgio, and G. Smith, *Flammability of Clean Room Materials*, J.I. OYOE6.RC, Factory Mutual Research, Norwood, MA, 1999.
84. A. Tewarson and M. M. Khan, Technical Report J.I. 0M2E1.RC, Factory Mutual Research, Norwood, MA, January 1989.
85. M. M. Khan, Technical Report J.I. OT1E2.RC, Factory Mutual Research, Norwood, MA, June 1991.
86. V. B. Apte, R. W. Bilger, A. Tewarson, G. J. Browning, R. D. Pearson, and A. Fidler, Grant Report ACARP-C5033, Workcover NSW, Londonderry Occupational Safety Center, Londonderry, NSW, Australia, December 1997.

NOMENCLATURE

A	Total exposed surface area of the polymer (m^2)
a	Surface absorptivity
c	Heat capacity (J/g·K)
C	Carbon atom
Cl	Chlorine atom
D_s	Specific optical density, $(V/AL)\log_{10}(100/T)$
f_j	Generation efficiency of product j
F	Fluorine atom
$\dot{G}''_j$	Generation rate of product j per unit polymer surface area (g/m^2·s)
H	Hydrogen atom
ΔH_{ch}	Chemical heat of combustion (kJ/g)
ΔH_d	Heat of decomposition of the polymer (kJ/g)
ΔH_g	Heat of gasification of the polymer (kJ/g)
ΔH_m	Heat of melting (kJ/g)

NOMENCLATURE

ΔH_T	Net heat of complete combustion (kJ/g)
ΔH_v	Heat of vaporization of the polymer (kJ/g)
k	thermal conductivity (W/m·K)
L	Optical pathlength through smoke (m)
$\dot{m}''$	Generation rate of polymer vapors (g/m²·s)
M	Molecular weight of the monomer (g/mol)
N	Nitrogen atom
O	Oxygen atom
$\dot{q}_e''$	External heat flux per unit surface area of the polymer (kW/m²)
$\dot{q}_{cr}''$	Critical heat flux per unit surface area of the polymer (kW/m²)
$\dot{q}_f''$	Flame heat flux per unit surface area of the polymer (kW/m²)
$\dot{q}_{rr}''$	Surface re-radiation loss per unit surface area of the polymer (kW/m²)
$\dot{Q}_{ch}''$	Chemical heat release rate per unit surface area of the polymer (kW/m²)
$\dot{Q}$	Heat release rate (kW)
s	Stoichiometric mass air-to-fuel ratio (g/g)
S	Sulfur atom
t	Time (s)
T	Percent light transmission through smoke
T_a	Ambient temperature (°C)
T_b	Boiling point (°C)
T_d	Decomposition temperature (°C)
T_{gl}	Glass transition temperature (°C)
T_{ig}	Ignition temperature (°C)
T_m	Melting temperature (°C)
T_v	Vaporization temperature (°C)
V	Volume of closed chamber (m³)
y_j	Yield of product j (g/g)

Greek

α	Thermal diffusivity (mm/s)
δ	Thermal penetration depth (mm)
χ	Combustion efficiency
ρ	Density (g/cm³)
Ψ_j	Maximum possible mass stoichiometric yield of product j (g/g)
Φ	Total mass of oxygen available for combustion per unit total mass of polymer vapors burned normalized by the mass stoichiometric oxygen-to-fuel ratio
Ω	Flame heating parameter (kW²/m³)

Subscripts

a	original, ambient
ch	Chemical
cal	Calculated
d	Decomposition

488 FLAMMABILITY OF POLYMERS

f	Flame
g	Gasification
ig	Ignition
r	Radiative
l	Molten polymer
meas	Measured
n	Net
s	Solid polymer
sm	Smoke
T	Total or complete
V	Vaporization

Superscripts

| . | per unit of time (1/s) |
| ″ | per unit area (1/m^2) |

General Abbreviations

CHF	Critical heat flux (kW/m^2)
CRF	Critical radiant flux (kW/m^2)
FR	Fire retarded
FPI	Fire propagation index (m/s$^{1/2}$)/(kW/m)$^{2/}$
HRP	Heat release parameter ($\Delta H_{ch}/\Delta H_g$; kJ/kJ)
PGP	Product generation parameter ($y_j/\Delta H_g$; g/kJ)
TRP	Thermal response parameter (kW·s$^{1/2}$/m^2)

Abbreviations for Polymers

ABS	Acrylonitrile–butadiene–styrene copolymer
CPVC	Chlorinated polyvinyl chloride
ECTFE	Ethylenechlorotrifluoroethylene
EPDM	Ethylene–propylene–diene rubber copolymer
EVA	Ethylene–vinylacete copolymer
EVAL	Ethylene–vinyl alcohol copolymer
ETFE	Ethylene–tetrafluoroethylene copolymer
FEP	Perfluoroethylene–propylene copolymer
HFP	Hexafluoropropylene
MFA	Tetrafluoroethylene–perfluoromethyl vinyl ether copolymer
PA	Polyamide
PAI	Polyamide imide
PAN	Polyacrylonitrile
PBT	Polybutylene terephthalate
PC	Polycarbonate
PCTFE	Polychlorotrifluoroethylene
PE	Polyethylene
PE-Cl	Chlorinated polyethylene
PE-HD	Polyethylene–high density
PE-LD	Polyethylene–low density
PEI	Polyetherimide

PEEK	Polyetheretherketone
PEK	Polyetherketone
PES	Polyethersulfone
PET	Polyethylene terephthalate
PEU	Polyetherurethane
PESPU	Polyester polyurethane
PF	Phenol-formaldehyde
PFA	Perfluoro-alkoxyalkane copolymer
PI	Polyimide
PIB	Polyisobutylene
PMMA	Polymethylmethacrylate
PMVE	Perfluoromethyl vinyl ether
POM	Polyoxymethylene, Polyformaldhyde, Polyacetal
PP	Polypropylene
PP-Cl	Polypropylene-chlorinated
PPO	Polyphenylene oxide
PPS	Polyphenylenesulfide
PS	Polystyrene
PPVE	Perfluoropropyl vinyl ether
PSF	Polysulfone
PTFE	Polytetrafluoroethylene
PU	Polyurethane
PVAC	Polyvinyl acetate
PVAL	Polyvinyl alchol
PVC	Polyvinyl chloride
PVDC	Polyvinylidene chloride
PVDF	Polyvinylidene fluoride
PVF	Polyvinyl fluoride
SAN	Styrene–acrylonitrile copolymer
SB	Styrene–butadiene copolymer
SMC	Sheet molding compound
Si	Silicone
TFE	Tetrafluoroethylene
TPO	Thermoplastic polyolefin
VDF	Vinylidene fluoride
XLPE	Crosslinked polyethylene
XLPO	Crosslinked polyolefin

CHAPTER 12

BIODEGRADABLE WATER-SOLUBLE POLYMERS

GRAHAM SWIFT
GS Polymer Consultants, Chapel Hill, North Carolina

12.1. INTRODUCTION

The scope of this chapter is the synthesis of water-soluble polymers designed to biodegrade in the environment. It includes definitions and testing protocols and many of the synthesis approaches evaluated for biodegradable water-soluble polymers regardless of their degree of ultimate success. Structurally similar water-soluble biodegradable polymers designed for biomedical and controlled drug delivery applications are not included. There are some necessary comparisons and correlations with the extensive work and literature on biodegradable plastics that are essential for the understanding of biodegradable water-soluble polymers and their development.

Interest in biodegradable polymers began over 40 years ago with the recognition that common commodity packaging plastics including polyethylene (PE), polypropylene (PP), polystyrene (PS), poly(vinyl chloride) (PVC), and poly(ethylene terephthalate) (PET) were accumulating in the environment. Their carefully designed and well-established resistance to environmental degradation was observed to be contributing to landfill depletion and litter problems due to careless disposal after use. At that time, the major focus was on replacing these synthetic packaging plastics with environmentally biodegradable substitutes.

More recently, in the last 25 years, it has become increasingly apparent that, in addition to the major commodity synthetic plastics, water-soluble commodity and specialty polymers and plastics, such as poly(acrylic acids), polyacrylamide, poly(vinyl alcohol), poly(alkylene oxides), and even some modified natural polymers, for example, cellulosics and starch, may potentially contribute to environmental problems and should also be targets for biodegradable substitutes.

Plastics and the Environment, Edited by Anthony L. Andrady.
ISBN 0-471-09520-6 © 2003 John Wiley & Sons, Inc.

Water-soluble polymers are widely used as coatings additives, temporary packaging, temporary coatings, pigment dispersants, mining, water treatment, detergents, and the like. The fact that they are water soluble and pass unseen into the environment is not an indicator that they are biodegradable, which for so long seemed to be the accepted reasoning. Just the opposite pertains, water-soluble polymers are capable of moving freely through the aqueous environment with the potential for many deleterious effects unless proven otherwise. Demonstrated biodegradation or proof of no significant fate and effects are essential for these polymers to be considered environmentally benign. The surest way of confirming acceptability is to establish biodegradability on disposal in any aqueous compartment they are likely to contact and preferably in the primary disposal site.

There are many well-documented similarities in the chemistries, biodegradation requirements, and synthetic routes evaluated in the development of biodegradable plastics and water-soluble polymers [1–35], which are recommended reading even though this chapter focuses only on approaches to the latter. Biodegradable water-soluble polymer design and synthesis have been widely based on natural renewable resources, nonrenewable petrochemical sources, and combinations of the two. They are, primarily, with few exceptions such as temporary coatings and water-soluble packaging polymers and plastics, designed to be biodegradable on disposal in aqueous environments after use, such as municipal wastewater treatment plants, lakes, and rivers. Ideally, as most of these polymers are discarded in sewage streams, they should biodegrade completely in industrial or municipal wastewater treatment plants. This would ensure that they do not enter rivers and streams where they can readily move around the environment with unknown fate and effects, unless biodegradation can be established in these subsequent compartments.

Water-soluble biodegradable polymers may degrade in the environment by any of the well-established and accepted degradation pathways for biodegradable plastics, photodegradation, biodegradation, and chemical degradation, which includes hydrolytic and oxidative degradation. However, it should be recognized that, of these pathways, only biodegradation can lead to complete removal from the environment, and accordingly it has justifiably received the highest attention. The other degradation pathways are appropriately described as environmental degradation, deterioration, or disintegration to produce fragments or other chemicals that may or may not biodegrade.

There are alternative disposal technologies competing with biodegradation for the waste management of plastics such as recycling (which includes recycling of plastics), recycling of plastics to monomers and subsequent repolymerization to the same or new polymers, recycling to olefinic feedstocks by pyrolysis, burial in landfill sites, and incineration. In contrast, with the exception of incineration, none of these alternatives is readily applicable to water-soluble polymers due to the difficulty of recovering them from a dilute aqueous disposal stream after use. Incineration of water-soluble polymers that are adsorbed on wastewater treatment plant sludge is practiced in some locations.

Acceptable definitions for environmentally degradable and biodegradable polymers and meaningful laboratory testing protocols for quantitatively measuring biodegradation and environmental fate and effects are essential for progress in the design and development of biodegradable polymers. It is critically important to establish that biodegradation test protocols correlate closely with real-world exposures on disposal in order to have confidence in their predictability for new polymers. Consequently, definitions and test methods will be addressed early in this review, prior to describing many of the important approaches evaluated for the synthesis of biodegradable water-soluble polymers. Synthesis approaches mainly fall into the two broad categories indicated below:

- Synthetic polymers
  - Carbon–carbon chain polymers
  - Heteroatom chain polymers
- Modified natural polymers
  - Graft polymers
  - Chemically modified polymers

The chapter concludes with some projections for future directions for research and commercial opportunities for environmentally biodegradable water-soluble polymers.

12.2. DEFINITIONS

There have been numerous definitions proposed for environmentally degradable and biodegradable water-soluble polymers and plastics; almost every article written on the subject includes the author's favorite definitions. All tend to encompass the same broad concepts but are slightly different in phraseology depending on the author's perspective and discipline, that is, chemist, biochemist, layman, lawyer, legislator, and so forth. This is both fortunate and unfortunate since, on the one hand, it should ensure that there is broad understanding of the problem but, on the other hand, it may also mean that we will never reduce to words an acceptable definition that has worldwide consensus. Consensus may be better reached by addressing the specifications for polymers in various degradation pathways as established by the testing protocols to be discussed later. However, definitions are important because they are indicative of general expectations for the acceptance of environmentally degradable and biodegradable polymers and of the types of testing protocols that are needed to establish the acceptability of polymers in a particular environment. At this time, the definitions developed by the American Society for Testing and Materials (ASTM) [1] for degradable, biodegradable, photodegradable, hydrolytically degradable, and oxidatively degradable plastics, indicated below, are probably the most widely accepted as written or in some slightly modified form. They are equally applicable to water-soluble polymers and are used as such in this chapter.

Degradable plastic, a plastic designed to undergo a significant change in its chemical structure under specific environmental conditions resulting in a loss of some properties as measured by standard test methods appropriate to the polymer

Biodegradable plastic, a degradable plastic in which the degradation results from the action of naturally occurring microorganisms such as bacteria, fungi, and algae

Hydrolytically degradable plastic, a degradable plastic in which the degradation results from hydrolysis

Oxidatively degradable plastic, a degradable plastic in which the degradation results from oxidation

Photodegradable plastic, a degradable plastic in which the degradation results from the action of natural sunlight

It is immediately apparent that these definitions do not quantify the extent of degradation in any of the pathways defined; they are only indicative of the mechanism that is operating to promote degradation. While this is acceptable in a scientific sense to define a process, the definitions do not satisfy the requirement for environmentally acceptable degradable and biodegradable polymers, which, in the minds of legislators and laypeople, is the key issue. As indicated above, specifications for acceptability have to be set and then monitored by the standard testing protocols to be discussed later. The environmental degradation processes are interrelated, as shown schematically in Figure 12.1.

All four environmental degradation pathways for polymers, biodegradation, oxidation, hydrolysis, and photodegradation initially give intermediate products or fragments. These may (bio)degrade further to some other residue, biodegrade completely and be removed from the environment entirely, ultimately mineralized as indicated in Figure 12.1, or remain unchanged in the environment. It should be noted that mineralization is a slow process that refers to complete conversion of a

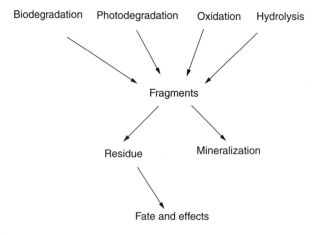

Figure 12.1. Interrelationships of environmental degradation pathways for polymers.

polymer (or any organic compound) to carbon dioxide and/or methane (depending on aerobic or anaerobic environment), water, and salts. It is used here loosely to indicate complete or total removal from the environment to carbon dioxide or methane, water, and biomass. In the instances where residues remain, these must be established as harmless in the environment by suitably rigorous fate-and-effect evaluations. Clearly, only biodegradation has the potential to remove polymers completely from the environment, and this should be recognized when developing and designing polymers for degradation by any of the other pathways. The final degradation stage should preferably be complete biodegradation and removal from the environment with ultimate mineralization. In this way, the polymers are recycled through nature into microbial cells, plants, and higher animals [2] and then back into renewable resource chemical feedstocks.

Based on the above arguments, an acceptable environmentally degradable polymer may be defined as one that degrades by any of several, defined mechanisms — biodegradation, photodegradation, oxidation, or hydrolysis — to leave no harmful residues in the environment. This definition has the advantage of not limiting the rate or degree of degradation for a particular polymer but requiring sufficient testing of fragments and degradation products that are incompletely removed from the environment to ensure no long-term damage or adverse effects to the ecological system. Polymers meeting this definition should be acceptable for disposal in the appropriate environment anywhere in the world. But, it should be emphasized that the most significant goal for water-soluble polymers should be environmental biodegradation, regardless of the mechanisms involved prior to the biodegradation stage, with complete removal from the disposal environment. This categorically precludes any unwanted adverse environmental impacts and prolonged fate-and-effects evaluations.

12.3. OPPORTUNITIES FOR BIODEGRADABLE WATER-SOLUBLE POLYMERS

Opportunities are the drivers for research into environmentally biodegradable polymers and for the development of laboratory testing protocols. It is opportunities that indicate the property and disposal requirements for the polymer. Disposal methods and locations identify the testing protocols that must be established for polymers in order to evaluate their environmental degradation under laboratory-simulated environmental exposure conditions. As already mentioned, the drivers for environmentally biodegradable polymers are waste management programs to avoid environmental contamination from the use for nondegradable polymers. Figure 12.2 indicates likely disposal pathways for water-soluble polymers.

Water-soluble polymers, after use, are usually in very dilute aqueous solutions and are preferably disposed of through wastewater treatment facilities or sometimes directly into the aquatic environment, as shown schematically. On entering a wastewater treatment plant, a polymer may pass straight through, remaining soluble, into streams, rivers, lakes, and other aquatic environments, or it may be adsorbed onto the suspended solids in the treatment plant. If it

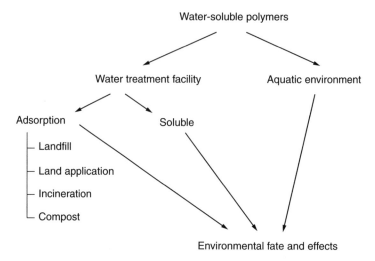

Figure 12.2. Environmental disposal options for water-soluble polymers.

passes straight through, it is no different from direct disposal into those aqueous environments, and both raise similar questions as to the fate and effects of the polymers involved. On the other hand, adsorption of a polymer onto sewage sludge results in the possibility of the polymer being land filled, incinerated, composted, or land applied as fertilizer or for soil amendment, depending on the local options available. However, here again, environmental fate and effects need to be addressed, it must be established how and where these polymers move in their new environmental compartments and where incineration products are ultimately deposited. Clearly, for water-soluble polymers there is a preference for complete biodegradation in wastewater treatment facilities. Polymers should be designed to be completely biodegradable and removed in the immediate disposal compartment, generally the sewage treatment facility. If biodegradation is not complete in the disposal environment, it must be assessed, often with difficulty, in subsequent compartments. Once the biodegradation of a polymer has been confirmed in any environmental compartment, no uncertainty remains as to its fate and effects. The advantage of complete biodegradation is that it is a positive result that can be established, whereas the assessment of environmental fate and effects are risk assessments and are negative results, based on limited studies with a few aquatic species. These results may not be representative of all the species that a water-soluble polymer, or its partial degradation fragments, may contact as it moves through many aqueous environment compartments.

A major difference in the waste management of water-soluble polymers and plastics using biodegradable substitutes is that there is a well-established widely available disposal infrastructure of wastewater treatment plants for the former, whereas plastics being developed for composting will require the large-scale implementation of a composting infrastructure; and this will slow their acceptance.

Hence, given the need for biodegradable water-soluble polymers to protect the environment and the ready availability of a disposal infrastructure in almost every country in the world, the opportunity is obvious and waiting a solution.

12.4. TEST METHODS FOR BIODEGRADABLE WATER-SOLUBLE POLYMERS

As indicated earlier, environmentally biodegradable polymers fall into degradation categories such as biodegradation, photodegradation, oxidation, and hydrolysis. Biodegradation has received practically all the attention in the case of water-soluble polymers in the environment, as the other modes of degradation are less likely to occur, although a few reported studies have shown that some water-soluble polymers degrade by hydrolysis, ozonolysis, and free-radical oxidation to low-molecular-weight fragments, which may or may not subsequently biodegrade. However, no standard test methods correlate with these observations at the present time.

Test methods attempt to define and develop laboratory simulations of the real-world conditions to which a particular polymer will be exposed on disposal. Results are expected to indicate the rate and extent of biodegradation under test conditions, which should correlate with real-world environmental exposure. The ultimate goal of laboratory test methods, therefore, is to be predictable of environmental response to new polymers. This goal is difficult since the environments for biodegradation differ widely in microbial composition, pH, temperature, moisture, and the like and are not easily reproduced in the laboratory. Once an environment has been sampled and placed within the confinements of a laboratory vessel, it no longer interacts with the greater environment in response to an added xenobiotic, and results may not always be representative of real-world exposures. Consequently, biodegradation observed in the laboratory is generally consistent with biodegradation in the environment, but lack of biodegradation in the laboratory is not conclusive of recalcitrance in the environment. This should be clearly understood by all in the field of designing and developing biodegradable polymers. Repeated laboratory testing with a variety of environmental samples is a recommended practice before deciding any approach should be terminated.

Biodegradation may occur either aerobically or anaerobically, the former environment has received the most attention, as oxygen is usually present in disposal environments for water-soluble polymers. However, anaerobic conditions do occur, for example, in anaerobic digesters in sewage treatment plants. The chemistries of the two different biodegradation pathways are expressed by the equations below:

Aerobic Environments

$$\text{Polymer} + O_2 \longrightarrow CO_2 + H_2O + \text{Biomass} + \text{Residue}$$

Anaerobic Environments

$$\text{Polymer} \longrightarrow CO_2 + CH_4 + \text{Biomass} + \text{Residue}$$

To quantitatively assess the degree of biodegradation, analytical techniques are needed for all the reactants and products. Usually, test methods measure oxygen uptake, biochemical oxygen demand (BOD), or CO_2 production for aerobic degradation and CO_2 and CH_4 production for anaerobic degradation. Analysis for residue should also be done where possible as this is a good indicator of biodegradation also. It is, also, usual to run blank tests, with killed inoculums, in concert with the authentic tests with a living environment to differentiate abiotic and biotic degradation.

Water-soluble polymers are readily tested for biodegradation in simple tests, but results may often be difficult to interpret because of variability due to issues mentioned above. Hence, several tests of increasing complexity are often coupled to determine whether a polymer is biodegradable. Simple tests such as biochemical oxygen demand (BOD) as a fraction of theoretical oxygen demand (ThOD) or chemical oxygen demand (COD) in aerobic environments may be misleading as the result may be due to competing abiotic oxidation. Advanced tests such as the measurement of CO_2 (CO_2 and CH_4 for anaerobic environments) and removal of soluble residues or total organic content (TOC) in laboratory-simulated sewage plants are more definitive. However, it must be ascertained whether removal of TOC is by biodegradation or adsorption on solids. Ultimately, the total fate of the polymer needs to be established to claim biodegradability. To do this categorically, isotopic labeling of the polymer with, for example, radiocarbon or tritium, is often used and recommended for complete and thorough confirmative testing. A very comprehensive and authoritative account of biodegradation testing for water-soluble polymers and organic chemicals is given by Swisher [7]. Additionally, work by Larson and Swift and their co-workers, especially for water-soluble polycarboxylates [36, 39], and others [37, 38, 40], are recommended reading as they are aimed at improving test methodologies.

In addition to the above test methods, microbial activity in polymer solutions has been used to assess biodegradation by measuring population growth using optical density as an indicator of activity. This approach is amenable to the quantification of metabolites, which is useful to establish biodegradation mechanisms. Analytical tools such as mass spectroscopy and gas chromatography are widely used. Gel permeation chromatography is also used to monitor changes in polymer molecular weight and molecular weight distribution as an indicator of biodegradation. Careful attention to molecular weight and distribution changes can be used to establish endo-biodegradation (chain random scission) or exo-biodegradation (terminal scissions) mechanisms.

As a precautionary note, when working with biodegradable water-soluble polymers, as with any polymers, it is important to recognize that low-molecular-weight fractions and impurities are usually present. These may be more biodegradable than higher molecular weight fractions, and this may give rise to false-positive results, particularly in short-term tests such as 5-day BOD tests sometimes run as rapid screens for biodegradation.

In many of the examples to be considered in this chapter, the reader is warned that often results claimed by the original authors are not always substantiated due to difficulty in estimating test reliability. However, the references are included as the polymers are often anticipated to be biodegradable to some extent and hence are a good indicator for future research directions and opportunities.

12.5. SYNTHESIS OF BIODEGRADABLE WATER-SOLUBLE POLYMERS

Water-soluble polymers may be anionic, cationic, or nonionic in character, which influences both their applications and their biodegradability. They may be totally synthetic, totally natural, or synthetically modified natural polymers. The synthetics may be subdivided into addition polymers and condensation polymers, the former being mostly carbon chain backbone polymers and the latter being chiefly heteroatom chain backbone polymers. Poly(acrylic acid) and poly(vinyl alcohol) are representative of carbon chain polymers, which may be considered functional polyolefins. Polyesters, polyamides, and polyethers are examples of condensation polymers, which are structurally similar to many naturally occurring polymers. Hence these polymers are good structural models for the development of synthetic water-soluble biodegradable polymers. The predominant commercially successful synthetic water-soluble polymers, on the other hand, are largely carbon chain polymers, and these are usually resistant to fast or measurable biodegradation as are their counterparts in nature such as lignins, rubbers, and the like. Nevertheless, cost drives commercial success, and much effort has been expended, as we shall see, on trying to develop carbon chain biodegradable water-soluble polymers, especially carboxylic acids.

All natural polymers are considered to be biodegradable or at least not harmful to the environment. However, modification of natural polymers either by grafting synthetic polymers or by chemical conversions such as oxidation and esterification, changes their properties and biodegradation characteristics significantly. Therefore, polymers produced by any of these modifications must be evaluated for biodegradability in the same manner as purely synthetic polymers.

12.5.1. Carbon Chain Polymers

Functional carbon chain backbone polymers show limited propensity to biodegrade at useful molecular weight (M_n), which are usually >1000. Suzuki et al. [41] demonstrated this limitation very elegantly by ozonizing higher molecular weight polymers such as polyacrylic acid, polyvinyl pyrrolidone, and polyacrylamide to oligomers of molecular weight <1000, which were much more biodegradable, though polyacrylamide was slower to biodegrade than the other two polymers. Suzuki et al. similarly showed that polyethylene oxide and polyvinyl alcohol, both of which are established to be biodegradable over low- to high-molecular-weight ranges, biodegraded more rapidly at lower molecular weight.

The biodegradability of functional derivatives of polyethylene, particularly polyvinyl alcohol and polyacrylic acid and derivatives have received attention because of their water solubility, high-volume use, and disposal into the aqueous environment. Polyvinyl alcohol is used in a wide variety of applications, including textiles, paper, plastic films, and temporary packaging, and polyacrylic acid is widely used in detergents as a builder, super absorbent for diapers and feminine hygiene products, water treatment, thickeners, pigment dispersant, and the like.

Poly(Vinyl Alcohol) Poly(vinyl alcohol), obtained by the hydrolysis of polyvinyl acetate is probably the only carbon chain polymer widely accepted to be fully biodegradable in the environment and confirmed in current standard tests. But, even in these laboratory tests, acclimation is usually essential to effect rapid biodegradation, indicating the complexity of laboratory testing mentioned earlier. The biodegradation mechanism is considered to be a random chain cleavage of 1,3 diketones formed by an enzyme-catalyzed oxidation of the polymeric secondary alcohol functional groups. Biodegradation was first observed by Yamamoto et al. [42] as a reduction in aqueous viscosity of the polymer in the presence of soil bacteria. Subsequently, Suzuki et al. [43] identified a *Pseudomonas* species as the soil bacteria responsible for the degradation over a degree of polymerization range of 500–2000 utilizing the polymer as a sole carbon source. An aqueous polymer solution at a concentration of 2700 ppm was reduced to 250–300 ppm concentration in 7–10 days at pH 7.5–8.5 and 35–45° C.

An oxidative endo mechanism was proposed based on the data shown in Table 12.1 and later substantiated [44] by quantifying the oxygen uptake at one mole for every mole of hydrogen peroxide produced and identifying the degradation products as ketones and carboxylic acids, as shown in Scheme 1. Included in the diagram is the alternative mechanism proposed by Watanabe et al. [45, 46] in

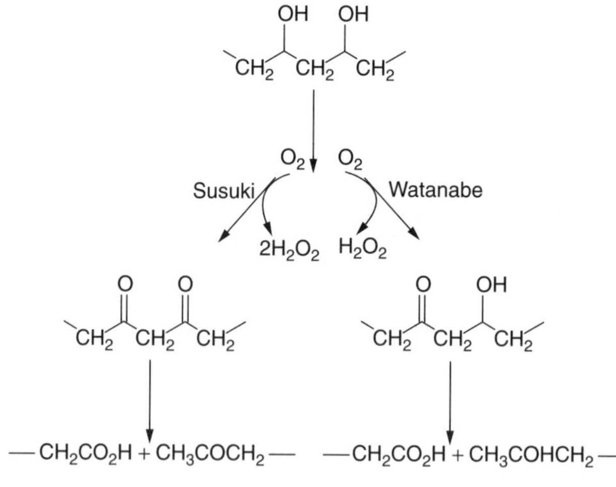

Scheme 1. Biodegradation pathways for polyvinyl alcohol.

Table 12.1 Consumption of Oxygen and Stoichiometry of Metabolites in the Biodegradation of Polyvinyl Alcohol

Reaction time (h)	Oxygen Consumed (μM)	Hydrogen Peroxide (μM)	Ketones (μM)	Carboxylic Acids (μM)
1.0	2.92	2.94	ND[a]	ND
2.0	5.40	5.15	ND	ND
6.0	ND	4.0	1.8	1.8
24.0	ND	7.2	3.5	3.1

[a]ND Not determined.

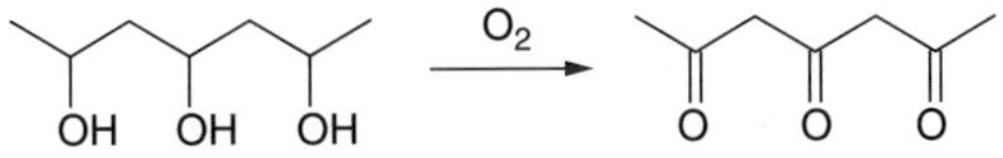

Scheme 2. Oxidation of polyvinyl alcohol.

which the products were identified as an alcohol and a carboxylic acid. This was subsequently proved in error and the Suzuki mechanism is now widely accepted for those microbes evaluated in his work.

The proposed biodegradation mechanism is supported by the rapid biodegradation of the polyketone obtained by the chemical oxidation of polyvinyl alcohol [47–49] shown in Scheme 2.

Other bacterial strains identified as biodegrading polyvinyl alcohol include *Flavobacterium* [50], *Acinetobacter* [51], and many others as well as fungi, molds, and yeasts [52]. Industrial evaluations at DuPont [53] and Air Products [54] indicate that over 90% of polyvinyl alcohol entering wastewater treatment plants is removed in those locations and hence no environmental pollution is likely.

Polyvinyl acetate, the precursor for poly(vinyl alcohol), hydrolyzed to less than 70%, is claimed to be nonbiodegradable under conditions similar to those that biodegrade the fully hydrolyzed polymer [55]. There are several packaging applications of partially hydrolyzed polyvinyl acetate whereby the rate of biodegradation is retarded deliberately to give an acceptable use life for the package. These include hot water disposable films and fabrics [56], body wrappings for mortuary use [57], moldings [58], and microfibrils for synthetic paper pulp [59].

Blends of poly(vinyl alcohol) with polycaprolactone [60] and ethylene-maleic anhydride copolymers [61] are also used in biodegradable packaging, though the latter blend is unlikely to be biodegradable.

Other interesting articles and reviews include polyvinyl alcohol composites, generally [62], with proteins [63–65], starch [66], chitosan [67], simple sugars [68], and a Baeyer Villiger oxidation [95] (Scheme 3) of a an ethylene–vinyl alcohol copolymer to produce a biodegradable polyhydroxypolyester.

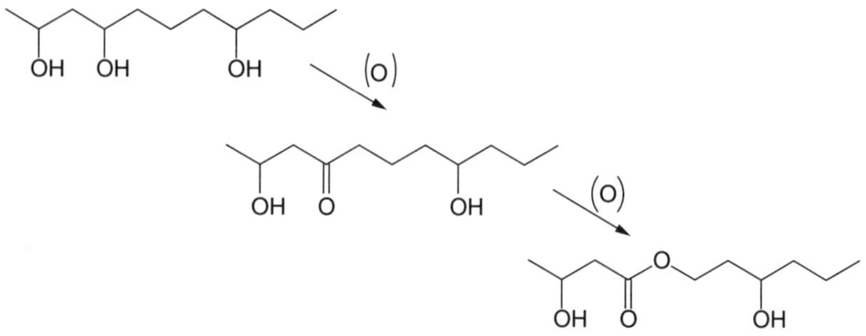

Scheme 3. Baeyer Villiger oxidation of ethylene–vinyl alcohol copolymer.

Polycarboxylates Carboxylate derivatives of poly(vinyl alcohol) are biodegradable and functional in detergents as co-builders, although too costly to be practical replacements for polyacrylic acid at this time. Matsumura et al. polymerized vinyloxyacetic acid [69, 70] and Lever has patented polymers based on vinyl carbamates obtained from the reaction of vinyl chloroformates and amino acids such as aspartic and glutamic acids [71]. Both hydrolyze (Scheme 4), to polyvinyl alcohol, which is biodegradable.

Copolymers of vinyl alcohol with acrylic and/or maleic acid have been evaluated in detergents as potentially biodegradable co-builders in a number of laboratories [72–74], but the results were not encouraging for balancing biodegradation and performance. Higher than 80 mol % of vinyl alcohol is required for biodegradation, and less than 20 mol % for acceptable performance.

The use of polycarboxylic acids in detergents, particularly poly(acrylic acids), and the search for biodegradable alternatives is well established and has been well reviewed in many articles mentioned earlier [12, 18–21] and by Hunter

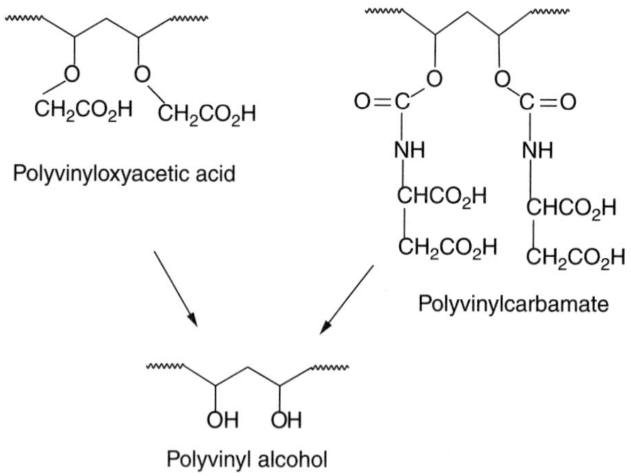

Scheme 4. Hydrolysis of carboxylated poly(vinyl alcohol) derivatives.

et al. [75]. The current polymers have little biodegradability at preferred performance molecular weights, ca. 5000 Dal for polyacrylic acid and 70,000 Dal for copolyacrylic–maleic acids. This, even though there are data to indicate no harmful environmental effects, has resulted in a massive search for biodegradable replacements. The value of negative results mentioned earlier is strongly reflected here.

Many efforts to radically copolymerize acrylic and maleic acids with a whole range of vinyl monomers to produce biodegradable polymers [76–83], and graft substrates, including polysaccharides [84], have failed to produce completely biodegradable water-soluble polycarboxylates. Therefore, it must be considered that the pioneering biodegradation work of Suzuki et al. [41] with carbon chain backbone functional polymers is probably correct and only low-molecular-weight oligomeric carbon chain polymers are likely to be biodegradable, at least at any reasonable rate. Recent confirmation of Suzuki's work has come from Kawai [85] and research in NSKK and Idemitsu Laboratories in Japan [86, 87].

As indicated earlier, Suzuki et al. [41] ozonized high-molecular-weight polyacrylic acid, polyacrylamide, and polyvinyl pyrrolidone to oligomers with molecular weights less than 1000 Dal and observed a marked increase in their biodegradability. The work of Kawai [85] and Tani [86] (NSKK) was based on carefully synthesized oligomers. The results all indicate that polyacrylic acids are not completely biodegradable above about a degree of polymerization of 6–8 (400–600 Dal). The work reported by Larson and Swift in their methodology [39] development also supports this result.

Other efforts to use radical polymerization to synthesize carbon chain biodegradable carboxylated polymers have been based on combining low-molecular-weight oligomers through degradable linkages and by introducing weak links into the polymer backbone. BASF [88] and NSKK [89] have patented acrylic oligomers chain branched with degradable linkages (X), as in Scheme 5. More recently, Idemitsu has patented variations on these themes [90]. Grillo Werke has patented copolymers of acrylic acid and enol sugars [91]. The degradability of all these polymers has not been categorically established, clearly related to the molecular weight control of the fragmented acrylic oligomers portion.

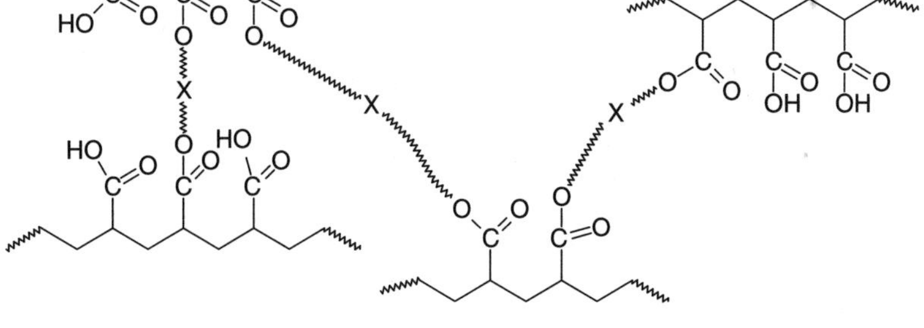

Scheme 5. Branched acrylic oligomers.

Several miscellaneous carbon chain polymers have been claimed as biodegradable without clear evidence. These include copolymers of methyl methacrylate and vinyl pyridinium salts [92–94], where the pyridinium salt is hypothesized as a "magnet" for bacteria that then cleave the chain into small fragments that ultimately biodegrade completely. Careful analysis indicates absorption on sewage sludge present for the laboratory test; and, since biodegradation was measured by TOC loss, the results were misleading and misinterpreted. The work at Deutsche Gold-und Silber [96] with copolymers of acrolein and acrylic acid post converted by a Cannizzaro reaction to copolymers containing hydroxyl and carboxyl functionality (Scheme 6), indicated high levels of biodegradability by BOD. However, the low-molecular-weight fragments and other organics present certainly added to this speculation and no further work has been reported.

Polymers of α-hydroxyacrylic acid [97] are prepared as indicated the Scheme 7 from the polymerization of alpha chloroacrylic acid and subsequent hydrolysis. The claims of biodegradability for these polymers [98–100] were not substantiated by Mulders and Gilain [97] using radiolabeled carbon polymers. There was no clear differentiation of absorption on sludge versus bacterial cellular absorption in the sewage testing protocol.

Mitsubishi [101] claimed a unique biodegradable polycarboxylate (Scheme 8) containing ethylene, carbon monoxide, and maleic anhydride monomers. The initial degradation step is photoactivation to yield low-molecular-weight fragments as indicated in Table 12.2. Unfortunately, no biodegradation data were reported on these fragments.

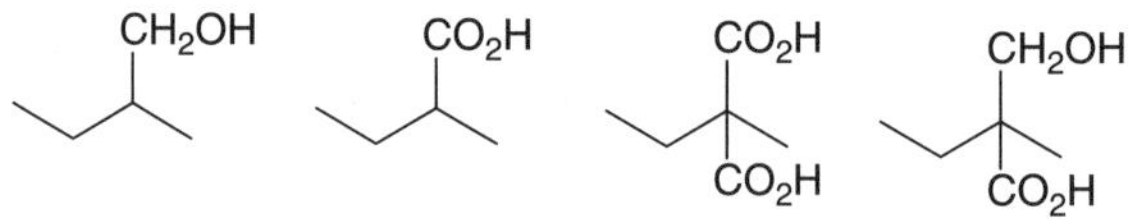

Scheme 6. Cannizzaro products from acrylic/acrolein polymers.

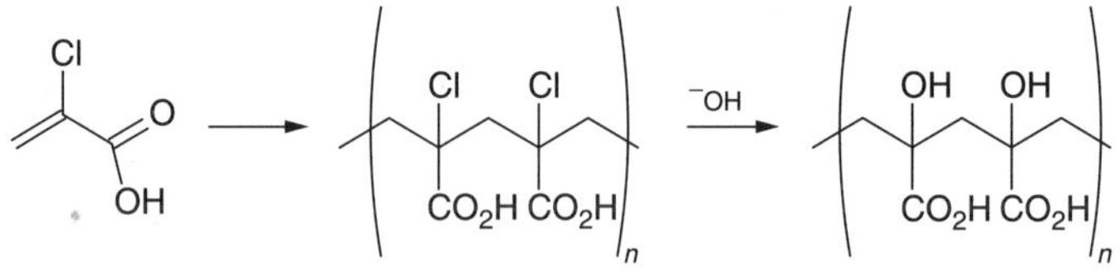

Scheme 7. Synthesis of poly α hydroxy acrylic acid.

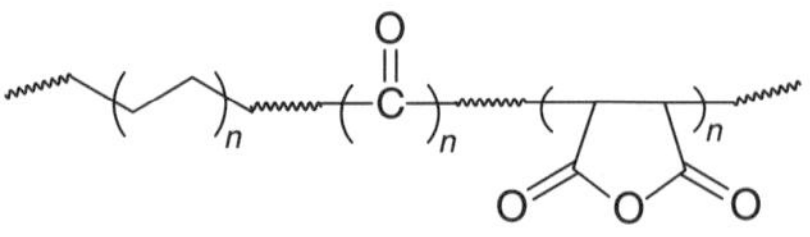

Scheme 8. Photo/biodegradable copolymeric carboxylates.

Table 12.2 Photodegradation of Photo/Biodegradable Carboxylates

CO (mol %)	MW Initial	MW 2 h	MW 4 h	MW 5 h
0	94,000	83,100	73,500	57,200
8	120,500	7,200	4,500	2,300
12	202,400	4,600	2,300	7,800

12.5.2. Heteroatom Chain Polymers

This class of polymers includes polyesters, which have been widely studied from the initiation of research on biodegradable polymers, polyamides, polyethers, and other condensation polymers. Their chemical linkages are quite frequently found in nature, and these polymers are considered more likely to biodegrade than the carbon chain hydrocarbon-based polymers discussed in the previous section.

Polyesters and Polyamides Bailey and co-workers developed a very clever free-radical route to polyesters that they used to introduce weak linkages into the backbones of hydrocarbon polymers and render them susceptible to biodegradability [101–106]. Copolymerization of ketene acetals with vinyl monomers incorporates an ester linkage into the polymer backbone by rearrangement of the ketene acetal radical as it incorporates into the polymer chain. The ester bond produced is a potential site for biological abiotic attack to yield low-molecular-weight fragments likely to biodegrade. Bailey demonstrated the chemistry with ethylene; see Scheme 9, and it has been extended to copolymers of acrylic acid [107, 108]. The biodegradation of the resulting copolymers has not been demonstrated, just claimed.

Water-soluble polyesters and polyamides containing carboxyl functionality are reported to be biodegradable detergent polymers by BASF and may be obtained by condensation polymerization of monomeric polycarboxylic acids such as citric acid, butane-1,2,3,4-tetracarboxylic acid, tartaric acid, and malic acid with polyols [109], amino compounds, including amino acids [110], and polysaccharides [111]. Earlier work by Abe et al. [112] and Lenz and Vert [113] demonstrated the self-condensation of malic acid to biodegradable carboxylated polyesters regardless of the ester linkage, alpha or beta, formed. Procter & Gamble has patented succinylated polyvinyl alcohol [114] as a detergent polymer.

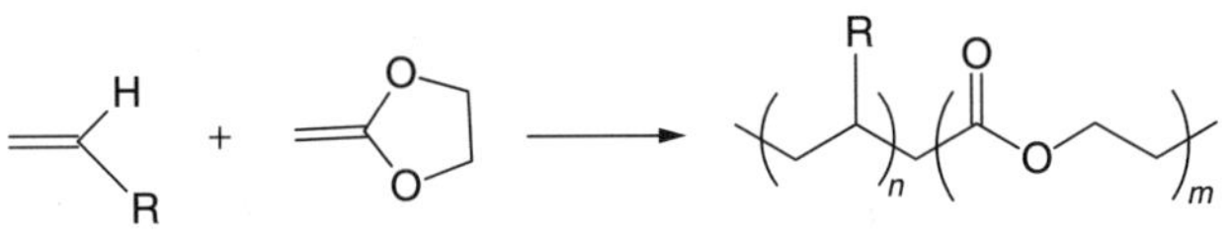

Scheme 9. Copolymerization with ketene acetals.

Nonionic water-soluble biodegradable polyamides are reported by Bailey [115], Chiellini et al. [116], and Ahmed for disposable fibers and webs [117].

Polyanionic polyamides are available by the condensation of polycarboxyamino acids such as glutamic acid and aspartic acid. Though both homopolymers are known and claimed as biodegradable, aspartic acid is more amenable to a practical industrial thermal polymerization since it has no tendency to form an internal N-anhydride as is the case with glutamic acid. An alternative synthesis for polyaspartic acid is from ammonia and maleic acid.

Acid-catalyzed condensation of L-aspartic acid yields an authenticated biodegradable polymer [118] by standard biodegradation test methodology. The noncatalyzed polymerization and the ammonia/maleic acid processes give partially (ca 30 wt % residue remains in the Sturm test) biodegradable polymers due to the molecules being highly branched and more resistant to enzymatic attack.

The two structures starting from L-aspartic acid are shown in Scheme 10. Pathway A is the acid-catalyzed thermal condensation, and B is the noncatalyzed thermal condensation. The polysuccinimide intermediates hydrolyze at the points indicated to give mixtures of α, β poly (D,L- aspartic acid) salts. Regardless of the stereochemistry of the starting aspartic acid, D or L, the final polymeric product is always the DL racemate.

Many synthesis patents and publications for polyaspartic acids from aspartic acid [119–124] and ammonia/maleic [125, 126] processes have issued, and the products find use in many applications including dispersants [127, 128] and detergents [129, 130]. BASF has an aspartic acid copolymer patent with carbohydrates and polyols [131], and Procter & Gamble [132] has a patent for poly(glutamic acid), both for biodegradable detergent co-builders.

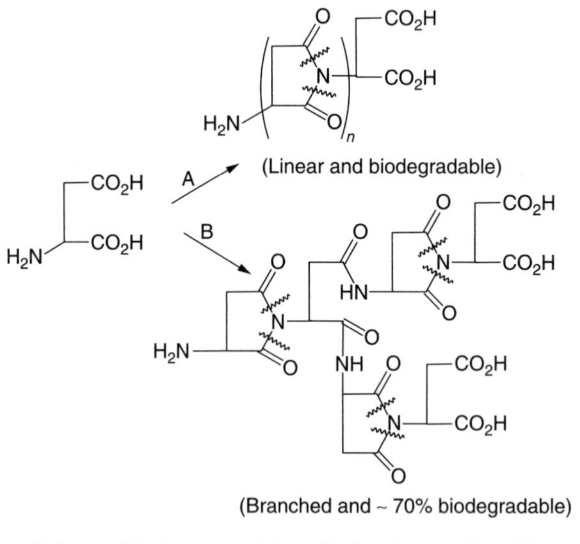

Scheme 10. Linear and branched polyaspartic acids.

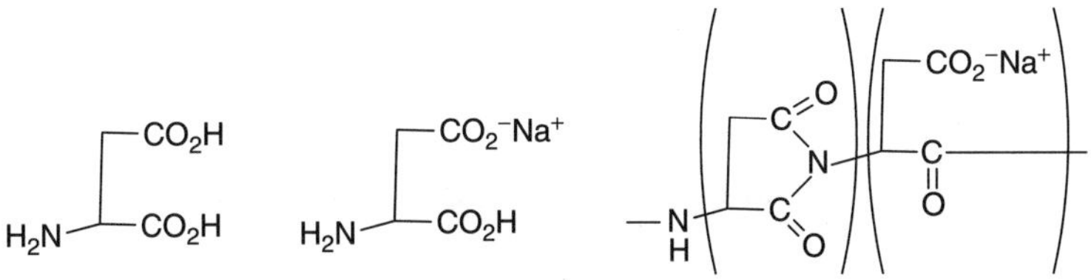

Scheme 11. Copolymerization of aspartic acid and monosodium aspartate.

This exciting field of biodegradable poly(aspartic acids) has generated many new opportunities for anionic polymers. Some examples of an ever-expanding list include dispersants for oil production [133], tobacco filters [134], cosmetics [135], hydrophobic associating thickeners [136], adhesives [137], and crosslinked superabsorbents [138–140]. A recent review on polyaspartic acids up to 1997 includes a large body of information [141].

A recent huge advance in this chemistry is reported by Sikes [142] and shown in Scheme 11. This advance allows the synthesis of water-soluble reactive intermediates by copolymerizing monosodium aspartate with aspartic acid in a preselected ratio. The intractable polysuccinimide intermediate obtained in other polymerizations mentioned earlier in other approaches is avoided, and the copolymer is soluble in water for further functionalization. Easier handling is very important for future characterization and development of applications.

Polyethers The biodegradation of polyethers has been investigated since about 1962, especially poly(ethylene oxides), which are water soluble and widely used in detergents and shampoos, and as a synthesis intermediate in polyurethanes, for example. This whole field has been very comprehensively review by Kawai [143] who established a symbiotic nature for the degradation of poly(ethylene oxides) with molecular weights higher than 6000 Da, while below a molecular weight of 1000 Da the polymer is biodegraded by many individual bacteria. The enzymatic exo-biodegradation pathway described by Kawai is shown in Scheme 12: the first stage is a dehydrogenation, the second stage is an oxidation, the third stage is an oxidation followed by a hydrolysis to remove a two-carbon fragment as glyoxylic acid. Biodegradation of polyethylene oxides at molecular weights as high as 20,000 Da has been reported.

Anaerobically, poly(ethylene oxides) degrade slowly, although molecular weights up to 2000 Da have been reported to completely biodegrade [143, 144, 145]. The biodegradation of higher poly(alkylene oxides) is hindered by their lack of water solubility. Only the low oligomers of polypropylene oxides are biodegradable with any certainty [146–148], as are those of poly(tetramethylene oxides) [149]. A similar exo-oxidation mechanism to that described for poly(ethylene oxides) is proposed.

Polyether carboxylates have been evaluated as biodegradable detergent polymers, initially by Crutchfield [150] and later by Procter & Gamble [151], and Matsumura et al. [152]. All these polymers fit the general structure of the series

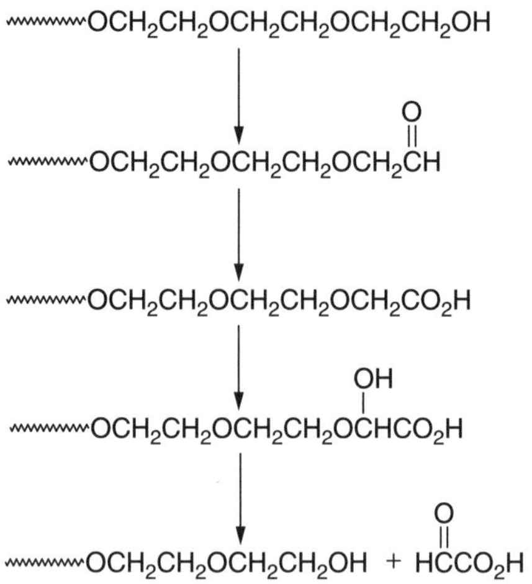

$\sim\!\sim\!\sim\!\sim\!OCH_2CH_2OCH_2CH_2OCH_2CH_2OH$

↓

$\sim\!\sim\!\sim\!\sim\!OCH_2CH_2OCH_2CH_2OCH_2\overset{O}{\overset{\|}{C}}H$

↓

$\sim\!\sim\!\sim\!\sim\!OCH_2CH_2OCH_2CH_2OCH_2CO_2H$

↓

$\sim\!\sim\!\sim\!\sim\!OCH_2CH_2OCH_2CH_2O\overset{OH}{\overset{|}{C}}HCO_2H$

↓

$\sim\!\sim\!\sim\!\sim\!OCH_2CH_2OCH_2CH_2OH + H\overset{O}{\overset{\|}{C}}CO_2H$

Scheme 12. Exo-biodegradation of polyethylene oxides.

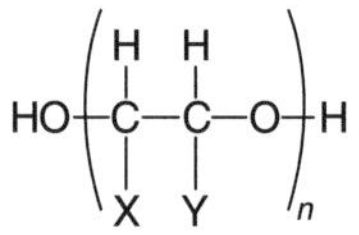

$$HO\!-\!\!\left(\!\!\begin{array}{cc} H & H \\ | & | \\ C\!-\!C\!-\!O \\ | & | \\ X & Y \end{array}\!\!\right)_{\!\!n}\!\!\!-\!H$$

Scheme 13. Functional polyethers.

made in Matsumura's extensive evaluation of anionic and cationic polymerized epoxy compounds shown in Scheme 13. Polymers with a molecular weight range of several hundreds to a few thousands, where X or Y may be carboxyl functionality and X or Y may be hydrogen or a substituent bearing a carboxyl functionality were evaluated. The extent of biodegradability, based on biochemical oxygen demand (BOD), is structure dependent.

Water-soluble biodegradable polycarboxylates with an acetal weak link were the clever invention of Monsanto scientists [153] in their search for biodegradable detergent polymers. Clever science, however, was no match for economic necessity, and the polymers never reached commercial activity. The polymers are based on the anionic or cationic polymerization of glyoxylic esters at low temperature (molecular weight is inversely proportional to the polymerization temperature) and subsequent hydrolysis to the salt form of the polyacids. The salt is reasonably stable under basic conditions being a hemi acetal, or ketal when methylglyoxylic acid is used. Hydrolytic instability results from the pH drop a detergent polymer solution experiences as it leaves the alkaline laundry environment, pH ca. 10, and enters the sewage or groundwater environment where pH is close to neutral. The polymer is unstable and hydrolyzes to monomer, which rapidly biodegrades. The chemistry is outlined in Scheme 14 and is reported in many patents [153] and several publications [154, 155].

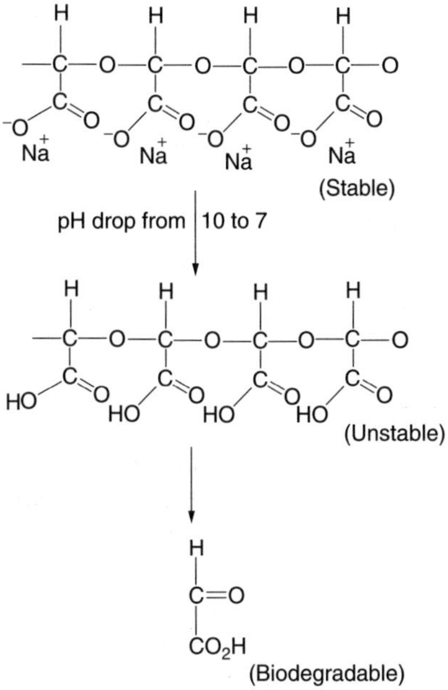

Scheme 14. Hydrolysis/biodegradation of polyacetal carboxylates.

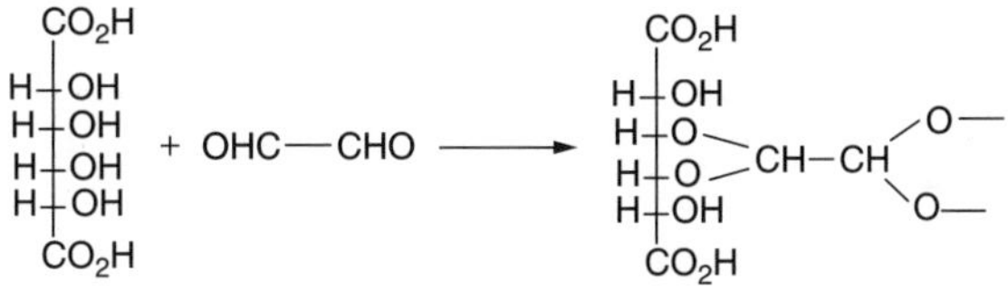

Scheme 15. Polyacetal formation from carboxysugars and dialdehydes.

Similar polyacetal structures were prepared by BASF scientists from general dialdehydes and aliphatic hydroxycarboxylic acids derived from sugars [156, 157], shown in Scheme 15. Alternatively, carboxypolyacetals are available by the addition of polyhydroxy carboxylic acids, tartaric acid, for example, to divinyl ethers [158].

Miscellaneous recent acetal polymer chemistry to produce biodegradable water-soluble polymers includes superabsorbents [159], the use of hydroxycarboxylic acid to produce lubricants [160], scale inhibitors [161], and water-soluble packaging materials [162].

12.6. MODIFIED NATURAL POLYMERS

Modification of natural polymers such as starch, cellulose, and proteins is a way of capitalizing on the well-accepted biodegradability of the base material with the intention of developing polymers that might be biodegradable or environmentally acceptable and functional in a variety of commercial applications. This, of course, will only be true if the modification is shown not to interfere with the biodegradation process and the product meets the guidelines listed earlier for environmental acceptability: either demonstrated to be totally biodegraded and be removed from the environment or biodegradable to the extent that no environmentally harmful residues remain. With this in mind, the approaches that have received the major attention are grafting of another polymeric composition and chemical modification to introduce some desirable functional group by oxidation or other simple chemical reaction such as esterification or etherification.

12.6.1. Graft Polymers

Starch, as mentioned earlier in the work of Sanyo [84] for detergent additives, has been a substrate of choice for biodegradable water-soluble polymers by grafting with synthetic polymers to achieve property improvement and new properties such as carboxyl functionality not available in starch and retain as much biodegradability as possible. As noted earlier, this was not successful due to the high molecular weight of the carbon chain graft. Further attempts have been made to meet these molecular weight limitations mentioned earlier (less than DP of ca. 6–8). But, acrylic acid grafts onto polysaccharides in the presence of alcohol chain transfer agent [163] were not completely biodegradable, nor were the ones based on the use of metal initiators [164] such as Ce^{4+} and mercaptans as chain transfer agents [165, 166]. Other efforts include methanol chain transferred graft of methyl acrylate onto starch [167], sodium acrylate grafts [168], maleic anhydride grafts [169], polyvinyl alcohol grafts [170], and polysuccinimide grafts [171].

Protein substrates [172] are expected to be similar to the starch grafts; the fundamental problem remains the control acrylic acid polymerization to the oligomers range, in order to have complete biodegradability.

Other grafts to natural materials are exemplified by Dordick's work [173] in which he produced polyesters from sugars and polycarboxylates by enzyme catalysis of the condensation polymerization. These polymers and the method of synthesis may well be the future of renewable resource chemistry.

12.6.2. Chemical Modification of Polysaccharides

Simple chemical reactions on natural polymers are well known to produce polymers such as hydroxyethyl cellulose, hydroxypropyl cellulose, carboxymethyl cellulose, cellulose acetates and propionates, and many others that have been in commerce for many years. Their biodegradability is often taken for granted, but, in many cases, is not at all well established. Carboxymethyl cellulose, for example, has been claimed as biodegradable below a degree of substitution of about 2, which is similar to the claim for cellulose acetate. More recently, there has been attempts to more rigorously quantify the biodegradation of the cellulose acetates [174,175] and to establish a property–biodegradation relationship.

Rhone-Poulenc indicates that cellulose acetate with a degree of substitution of about 2 is biodegradable, in agreement with its earlier reference [176]. Cellulose has been discussed as a renewable resource [177]. A recent publication [178] on chitosan reacted with citric acid indicates that the ampholytic product is biodegradable. Chitosan acetate liquid crystals [179], hydrophobic amide derivatives [180], and crosslinked chitosan [181] are also claimed to be biodegradable.

Carboxylated natural polymers have been known for many years with the introduction of carboxymethyl cellulose, as noted above. This product has wide use in detergents and household cleaning formulations, even though of questionable biodegradability at the level of substitution required for performance. Nevertheless, carboxylated polysaccharides are a desirable goal for many applications, and the balance of biodegradation with performance has been recognized as an attractive target with a high probability of success by many people. Three approaches have been employed: esterification, oxidation, or Michael addition of the hydroxyl groups with a suitable vinyl receptor. Attempts have also been made to react specifically at the primary hydroxyl, the 6 position, or secondary sites at the 2,3 positions of polysaccharides.

Esterification with poly(carboxylic anhydrides) can be controlled to minimize diesterification and crosslinking to produce carboxylated cellulosic esters. Eastman Kodak in a recent patent claimed the succinylation of cellulose to different degrees: 1 per 3 anhydroglucose rings [182] and 1 per 2 rings [183]. Henkel [184] also has a patent for a surfactant by the esterification of cellulose with alkenylsuccinic anhydride, presumably substitution degree governs the hydrophile–hydrophobe balance of the product and its surfactant properties.

Oxidation of polysaccharides is a far more attractive route to polycarboxylates and potentially cheaper and cleaner than esterification. Selectivity at the 2,3-secondary hydroxyls and the 6-primary is claimed possible. Total biodegradation with acceptable property balance has not yet been achieved, though. For the most part, oxidations have been with hypochlorite/periodate under alkaline conditions, but more recently catalytic oxidation has appeared as a possibility, and chemical oxidations have also been developed that are specific for the 6-hydroxyl group.

Matsumura [185–188] has oxidized a wide range of polysaccharides, starch, xyloses, amyloses, pectins, and the like with hypochlorite/periodate. The products are either biodegradable at low oxidation levels or functional at high oxidation levels; the balance has not yet been established for commercial success. Other than Matsumura, van Bekkum and co-workers, at Delft University, has been the major player in the search to control the hypochlorite/periodate liquid-phase oxidations of starches [189–191]. He has been searching for catalytic processes to speed up the oxidation with hypochlorite. Hypobromite is a more active oxidant than hypochlorite but more expensive, however, it may be generated in situ from the cheap hypochlorite and bromide ion in one solution [191, 192]. This is shown in Scheme 16.

deNooy et al. [193] has also published a method for oxidizing specifically the 6-hydroxyl group (primary) of starch by using TEMPO and bromide/hypochlorite, as shown below in Scheme 17.

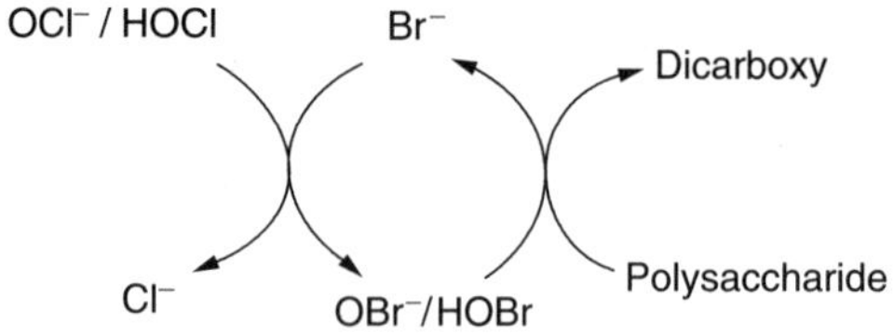

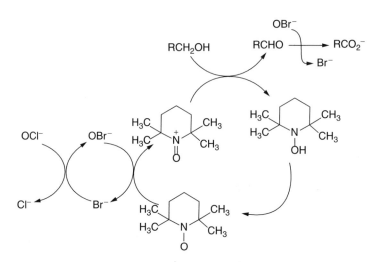

Scheme 16. Catalytic oxidation of polysaccharides with Br⁻/OCl⁻.

Scheme 17. Catalytic oxidation of C-6 hydroxyl of polysaccharides.

Chemical oxidation of polysaccharides with strong acid is reportedly selective at the 6-hydroxyl, with a mixture of nitric acid/sulfuric acid/vanadium salts [194], which is claimed as specific for up to 40% conversion. Alternatively, dinitrogen tetroxide in carbon tetrachloride has similar specificity up to 25% conversion [195].

Catalytic oxidation in the presence of metals and oxygen is claimed as both nonspecific and specific for the 6-hydroxyl oxidation depending on the metals used and the conditions employed for the oxidation. Nonspecific oxidation is achieved with silver or copper and oxygen [196] and with noble metals with bismuth and oxygen [197]. Specificity results with platinum catalyst at pH 6–10 in water in the presence of oxygen [198]. A related patent to produce a water-soluble carboxylated derivative of starch is Hoechst's on the oxidation of ethoxylated starch and another on the oxidation of sucrose to a tricarboxylic acid; all the oxidations are specific for primary hydroxyls and use a platinum catalyst at a pH near neutrality in the presence of oxygen [199, 200].

For further reading on polysaccharides as raw materials in the detergent industry may be found in articles by Swift et al. [201, 202].

12.7. CONCLUSIONS

Although there has been much activity to produce totally biodegradable water-soluble polymers for a variety of applications, especially for detergents, few complete successes have been registered beyond poly(vinyl alcohol), poly(ethylene oxides), and poly(aspartic acids). Almost all efforts to obtain carbon chain polymers that are completely biodegradable have failed, at least in the short-term testing protocols currently in use. Some promising leads are noted in condensation polymers, polyaspartic acids, and acetals and in the modification of natural polysaccharides.

This suggests that future research is in the direction of condensation polymers and modified natural polymers, especially starch as a very cheap and widely available natural resource. The research will require close attention to cost–performance characteristics to develop competitive products since current nonbiodegradable products are not apparently creating environmental problems, though there is a nagging doubt since all the evidence is based on the absence of negative responses to date.

REFERENCES

1. ASTM D 883-93, *Terminology Relating to Plastics*, American Society for Testing and Materials, West Conshohocken, New York.
2. R. Narayan, Annual Meeting of the Air and Waste Management Assoc., June 24–29, 1990, 40.
3. *ASTM Standards on Environmentally Degradable Plastics*, ASTM Publication Code Number (PCN): 03-420093-19, American Society for Testing and Materials, West Conshohocken, PA, 1993.
4. J. E. Potts et al., *Polymers and Ecological Problems*, Plenum, Newyork, 1973; EPA Contract, CPE-70-124, 1972.
5. R. A. Clendinning, S. Cohen, and J. E. Potts, *Great Plains Agric. Council Pub.* **68**, 244 (1974).
6. R. D. Fields, F. Rodriquez, and R. K. Finn, *J. Am. Chem. Soc., Divn. Poly., Chem.* **14**, 2411 (1973).
7. R. D. Swisher, *Surfactant Biodegradation*, 2nd. ed., Marcel Dekker, New York, 1987.
8. *OECD Guidelines for Testing Methods, Degradation and Accumulation Section*, OECD, Washington, DC, Nos. 301A-E, 302A-C, 303A, and 304A, 1981.
9. R. J. Tanna, R. Gross, and S. P. McCarthy, *Poly. Mater. Sci. Eng.* **67**, 230–231 (1992).
10. J. D. Gu, S. P. McCarthy, G. P. Smith, D. Eberiel, and R. Gross, *Polym. Mater. Sci. Eng.* **67**, 294–295 (1992).
11. A. C. Palmisano and C. A. Pettigrew, *Bioscience* **42**(9), 680–685 (1992).
12. G. Swift, *Poly. News* **19**, 102–106 (1994).
13. G. Swift, International Biodegradable Polymer Workshop, Osaka, Japan, November, 1993.

14. A. L. Andrady, *JMS-Rev. Macromol. Chem. Phys.* **C34**(1), 25–76 (1994).
15. G. Swift, *Polym. Degrad. Stab.* **45**, 215 (1994).
16. G. Swift, *Polym. Degrad. Stab.* **59**, 19 (1998).
17. G. Swift, in *Kirk-Othmer of Chemical Technology*, 4th ed., Vol. 19, 1996.
18. G. Swift, in J. E. Glass and G. Swift, ed., *Agricultural and Synthetic Polymers*, ACS Symposium Series Number No. 433, 1990.
19. G. Swift, in *ACS Advances in Chemistry*, J. E. Glass, ed., series No. 248, American Chemical Society, Washington, DC, 1996.
20. G. Swift, *Acc. Chem. Res.* **26**, 105 (1993).
21. G. Swift, *Macromol. Symposia* **130**, 379 (1998).
22. F. Lo, J. Petchonka, and J. Hanly, *Chem. Eng. Prog.* **89**(7), 55 (1993).
23. E. Chiellini and R. Solari, *Adv. Mater.* **8**(4), 305 (1996).
24. Y. Tian, L. Hou, X. Yu, and Y. Tang, *Shanghai Huagong* **24**(6), 28 (1999).
25. S. Matsumura, *Yukagaku* **44**(2), 97 (1995).
26. S. Matsumura, *Petrotech (Tokyo)* **18**(3), 207 (1995).
27. S. Matsumura, *Zairyo Gijutsu* **8**(5), 164 (1990).
28. G. Swift, 213th ACS National Meeting, San Francisco, PMSE Divn., 283 (1997).
29. G. Swift, *Polym. Mater. Sci. Eng.* **63**, 846 (1990).
30. F. Kawai, *Kobunshi Kako* **42**(4), 176 (1993).
31. F. Kawai, *Macromol. Symp.* **123** (37th Microsymposium on Macromolecules, Biodegradable Polymers Chemical, Biological and Environmental Aspects, 1996) 177 (1996).
32. S. Matsumura, *Sen'I Gakkaishi* **52**(5), 209 (1996).
33. F. Kawai, *Adv. in Biochem. Eng./Biotech.* **52**, 151 (1995).
34. E. A. Smith and F. W. Oehme, *Rev. Environ. Health* **9**(4), 215 (1991).
35. S. Takahashi, *Idemitsu Giho* **38**(5), 480 (1995).
36. R. J. Larson, R. Williams, and G. Swift, *Polym. Mat. Sci. and Engr.* **67**, 348 (1992).
37. J. Doi, *Ecol. Assess. Polym.* 33 (1997).
38. T. Suzuki, *Kobunshi* **24**(6), 384 (1975).
39. R. J. Larson, E. A. Bookland, R. T. Williams, K. M. Yocom, D. A. Saucy, M. B. Freeman, and G. Swift, *Environ. Polym. Degrad.* **5**(1), 41 (1997).
40. N. Scholz, *Tenside, Surfactants, Detergents* **28**, 277–281 (1991).
41. T. Suzuki, K. Hukushima, and S. Suzuki, *Environ. Sci. Technol.* **12**(10), 1180 (1978).
42. T. Yamamoto, H. Inagaki, J. Yagu, and T. Osumi, *Abst. Annual Meeting Agric. Chem. Soc. Jpn.* 133 (1966).
43. T. Suzuki, Y. Ichihara, M. Yamada, and K. Tonomura, *Agric Biol. Chem.* **34**(4), 747 (1973).
44. T. Suzuki, Y. Ichihara, M. Yamada, and K. Tonomura, *J. Appl. Poly. Sci., Appl. Poly. Symp.* **35**, 431 (1979).
45. Y. Watanabe, M. Morita, N. Hamada, and Y. Tsujisaka, *Agric. Biol. Chem.* **39**(12), 2448 (1975).
46. Y. Watanabe, M. Morita, N. Hamada, and Y. Tsujisaka, *Arch. Biochem. Biophys.* **174**, 575 (1976).

47. S. J. Huang, E. Quingua, and I. F. Wang, *Org. Coat. Appl. Polym. Sci. Proc.* **46**, 345 (1982).
48. J. C. Huang, A. S. Shetty, and M. S. Wang, *Adv. Polym. Tech.* **10**, 23 (1990).
49. Kuraray, JP 03263406 and JP 03263407.
50. F. Fukanaga, et al., Japan Kokai 7794471 (1977).
51. F. Fukanaga, et al., Japan Kokai, 76125786 (1976).
52. M. Shimao and N. Kato, *Int. Symp. Biodegradable Polym. Abstr.* **80** (1990).
53. J. P. Casey and D. G. Manley, Proc. 3rd. International Biodeg. Symp. Appl. Sci. Publ. 1976, pp. 731–741.
54. O. D. Wheatley and C. F. Baines, *Text. Chem. Color* **8**(2), 28–33 (1976).
55. S. Matsumura, S. Maeda, J. Takahashi, and S. Yoshikawa, *Kobunshi Robunshu* **45**(4), 317 (1988).
56. Planet Polymers, U.S. Patent 5658977 (1996).
57. Negoce et Distribution S.a.r. (NEDI), French Patents 2553423 (1985).
58. Unitika Chem Kk, Japan Kokai, 05295210 (1993).
59. L. M. Robeson, R. J. Axelrod, M. R. Kittek, and T. L. Pickering, *Polym. News* **19**(6), 167 (1994).
60. H. Haschke, I. Tomk, and A. Keilbach, *Monatsh. Chem.* **129**(4), 365 (1998).
61. S. Coombs, Y. B. G. Christie, and L. Yu, PCT Int. Applic. WO 2000036006 (1999).
62. E. Chiellini, P. Cinelli, A. Corti, E. R. Kenawy, E. G. Fernandez, and R. Solaro, *Macromol. Symp.* **152**, 83 (2000).
63. A. Breitenbach, G. Nykamp, and T. Kissel, *Proc. Int. Symp. Controlled Release Bioact. Mater.*, 1999, p. 150.
64. A. Breitenbach and T. Kissel, *Polymer* **39** (14), 3261 (1998).
65. P. Alexy, D. Bakos, K. Kolomaznik, M. Sedlak, and E. Sedlakova, Int. Pat., PCT Int. Appl. WO 20001019 (2000).
66. C. Bastioli, V. Belloti, M. Camia, D. L. Giudice, and A. Rallis, *Stud. Polym. Sci.* **12**, 200 (1994).
67. D. K. Kweon, W. D. Kang, and J. N. Chung, *Polymer (Korea)* **22**(5), 786 (1998).
68. R. H. Haschke, R. Rauch, F. Reiterer, and F. Wehrmann, PCT Int. Appl. 9603443 (1996).
69. S. Matsumura, J. Takahashi, S. Maeda, and S. Yoshikawa, *Macromol. Chem. Rapid Commun.* **9**(1), 1–5 (1988).
70. S. Matsumura, J. Takahashi, S. Maeda, and S. Yoshikawa, *Kobunshi Ronbunshi* **45**(4), 325 (1988).
71. Lever, U.S. Patent 5062995.
72. S. Matsumura, et al., *J. Am. Oil. Chem. Soc.* **70**, 659–665 (1993), and ACS Symposium Series, 627 Hydrogels and Biodegradable Polymers for Bioapplications, 1996, p. 137.
73. S. Matsumura, s. Ii, H. Shigeno, T. Tanaka, F. Okuda, Y. Shimura, and T. Toshima, *Makromol. Chem.* **194**(12), 3237 (1993).
74. Rohm and Haas, US Patent 5191048 (Swift and Weinstein).
75. M. Hunter, D. M. L. daMotta Marques, J. N. Lester, and R. Perry, *Environ. Technol. Lett.* **9**, 1–22 (1988).

76. S. Matsumura, et al., *Yukagaku* **33**, 211, 228 (1984).
77. S. Matsumura, et al., *Yukagaku* **34**, 202, 456 (1985).
78. S. Matsumura, et al., *Yukagaku* **35**, 167 (1985).
79. S. Matsumura, et al., *Yukagaku* **35**, 937 (1986).
80. S. Matsumura, et al., *Yukagaku* **30**, 31 (1980).
82. S. Matsumura, et al., *Yukagaku* **30**, 757 (1981).
83. T. Yamamoto and K. Itakura, JP Kokai 04304216 (1992).
84. Sanyo, JP 6131498.
85. F. Kawai, JP 05237200-A, and *Appl. Microbiol. Biotech.* **39**(3), 382–385 (1993).
86. Y. Tani, et al., *Appl. Environ. Microbiol.* 1555–1559 (1993).
87. S. Matsuo, JP Kokai, 11172039 (1999).
88. BASF, DE 373348A, EP 289827A (1988); DE 3716543A, EP 291808A (1988); DE 3716544, EP 292766A (1988); EP 289787A (1988), EP 289788A (1988), DE 381426 (1989), U.S. Patent 4952655 (1990), DE 4319934 (1994).
89. NSKK, EP 529910A.
90. S. Matsuo (Idemitsu), JP Kokai 2000281801 (2000).
91. Grillo Werke, EP 289895A (1988).
92. N. Kawabata, *Prog. Polym. Sci.* **17**(1), 1 (1992); *Nippon Gomu Kyokaishi* **66**(2), 80–87 (1993).
93. X. Peng and J. Shen, *J. Appl. Polym. Sci.* **71**(12), 1953 (1999).
94. T. Yamamoto, O. Kakajima, and K. Katsuhiko, JP Kokai 05059130 (1993), (Mitsubishi Petro.)
95. Quantum Chemicals, U.S. Patent, 5219930.
96. Deutsche Golde-und Silber, U.S. Patent 3686145 (1972); U.S. Patent 3896086 (1975); U.S. Patent 3923742 (1975).
97. J. Mulders and J. Gilain, *Water Res.* **11**(7), 571–574 (1977).
98. Henkel et Cie, DE2061584 (1972).
99. Solvey et Cie, Belg. Patent 786464 (1972); US Patents 4107411 (1978) and 4182806 (1980).
100. Hoechst, A. G., U.S. Patent 3890288 (1975).
101. Mitsubishi Petrochemicals, EP 0281139 (1988) and JP Kokai 63284296 (1989).
102. W. J. Bailey, et al., *Contemp. Topics Polym. Sci.* **3**, 29 (1979).
103. W. J. Bailey, et al., *J. Polym. Sci., Polym. Lett. Edn.* **13**, 193 (1975).
104. W. J. Bailey, W. J. Gu, Y. Lin, and Z. Zheng, *Makromol. Chem. Makromol. Symp.* **42/43**, 195 (1991).
105. W. J. Bailey and B. Gapud, *Polym. Stab. Degrad.* **280**, 423 (1985).
106. W. J. Bailey and L. L. Zhou, *Macromolecules* **25**(1), 3 (1992).
107. American Cyanamid, U.S. Patent 4923941 (1990).
108. Toa Gosei Chemical Industry Co., JP Kokai, 2000038595 (2000).
109. BASF, EP 4223807; U.S. Patent 5217642; WO 9216493-A1; DE 4108626-A1.
110. BASF, DE 4225620-A1, DE 4213282-A1.
111. BASF, DE 4108626-A1, DE 4034334-A1.
112. Y. Abe, S. Matsumura, and K. Imai, *Yukagaku* **35**(11), 937 (1986).

113. R. W. Lenz and M. Vert, *ACS Polym. Preprints* **20**, 608 (1978).
114. Procter & Gamble, U.S. Patent 5093170.
115. W. J. Bailey, Y. Okamoto, W.-C. Kuo, and T. Narita, *Proc. Int. Biodegradation Symp.* 765 (1976).
116. E. Chiellini, R. Bizzarri, and R. Solari, *J. Bioact. Compat. Polym.* **14**(6), 504 (1999).
117. S. U. Ahmed, US Patents 5869596, 5869597 (1999)
118. G. Swift, M. B. Freeman, Y. H. Paik, S. Wolk, and K. M. Yocom, ACS Biotech. Secretariat Abstr., San Diego, Spring, 1994 and, 6th International Conference on Polymer Supported Reactions in Organic Chemistry (POC), Venice, June 19–23, 1994, Abstr. p. 21.13. and, 35th, IUPAC International Symposium on Macromolecules, Akron, Ohio, July 11–15, 1994, Abstr. 0-4.4-13th, p. 615; and *Macromol. Symp. 123, 195 (1997).*
119. Cygnus, U.S. Patent, 5219952.
120. Rohm and Haas, EP 578448-A1.
121. Rhone-Poulenc, EP 511037-A1.
122. Donlar Corporation, U.S. Patent 5221733.
123. V. S. Rao, *Makromol Chem.* **194**, 1095 (1993).
124. Donlar Corporation, WO 9214753.
125. SR Chem., U.S. Patent 5288783.
126. SR Chem., WO 9323452.
127. S. Sikes, University of South Alabama, US Patent 5260272.
128. Donlar Corporation, U.S. Patent 5116513.
129. Montedipe, EP 454125.
130. Lever, EP 561464, and EP 561452.
131. BASF, DE 4221875-A1.
132. Procter and Gamble, WO 93 06202.
133. G. Fan, L. P. Kokan, and R. J. Ross, 215th ACS Abstracts. National Meeting, ENVR-029 (1998).
134. H. Taniguchi and K. Nishimura, JP Kokai 95–200764 (1995).
135. BASF, DE 19631380 (1998).
136. T. Nakato, M. Tomida, M. Suwa, Y. Morishima, A. Kusuno, and T. Kakuchi, *Polym. Bull. (Berlin)* **44**(4), 385 (2000).
137. Bayer Corp., EP Appl., 1085072 (2001).
138. Mitsui Chemicals, EP Appl., 856539 (1998).
139. Mitsui Chemicals., JP Kokai, 11217436 (1998).
140. NSKK, JP Kokai, 08059820 (1996).
141. S. Roweton, S. J. Huang, and G. Swift, *J. Environ. Polym. Degrad.* **5**(3), 175, (1997).
142. Sikes, US Patent 5981691 (1999).
143. F. Kawai, *CRC Critical Reviews in Biotechnology*, Vol. 6, C.R.C. Press, Boca Raton, FL, 1987, p. 273.
144. B. Schink and H. Strab, *Appl. Environ. Microbiol.* **45**, 1905 (1983).
145. B. Schink and H. Strab, *Appl. Microbiol. Biotech.* **25**, 37 (1986).
146. F. Kawai, K. Hanada, Y. Tani, and K. Ogata, *J. Ferment. Tech.* **55**, 89 (1977).

147. F. Kawai, T. Okamoto, and T. Suzuki, *J. Ferment. Tech.* **63**, 239 (1985).
148. F. Kawai, *J. Kobe Univ. Commerce* **18**(1–2), 23 (1982).
149. F. Kawai, and H. Yamanaka, *Ann. Meeting Agric. Chem. Sic. Japan*, Kyoto, 1986.
150. M. M. Crutchfield, *J. Am. Oil Chem. Soc.* **55**, 58 (1978).
151. Procter & Gamble, U.S. Patents 4654159 (1987), 4663071 (1987), 4689167 (1987), and EP 192441 (1986), 192442 (1986), 236007 (1987), 264977 (1987).
152. S. Matsumura, K. Hashimoto, and S. Yashikawa, *Yukagaku* **36**(110), 874–881 (1987).
153. Monsanto, U.S. Patents, 4144226, 4146495, 4204052, 4233422, 4233423.
154. W. E. Gledhill and V. W. Saeger, *J. Ind. Microbiol.* **2**(2), 97 (1987).
155. W. E. Gledhill, *Appl. Environ. Microbiol.* **12**, 591 (1978).
156. BASF, DE 4204808-A1.
157. BASF, DE 4106354-A!, WO 9215629-A1.
158. BASF, DE 4142130-A1.
159. NSKK, JP Kokai, 09124754 (1997).
160. Clariant G.m.b.h., DE 19636688 (1998).
161. Mitsubishi Gas Chemical Company, EP Appl., 764675 (1997).
162. NSKK., JP Kokai, 10081817 (1998).
163. Taechang Moolsan Co. Ltd., WO 9302118-A1.
164. Rhone-Poulenc, EP 465286, and 465287.
165. BASF, DE 4003172.
166. Stockhausen, WO 9401476-A1.
167. X. Q. Xu, M. L. Duan, X. H. Dong, and J. X. Feng, *Huaxue Gongye Gongcheng (Tianjin)* **17**(5), 307 (2000).
168. Mitsubishi, JP Kokai, 04055411 (1992); Mitsubishi, JP Kokai, 04055412 (1992).
169. NSKK., JP Kokai, 06298866 (1994).
170. Y. Tokiwa and M. Kitagawa, *Sci. Technol. Polym. Adv. Mater.* 447 (1998).
171. Dainippon Ink and Chemicals, JP Kokai, 2000290502 (2000).
172. BASF, U.S. Patent 5027941, and DE 4029348.
173. J. Dordick, University Iowa, State Res. Found., WO 92221765.
174. S. Kim, V. T. Stannett, and R. D. Gilbert, *J. Polym. Sci. Lett. Edn.* **11**(12), 731 (1973).
175. C. M. Buchanan. R. Komanek, D. Dorschel, C. Boggs, and A. W. White, *J. Polym. Sci.* **52**(10), 1477 (1994); and, J. D. Gu, D. T. Ebereiel, S. P. McCarthy, and R. A. Gross, *J. Environ. Polym. Degrad.* **1**(2), 143 (1994).
176. Rhone-Poulenc Announcement, Eur. Plastics News, #20,16 (1993).
177. A. Arch, *J. Macromol. Sci.* **A30**(9/10), 733 (1993).
178. W. A. Monal and C. P. Covac, *Macromol. Chem. Rapid Commun.* **14**, 735 (1993).
179. K. K. Canon, JP Kokai, 06329966 (1994).
180. C. L. Yu, R. Kumar, J. Pu, and S. Mccarthy, 216th Acs Book of Abstracts, Poly-251 (1998).
181. J. M. Mayer and D. L. Kaplan, U.S. Patent 5015293 (1991).
182. Eastman Kodak, WO 92210521-A1.

183. Eastman Chemical, EP 560891-A1.
184. Henkel, EP 254025-B.
185. S. Masumura, et al., *Angew. Makromol. Chem.* **205**, 117 (1993).
186. S. Matsumura, S., Maeda, and S. Yoshikawa, *Macromol. Chem.* **191**(6), 1269 (1993).
187. S. Matsumura, et al., *Poly. Preprints Jpn*, **41**(7), 2394 (1992).
188. S. Matsumura, K. Aoki, and K. Toshima, *J. Am. Oil. Chem. Soc.* **71**(7), 755 (1994).
189. H. van Bekkum, et al., *Prog. Biotech.* **3**, 157 (1987).
190. H. van Bekkum, *Starch-Starke* 192 (1988).
191. A. C. Besemer, Thesis (Delft), 1993, and EP 4273459-A2, WO 9117189.
192. A. C. Bessemer, and H. van Bekkum, *Starch-Starke* **46**, 95–100 (1994); **46**, 101–106 (1994).
193. A. E. J. deNooy, A. C. Bessemer, and H. van Bekkum, *Rec. Trav. Chim.* **113**(3), 165–166 (1994).
194. Procter & Gamble, EP 542496-A1.
195. Henkel, DE 4203923-A1; WO 9308251-A1.
196. Novamont, EP 548399-A1, WO 9218542-A1, WO 9238205.
197. Roquette Freres, EP 455522-A, U.S. Patent 4985553.
198. Mercian Corp., JP 05017502.
199. Hoechst, U.S. Patent 5223642, WO 9102712.
200. Hoechst, U.S. Patent 5238597.
201. G. Swift, Y. H. Paik, and E. S. Simon, *ACS Polym. Mater. Sci. Eng. Abstr.* **69**, 496 (1993); *Chem. and Ind.* Jan 16th, 55 (1995).
202. G. Swift, *Book of Abstracts*, 216th ACS National Meeting, Poly-002, 1998.

PART 4

CHAPTER 13

POLYMERS, POLYMER RECYCLING, AND SUSTAINABILITY

JOHANNES BRANDRUP, WIESBADEN

13.1. INTRODUCTION: WHAT IS SUSTAINABILITY

The limits and problems of our globe were the content of the famous Brundtland report to the World Commission on Environment and Development in 1987. The term *sustainable development* was coined, meaning a development satisfying the needs of the present generation by considering the needs for a sufficient supply for future generations.

Five years later at the conference in Rio members of more than 100 countries formulated an action program along this basic principle and made it known worldwide. For a long time different definitions and contents for this basic principle have been discussed. It is publically accepted, meanwhile, that a sustainable development has to consider social, economic, and ecological aspects because it is a development goal for the human species on this globe (Fig. 13.1).

$$\text{Sustainability} = \text{Social welfare} + \text{Economy} + \text{Ecology}$$

Figure 13.1. Components of sustainable development.

Environmental impact is generated solely by the existence of human beings in this world—this part is governed by social and even religious or theological aspects. Environmental impact is certainly caused by our standard of living

---

Dedicated to Dr. W. Glenz on his 60th birthday.

Plastics and the Environment, Edited by Anthony L. Andrady.
ISBN 0-471-09520-6 © 2003 John Wiley & Sons, Inc.

and our use of technology to improve it. This certainly depends to a large extent on economic considerations. But there are different ways to realize this standard of living. You may use a lot of resources (abuse/exploitation) or you may try to save resources knowing that they do not last forever. Since the first two components which contribute to environmental impact are bound to increase, only the last aspect can try to counteract — at least to a certain extent — this development that we then call sustainable (Fig. 13.2).

$$\text{Environmental impact} = \text{Population} \times \frac{\text{Wealth}}{\text{population}} \times \frac{\text{Environmental impact}}{\text{wealth}}$$

Figure 13.2. Components of environmental impact.

Environmental impact may mean the saving of resources, the minimization of emissions including disposal, or the avoidance of hazardous substances. The public emphasis concentrates presently on minimization of emissions if we consider the debate on global warming or on dioxins. But saving of resources will obtain equal importance in the future if the limited availability of fuel will get more public awareness again.

The positioning of plastics within this framework is the topic of this contribution. It will try to give answers to certain questions raised today but obviously will at the same time raise additional questions because sustainability is a complicated frame due to the very different aspects contributing to it.

While we are used to dealing with economic questions for a long period of time, already the treatment of ecological questions in comparison is still not mature, and the combination of several compartments such as economy and ecology or even social welfare is still in its infancy.

13.2. POLYMERS AND SUSTAINABILITY

The positioning of a product depends mainly on its efficient use of resources, on the environmental impact of its production, on the environmental savings by using it in the marketplace, and by the amount of final waste left at the end of its lifetime — in other words, on its ability to be recycled. Plastics make a positive contribution in most of these sectors.

13.2.1. Production of Polymers

Polymers use their main feedstock — the crude oil — more efficiently than the direct incineration of it which is true today for more than 82% of fuel consumption. Plastics — which need ca. 6% of the total fuel demand in Germany — can be considered as the first recycling loop of crude oil since all the energy content of the fuel is preserved and can still be used after disposal of plastic products.

Plastics rely on a feedstock used for energy production, but at the same time they need less of it per ton of product than other materials. The European Association of Plastic Manufacturers (APME) in Brussel has made a big effort over the last decade to develop a life-cycle analysis approach for an environmental positioning of all major plastic materials [1]. Meanwhile, the data for most polymers for resource demand and major emissions have been evaluated and published based on the average data of all European production sites.

The issue of hazardous substances is very complex — if you only consider, for example, the issue of possible endocrine effects of substances. It is a topic of its own and will not be dealt with in this chapter.

The results for resource demand of plastic materials are given in Figure 13.3. If you compare these results with the energy demand of other materials you obtain the diagram shown in Figure 13.4. While the comparison per weight is only of partial interest, the comparison by volume illustrates the reality better since final products are mainly used by volume. Thus, the lightness of plastics compared to the specific weight of other materials is an environmental benefit.

But there are more arguments for this positioning of plastic material. Production and processing require little energy in general since these are low-temperature processes in comparison to other materials production processes. Emissions are low and the waste generated during production and processing can be reused immediately, thus the overall waste rate at this stage is low. In addition, the utilization value of plastics is high since plastics help to save resources in different important aspects of our life.

13.2.2. Use of Plastics — Examples for Saving Total Primary Energy Demand

Insulation of Houses Primary energy is needed for heating or cooling our houses. The amount can be reduced by efficient insulation. One way of doing it effectively is by using polystyrene foam. We need energy and resources for the production of this foam. But the amount needed is saved within less than one year heating a house in wintertime or cooling it during the hot season.

Poly (ethylene)	85.8 (83.0 – 88.5)
Poly (propylene)	80.0
Poly (styrene)	102.2 (96.2 – 105.3)
ABS	95.2
Poly (vinyl chloride)	66.8 (65.0 – 74.9)
Poly (ethylene terephthalate)	83.8
Poly (butadiene)	84.5
Poly (urethan)	104.9
Poly (methyl methacrylate)	111.7/133.4 (beads/sheet)
Poly (carbonate)	116.3
Poly (amide 66)	143.6
Epoxy resins	141.7

Figure 13.3. Production of plastics; Gross energy requirements (MJ/kg).

Specific energy for the production of different materials

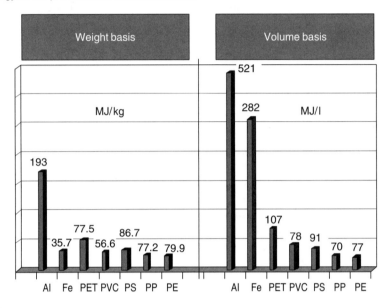

References:
Al, Fe: life cycle inventories for packagings. Environmental series no. 250/I.
Swiss agency for the environment, forset and landscape. Berne, 1998
Plastics: actual data sets of APME (http://lca.apme.org. date: 31.05.2000)

Figure 13.4. Specific energy for the production of different materials (from R. Wittlinger, BASF, 2000).

An investigation by Fraunhofer Institute Karlsruhe in 1998 has shown that 2.0 GJ/cbm of foam energy are needed and 60 kg CO_2/cbm are emitted when producing the foam, but the annual savings when used for insulation amount to 3.7 GJ/cbm energy and 243 kg CO_2/cbm. Thus, the amortization of the environmental loads occurs within less than one year [2].

Car Industry The fuel demand of a car depends among other factors also on the car weight. This can be reduced drastically by using lightweight materials such as plastics. An investigation by Fraunhofer Institute Freising in 1999 has given the following results: An average German medium-sized car uses ca. 165 kg of different plastic materials (Fig. 13.5). Out of this only 75 kg substitute alternative materials and thus save car weight. The remaining part has functions that cannot be fulfilled otherwise (airbag, seat construction, etc.).

The amount of fuel saved by reducing car weight certainly depends on the mileage of the car over its lifetime and on the efficiency of the fuel mileage. For a low car mileage of 100,000 km over the lifetime of the car there are:

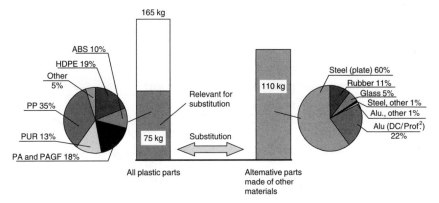

Figure 13.5. Weight-relevant plastic parts and relevant alternative materials in a medium-sized car in Germany [3].

- Primary energy savings per kilogram plastic of between ca. 10 and 60 MJ/kg plastic
- Reductions in CO_2 emissions per kilogram plastic of between ca. 2 and 4 kg CO_2/ kg plastic

For a higher car mileage (200,000 km), the corresponding savings are between:

- 60 and 120 MJ/kg plastic for primary energy
- 6 and 10 kg CO_2/ kg plastic for CO_2 emissions

These savings during the lifetime of the car have to be compared with the environmental load of producing the corresponding plastic parts and the environmental load or benefit of disposal after the use phase (Figs. 13.6 and 13.7). They were calculated together with 105 MJ/kg part for the energy and with ca. 4.5 kg CO_2/ kg part. Thus the use of weight-relevant plastics in a car replacing heavier alternative materials results in one or two times the primary energy savings and CO_2 emission reductions compared to the values arising from their manufacture [3].

Packaging Even packaging plastics can be resource minded. Manufacturers have substituted other materials such as paper, glass, and metals for 45% of all household packagings (Fig. 13.8). But they represent only 10% of the weight of all packaging materials. This saves four times the weight or two times the volume of alternative packaging materials as was calculated by a German Packaging Institute using the data for 1991 [4].

Another interesting aspect of packaging are the consequences for the packed material. For example, 54% of all goods in the household area is food — much

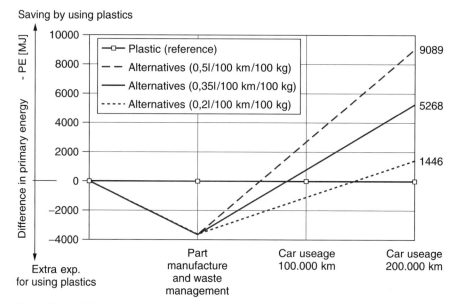

Figure 13.6. Primary energy saving by using plastic products in medium-sized cars in Germany 1995 [3].

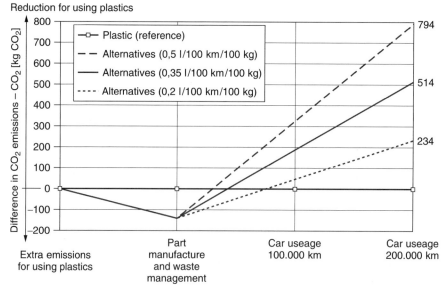

Figure 13.7. Reduction in CO_2 emissions by using plastic products in medium-sized cars in Germany 1995 [3].

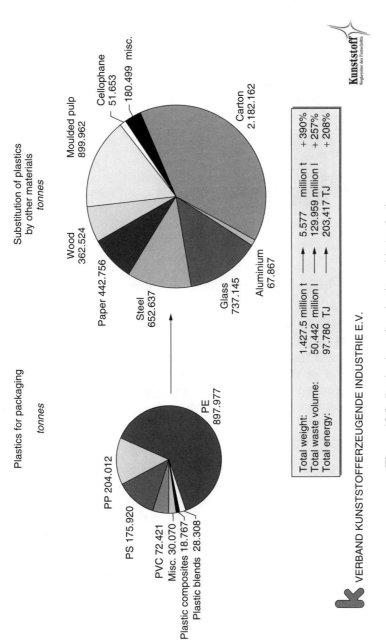

Figure 13.8. Production of packaging without plastics.

530 POLYMERS, POLYMER RECYCLING, AND SUSTAINABILITY

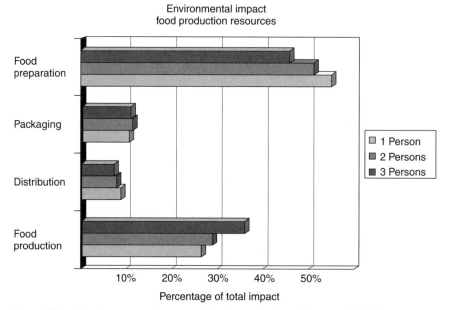

Figure 13.9. Environmental impact of food production. Share of individual process steps [5].

of it perishable. The largest environmental impact of food is observed for food production and food preparation where the differences for various household sizes are very small [5]. The share of environmental impact of distribution and of plastics packaging is small in comparison to the other influences (Fig. 13.9).

Thus preserving food effectively represents a larger environmental impact than the question of recyclability of the packaging material. This especially impacts small packaging that protecting small amounts of food for single households [6].

Prerequisites for Other Technologies But plastics are also prerequisite for numerous modern technologies that would not exist without plastics. The transport of electricity is the most famous example. What alternative materials exist for a flexible, longlasting insulation? The storing of information on polyethylene terephthalate (PET) tapes or on polycarbonate disks is another example. A so-called welfare cost–benefit analysis for a huge number of different plastics applications has been carried out by GUA.Geselllschaft für umfassende Analysen, Wien, initiated by APME [7].

13.3. POLYMER RECYCLING AND SUSTAINABILITY

While these contributions of plastics toward sustainability are seldom questioned, it was a common believe in the early nineties in Germany that this material could not be recycled. This provoked a critical public position toward plastics in general.

This was especially true for household plastic packaging and resulted in a rather restrictive legal situation on collection and recycling of packaging materials.

Meanwhile a lot happened in Germany, certainly stipulated by these governmental legal restrictions:

Technologies for recycling of plastic materials have been developed.

Financial aspects of this ecological-driven recycling have been solved.

Thus we can make the statement that plastics are recyclable by several routes.

Mechanical Recycling If you are able to collect rather clean, homogeneous plastic parts

Feedstock Recycling Like pyrolysis, hydrogenation ceased meanwhile syngas production, or use in a blast furnace, if you have to deal with rather small, lightweight, contaminated mixtures that fulfill certain process specifications.

Energy recovery of the fuel value if other routes fail or if you deal with hazardous contents

The German situation for plastics production, consumption, and recycling is shown in Figure 13.10.

13.3.1. Environmental Aspects

Recycling of Household Plastic Packaging Material The situation for recycling of household packaging in 1997 is shown in Figure 13.12. It looks

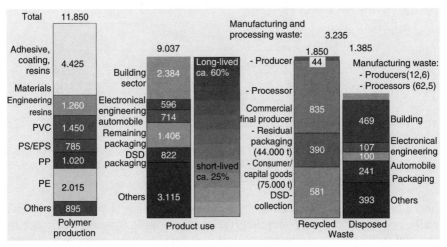

Figure 13.10. Situation of plastics in Germany in 1997 kt/a—production, consumption, and waste [5]. (Courtesy of Association of Plastics Producers Germany (VKE), Frankfurt/M.)

532 POLYMERS, POLYMER RECYCLING, AND SUSTAINABILITY

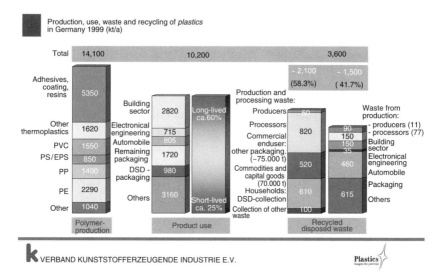

Figure 13.11. Plastics production, consumption, and waste in Germany (1999).

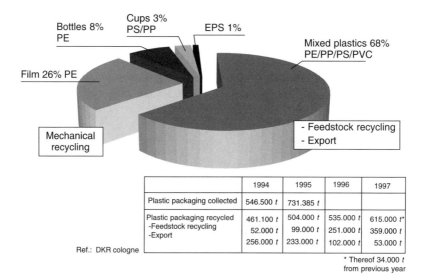

Figure 13.12. Plastics packaging waste from yellow bag collections.[1]

like a big success for a sustainable development. But is it really so? Are we really saving resources by recycling the smallest, most contaminated piece by material recycling or even by mechanical recycling? Is incineration an ecological

---

[1] The amount of plastic waste collected by DSD was 610 kt/y in 1999 and 570 kt/y in 2000.

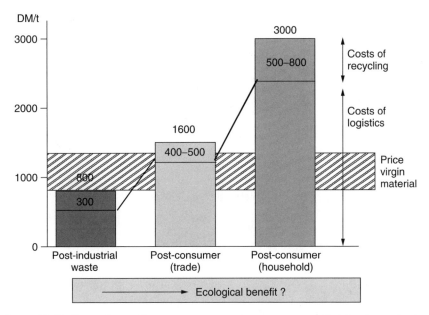

Figure 13.13. Dependence of waste management costs on waste quality plastics packaging.

alternative or not? A lot of questions, but no answers. What was realized rather soon was a tremendous increase in overall costs of recycling for society, without information about the ecological benefit obtained from these economic expenses (Fig. 13.13).

Typically plastic packagings are lightweight: 60% of all plastic packagings weigh less than 10 g, and their light weight reflects their outstanding performance. Only 10 wt% of the total packaging consumption are plastic packagings, while these packagings protect 42% of all goods in private households (of which 54% are food packagings) (Fig. 13.14).

The stringent requirements of the German Packaging Ordinance of those days, that is, to collect 80%, to sort 80%, and to channel 64% into material recycling forced the Dual System from the very start to sort, unlike in other countries, not only large plastic packagings (such as bottles and huge bags) but also small packagings (e.g., bags for chips, yoghurt cups) and utilize those packagings in material recycling, because large packagings (weight >10 g) merely account for 37% of the total market input of plastic packagings. Consequently, unlike in other countries, no individual return system was possible in Germany, but a high-cost cerbside collection system had to be implemented. Since it would have been too exacting for private households to cope with several containers for waste collection, the idea of the yellow bag — intended not only for plastics but also for composite packagings and metal tins — was born. However, the frequent abuse of the yellow bag with ordinary waste disposal was overlooked so that the actual contents of the Yellow Bag were as shown in Figure 13.15.

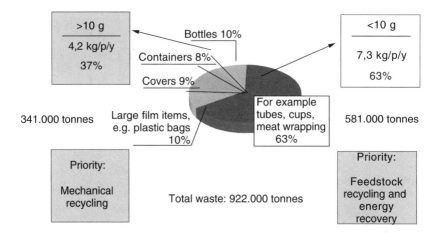

Figure 13.14. Plastics Packaging—Domestic Household Use in 1991, Germany (kt/y).

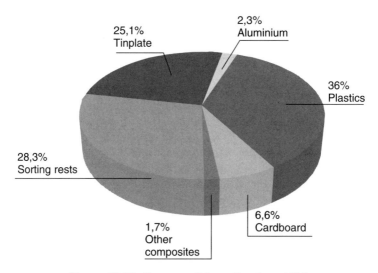

Figure 13.15. Contents of the yellow bag, 1995.

Regional analyses (Dresden, Lahn-Dill district) repeatedly confirmed this picture. Given such framework conditions, it becomes obvious why some two thirds of the total costs of plastics collection and recycling arise in the collection and sorting of lightweight and contaminated—and therefore difficult to utilize—waste, while just one third of the total cost falls to the share of recycling (see Fig. 13.13).

Total costs in the cycle management of plastic packagings from private households exceeded 2 billion DM per year in 1995, that is, more than 50% of the total costs of the Dual System. Referred to the final outcome—that is, 535,000 tons of recycled plastics—this results in a price of roughly 4 DM/kg of plastic. The total amount is covered by the payment of 27 DM per person, that is, a private household consisting of 4 persons pays annually some 100 DM alone for the disposal of 40 kg/household of purchased, mainly small-size plastic packagings. By contrast the annual fee payable for the remaining waste (4×283 kg $= 1, 132$ kg/household) varied around 520 DM per year, depending on the municipality. The plastic materials thus obtained are three times as expensive as the purchase of new products (1.20 DM/kg approx.). Therefore the following questions are obvious:

- What are the advantages to the environment to justify such extra cost?
- Is there a savings potential that makes it possible to reach the same or almost the same ecological result at lower cost?

Opinions differ about the process with more ecological benefit. The public preference is for mechanical recycling since supposedly you save part of the processing energy. Industry emphasizes the low quality and the limited market of the products obtained from postconsumer plastics packaging and favors instead the most economical process of recycling in order to keep the cost of recycling low. In fact the most economical process is very often also the best ecological process since it uses generally at least less energy, but this could not be proven for the recycling of plastics packaging from household waste.

Life-Cycle Analysis of Recycling of Plastic Packaging Material in Germany In order to give a more specific answer to this problem German plastics industry initiated a life-cycle analysis for all presently known routes of recycling and recovery and gained financial support from the Association of Plastics Manufacturers in Europe (APME), Verband der Chemischen Industrie e.V. (VCI), and Duales System Deutschland GmbH (DSD).

The intent was to evaluate different established routes using different technologies of plastics recycling from an environmental point of view using life-cycle analysis (LCA) (Fig. 13.16). Approximately 20 companies participated. This means they were prepared to describe their process in detail and to inform on input of resources such as energy, feedstock, and the like and on output such as airborne and waterborne emissions, solid waste, heavy metals, and other environmentally important products. A secrecy agreement assured the companies protection of any proprietary information.

The German Agency on Environment participated as guest and the German TÜV validated the individual data and their correct interpretation. The then present state of ISO rules on LCA was followed as closely as possible. The final report underwent a critical assessment as requested by ISO rules. The study was awarded the Oce' van der Grinten prize in 1997 and its findings were published [8].

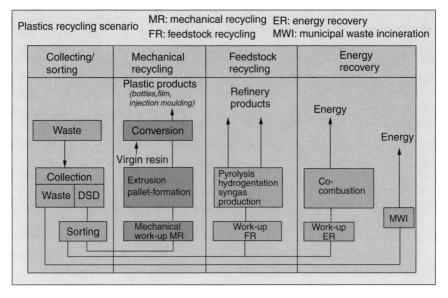

Figure 13.16. Plastics recycling scenario.

The LCA was carried out by four institutes, which had the following tasks

- Mechanical recycling replacing plastic products (Technical University Berlin)
  - Bottle to bottle (detergent and fertilizer bottles from the DS bottles fraction)
  - Film to film(refuse bags from the DS film fraction)
  - Film to cable conduit(electric cable conduit from DS film fraction)
- Mechanical recycling replacing wood or concrete products (Fraunhofer Institute, Freising and Research Institute for Plastics and Recycling, Willich)
  - Mixed plastic waste to products normally made from wood or concrete, such as palisades, fence post, and the like
- Feedstock recycling (University Kaiserslautern)
  - Thermolysis of products into petrochemical products[2]
  - Hydrogenation[3]
  - Syngas production (fixed bed, fluidized bed) together with lignite or vacuum residue oils
  - Blast furnace — use as reducing agent
- Energy recovery (Fraunhofer Institute, Freising)
  - Monocombustion in a fluidized bed

---

[2] Pilot plant no longer in use.
[3] The process was discontinued for economic reasons.

- Municipal waste incineration with partial recovery of useful energy
- Dominance and sensitivity analysis (Fraunhofer Institute, Freising)

The technical background for all processes is described in detail in the literature [9]. Over a 2-year period more than 100 persons participated in the project collecting data, evaluating, and discussing it in various working groups. The first topic to be agreed upon was to develop a method enabling the comparison of something uncomparable such as products and energy. For example, 1-kg plastic waste can in theory either be recycled into 1 kg of resin again via mechanical recycling or can be converted into 26 MJ of energy via optimum incineration. What is the ecological preferred route? How to compare them?

Earlier approaches were based on process efficiencies, energy equivalents, and so on. All of them lack a satisfying scientific logic. The underlying public demand for products was chosen to be the common denominator. This is the reason why plastic resins, naphtha, or energy are produced and sold. All these products are needed — they have a market — and if these products are not obtained via recycling from waste, they have to be produced via conventional processes using primary resources only.

In the example above it would mean to draw energy from natural resources if 1 kg of plastic waste was turned into resin again. Alternatively, 1 kg of virgin resin has to be produced from crude oil if 1 kg of waste had been incinerated. This leads to the development of the so-called basket of goods (Fig. 13.17).

This basket of goods always has the same contents and one needs only compare the different routes of filling this standard basket. Thus everything put into the

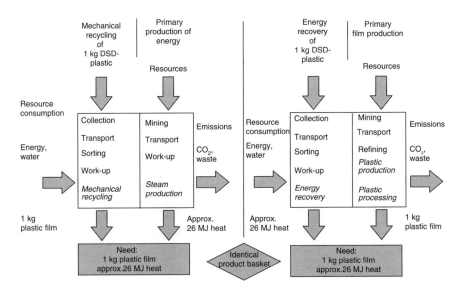

Figure 13.17. Comparison of different production routes based on an identical basket of goods.

basket via recycling routes must not be drawn from primary resources. On the other hand, anything that cannot be provided by recycling has to come from primary resources.

This procedure enables us to determine the most ecological route to offer the sum of all the required products in the required amount to fill public demand. There exist many different routes to make use of plastic waste. In other words, plastic in recycling is as versatile as plastic in original applications. These different processes yield very different products that all have a market value.

Thus the real basket of goods corresponding to the plastics recycling scenario shown in an earlier figure consists of about 12 different products (Fig. 13.18). These products are produced by different technologies that by using plastic waste as resource save the corresponding primary resources. All processes were grouped in one of the following categories: *mechanical recycling* or *feedstock recycling* or *energy recovery*. These categories than were compared with a reference scenario where the waste was disposed of by landfill, and all products of the basket of goods had to be produced from primary processes.

Each scenario (Fig. 13.19) produces the full complement of products of all recovery processes in a given proportion. This approach makes the scenarios directly comparable on a given criterion (e.g., CO_2 emissions), as their waste input and product output are the same. When, as in the simplified example, the various scenarios are compared with a reference scenario that produces all products by conventional means, the effects of the complementary processes are eliminated by subtraction. This considerably simplifies the calculation. As illustrated in Figure 13.20, comparing a given scenario with a conventional reference scenario reduces the comparison to the sum of:

The net difference between a recycling or recovery process and its complimentary process and

Naphtha	12 kg
Benzene	7 kg
Syncrude	82 kg
Syngas	60 kg
Blast furnace gas	1255 kg
Iron	862 kg
Coks	4.8 kg
Electricity	406 kWh
Heat	1618 MJ
Bottle	5.5 kg
Film	2.4 kg
Cable conduit	3.4 kg

Figure 13.18. Basket of goods — product content.

POLYMER RECYCLING AND SUSTAINABILITY 539

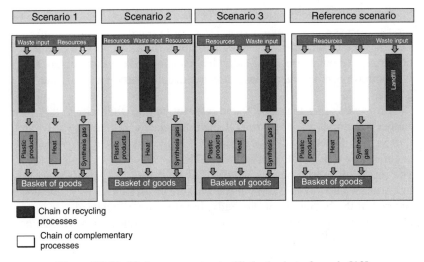

Figure 13.19. Various scenarios to fill the basket of goods [10].

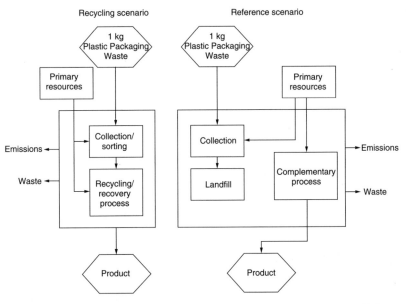

Figure 13.20. Recycling scenario and complementary process route [10].

The net difference between collecting and sorting waste plastics on the one hand and collecting and landfilling them as a component of household waste

The choice of the complementary process and its efficiency are certainly of utmost importance, which is criticized by some. Examples for this are given later in the results.

Ecological data have been collected for all the products in this project as well as for all process steps required (see Table 13.1). First a "life-cycle inventory" was established for all relevant ecological inputs. In a second step a data assessment was carried out aggregating individual data into 10 environmental categories such as anthropogenic green house effect, eutrophication potential, and the like (Fig. 13.21).

Only five categories that were found relevant will be dealt with in this chapter. The other categories such as ozone depletion, mineral resources, total water, or total mass of radioactivity did not show any or only very little differences between different routes of filling the basket of goods.

The report, discussed here, was criticized by the expert panel for not having included an evaluation of the influence on human toxicity. This evaluation was considered in the beginning of the study, but it soon was realized that very little data existed or were available. A second reason for excluding it was the nonexistence of a scientific approved and accepted way of assessing these category at the time of the study. Meanwhile, some proposals exist [11].

Table 13.1 Ecological Inputs and Outputs

	Environmental Impact Categories
Resources (input factors)	
Coal	
Crude oil	Energy equivalent of nonrenewable energy resources
Natural gas	
Minerals	Energy equivalent of renewable energy resources
Hydroelectric power	
Wood	Total mass of mineral resources
Water appropriated	Total water appropriated
Air appropriated	
Secondary raw materials	
Sum of resources	
Inventory analysis	
Emissions (output factors)	
Solid waste	Mass of household-type solid waste
Airborne emissions	Mass of hazardous solid waste (from
Waterborne emissions	power generation) and other hazardous waste
Water returned	
Air returned	Material flows contributing to the anthropogenic greenhouse effect
	Material flows contributing to catalytic stratospheric ozone depletion
	Material flows with acidification potential
	Material flows with eutrophication potential
Sum of emissions	

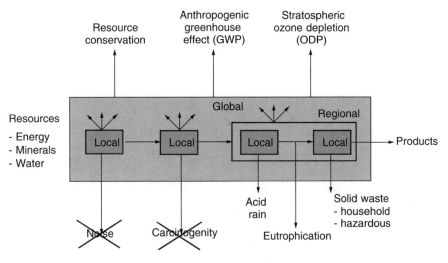

Figure 13.21. LCA — data assessment.

Furthermore, recycling routes are categorized as follows:

- Recycling routes that are able to use any kind of plastic waste — certainly after some kind of pretreatment such as agglomeration or dehalogenation, for instance. They are called universal processes. A second prerequisite for this definition is the existence of a very large market for the resulting products in comparison to the recycled amount.
- In mechanical recycling, which requires rather homogeneous plastic waste. Reality has demonstrated, that only 20–30% of household waste can be separated, purified, and worked up for mechanical recycling for economical reasons and due to insufficient market potential. This practical finding corresponds with the fact that more than 60% of plastics packaging weigh less than 10 g. The smaller the item the more uneconomical the sorting process — manually or technically.

The difference in waste input — required for technical reasons — certainly affects the environmental impacts so that only scenarios involving identical waste input can be subjected to a quantitative comparison. In practice, however, the differences in the investigated input mix are sufficiently minor (higher or lower polyolefin content) to permit qualitative comparison. The different waste input categories are indicated by separator lines in the bar charts in Figures 13.22 and 13.23.

INFLUENCE ON ENERGY DEMAND AND GLOBAL WARMING POTENTIAL The most relevant differences between different routes are found for saving in energy consumption or for energy-dependent causes such as global warming potential

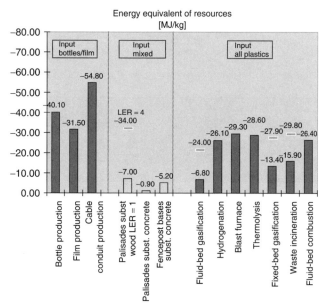

Figure 13.22. Saving of energy resources of different routes of recycling in comparison to landfill (MJ/kg of recycled plastic) (LER = life expectancy ratio) [10].

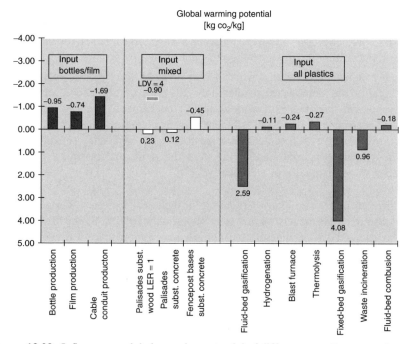

Figure 13.23. Influence on global warming potential of different recycling routes in comparison to landfill (kg CO_2/kg recycled plastic) (LDV = LER) [10].

FEEDSTOCK RECYCLING AND ENERGY RECOVERY   When we discuss the processes of feedstock recycling and energy recovery first, we realize that they have the potential to save ca. 30 MJ/kg of primary energy in comparison with landfill and thus use most of the energy content of waste plastics. The blast furnace and thermolysis process perform best, closely followed by fluid-bed combustion and hydrogenation. The waste incineration and gasification processes perform significantly worse. In the case of waste incineration this is primarily because only 34% of the generated steam was assumed to be utilized as energy, which corresponds to the present average of German waste incineration plants. The poor performance of the gasification processes is attributable to the choice of the complimentary process — a European mix in which 73% of the synthesis gas is produced from natural gas (far more energy efficient), 22% from vacuum residue oil, and only 5% from lignite

If energy recovery could be optimized in waste incineration, for example, by fully utilizing the generated steam[4] or if for the gasification processes a complimentary process is chosen that gasifies lignite only,[5] all investigated feedstock recycling and energy recovery processes would save 24–30 MJ of primary energy as indicated by the additional bars in Figures 13.22 and 13.23.

The impact on the global warming potential depends upon the use of energy resources of each process. The main greenhouse factors involved are gaseous emissions of CO_2 and methane. Fluid-bed and fixed-bed gasification and waste incineration yield increased gas emissions in comparison to landfill because they release them immediately and not in more than 100 years, as assumed for the landfill.

Energy recovery in a cement kiln was not investigated in the main study, whose results are reported here, but in a separate investigation initiated by APME [12]. Here plastic waste substituted coal as fuel for the cement kiln. The investigation showed a significantly better resource saving potential than feedstock recycling (29–33 MJ/kg vs. 26–30 MJ/kg); the saving potential for eutrophication and for acidification was also improved.

MECHANICAL RECYCLING REPLACING VIRGIN PLASTIC MATERIALS   We turn now to mechanical recycling and compare this route of filling the basket of goods with the three best feedstock recycling routes. As can be seen mechanical recycling enables in certain cases to save part of the processing energy of plastic waste and is in these cases to be ecologically the preferred process. But, there are also examples of mechanical recycling that save comparable or even less energy than the best universal processes. In these cases feedstock recycling is the ecologically preferred process.

The positioning of this "break-even point" depends on the efficiency by which recyclate is able to replace virgin resin (Fig. 13.24). If 1 kg of recyclate is able to substitute 1 kg of virgin resin and the final products fulfill identical practical

---

[4] Several plants of this type exist in Germany where the steam is used in industrial plants close by.
[5] This would be the case for the Schwarze Pumpe plant in Germany.

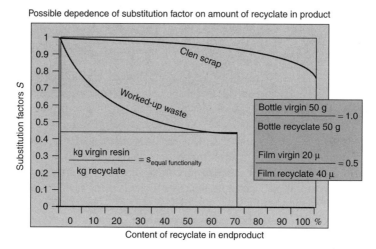

Figure 13.24. Possible dependence of substitution factor on amount of recyclate in product.

functions, this would mean dealing with a substitution factor equals 1. If, on the other hand, a bag has to be 30% thicker in order to have the same impact strength as a bag made from virgin material, the substitution factor equals only 0.77. This substitution factor certainly varies with the ratio between recyclate and virgin material in the final product. It will mostly be 1 at low percentage of recyclate and drop drastically with increasing percentage of recyclate in the final product. Two limiting cases can be assumed in theory, one where clean scrap substitutes virgin resins up to 100% with only a certain drop in properties, and one where worked up waste substitutes virgin resin only up to a certain content, and this with a much larger drop in properties.

The critical substitution factors, or break-even points, between mechanical recycling and feedstock recycling (i.e., when the ecological effects of mechanical recycling are identical to those of the best feedstock processes) have been calculated for the ecological categories considered for the investigated here routes of mechanical recycling (Fig. 13.25).

The conclusion is that mechanical recycling substituting virgin resins in its application has the potential of saving significantly more primary resources than the other routes of recycling because both the energy content (some = 40 MJ/kg) and the process energy used to produce them (another 40 MJ/kg) are largely conserved. But each mechanical recycling route is different as shown with three examples only. This means a general route of mechanical recycling does not exist. In addition, different polymers such as polypropylene (PP), polystyrene (PS), PET, and polyvinyl chloride (PVC) can be recycled each in a very special process.

Therefore, individual cases have to be considered. A general environmental positioning in comparison to other routes is impossible. The procedure developed in this project enables us to position individual processes if the environmental data are available.

Input	Best feedstock process	Bottle	Film	Cable conduit
Energy total (MJ)	Blast furnace	0.9	0.7	0.6
Energy (MJ) renewable	Municipal waste incineration	X	0.6	0.6
nonrenewable	Blast furnace	0.9	0.7	0.6
Water (m^3)	Hydrogenation	0.8	0.1	0.2
Output				
Waste (kg)	Municipal waste incineration	X	X	X
Special waste (kg)	Blast furnace	0.6	X	0.1
Eutrophication (mol PO$_4$)	Thermolysis(BASF)	0.7	0.3	0.3
Acidification (mol H$^+$)	Thermolysis(BASF)	0.7	0.3	0.3
GWP (kg CO$_2$)	Blast furnace	0.9	0.5	0.6

X = worse than feedstock recycling   0.6-better than feedstock recycling above substitution factor given

Figure 13.25. Ecological break-even points: mechanical recycling vs. feedstock recycling.

MECHANICAL RECYCLING REPLACING PRODUCTS MADE FROM WOOD OR CONCRETE
End products in the mechanical recycling of mixed plastics are thick-walled sections or preformed parts usually not intended to substitute end products made of primary plastics but end products made of wood or concrete. Here thick-walled sections for the production of various articles (park benches, landing stages, bank stabilizations) are a good example. Roughly one third of the quantities utilized in 1997 in Germany by DKR in mechanical recycling provided such target products. Merely two thirds (i.e., approx. 180,000 tons) substituted plastics as primary commodities (Table 13.2).

The decisive criterion in the case of wood substitution is the life expectancy ratio (LER) of the compared products made of mixed plastics or wood, respectively. Only above a certain limit value (in the examined case LER = >3.3) — that is, a clearly longer life cycle of the plastic product as compared with the wooden product — is the level of resource reduction in feedstock recycling/energy recovery reached. The situation is similar as far as acidification and eutrophication are concerned. Advantages are found only with regard to waste minimization.

In the case of concrete substitution, assessment criteria are the life expectancy ratio as well as the weight ratio between concrete and plastic recyclate. However, ecological savings found in the examined cases are invariably much lower than savings possible through feedstock recycling/energy recovery, so that here mechanical recycling is obviously unsuitable to reach ecological objectives.

Table 13.2 Situation of Mechanical Recycling of DSD/DKR Waste in 1997 (tons/year)[a]

Substitution of	Germany		Export	
	Virgin Resin	Wood Concrete	Virgin Resin	Wood Concrete
Film	71,707	18,379	28,380	2,855
Bottles	53,501	710	0	0
Cups	9,766	95	0	0
Mixed plastic	11,483	18,258	2,306	1,600
Agglomerate[a]	7,516	9,724	1,311	10,265
Total	153,973	47,166	31,997	14,720
Grand total	201,139		46,717	

[a]Agglomerates are used for feedstock recycling. Two hundred thousand tonnes for instance.

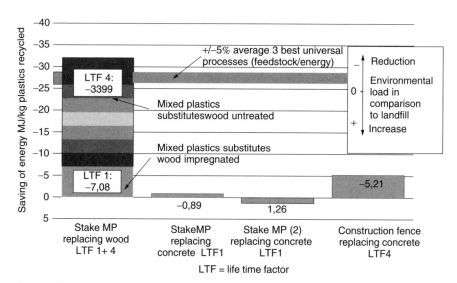

Figure 13.26. Mechanical recycling of mixed plastics (MP) (DSD system). Saving of energy consumption in comparison to landfill (LTF = LER) [8].

Based on this study it is concluded that the mechanical recycling of mixed plastics is doubtful from an environmental point of view. It reaches the ecological level of the universal processes in feedstock recycling/energy recovery only in exceptional cases (LER > 3.3) (Fig. 13.26). While in the case of plastics substitution the substitution factor is important, it is the life expectancy ratio for the substitution of wood and the weight ratio for the substitution of concrete. The respective ecological break-even points for each recycling route are given in Table 13.3.

Table 13.3 Analysis of Substitution of Materials

Products of Mechanical Recycling Substituting	Influence Parameters	Ecological Break-even Point Mechanical vs. Feedstock/Energy Recovery
Plastic	Substitution factor	Ca 0.7–1.0
Wood	Life expectancy ratio	Ca 3.3 (at weight ratio 0.75)
Concrete	Weight ratio	Ca 7–28 (at life expectancy ratio 4–1)

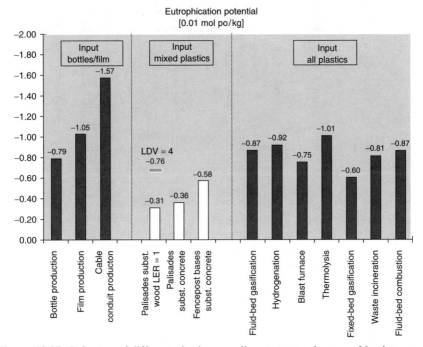

Figure 13.27. Influence of different plastics recycling routes on the eutrophication potential [10].

SPECIAL ASPECTS

Eutrophication and Acidification Eutrophication potential comprise airborne and waterborne emissions of nitrogen and phosphor compounds that promote excessive plant growth (Fig. 13.27). Acidification potential takes into account emissions of some nitrogen and sulfur compounds, hydrogen fluoride, and other

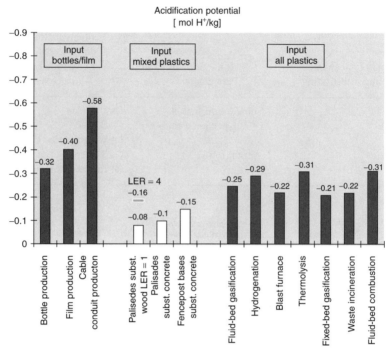

Figure 13.28. Influence of different plastics recycling routes on the acidification potential [10].

acids (Fig. 13.28). All these emissions are associated with energy-consuming processes. The results are, therefore, in principal similar to the results of the global warming effect. The mechanical processes substituting virgin plastics perform best; the feedstock recycling and energy recovery processes come in the middle, and the mechanical processes substituting wood or concrete perform worst.

Municipal Solid Waste and Hazardous Waste Waste incineration is the most effective method of waste reduction. All other recycling routes involve more waste in the total recycling chain (Fig. 13.29). The special case of mechanical recycling substituting wood or concrete, apparently saving more waste than they replace, is due to the assumptions made. In the case of concrete substitution, the portion in excess of 1 kg/kg of recycled plastic results from the low density of mixed plastics compared with the far heavier concrete product. The wooden product generates more MSW on final disposal than the mixed plastic solely because of the assumed life expectancy ratio of 4.

As no hazardous waste arises in the production of secondary raw materials or products made of virgin materials, the scenarios for mechanical recycling substituting plastics show no effect at all (Fig. 13.30). The increased quantities of hazardous waste in the case of processes that substitute wood or concrete result largely from the final disposal by incineration (filter dust).

POLYMER RECYCLING AND SUSTAINABILITY 549

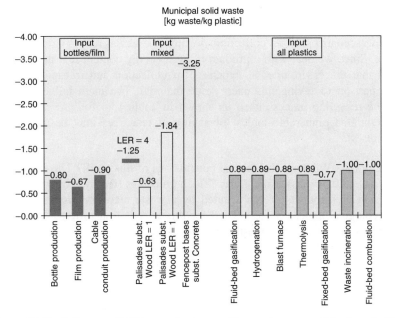

Figure 13.29. Influence of different plastic recycling routes on municipal waste [10].

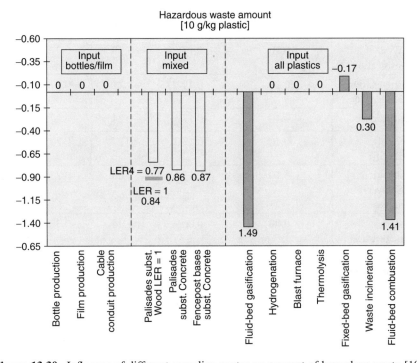

Figure 13.30. Influence of different recycling routes on amount of hazardous waste [10].

CONTRIBUTIONS FROM INDIVIDUAL STEPS IN THE RECYCLING CHAIN   An interesting question is what are the environmental contributions of individual steps in the recycling scenario such as collection, sorting, the actual recycling process, and the like. Since we deal with a product of high energy content it is not surprising to find that the environmental expenses for collection and transport are small in comparison to saving this energy-rich material. The main influences derive from the recycling process itself as shown in Table 13.4 for the global warming potential. Comparable tables for saving of resources may be found in the literature [8].

CONCLUSIONS   What are the general conclusions of this work? The decisive factors in an ecological assessment are: Which processes (fed by primary resources) and which pre-products are substituted by a waste recovery process chain and what ecological expenses are needed for this substitution process (called complementary process in the investigation).

For mechanical recycling this means substituting virgin plastic materials, for instance; this is an ecologically more efficient process than feedstock recycling if

Table 13.34 Contributions of Individual Steps to the Reduction (−) or Increase (+) of the Global Warming Potential in Comparison to Landfill (kg CO_2 Equivalent/kg Recycled Plastic)

	Collection/sorting	Treatment	Recycling process	Landfill	Total
HTW gasification	0.10	0.17	2.59	−0.27	2.59
Hydrogenation (KAB)	0.10	0.17	−0.11	−0.27	−0.11
Blast furnace	0.10	0.17	−0.24	−0.27	−0.24
Thermolysis[1] (BASF)	0.10	0.17	−0.27	0.27	−0.27
Gasification (Schwarze pumpe)	0.10	0.17	4.08	−0.27	4.08
Waste incineration	0.00	0.00	1.34	−0.38	0.96
Mono-combustion	0.10	0.17	−0.07	−0.38	−0.18
Bottle recycling	0.10	0.54	−1.27	−0.31	−0.95
Film recycling[2]	0.10	2	−0.48	−0.36	−0.74
Cable conduit	0.10	0.41	−1.84	−0.36	−1.69

the expenses for mechanical recycling are low. If they are excessive — and this investigation has provided examples for this — feedstock recycling or energy recovery may be ecological comparable or even the more preferred processes.

Mechanical recycling is suitable for single waste fractions, while feedstock recycling and energy recovery are needed for recycling mixed waste fractions. The mechanical recycling of mixed plastics into products substituting wood or concrete is of little environmental benefit. This leads to the final conclusion of thin LCA analysis described here in detail that there is not only one preferred process for plastics recycling from household packaging, but an ecologically sensible recovery strategy has to combine mechanical and feedstock recycling as well as energy recovery routes taking into account ecological as well as economical aspects.

Life-Cycle Analysis of Recycling of Household Packaging Material in Other Countries of Europe The results of the LCA analysis described in detail have been confirmed by several studies published. An investigation by the Centre for Energy Conservation and Environmental Technology in Delft has shown that a combination of feedstock recycling (Texaco process of gasification[6]) and mechanical recycling will compensate ca. 21% of the environmental impact related to Dutch production of high-density polyethylene (HDPE), low-density polyethylene (LDPE), and PET or ca. 1% of the environmental impact related to total Dutch passenger travel by car. The investigated environmental impact categories (greenhouse effect, eutrophication, acidification, etc.) were weighed and aggregated into one environmental parameter. Weighing was based on the distance to target method. Target was the national goal for each environmental parameters [13]. This enabled the investigators to make a correlation to costs as discussed later.

A further investigation dealt with the importance of the complementary process. CE/Delft has compared the environmental effect using a separated plastic/paper fraction from household waste as fuel in a coal-fired power plant, co-firing it with biomass, using it as fuel in a cement kiln, or incinerating it in a municipal waste incineration plant. The largest environmental benefit was obtained substituting coal in a power plant [14].

A study in Norway investigated the influence of mechanical recycling and of energy recovery by an LCA approach in 1999 [15]. The study used also a comparative, that is, a basket of goods approach. The environmental benefits of mechanical recycling were clearly shown, but at the same time the importance of the complementary process for the positioning of energy recovery was demonstrated. The study recommended replacing conventional energy resources with high environmental loads by energy recovery of plastic waste incineration.

A Swedish report investigated the environmental impact of the Swedish system of collection and recycling of plastic waste in 1999 by an LCA approach and confirmed the findings that mechanical recycling has the highest environmental benefit and that energy recovery from plastic waste is better than landfill [16].

---

[6] Has been discontinued.

Recycling of Plastic Waste from Other Sources Recycling of plastic waste today is not a question of technologies — they exist and are developed further [9] — but a question of logistics (proper collection and sorting) and especially a question of financing. While the recycling of industrial scrap or the recycling of higher prized postconsumer waste (e.g., PET bottles) is profitable in most of the cases, the recycling of postconsumer waste generally is not. An accompanying system of financing this activity is necessary. It is the DSD levy system in Germany or other systems in other countries that finance the recycling of household packaging.

Other ways of financing are necessary when recycling postconsumer plastic waste from sources such as cars, electro/electronics, and the like because here the organization of the markets are different. Packaging material are short-lived materials (less than 1–2 years on average). The consumer paying the levy obtains the recycling compensation soon after. But, in the case of the other waste sources, we deal with longer life applications — cars, for instance, around 10 years or buildings more than 50 years. Who is going to be responsible for the recycling after such a long time? Who is going to finance it finally? The consumer in the beginning or the producer of the product at the end? How many items are left after this time period? A lot of open questions. They are presently being discussed in Europe.

In addition, similar questions will be raised here later that we have discussed for plastics packaging. How much subsidy makes sense to obtain a sizable environmental gain and where is the break-even point beyond which the environmental gain is too small for the money spent, as in the German DSD system. These questions lay ahead for society. The likelihood of a positive overall scenario seems larger according to a recent investigation in Austria [17].

13.3.2. Correlation of Ecological and Economic Aspects of Sustainability

Life-cycle assessment when carried out according to the ISO rules has shown its ability to deliver data for certain more global environmental compartments like the impact potential on saving of resources, global warming potential, acidification, ozone depletion, and the like. It usually does not cover local effects such as noise or smell and hazardous substances. Here risk assessment or other methodologies are needed. The evaluation of effects regarding human toxicity is hampered by a lack of sufficient data and by a still undecided question of data evaluation. Thus, life-cycle analysis is a useful tool but not the only answer to all environmental aspects.

The question gets even more complicated if we try to combine only two of the three basic compartments of sustainability, economic costs and environmental benefit. This combined assessment is very essential if we want to find an optimal balance between both aspects of life. The sole statement that mechanical recycling is better than other recycling routes is of little help if not combined with a corresponding money tag. It even may cause the introduction of very expensive waste treatment solutions without any environmental gain. Thus, an increase of recycling rates from 15 to 50% will increase the cost by a factor of 3 while environmental impacts remains broadly similar according to an investigation by TNO, Netherlands [18].

While the economic compartment is easily defined by a cost–profit relationship, the correlation with the ecological compartment is difficult because we deal with several very different aspects, such as global warming potential, and different aspects may have different importance under different situations (smog in California or eutrophication of the Caspian sea).

The LCA rules do not recommend the combination of different environmental parameters into one number and request a separate treatment of each environmental aspect.

Comparison of Environmental Benefits Versus Cost for Recycling of Plastic Household Packaging in Germany in 1996

Ecological Aspects One first attempt to correlate economic and environmental aspects according to these rules were made by VKE/Germany [19]. The ecological results of the study described (8, 10) were used to evaluate the German situation of recycling plastic household packaging. The saving of energy resources and the global warming potential — the main parameters according to the findings — were chosen as environmental parameters. In order to do this, the ecological data for resource saving obtained by this investigation have been applied to the annual amount of plastic waste collected by DSD and afterwards correlated with the corresponding costs for four different recycling scenarios.

In case A it is assumed that all plastic goods are made from virgin material and plastic waste remains part of the household waste and is channeled — together with residual waste and in compliance with the German technical instruction on residential waste (TA Siedlungsabfall) — into thermal treatment. In this case the entire collected quantity (750,000 tons) undergoes thermal treatment. Using the ecological savings of the study for each recycling route, this leads to resource savings in comparison to landfill of:

$$-300{,}000 \text{ tons/year fuel oil}[8]$$

At the same time the potential effect on global warming goes up by +643,000 tons/year in CO_2 equivalents.

In case B it is assumed that waste is collected separately, but very little sorting is done because in this case the entire quantity is channeled into feedstock recycling. The following savings potential is calculated from ecological mean values obtained in the three best processes available for feedstock recycling:

---

[7] Example: 750,000 tons/year waste × −15.95 MJ/kg energy savings: 40 MJ/kg (H_u fuel) = 300,000 tons/year fuel.

Fuel oil: −382,000 tons/year or

Greenhouse potential: −159,000 tons/year in CO_2 equivalents

The scenario of an optimized energy recovery is almost fully congruent as far as resource savings are concerned (−360,000 t/y of fuel oil). However, the situation is more complex with regard to the potential effect on global warming and depends on individual framework conditions.

In case C it is assumed that the same quantity channeled into recycling is entirely (100%) utilized in the ecologically most favorable method available in mechanical recycling (conduit production), resulting in the following theoretical savings potential:

Fuel oil: −715,000 tons/year, or

Greenhouse potential −546,000 tons/year in CO_2 equivalents

Since the practical implementation of case C is impossible for numerous reasons outlined earlier, in case D a more realistic scenario is assumed with 40% mechanical recycling and 60% feedstock recycling. In mechanical recycling two thirds of the amount is intended to substitute virgin resins, and roughly one third is intended to substitute wood and concrete. For the sake of simplicity, the ecological effect of the latter is equaled with feedstock recycling — a rather positive view as previously described.

The following savings potential is calculated in this scenario:

If a degree of substitution of 1:1 is achieved for the primary commodity:

−470,000 tons/year of fuel oil, or

−262,000 tons/year in CO_2 equivalents

If a degree of substitution of 0.7 is reached:

−413,000 tons/year of fuel oil, or

−218,000 tons/year in CO_2 equivalents

In scenario E it is assumed that — complying with the demand voiced by some politicians — the share of mechanical recycling is raised to 60%. The following savings potential is calculated based on assumptions analogous to scenario D:

If a degree of substitution of 1:1 is achieved for the primary commodity:

−514,300 tons/year of fuel oil, or

−313,400 tons/year in CO_2 equivalents

If a degree of substitution of 0.7 is reached:

−428,700 tons/year of fuel oil, or

−248,000 tons/year in CO_2 equivalents

The differences in the savings potential between case D and E are +15,000 tons/year of fuel oil in the lower borderline case and +44,000 tons/year of fuel oil in the upper borderline case, that is, differences are in the vague zone of these calculations (Fig. 13.31).

The discussions about those differences during the last amendment to the German Packaging Ordinance had the character of religious wars. Looking at the real differences, especially in scenarios D and E, one may wonder whether always the right priorities have guided public discussion. This question is even more justified when relating the values calculated here to the present total emissions in Germany (Fig. 13.32).

Economic Consequences of Plastics Recycling An evaluation of scenarios A–E with known average costs of collection, sorting, and recycling or disposal[8] in 1995 resulted in the diagram shown in Figure 13.33. Cost differences are large — varying between 0.5 and 2 billion — while added ecological benefits are small — around 150–200,000 tons/year of fuel as described.

Analyzing the costs of the Dual System, it emerges that more than two thirds of the total costs are caused by logistics, that is, collection and sorting. This situation results from the joint collection of several material flows so that we

	Saving potential resources, measured in t/y fuel oil	Saving potential greenhouse effect, measured in t/y CO_2 equivalents
Case A: Traditional waste management/ waste in MWI	−300.000	+643.000
Case B: 100% Feedstock recycling	−382.000	−159.000
(Case C: 100% Mechanical recycling) hypothetic, as beyond means	(−715.000)	(−546.000)
Case D: 40% Mechanical recycling 60% Feedstock recycling	−413 – 470.000 *	−218 – 262.000 *
Case E: 60% Mechanical recycling 40% Feedstock recycling	−428 – 514.000 *	−248 – 313.000 *

*according to efficiency of virgin resin substitution

Figure 13.31. Ecological comparison on different recycling scenarios for plastic waste in the household area (based on 750,000 tons of waste collected with −535,000 tons recycled).

---

[8] Municipal waste collection 300 DM/ton
   municipal waste incineration 500 DM/ton
   DSD collection 1000 DM/ton
   DSD collection for feedstock recycling 700 DM/ton
   DSD sorting 25% 600 DM/ton
   DSD sorting 50% 1200 DM/ton
   DKR gate fee for recycling 700 DM/ton

	kg CO_2/kg	For 535.000 t/y recycled plastic packaging waste (mio. t CO_2/y)
Scenario A: 100% municipal waste incinerator	+ 0,858 *	+ 0,643 *
Scenario B: 100% blast furnace	– 0,348 *	– 0,159 *
Scenario C: 100% best mechanical recycling (pipe processing)	– 1,032 *	– 0,546 *

*Deviation in comparison to scenario landfill
+ = Increase / – = Decrease

Comparison		
Energetical use of oil	104 mio. t/y	326,0
Total emissions Germany		888,0 *
CO_2-formation by respiration 80 mio. persons		20,0

* OECD-calculation (FAZ 25.06.97)

Figure 13.32. Plastic waste scenarios — influence on global warming potential.

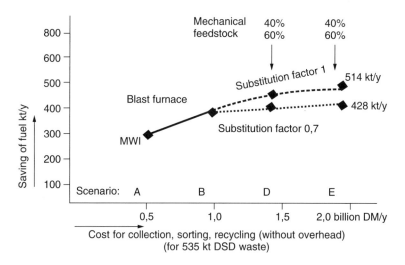

Figure 13.33. Ecology vs. economy for different strategies of recycling and recovery [19].

practically have a second waste, not raw material, collection system as far as plastic materials are concerned.

There are more efficient routes to saving comparable amounts of resources. Therefore, the actual conclusion is that also in ecological evaluations the economic aspect has to be considered. The money spent here cannot be used for potentially higher ecological benefits somewhere else.

Development of "Ecoefficiency" Method and Its Application to Recycling of Plastics Packaging in Europe *Sustainability* is a word coined for the use in general society and not for science only. Argumentations in the political world are often black and white in order to clarify a position. The if's and when's are left to the experts. For these kind of discussions the method of ecoefficiency was recently developed by BASF for an evaluation of its products and applied to plastics packaging recycling in Europe by TNO recently [20]. The method combines information on economical aspects as well as environmental aspects and presents the data in a standardized, that is, dimensionless, diagram of cost versus environment in the form of a portfolio analysis. Only the differences between different processes are shown, not the absolute values. All costs of one scenario are summarized into one number. In order to enable a similar procedure for the ecologic aspects, several different environmental aspects (in this case eight parameters) have to be summarized into one figure. This can be done by weighing the individual parameters. The weighing occurred by using normalization factors (effect per capita per year), by turning it into relative factors (fraction of the total score of that environmental effect in Europe) and by summarizing them with equal weight[9] into one number for the diagram. Weighing is a very subjective step, and no weighing or ranking method has broad support by society. Therefore, ISO guidelines (ISO14042) do not recommend using this weighing procedure, although LCA studies on weighing exist [13].

There is no consensus yet on how to proceed, and different parts of society may use different ranking procedures. Therefore, the study has taken great care to show the influence of different weighing procedures and could show that the basic results are valid under all circumstances.

Different recycling scenarios for plastic industrial waste and for plastic household waste with bottle banks, separate collection in a yellow bag as in Germany or with a less elaborated system (gray bag[10]) for different recycling quotas (15, 25, 35, and 50%) have been investigated with this method. Increasing recycling quotas cannot be fulfilled by mechanical recycling only due to limited markets for the products, the difficulties in collecting dirty and light packaging, and the technical developments on the virgin market tend to make products thinner, lighter, and technically of higher standard as was shown by an earlier TNO investigation [18]. The investigation compared therefore the following scenarios, which included mixed plastic recycling and feedstock recycling in order to achieve the targets (Table 13.5).

The basic result is, in principal, an agreement with the German investigation described earlier [19]. The reference scenario landfill has the lowest cost but the greatest environmental load. A recycling ratio of 15% (R15 in Fig. 13.34) gives an obvious decrease of the environmental load without a significant increase

---

[9] A sensitivity analysis with different weighing ratios and methods shows the influence of this procedure but also shows that the basic results are not influenced by this.

[10] The gray bag system consists of two bins, one for collection of dry items and one for wet items; 98% of all plastic waste is found in the dry compartment.

Table 13.5 Recycling Targets for Various Scenarios [13]

Scenario	Code	Mechanical Recycling (%)	Mixed Plastic Recycling (%)	Feedstock Recycling (%)	Energy Recovery in MWI (%)	Landfill (%)
Reference 1	Landfill					100
Reference 2	NOW	10	2	3	15	70
Scenario I	R15	15			85	
Scenario II	R25y R25g	25		10	75	
Scenario III	R35y R35g	15	10	10	65	
Scenario IV	R50y R50g	15	20	15	50	

For recycling rates up to 15% it was assumed that this target can be achieved by collection of industrial monostreams and by bottle bank collection. For higher recycling rates it was assumed that more comprehensive routes such as gray bag or yellow bag systems are required.

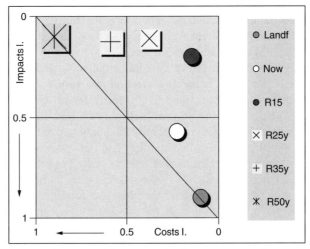

Figure 13.34. Ecoefficiency portfolio, comparing increasing recycling targets with landfill using the yellow bag system (Germany) [18].

of costs. An increasing recycling ratio of 25, 35, or 50% (R25, etc.) shows an increase in costs without an obvious further reduction of environmental impacts.

When comparing the yellow bag system with the gray bag system, it was shown that the yellow bag has higher costs while the gray bag system has slightly more environmental load (Fig. 13.35). The main conclusion thus is that

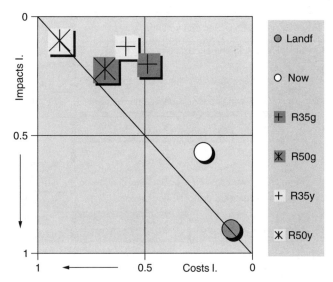

Figure 13.35. Ecoefficiency portfolio, comparing increasing recycling targets with landfill using the gray bag system [18].

the diversion of plastic waste from landfill, rather than the choice of recovery option, is the most important factor influencing the ecoefficiency of waste recovery and that recycling rates above 15% have no major environmental benefit due to the market limitations for mechanical recycling and achieving higher targets in reality only via feedstock and mixed plastics recycling. This approach is useful as a first step to identify trends; it should be supported by further investigation of the details. The critical review of this investigation — as requested by ISO rules — has been carried out meanwhile and the results confirmed in principle.

Cost–Benefit Analysis and Its Application to Recycling of Plastic Packaging Material in Austria Another approach to combine environmental data with economic costs was taken by Hutterer and Pilz in Austria [21]. They attempted to monetize environmental effects of different recycling and recovery routes by using avoidance costs. Avoidance costs are costs that could be invested elsewhere in the economy to prevent a certain amount of emissions (e.g., costs for installing a filter to clean the emissions of a factory or costs of insulation to reduce CO_2 emissions). The level of avoidance costs depends, of course, on the assumed target for the reduction of emissions. The investigation takes this into account by varying avoidance costs (e.g., for CO_2) between current marginal costs (17 Euro/t CO_2) and marginal costs for a reduction of 20% (84 Euro/t CO_2) in Austria.

The recycling situation for plastics waste existing in Austria in 1996 was investigated and the microeconomic (waste management), macroeconomic, and environmental costs were investigated using cost–benefit analysis. Ecological parameters chosen were the primary energy consumed, the CO_2 and CH_4

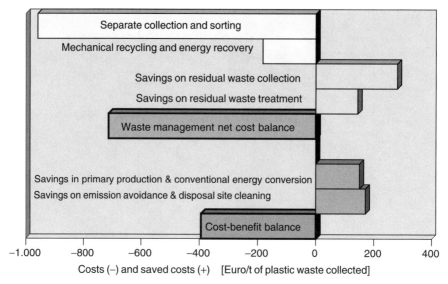

Figure 13.36. Cost–benefit analysis for mechanical recycling of plastics packaging from households in Austria 1996 (/t collected material [21]).

emissions, the TOC emissions, and the volume of waste for landfill. The individual parameters shown in Figure 13.36 were taken into account.

While the analysis of the recycling of plastic household packaging showed an overall negative result, leading to a reduction of recycling goals for this sector (collection limited to easily collectable and sortable items), alternative analysis of other sectors, like commercial plastic packaging and certain plastic waste from buildings, cars, electro/electronics, and agriculture showed an overall positive result [17].

13.4. CONCLUSIONS

While the methods to combine ecological and environmental aspects of sustainability differ, the results of the different approaches are more or less compatible. All routes diverting plastic waste from landfill show a positive environmental contribution. Mechanical recycling has the chance — in addition to saving the energy content of the product — to save part of the processing energy, but the realization is limited to rather clean waste such as industrial scrap or selected clean fraction of household waste, such as bottles due to market limitations and excessive cost of collection and sorting. The difference between different routes of recycling, including routes of energy recovery, is rather small, especially if compared to the total environmental benefit of plastics. Plastics are recyclable in principle, but the amount depends on the complexity of the product that was turned into waste, the costs involved in separating it into a clean plastic fraction, and the recycling route chosen. Thus, all requirements cited in the beginning for being a sustainable product are fulfilled for plastics.

REFERENCES

1. Further information may be obtained from Association of Plastics Manufacturers Europe, (APME), Brussels, or on the Internet at *http://lca.apme.org/reports/htm/14.htm*.
2. Einfluß des Einsatzes von Kunststoffen auf den Energiebedarf und die energiebedingten CO_2 — Emissionen im Bereich der Wärmedämmung, Fraunhofer Institut für Systemtechnik und Innovationsforschung, Karlsruhe June 1999 or on the internet at *http://www.vke.de/text/deutsch/download.htm*.
3. M. Heyde and T. Nuerrenbach, *Use of Plastics in the Car Manufacturing Industry*, Fraunhofer Institute for Process Engineering and Packaging, Freising, Germany, July 1999.
4. *Gesellschaft für Verpackungs-Marktforschung*, Wiesbaden, Germany, 1997.
5. J. M. Kooijmann, Environmental Assessment of Food Packaging: Impact and Improvement, *Packaging Technology and Science* 7, 111 (1994).
6. Untersuchung zu möglichen ökologischen Effekten bei der Substitution von Kleinverpackungen, Fraunhofer Institute Freising 2000, obtainable from Association of Plastics Producers Germany (VKE), Frankfurt/M. Kunststoffkleinverpackungen im Spiegel von Konsumverhalten und Ökologie, Memorandum Industrieverband Kunststoffverpackungen (IK), Bad Homburg 2000.
7. The Benefits of Using Plastics, GUA Gesellschaft für umfassende Analysen, Wien, 2000; e-mail: *office@gua-group.com* Information available from Association of Plastics Manufacturers Europe (APME), Brussel.
8. M. Heyde and M. Kremer, Recycling and Recovery of Plastics from Packagings in Domestic Waste, *LCA Documents* Vol. 5, 1999.
9. J. Brandrup, M. Bittner, W. Michaeli, and G. Menges, eds., *Recycling and Recovery of Plastics*, Hanser, New York, 1996.

    A. L. Bisio, and M. Xanthos, eds., *How to Manage Plastic Waste, Technology and Market Opportunities*, Hanser, New York, 1994.

    H. Sutter, *Erfassung und Verwertung von Kunststoff*, EF, Berlin, 1993.
10. Recycling and Recovery of Plastics from Packaging in Domestic Waste — LCA-type Analysis of Different Strategies, Summary Report, December 1998, obtainable from VKE, Frankfurt, or DSD, Cologne.
11. M. A. J. Huijbrechts, *Priority Assessment of Toxic Substances in the Frame of LCA*, Interfaculty Department of Environmental Science, University of Amsterdam, 1999.
12. M. Heyde and M. Kremer, *Energy Recovery from Plastic Waste as an Alternative Fuel in the Cement Industry*, Fraunhofer Institute, Freising, Germany, 1998; Information available from Association of Plastics Manufacturers, Europe (APME), Brussel.
13. H. Croezen and H. Sas, *Evaluation of the Texaco-Gasification Process for Treatment of Mixed Plastic Household Waste*, Centre for Energy Conservation and Environmental Technology, Delft, The Netherlands, 1997.
14. Subcoal: An environmental assessment, CE, Delft 2000. Information available from Association of Plastics Manufacturers Europe (APME), Brussel.

15. *Recycling of Plastic Packaging Waste in the Drammen Region of Norway*, Ostfold Research Foundation, 1999.
16. *Recycling, Incineration or Landfill? A Life Cycle Assessment of Three Different Waste Treatment Methods*, Chalmers Industriteknik Ekologik (CIT), Göteborg, Sweden, 1999.
17. H. Hutterer, H. Pilz, G. Angst, and M. Musical-Mencik, *Stoffliche Verwertung von Nichtverpackungskunststoffabfällen. Kosten-Nutzen-Analyse*, Bundesministerium für Umwelt, Jugend und Familie, *Monographien* Vol. XX, Wien, 1999.
18. *Potential for Post-User Plastic Waste Recycling*, TNO, Netherlands, 1998, available through Association of Plastics Manufacturers Europe (APME), Brussel.
19. J. Brandrup, *Müll und Abfall 30* **8**, 492 (1998).
20. Eco-efficiency of recovery scenarios of plastic packaging TNO report R 2000/119, Apeldorn 2000. Information available from Association of Plastics Manufacturers Europe (APME), Brussel.
21. H. Hutterer and H. Pilz, *Kosten-Nutzen-Analyse der Kunststoffverwertung, Bundesministerium für Umwelt, Jugend und Familie, Monographien* Vol. 98, Wien, 1998.

CHAPTER 14

PLASTICS RECYCLING

MICHAEL M. FISHER
Director of Technology, American Plastics Council

14.1. INTRODUCTION

Polymer recycling can encompass a broad range of materials and products. These could include natural cellulose from paper fibers, high-density polyethylene (HDPE) from detergent bottles, and rubber from automobile tires. These examples cover biopolymers, synthetic polymers in the form of plastics, and natural and synthetic polymers in the form of rubber. The emphasis of this chapter is on plastics recycling since this is the field of polymer recycling that is broadest in scope, highest in visibility, and the most heuristic. Paper recycling is a specialized area, and the corresponding technical issues are quite unique to paper and wood-based products. A discussion of plastics recycling covers the broad field of synthetic polymers, and the basic technologies can be applied to most rubber and elastomers. This chapter begins with a brief discussion of definitional issues in order to provide a context and framework for the technical material that follows. Plastics recycling is covered from a "generic" technology perspective and a product-focused perspective. The chapter includes statistical data on plastics recycling in the United States, Europe, and Japan. A final section looks at the future for plastics recycling. The subject of polymer recycling is interesting and significant not only from a technology perspective but from a political, economic, and social perspective as well. The coverage of these subjects in detail is beyond the scope of this chapter.

Disclaimer: The views expressed in this article are those of the author and do not necessarily represent the views or policies of the American Plastics Council, a part of the American Chemistry Council, Inc. Reference to specific manufacturers and trade names is for informational purposes only and does not constitute an endorsement of either the company or the product(s).

Plastics and the Environment, Edited by Anthony L. Andrady.
ISBN 0-471-09520-6 © 2003 John Wiley & Sons, Inc.

Bibliographical Note: The references at the end of this chapter includes a list of plastics recycling related monographs [1–7]. By far the most comprehensive is Brandrup and co-workers [1]. Also of interest for its extensive list of references is Bisio and Xanthos [2]. In general, much of the polymer recycling technology developed over the past 15 years is described in industry reports not readily available to the public. A list of industry associations that have sponsored numerous studies is included in the bibliography. Fortunately, much of this work did make it into conference proceedings. Many of the reference in this chapter are to such proceedings. Conferences sponsored by the Society of Plastics Engineers (SPE, Brookfield, CT), SAE International (Warrendale, PA), IEEE (Piscataway, NJ), the Association of Plastics Manufacturers in Europe (Brussels), Maack Business Services (Zurich), and Recovery Recycling Reintegration International Conferences R'95, R'97, and R'2000 (EMPA, Saint Gall, Switzerland) are especially fruitful. In comparison to the tremendous amount of information presented in conference papers, relatively little has made it to the more readily available scientific, engineering, and technical journals.

14.2. BASIC PLASTICS RECYCLING DEFINITIONS AND NOMENCLATURE

As in any technical field, when communicating about the technology of plastics recycling, it is important to use common terminology and definitions. Early (pre-1970) terminology for commercial, industrial, and postconsumer recycling of materials developed around the needs of the paper and textile (fiber) recycling industries and the metal and glass recycling industries. After the first oil price shock of the 1970s, plastic recycling gained attention, and it soon became apparent that the recycling definitions and nomenclature for the other commodities did not apply adequately to the collection and recycling of plastics. The problem reflected both the unique properties of plastics along with their versatility and limitations in terms of some processing technologies. Over the years, the Institute of Scrap Recycling Industries (ISRI, Washington, DC) has developed an extensive terminology and a list of definitions applicable to paper, textiles, glass, and metals. An authoritative source of terminology and definitions applicable to plastics recycling is ASTM International's (West Conshohocken, PA) *International Guide for the Development of Standards Relating to the Proper Use of Recycled Plastics, D5033-00*. This guidance document was revised in 2000. The following discussion of plastics recycling definitions and nomenclature reflects the information that can be found in this guide.

Primary metals for the most part can be recycled into secondary metals that either have the purity to compete directly with primary metals in high-performance applications or have somewhat diminished properties and restricted applications. In all cases, the practical form of recycling is metal back to metal. The same is essentially true for glass-to-glass and fiber-to-fiber recycling. Plastics on the other hand have high energy content and can also be burned as fuel to recover their feedstock energy. The heat content of most commodity plastics is significantly greater than that of wood or coal. Many plastics can also be thermally or chemically depolymerized into monomer(s), petrochemical feedstocks, and fuels. The end result is that several forms of plastics recycling can be defined. D5033-00 describes the following:

Primary recycling is the "processing of scrap plastic product into a product with characteristics similar to those of the original product."

Secondary recycling is the "processing of scrap plastic into a product that has characteristics different from those of the original product."

Tertiary recycling is the "production of basic chemicals or fuels from segregated plastic scrap or plastic material that is part of a municipal waste stream or other source."

Quaternary recycling is the "useful retrieval of the energy content of scrap plastic by its use as a fuel to produce products such as steam, electricity, and so forth."

Primary and secondary recycling are often referred to as *mechanical recycling* since the principal recycling processes involve cleaning and separation of the plastic by mechanical rather than chemical or thermal means. Quaternary plastics recycling is not universally recognized by government agencies in the United States or Europe as a form of plastics recycling. Energy recovery in the form of steam, hot water, or electricity by direct, controlled combustion of plastics as well as via the intermediate production of liquid, gaseous, or solid fuels from scrap plastic by thermal or chemical methods are often referred to as *resource recovery* processes rather than recycling. In Japan, quaternary recycling is often referred to as thermal recycling and is included in the broad definition of plastics recycling.

There is also uncertainty in the regulatory status of tertiary recycling when it does not result in the direct production of monomers suitable for polymerization into new plastic. The European Commission has at times supported the chemical recycling (depolymerization) of condensation polymers such as polyethylene terephthalate back to monomer (e.g., dimethyl terephthalate) as recycling for the purpose of government-mandated plastics recycling rate calculations, but not the liquefaction of polyolefin plastics back to petrochemical feedstocks for reprocessing in a refinery. Discussions around these types of definitional issues, and their environmental and economic implications, are likely to continue for many years to come.

Lastly is the subject of *postconsumer* and *preconsumer* plastics. In the United States for the purposes of making recyclability or recycling content claims for products, it is important to differentiate postconsumer from preconsumer plastics (16 CFR Part 260 *Guides for the Use of Environmental Marketing Claims*, available from U.S. Federal Trade Commission, *www.ftc.gov*). The paper industry now widely differentiates between postconsumer fiber and preconsumer fiber. The following definitions of postconsumer and preconsumer plastics are contained in ASTM D5033-00:

Postconsumer plastic: "Plastic material or finished product that has served its intended use and has been diverted or recovered from waste destined for disposal, having completed its life as a consumer item."

Preconsumer plastic: "Plastic material diverted from the waste stream following an industrial process, but excluding reutilization of material such as rework, regrind, or scrap generated in a process and capable of being reclaimed within the same process."

The term *postuse plastic* is sometimes used to cover both postconsumer or preconsumer plastics. The above definitions are in line with the recycling industry's position that scrap destined for recycling is not waste.

14.3. POLYMER RECOVERY, RECYCLING, RESOURCE CONSERVATION, AND INTEGRATED RESOURCE MANAGEMENT AS GLOBAL CONCEPTS

The previous section highlighted the different forms of plastics recycling and noted that some parties differentiate plastics recycling for material recovery from plastics recycling for fuel or energy recovery by designating energy recover and sometimes fuel recovery as resource recovery rather than recycling. As noted above, Japan usually includes thermal recycling in its definition of plastics recycling, while Germany does not. Responsible arguments can be made on both sides of this discussion. In the final analysis, what is most important globally is resource conservation. Figure 14.1 shows a general classification of resource management options for postuse plastics. Depending on the particular region of the world, some options, and not others, may provide a better balance of environmental benefit and economic sustainability.

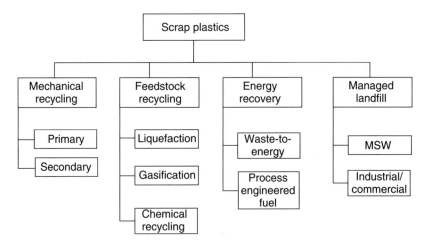

Figure 14.1. Resource management options for postuse plastics.

In the final analysis, plastics recycling should not be looked at in isolation but in the broad context of overall societal needs and ecological benefits. Because polymers can be recovered in so many ways compared to glass or metals—as fiber, fuel, energy, plastic resins, monomers, other chemical feedstocks, or syngas, for example, they have the ability to integrate into many industrial and commercial processes [1]. Integrated resource management (IRM) is the utilization of a range of recovery methods and disposal options that best meets local requirements. When postuse plastic serves the chemical, energy, or material needs of another industry, the end result is called integrated resource management across industries. This is becoming a long-term goal in many regions of the world. One example is the use of postconsumer and postindustrial plastics as a reductant and fuel in steel mill blast furnaces discussed briefly in a later section of this chapter. The goals of IRM are resource conservation, waste minimization, and progress toward sustainable development.

14.4. EARLY HISTORY OF PLASTICS RECYCLING (PRE-1990)

Four events focused attention of plastics recycling prior to 1990 (see Table 14.1). The first was the oil crisis of the early 1970s. The synthetic rubber and plastics industries largely depend on petroleum and natural gas for both their energy needs and feedstock requirements. When the price of a barrel of oil began to skyrocket in the early 1970s, efforts turned to reusing the energy content or material content of plastics as an alternative to purchasing oil from oversees markets. This stimulated a number of industry-led and government-led R&D initiatives from about 1975 until 1985 to explore technology for the recovery and recycling of scrap plastics. As one might expect, given the strong and lightweight properties of plastics, the automotive industry was the focus of much of this early R&D. It was broadly recognized that the expanded use of lightweight plastics would save gasoline through improved automobile fuel economy as well as provide additional energy and material conservation benefits if recovered and recycled.

Because of the concerns about an adequate and reliable supply of petroleum due to the pricing actions of the Organization of Petroleum Exporting Countries (OPEC), the 1970s also saw an interest in the development of bio-based feedstocks for the production of plastics and elastomers. In recent years this area has seen renewed attention as companies pursue their sustainability objectives and look to replace nonrenewable feedstocks with renewable feedstocks where economically and environmentally sustainable.

The second event that increased attention on plastics recycling was the establishment of a deposit fee on bottles for carbonated beverages. Although at the time the fee was imposed by several states glass bottles predominated, as product manufacturers and bottle fillers increasingly turned to plastic bottles, a recycling infrastructure for the plastic was established. This infrastructure grew significantly during the 1980s. The plastic that predominated in the carbonated beverage market, and still does, was polyethylene terephthalate (PET).

Table 14.1 Significant Events in the History of Plastics Recycling

Event	Result
Oil crisis of the 1970s	R&D on plastics recycling in the packaging and automotive markets. R&D on biopolymers and agricultural-based polymer feedstocks
Bottle-bill legislation passes in several states in the United States (1972–1985)	Plastic bottle recycling infrastructure begins to be established to complement that for glass as the market for plastic carbonated beverage bottles grows
Mobro 400 barge results in U.S. focus on municipal solid waste management, a perceived landfills crisis, and product bans (1987–1995)	Major new R&D initiatives by government and private industry to demonstrate responsible resource management strategies for plastics packaging; formation of CSWS and APC by the plastics industry; hundreds of state laws proposed
European governments (Germany and EU) focus on postconsumer solid waste and recycling (1990 to present)	Landfill bans; major new R&D initiatives focused on integrated resource management (IRM) by government and private industry in Europe, Japan, and the United States; all markets for plastics affected—packaging, other nondurables, automotive, electrical and electronic products, building and construction, and furnishings

The third event was the publicity surrounding the barge *Mobro 400* that sailed the east coast of the United States for months in 1987 looking for some willing location to take its load of trash. That one event turned the country's attention to solid waste and led to a growing perception that there soon would be no landfill space available for society's trash. This became known as the solid waste or landfill crisis of the late 1980s and early 1990s. Many individuals, groups, and institutions began to argue that we could recycle our way out of the solid waste crisis. As time progressed, the economics of municipal solid waste management began to be better understood. New and often very large landfills (often referred to as mega landfills) were built to revised Environmental Protection Agency (EPA) standards to replace the hundreds of small dumps that were closed, and waste-to-energy plants became highly regulated and expensive to build and operate. The true costs for collecting and sorting municipal solid waste for the purpose of commodity recycling were observed first hand throughout the country, and landfilling continued to be the waste disposal option of choice throughout most of the United States. However, significant efforts at the local, state, and federal levels went into trying to expand the level of cost-effective recycling. Since lightweight plastics were targeted, albeit incorrectly, as the principal contributor to the perceived solid waste crisis, the plastics industry became actively engaged in addressing plastics recycling issues beginning in 1988 with the founding of the Council for Solid Waste Solutions (CSWS), the precursor to the American Plastics Council (APC). CSWS/APC carried out extensive research on the recycling of plastics from packaging, other nondurables, and durable goods such as automobiles and consumer electronics. A similar industry-led initiative, formation of a Plastics Waste Management Institute (PWMI), was established in Europe under the Association of Plastics Manufacturers in Europe (APME) and earlier in Japan (PWMI).

The fourth event was the high level of attention that began to be paid to recycling in Europe, especially in Germany, beginning in the late 1980s. This took the form of both a solid waste management strategy as well as a high-profile environmental policy initiative. Early focus on plastic beverage bottles [PET and polyvinyl chloride (PVC)] soon turned to plastics from end-of-life automobiles, packaging film, and end-of-life electrical and electronic equipment. Germany passed demanding recycling quotas for plastics packaging in 1992, and soon after the European Union began to develop a series of legislative directives that laid out specific waste management policies and recycling targets for both products and materials. Although Europe and the United States largely went down separate paths in addressing packaging wastes, the recycling of durable goods was another matter. Since durable goods markets are largely global, early developments in Europe had a profound effect on how the United States and Japan addressed recycling issues. One result was that major investments in the development of plastics recycling technology for end-of-life durables such as automobiles, refrigerators, and computers, began as a coordinated effort among original equipment manufacturers and their material suppliers. Some of the earliest technology development efforts began in the United States and emphasized mechanical recycling [8–10]. Reflecting legislative challenges, Europe and Japan applied a larger proportion of resources to feedstock recycling and energy recovery [1, 11].

14.5. POLYMER RECYCLING STATISTICS

14.5.1. Plastics Recovery in the United States

In the United States, three organizations are primarily responsible for maintaining plastics recycling statistics: the American Plastics Council (APC), the National Association for PET Container Resources (NAPCOR), and the U.S. Environmental Protection Agency. APC and NAPCOR provide recycling rate data based on actual surveys of plastics recyclers. The EPA estimates recycling rates for specific commodities based on broad market data for products and an evaluation of municipal solid waste management data from the states [12]. The APC statistics are the most complete for plastic bottles [13]. Figure 14.2 shows the growth in postconsumer plastic bottle recycling from 1990 to 2000. In 2000, over 1.5 billion

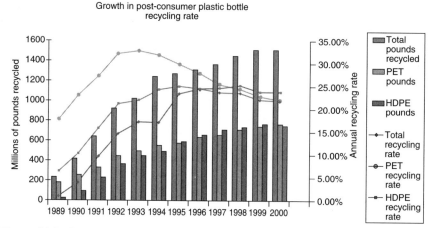

Figure 14.2. Growth in plastics bottle recycling in the United States between 1990 and 2000. (Source: R. W. Beck, Inc. for the American Plastics Council.)

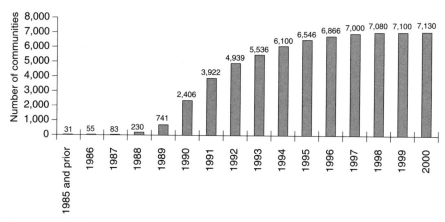

Figure 14.3. Historical increase in curbside collection programs in the United State (1995–2000). (Source: R. W. Beck, Inc. for the American Plastics Council.)

pounds of plastic bottles collected and recycled were made up almost entirely of PET and HDPE. Figure 14.3 shows the historical increase in curbside bottle collection programs in the United States. Table 14.2 provides detailed recycling rate data for PET, HDPE, and miscellaneous plastic bottles for 1999 compared with 2000. The total plastic bottle recycling rate for 2000 was 21.8%, down slightly from the 22.1% rate in 1999.

Looking beyond bottle recycling, over 200 million pounds of polypropylene are recycled each year from automotive batteries [14]. The lead and acid are also recovered. The recycling rate for lead-acid automotive batteries is greater than 90% according to Battery Council International (*www.batterycouncil.org*).

Table 14.2 Postconsumer Plastic Bottle Recycling Rates in United States for 1999 and 2000[a]

Plastic Bottle Type	Calendar Year 1999			Calendar Year 2000		
	Plastic Recycled	Resin Sales	Recycling Rate	Plastic Recycled	Resin Sales	Recycling Rate
PET soft drink	581	1,628	35.7	585	1,670	35
PET custom	159	1,622	9.8	173.3	1,774	9.8
Total PET bottles	*740*	*3250*	*22.8*	*758.3*	*3,445*	*22*
HDPE natural	446.7	1,503	29.7	416	1,402	29.7
HDPE pigmented	315.2	1,702	18.5	329.4	1,732	19
Total HDPE bottles	*761.9*	*3,205*	*23.8*	*745.4*	*3,134*	*23.8*
PVC	0.8	165	0.5	0.8	161	0.5
LDPE/LLDPE	0.2	41	0.5	0.2	34	0.6
PP	5.6	164	3.4	5.7	131	4.3
PS	0.2	10	0.1	0.2	10	0.1
Total bottles	1,508.70	6,835	22.1	1,510.60	6,915	21.8

[a] Volumes in millions of pounds per year. (Source: R. W. Beck, Inc. for the American Plastics Council.)

NAPCOR reported that 771 million pounds of PET was recycled from bottles in 1999 at a recycling rate of 23.7% compared to the APC numbers of 740 million pounds and 22.8%, respectively [15]. In part, these differences reflect the slightly different methodologies used to account for recycling efficiencies and yields.

The U.S. EPA reported that the estimated recycling rate for all plastics in municipal solid waste (MSW) was 5.6% in 1999 [12]. The corresponding percentages for specific plastic product categories were: durable goods, 3.8%; nondurable goods, negligible; containers and packaging, 9.7%. It is known that a significant quantity of plastic is used in products that, at end of their useful life, do not end up in MSW. Examples are automotive plastics (other than lead-acid batteries) and building and construction plastics. Little postconsumer plastic from these markets, however, is presently recovered and recycled.

An additional recycling statistic of interest is the growth in the number of companies recycling plastics. According to APC data, in 1989 the number of companies was only about 230 but increased to almost 1400 companies by 1998.

14.5.2. Plastics Recovery in Europe

The Association of Plastics Manufacturers in Europe (APME) tracks plastics recycling statistics for the overall European market. Table 14.3 shows historical data from APME for the recovery and recycling of postuse plastics in Europe [16]. The full report can be found on the APME website *www.apme.org*. The total percentage of plastics diverted from landfill in 1990 was 23%, increasing to 32% in 1999. Energy recovery was the predominate method of recovery followed

Table 14.3 Plastics Recovery and Recycling Rates in Europe (1990–1999)

	1990	1991	1992	1993	1994	1995	1996	1997	1998	1999
Total plastic waste	13594	14637	15651	16211	17505	16056	16871	16975	18457	19166
Mechanical recycling	958	1080	1129	915	1057	1222	1320	1455	1614	1800
Feedstock recycling	0	0	0	0	51	99	251	334	361	346
Energy recovery	2108	2138	2599	2425	2348	2698	2496	2575	3834	3949
Total plastics waste recovered	23%	22%	24%	21%	20%	26%	25%	26%	31%	32%

Source: APME.

by mechanical recycling. The mechanical recycling rate increased from 7% in 1990 to 9% in 1999. According to APME the diversion of some postuse plastics to feedstock recycling began in 1994. It is important to keep in mind that the APME data is for all postuse plastics in all markets, while in the United States only plastic bottle recycling is closely monitored.

14.5.3. Plastics Recovery in Japan

In Japan, the Plastics Waste Management Institute (PWMI) publishes plastics recovery and recycling data. Statistical data for 1999 can be found on its website *www.pwmi.or.jp*. It is difficult to compare this data directly with U.S. and European data due to differences in the way that preconsumer and postconsumer plastics are categorized and recorded. The total plastics waste figure given by PWMI for 1999 is 9,760,000 tons of which 4,900,000 tons is assigned to the industrial sector and 4,860,000 is assigned to general consumer waste. The overall recovery rate for general waste is reported to be 44%. Table 14.4 shows the summary data for disposal and recovery. Mechanical recycling is only 2% for general postconsumer waste but 25% for industrial plastic waste. Approximately 50% of the combined plastics are disposed of in landfills or incineration plants

Table 14.4 Plastics Recovery and Recycling in Japan (1999)

	Industrial Scrap Plastic (%) 4,900,000 tons	General Scrap Plastic (%) 4,860,000 tons
Mechanical recycling	25	2
Feedstock recycling	1	0
Energy recovery	23	42
Incineration	8	35
Landfill	43	21

Source: PWMI, Japan.

without energy recovery. In contrast to the United States where only about 15% of plastics in MSW are recovered for their energy value, in Japan the corresponding figure is over 40%.

14.5.4. Rubber and Elastomer Recovery

The recycling of industrial scrap rubber is a relatively mature industry throughout much of the world, but reliable and reasonably comprehensive statistical data is hard to come by. Meaningful recycling statistics for postconsumer rubber and elastomers exists only in the case of scrap tires. The Rubber Manufacturers Association (Washington, DC) maintains a database on the recovery, recycling, and reuse of automotive tires. The website is *www.rma.org/scraptires*. Each year approximately 270 million scrap tires are generated. Although efforts continue to increase the recovery and recycling of nontire rubber (e.g., synthetic elastomers such as ethylene-propylene-diene-monomer elastomer or EPDM), it has been difficult to achieve critical mass from a collection standpoint and find viable markets for the recovered rubber. Most rubber is used in crosslinked form, which limits mechanical recycling options to size reduction to produce chips, granulate, or fine powder. Although chemical devulcanization processes for crosslinked rubber have been developed, their use is limited for economic reasons. The following information on scrap tire recovery and recycling statistics is from the Rubber Manufacturers Association (*www.rma.org*) [17].

There are three general markets for scrap tires: tire-derived fuel (TDF), civil engineering applications, and ground rubber applications. The use of scrap tire materials in each of these markets has been gradually increasing. In 2001, an estimated 273 million scrap tires were generated in the United States. Additionally, the efforts of state abatement programs removed approximately 30 million scrap tires from stockpiles. From the total, 118 million tires were used as fuel in cement kilns, pulp and paper mill boiler, and other furnaces. Also 40 million tires were used in civil engineering applications, such as lightweight back fill, road material, septic field drainage medium, and in landfill construction. Some 15 million tires were exported, while another 33 million were processed in fine powder, referred to as ground rubber (aka crumb rubber). Ground tire rubber is mixed with asphalt binder, used in tire manufacturing and in molded or extruded products. Eight million tires, bias ply tires in general, were punched/stamped into new products. Finally, the last market for scrap tires, one of the newest markets, is the use of a larger sized ground rubber, one-eighth- to one half-inch particles. These particles are being used in playgrounds, as a soil amendment, in running tracts and athletic fields, as well as for mulch and other horticultural materials. About 7 million tires are used in agricultural and miscellaneous applications. There were approximately 27 million tires that were legally landfilled. While not a market for tires, landfilling remains an acceptable management option in at least 10 states. Figure 14.4 shows this application data in graphical form.

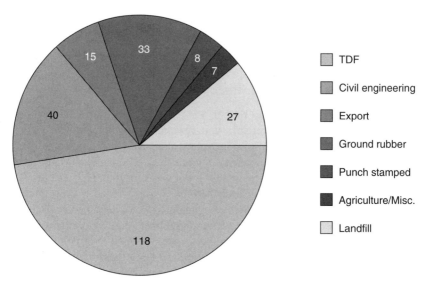

Figure 14.4. Markets for scrap tires and rubber derived from scrap tires. (Source: Rubber Manufacturers Association.)

14.6. MARKING OF PLASTIC PACKAGES AND PRODUCTS

14.6.1. Physical Marking

To facilitate the manual sorting of plastic packages and products for recycling, various methodologies have been developed to place an identifying mark on the package or product to identify the type of plastic. In practice, the development of fast automated sorting of plastic bottles and rapid flake sorting of commodity and engineering plastics has limited the value of these types of markings. Nevertheless, since some curbside bottle collection programs still accept some but not all types of plastics, consumers often rely on the markings to help separate out "acceptable" from "nonacceptable" packages. In commercial operations, time is money, and taking the time to examine an individual plastic bottle to find the plastic resin identifying mark is expensive. In the case of PET soft drink bottles and many pigmented and unpigmented high-density polyethylene bottles, simple sorting by bottle type provides sufficient resin purity for downstream plastics recycling operations.

Four organizations have developed and publicized marking systems for plastic packages and products (Table 14.5). The original chasing arrow system developed by The Society of the Plastics Industry (Washington, DC) covers seven plastic resin categories and was developed for the marking of plastic bottles only. In practice the chasing arrow symbol is often found on plastic film and other products. It is still widely used to mark bottles, jars, and containers in the United States. The system (Fig. 14.5) covers the six principal resins used in packaging applications: polyethylene terephthalate (PETE or PET), high–density polyethylene (HDPE), polyvinyl chloride (V), low-density polyethylene (LDPE/LLDPE),

Table 14.5 Plastic Marking Guidelines and Standards

Guideline/Standard	Responsible Organization	Date of Publication	Contact Information
Technical Bulletin RPCD-13-1989 Revision 1 Voluntary Guidelines Rigid Container Material Code System Mold Modification Drawings	Society of the Plastics Industry	1990	Society of the Plastics Industry 1801 K Street, NW, Suite 600K Washington, DC 20006-1301 Tel: 202-974-5200 Fax: 202-296-7005 http://www.plasticsindustry.org
Surface Vehicle Recommended Practice SAE J1344 Marking of Plastic Parts	SAE International	Rev. July 1997	SAE International 400 Commonwealth Drive Warrendale, PA 15096-0001 Tel: 724-776-4841 Fax: 724-776-5760 http://www.sae.org
International Standard ISO 1043-1 Plastics—Symbols and abbreviated terms—Part 1: Basic polymers and their special characteristics	International Organization for Standardization (ISO)	1043-1 Second Edition 1-3-97	International Organization for Standardization (ISO) 1, rue de Varembe, Case postale 56 CH-1211 Geneva 20, Switzerland Tel: +41 22 749 01 11 Fax: +41 22 733 34 30 http://www.iso.org
International Standard ISO 1043-2		1043-2 Second Edition 7-15-02	

(*continued overleaf*)

Table 14.5 (*continued*)

Guideline/Standard	Responsible Organization	Date of Publication	Contact Information
Plastics—Symbols and abbreviated terms—Part 2: Fillers and reinforcing materials			
International Standard ISO 1043-3		1043-3 Second Edition 4–15–96	
Plastics—Symbols and abbreviated terms—Part 3: Plasticizers			
International Standard ISO 1043-4		1043-4 First Edition 2–15–98	
Plastics—Symbols and abbreviated terms—Part 4: Flame retardants			
International Standard ISO 11469		11469 Second Edition 5–15–00	
Plastics—Generic identification and marking of plastics products			
Standard Practice for Generic Marking of Plastic Products	ASTM International	1992–1997	ASTM International 100 Barr Harbor Drive PO Box C700 W. Conshohocken, PA 19428-2959 Tel: 610-832-9585 Fax: 610-832-9555 *http://www.astm.org*

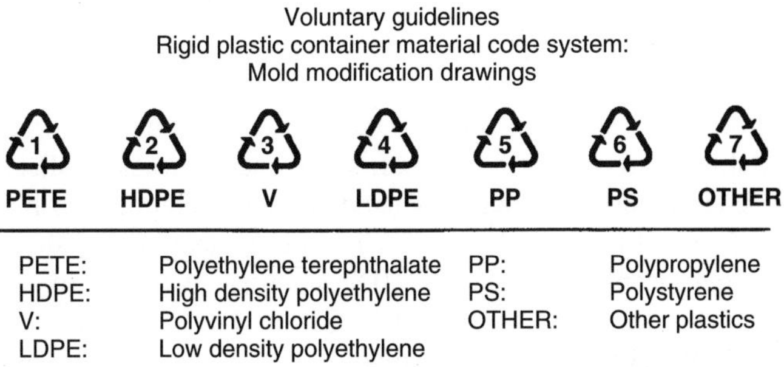

**Voluntary guidelines
Rigid plastic container material code system:
Mold modification drawings**

♳ ♴ ♵ ♶ ♷ ♸ ♹
PETE HDPE V LDPE PP PS OTHER

PETE:	Polyethylene terephthalate	PP:	Polypropylene
HDPE:	High density polyethylene	PS:	Polystyrene
V:	Polyvinyl chloride	OTHER:	Other plastics
LDPE:	Low density polyethylene		

Figure 14.5. Marking system for plastic bottles developed by The Society of the Plastics Industry.

polypropylene (PP), polystyrene (PS), and an "other" category. In line with Federal Trade Commission environmental labeling guidelines, the system should not be used to denote or advertise actual recyclability of the package.

The remaining three methodologies were developed by international standards organizations (ASTM International, SAE International, and the International Organization for Standardization, or ISO) to serve the needs of all product manufacturers. In contrast to the six specific commodity resins (seven when linear low-density polyethylene is included in the LDPE category) covered by the SPI bottle marking system, these other systems cover in excess of 200 different polymer resins, alloys, and blends. Both thermoplastic and thermoset resins are included. In addition, since these marking systems were designed to cover engineering resin formulations, they can accommodate the use of fillers, reinforcements, and some chemical additives such as flame retardants. The following marking is in agreement with the ASTM, SAE, and ISO guidelines and designates a polyurethane plastic reinforced with 20% by weight glass fiber: **>PU + 20GF<**. The standards themselves should be consulted for details and can be ordered online from all three organizations.

14.6.2. Chemical Marking

The above discussion related to the physical marking of plastic packages and products for resin identification. During the 1990s several groups in Europe and the United States investigated the use of chemical marking methods based on the addition of fluorescent dye markers to the plastics [18, 19]. While technically feasible for the identification of virgin resins, the practical issues of cross contamination during recycling and the complexity cost, and possible legal issues surrounding worldwide implementation by the plastics and product manufacturing industries have stymied commercialization.

14.7. COLLECTION OF PLASTICS FOR RECYCLING

14.7.1. Consumer Packaging

In the United States, the collection of plastic postconsumer packaging from households is mostly confined to HDPE and PET bottles [13, 20]. This is reasonable since HDPE and PET bottles make up over 95% of all bottles on the market, and much of the remaining plastics packaging consists of lightweight film and foam items that are difficult to collect, clean, and sort economically [21–23]. Households are often asked to set out HDPE and PET bottles commingled with other recyclables. This is not a problem since special trucks collect the bottles and sorting equipment is available at handlers and reclaimers to make the necessary separations often at high speed. Curbside collection and drop-off centers are the principal means for collecting HDPE bottles such as natural HDPE milk jugs and water bottles and pigmented HDPE liquid detergent bottles.

The same collection routes are available to PET bottles, but PET soft drink bottles are also collected through refund centers and reverse vending machines in "bottle-bill" states where a deposit fee is placed on the bottle at the time of purchase. As shown in Figure 14.3, over 7000 curbside collection programs are presently operating in the United States [13]. In 2000, the American Plastics Council began promoting all-bottle collection programs to increase the amount of plastic bottles available to recyclers [24]. Pilot studies demonstrated that consumers put out more HDPE and PET bottles when they are allowed to put all bottles into a collection bin rather than just 1's and 2's.

Many communities do not ask households to do sorting but collect all municipal solid waste together and send the mixed waste to a material recovery facility (MRF) for processing. MRFs are capable of producing separate streams of PET soft drink bottles, PET custom, natural HDPE milk jugs, and mixed color HDPE bottles. Most mixed waste MRFs in the United States use manual sorting. PET sorted at a mixed waste MRF is not as clean as that obtained through a curbside bottle collection program.

Very little film plastic is collected from households due to high costs [21, 23]. Retail grocery stores and dry cleaning establishments often provide drop off centers for postconsumer plastic bags. Significant collection and recycling of plastic film such as shrink wrap from commercial and industrial sources takes place [25]. Plastic lumber is an important market for film plastic [26]. Most plastic packaging that is collected goes first to handlers who sort the material for reclaimers. Reclaimers process the plastic products to produce marketable clean plastic flake or granulate. Some reclaimers have extrusion and pelletizing equipment, as well. A list of handlers and reclaimers operating throughout the United States is available from APC on the Internet at *www.plasticsresource.org* (Recycled Plastic Products and Markets Databases).

In Europe the collection of packaging is legislated throughout the European Union. Various collection schemes have been organized country by country. The most publicized system is Germany's Duales System Deutschland (DSD). The principal purpose of this system is to collect all packaging for which a recycling fee has been paid. Product manufacturers who use plastics packaging have come together to fund a workable system. Subsidies are quite high since collection of the plastic packaging at legislated levels is not economically sustainable in the absence of subsidies. Not all of the plastic packaging collected in Germany goes to mechanical recycling. Much is used for energy recovery and some for feedstock recycling as noted earlier in the chapter. The German collection system has stimulated development of relatively sophisticated automated sorting equipment for mixed packaging collected at curbside [27].

14.7.2. Consumer Electronics

Perhaps the fastest growing type of residential collection program in North America is the collection of consumer electronics—TVs, computers, audio equipment, VCRs, fax machines, and small appliances (Fig. 14.6). These consumer products are included in the definition of municipal solid waste in the United States [12], and on average, plastics make up about 20% by weight of these products [28, 29].

A recycling infrastructure to deal with computers and other major office equipment from commercial and institutional sources has been in place for several years. Several pilot studies were initiated in the United States in the late 1990s to evaluate different collection schemes such as curbside, drop-off center, retail store drop-off, and so forth for postconsumer electrical and electronic products [30–35]. New programs are starting up almost monthly. Figure 14.7 shows the types of products that have been collected from residential sources in a multi-year program in Henvipin Country, Minnesota [36] studies. In all of the pilot

Figure 14.6. Collection events for postconsumer electrical and electronic equipment. Ref. 37.

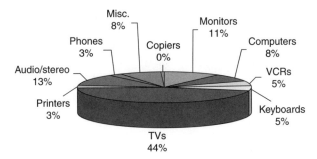

Figure 14.7. Types of end-of-life electronic equipment collected predominately from residential sources. Ref. 36.

studies, TVs predominate, ranging from 36 to 69% by weight. Evidence to date is that the collection of these very heterogeneous and bulking products is costly. Collection volumes in 2001 were still too small to develop a detailed understanding of the true value of the recyclable secondary materials, including the plastics available from consumer electronics. All evidence points to extensive collection of consumer electronics in the future in North America, Europe, and Japan, indeed throughout the world, and significant volumes of plastics will likely be available for recovery beginning in 3–5 years. General information on the collection of end-of-life consumer electronic equipment for recycling is available from the Electronic Industries Alliance (Arlington, VA), *www.eia.org*, and from the International Association of Electronics Recyclers (Albany, NY), *www.iaer.org*. Tables 14.6 and 14.7 show the types of plastics available from

Table 14.6 Types of Plastics Found in End-of-Life Electronic Products Collected in Hennipin County, MN in 1999. Ref. 36

Plastic Resin	Television Plastics	Computer Plastics	Misc. Plastics	Percent of Total Sample
HIPS	75%	5%	50%	59
ABS[a]	8%	57%	24%	20
PPE	12%	36%	11%	16
PP	3%	—	3%	2
Other	2%	>1%	2%	2
PE	—	—	6%	1
PC/ABS	—	2%	—	>1
PC	—	—	2%	>1
PVC[a]	—	—	2%	>1

[a]This category includes SAN (styrene acrylonitrile).

Table 14.7 Types of Plastics Found in End-of-Life Electronic Products Collected in a MultiCounty Collection Pilot Program Organized by MOEA in 2000. Ref. 39

Plastic Resin	Television Plastics (%)	Computer Plastics (%)	Miscellaneous Plastics (%)	Percent of Total Sample
HIPS	82	25	22	56
ABS	5	39	41	20
PPE	7	17	4	11
PVC	<1	5	15	3
PC/ABS	0	6	7	3
PP or PE	0	3	8	2
PC	1	4	1	2
Other	<1	<1	2	<1
Unidentified	5	0	0	3

end-of-life electrical and electronic equipment based on two recent pilot collection projects in the United States. The Minnesota Office of Environmental Assistance (MOEA) study was the largest and demonstrated that three resins (HIPS, ABS, and PPE) make up over 80% of the available plastics [36–39].

14.7.3. Major Appliances (White Goods)

Major appliances, often referred to as white goods, consist of refrigerators, stoves, dishwashers, washing machines, clothes dryers, air conditioners, microwave ovens, and hot water heaters. Older white goods available for recycling contain large quantities of metal and can be profitably recycled by scrap metal processors using the existing metal recycling infrastructure. The normal procedure is to co-process major appliances with automobiles at automobile shredders. The plastics in white goods end up in shredder residue (see next section). Collection of white goods for recycling occurs by retailers, remanufacturers, special curbside collection programs, and drop-off.

There has been significant research exploring recovery methods for ABS and HIPS plastic liners and polyurethane foam insulation from refrigerators [40–42]. Some recovery of the plastics is underway in Japan and Europe, driven by legislation requiring special processing of scrap refrigerators to recover the chlorofluorocarbon refrigerants and polyurethane foam blowing agents. In the United States, special handling of scrap refrigerators for the purpose of plastics recycling has not proved economically viable compared to whole refrigerator shredding to maximize metals recovery. General information on the collection of end-of-life major appliances for recycling can be obtained from the Association of Home Appliance Manufacturers (Washington, DC), website *www.aham.org*. Reports are available discussing plastics recovery from major appliances, especially ABS, HIPS, and PUR from refrigerators and freezers [43, 44].

14.7.4. End-of-Life Vehicles

End-of-life vehicles (ELVs) are not categorized as part of municipal solid waste by the EPA [12]. A separate recycling infrastructure has developed for ELVs. This recycling infrastructure is highly efficient, and in the United States close to 95% of all ELVs are recycled for scrap metal. Recycling rates are also high in Europe and Japan. However, many older cars are exported from the European Union countries to the eastern part of Europe for reuse, and Japan exports many older cars to Asia. Since the recycling of ELVs is profitable today, ELVs are collected by recyclers either for highly valuable used parts or for their scrap metal value. Each year over 10 million ELVs are collected for recycling in North America. Information on the collection and dismantling of scrap vehicles for used parts can be obtained from the Automobile Recyclers Association (Fairfax, VA), *www.autorecyc.org*. Once processed for resalable parts, ELV hulks are sent to shredders where giant hammer mills tear the cars into fist-sized chunks of metal. The metal scrap is sold to steel mills for recycling. The principal trade association of the automobile shredding industry is the Institute of Scrap Recycling Industries (Washington, DC), *www.isri.org*.

The automotive plastics that are not reused in the form of resalable parts end up in an industrial waste stream from the automobile shredders called automotive shredder residue (ASR) [45]. ASR contains residual ferrous and nonferrous

Table 14.8 Representative Composition of Automotive Shredder Residue

Material	Composition Range Percent by Weight
Residual metals	2–10
Plastics (nonfoam)	15–20
Rubber and elastomers	10–20
Foam plastics	4–6
Wood and paper	1–5
Textiles and leather	1–5
Dirt and rocks	5–15
Moisture	2–25
Fines (less than 3/8-inch)	40–50
Other (automotive fluids, etc.)	1–3

After Refs [44–46, 74, 134].

Table 14.9 Polymer Composition of Automotive Shredder Residue Lights Fraction Collected in the United States and Processed to Recover Nonfoam Plastics [52]

Polymer	Approximate Composition (wt %)
Unidentified polymers plus size-reduced PUR flexible foam (aspirated fraction)	30–40
Plastic/rubber-rich Product	30–40
Polymer Composition	
Rubber	~33
PP	~19
PE	~9
ABS	~8
PUR	~6
Nylons	~4
PVC	~2
PC and PC/ABS	~2
Wood	~2
PPE	~1
SMA	~1
PMMA	~1

metals, rust, glass, fluids, wood, rubber, plastics, stones, and dirt. Approximate compositions are shown in Table 14.8 [45, 46]. Detailed discussions of ASR composition are available [45]. Each year approximately 4 billion pounds of rigid, flexible, and foam plastics are disposed of in ASR. Almost all ASR is landfilled in the United States, Europe, and Japan. A minor amount of ASR is processed in waste-to-energy plants. Co-combustion with MSW is presently the recovery option of choice for ASR in several areas of Germany, France, Sweden, and Switzerland, and experience in recent years has shown the technology to be both economically and environmentally sound [47, 48]. Nevertheless, an active area of research continues to be ASR processing for the purpose of recovering rubber, elastomers, and especially plastics [44]. Limited quantities of plastics from ELVs are beginning to be collected at automobile dismantling yards in both the United States and Europe [49–51]. The emphasis is on collecting large plastics-intensive parts such as bumpers and fascia. Table 14.9 shows experimental data on the polymer composition of ASR [52, 53].

14.8. OVERVIEW OF PLASTICS RECYCLING TECHNOLOGY

14.8.1. Coarse Manual Sorting

In industrialized countries, cost-effective plastics recycling from postconsumer packaging or products depends on the application of sophisticated automated sorting equipment rather than piece-by-piece manual sorting that by its very nature is highly labor intensive. This is true for both bottle recycling and recycling of plastics from end-of-life durable goods such as computers and small household appliances. In the case of durables, some dismantling of end-of-life products is usually carried out in order to segregate potentially hazardous materials (e.g., lead-containing cathode ray tubes) or contaminated plastic parts (e.g., those containing very high levels of metallic paint, decorative film laminates, or labels) and to reclaim parts for remanufacture and/or resale. Although according to industry sources work is underway to automate the dismantling of some relatively homogeneous streams such as TVs, dismantling largely remains a labor-intensive process. During the dismantling stage, some coarse sorting of plastics, computer housings separate from TV housings, for example, can be carried out at minimal added cost. Downstream processing of the collected plastic housings can be facilitated by this coarse presorting (see Tables 14.6 and 14.7). However, extensive manual sorting of plastic parts, bottle, and the like is counterproductive since accurate, high-speed flake-sorting technology exists to separate one plastic from another in many situations. This technique is discussed later in the chapter. Coarse sorting is usually done on a product-by-product basis without the use of plastics ID technology or a search for plastic labeling marks [10, 36, 39].

As a general principle, coarse manual sorting of postconsumer plastics packaging or durable goods containing significant plastics is a necessary and sufficient condition to kick-start a viable plastics recycling process. All automated plastics cleaning and sorting technologies assume that some manual sorting has occurred.

What is important to realize is that the plastics recycling technologies that have developed over the past decade do not require the piece-by-piece manual sorting of plastic packages and products by type of resin. This can be done using automated equipment if the plastics recycling plant is appropriately designed.

14.8.2. Automated and Semiautomated Plastics Identification Technologies

One of the true breakthroughs during the past decade that has facilitated the commercial recycling of plastics from both bottles and durable goods is the successful development of fast and accurate spectroscopic methods to identify plastics by resin type [19, 54–57]. In bottle recycling, the focus has been on identifying PET, HDPE, PP, and PVC for separation. Sophisticated technologies from several vendors not only separate bottles by resin type but by color as well [57–61]. In the case of durable goods, identification of a broader range of plastics is required [10, 19, 28, 62, 63]. The list of plastics applicable to electrical and electronic (E&E) equipment includes ABS, HIPS, PA, PC, PC/ABS, PBT, PMMA, PP, PPE, PUR, and PVC. As noted earlier, coarse sorting is needed to separate out plastic parts and housings that have extensive contamination due to laminates and metal coatings. The variety of spectroscopic methods that have been developed for rapid plastics identification and sorting are summarized in Table 14.10. One or more of these techniques can prove extremely useful to a recycler for field appraisals of potential plastic feedstocks, incoming inspection of plastics, in-process control during downstream processing of mixed plastic streams, and as part of the standard quality control (QC) procedures on final products. The ideal plastics ID equipment is fast, accurate, reliable, affordable, robust, and easy to use.

The following types of commercial plastics identification equipment have been developed and are in use:

Hand-held devices	In field use to provide coarse sorting capabilities for specific durable goods such as refrigerators, automobiles, and computers
Portable devices	Useful in dismantling or selected field applications where enough plastics of sufficient variety are present to justify cost and need for additional supporting capabilities
Bench top	Useful in dismantling operations where enough plastics of sufficient variety are present to justify cost and need for additional capabilities; also useful in recycling, compounding, and plastic supplier QC laboratories for rapid ID capabilities
Automated — Parts	In large recycling operations where the costs can be justified; for near-term systems, parts or pieces need to be rather large (probably at least 200 g) and have limited requirements for sorting (approximately six or less plastics)

Table 14.10 Technologies Developed for Rapid Identification of Postconsumer Plastics by Resin Type [10]

Technology	Advantages	Disadvantages
MIR	• Fundamental vibrations yield "fingerprints"—increased accuracy and information • Can measure black plastics • Proven technology	• Very surface sensitive • MIR fiber optics are limited in range, expensive, and fragile • Remote sensing difficult • Commercial MIR instruments slower than NIR instruments
NIR	• Commercial units available • Can use "normal" fiber optics • "Portable" units already used for QC • Fast and can be done without contact • Some have no moving parts (rugged)	• Limited information in this range—overtone vs. fundamental peaks • Carbon black absorbs and scatters highly at NIR frequencies, making dark plastics difficult to probe
SWNIR	• Low-cost equipment • Very small instrument with fiber optics • No moving parts (rugged)	• Only limited polymers (and colors) can be detected • Still somewhat developmental
Raman	• Can be fast and remote is possible • Fiber-optic probes possible • Spectral detail similar to MIR	• Fluorescence of black pigments • Lasers expensive
Pyrolysis and plasma techniques	• Can obtain very accurate identifications • Can be very fast • Additive ID possible	• Sampling could be difficult • Polymer degradation questions • Still in laboratory stage
Triboelectric	• Only known true hand-held device • Completely portable and easy to use • Fast response • Inexpensive	• Very limited in number of polymers • Can be sensitive to moisture and surface contamination • Still somewhat developmental
Thermography	• Remote probing possible • Some coatings may not be a problem • Can be very fast	• "Signatures" of many polymers very similar • Still developmental
X-ray	• Can detect heavy atom additives and components, like Cl, Br, Cd, Pb, etc. • Fast and remote • Proven technology	• Can't distinguish between different polymers • Expensive • Radiation safety issues
Automated—Flakes	Can be used for colors today, and potentially for some material ID in the future; usually limited to removing one material present in small quantity from major material	

Figure 14.8 shows a mixed plastic bottle sorting line that employs a combination of near-infrared, visible, and X-ray methods to separate PET, HDPE, and PVC bottles. Throughputs above 1500 pounds per hour per line are possible and sorting accuracy and yields are excellent. In bottle sorting, achieving a singulated

Figure 14.8. Automated bottle sorting line. Photos courtesy of MSS, Inc.

stream of bottles for presentation to the detector and air ejectors (a blast of air is usually used to eject an identified bottle off of a moving belt downstream from the detector) is an important part of the overall ID and sorting technology.

The ability to ID black plastics is of secondary importance in the case of bottles but becomes more important for some electrical and electronic plastics and for many automotive plastics. Mid-infrared (MIR) can identify many black plastics but not near-infrared (NIR), where the carbon black interferes [19, 64, 65]. Recently, a company has reported success in applying laser Raman techniques to the rapid identification of black plastics [66].

Many companies have been instrumental in the development of technology and equipment for the rapid identification of plastic bottles [54] and durable plastic parts [10, 19, 63] for the purpose of recycling. Three companies that market plastics ID and related sorting equipment internationally are referenced here:

Magnetic Separation
  Systems, Inc. (MSS)
3738 Keystone Avenue
Nashville, TN 37211
Tel +1 (615) 781 2669
Fax +1 (615) 781 2923
magsep@magsep.com
www.magsep.com

National Recovery
  Technologies, Inc. (NRT)
566 Mainstream Drive
Suite 300
Nashville, TN 37228
Tel +1 (615) 734 6400
Fax +1 (615) 734 6410
nrtinfo@nrt-inc.com
www.nrt-inc.com

LLA Instruments GmbH
  Schwarzschildsr. 10
D-12489 Berlin-Adlershof
Germany
Tel +49 30 6719 8376
Fax +49 30 6392 4766
mail@lla.de
www.lla.de

In some situations, manual sortation of plastics recovered from durable goods is feasible because of the size and value of each item and compatibility with present disassembly practices. Figure 14.9 shows the use of an MIR instrument to ID dismantled plastics parts from E&E equipment or automobiles. If the durable plastic part has been painted, the identification needs to employ an unpainted surface if available or the paint film must be physically removed first.

In most situations, the manual identification and sorting of plastics parts from automobiles or computer and business equipment housings using fast infrared techniques has significant throughput limitations. Overall ID and sorting rates of

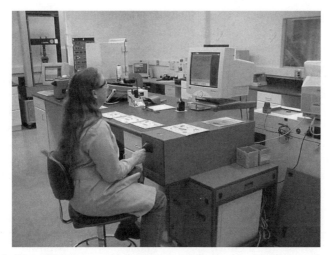

Figure 14.9. Use of bench-top mid-infrared rapid plastics identification equipment. Photo courtesy of MBA Polymers, Inc.

about 250 lb per hour for automobile interior trim plastics and near 900 lb per hour for computer and business equipment housings can be expected [49, 67]. This is not very impressive when compared to the fully automated flake-sorting rates near 10,000 lb per hour achieved using state-of-the-art density separation methods as discussed later in this chapter.

14.8.3. Identification of Additives

Several techniques have been developed to rapidly identify additives in plastics as part of an overall plastics recycling operation. Some techniques such as Fourier transform infrared (FTIR) can combine resin identification with information of the presence or absence of additives such as flame retardants or talc fillers [49, 62, 65]. The sliding spark technique developed at the University of Duisburg in Germany can identify a range of heavy metals along with the type of resin [68, 69].

A reliable and reasonably fast method to identify a broad range of metals, halogens, and other heteroatoms in plastics is X-ray fluorescence spectroscopy [70]. X-ray fluorescence (XRF), which is based on energy level changes of core electrons of atoms, can be used to detect most elements in plastics except for H, C, N, and O. Since these atoms are usually present in relatively large amounts as part of functional groups observable by other techniques (such as infrared spectroscopy), XRF is a complementary technique for the analysis of other elements. Many of these other elements are found in additives and can appear in plastics in concentrations from parts per million to near 50%. XRF has the ability to cover this entire range for many elements. Though often less quantitative than other techniques, XRF has several advantages such as operator ease of use, quick sample preparation, quick sample analysis, minimal recalibration, a broad range of elements covered, and a broad concentration range for analysis. XRF is thus ideally suited for real-world plastics recycling environments where analytical chemistry may not be available. XRF is used to look for elements of particular interest in the recovery of engineering thermoplastics. Quantitative analysis can be performed on elements frequently found in pigments and flame retardant additives (Cl, Br, Sb, and Ti). Qualitative or semiquantitative analysis for other elements (e.g., Pb, Cd, Zn, Si, S, P, Ca, and Fe) can also be performed on a variety of plastics using various XRF instruments. Figure 14.10 shows a schematic of the XRF technique, and Figure 14.11 shows an X-ray fluorescence instrument in a plastics recycling QC laboratory environment. Contact information for seven manufacturers of XRF instruments follows:

Examples of XRF Instrument Manufacturers

Company	U.S. Office	Phone	Website
Jordan Valley AR	Austin, TX	(512) 973–9229	jordanvalley.com
Kratos Analytical	Chestnut Ridge, NY	(914) 426–6700	kratos.com
Niton	Bedford, MA	(800) 875–1578	niton.com
Oxford Instruments	Concord, MA	(800) 447–4717	oxinst.com
Philips Analytical	Natick, MA	(508) 655–1222	analytical.philips.com

14.8.4. Automated Sorting of Plastic Bottles and Electronic Parts Using NIR

As previously noted, many commercial plastic bottle sorting lines use near-infrared spectroscopy to accurately ID bottles by resin type at very high speed.

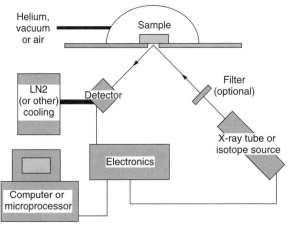

Figure 14.10. Schematic of X-ray fluorescence method for the identification of heteroatoms/additives in postuse plastics [70].

Figure 14.11. X-ray fluorescence instrument used in plastics recycling. Photo courtesy of MBA Polymers, Inc.

Compared to MIR, in addition to speed, NIR has the advantage that direct or close contact between the detector and the sample is not necessary [67]. NIR instruments are also compatible with flexible fiber-optic probes [19].

NIR spectroscopy has been applied to the fully automated sorting of dismantled plastic parts on a conveyor belt (Fig. 14.12). An interesting evaluation of the technology can be found in a recent International Symposium on Electronics and the Environment (ISEE) conference paper [71]. The point raised in the paper is that under optimum conditions sorting accuracy is at best about 98% and that 98% resin purity is unacceptable for higher value engineering polymers. Since the NIR-sorted parts would need to go through another separation and purification step, the paper points out that it may be more cost effective to go directly to this type of processing step if it can be run with sufficient accuracy and throughput. The point may be valid under some situations and should be part of any overall recycling technology assessment, especially in the case of mixed plastics obtained from end-of-life vehicles and electrical and electronic equipment. Such an assessment leads to consideration of whether plastics sorting should take place at the parts dismantling stage using instrumented ID methods to separate by resin type [49] or at the granulate/flake stage after size reduction [10, 49, 71, 72] where higher throughputs are in principle possible.

Figure 14.12. Near-infrared plastics ID equipment for the rapid identification of dismantled plastic parts on a moving belt [71].

A related study compared recyclate quality and plastic separation costs for three alternative processing schemes [67]. The first scheme was called a manual sort and involved bench-top spectroscopic (MIR or NIR) identification and manual presorting of plastic electronic product housings followed by dry and wet processing of the manually sorted plastic to obtain clean granulate. The second scheme was called automated ID and sort and used an automated NIR process to identify and sort the plastic housings on a conveyor belt followed by dry and wet processing. The third method used dry and wet processing to separate the plastics from a stream of mixed housings without going through the manual or automated presort of the housings. The conclusion of the study was that under most conditions using real-world feedstocks neither the manual or automated presort produced a final recyclate stream with adequate purity (99 + percent). In all cases, dry and wet processing that included a sophisticated flake sorting step was required to achieve high purity. Table 14.11 shows the comparative economics developed in the study. Information such as this will become more important as more and more end-of-life electronics are collected for recycling and both the public and private sectors work to minimize overall system costs and maximize material value.

14.8.5. Mechanical Recycling

Size Reduction Once plastic packages, parts, or products have been collected, undergone incoming QC, and been presorted, size reduction is the next critical

Table 14.11 Comparative Sorting Costs for Plastics from End-of-Life Electrical and Electronic Products for Different Part and Flake Sorting Strategies. Adapted from Ref. 67

	Costs ($/lb)		
	Manual ID & Sort	Automated ID & Sort	Flake Sorting
Sourcing	0.01	0.01	0.01
Material costs	0.00–0.05	0.00–0.05	0.00–0.05
Shipping	0.03–0.10	0.03–0.10	0.01–0.06
Manual ID & & presorting	0.042	—	—
Automated ID & presorting	—	0.023	—
Particle sorting	0.15–0.20	0.15–0.20	0.15–0.20
Handling, tracking, packaging, storage	0.06–0.12	0.06–0.12	0.06
Total	0.292–0.522	0.273–0.503	0.23–0.38

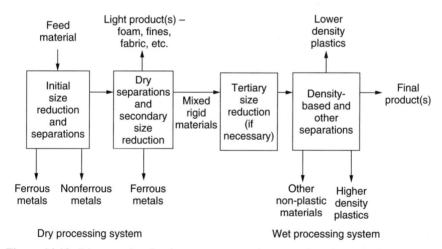

Figure 14.13. Diagram showing important processing operations in a plastics recycling plant designed to handle complex multimaterial streams. Source: American Plastics Council.

step in the plastics recycling operation. This is the first critical step in a plastics recycling plant and especially important in a plant designed to handle plastics from end-of life durables. Figure 14.13 shows key processing steps in a plastics recycling plant capable of handling complex, contaminated, mixed plastic feedstreams. For complex material steams, more than one size reduction step may be required. Depending on the degree of contamination, a coarser, more robust shredding operation might be needed, followed by contaminate removal, final cleaning, and then granulation. In the case of plastics from end-of-life durables, due to extensive multimaterial contamination that often occurs, size reduction becomes perhaps the most critical step in the entire plastics recycling operation. In the case of plastic bottle recycling one can often turn directly to granulation to produce a size-reduced flake. Table 14.12 provides a summary of several size reduction technologies for different plastics recycling applications.

Size reduction is used to increase bulk density, lowering storage requirements and shipping/transport costs, ease material handling and conveying, and liberate foreign materials. There are many specific challenges associated with durables recycling: large and widely varied parts, often significant amounts and sizes of metal, thick wall sections, tough engineering plastics, high modulus but brittle plastics, high rubber content sometimes present, film plastics sometime present, wide range of cutting and fracture behaviors, extremely wide range of foreign materials, well-adhered foreign materials (labels, foams, fabrics, laminates, metal foils, etc.), hardened metals, high dust/fluff loadings, and numerous different material types used in many different applications leading to more equipment cleanouts.

Desired attributes of size reduction equipment for the recycling of plastics from end-of-life durable goods include: accommodates large amounts of metal, handles tough engineering plastics at reasonable throughputs, liberates molded-in and well-adhered materials, does not imbed or encapsulate foreign materials, produces uniform particle shapes and sizes, is safe for wide range of operators, accommodates very large parts or bales of materials, provides high throughput to power requirement ratio, minimizes fines generation, can be enclosed or evacuated, is reasonably priced, needs low maintenance, is easy to clean for material switch-overs, produces low noise, and has reasonable power requirements [10]. Figure 14.14 summarizes the many factors that must be considered in selecting a size reduction system.

As summarized in Table 14.12, the types of equipment available to meet these challenges include: hammer mills, ring mills, sheer shredders with screens, four-shaft shear shredders with screens, rotary grinders, and granulators. Depending on the types of feedstocks coming into a plastics recycling plant, there may be a need for several different stages of size reduction and liberation, each involving a different type of equipment. In some applications, wet or underwater grinding may be desirable.

Rotary grinders deserve special mention. In early APC-sponsored studies in collaboration with MBA Polymers and wTe Corporation, rotary grinders (originally developed in Europe for the wood chipping industry) were found to work for a wide variety and size of plastic parts, tolerate moderate amounts of metal, and provide acceptable throughputs, especially when a ram was used to help push and hold the parts against the rotor and teeth [10, 49, 73]. Rotary grinders employ a rotating drum that contains numerous square teeth arranged such that the diagonal axis is parallel with the direction of rotation of the rotor. Figure 14.15 shows a closeup of a rotary grinder showing one type of tooth configuration. Unlike a shredder that shears material into strips as it is pulled through stacks of

Table 14.12 Summary of Various Size Reduction Equipment Used in Plastics Recycling [10]*

Machine	Advantages	Disadvantages
Traditional Granulators	• Can produce fine particle sizes (<1/8-inch mesh) • Excellent liberation of materials • Well-known technology • Can have high throughputs	• Cannot handle metals • Maintenance costs rather high • High speed and can be noisy
Granulators-Modified	• Can handle small amounts of metals • Can produce fine particle sizes (<1/8-inch mesh) • Excellent liberation of materials	• Newer technology and not widely available • Can't handle large amounts of metal or thick metal
Traditional Shear Shredders	• Can handle large and heavy metal • Large feed hopper • Well-known technology	• Poor particle size uniformity • Feed rams not a stock item • Difficult to replace blades • Typically no screens provided
Four Shaft Shredders and Modified Two Shaft Shear Shredders	• Can handle metal • Relatively low power requirements • Good liberation of materials • Better size control	• Not widely available • Much lower throughput rates compared to shredder with no screen • Feed rams not a stock item • Difficult to replace blades
Hammer Mills	• Can handle significant amounts of metals • High throughputs • Very robust designs • Well-known technology	• Poor particle uniformity • Relatively high power requirements • Noise can be high
Rotary Grinders	• Can handle moderate metal • High throughputs • Blades easily replaced & sharpened • Relatively low power requirements • Automatic ram feed • Comes with screens • Very good liberation of materials • Reasonably easy to clean	• Cannot handle large amounts of hard metals • Cannot easily produce small particle sizes (<1/4-inch mesh) • Higher cost
Cryogenic Grinding	• Very fine partial sizes possible (<60 mesh) • Excellent liberation of materials	• High Cost • Low throughputs • Potentially high operating cost due to liquid nitrogen needs • Cannot handle metals

*Advantages and disadvantages are considered representative for each class of equipment but may not be valid in all cases as technology is steadily evolving.

594   PLASTICS RECYCLING

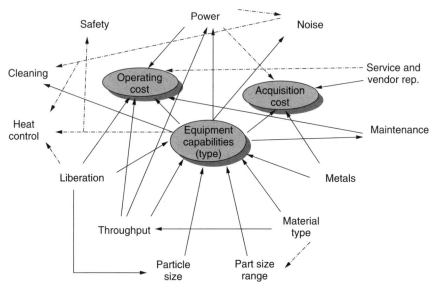

Figure 14.14. Factors to consider in selecting size reduction equipment for a plastics recycling plant. Adapted from reference 10.

Figure 14.15. Closeup of a rotary grinder showing placement of cutters.

counterrotating knives, the rotary grinder's teeth chip away the material, almost like a punch-and-die operation. As a result, the first cut produces a much smaller particle size, approximately the size of the teeth, and little material is recycled back around the rotor. This approach produces fewer fines and much more uniform particles in a more efficient manner than many other traditional size reduction techniques.

Because the cutting tooth design used on this machine is similar to the tooth design used on metal milling machines, it is effective in cutting through some amount of metal typically found attached to plastics in many durable goods. Because each tooth has four independent cutting surfaces, the machine can be refitted with new cutting surfaces four times before a new set of cutters is needed; cutters can be rotated and flipped without dismantling the machine by removing one socket head screw. Both hardened steel and carbide cutters are available.

One further, and significant, advantage of rotary grinders is the hopper and ram design. Most rotary grinders have very large hoppers that can accommodate large parts and even bales of materials. More unique, however, is the horizontal ram that is used to push material into the rotor. The ram is usually driven by hydraulics, and the load is monitored such that the ram keeps a constant force of material on the rotating shaft. It also senses when the load drops too low, and it should reverse its direction to allow more material to fall in front of it for another cycle. Rotary grinders are equipped with screens to allow recirculation of oversized particles. Today, rotary grinders have found an important place in plastics recycling of all types.

Removal of Nonplastic Materials Nonplastic materials or low-volume, nonrigid plastic materials that can be present in plastics recycling streams before and after the primary size reduction step include ferrous metals, nonferrous metals, foams, film, rubber, labels, paint and coatings, metallic foils, glass, rocks, sand, dirt, and the like. The removal of all of these materials is required if the objective of the recycling operation is to produce recyclates capable of competing with virgin resins in value-added applications. Size-reduced plastics must therefore be subjected to several additional processing steps in order to prepare plastics free of nonplastic contaminants (see Fig. 14.13). Although for simplicity only dry processing may be desirable, in practice some separations require the use of a fluid medium (wet processing) for reasons of effectiveness and efficiency.

Metals Removal Ferrous metals, including low-grade stainless steel and nickel alloys, can be removed using various types of magnets. Overhead belt magnets, magnetic pulleys, and drum magnets are common. Specialized devices are used to remove final traces of magnetic materials as part of a final polishing step. Nonferrous metals are removed using eddy current separators and/or electrostatic separators. These are fully commercial technologies that have the advantage of being dry processes. Neither technology provides a perfect separation, and both eddy current and electrostatic separators can be expensive ($100,000+). Fluidized beds have also been used to separate nonferrous metals as part of a plastics recycling operation, but experience is still limited compared to eddy current and electrostatic methods. With added complication due to water treatment requirements, wet processes such as elutriation, heavy media tanks, and mineral jigs are also used to remove nonferrous metals and glass. Some recycling processes starting with very contaminated feedstocks such as automotive shredder residue place a heavy media tank very early in the process in order to drop out most metal, rocks, glass, and sand that could damage size reduction equipment [74].

Lights Removal Air classification methods are used to remove light contaminants such as dust, film and foam fragments, and paper glass powder in a recycling operation in the absence of water [1, 10]. Two common types of air classifier systems, cyclone separators and multiaspirators, are shown in Figures 14.16 and 14.17. Other types of air classifiers are air knives, elutriators, zig-zag classifiers, and air tables. Air classifiers are rather simple equipment where control is often more art than science, and the equipment must be "tuned" for each stream of material. Separations of materials are based on differences in terminal velocities in an airstream and are highly dependent on particle size and shape.

The Kongskilde aspirator operates like a cyclone in that the heavy particles in a rotating airstream are thrown outward against the wall of the aspirator chamber and fall out the bottom. The lighter particles are drawn up and out of the aspirator by the moving air stream. The "light" particles are not necessarily the lightest in strict particle mass. Rather, they have the lowest terminal velocity. Therefore, a more dense particle with a large surface area-to-mass ratio, and therefore more drag, can be carried into the light stream, while a less dense particle with a small surface area-to-mass ratio, and less drag, can report to the heavy stream. To help break up and evenly distribute the commingled feed stream into the airstream, Kongskilde incorporates an axial fan inside its aspirator. This lower fan design helps to disperse the feed stream, allowing the light materials to be vacuumed up quickly by the external fan. Although this does help break up and distribute the feed, it is also prone to plugging during operation, especially when using particles larger than 1 or 2 inches.

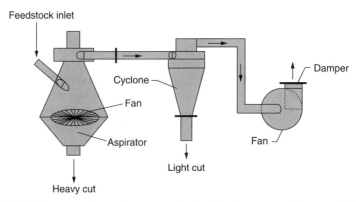

Figure 14.16. Diagram of a cyclone-type air classification system. Source: MBA Polymers, Inc. for the American Plastics Council.

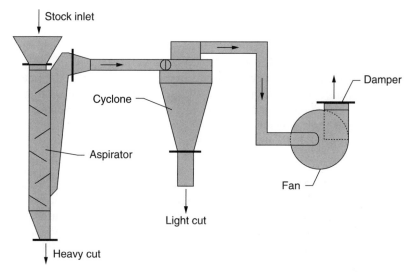

Figure 14.17. Diagram of a Kice multiaspirator-type air classification system. Source: MBA Polymers, Inc. for the American Plastics Council.

The Kice multiaspirator can be built to many different size specifications and is very effective except for especially high foam or fluff loadings. Air is introduced at each separation step, increasing the number of times that materials can be cut out of the stream.

Air classification is indispensable in a plastics recycling operation that takes in complex plastic-rich parts from end-of-life durables such as automobiles, computers, televisions, business equipment, and small household appliances [8, 10, 49, 73]. Air classification process steps can also be staged to drop out various materials depending on particle size, shape, and buoyancy. Some advantages and disadvantages of air classification processing are shown next:

Air Classification

Advantages	Disadvantages
• Dry process • Simple to use • Rather inexpensive • Uses only air • Minimum floor space requirements	• Noisy • Frequent material plugging, especially with fluffy materials • Static electricity can introduce problems • Throughputs highly variable • Highly particle size and shape dependent • Seldom 100% effective

Plastic–Plastic Separation Both dry and wet processes can be used to separate one plastic from another in a recycling operation. The most common wet separation methods are based on differences in density between particles of different plastics [1, 2, 10, 49]. Density is a bulk property of the plastic. Unfortunately, many factors beyond polymer type affect density. Plastic particles can also show differences in surface properties. Plastic–plastic separation methods based on surface property differences include triboelectric separation [75] (a dry process) and froth flotation [75] (a wet process). Density-based and surface-property-based methods for separating a stream of mixed plastics are covered in the following sections.

Density-Based Wet Methods The most widely used and accepted techniques to separate mixed plastics are based on density because many plastics differ sufficiently in this material property [10, 49, 76, 77]. It can be used to separate polymers within the same family containing different additives in many cases. As these techniques are based on differences in a bulk material property, they are more predictable and reliable than separation techniques exploiting only surface-sensitive properties that can more easily vary with surface contamination such as dirt, oils, coatings, and the like and with environmental exposure such as ultraviolet (UV) degradation or oxidation. Plastic resins cover a very wide density range from less than 1 g/cm^3 to over 1.5 g/cm^3. Some representative values are shown in Table 14.13. The standard procedure for density-based separation is to place the mixture in a medium having a density between that of the targeted material and the other materials. If there is a sufficient density difference between target plastic and the remaining plastic(s), the material less dense than the media will float and the more dense material will sink. Three types of equipment have been used in commercial plastics recycling operations to accomplish plastic–plastic separation using differences in density. These are sink–float

Table 14.13 Densities of Some Common Neat Plastic Resins Used in Packaging and Durable Applications

Polymer	Specific Gravity
Polypropylene (PP) copolymer	0.890–0.905
Polypropylene (PP) homopolymer	0.900–0.910
High-density polyethylene (HDPE)	0.952–0.965
Polystyrene (PS)	1.04–1.05
ABS	1.01–1.06
Nylon 66 (PA 66)	1.13–1.15
ABS ignition resistant	1.16–1.21
Acrylic	1.17–1.2
Polycarbonate (PC)	1.2
Polyethylene terephthalate (PET)	1.34–1.39
Polyvinyl chloride (PVC), rigid	1.3–1.58

After Refs. [51] and [76].

tanks, hydrocyclones, and centrifuges [10, 49, 75, 77]. Each has advantages and disadvantages.

SINK–FLOAT TANKS OR CLASSIFIERS   The most simple density separations use large sink–float tanks and water as the media. Water, for example, allows the separation of HDPE bottle resin (specific gravity less than 1.00 g/cm^3) and PET bottle resin (amorphous PET specific gravity of 1.34 g/cm^3). Simple augers can be used to transport resins between tanks. To separate two plastics with densities greater than that of water, a media within the appropriate density range must be used. Examples of heavier than water media are shown in Table 14.14. Alcohols and aerated liquids have been explored to obtain separation media with densities below that of water [2]. A relatively esoteric method to separate mixed plastics by density involves the use of near-critical or supercritical carbon dioxide and sulfur hexafluoride in which density of the fluid separation medium can be varied within useful ranges by slight variations in temperature and pressure [78].

Sink–float tanks have been extensively evaluated for the separation of plastics from end-of-life durables [49, 73, 79, 80]. Engineering plastics from end-of-life durable goods cover an exceptionally broad range of densities (less than 0.9 g/cm^3 to greater than 1.7 g/cm^3). There are many factors that influence the actual density of the compounded plastic besides the density of the base resin. Several of these are listed here:

Chemical structure of base polymer	Colorants	Plasticizers
Degree of crystallinity	Type of processing	Impact modifiers
Degree of foaming	Antioxidants	Pigments
Fillers	Flame retardants	Lubricants
Reinforcements	Smoke suppressants	

The end result is that there can be density overlap between two resins even though the densities of the base resins differ significantly (Fig. 14.18). A classic example is common talc-filled polypropylene (PP) that can have a density very similar to that of unfilled acrylonitrile–butadiene–styrene (ABS) resin. These

Table 14.14 Common Heavy Media Selection Options for Separating Plastics by Density

Media	Density Range							
Aqueous calcium nitrate solution	1.05	1.10	1.15	1.20	1.25	1.30	1.35?	
Aqueous calcium chloride solution		1.05	1.10	1.15	1.20?			
Aqueous sodium chloride solution		1.03	1.08					
Aqueous calcium carbonate suspension	1.05	1.10	1.15	1.20				
Potassium carbonate								1.4/1.5
Magnetite						1.3?	1.4	1.5/1.6
Ferrous silicate								>1.6

Source: American Plastics Council.

Figure 14.18. Density ranges and overlap for compounded plastic resins used in durable applications. Source: American Plastics Council.

resins have been commonly used in automotive interiors, for example, complicating their straightforward separation using density methods alone [49]. Some of the advantages and disadvantages of sink–float classifiers are summarized below.

Sink–Float Tanks or Classifiers

Advantages	Disadvantages
• Most commonly practiced technology — well known • Can be very inexpensive • Few moving parts • Discrete separations are possible (if particles are well wetted	• Plastics have density distributions rather than discrete values • Must use some media in water — salts are commonly used • Some media loss will occur • Water treatment issues related to media • Reliable wetting of particles can be a problem • Must control density of bath

HYDROCYCLONES Hydrocyclones are an economical and effective tool for separating mixed plastics and for removing many contaminants from a target plastic [75, 77]. The basic components of a hydrocyclone are shown in Figure 14.19, and Figure 14.20 shows a schematic demonstrating the working principle behind the separation. The motive force for effecting hydrocyclone separations is again density differential, and the greater the difference in density, the higher probability of separating two dissimilar components. The shape of the particles to be

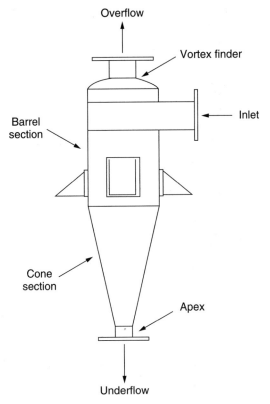

Figure 14.19. General arrangement of a hydrocyclone useful in high-throughput plastic–plastic separations [77].

separated in a hydrocyclone is also an important consideration. Since one hydrocyclone cannot guarantee close-tolerance separations, it is common practice to install hydrocyclones in series.

Hydrocyclones lie between sink–float tanks and centrifuges with respect to cost and complexity [77]. They represent a realistic density separation tool because they are relatively inexpensive, require very little space, suffer from fewer material-wetting problems than sink–float systems, and most importantly can be operated at extremely high throughputs. A single 10-inch cyclone can theoretically sort materials at rates in excess of 5000 lb/h. Though multiple stages and pumps are usually required for challenging separations, the overall capital costs can be very reasonable particularly if water alone is used as a separation medium.

Extensive testing of the fundamental performance of hydrocyclones for the separation of engineering plastics has been reported [77]. Initial testing was performed on a hydrocyclone test stand. The test stand could accommodate up to three hydrocyclones, was used to demonstrate the cyclone capabilities,

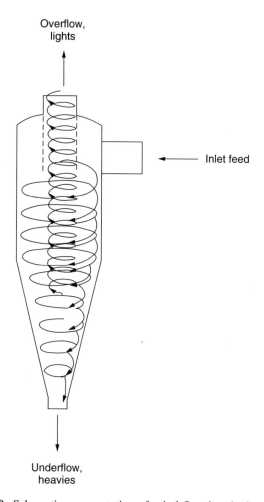

Figure 14.20. Schematic representation of spiral flow in a hydrocyclone [77].

and experiment with new concepts. A next-generation design was then built to test the most promising approaches suggested from the test stand studies. The research elucidated the fundamental parameters of the cyclone and demonstrated new configurations that are more efficient. Areas of study included the use of variable-flow/pressure control, variable apex and vortex combinations, multistep cyclones, elevated cyclones, separation curves, and statistical analysis.

This research led to a greater understanding of the capabilities and fundamental operating parameters for hydrocyclones and was used to help design a pilot separation line, the purpose of which was to demonstrate the application of hydrocyclone separations to a wide variety of plastic recycle streams [10]. Some of the advantages and disadvantages of hydrocyclones are summarized here:

Hydrocyclones

Advantages	Disadvantages
• High throughputs are possible • Moderately simple in design • Much less expensive than centrifuges • Particle wetting less of a problem compared to sink–float	• Must use some media in water • Some medial loss will occur • Water treatment issues (media) • Need to pump to drive cyclone • particle size and shape important

Table 14.15 lists several manufacturers of hydrocyclones in the United States.

CENTRIFUGES Centrifuges represent perhaps one of the most sophisticated density separation tools and can make very accurate separations at reasonably high throughputs [75, 81]. The primary drawback to centrifuges is their relatively high purchase and maintenance costs. Furthermore, only rather small particles can be fed to most centrifuges, so size reduction expenses can be high as well. However, the high centrifugal forces can overcome particle shape effects. Thus centrifuges have been shown to have utility for the separation of plastics in the form of rigids, films, and fibers. Some of the advantages and disadvantages of centrifuges are summarized here:

Centrifuges

Advantages	Disadvantages
• Most discrete density separations are possible due to high g-factors • Particle shape of little importance (film and rigids can be separated in same system) • Fairly compact footprint	• Must use some media in water (most use some type of salt) • Some media loss will occur • Water treatment issues (media) • Much more expensive than sink–float or hydrocyclone

A rough comparative estimate of cost and throughput capacity for hydrocyclones, centrifuges, and sink–float systems is shown here [77]:

	Hydrocyclone System	Centrifuge System	Sink–Float Tanks
Cost ($K)	30–60	450–650	15–50
Throughput capacity (lb/h)	5,000–10,000	1,000–3,000	1,000–2,000

Table 14.15 U.S. Hydrocyclone Manufacturers

Bird Manufacturing Company (South Walpole, MA)	Lacos Separators International (Fresno, CA)
Carpco Inc. (Jacksonville, FL)	Linatex Corporation of America (Stafford Springs, CT)
Demco Incorporated (Oklahoma City, OK)	
Donaldson Company Inc. (Minneapolis, MN)	Ohio Rubber (Denton, TX)
Dorr-Oliver Inc. (Stamford, CT)	Townley Engineering and Manufacturing Co
Heyl & Paterson (Pittsburgh, PA)	Inc. (Candler, FL)
Krebs Engineers (Menlo Park, CA)	Wemco Division, Envirotech Corporation (Sacramento, CA)

Only costs associated with the separation units and necessary associated equipment such as pumps, piping, and slurry tanks are shown. Other items are not included, such as water treatment, dewatering, drying, screening, conveying, electrical, and mechanical installation costs. Each of the systems discussed has some variability in capacity that naturally changes the cost of the system, and the capacities can change dramatically depending on the form of the material being processed. The costs shown reflect the capacity given.

Density-Based Dry Method The most common density-based dry separation methods include air classification discussed above and several technologies drawn from mineral processing and metal recycling industries such as air table classifiers and gravity table separators. Polymer melt density can also be used. Most polymers are incompatible in the solid or liquid state and liquefy at different temperatures [82]. This has resulted in the exploration of a developmental melt centrifuge technique to separate individual plastics from mixtures of plastics. The technique has been applied to the separation of plastics used in several automotive applications [83].

Surface-Property-Based Separation Methods

FROTH FLOTATION   Froth flotation was originally patented in 1906 and has allowed the effective and efficient separation of low-grade and complex minerals that were once unrecoverable. Froth flotation is based on the differences in physicochemical surface properties of target materials and therefore can work for two plastics with similar densities but different surface properties [1, 2]. After treatment with appropriate chemicals, such differences in surface properties can be significantly enhanced in a flotation slurry. To enable flotation, air (or gas) bubbles must be added and must attach themselves to the particles. The bubble–particle clusters (froth) float to the slurry surface. Usually, the process can only be utilized on relatively fine materials because the adhesion between the target particles and bubbles are limited. It required far more bubble attachment to lift large particles, and large particles have a lower surface/mass ratio, making it more difficult to get sufficient buoyancy.

A routine flotation process may include the following steps:

1. Grinding the feed material to a size suitable for froth flotation.
2. Conditioning the slurry in a mixing tank by adding feeding material, water, and some selected chemicals. The order for adding chemicals and mixing time required for each chemical material must be predetermined in testing. The objective of this step is to prepare a flotation slurry with a desirable solids content and to ensure that the chemicals attach to the surface of target materials.
3. Carrying out flotation in a flotation machine by adding air (or gas). Hydrophobic particles adhere to air bubbles generated in the machine, float to the slurry surface, and are collected as a froth fraction, while the hydrophilic particles remain in the cell and discharge as a cell (sink) fraction.

The overall flotation performance depends on the following two groups of parameters: (1) interaction between the material and chemicals, which is controlled by the selected chemicals, and may be optimized by selecting highly effective chemicals; and (2) interaction between the material and air bubbles, which is primarily controlled by the flotation machine. Both are also influenced by the cleanliness of the material's surface. The application of froth flotation to separate mixtures of plastic particles has been described in the patent literature by several research groups (Table 14.16) and in several publications and conference proceedings [1, 2, 41, 53, 84–86]. Some of the advantages and disadvantages of froth flotation technologies are summarized here:

Froth Flotation

Advantages	Disadvantages
• Usually water based • Reasonably known technology • Rather inexpensive • Few moving parts • Demonstrated for binary mixtures of bottle resins and plastic refrigerator liner resins	• "Purifier," not a separator • Surface technique — cleanliness very important • Difficult stand-alone strategy for complex mixtures • Need to find and add correct "agents" to water

ELECTROSTATICS Plastic separation technologies that use electrostatic charging of the particle surface are in commercial use [87, 88]. Methods based on triboelectric

Table 14.16 Examples of Plastics Recycling Froth Flotation Patents

Separation	Author	Patent Number	Year
PE/PS/Ebonite	K. Saitoh	3,985,650	1976
R-PVC/S-PVC	Mitsui	4,046,677	1977
PVC/PET; PS	K. Saitoh	4,119,533	1978
ABS/PP	Mitsui	4,132,633	1979
PET/PVC	E. A. Sisson	5,120,768	1992
PET/PVC	R. W. Kobler	5,234,110	1993
PET/PVC	Deiringer et al.	5,248,041	1993
Diverse	J. Y. Hwang	5,377,844	1995
ABS/HIPS	B. Jody et al.	5,653,867	1997

charging are of particular interest and utilize the simple principle that when dissimilar materials, for example, particles of two different plastics, are rubbed together they transfer electrical charge and the resulting surface electrical charge differences can be used to separate the two plastics in an electric field [42, 75, 87, 88]. Table 14.17 shows a triboelectric series for some common polymers. Figure 14.21 shows an example of an experimental device that has been used to evaluate the separation of numerous mixed plastic streams. The operating principle is shown schematically in Figure 14.22. Many plastic separations are possible using electrostatics [42, 87–91] including the separation of ABS and HIPS from end-of-life electronics, ABS and PMMA from automotive taillights, PP and PE, PET and nylon, PET and PVC, PVC and PE from wire and cable scrap, and the separation of water bottle PVC and PC. Consistent results require careful cleaning and preconditioning of the mixed plastic feedstocks. Some of the advantages and disadvantages of electrostatic separation methods are summarized here:

Electrostatic Separators

Advantages	Disadvantages
• Dry process • Rather known technology from the mineral processing industry • Very few moving parts • Rather simple design • Demonstrated for several binary mixtures	• "Purifier" more than a separator • Particles must be small but not too small • Particle shape can be important • Must control humidity and particle moisture level • Surface-sensitive technique: surface cleanliness important • Somewhat expensive ($100,000 + range)

OVERVIEW OF PLASTICS RECYCLING TECHNOLOGY

Table 14.17 Representative Triboelectric Series for Plastics

PTFE, FEP, etc. (fluorinated polymers)	Negatively charging
Polyvinyl chloride (PVC),	
Polyethylene terephthalate (PET)	
Polypropylene (PP)	
Polyethylene (PE)	
Polystyrene (PS)	
Acrylonitrile–butadiene–styrene (ABS)	
PA 66	
Polycarbonate (PC)	
Polymethylmethracrylate (PMMA)	
Polyurethane (PUR)	
Polyethylene oxide (PEO)	Positively charging

Figure 14.21. Triboelectric separator built by Plas-Sep and used by APC and MBA polymers for R&D on mixed plastics. Photo courtesy of MBA Polymers, Inc.

Three companies that supply electrostatic separation equipment to the plastics recycling industry are listed here:

Plas-Sep Limited	Outokumpu Technology, Inc.	Hamos GmbH
293 Dawlish Ave	Carpco Division	Im Thal 17
Toronto, Ontario	1310-1 Tradeport Drive	82377 Penzberg
Canada	Jacksonville, Florida 32218	Germany
Tel: 416-481-6560	Tel: +1-904-353-3681	Tel: +49-8856-9261-11
Fax: 416-481-1059	Fax: +1-904-353-3681	Fax: +49-8856-9261-99
	www.carpco.com	hamos@hamos.com
		www.hamos.com

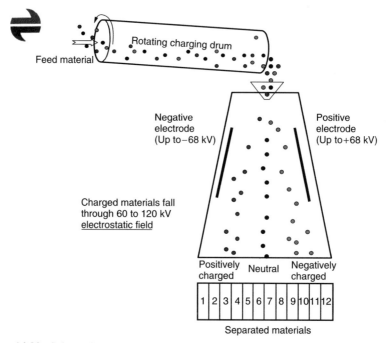

Figure 14.22. Schematic showing the operating principles for a triboelectric separator [42].

SOFTENING POINT   Most amorphous thermoplastics become tacky at elevated temperatures, and the onset of tackiness occurs at different temperatures for different polymers [76]. This phenomenon has been explored to achieve separation of plastics [2, 75, 92]. Recently, the controlled softening of mixed thermoplastics using heat lamps followed by mechanical pickup of the softened resin on a specially designed rotating surface-modified drum has been reported by a Belgian company [92]. The commercial application of a technology may be integrated with some form of density separation to prepare an acceptable mixed plastic feedstock. Nondensity-based separation techniques will continue to be explored for the efficient recycling of the broadest possible range of engineering plastics from postconsumer durable goods.

14.8.6. Selective Dissolution

The solubility of polymers is highly solvent and solution temperature dependent [76]. As a result, some of the early pioneering work on plastics recycling examined selective dissolution of polymers as a means to separate different plastics from mixtures [2, 75, 93–95]. Various plastic–solvent combinations are shown here:

   PVC: esters, ketones, chlorinated hydrocarbons, tetrahydrofuran
   ABS: aromatics, ketones, chlorinated hydrocarbons, tetrahydrofuran

PA: phenols
PC: chlorinated hydrocarbons, dimethylformamide
PS: aromatics, chlorinated hydrocarbons
PE: aromatics (hot), chlorinated hydrocarbons
PP: aromatics (hot), chlorinated hydrocarbons

One approach to selective dissolution is based on temperature-dependent solubility of different plastics in a single solvent, for example, xylene [2, 93]. A second approach is to use solvent-dependent solubility of different plastics at a specified temperature [2, 46, 75]. The following is a simple flow diagram for a selective dissolution process:

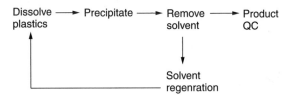

Recently, a solvent extraction method to recover pure PVC compounds from a variety of PVC-based composite products was developed and commercialized [96–98]. Selective dissolution using supercritical solvents such as supercritical CO_2 to extract additives such as flame retardants [99] or contaminants such as motor oil [100] from plastics have also been evaluated with some success.

It is likely that solvent-based processing of recovered plastics will grow in the future as part of the overall global infrastructure for plastics recovery. Economies of scale will help to cover the overhead costs required to meet demanding environmental and safety requirements for chemical processing. Some advantages and disadvantages of solvent-based separation processed are summarized here:

Selective Dissolution

Advantages	Disadvantages
• Potential to handle complex mixtures of plastics	• Expensive
• Ability to separate many different plastics	• Siting difficulties
	• "Chemical plant" permitting
	• Limited scale-up experience
• Potential to purify base resins	• Extensive solvent handling
• 98+% product purity	• Flammability issues
• Contaminate removal by filtration	• Hazardous waste potential
	• Product purity limitations
• Integral compounding possibilities	

14.8.7. Mechanical Recycling of Thermosets

Many of the recycling technologies covered in previous sections are applicable only to thermoplastics. Because thermoset plastics cannot be readily dissolved, melted, recompounded, and reshaped like thermoplastics, different mechanical recycling strategies are required [101]. Polyurethane plastics [5, 6, 102–104] and sheet molding compound (SMC) [105–107] have been studied extensively. During the past decade, the following have been extensively examined with significant success: (1) grinding thermoset plastics to fine powders that can be used as fillers in new thermoset resins or thermoplastic compositions and (2) similar to (1) with the additional or alternative strategy of using controlled size reduction to liberate and recover the original fillers or reinforcement fibers in the thermoset resin that may, in fact, be of higher value than the original matrix resin [108]. Mechanical recycling approaches such as these complement chemical recycling [1], thermal feedstock recycling processes such as gasification and liquefaction [1, 109], and energy recovery processes [110, 111] that remain practical for a broad range of thermoset materials [1, 2].

14.8.8. Color Sorting

The ability to sort plastics by color has been an important component of HDPE and PET bottle sorting technology in order to produce final recyclates of highest value [56–58]. Many bottle sorting lines offer the capability to separate bottles by color as well as by type of resin. In bottle recycling operations, high throughput flake/particle color sorters have also been used to accomplish a final polishing step, for example, the removal of traces of colored HDPE flake from a primary stream of natural HDPE resin [112].

Recently the application of particle color sorting to the value-added recycling of plastics from end-of-life durables has been explored [112]. Figure 14.23 shows a picture of a commercial particle color sorter at a plastics recycler. Figure 14.24 is a schematic of the color sorting operation. Air nozzles are used to remove optically identified particles of one color from multiple falling streams of mixed color resins. It is anticipated that color sorters of various types will become an integral part of plastics recycling operations targeting highest value recyclates for high value end-use markets.

14.8.9. Final QC

When postconsumer or postindustrial plastics are being recycled back into high-value applications to compete with virgin resins, it is an absolute requirement that the highest standards of final product QC be part of the overall operation. The establishment of a properly equipped quality control laboratory for the analysis and testing of plastics will have to be an integral part of the plastics recycling manufacturing operation (Fig. 14.25). Product consistency is becoming more and more an absolute requirement in the marketplace. In order to ensure product consistency, compounding and pelletizing operations will often be included [1, 7, 113, 114]. In order to supply recycled plastics to some markets, for example, the

Figure 14.23. Color sorter installation at a plastics recycler. (Photo courtesy of MBA Polymers, Inc.).

information technology market, safety-related specifications developed by groups such as Underwriters Laboratories Inc. (UL) also need to be considered [115]. This can put significant additional demands on final product quality control in a plastics recycling operation.

14.8.10. Feedstock/Chemical Recycling

Chemical recycling and feedstock recycling of plastics describe a family of plastics recycling processes that convert solid plastic materials through the use of heat or heat and catalysts into useful smaller molecules, usually liquid or gases, but sometimes solids or waxes [1, 116, 117]. Today, with few exceptions, these tertiary recycling technologies remain developmental but are of considerable interest for their longer term potential. For environmental as well as economic reasons, they will likely play a significant role in the overall recovery and recycling of plastics in the future. Figure 14.26 shows the range of tertiary recycling processes for plastics that have been under development at one time or another over the past decade and a half.

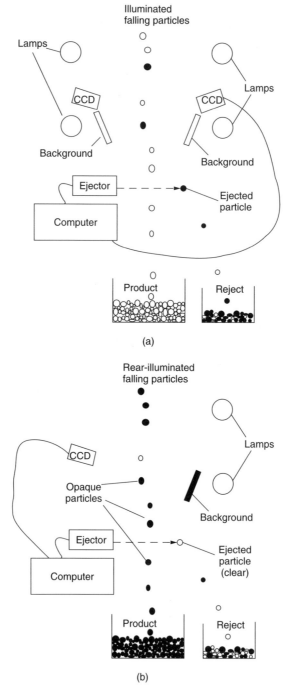

Figure 14.24. Schematic showing the operating principles of a commercial flake color sorter [112].

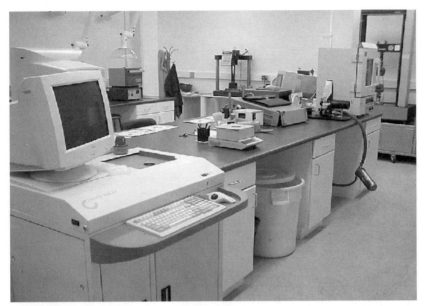

Figure 14.25. Quality control laboratory supporting a plastics recycling plant. (Photo courtesy of MBA Polymers, Inc.).

Feedstock Recycling The term feedstock recycling broadly defined encompasses chemical recycling but is most often applied to the thermal cracking of polyolefins (polyethylenes and polypropylene) and substituted polyolefins into a variety of smaller hydrocarbon intermediates. In some processes, the yields of ethylene, propylene, and butylene can be significant [1]. In some cases, other addition polymers can be thermally depolymerized directly back into monomers in reasonably high yield. Examples are polystyrene to styrene [118] and polymethyl methacrylate to methyl methacrylate [119, 120]. A special class of feedstock recycling processes yields an important raw material called synthesis gas, a mixture of hydrogen and carbon monoxide [121, 122]. Hydrocarbons and synthesis gas can all be used as chemical feedstocks for further upgrading to commercially important products at oil refineries and chemical plants. The production of methanol can be mentioned as a specific example [122]. Feedstock recycling has been extensively applied to the recycling of mixed postconsumer packaging plastics, but feedstock recycling of plastics from end-of-life electrical and electronic equipment and end-of-life automobiles and appliances has also received attention [1, 123–126]. The U.S. Department of Energy has extensively investigated both the co-liquefaction and co-gasification of postuse plastics with coal to produce petrochemical feedstocks and transportation fuels [127, 128].

Chemical Recycling The term chemical recycling is most often applied to the depolymerization of certain condensation and addition polymers back to monomers using chemical reagents and/or solvents [1, 116]. Polyethylene

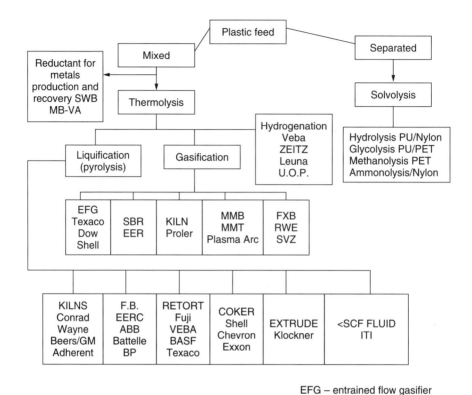

Figure 14.26. Overview of range of feedstock recycling technologies that have been investigated for the recycling of postuse plastics. Source: American Plastics Council.

terephthalate (PET), certain polyamides (nylon 6 and 66), and polyurethanes (PURs) can be efficiently depolymerized. The resulting chemicals can then be used to make new plastics that can be indistinguishable from the initial virgin polymers.

Several chemical recycling processes for postconsumer PET have been scaled up [1, 116]. For example, reacting PET with ethylene glycol forms bis-hydroxyethyl terephthalate, a starting monomer for the synthesis of PET. Hydrolysis of PET yields terephthalic acid, another monomer. A third process uses methanol to depolymerize PET to the dimethyl ester of terephthalic, which can also serve as a monomer for the manufacture of PET.

Nylon 6 can be cracked back to caprolactam in the presence of phosphoric acid catalyst and superheated steam, and one nylon manufacturer has used the process to recycling the nylon face fiber on carpets [1, 116, 129]. Mixtures of nylon 6 and 66 can be depolymerized to useful monomers by ammonolysis at above 300°C by combining nylons and ammonia in the presence of catalysts [130].

Commercial polyurethanes depolymerization has been by glycolysis. By reacting polyurethanes with glycols, new polyols can be made. Following purification, the new polyols can be used for new polyurethane manufacture [1, 5, 6].

14.8.11. Plastics as a Reductant and Fuel for Steel Blast Furnaces

Recovered plastics from packaging and from end-of-life durable goods have been evaluated as a replacement for coke in steel mill blast furnaces [131]. Coke serves as both a reductant and fuel according to the following basic blast furnace chemistry:

Blast Furnace Chemistry

Reduction of Iron Oxide	Cracking of Hydrocarbons (Plastics)
$Fe_2O_3 + 3CO = 2Fe + 3CO_2$	$C_2H_4 = 2C + 2H_2$
$Fe_2O_3 + 3H_2 = 2Fe + 3H_2O$	$2C + O_2 = 2CO$

Figure 14.27 shows a schematic of a steel mill blast furnace. Plastics, like coke, oil, and coal, can be an effective source of carbon and hydrogen, and their use in the steel-making process is an excellent example of integrated resource management. The technology is still under development in Japan and in Germany where some commercial use has already occurred [132, 133]. A recent paper study in the Unites State and Canada thoroughly evaluated the opportunity to use polymeric materials from automotive shredder residue and postconsumer packaging in North American blast furnaces [132, 134–136].

In the future, feedstock recycling of mixed plastics to synthesis gas, petrochemical feedstocks, coal and coke replacements, and monomers has the potential to complement mechanical recycling processes at a significant scale worldwide. Many of the plastic feedstock recycling processes appear to be technically feasible and robust enough to warrant further development. Much more needs to be understood about infrastructure requirements, feedstock quality requirements, processing, and economics.

14.8.12. Energy Recovery and Fuel Recovery

Recognizing that the broad mechanical recycling of postconsumer plastics has technical or economic limitations that are only now beginning to be overcome in specific cases, the Association of Plastics Manufacturers in Europe (APME) embarked on a ground-breaking series of studies beginning in the early 1990s to evaluate state-of-the-art municipal solid waste combustors for the co-combustion of packaging plastics to produce heat and electricity. Gaseous emissions and solid residues from the combustion process were evaluated and showed to be in compliance with strict German regulations. These studies have been well documented in the literature and in APME Technical Reports [111, 137–140].

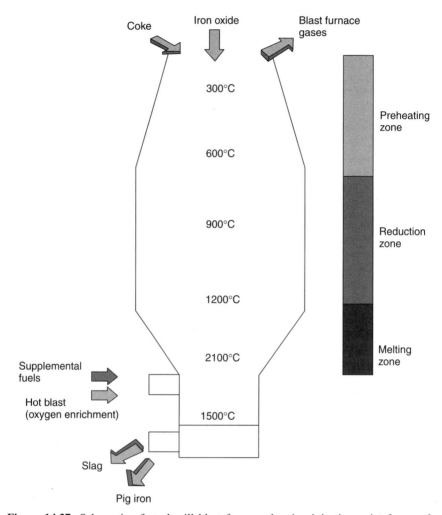

Figure 14.27. Schematic of steel mill blast furnace showing injection point for supplemental fuels and reductants such as postconsumer plastics [134].

Additional studies have examined the co-combustion of automotive shredder residue [47, 48, 111], electrical and electronic plastics [141–144], and building insulation foam [144] with municipal solid waste. Co-combustion of these plastics for energy production can also be carried out in compliance with U.S. and German waste-to-energy plant regulations in modern municipal solid waste combustors.

As an alternative to using postconsumer plastics as a fuel in municipal waste combustors, several groups have examined processed plastics or mixed paper and plastics as a fuel for cement kilns [1, 145, 146] and industrial and utility boilers [147–151]. In this application, the term process engineered fuel (PEF) is often used. Solid PEF can be produced in densified or shredded form. In contrast to municipal solid waste, PEF would be developed and marketed in compliance with accepted fuel standards. A recent full-scale trial successfully demonstrated the use of postcommercial greenhouse polyethylene film as a co-fuel in a coal-fired power plant [152]. For technical, economic, and environmental reasons, it can be anticipated that waste-to-energy and process engineered fuel, along with chemical and feedstock recycling, will contribute significantly to the environmentally and economically sound integrated resource management of plastics and fibers [153] alongside the more well-known mechanical recycling processes.

14.9. LOOKING AHEAD

Plastics are inherently strong, lightweight, durable, versatile, and resource-efficient materials with significant life-cycle environmental benefits. Polymer-based products will continue to provide unique technical solutions in virtually all markets [154, 155]. Although many challenges to the cost-effective and environmentally sound recovery and recycling of polymers remain, significant advances in science and technology have been made over the past decade, major technical and regulatory barriers have been addressed, and the future for increased recovery of polymers, especially plastic packaging and engineering plastics looks extremely promising. Integrated resource management across industries will play an important role combining mechanical recycling, feedstock/chemical recycling, and fuel/energy recovery. Although the mix of resource management options will likely vary by region, postuse plastics recovery will provide material and energy resources of increasing importance to industry and society. Over the next 20 years, a significantly expanded plastics recycling infrastructure will develop globally providing new economies of scale.

Some recovery challenges are unique to low-cost commodity resins used in a wide variety of packaging and nondurable applications. Beyond economics, collection, sorting, and cleaning of relatively lightweight plastic packaging can be thermodynamically constrained relative to other materials that are both heavy and stable to very high temperatures. If sorting and cleaning are not adequate, downstream markets for secondary polymers will be limited, especially at the higher value end of the spectrum. Nevertheless, it can reasonably be predicted that the collection of plastic bottles will grow and recycling rates for plastics will increase. In the case of the relatively new engineering polymers and composite materials, lack of large, readily accessible volumes and the lack of a proven recycling infrastructure comparable to that developed over the past century for metals are significant barriers. However, cost reduction will remain a powerful driver for the recycling of engineering polymers, and economics will drive further advances in more cost-effective collection methods and in sorting and cleaning technologies.

Acknowledgments The author would like to give special acknowledgement to the following individuals and organizations for their valuable contributions to advancing the field of polymer recycling through ground-breaking fundamental research, process and product development, communication and educational activities, and technology transfer: Mr. Peter W. Dinger, Director of Technology at the American Plastics Council and, since 1989, a catalyst for advancing the recycling of plastics packaging; Mr. Chris Ryan, manager of the first durables plastics recycling R&D facility operated by wTe Corporation; Dr. Michael B. Biddle, President of MBA Polymers, Inc. and the entire MBA technical staff under Trip Allen for their pioneering work and leadership in evaluating, developing, and promoting technologies for the recycling of plastics from end-of-life durables; the Association of Plastics Manufacturers in Europe (APME) for its major investment over the past decade in the development and communication of energy recovery and feedstock recycling technologies for plastics; Dr. Frank E. Mark of Dow Europe and Martin Frankenhaeuser of Borealis Polymers Oy have made major contributions in the areas of energy recovery; the Vehicle Recycling Partnership (VRP) under the United States Council for Automotive Research (USCAR) which, since 1991, has actively worked to overcome major technical barriers to the cost-effective recovery of plastics from end-of-life vehicles; the many universities and national laboratories that have taken on plastics recycling R&D challenges and made important contributions both from a scientific and engineering perspective; and a very special thanks must go to the American Plastics Council, its staff and individual member companies for their support to the development of new, enabling technologies for the mechanical recycling of plastics in both the packaging and durables markets and to technology development in the areas of feedstock recycling, waste-to-energy, and process engineered fuel.

BIBLIOGRAPHY

Plastics Industry Trade Associations Active in Plastics Recycling

American Plastics Council (APC), 1300 Wilson Boulevard, Arlington, VA 22209, Tel 703-741-5000, Fax 703-741-6095, *www.plastics.org*.

The Society of the Plastics Industry (SPI), 1801 K Street, NW, Suite 600K, Washington, DC 20006, *www.plasticsindustry.org*.

Canadian Plastics Industry Association Environment and Plastics Industry Council (CPIA/EPIC), 5925 Airport Road, Suite 500, Mississauga, Ontario L4V 1 W1, Tel 905-678-7405, *www.plastics.ca/epic*.

Association of Plastics Manufacturers in Europe (APME), Avenue E. Van Nieuwenhuyse 4, Box 3, B-1160 Brussels, Belgium, Tel (32-2) 672 82 59, Facsimile (32-2) 675 39 35, *www.apme.org*.

Plastics Waste Management Institute of Japan (PWMI), Fukide Bldg., 1–13, Toranomon 4-chome, Minato-ku, Tokyo, Japan T105-0001, Tel 81-3-3437-2251, Fax 81-3-3437-5270, *http://www.pwmi.or.jp*.

Other Associations/Organizations Active in Plastics Recycling

Society of Plastics Engineers (SPE), 14 Fairfield Drive, PO Box 403, Brookfield, CT 06804, Tel 203-775-0471, *www.4spe.org*.

Association of Post Consumer Plastics Recyclers (APR), 1300 Wilson Boulevard, Arlington, VA 22209, *www.plasticsrecycling.org.*

National Association of PET Container Resources (NAPCOR), 2105 Water Ridge Parkway, Suite 570, Charlotte, NC 28217, Tel 704-423-9400, Fax 704-423-9500, *www.napcor.com.*

Institute of Scrap Recycling Industries (ISRI), 1325 G St., N.W., Suite 1000, Washington, DC 20005, *www.isri.org.*

Automotive Recyclers Association (ARA), 3975 Fair Ridge Drive, Suite 20, Terrace Level North, Fairfax, VA, *www.automrecyc.org.*

Vehicle Recycling Partnership of USCAR, 1000 Town Center Building, Suite 300, Southfield, MI 48075, *www.uscar.org.*

Association of Home Appliance Manufacturers (AHAM), 1111 19th St., N.W., Suite 402, Washington, DC 20036, *www.aham.org.*

Electronic Industries Alliance (EIA), 2500 Wilson Boulevard, Arlington, VA 22201, *www.eia.org.*

Directories of Plastics Recyclers and Recycled Plastic Product Manufacturers

2002 Scrap Plastics Markets Directory Including Plastics Recycling Equipment Listings
Resource Recycling, Inc.
P.O. Box 42270
Portland, OR 97242-0270
Tel 503-233-1305 Fax 503-233-1356
www.resource-recycling.com

United States & Canada Recycled Plastic Markets Database
American Plastics Council
1300 Wilson Boulevard, Suite 800
Arlington, VA 22209
Tel 703-253-0700 Fax 703-253-0701
http://markets.plasticsresource.com

Directory of U.S. and Canadian Companies Involved in the Recycling of Vinyl Plastics and Manufacturing Products from Recycled Vinyl, August 2000
The Vinyl Institute
1300 Wilson Boulevard, Suite 800
Arlington, VA 22209
Tel 703-253-0700 Fax 703-253-0701
www.vinylinfo.org

Market Data Book
Plastics News, December 2001
1725 Merriman Road Akron, Ohio 44313-5283

Tel 330-836-9180 Fax 330-836-2322
www.plasticsnews.com

Recycled Plastic Products for Building and Construction Sourcebook, American Plastics Council, 2000

U.S and Canadian Recycled Plastic Products, American Plastics Council and Environment and Plastics Industry Council of Canada, 2001

REFERENCES

1. J. Brandrup, M. Bittner, W. Michaeli, and G. Menges, eds., *Recycling and Recovery of Plastics*, Hanser/Gardner Publications, Cincinnati, OH, 1996.
2. A. L. Bisio and M. Xanthos, eds., *How to Manage Plastics Waste Technology and Market Opportunities*, Hanser, New York, 1995.
3. R. J. Ehrig, ed., *Plastics Recycling Products and Processes*, Hanser Publishers, New York, 1992.
4. F. P. La Mantia, *Recycling of PVC & Mixed Plastic Waste*, ChemTec, Toronto, Canada, 1996.
5. K. C. Frisch, D. Klempner, and G. Prentice, *Recycling of Polyurethanes, Advances in Plastics Recycling*, Vol. 1, Technomic, Lancaster, PA, 1999.
6. W. Rashofer and E. Weigand, *Automotive Polyurethanes, Advances in Plastics Recycling*, Vol. 2, Technomic, Lancaster, PA, 2001.
7. Recycling of Polymers, *Die Makromolekulare Chemie, Macromolecular Symposia*, IUPAC, **57**, 1–395 (1992).
8. M. B. Biddle and M. M. Fisher, Proceedings of the SPI Structural Plastics Division 22nd Annual Conference, 119–130 Washington, DC, 1994.
9. M. M. Fisher and M. A. Maten, Proceedings of AutoRecycle '95, 39–68 Schotland Business Research, Inc., Dearborn, MI, November 15–16, 1995.
10. M. B. Biddle, P. Dinger, and M. M. Fisher, IdentiPlast II Conference Proceedings, APME, Brussels, April 1999.
11. F. E. Mark and J. Vehlow, *VGB Power Tech.* 46–50 (2000).
12. U.S. EPA, "Municipal Solid Waste in the United States 1999 Facts and Figures," EPA 530-R-01-014, U.S. EPA, July 2001.
13. 2000 National Post-Consumer Plastics Recycling Report, available from the American Plastics Council website, Arlington, VA, 2001.
14. Battery Council International Battery Recycling Report, Chicago, IL, 2001.
15. *Report on Post Consumer PET Container Recycling Activity*, NAPCOR, Charlotte, NC, 2000.
16. *An Analysis of Plastics Consumption and Recovery in Western Europe 1999*, APME, Brussels, 2001.
17. M. Blumenthal, Proceedings of SPE GPEC 2002, February 13–14, Detroit, MI, 185–192, (2002).
18. B. Overton, *IdentiPlast II Proceedings*, APME, Brussels, April 1999.

19. B. Krummenacher, P. Peuch, M. Fisher, and M. Biddle, APME Technical Report 8027, Association of Plastics Manufacturers of Europe, Brussels, November 1998.
20. *How to Collect Plastics for Recycling*, American Plastics Council, Arlington, VA, 1995.
21. S. Apotheker, *Resource Recycling*, May 1995, p. 35.
22. *Understanding Plastic Film: Its Use, Benefits and Waste Management Options*, American Plastics Council, Arlington, VA, December 1966.
23. *Plastic Film Recovery Guide*, American Plastics Council, Arlington, VA, 1999.
24. *Simplicity = Success: Answers to Six Common Questions About "All Plastic Bottle" Collection Programs*, American Plastics Council, Arlington, VA, 2000.
25. *Stretch Wrap Recycling: A How-To Guide*, American Plastics Council, Arlington, VA, 1994.
26. K. Sparks, *Resource Recycling*, 1999, pp. 33–37. See also P. Krishnaswamy and R. Lampo, *J. Stand. Engr. Soc.* **53**(5), September/October (2001).
27. I. Bremerstein, Proceedings of IdentiPlast III, Brussels, April 2001.
28. M. M. Fisher and P. Peuch, Society of Plastics Engineers, Annual Recycling Conference, ARC98, Chicago, November 65–72 (1998).
29. *Plastics, A Material of Choice for the Electrical and Electronic Industry—Plastics Consumption and Recovery in Western Europe*, Association of Plastics Manufacturers of Europe, Brussels, 1995.
30. L. B. Jung (Vista Environmental), Environmental Protection Agency's Common Sense Initiative, Computer and Electronics Overcoming Barriers Workgroup, EPA contract No. 7W-3901-TASA, July 1998.
31. P. Boyles and S. Bennett (Northeast Resource Recovery Association), Environmental Protection Agency's Common Sense Initiative, EPA Grant No. X991642-01-0, February 1998.
32. Analysis of Five Community Consumer/Residential Collections: End-of-Life Electronic and Electrical Equipment, U.S. EPA, EPA-901-R-98-003, April 1999.
33. Household End-of-Life Electrical and Electronic Equipment Pilot Collection Project, U.S. EPA, February 1998.
34. *Electronic Product Recovery and Recycling Baseline Report: Recycling of Selected Electronic Products in the United States*, National Safety Council's Environmental Health Center, Itasca, ILL, May 1999.
35. Recovery of Plastics from Municipally Collected Electrical and Electronic Goods, a summary report of research sponsored by the American Plastics Council and The Materials for the Future Foundation, American Plastics Council, Arlington, VA, March 1999.
36. Plastics from Residential Electronics Recycling Report 2000, American Plastics Council, Arlington, VA, April 2000.
37. Recycling Used Electronics: Report on Minnesota's Demonstration Project, Minnesota Office of Environmental Assistance, *www.moea.state.mn.us*, 2001.
38. T. Hainault, D. Smith, D. J. Cauchi, D. A. Thompson, M. M. Fisher, and C. Hetzel, ISEE Proceedings, San Francisco, May 8–10, 2000, pp. 310–317.
39. M. M. Fisher, M. B. Biddle, T. Hainault, D. S. Smith, D. J. Cauchi, and D. A. Thompson, Proceedings of SPE Annual Recycling Conference ARC'2000, 177–182 Dearborn, MI, November 8–9, 2000.

40. D. E. Karvelas, B. J. Jody, B. Arman, J. A. Pomykala Jr., and E. J. Daniels, Proceedings EPD Congress 1996, Garry W. Warren, ed., TMS, Warrendale, PA (1997).
41. J. Karvelas, B. Jody, J. Pomykala, and E. J. Daniels, Proceedings, SPE 6th Annual Recycling Conference ARC'99, Detroit, Michigan, November 9–11, pp. 231–237.
42. C. Xiao, M. B. Biddle, and M. M. Fisher, Proceedings, SPE 6th Annual Recycling Conference ARC'99, Detroit, Michigan, November 9–11, 1999, pp. 219–230.
43. *Composition, Properties and Economic Study of Recycled Refrigerators*, American Plastics Council, Arlington, VA, April 1994.
44. B. J. Jody and E. J. Daniels, Report ANL/ESD/TM-152, Energy Systems Division, Argonne National Laboratory, Argonne, IL, June 1999.
45. M. M. Fisher and F. E. Mark, SAE Technical Paper 1999-01-0664, SAE World Congress, Detroit, MI, March 1999.
46. B. J. Jody, E. J. Daniels, P. V. Bonsignore, and N. F. Brockmeier *JOM*, **46**(2), 40–43 (1994).
47. F. E. Mark and M. M. Fisher, SAE Technical Paper 1999-01-0990, SAE World Congress, Detroit, MI, March 1999.
48. A. A. Jean, *Polym. Recycling*, **2**(4), 291–297 (1996).
49. M. M. Fisher, M. B. Biddle, and C. Ryan, SAE Technical Paper 2001-01-0697, SAE International, Warrendale, PA, 2001.
50. H. Hock and M. A. Maten, SAE Technical Paper 930561, SAE International, Warrendale, PA, 1993.
51. J. Fosnaugh and M. Biddle, in R. A. Myers, ed., *Encyclopedia of Environmental Analysis and Remediation*, Wiley, New York, 1998.
52. MBA Polymers, Inc., American Plastics Council, *Vehicle Recycling Partnership*, unpublished results.
53. G. R. Winslow, S. X. Liu, and S. G. Yester, SAE Technical Paper 980093, SAE International, Warrendale, PA, 1998.
54. P. Dinger, in *Modern Plastics Encyclopedia*, McGraw-Hill, Inc., New York, November 1995, p. 30.
55. P. Dinger, *BioCycle*, March, 1992.
56. P. Dinger, *Mod. Plastics*, November, 1994.
57. J. H. Schut, *Plastics Tech.*, September, 1992.
58. R. Massen, IdentiPlast II Proceedings, APME, Brussels, April 1999.
59. G. R. Kenny and F. A. Hottenstein, Proceedings, SPE 6th Annual Recycling Conference, Detroit, Michigan, November 9–11, 1999, pp. 181–186.
60. F. Hottenstein and G. R. Kenny, Proceedings of IdentiPlast 2001, Brussels, Belgium, April 23–24, 2001.
61. N. Morley, *Polym. Recycling* **3**(3), 217–226 (1997/1998).
62. M. R. Costello, Proceedings, SPE Annual Recycling Conference ARC'98, Chicago, IL, 1998, pp. 95–102.
63. H. Lucht, L. Kreuchwig, and A. Uhl, Proceeding of SPE Global Plastics Environmental Conference GPEC 2002, Detroit, MI, February 13–14, 2002, pp. 279–284.
64. G. Zachman, *J. Mol. Struct.* **348**, 453–456 (1995).
65. T. Imai, Proceedings of IdentiPlast 2001, Brussels, Belgium, April 23–24, 2001.

66. S. Kumer, E. R. Grant, and C. M. Duranceau, SAE Technical Paper 2000-01-0739, SAE 2000 World Congress, Detroit, MI, March 6–9, 2000; see also *http://news.uns. purdue.edu/html4ever/010119.SpectraCode.recycle.html*.
67. D. F. Arola, L. E. Allen, M. B. Biddle, and M. M. Fisher, Proceedings of SPE 6th Annual Recycling Conference ARC'99, Detroit, MI, November 9–11, 1999, pp. 241–249.
68. T. Seidel, IdentiPlast, Brussels, October 27–28, 1997.
69. A. Gollach, A. Moormann, T. Seidel, and D. Siegmund, Proceedings of the Pittsburgh Conference, #587, March 1995.
70. B. L. Riise, M. B. Biddle, and M. M. Fisher, R'2000 Proceedings, 5th World Congress, Toronto, Canada, June 5–9, 2000.
71. D. F. Arola, L. E. Allen, and M. B. Biddle, 1999 ISEE Conference Proceedings, Danvers, MA, May 11–13, 187–191 (1999).
72. E. Langerak, ISEE Proceedings, San Francisco, CA, May 5–7, 1997.
73. M. M. Fisher, Proceedings GLOBEC '96, Mach Business Services, Davos, Switzerland, March 18–22, 12–3.1–9 (1995).
74. R. Kobler, G. Winslow, C. J. Bedell, and P. Christopher, Proceedings of SPE GPEC2002, Detroit, MI, February 13–14, 333–340 (2002).
75. W. Michaeli, M. Bittner, K. H. Unkelbach, Ingo Stahl, U. Kleine-Kleffmann, and S. Bletsch, "Sorting Techniques" in J. Brandrup, M. Bittner, W. Michaeli, and G. Menges, eds., *Recycling and Recovery of Plastics*, Hanser/Gardner, Cincinnati, 254–286 (1996).
76. J. Brandrup, E. H. Immergut, E. A. Grulke, A. Abbe, and D. Bloch, eds., *Polymer Handbook*, 4th ed., Wiley, New York, 1999.
77. *Development of Hydrocyclones for Use in Plastics Recycling*, American Plastics Council, Arlington, VA, February 1999.
78. M. S. Super, R. M. Enick, and E. Beckman, Davos Recycle'92 Conference Proceedings, Maack Business Services, Zurich, April 7–10, 1992.
79. C. Ryan, M. M. Fisher, and M. M. Fisher, Proceedings of AutoRecycle '95 Conference, 99–117 Schotland Business Research, Dearborn, MI, November 15–16, 1995.
80. M. S. Corbett and C. Ryan, ISEE Proceedings, San Francisco, May 5–7, 161–166 (1997).
81. H. Neureither and K. H. Unkelbach, R'97 Congress Proceedings, V. III, Geneva, 117–125 (1997).
82. A. K. Bledzki, P. Orth, P. Tappe, M. Rink, and K. Pawlaczy, *Polym. Recycling* **4**(2), 85–92 (1999).
83. W. Diegmann, F. H. Adam, R. Tiedeck, and W. Hoffmanns, SAE Technical Paper 2000-01-0736, 2000, pp. 133–138.
84. R. Buchan and B. Yarar, *JOM*, February, 52–55 (1995).
85. S. J. Read, G. C. Lees, S. J. Hurst, R. S. Whitehouse, and M. Barrel, *Polym. Recycling* **1**(3), 197–205 (1995).
86. S. J. Read, G. C. Lees, S. J. Hurst, R. S. Whitehouse, and M. Barrell, *Polym. Recycling* **2**(1), 49–56 (1996).
87. R. MeierStaude and R. Koehnlechner, 7th Annual Recycling Conference ARC 2000 Proceedings, Society of Plastics Engineers, Dearborn, MI, November 8–9, 199 (2000).

88. J. D. Brown, 7th Annual Recycling Conference ARC 2000 Proceedings, Society of Plastics Engineers, Dearborn, MI, November 8–9, 201–207 (2000).
89. R. Koehnlechner, Proceedings of IdentiPlast 2001, Brussels, Belgium, April 23–24, 2001.
90. I. Inculet, G. S. Castle, and J. D. Brown, Davos Recycle'92 Conference Proceedings, April 7–10, Maack Business Services, Zurich, 1992.
91. U. Kleine-Kleffmann, A. Hollstein, and I. Stahl, Davos Recycle'92 Conference Proceedings, April 7–10, Maack Business Services, Zurich, 1992.
92. See, for example, *http://www.salypnet.com*.
93. M. Matsuda, S. Asada, K. Webber, J. Lynch, and E. B. Nauman, SAE Technical Paper 941023, pp. 47–62, SAE International, Warrendale, PA, 1994.
94. H. Berthold, Recycle'91 Proceedings, Davos, Switzerland, Maack Business Services, Zurich, April 3–5, 1991.
95. M. Schneider, C. Ducommun, U. Schurr, and E. Pohl, SAE Technical Paper 930036, SAE World Congress, SAE International, Warrendale, PA, 1993.
96. J.-M. Yernaux, Proceedings, SPE 6th Annual Recycling Conference ARC'99, Detroit, MI, November 9–11, 1999, pp. 263–267.
97. J. Milgrom, Proceedings of SPE 7th Annual Recycling Conference ARC'2000, Dearborn, MI, November 8–9, 2000, pp. 93–99.
98. J.-M. Yernaux and P. Crucifix, Proceedings of SPE Global Plastics Environmental Conference GPEC 2002, Detroit, MI, February 13–14, 2002, pp. 193–199.
99. E. Marioth, G. Bunte, and Th. Hardle, *Polym. Recycling* **2**(4), 303–308 (1996).
100. ITec International Corporation, a subsidiary of Beechport Capital Corp., Oakdale, CA, 2002.
101. W. J. Farrissey, in R. J. Ehrig, ed., *Plastics Recycling Products and Processes*, Hanser, New York, 233–262 (1992).
102. E. Weigand, "Properties and Applications of Recycled Polyurethanes" in J. Brandrup, M. Bittner, W. Michaeli, and G. Menges, eds., *Recycling and Recovery of Plastics*, Hanser/Gardner, Cincinnati, 683–701 (1996).
103. W. K. Law, T. Patel, K. Swisher, and F. Shutov, *Polym. Recycling*, **3**, 269–274 (1997/98).
104. E. Vilgili, H. Arastoopour, and B. Bernstein, Proceedings of the SPE Annual Recycling Conference ARC'98, Chicago, ILL, November 11–13, 1998, pp. 320–326.
105. P. Schaefer, in J. Brandrup, M. Bittner, W. Michaeli, and G. Menges, eds., *Recycling and Recovery of Plastics*, Hanser/Gardner, Cincinnati, OH, 702–716 (1996).
106. C. N. Cucuras, A. M. Flax, W. D. Graham, and G. Hartt, SAE International Congress and Exposition, Paper No. 910387, February 1991.
107. H. Kelderman and H. Kluczka-Koss, Proceedings of IdentiPlast 2001, Brussels, Belgium, April 23–24, 2001.
108. W. D. Graham, *Polym. Recycling* **1**(2), 87–97 (1995).
109. G. Ramlow and M. Christill, *Polym. Recycling* **4**(1), 41–55 (1999).
110. M. Fisher, SPE 3rd Annual Recycling Conference ARC'96, Chicago, IL, November 7–8, 1996, pp. 385–394.
111. F. E. Mark, M. M. Fisher, and K. A. Smith, APME/APC Technical Report 8026, Brussels, September 1998.

112. B. Riise, T. Allen, M. B. Biddle, and M. M. Fisher, Proceedings of the International Symposium on Electronics and the Environment ISEE 2001, Denver, CO, May 7–9, 223–228 (2001).
113. H. Herbst and R. Pfaendner, *Polym. Recycling* **4**(2), 75–83 (1999).
114. H. Guo and A. Merrington, Proceeding of SPE 3rd Annual Recycling Conference ARC'96, November 7–8, 1996, Chicago, IL, pp. 147–154.
115. G. J. Fechtmann, Davos Recycle'95 Conference Proceedings, Davos, Switzerland, Maack Business Services, Zurich, 1995.
116. See *www.plasticsresource.com* (APC website) and *www.apme.org* (APME website) for a broad range of recent reviews, conference summaries, and reports on chemical and feedstock recycling.
117. J. M. Forgac, Proceedings, SPE 3rd Annual Recycling Conference ARC'96, Chicago, IL, November 7–8, 1996, pp. 359–384.
118. K. S. Strode and D. G. Demianiw, *Advanced Recycling of Plastics: A Parametric Study of the Thermal Depolymerization of Plastics*, Final Report with the American Plastics Council, Arlington, VA, 1995.
119. J. B. Schneider, in R. J. Ehrig, ed., *Plastics Recycling Products and Processes*, Hanser, New York, 171–186 (1992).
120. N. Brand, "Depolymerization of Polymethyl Methacrylate (PMMA)" in J. Brandrup, M. Bittner, W. Michaeli, and G. Menges, eds., *Recycling and Recovery of Plastics*, Hanser/Gardner, Cincinnati, OH, 488–493 (1996).
121. M. Gebauer and D. Stannard, "Gasification of Plastics Wastes" in J. Brandrup, M. Bittner, W. Michaeli, and G. Menges, eds., *Recycling and Recovery of Plastics*, Hanser/Gardner, Cincinnati, OH, 455–480 (1996).
122. T. Obermeier, Proceedings of IdentiPlast 2001, Association of Plastics Manufacturers in Europe, Brussels, Belgium, April 23–24, 2001.
123. T. Lehner, Proceedings of IndentiPlast Conference 2001, Association of Plastics Manufacturers in Europe, Brussels, Belgium, April 23–24, 2001.
124. F. E. Mark and T. Lehner, APME Technical Report 8036, Association of Plastics Manufacturers in Europe, Brussels, July 2000.
125. Feedstock Recycling of Electrical and Electronic Plastics Waste, APME Technical Report 8024, Association of Plastics Manufacturers in Europe, Brussels, 1997.
126. L. D. Busselle, T. A. Moore, J. M. Shoemaker, and R. E. Allred, International Symposium on Electronics and the Environment ISEE 1999 Proceedings, IEEE, Danvers, MA, May 11–13, 192–197 (1999).
127. G. P. Huffman and L. L. Anderson, *Fuel Process. Tech.*, Special Issue, **49** (1996).
128. G. P. Huffman and N. Shah, Report prepared for DOE Contract No DE-FC22-93-PC93053, September 1998, *www. cffls.uky.edu*.
129. H. Sheehan, SPE Annual Recycling Conference, ARC'99, Detroit, MI, Nov. 9–11, 381–387 (1999).
130. R. A. Smith and B. E. Gracon, Proceedings of Recycle'95, Davos Switzerland, May 15–19, 1995.
131. B. Niemoller, "Reduction in Blast Furnaces" in J. Brandrup, M. Bittner, W. Michaeli, and G. Menges, eds., *Recycling and Recovery of Plastics*, Hanser/Gardner, Cincinnati, OH, 481–487 (1996).

132. *Review of ASR-Blast Furnace Concept with European Interests*, Competitive Analysis Centre, Inc., Ontario, Canada, March 1998.
133. Blast Furnaces: Plastic as Ironmaking Fuel at NKK, May 1998, *http://www.newsteel.com/features/NS9805f5.htm*.
134. *Automotive Shredder Residue: Its Application in Steel Mill Blast Furnaces—A Preliminary Analysis Prepared for the American Plastics Council and the Environmental and Plastics Industry Council of CPIA*, Competitive Analysis Centre, Inc., Ontario, Canada, October 1997.
135. *Automotive Shredder Residue: Its Application in Steel Mill Blast Furnaces—Review of the Concept with North American Stakeholders*, Competitive Analysis Centre, Inc., Ontario, Canada, March 1999.
136. M. M. Fisher, D. Barnett, Fred Edgecombe, D. G. Tate, and A. Underdown, Proceedings, SPE 6th Annual Recycling Conference ARC'99, Detroit, MI, November 9–11, 1999, pp. 299–306.
137. F. E. Mark, APME Technical Report 8004, Brussels, January 1994.
138. F. E. Mark, A. H. M. Kayen, and J. L. Lescuyer, APME Technical Report 8008, Brussels, December 1994.
139. F. E. Mark, *Polym. Recycling* **1**(2), 115–123 (1995). See also F. E. Mark and R. Martin, APME Technical Report 8010, Brussels, February 1995.
140. F. E. Mark, APME Technical Report 8019, Brussels, November 1996.
141. F. E. Mark and J. Vehlow, APME Technical Report 8020, Brussels, February 1997.
142. J. Vehlow, Proceedings of IdentiPlast 2001, Association of Plastics Manufacturers in Europe, Brussels, Belgium, April 23–24, 2001.
143. J. Vehlow, B. Bergfeldt, H. Hunsinger, K. Jay, F. E. Mark, L. Tange, D. Drohmann, and H. Fisch, APME Technical Report 8040, Brussels, February 2002.
144. J. Vehlow and F. E. Mark, APME Technical Report 8012, Brussels, October 1995.
145. A. Caluori, F. E. Mark, M. Moser, and A. Prisse, APME Technical Report, 8021, Brussels, October 1997.
146. F. E. Mark and A. Caluori, APME Technical Report 8028, Brussels, September 1998.
147. L. Tomczyk and M. M. Fisher, SPE Annual Recycling Conference, ARC'99, Detroit, MI, November 9–11, 1999.
148. L. Tomczyk, *Solid Waste Tech.* **11**(1), 25–30 (1997).
149. M. Frankenhaeuser, APME Technical Report 8002, Brussels, 1992.
150. G. W. Krajenbrink, H. M. G. Temmink, J. A. Zeevalkink, and M. Frankenhaeuser, Consortium Report TNO-MEP R 98/220, January 1999.
151. L. A. A. Schoen, M. L. Beckes, J. van Tubergen, and C. H. Korevaar, APME Technical Report 8035, Brussels, 2000.
152. F. E. Mark and J. Rodriguez, APME Technical Report 8031, Brussels, July 1999.
153. M. K. Mishra, ed., Polymer-Plastics Technology and Engineering, Special Issue, *Polymer and Fiber Recycling* **38**(3) (1999).
154. See the APC and APME websites, *www.plastics.org* and *www.apme.org*, respectively, for summary reports on the current and future role of plastics in diverse

applications such as space, packaging, automotive, sports, health, consumer electronics, and information technology.

155. Committee on Polymer Science and Engineering, *Polymer Science and Engineering, The Shifting Research Frontiers*, National Research Council, National Academy Press, Washington, DC, 1994.

CHAPTER 15

THERMAL DESTRUCTION OF WASTES AND PLASTICS

ASHWANI K. GUPTA
Department of Mechanical Engineering, University of Maryland

DAVID G. LILLEY
School of Mechanical & Aerospace Engineering, Oklahoma State University

15.1. INTRODUCTION

In recent years, much emphasis has been placed on the thermal destruction of plastics and other hydrocarbon wastes, together with associated challenges on the environmentally benign energy recovery. The main problem for combustion engineers has been how to recover maximum energy from waste (fuel) with minimum pollution, using existing, modified, or newly designed equipment. The thermal destruction opportunities from wastes discussed here provide unique challenges to combustion engineers for environmentally benign energy recovery. The amount of plastic and other waste generated is projected to grow with increase in productivity. Almost all waste has significant energy potential, and its utilization can save millions of dollars on the national scale. The old practice of waste disposal has been to dump in open landfills, which results in *garbage in and garbage remains*. The landfill results in odor, bacteria growth, and slow release of greenhouse gases. Some materials in the waste (such as plastics, metals, and ceramics) are not biodegradable. This exacerbates the offensiveness of the existence of the landfill waste sites. The goal for the new millennium must be *garbage in and energy out* in an environmentally acceptable manner. It is anticipated that the amount of waste generated will continue to grow from many of the developing countries, along with associated changes in its composition.

Plastics and the Environment, Edited by Anthony L. Andrady.
ISBN 0-471-09520-6 © 2003 John Wiley & Sons, Inc.

Thermal destruction of wastes to the molecular and atomic level allows one to more cleanly convert waste into usable energy. The challenges provide opportunities for combustion engineers, whose research is now becoming even more important with new emphasis. Research in combustion-related areas has expanded significantly in recent years, particularly in areas related to low-quality fuels and wastes being used as fuels. The field of combustion is further diversified by the complex nature of most reaction processes. Fuel chemistry, fluid mechanics, mixing, convective and radiative heat transfer, gas-phase elementary reactions, turbulence, and particle kinetics and dynamics are relevant processes that often have a direct, and sometimes controlling, influence on the behavior of a particular combustion system. Sensors, diagnostics, and miniaturization of the system continue to be of major importance for successful implementation of this new technology. Recent developments in optical diagnostics has finally provided the prospects for obtaining this needed information about variables, such as size, velocity, species concentration, temperature, and the correlations between them.

In thermal destruction systems, the designer must accomplish efficient and environmentally benign thermal treatment and/or energy conversion. The thermal and chemical behavior of the flow in most power and propulsion systems is complex. In design situations, experimental data is used to verify and develop models. These mathematical models then supplement and reduce the amount of costly and time-consuming experimental procedures. These models bring benefits and entail costs. Benefits include knowing quantitatively, in advance, what will be the performance of equipment that has not yet been built or that has not yet been operated in the manner under investigation. Most mathematical models simulate the physical processes (e.g., turbulence, radiation, combustion, pollutant formation, and multiphase effects) by solving an associated set of coupled partial differential equations. A rational approach of parallel efforts between modeling and diagnostics has the real true hope for the cost-effective solution to the successful design and development of combustion systems.

Almost all thermal destruction and incineration problems of today involve multiphase situations. In two-phase flows, accurate size and velocity measurements of particles are important for a broad spectrum of applications. Solid particles are not spherical in shape. Their shape is quite complex and far from being uniform. Complete characterization of local properties includes such attributes as particle size and number density distribution, a velocity distribution function related to particles of different size, mean velocity of the gas phase, local gas and particle temperatures, composition of both the gas and particle phases, turbulence properties, and the like. In particle-laden flows the influence of turbulence on the particles and vice versa is important. The reliable and precise measurement of any one of the above properties is a nontrivial task.

15.2. MAGNITUDE OF THE GENERAL INCINERATION PROBLEM

The United States generates about 1.8 kg (approx. 4.0 lb) per person per day of solid waste; see Chopra et al. [1]. This translates to over one billion tons of municipal solid waste generated annually. In addition to this, the United States also generates about 300 million metric tons (where 1 ton is 1000 kg) of hazardous wastes, and over the past one or two decades it has been growing at a rate of about 5–10% annually. Oppelt [2] reviews the information. Medical waste (heating value approx. 14 MJ/kg or 6000 Btu/lb) generated in the United States varies between 4,180 and 14,000 tons/day. Hazardous wastes are produced by many industries and according to the Environmental Protection Agency (EPA) approximately 10–17% of all chemical wastes generated are hazardous.

A number of surveys have been made to evaluate the magnitude of the problem; see, for example, Gupta [3]. The 1985 EPA survey identified 2959 facilities, regulated under the Resource Conservation and Recovery Act (RCRA), which managed a total of 247 million tons of wastes per year. The Chemical Manufacturers Association (CMA) conducted a survey to determine hazardous waste generation by its member companies. The total waste generated was about 213 million tons per year. The Congressional Budget Office estimated the hazardous waste generation in 1982 to be between 223 and 308 million tons. The Office of Technology Assessment arrived at between 255 and 275 million tons of hazardous waste in 1983.

Whatever the source of information, the magnitude of the problem is very large. For example, the U.S. Navy generates about 1.5 kg (approx. 3.5 lb) per person per day of solid waste. Japan generates only about 1 kg per person per day. In addition, liquid wastes (black water, gray water, and bilge water) generated on board mission ships are very large, and this must be processed before discharging to sea, especially near to coastal areas. The disposal of both solid and liquid waste into ocean waters is becoming of increasing importance. The problem involved with the proper disposal of this waste in a densely populated ship city, that is environmentally acceptable, is very challenging. According to an EPA estimate, 80–90 percent of this waste is disposed of improperly. It is estimated that there are over 760,000 generators of hazardous wastes and about 40,000 produce less than 12 metric tons per year.

The composition of the solid waste varies significantly from source to source and also from season to season. The composition of the waste stream is therefore heterogeneous. Typical composition of the dry municipal solid waste (MSW) is given in Table 15.1.

The presence of moisture (which can vary from 10 to 66%) can have a significant influence on heating value of the waste. The heating value of dry MSW is given in Table 15.2 and is approximately 16.2 MJ/kg (or equivalently 6968 Btu/lb). As an example 10% increase in moisture will reduce the heating value of waste by about 1.67 MJ/kg (or equivalently 717 Btu/lb). The decrease in heating value (in Btu/lb) with increase in moisture content can reasonably be estimated from $6968 - 71.7x$, where x is the moisture content in weight percent. The average moisture content in material depends on the material type. Food waste may have moisture content of about 70%, while plastic and leather have very low moisture content of about 2%. There are seasonal variations of the moisture and energy content of the wastes. Therefore, the sorting of the material

Table 15.1 Characteristic Composition of the Dry Municipal Solid Waste [4]

Component	Average Content (% by weight)	Heating Value Dry (Btu/lb)
Food waste	33.5	6,528
Paper, cardboard	33.5	7,500
Plastics	9.2	14,000
Ferrous metal	7.8	300
Glass	5.8	250
Leather and rubber	4.8	13,000
Textiles and rags	3.7	7,652
Stones and ceramics	1.5	652
Nonferrous metals	0.2	13,000

Table 15.2 Comparison of Heating Values of Municipal Solid Waste with Other Wastes, Refuse-Derived Fuels, Biomass, and Fossil Fuels [4]

Waste Type (Fuel)	Heating Value (Btu/lb)	Heating Value (MJ/kg)
Cellulose	7,300	17.00
Lignin	9,111	21.20
Wood (pine)	9,600	22.30
Wood (oak)	8,296	19.30
Coal (subbituminous)	11,729	27.30
Peat	8,237	19.20
Municipal solid waste (dry)	6,968	16.20
Municipal solid waste (50% moisture)	3,380	7.90
Refuse-derived fuel (RDF)	7,942	18.50

cannot only provide near uniform chemical composition but also energy content. A comparison of heating values of municipal waste with various other waste fuels and biomass is given in Table 15.2. The data shows that the energy content in the MSW is comparable with some wood material and wastes. The energy associated with plastics is very high as compared to nonplastic wastes.

The composition of the solid waste has been changing over the past two decades. This is because of the changing needs of the society as well as some of the process and manufacturing needs and refinements. Table 15.3 shows the immediate past, present, and near-term projected refuse composition.

Irrespective of the type and amount of waste, the amount of wastes generated has been growing. From the energy availability there are significant opportunities.

Table 15.3 Average Composition of the Past, Present, and Near-Term Projected Wastes [5]

Composition (wt %, as discarded)	1970	1980	1990	2000
Paper	37.4	40.1	43.4	48.0
Yard wastes	13.9	12.9	12.3	11.9
Food wastes	20.0	16.1	14.0	12.1
Glass	9.0	10.2	9.5	8.1
Metal	8.4	8.9	8.6	7.1
Wood	3.1	2.4	2.0	1.6
Textiles	2.2	2.3	2.7	3.1
Leather, rubber	1.2	1.2	1.2	1.3
Plastics	1.4	3.0	3.9	4.7
Miscellaneous	3.4	2.7	2.4	2.1
Total	100.0	100.0	100.0	100.0

In rural areas the waste energy can be converted to useful electrical or mechanical energy. The environmental issues are of concern as the amount of pollutants emitted from wastes are often higher than that generated from clean fossil fuels, which also have high energy content. The emission of dioxins and furans are among the prime issues of concern from the thermal destruction of plastics. In the urban areas the energy recovery from wastes must be carried out in an environmentally benign manner. Many of the waste-to-energy conversion systems utilize extensive cleanup devices to the exhaust system. The emission of all the criteria pollutants must be met to the federal, state, and local emission standards. In some locations the local standards are more stringent than the federal levels. The allowed emission levels depend on the type and capacity of the system, year in which installed, and the type of fuel burnt. Table 15.4 gives some guidelines on the emission standards. It must be emphasized that the table may be used only as a guide. Many of the pollutants given here may have more stringent emission levels, depending on the size, type, and location of the plant, than the values given in the table. In addition the local and state governments may have additional emission regulations?

The energy recovery from wastes, although a challenge, is not really a major problem as compared to the environmental issues. Any waste recovery or conversion of waste to some other form of useful energy must be environmentally benign. The real challenge lies in making the system environmentally acceptable at reasonable costs so that one cannot only provide permanent disposal of the waste but also recover energy. The environmental issues are expected to become even more important in the foreseeable future so that opportunities from wastes will only continue to grow. We also envision significant opportunities from other types of wastes such as liquid wastes, plastics, chemical wastes, mixed waste,

Table 15.4 Pollution Emission Standards (at 7% O_2 and 293 K)

	Pollutant	Standard
Gas	Carbon monoxide (CO)	50 ppm
	Hydrocarbons (HC)	6.8
	Oxides of nitrogen (NO_x)	180
	Hydrogen chloride (HCl)	25
	Sulfur dioxide (SO_2)	30
Solid	Particulate	15 mg/dscm
Metal	Lead (Pb)	0.1 mg/dscm
	Mercury (Hg)	0.1 mg/dscm
	Cadmium (Cd)	0.01 mg/dscm
Organic	Dioxins/Furans	0.2 ng/dscm
Opacity		10%

and industrial wastes. Since the chemical and physical properties of each type of waste is often significantly different, research and technology efforts must be made for various kinds of wastes so as to achieve clean energy conversion.

15.3. SOURCES, DISPOSAL, AND RECYCLING OF PLASTIC AND OTHER WASTES

15.3.1. The Problem

In recent decades the increased use of plastics in domestic and industrial sectors including automobiles, homes, boats, and aircraft has caused serious concern on their proposed disposal. Plastics in automobiles play a major role in enhancing fuel efficiency through weight reduction and aerodynamic vehicle improvements, in addition to improving safety, durability, reduced cost, and design flexibility. As an example, plastics use increased from about 86 kg (approx. 190 lb) per vehicle in 1980 to over 150 kg (approx. 350 lb) in 1991. Increased amounts of vehicle plastics will increase automotive shredder residue (ASR) and the proportion of plastics in ASR. It is estimated that 2 billion old tires are in America's dumps and on its roadsides, with more than 200 million added every year. Tires have high energy content (about 30 MJ/kg or 13,000 Btu/lb) and are environmentally unfriendly due to unacceptable levels of pollutants emitted during their combustion. They are nonbiodegradable and under normal ambient condition they are indestructible. They attract mosquitoes and rats and cause fires that can smolder.

Without recycling options for automotive plastics and proper disposal of industrial, commercial, and residential wastes, the old practice of landfilling will see an increasing amount of materials that could otherwise be put to use. A reduction in the number of municipal landfills is escalating technology development in waste recycling and waste treatment. Improper disposal of household wastes, runoff of lawn and garden chemicals, and misuse and improper disposal of chemicals by businesses and institutions are important contributions to non-point-source discharges in many states. Some of the waste generated may also be hazardous. The users and disposers in many situations are unaware of the environmental and economic benefits to be derived from source reduction activities. Although many communities in the United States sponsor periodic waste collection programs and a few have permanent collection stations, only a small portion of the population is served in this way. The high costs of conducting household waste collection days have deterred wider implementation of such programs. Since household waste (both nonhazardous and hazardous) is not regulated, public education is a key ingredient in achieving proper management and reduction in the generation of these wastes. In many communities the household waste is presorted out at source into different categories, such as glass, cans, paper, plastics, and general waste. Hazardous wastes exhibit characteristics of reactivity, ignitability, toxicity, and corrosiveness and include many household cleaners, paint and related products, automobile-related wastes, pesticides, batteries, hobby items, and the like. Unfortunately, these wastes too often find their way into storm water and freshwater collection systems and eventually into streams and lakes. If these products are placed in municipal solid waste landfills or disposed of on landfills or disposed of on land, the potential of leachability and groundwater contamination, respectively, exists. If incinerated, these wastes may contribute to the hazardous emissions from stack and ash.

Regarding automobiles, engineers have done an outstanding job of meeting federal and local emissions standards, safety, and efficiency. However, today's vehicles are not designed for ease of disassembly and recyclability. Industry figures show that design for recycling initiatives implemented today will take about 12 years to impact on recyclability. Recovery economics for plastic parts/materials are directly related to ease of disassembly, volume of material recovered per part, time to decontaminate, and value of the material. The plastics recycling industry is in its infancy, with minimal sorting, washing, pelletizing, and compounding capabilities. Current plastics recycling processes are highly sensitive to contamination. This makes intermediate processing to remove contaminants absolutely necessary. Recycling of engineering-type resins, used in automotive applications, has been limited to clean, postindustrial scrap and postuse battery cases. Costs of dismantling and removing contaminants can be up to about $1.05/ kg of waste. The total cost associated with the recovery and reclamation, including the costs of transportation, distribution, and collection, can be up to $2.50/ kg of plastic.

15.3.2. Waste Disposal Options

Several waste disposal options used to date include direct landfilling, storage in surface impoundments, physical/chemical/biological treatment, chemical stabilization, and thermal destruction. The landfilling, storage, and mass-burn incineration are the options of the past, so that we can only expect to see an increase in rejection of these practices. The thermal destruction option offers an attractive permanent disposal option for waste because of maximum volume reduction (up to 99.9% of the original waste), energy recovery (heating value of compounds in the waste is in the range of about 12–24 MJ/kg or 5000–10,000 Btu/lb), and by-products may be used in a variety of ways. For the case of solid waste, presorting the waste and removing metals and glass can enhance the heating value. Pyrolysis, gasification, combustion, or some combination of these processes can describe almost all types of thermal destruction processes. Chopra et al. [1] and Oppelt [2] expand on these waste disposal options. Recent texts by Brunner [6] and Niessen [5] deliberate on these incineration problems.

Based on the type of thermal destruction process selected, there are several different commercial designs and configurations of the reactor that have been utilized for a particular application. Some of the most commonly used technologies include rotary kilns, starved air incinerators, fluidized beds, mass-burn incinerators, electrically heated reactors, microwave reactors, plasma, and other high-temperature thermal destruction systems. Recent advances include gasification and very high temperature steam reforming.

Low-temperature oxygen/air-enriched systems include rotary kiln, fluidized-bed, and mass-burn incinerators. These types of systems are sometimes suitable when there is no preprocessing of waste required and noncombustibles are removed prior to combustion of the waste. Fluidized beds require complex shredding and sorting of waste due to difficulty in the removal of waste from the bed material. They offer advantages for low NO_x pollution due to their low operational temperature, around 800–850°C (approx. 1472–1562°F). At this temperature the oxidation of atmospheric nitrogen is negligible compared to high-temperature operating systems. However, the thermal destruction efficiency is not as high as for systems operating at higher temperatures. In size the system is large, as is the expensive postprocessing of effluents.

Low-temperature starved air systems (e.g., rotary kiln, fluidized bed, and other reactors) have been used with varying degree of success both in the combustion and pyrolysis modes. These systems are less effective and require longer residence time, on the order of hours. This translates into larger size of the system. A secondary combustion chamber is required for the oxidation of gases formed from the pyrolysis reactions in the first stage. The advantages of easy recovery of metals and glass from the residue are often not enough to offset the increased size of the system, initial cost, and postprocessing of the effluent gases. High-temperature operating systems (such as plasma arc system) offer the advantage of reduced size of system and less pollution control devices, increased processing rate of waste, and reduced volume of the by-products.

The effectiveness and reliability of the thermal destruction system depends not only on the design and type but is critically dependent upon the operational parameters of the system, such as, operating temperature, heating rate, residence time, waste chemical composition, excess air, and chemical and thermal environment surrounding the waste. Lower pollution and higher thermal destruction efficiency requires controlled conditions throughout the waste heat up, pyrolysis, and combustion periods. Each of the thermal destruction processes involves a variety of equipment types and methods. Major differences in the processes arise from factors such as heat source, temperature, physical configuration, heat transfer medium, quantity of air used, and feed preparation.

Pyrolysis is controlled thermal decomposition of an organic material into one or more recoverable substances through the application of heat in an oxygen-free environment. The process is used to decompose cellulose, plastic, and rubber types of material into chemical products that can be reused while preserving their energy content. Pyrolysis should not be confused with incineration, which is the total oxidation of all the organic material. The end result of incineration is usually ash and metals, which may be used for landfill or construction. Pyrolysis is not new and has been used to reduce the stockpile of scrap tires in the United States and plastic and rubber waste in Europe. The plastics used in automobiles is roughly 22% polyurethane, 15% polypropylene, 13% polystyrene, 11% vinyl, 10% nylon, 5% polyester, and the remaining 24% resins. All these materials can be pyrolyzed for chemical and energy recovery. The materials from pyrolysis decompose into three major substances of pyro-gas, pyro-oil, and solid by-products. The gas is similar to natural gas in performance and can be further processed into thermoplastics and asphalt applications such as shingles or paring.

Gasification is a thermochemical process in which the organic substances are converted into combustible gases with the aid of some agent, such as air, oxygen, carbon dioxide, steam, hydrogen, or some mixture of these gases. Gasification has an advantage over pyrolysis since it promotes the char (residue) conversion to gas and therefore reducing the volume of solid residue and releasing chemical energy contained in the char. Gasification is a heterogeneous solid-gas reaction. The factors that influence the kinetics of reaction in pyrolysis also influence gasification. The gasification of municipal solid waste (MSW) gives rise to low to medium heating value gases as compared to high heating value gases in pyrolysis. However, the volume of gases evolved is much higher with gasification than pyrolysis. The heating value can be increased if the material is gasified in an oxygen-enriched environment. The gasification with air or oxygen has a significant effect on volume of gas produced and their associated heating value and the amount of remaining residue.

Steam reforming is also a thermochemical process in which the organic compound is converted into gas of very high chemical energy, such as hydrogen, using very high temperature air and steam. Steam assists in the production of hydrogen at very high temperatures. The amount and quality of high-energy gas produced depends upon the temperature of the reactants as well as the surrounding environment. It is expected that this process will lead to significant development and use in the near future as this process offers cleaner and highly desirable energy conversion.

15.3.3. Thermal Destruction (Incineration) of Wastes

Combustion of any moisture-containing organic material takes place in the four stages of drying, pyrolysis, gasification, and combustion of the volatiles and char produced. No comprehensive scientific kinetic or burning rate data are currently available for MSW. Most of the available data is on operational experiences with a particular system. The system is often custom made or developed by trial and error method. Significant amounts of aerospace-related research and development efforts have assisted in designing better thermal destruction systems for different kinds of solid, liquid, and gaseous wastes but much more is needed. Studies carried out on the wood, lignin, biomass, and biomass-related materials revealed that operating temperature, oxygen concentration, moisture content, size, and shape of feed particles have a significant effect on the burning rate (analogous to that found for various propellants). High operating temperatures or excess enthalpy combustion methods are favorable for devolatilization and evolution of gaseous products, which in turn assist in establishing the flame around the solid material. Available data show that the volatile burn time reduces by about 70% (i.e., from 40 to 12.5 s) with an increase in temperature from 900 to 1200°C (approx. 1650–2190°F temperature increase). It is expected that high operating temperatures (or ultra high temperatures within materials' limit of the incinerator) will reduce the residence time of the feed particles, and hence the size of the reaction chamber, for the same feed rate; see Chopra et al. [1].

Additional information regarding the effects of chemical composition of waste, surrounding gas composition, flow aerodynamics, turbulence levels, temperature, heating rate, residence of feed particles, radiation, and energy feedback (using, e.g., excess enthalpy flames) will assist in the design of advanced incineration systems. Generally speaking, if the waste can be incinerated, energy can be recovered from it by a properly designed system. However, some wastes contain high amounts of nitrogen (4–8% for polyurethane plastics), sulfur (from rubber), and chlorine (from plastics). This leads to excessive levels of NO_x, SO_x, chlorine, and HCl in addition to the formation of soot, dioxins, furans, and polycycloaromatic compounds, which are known to be carcinogenic and mutagenic. In addition, some waste also contains glass and metals, which could produce particulates, ash, and slag. The emission of some heavy metals can be health hazardous; see Oppelt [2].

15.3.4. Diagnostic Tools

The text by Gupta and Lilley [7] provides an extensive review of the flowfield modeling and diagnostic situation, with many citations to specialized studies. Eckbreth [8] and Chigier [9] also describe the measurement techniques. Basic flow visualization studies are undertaken so as to enable the main features of a flowfield to be characterized, including streamline patterns and recirculation zones. Usually a major portion of the flow domain is captured at once on film, and regions requiring more detailed study are quickly identified and subjected to more intense scrutiny. Recirculation zones are important to combustor designers, and the size and location of these regions in isothermal flows are readily deduced from flow visualization photographs of tufts, smoke, bubbles, and sparks responding to the experimental flowfield patterns. These techniques provide experimental characterization of corner and central recirculation zones.

Laser velocimetry (LV) and particle image velocimetry (PIV), for example, may be used for local particle and gas velocity measurements, particle size, and concentration fluctuation measurements. The diagnostic techniques for studying turbulence effects on pulverized coal combustion are fairly well developed. Measurement of particle and gas temperature in two-phase flows is a problem. Recent attempts on the measurement of mean and fluctuation particle temperatures, using two-color optical pyrometer, show good promise for measuring mean particle temperature greater than about 1000 K. Measuring particle temperature below 1000 K using a two-color pyrometer is a problem area because of poor signal–noise ratio. As long as the flow is optically thin, which is usually the case for model experiments, flow visualization (direct high-speed photography, shadowgraph, Schlieren interferometry, holography, speckle metrology, etc.) is not more difficult than for gas flows. Some techniques, such as direct high-speed photography, are even much simpler for two-phase combustion than for gas flames because of the strong radiation from flames. Particle sizing using photographic techniques requires the use of very bright light sources to pass through the luminous flame zone. These investigations are extremely time consuming and difficult. Multipoint conditionally sampled measurements have to be done for flame structure studies, combined probe and visualization measurements developed, and advanced data reduction techniques applied for evaluation and interpretation purposes in order to be able to extract information about individual events. Especially interesting in this context is the development of controlled excitation studies that provide the convenience of investigation of the details of coherent structures via phase lock on the induced perturbations.

Developments are expected to continue in the future in laser-based probes for flames and practical environments. In order to complete the flowfield description, temperature, major species densities, and fluctuating mass flux of the flame gases are required. In large-scale flames, optical access is a particularly difficult problem involving a full-scale dirty, vibrating, and confined environment. The development of compensating optics for aberrations introduced into the light beam is well overdue. The application of fiber optics and miniaturization of optics should be encouraged. Research activity on the miniaturization and their application to harsh and hostile environments are worthwhile.

The laser-based techniques currently being used and developed include vibrational Raman scattering, coherent anti-Stokes Raman scattering (CARS), Rayleigh scattering, laser-induced fluorescence, and planar laser-induced fluorescence (PLIF). This is an active research field at various research establishments. Laser-induced fluorescence spectroscopy has been used to measure several combustion intermediates, for example, CH, C_2, HCH, OH, NO, NO_2, HNO, CO, halogenated hydrocarbons, and polycyclic aromatic hydrocarbons.

15.3.5. The Combustion Aerodynamic Problem

Standard texts including the application of computational fluid dynamics to chemically reacting flow situations are available; see, for example, Gupta and

Lilley [7]. The discussion here is limited primarily to the fluid dynamics aspects and aimed at alternative fuels with different physical and chemical properties than conventional fossil fuels. Here, the following three flow domains are of particular importance and can be distinguished in any multiphase combustion system: flowfields connected with injection of fuel and air, flow regions dominated by free convection currents located far away from burners, and flow along the cooled walls of the combustion zone. Extensive reviews of the application of the classical turbulent combustion modeling methods have been presented for fossil fuels. For alternative fuels, it has been pointed out that there are major differences in the combustion of coal-derived liquids and shale oil in gas turbine combustors. Texts include Bartok and Sarofim [10] and Keating [11].

15.3.6. Computational Problem of Turbulent Reacting Flows

A complete and comprehensive mapping of a turbulent flame [e.g., mean and root-mean-square (rms) velocity, temperature, species concentrations, and their correlations] has, as yet, not been accomplished. The developments of correlations, such as the velocity–temperature correlation $u'T'$ (using, e.g., simultaneous LV or PIV and Raman scattering) should be encouraged and is now becoming a reality. This capability needs to be developed first for benchtop measurements prior to extending to scale-model measurements. Because of the highly nonlinear dependence of radiation energy flow on temperature, the average temperature is almost useless in computing local radiant heat flux. Techniques for measuring the mean and fluctuating local concentrations in large-size flames nonintrusively are also urgently required. A comprehensive set of such detailed measurements could allow the testing and development of mathematical models. Achievement of this goal is in sight and several investigators have demonstrated the feasibility of measuring many of the quantities and their correlations.

The development of model laws is urgently needed to account for the interaction between fluid dynamics and combustion. Both experimental and theoretical studies are needed. Extensive measurements should cover a wide range of dimensionless variables and include mean and fluctuating data and their correlations so as to aid the modeler. Empirical and physical hypotheses and laws need to be further investigated, verified, and refined. Computational studies should include macro- and micro-approaches, calibrate model laws and parameters with available data, and predict phenomena outside the range of previous experimental study. Here the efficacy of the modeling approach reveals itself. Regarding combustion studies, after success with fluid dynamic model laws with a given fuel, extension to a wider range of fuels and operating conditions are required.

There is a need to define and incorporate new finite-difference schemes and/or error reduction techniques into current computer code models. Improvements in accuracy of representation (differencing schemes) and rapidity of convergence (solution algorithms) are being sought before computer codes can give quantitative, rather than qualitative, predictions. Better numerics, improved understanding of inlet flow, improved turbulence modeling, and improved understanding of unsteady problems are required. Techniques for improved accuracy (including several bounded versions of skewed upwind differencing), improved versions of the pressure–velocity coupling technique, and methods for speeding convergence of iterative procedures are needed. Further information about turbulent flame understanding may be found in the literature [12–20].

15.4. INCINERATION OF CELLULOSE AND SURROGATE SOLID WASTES

15.4.1. Preamble

Experimental and numerical studies presented here are on the thermal destruction behavior of cellulose and surrogate solid waste. An experimental study was conducted in a controlled mixing history reactor (CMHR) using plasma gas as the heating device [21]. The effect of pyrolysis temperature, waste properties, residence time, and gaseous environment surrounding the waste has been examined. Gas generation rate, chemical composition, and heating value of evolved gases as well as the solid residue remaining after exposing the waste to prescribed environment during pyrolysis have also been examined.

Equilibrium thermochemical calculations were carried out using STANJAN and SOLGASMIX computer codes to provide information on the thermal destruction behavior of samples of surrogate solid waste and its excursions at different temperatures. Calculated results show good trends with the experimental data and can therefore be used as a guideline to describe the experimentally observed results on the thermal destruction behavior of the wastes. The results show that temperature and chemical composition surrounding the waste are important parameters for the pyrolysis process. These parameters affect destruction rate, solid residue remaining after pyrolysis, gas yield and its chemical properties, and pollutants and metals emission. The results also show a significant influence of waste properties and operational conditions on the waste destruction characteristics and products formation. Cellulose and chosen surrogate waste showed significant differences in the thermal decomposition behavior and products formation.

15.4.2. The Problem

Environmental catastrophes resulting from improper treatment/disposal of different types of wastes have caused increased public awareness of the growing problem of waste generated in all sectors of the public, industry, and government. Waste minimization and recycling can only provide a partial solution to the growing problem. The United States generates approximately 200 million tons of solid waste every year (about 4 lb/person/day) and this amount are projected to increase at a rate of 1% annually [21]. Therefore, stringent measures must be taken in order to provide a better and permanent solution to the problem. Thermal destruction of wastes can provide an increasingly important role in this area. This includes the application of three fundamental reacting processes of pyrolysis, gasification, and combustion. The existing destruction technologies that have been used include mass-burn-type incineration, fluidized bed, rotary kiln, molten salt bed, low- or high-temperature oxygen/air enriched systems, and low- or high-temperature starved air systems. More recently electric heating, microwave, and plasma-assisted systems have also appeared [22]. Thermal destruction offers distinct advantages over the other methods since it provides maximum volume reduction, permanent disposal, and energy recovery while the by-products can be used in several ways, such as material for building and roadbed construction [23]. For certain waste streams, under certain conditions, the by-product material can be very hard. As an example, the titanium and nitrates present in the waste material can form titanium nitrate at high temperatures. This is a very tough and strong material. This can be possible only with a controlled process so that the compound formed may be isolated from the other compounds in the by-products. Of all the permanent treatment technologies, thermal destruction provides the highest overall degree of destruction. In addition it provides maximum volume and mass reduction, maximum energy recovery, and the by-products can be nonleachable.

The disposal of municipal solid wastes (MSW) has traditionally been via landfills (about 83% of the waste generated) since the method is most convenient. Some of the gases released by this method [e.g., greenhouse gases and volatile organic compounds (VOCs)] are high and unacceptable. In addition the odors released into the environment are unacceptable. Other methods for waste disposal include incineration (6%) and recycling (11%) [24, 25]. The most common practice for the disposal of MSW has been landfill. The landfill disposal, therefore, creates the problem of odor, generation of toxic and other gases (e.g., methane and carbon dioxide), and intrusion of leachate generated from landfill site into soil and groundwater [26]. With the land prices continuously escalating and land becoming scarce, this method is unsatisfactory in addition to the environmental needs. As for the thermal destruction, special interest in air toxic organic pollutants and trace metals emission from incinerators came after the risk assessment findings toward human life [24]. Some of these metals (e.g., arsenic, cadmium, chromium, and beryllium) are very hazardous to humans in addition to being carcinogenic. In addition to the concern over pollutants such as NO_x, SO_2, HCl, CO, CO_2, unburned hydrocarbons, soot and particulates, the emission of dioxins, furans, volatile organic compounds, and metals have received increased attention from many countries around the world. The concern over pollutants, produced as by-products from direct result of combustion process [27], is common to all incineration systems.

While several methods are being used to treat the wastes, incineration has been widely used to provide the highest degree of destruction [21, 22, 24] for a broad range of waste streams, even though combustion contributes to pollution. Thermochemical behavior of cellulose and surrogate solid wastes is provided here. The goal is to provide the further knowledge and tools to destroy solid wastes while simultaneously providing energy recovery and reduction of toxic by-products.

The previous studies [21, 22, 27, 28] have shown that the formation of low-molecular-weight gases at elevated temperatures leads to a reduction in the molal mass of the product gas mixtures (e.g., by more than 100% at temperatures approaching 6000 K as compared to 1000 K). At these temperatures, destruction of the waste to molecular level occurs. Pyrolysis at elevated temperatures, using, for example, plasma gas, is most suitable to thermally destruct the solid wastes. Results also show that the gas composition from pyrolysis is significantly affected by the temperature and chemical properties of the solid waste material. In addition there is a dramatic increase in the volume of gas generated and heating value of the gas from pyrolysis at pyrolysis temperatures above 3000 K. There is a 250–300% increase in the volume of gas produced at temperatures approaching 6000 K as compared with 1000 K. The heating value of the gases generated is increased by about 225–350% over the same temperature range. The amount as well as heating value of the gases can be controlled via oxygen enrichment to air. Based on these observations, ultra-high-temperature operating systems, such as plasma arc systems, appear promising for the disposal of solid wastes, in particular when the space requirements are of concern. The volume of gas generated with combustion is much less with oxygen than normal air [29]. However, high-temperature chemistry and chemical kinetics are not fully understood, and comprehensive information on the high-temperature chemical kinetics is lacking. A thermochemical database for the solid wastes is required for the advanced high-temperature thermal destruction system. Therefore, fundamental studies must be carried out to obtain the basic information on the thermal destruction of solid wastes. Most studies in the literature have been on systems, so that any data at the fundamental level will assist in providing good understanding of the thermal destruction process. Comprehensive studies require details of chemistry and fluid dynamics for both organic and inorganic portions of the wastes. Information on cellulose and surrogate waste helps provide information on real wastes. It must be recognized that real wastes have poorly defined spatial and temporal composition so that scientific information can only be gained from surrogate wastes. Emphasis has been placed on the pyrolyzed gas characteristics as influenced by the waste composition, pyrolysis temperature, and gaseous environment surrounding the solid waste. Special interest has been on examining the effect of waste properties (as affected by presorting the waste) and operational parameters (surrounding temperature and chemical composition) on solid residue and product gas composition during pyrolysis using experimental and numerical studies.

15.4.3. Experimental Facility and Methodology

Experiments on the pyrolysis of cellulose were carried out in a controlled mixing history reactor (CMHR), which is a plasma-operated drop-tube furnace capable of operating at temperatures up to 2500 K. A schematic diagram of the CMHR is shown in Figure 15.1. The graphite core reactor tube is 2 inches in diameter and 60 inches long. The central test section of the reactor has two 24-inch long

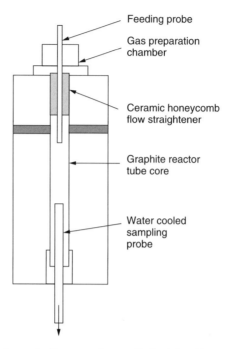

Figure 15.1. A schematic diagram of controlled mixing history reactor (CMHR).

$\times \frac{3}{4}$-inch wide diametrically opposite windows, which provide the desired optical access to the test section in the reactor. In addition it has two viewing ports of 1.25 inches in diameter located near the end of the test section. The reactedness of the graphite core under reducing conditions is negligible while that under oxidative conditions is less than 1% at the conditions examined. The reactor is heated with a nominal 40-kW nontransferred arc plasma torch. The desired gas temperature and composition within the reactor is achieved by diluting the high-temperature plasma gas with some dilution gas. The dilution gases used include argon, helium, hydrogen, nitrogen, air, carbon dioxide, or any mixtures of these depending on the type of environment (inert, oxidizing, or reducing) to be used. The temperature and chemical composition of gas in the test section of the reactor is therefore controlled. Particles of surrogate solid waste material are allowed to fall in the downward direction in the reactor. A transpiring wall water-cooled sampling probe, inserted from the bottom of the reactor, intercepts the particle stream after a desired residence time of particles in the high-temperature zone. Gaseous and particulate products are isokinetically sampled at various axial positions in the reactor. The reactor is allowed to move in a vertical direction relative to the fixed position of the sampling probe. The residence time of the particles in the high-temperature zone of the reactor is therefore controlled. The reactor has the capability to examine a wide range of materials and particles sizes exposed to different residence times, chemical environments, and temperatures.

A hot gas preparation chamber located at the top of the reactor allows for a thorough mixing of the plasma gases with inert gases to produce a uniform temperature and composition of the carrier gas. Varying the amount of gas (such as, N_2, Ar, O_2) to the gas preparation chamber provides control of gas composition inside the reactor. A water-cooled feed probe is used to feed the surrogate solid waste into the reaction chamber. This is located at the centerline of the hot gas preparation chamber. The waste particles are therefore kept isolated from the high-temperature environment until they are exposed in the test section for their thermal destruction. The particle feeder is essentially of a fluidized-bed type. The solid waste particles in the bed are elutriated with the carrier gas (argon) and flow through the bed. The design enables good control over the particle flow rate.

The transpiring wall water-cooled sampling probe design allows one to collect the gases and solid residue material under isokinetic conditions after the material has been exposed to high temperatures for a prescribed residence time. The solid products enter the probe where they are further quenched with argon gas flowing radially inward and then downward through the probe. This freezes the chemical composition of the incoming material instantly.

A cascade impactor is attached to the sampling probe exit to collect the solid residue for subsequent analysis. This provides information on weight loss as a function of residence time of the solid material in the high-temperature zone. The cascade impactor has six stages that collect solid material according to bin sizes down to 0.2 μm. The cascade impactor therefore separates out the particles into a number of size bins. The separation of the particles is accomplished by passing the collected gases (containing the particles) through orifices of successively smaller size diameter. Larger particles are inertially collected on the first collection plate while the smallest size on the last collection stage. The residue gases are pumped using EPA method 23 sampling apparatus and then analyzed using on line gas analyzers and a gas chromatograph.

The thermal destruction behavior of both cellulose and surrogate solid waste has been examined. The surrogate waste stream represents conditions wherein 90% of the food waste has been removed from the waste stream for pulping prior to thermal processing. The chemical constituency of the waste stream, although somewhat simplified, is quite realistic. The food waste entering the thermal destruction facility represents nonpulpable items such as corn cobs, bones, and food residue contaminated with metal, glass, and paper fraction of the waste. The steel component is mostly tin cans, and this has been simplified to pure iron. This simplification, although unacceptable for the slag chemistry, is reasonable for examining the thermal destruction behavior of the wastes. The glass fraction of the waste has also been simplified. The small amount of aluminum in the waste has also been omitted. The alkali content (Na and K) was generalized to Na_2O while the alkali earth content (Ca and Mg) was generalized to CaO.

After the bulk of the food waste (90%) has been removed from the given solid waste, the waste material will have the following composition: 6.75% food material (30% bone representing 75% hydroxyapatite and 25% organics, and 70% food representing 50% organics and 50% water), 62.22% paper (consisting

Table 15.5 Elemental Composition of Surrogate Solid Waste (SSW) and Its Excursions

	SSW	Without Paper	Without Food	Without Steel	Without Aluminum	Without Glass
C	25.02	4.00	25.20	29.90	26.87	27.11
H	4.83	1.31	4.65	5.77	5.18	5.23
O	41.71	19.40	40.72	49.89	44.80	41.30
N	0.13	0.33	0.00	0.15	0.14	0.14
Ca	1.24	3.29	0.68	1.49	1.34	0.66
P	0.28	0.74	0.00	0.34	0.30	0.30
Si	2.68	7.10	2.88	3.21	2.88	0.00
Na	0.81	2.13	0.86	0.96	0.87	0.00
Fe	16.40	43.40	17.59	0.00	17.62	17.77
Al	6.91	18.29	7.41	8.27	0.00	7.49

of 85% cellulose and 15% water), 16.4% steel (100% pure iron), 6.91% aluminum (100% pure), and 7.72% glass (consisting of 74.19% silica, 14.21% sodium oxide, and 11.58% calcia). The above surrogate waste has been examined here as complete surrogate or some systematic removal of some components from the waste. The elemental composition of the surrogate solid waste and its excursions are calculated. The thermal destruction characteristics of this waste have been examined using equilibrium codes (STANJAN and SOLGASMIX) and the plasma-assisted controlled mixing reactor. The results, shown in Table 15.5, are used to calculate the thermochemical equilibrium conditions.

15.4.4. Some Results

The organic portion of solid wastes has a carbon–hydrogen–oxygen ratio similar to that of cellulose $(C_6H_{10}O_5)_n$, a polymer that is widely present in nature [21, 25, 26, 30]. It is therefore expected that the thermal destruction behavior of cellulose will provide useful information on the organic portion of the solid waste. A pyrolysis process can be represented as:

$$\text{Hydrocarbons (waste)} + \text{heat} \Rightarrow \text{gaseous products} + \text{solid char}$$

The pyrolysis products include both the gas and solid phases. For gas phase, it is mostly carbon monoxide (CO) and hydrocarbons (HC). Experimental and calculated results obtained here show that an increase in pyrolysis temperature results in the formation of lower molecular weight hydrocarbon gases as well as a decrease in the fraction of solid residue. In contrast the calorific value and the volume of pyrolyzed gases increases.

Pyrolysis experiments on cellulose have been performed in the CMHR using the plasma jet as heat source. The plasma flow rate was set at 25 liters/min while the plasma power was varied from 3.3 to 32 kW. The temperature in the test

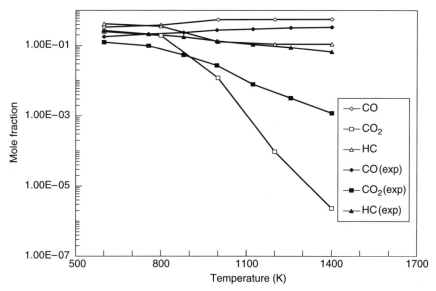

Figure 15.2. CO, CO_2, and HC concentration in gases produced during pyrolysis of cellulose.

section of the reactor was varied from 600 to 1400 K. The measurement results on the concentration of gaseous species (CO, CO_2, and HC) generated from the pyrolysis of cellulose in the controlled mixing history reactor are shown in Figure 15.2. The results are also compared with the calculated results. The results show an increase in CO and a decrease in CO_2 and unburned hydrocarbons with an increase in pyrolysis temperature. This is an indication of dissociation of the cellulose into low-molecular-weight products (e.g., CO, CO_2, and unburned hydrocarbons) at high temperatures. The calorific value of the gases at high temperatures is high, thus indicating a good source of energy recovery. Good trends were obtained between the experimental and calculated results except for the CO_2 that had very low concentrations at higher temperatures.

The solid residue from pyrolysis experiment was collected, measured, and further tested to examine its characteristics. The amount of solid residue collected at different temperatures is shown in Figure 15.3. It can be seen that the amount of solid residue collected decreases rapidly with an increase in pyrolysis temperature. This is due to the high heating rate that results in lower activation energy and a larger preexponential coefficient of the devolatilization reaction [31]. Further analysis showed that the solid residue from cellulose pyrolysis is primarily pure carbon.

The devolatilization reactions during pyrolysis will be completed only when the heat has fully penetrated the waste particle [32–34]. Therefore, the size and shape of the solid waste particles are important parameters that affect the pyrolysis reactions. The fed particles must have a characteristic size that is designed for a particular operating temperature, heating rate, and residence time.

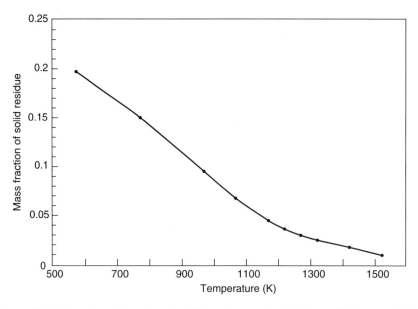

Figure 15.3. Experimental results of solid residue produced during cellulose pyrolysis.

Particle size and density are two important factors that affect pyrolysis since they determine the particle velocity and residence time in the CMHR. The effect of particle size and density was studied here using flow visualization. The flow visualization tests were performed on cellulose and surrogate solid waste particles as they travel downstream in the reaction zone of the CMHR. These tests provided the direct effect of temperature on residence time of particles in the reactor. Particle imaging was performed in the reactor with a charge coupled device (CCD) camera and frame grabber using Global Lab Image software and a computer. Particle images were taken under both nonburning and burning conditions. The flow inside the reactor was laminar. Near single particles were introduced into the furnace so that the equivalence ratio was low (fuel-lean). Sample images are shown in Figure 15.4. These images allow one to analyze the evolutionary behavior of solid particles, both temporally and spatially, as they travel through the test section of the reactor. The effect of particle size and density on the particle velocity, residence time, and transport behavior can be extracted from these images. It is found that, in general, larger cellulose particles move faster in the reactor and therefore have a shorter residence time as compared to the smaller size particles. The images also show a decrease in velocity as the particles travel downstream, in particular under the high-temperature environment of the reactor. As an example, at a temperature of 1286 K, the velocity of a 440-μm cellulose particle changes from 0.46 to 0.39 m/s within the viewing area of the camera. However, the velocity of a 400-μm particle is almost constant in the same viewing area at low ambient temperatures. Some scatter in the data (e.g., change in diameter for a given temperature or velocity for a given size) is attributed to the

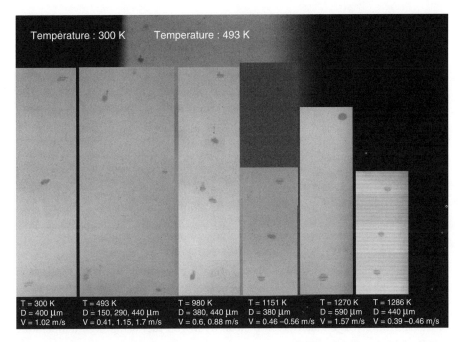

Figure 15.4. Sample images of cellulose particles in the CMHR at different temperatures.

nonspherical nature of the particle. The solid particles are nonspherical so that the recorded image plane of the particle does not provide the true information on the particle size. Furthermore there may be some variation in particle density that will then affect the particle velocity. Significant differences are expected to exist between the particle and the surrounding gas that will alter the particle kinetics. In this study no attempt was made to determine the particle temperature. This effect is important for the kinetics and should be examined in the future.

The effect of moisture contained in the cellulose sample on gas product generation during pyrolysis was examined. A sample containing 75% cellulose and 25% water by weight was pyrolyzed in the CMHR. The results showed a decrease in CO production and an increase in CO_2 and unburned hydrocarbons as seen from the results presented in Figure 15.5. This can be explained by the following water–gas shift reaction that enhances at elevated temperatures (above 1000 K):

$$H_2O + CO \Longleftrightarrow CO_2 + H_2$$

This reaction affects the formation of CO and CO_2 during pyrolysis since the carbon monoxide reacts with water to form hydrogen and carbon dioxide. The influence of intermediate radical species (such as OH) can have an influence on the above reaction. This influence as well as the influence of other species was not considered due to the complex interaction of various intermediate species. At high temperatures production of hydrogen occurs. This leads to new possibilities

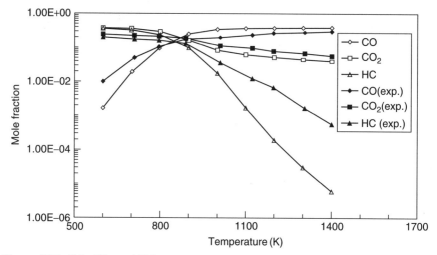

Figure 15.5. CO, CO_2 and HC concentration in the gases produced during pyrolysis of 1 mol of cellulose and 3 mol of water in CMHR.

of transforming waste materials into clean hydrogen fuel using high-temperature steam and air.

Oxidative pyrolysis of cellulose containing moisture was examined in the CMHR in order to determine the presence of air, oxygen, and moisture during the pyrolysis process. A test sample containing 75% cellulose and 25% moisture by weight was pyrolyzed in a gaseous environment that contained 1.428 mol of air for each mole of cellulose. Numerical calculations were performed according to:

$$C_6H_{10}O_5 + 3H_2O + 0.3(O_2 + 3.76N_2) \Longrightarrow \text{Products}$$

The results show a reduction in CO and unburned hydrocarbon concentration and a corresponding increase in CO_2 production with the addition of air and moisture, as compared to the pure pyrolysis case. The calculated results showed similar trends; see Figure 15.6.

The amount of pyrolyzed gases generated as well as their composition and heating value are important for the design and development of waste destruction systems and to control the emission of pollutants. Significant increase in both the heating value and volume of gases produced during cellulose pyrolysis can be seen with an increase in pyrolysis temperature; see Figure 15.7. At temperatures below 3000 K, large amounts of H_2 and CH_4 are found. At temperatures greater than 3000 K there is a sharp increase in the H species (atomic hydrogen) and C species (carbon gas) formation due to the thermal decomposition of CH_4, CO_2, CO, and other higher molecular weight gases; see Table 15.6.

In addition, it is also to be noted that no solid char residue remains at such high temperatures. The increase of low-molecular-weight gases at temperatures greater than 3000 K causes a dramatic increase in the heating value of the gas and overall

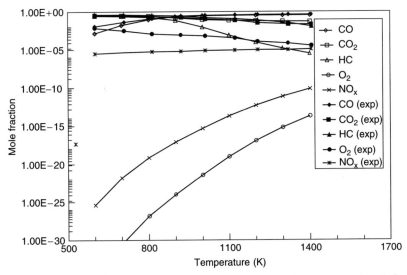

Figure 15.6. Concentration of CO, CO_2, NO_x and HC in the gases produced during pyrolysis of cellulose with water and air addition in CMHR.

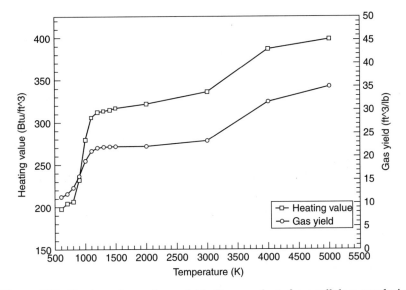

Figure 15.7. Heating value and gas yield of gas products from cellulose pyrolysis.

volume of the gas produced. Table 15.7 shows the molal mass of gases produced during the pyrolysis of cellulose at temperatures in the range of 800–6000 K. It can be seen that an increase in pyrolysis temperature significantly reduces the molal mass of the pyrolyzed gas mixture. The change in chemical composition at various temperatures is given in Table 15.6.

Table 15.6 Cellulose Pyrolysis Products at Different Temperatures

				Mol/Mol of Cellulose				
T (K)	CH_4	CO	CO_2	H_2	H_2O	C(s)	C(g)	H
1000	0.08	4.43	0.18	4.64	0.20	1.30	1.53×10^{-25}	1.71×10^{-7}
	6.22×10^{-3}	4.99	1.68×10^{-3}	4.98	4.32×10^{-3}	1.00	1.94×10^{-16}	
1500	9.74×10^{-4}	5.0		5.0	2.93×10^{-4}	0.99	3.33×10^{-10}	1.23×10^{-4}
	5.32×10^{-4}	5.0	6.49×10^{-5}	4.85	5.0×10^{-4}	0.99	1.28×10^{-6}	
2000	1.17×10^{-4}	5.0		4.46	1.82×10^{-5}	0.99	5.71×10^{-4}	1.14×10^{-2}
	3.43×10^{-6}	5.0	1.37×10^{-5}	1.41	1.20×10^{-6}	0.03	0.97	
2500	2.66×10^{-10}	5.0		0.14	4.13×10^{-6}	0.00	1.0	0.13
3000			2.83×10^{-6}					1.08
4000			5.02×10^{-7}					7.18
5000			1.57×10^{-5}					9.71

Table 15.7 Molal Mass of Product Gases from Pyrolysis of Cellulose

	Molal Mass (kg/kmol of cellulose)					
Pyrolysis temperature (K)	800	2000	3000	4000	5000	6000
Gas molal mass	25.268	18.008	14.392	11.110	10.221	9.860

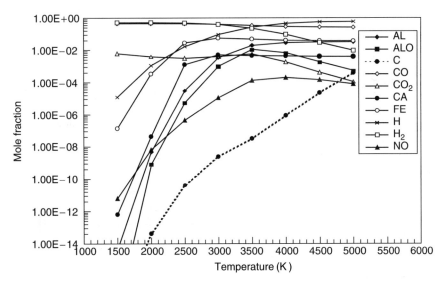

Figure 15.8. Gas species generated from pyrolysis of surrogate waste No. #2.

Chemical equilibrium calculations were carried out for the pyrolysis of surrogate solid waste and its excursions at different temperatures [35–37]. The number of species can be selected depending on the data file. However, in the present study only 20 species were considered for the STANJAN code. The calculated results are shown in Figures 15.8–15.10, which include pyrolyzed gas composition (Fig.15.8), gas heating value (or energy recovery; Fig. 15.9), and pyrolyzed gas generation rate (or solid residue fraction; Fig. 15.10) as affected by the pyrolysis temperature. The results have been compared with cellulose pyrolysis. The effect of temperature on gas composition, gas yield, and gas heating value during pyrolysis of surrogate solid waste are similar in trend to those obtained for cellulose pyrolysis. High heating value of the pyrolyzed gases is produced at high temperatures from the surrogate solid wastes as seen from Figure. 15.9. At temperatures below 3000 K there is a large percentage of H_2, CO, and CO_2 in the pyrolyzed gas. The increase of low-molecular-weight gases at temperatures greater than 3000 K causes a dramatic increase in the overall volume of the gas produced, as seen from Figure. 15.10. The results obtained have shown that

654 THERMAL DESTRUCTION OF WASTES AND PLASTICS

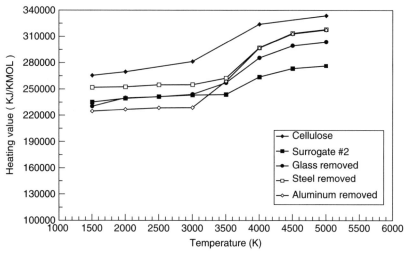

Figure 15.9. Heating value of the gases produced during pyrolysis of different surrogate wastes.

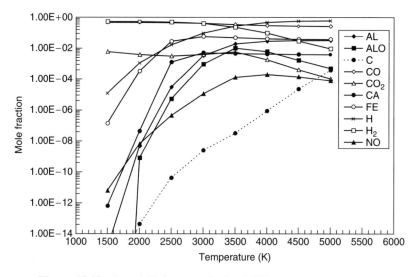

Figure 15.10. Gas yield from pyrolysis of different surrogate wastes.

the initial composition of waste material has a significant effect on the thermal destruction process. The removal of some constituents from the waste results in different products of pyrolysis as well as thermal and chemical characteristics. The results shown in Figures 15.9 and 15.10 show that the presence of paper in the waste provides a major role in both the gas yield and gas heating value from the pyrolysis of surrogate solid wastes. Removal of paper from the solid waste causes a large reduction in both the gas yield and gas heating value. However, with the steel removed from the waste, the gas yield is very close to that of cellulose, while the heating value of the gas is remarkably higher. This indicates that removal of steel from the surrogate solid waste as pretreatment approach provides significant measure for an efficient thermal destruction of the solid wastes.

15.4.5. Closure

Thermal destruction behavior of cellulose $(C_6H_{10}O_5)_n$ and surrogate solid waste has been examined both theoretically and experimentally. Theoretical calculations were conducted using STANJAN and SOLGASMIX equilibrium codes. Experimental studies were carried out using a plasma-gas-operated controlled mixing history reactor. There was a general qualitative agreement between the calculated results under equilibrium conditions and those obtained experimentally. Pyrolysis temperature, surrounding gas chemical composition, and waste composition affected the pyrolysis behavior of surrogate solid wastes. At high temperatures (>3000 K), a large amount of lower molecular weight gas species (mostly atomic species) were generated. Significant increase in gas yield and gas heating value is also found at high temperatures.

The amount of solid residue remaining also decreases quickly with increase in pyrolysis temperature. This suggests lower activation energy and a larger preexponential coefficient of the devolatilization reaction under high-temperature conditions. Removal of metals from the surrogate solid waste yields pyrolysis behavior similar to that obtained for cellulose. However, removal of paper from the waste results in a reduction in both gas yield and gas heating value. The results provided here are aimed at providing guidelines for the design and development of advanced thermal destruction systems.

15.5. THERMAL DESTRUCTION OF PLASTIC AND NONPLASTIC SOLID WASTE

15.5.1. Preamble

Experimental and theoretical studies are presented from a laboratory-scale thermal destruction facility on the destructive behavior of surrogate plastic and nonplastic solid wastes. The nonplastic waste was cellulosic while the plastic waste contained compounds such as polyethylene, polyvinyl chloride, polystyrene, polypropylene, nylon, rubber, and polyurethane or any of their desired mixtures. A series of combustion tests was performed with samples containing varying composition of plastic and nonplastic. Experimental results are presented on combustion parameters (CO, excess air, residence time) and toxic emissions (dioxin, furan, metals).

Equilibrium thermochemical calculations are presented on the thermal destruction behavior of samples under conditions of pyrolysis, combustion, and pyrolysis followed by combustion. Special interest is on the effect of waste properties and input operational parameters on chemistry and product composition. STAN-JAN and SOLGASMIX computer codes were used in the chemical equilibrium study [35, 36].

Analysis and interpretation of the data reveal the effect of waste feed composition on combustion parameters and dioxin, furan, and metals emission. Equilibrium calculation results are used to describe the experimentally observed trends for the thermal destruction behavior of these wastes. The results show significant influence of plastic on combustion characteristics, and dioxin, furan, and metals emission.

15.5.2. The Problem

Numerous environmental catastrophes resulting from the improper disposal practices of the wastes of different kinds have caused increased public awareness of the growing problem of waste generated in all sectors of the public, industry, and government. Waste minimization and recycling provides only a partial solution. Further stringent measures must be taken to solve the problem of waste disposal completely. The United States generates approximately 0.4×10^{12} lb of waste every year, and even with extensive waste minimization plans, this amount is projected to increase at a rate of 1% annually [26].

Thermal destruction in a broad scope includes the application of three fundamental reacting processes of pyrolysis, gasification, and combustion. Thermal destruction technologies include mass-burn-type incinerators, fluidized-bed, rotary kiln, molten salt bed, low- or high-temperature oxygen/air enriched systems, and low- or high-temperature starved air systems. More recently electric heating, microwave, and plasma-assisted systems have also appeared [38]. Thermal destruction offers distinct advantages over the other methods since it provides maximum volume and mass reduction, permanent disposal, energy recovery, and the by-products can be used in several ways in building material and roadbed construction [25]. Of all the waste treatment technologies, thermal destruction provides the highest overall degree of destruction. Any thermal destruction technique selected must be environmentally clean and accepted by the public since it is well known that the by-products can be health hazards and detrimental to our environment. As an example, emissions of toxic gases, metals, and dioxins are not perceived to be acceptable by the public.

Municipal solid wastes (MSW) presently use landfills (83%), incineration (6%), and recycling (11%) for waste disposal [23, 24]. The most common form for the disposal of MSW continues to be landfill. This form of landfill disposal creates problems of odor, methane, carbon dioxide, and toxic gases. In addition the leachate generated from landfill site can provide intrusion into soil and groundwater [39]. Special interest in air toxic organic pollutants and trace metals emission from incinerators came after their risk assessment findings toward human life [23]. As an example, some of these metals are very hazardous to humans (e.g., arsenic, cadmium, mercury, chromium, and beryllium were found to be carcinogenic). Carbon monoxide is poisonous gas and acid gases such as

NO_x, Cl_2 and SO_x cause corrosion. In addition to the above concerns, emission of pollutants such as HCl, CO_2, unburned hydrocarbons, soot, particulates, dioxins, furans, and volatile organic compounds and metals have received increased attention from many countries around the world. The concerns over pollutants, which are produced as by-products from the direct result of the combustion process [40], are common to all combustion sources.

The experimental and numerical investigation described here explore the thermal destruction behavior of different plastic and nonplastic mixtures in a laboratory-scale facility and its effect on carbon monoxide, particulates, dioxin, furan, and toxic metals emission. The results show that the composition of the waste has a significant influence on the emissions characteristics. The results also show that by using a suitable combination of various components in the waste, enhanced burning of waste occurs with reduced toxic emissions and solid residues.

15.5.3. Experimental Facility and Methodology

A schematic diagram of the experimental facility is shown in Figure 15.11. The laboratory-scale thermal destruction facility (TDF) consists of a destruction chamber, controls, and instrumentation for TDF, EPA method 23 sampling train, continuous gas monitoring system for gas analysis, scanning electron microscope (SEM) for fly ash metal analysis, and neutron activation method (NAM) for bottom ash analysis. The chamber is 12 × 12 × 18 inches in size. A natural gas burner is used for startup and auxiliary burning. Inside the chamber was lined with a 2-inch-thick insulating refractory cement, type ASTM C449, to minimize

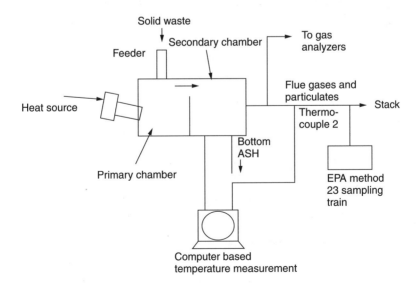

Figure 15.11. Schematic diagram of laboratory-scale thermal destruction facility.

heat transfer to the surroundings. A rotary feeder driven by an electric motor achieves continuous feeding. The fly ash and product gases leave through an exhaust pipe while the bottom ash is removed through the ash and slag pipe located at the bottom of the chamber. Gas-sampling and temperature-measuring ports are provided on the side and exit of the TDF for continuous monitoring and measurement of operating parameters. The TDF has all the necessary provisions to study the effect of operating parameters, such as fuel type and size, feed rate, excess air, residence time, and temperature on thermal destruction and products formation.

The combustion air and fuel gas flow to the gas burner were measured with flow meters. The flow rates were corrected for temperature and pressure pertaining to experimental conditions. The waste feed rate is calculated from the speed of the feeder motor. Two digital temperature indicators are used for continuously monitoring the temperatures in the furnace and exhaust duct, respectively. The furnace temperature is measured using a type R [Pt—Pt/13% Rhodium] thermocouple of wire size 0.01 inch while the exhaust gas temperature is measured using a type K (Cr—Al) thermocouple having 0.025 inch wires diameter. A microprocessor-based data acquisition and analysis system was used to measure the instantaneous and time-averaged temperature in the combustion chamber and exhaust duct.

Experiments were designed to study the effect of varying plastic and nonplastic fractions in the waste on emission characteristics. Seven different samples were examined, with chemical compositions of the nonplastic and plastic materials and their heating values being given in Table 15.8. The overall chemical

Table 15.8 Composition of Examined Nonplastic and Plastic Samples

Sample No.	Nonplastic (% volume)	Plastic (% volume)
1	100[a]	0
2	85	15
3	50	50
4	50	50[b]
5	50	50[c]
6	50	50[d]
7	0	100[e]

[a] Paper (75%) and cardboard (25%) by mass.
[b] Plastic A: mixture without polyurethane and latex.
[c] Plastic B: mixture without polyvinyl chloride.
[d] Plastic C: mixture without nylon, latex, polyurethane, and acetate.
[e] Plastic mixture of polyethylene (73%), PVC (2%), polystyrene (4%), polypropylene (8%), polyethylene (4%), nylon (2%), latex (2%), acetate (2%), and polyurethane (3%) by mass.

composition of the nonplastic sample 1 waste was $C_6H_{10}O_5$ while that for the 100% plastic waste was calculated to be $C_8H_{14}O_{14.5}N_2Cl$ based on the composition for sample 7. The bulk density of all the samples, determined by weighing a known volume of the sample, was found to be between 5.4 and 8.5 lb/ft^3.

The TDF chamber was heated up with a natural gas burner and kept in pilot running mode to achieve the desired temperature. The air flow rate was kept constant at 6.75 scfm and the mixture feed rate varied from 0.128 to 0.441 lb/min as a result of variation in the heating value of plastic and nonplastic in each sample. This procedure allowed TDF temperature to be maintained in the range of 1400–1600°F. As a result of this fuel feed rate variation, the equivalence ratio and gas residence time varied from 0.58 to 0.93 and 0.67 to 1.00 s, respectively. The combustion process was, therefore, held in a pseudo-steady-state with the TDF temperature maintained in the range of 1400–1600°F. During each test, fly ash was collected while the bottom ash was collected at the end of the test. The online monitoring of CO during all the tests was carried out using an infrared CO analyzer.

EPA method 23 (modified method 5) [41] was used to sample polychlorinated dibenzo dioxins (PCDD) and polychlorinated dibenzo furans (PCDF). The subsequent analysis was carried using a high-resolution gas chromatograph combined with a mass spectrometer. Two different ions for dioxin (304.2, 306.2) and furan (320.2, 322.2) groups were monitored to obtain the peaks using selective ion-monitoring technique. Fly ash samples were tested for unburned carbon loss by evaluating loss on ignition (LOI) using a high-temperature furnace. Fly ash samples were heated to a high temperature of 1700°F, and the weight loss measured provided an indication of unburned carbon present in the ash. Bottom ash analysis for toxic metals was carried out by neutron activation method. This was carried for short irradiation metals where the sample is made radioactive. A detector is used to count the isotopes coming out of the spectrum. However, this method will not detect some long irradiating metals such as cadmium, lead, and iron because of their longer waiting time. The semiquantitative analysis of fly ash samples was performed using SEM. The spectrum of different metals in the fly ash and their weight percentage were determined by an X-ray microanalysis system.

15.5.4. Some Results

Computational Results Equilibrium thermochemical calculations (using STANJAN [35] and SOLGASMIX [36] codes) of a mixture of nonplastic and plastic surrogate solids were carried out under conditions of pyrolysis and combustion; see Table 15.9. A large thermodynamic data file compiled from JANAF tables is used in these codes [37]. The nonplastic material is assumed to be cellulose while the plastic material may contain any or all of the following plastics: polyethylene, polyvinyl chloride, polystyrene, polypropylene, polyethylene teraphathalic, nylon, latex in the form of rubber, polyurethane, and acetate. Cellulose represents the organic portion of the waste such as paper and cardboard.

Table 15.9 Chemical Composition and Heating Value of Examined Materials

Material	Chemical Formula	Heating Value (Btu/lb)
Polyethylene	$(CH_2H_2)_n$	19,932
Polyvinyl chloride	$(C_2H_3CL)_n$	7,875
Polystyrene	$(C_8H_8)_n$	17,838
Polypropylene	$(C_3H_6)_n$	19,948
Polyethylene	$(C_{10}H_8O_4)_n$	12,700
Nylon	$(C_6H_{11}ON)_n$	13,640
Latex (rubber)	$(C_4H_8O_2S)$	19,465
Polyurethane	$(C_{12}H_{22}O_4N_4)_n$	11,203
Acetate paper cardboard	$(C_4H_6O_2)_n$	12,050
	$(C_6H_{10}O_5)_n$	7,200
	$(C_6H_{10}O_5)_n$	7,000

The adiabatic flame temperature and product composition under conditions of direct combustion was calculated over a range of conditions extending from fuel-lean to fuel-rich modes (up to 10 mol of air). Twenty major species considered in the present analysis were: C (gas), C (solid), C_2, CH, CH_4, CO, CO_2, C_2H_2, C_3H_8, H, OH, H_2, H_2O (vapor), H_2O (liquid), N, N_2, NO, NO_2, O, and O_2. Additional species (a total of up to 400) could be added using the JANAF tables, but the analysis time would increase with the increase in number of species considered. Results obtained with additional species showed minor changes in the results, but the overall trends remained unchanged. In order to provide a compromise between the extensive computational time and the quality of data, it was therefore decided to limit the calculations with the above species since the overall conclusions would remain unaffected. Results shown in Figure 15.12 for the seven samples reveal the importance of plastic on the distribution of adiabatic flame temperature. Increased plastic content in the waste yields higher temperature than 100% nonplastic, and the maximum temperature is obtained when burning 100% plastic. The adiabatic flame temperature shifts to correspond with lower number of moles of air with increase in plastic content. Inclusion of plastic with nonplastic materials has a significant influence on temperature whereas exclusion of certain plastic components within the mixture has negligible influence on the peak temperature. The higher predicted flame temperature with plastic is attributed to the direct result of higher heat content of plastic material.

The product mole distribution is calculated using SOLGASMIX [36] for different moles of air and the corresponding adiabatic flame temperature. Specifically, the products formed during the combustion of seven different samples in air were examined according to the following combustion equation:

$$\text{Plastic/nonplastic} + a(O_2 + 3.76N_2) \longrightarrow \text{products}$$

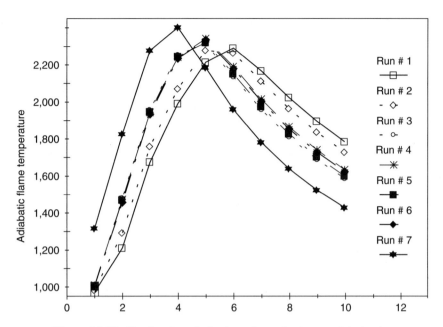

Figure 15.12. Combustion of plastic and nonplastic materials in air.

where the fuel (plastic/nonplastic) represents any of the seven samples examined here. The results shown in Figures 15.13a and 15.13b and 15.14a and 15.14b are for 100% nonplastic and 100% plastic, respectively. Yields of CO_2 and H_2O reach a maximum at stoichiometric conditions and then decrease as the number of moles of air increases for both the 100% nonplastic and 100% plastic (note the log scale for mole fraction in the figure). Concentrations of CO and H_2 decay more rapidly with 100% nonplastic than with plastic due to the variation in the reaction temperatures. Emission of NO from both the 100% nonplastic and plastic wastes first increases with the increase in moles of air (up to stoichiometric mixture) and then decreases. A similar trend was found for the NO_2 except for the nonplastic waste wherein the values kept increasing well into the fuel-lean region. The amount of NO and NO_2 produced from nonplastic-fueled flames was higher than plastic-fueled flames. This is attributed to some chemical interaction between the reactive chemical species produced in highly luminous plastic-fueled flames since the temperatures in nonplastic flames was determined to be lower than in plastic-fueled flames. It is also to be noted that plastic waste contained some chemically bound fuel-bound nitrogen, which should have provided higher NO_x. The results therefore show the importance of reaction chemistry on the formation of NO_x in flames. The exact mechanism on the formation of NO and NO_2 in these flames was outside of the scope of this study, but nevertheless this requires further examination. Formation of compounds such as H_2S and HCl from the combustion of plastic is due to the presence of sulfur and chlorine in different kinds of plastics (see Table 15.9).

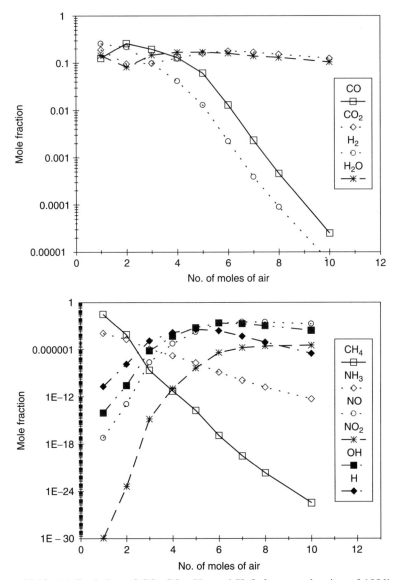

Figure 15.13. (*a*) Evolution of CO, CO_2, H_2, and H_2O from combustion of 100% nonplastic in air. (*b*) Evolution of CH_4, NH_3, NO, NO_2, OH, and H from combustion of 100% nonplastic.

Experimental Results The mean values of operating variables monitored during the experimental tests include furnace temperature, exhaust temperature, waste feed rate, air flow rate, and carbon monoxide concentration and are given in Table 15.10. The data show that 100% plastic feed rate is less than 100% nonplastic for the same inlet air and furnace temperature (compare samples 1

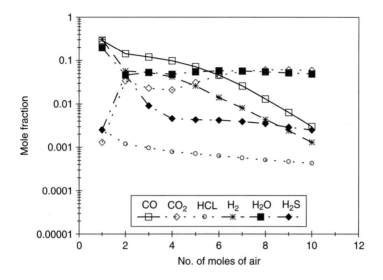

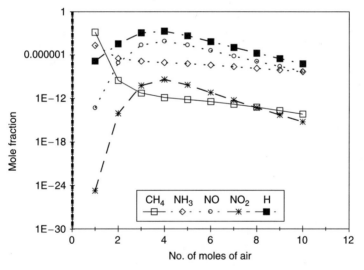

Figure 15.14. (*a*) Evolution of CO, CO_2, HCl, H_2, and H_2O from combustion of 100% plastic in air. (*b*) Evolution of CH_4, NH_3, NO, NO_2, and H from combustion of 100% plastic in air.

and 7). Higher heating value of the plastic and air stoichiometry requirements are the reasons for lower feed rate and air requirement. The furnace temperature was observed to be fluctuating by as much as 30°C at the measurement location as a result of large temperature gradient in the flame zone. The unique burning characteristics of plastic and nonplastic (ignition temperature, heat capacity) as

Table 15.10 Experimental Conditions and Measured CO Levels

Run No.	Mean Furnace Temp. (°F)	Mean Exhaust Temp. (°F)	Waste Flow Rate (lb/min)	Gas Flow Rate (scfm)	CO at 1500°F (vol%)	CO at 1350°F (vol%)
1	1500	1270	0.44	46.11	0.1	0.14
2	1400	1200	0.32	39.04	0.1	0.18
3	1600	1300	0.36	43.08	0	0.1
4	1400	1200	0.30	37.95	0.04	0.12
5	1500	1236	0.31	39.49	0.02	0.04
6	1400	1250	0.29	38.69	0.02	0.04
7	1600	1250	0.12	31.0	0	0.04

Table 15.11 Evaluated Flow Parameters

Run No.	Gas Velocity (ft/s)	Run Time (s)	Fuel/Air Ratio (actual)	Fuel/Air Ratio (theoretical)	Equivalence Ratio	Excess Air
1	1.73	0.67	0.145	0.17	0.874	1.14
2	1.46	0.79	0.119	0.18	0.658	1.51
3	1.62	0.72	0.185	0.20	0.925	1.08
4	1.42	0.82	0.158	0.20	0.789	1.26
5	1.48	0.79	0.163	0.20	0.815	1.22
6	1.45	0.80	0.154	0.20	0.774	1.29
7	1.16	1.00	0.147	0.25	0.589	1.69

well as the local furnace conditions (e.g., large-scale turbulence, flow motion, recirculation, temperature gradients) are possible sources for this contribution. These temperature fluctuations were the highest for the 100% plastic tests. This may well be due to the depolymerization of the plastic polymers under high-temperature conditions. However, in the present study, the exhaust temperature was maintained constant at 1250°F (within ±50°F) during all tests. From the test operating data, the fundamental combustion parameters such as excess air, equivalence ratio, and gas residence time were calculated and these are presented in Table 15.11.

The residence time was computed by dividing the gas volumetric flow by the furnace volume. The calculated residence time for all types of waste was somewhat lower (<1 s) than that often used in many commercial incinerators (1–2 s). The actual fuel–air ratio was evaluated from the waste and air flow rates. The theoretical fuel–air ratio was obtained from SOLGASMIX calculation for combustion and used in the calculation of equivalence ratio and excess air for the test operating conditions.

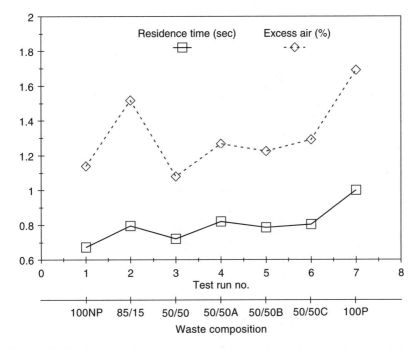

Figure 15.15. Variation of residence time and excess air with sample composition.

Figure 15.15 shows the variation of residence time and excess air level inside the reactor for varying waste composition. The excess air level was very low with 100% nonplastic, very high with 100% plastic, and in between for the mixtures. Consequently, the residence time associated with the 100% plastic was the highest, lowest with 100% nonplastic, and almost constant with mixtures. Excess air can therefore be related to the residence time as shown in this figure.

High excess air level was found to be necessary for the 100% plastic case. Maintaining a lower plastic waste feed rate was inadequate to achieve a constant furnace temperature at constant air flow rate. Furnace operation at lower feed rates provided unacceptable levels of temperature fluctuations in the furnace and in the exhaust section. The larger residence times with plastic wastes are also associated with the larger amounts of gas produced during combustion. This effect may be seen from the increase in both the residence time and excess air for the 100% plastic test case. The test operating conditions for all runs were fuel lean; see Figure 15.15. No significant variation in excess air level and residence time for the 50/50 plastic and nonplastic mixtures could be observed. This is due to the fact that the inclusion or exclusion of certain types of plastics within the plastic waste mixture has little effect on the overall waste composition.

Emission of Carbon Monoxide The concentrations of carbon monoxide (CO) in volume percentage, monitored for the lowest and highest furnace temperature

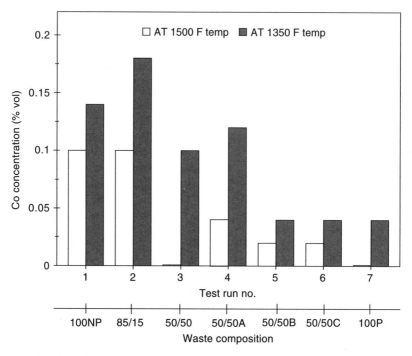

Figure 15.16. Variation of carbon monoxide concentration with sample composition.

condition are given in Table 15.4. The CO emission is an indicator of the combustion efficiency and performance of the combustion devices. The variation of CO concentration with sample composition is given in Figure 15.16. As expected, the reactor operation at higher temperatures decreases CO. The results show high CO concentration with 100 and 85% nonplastic test runs (samples 1 and 2). This incomplete combustion condition is attributed to the lower heating value and higher ash content of paper as compared to the plastic. The 50/50 mixture of plastic and nonplastic produces less CO than nonplastic alone. The presence of plastic in the mixture is therefore advantageous from the point of view of CO. Absence of one or more types of plastics in the mixture changes somewhat the CO concentration. The CO concentration during the test runs varied between 1000 to 1800 ppm.

A comparison of the mean value of CO with residence time shown in Figure 15.17 reveals higher CO concentration at lower residence time for all samples except in the case of sample 4 (50% nonplastic/50% plastic A).

Emission of Particulate Matter Particulate matter leaving the TDF is determined from the fly ash samples collected in the exhaust duct. From the gas flow rate exiting the chamber and gas temperature, the particulate loading was calculated in milligram per cubic meter. The results given in Table 15.12 form a basis to make a qualitative comparison between tested samples.

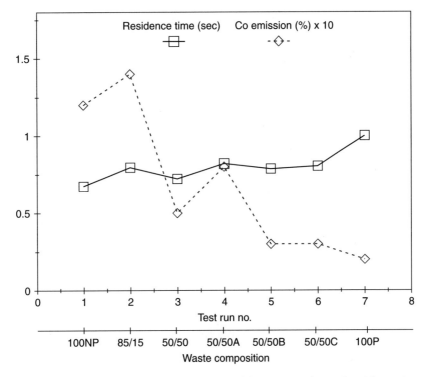

Figure 15.17. Variation of mean carbon monoxide concentration and residence time.

Table 15.12 Particulate Matter Emission for Different Samples

Run No.	Waste Feed (kg/h)	Firing Density (kg/h/m³)	Total Fly Ash (g)	Flue Gas Flow (scfm)	Particulate Matter (mg/m³)
1	12.0	801	16.1	46.1	88.3
2	8.88	592	10.4	39.0	105
3	9.84	656	4.89	43.0	66.8
4	8.22	548	3.13	37.9	44.8
5	8.64	576	4.39	39.4	60.4
6	7.98	532	5.53	38.6	91.7
7	3.5	333	0.32	31.0	10.5

Maximum nonplastic containing samples (1 and 2) have the maximum particulate emission, which is attributed to higher ash content of paper in addition to higher flue gas flow rate. The lowest particulate emission was from 100% plastic (sample 7). Other test samples provided results that bounded between samples 1 and 7. The particulate emission level decreases with increase in plastic content

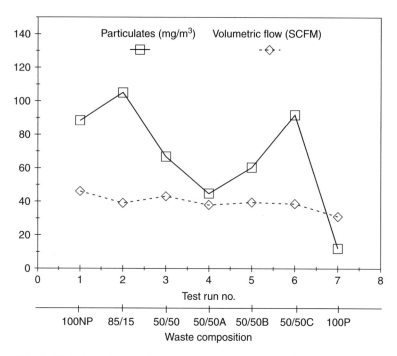

Figure 15.18. Variation of particulate matter and gas volume with sample concentration.

due to reduced ash content in the plastic fraction of the mixture. Higher particulate emission from sample 2 (85% nonplastic/15% plastic) is due to its high firing density. Figure 15.18 shows the relative particulate matter emission and the almost constant gas volumetric flow rate for the different samples. The particulate emission is very low in the case of 100% plastic (sample 7) and is attributed to the low flue gas volumetric flow rate and ash content in the waste. A lack of good correlation for mixtures is probably due to the complex burning characteristics associated with the inhomogeneous and complex fuel mixture.

Emissions of Dioxin and Furan The total dioxins and furans were determined from the measured concentration levels of PCDD obtained at ion 320.2 and 322.4 and PCDF measured at ion 304.2 and 304.6, respectively.

Experimental results show highest concentrations of dioxins and furans from 100% plastic (sample 7) and lowest from 100% nonplastic (sample 1); see Figure 15.19. The concentrations recorded from other samples are between those of pure plastic and nonplastic. Sample 5, which is a combination of 50% plastic B to 50% nonplastic has lower value of PCDD/PCDF than sample 2 (85% nonplastic and 15% plastic). This is mainly due to the absence of polyvinyl chloride in sample 5. The results reveal the important and complex chemistry associated with the chlorinated hydrocarbons and in producing the precursor compounds that are responsible for dioxin and furan formation.

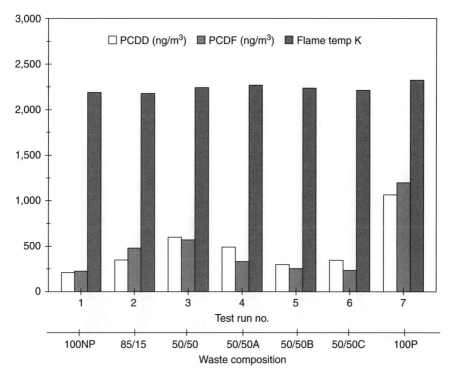

Figure 15.19. Dioxin and furan concentration levels with adiabatic flame temperature.

The small amount of PCDD/PCDF detected in 100% paper may be due to the trace amount of chlorine in the paper (from the bleaching process) and from the interaction with the refractory material, which had trace amounts of chlorine. The residual carbon collected in the region of 300–400°C is responsible for the formation of PCDD/PCDF by de nova synthesis. The presence of chlorine and copper in ash, in trace levels, implies the presence of copper chloride, which can play a vital role in the formation of PCDD and PCDF through de nova synthesis [42, 43].

Figure 15.19 also shows the absolute concentration levels of PCDD/PCDF from the various samples as well as the corresponding adiabatic flame temperatures. High temperature has no pronounced effect on the emission of these toxic compounds. The dioxin and furan formation is a weak function of the furnace temperature and strongly depends on waste composition [37]. The concentration of dioxin varies between the various samples tested. Out of the four 50/50 mixture of plastic and nonplastic, PCDD/PCDF emission was found to be lowest from the sample with no polyvinyl chloride (sample 5).

Figure 15.20 shows the emission of PCDD, PCDF, and CO as a function of the waste composition. A comparison of the results between sample 1 and sample 7 reveals that the dioxin/furan emission is inversely related to CO

670 THERMAL DESTRUCTION OF WASTES AND PLASTICS

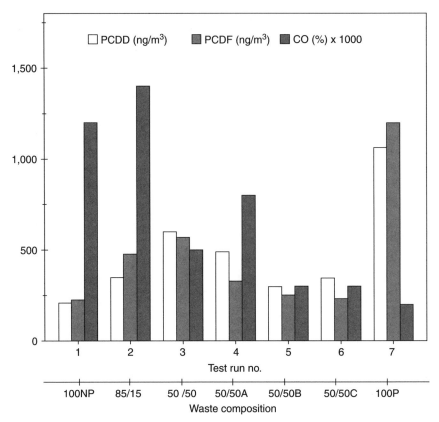

Figure 15.20. Variation of dioxin, furan, and CO concentration levels with waste composition.

emission. The relationship between dioxin or furan and CO emission for all of the 50/50 nonplastic and plastic mixtures was found to be poor and is attributed to the poor mixing between plastic and nonplastic in the mixture.

Metals in Fly Ash Particulates from waste incinerators typically contain oxides of silicon, iron, calcium, and aluminum. The other trace metals of importance are antimony, arsenic, barium, beryllium, cadmium, chromium, lead, mercury, silver, and thallium. Carcinogenic toxic metals include arsenic, cadmium, chromium, and beryllium. Arsenic is only present in samples 7 (100% plastic) and 6 (50% nonplastic/50% plastic C) and not present in other samples.

The noncarcinogenic toxic metals are antimony, barium, lead, mercury, silver, and thallium. Out of these only lead is invariably present in all the samples. Lead would have come as a contaminant through these samples. However, the source of lead contaminants is not clearly understood. The presence of chlorine species can affect metal species and volatility temperature [30]. The chlorine content is highest in 100% plastic due to the presence of polyvinyl chloride. The 100%

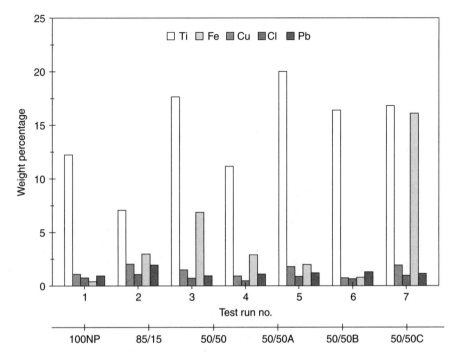

Figure 15.21. Some of toxic and rare metal elements detected by SEM.

nonplastic sample contains trace amounts of chlorine that may be left after the bleaching process in papermaking. The other metals found in the fly ash analysis are titanium, iron, and copper. In general metals are present more in plastic than nonplastic. Figure 15.21 shows the variation of some of the toxic and rare metal elements for all the test runs.

Metals in Bottom Ash The neutron activation method (NAM) has been used to analyze metals present in bottom ash. The accuracy is down to 100th of a percent. However, the limitation is with the number of metals that can be detected. Only short irradiation metal analysis was carried out, and metals with longer radioactive decay have to stay longer (months) in the reactor to complete the test. The presence of the rare metals vanadium, barium, indium, magnesium, and manganese was detected. The higher vanadium content is found in plastic/nonplastic mixtures than in 100% nonplastic. Barium, which is a noncarcinogenic toxic metal, was found in small percentages in samples 4 (50% nonplastic/50% plastic A) and 6 (50% nonplastic/50% plastic C).

A small fraction of indium is detected from the nonplastic/plastic samples 2, 3, 4, and 5. Strontium was detected with sample 1 (100% nonplastic) and some in sample 2 (85% nonplastic/15% plastic) and sample 6 (50% nonplastic/50% plastic C). Figure 15.22 shows some of the rare and toxic metals present in bottom ash and their variation with test runs.

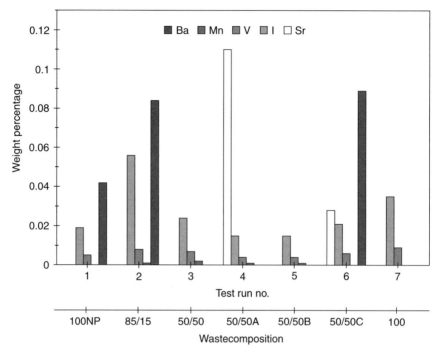

Figure 15.22. Some toxic and rare metal elements detected by NAM.

15.5.5. Closure

Results of equilibrium thermochemical calculations for the thermal destruction of nonplastic and plastic materials show the effect of material composition on the flame temperature, particulate emission, metals, dioxins, and product gas composition. The effect of waste composition has greater influence on adiabatic flame temperature, combustion air requirement, and the evolution of products and intermediate species. The combustion of waste in air produces higher flame temperature for 100% plastic than for nonplastic and mixtures. The 100% plastic requires lower number of moles of oxidant than 100% nonplastic and mixtures. Plastic produces HCl and H_2S with concentration levels ranging from 1000 to 10,000 ppm. Emission of NO and NO_2 from 100% nonplastic showed an increase with increase in moles of air while that from 100% plastic a slight decrease with increase in moles of air. The higher theoretical flame temperatures predicted with plastic waste corresponds to lower waste feed rate requirement of plastic at constant furnace temperature. This resulted in higher excess air operation with plastic waste and hence lower equivalence ratio. The gas residence time calculated for all the samples was found to be about 1 s. Variation of residence time more or less follows the same trend as excess air for all the samples.

The experimental data and calculated combustion parameters show different burning characteristics of plastic and nonplastic. Temperature dependency of

CO emission is in good agreement with the theoretical calculations. Higher CO emission was observed for both (100 and 85%) nonplastics than for 100% plastic. Incomplete combustion occurs with the paper burning because of lower heat release than plastic. Concentrations of dioxin vary between the various samples tested. Dioxin and furan levels were found to be higher from nonplastic/plastic samples than from 100% nonplastic. Nonplastic waste yields lowest levels of dioxins and furans. Out of four samples of various 50/50 plastic/nonplastic mixtures, PCDD/PCDF emission levels are lowest for the sample without polyvinyl chloride (sample 5). Metals in fly ash and bottom ash did not reveal any specific pattern for their variation; however, in general plastic increases metals emission. The presence of lead was found in all samples after combustion, although its source is not well understood. The highest amount of chlorine was found in a 100% plastic sample. Zinc and arsenic are present only in plastic samples. Sulfur is only detected in trace quantities from 100% nonplastic sample. Bottom ash analysis proved the presence of trace metal elements such as vanadium, barium, indium, magnesium, and manganese in ultra trace levels. The results provided here give the insights on the design guidelines for the thermal destruction of plastic containing wastes.

15.6. THERMAL DESTRUCTION OF POLYPROPYLENE, POLYSTYRENE, POLYETHYLENE, AND POLYVINYL CHLORIDE

15.6.1. Preamble

Results on the thermal decomposition behavior of polypropylene, polystyrene, polyethylene, and polyvinyl chloride under controlled environment are presented and compared with cellulose. Thermogravimetry (TGA) tests were conducted on the decomposition of cellulose, polypropylene, and polystyrene in inert (nitrogen) and oxidative (air) atmospheres. These results provide information on the importance of gasification and incineration of the wastes. These test were performed at a heating rate of 5, 10, 30, 50°C/min. Differential scanning calorimetry (DSC) was also used to measure the heat flow into and out of the sample during thermal decomposition of the material. The TGA results on the mass evolution of the materials studied versus temperature showed that the cellulose contained moisture whereas no moisture was found for the polypropylene (PP), polystyrene (PS), polyethylene (PE), and polyvinyl chloride (PVC). The DSC curve showed the heat flow into and out of the sample during the process of pyrolysis and oxidative pyrolysis. The results show that less than 10% of the energy is required to decompose cellulose. The gases evolved are much leaner to burn. The temperature dependence and mass loss characteristics of materials were used to evaluate the Arrhenius kinetic parameters (A and E). These kinetic parameters provide information on the rate of material decomposition. The surrounding chemical environment, heating rate, and material composition and properties affect the overall decomposition rates under defined conditions. The composition of these materials could be related to their thermal decomposition behavior. Experimental results show that decomposition process shifts to higher temperatures at higher heating rates as a result of the competing effects of heat and mass transfer to the material. The results on the Arrhenius chemical kinetic parameters, maximum decomposition temperature, and heat of pyrolysis obtained from the thermal decomposition of surrogate wastes show these three materials have considerably different features. These results assist in the design and development of advanced thermal destruction systems, such as gasification plants or incineration plants or their combination.

15.6.2. The Problem

Recently, interest in waste disposal has increased all over the world especially in the developed countries. The two distinct reasons for this are (1) reduction in the number of available sites for direct landfill and (2) emission of hazardous chemicals from disposition of wastes. The waste consists of 7–9% by weight or 25–30% by volume of plastics [24]. The use of plastics has dramatically increased during the last couple of decades and the problem of disposing them has become progressively acute. In addition the United States produced over 4 million tons of polypropylene and 2.5 million tons of polystyrene annually [43]. The development of measuring technique has revealed the presence of newer and harmful chemicals (that may be in a very small quantity) from chemical wastes. The biologists have warned of risk on some of the chemicals not only for the typical pollutants, such as, NO_x, SO_2, HCl, CO, CO_2, and unburned hydrocarbons but also the new chemicals called endocrine disruptors (bisphenol A, polychlorinated biphenyl (PCB), dichloro diphenyl trichloro ethane (DDT), dioxin, etc.) [44, 45]. In addition phosgene ($COCl_2$) can also form in chlorine containing wastes under certain operational conditions. Several waste disposal options used to date include direct landfilling, storage in surface impoundments, physical/chemical stabilization, and direct incineration. Although all these solutions seen viable, none is without problems. Of all the treatment technologies thermal destruction offers distinct advantage over the other methods since it provides maximum volume reduction (typically 80% of the original waste), its effectiveness to plastic, energy recovery (waste has higher heating value (HHV) value of about 5500 Btu/lb) and the by-products can be used in several ways, such as building material and roadbed construction. Presorting the waste, for example, by removing the metals, glass, and other materials from the waste, can further enhance the volume reduction. However, the disadvantage is that it must be environmentally clean and accepted by the public since the by-products from their determination can be health hazardous and detrimental to our environment [46]. It is to be noted that trace elements, volatile organic compounds (VOCs), and dioxins are worse among the pollutants emitted from waste destruction.

Even though the incineration is only a small fraction for the waste disposal, it is a permanent disposal option (less the environmental issues). Among the reasons that incineration is not widely accepted is that the public is rather sensitive to having an incineration plant as it may cause some short- or long-range harmful effects. Several efforts have been made for the better design of incineration systems. Examples include converting the waste to gas by using thermal energy, gasifying the waste in the presence of some suitable gas, high-temperature thermal destruction, and steam reforming. In all cases the objective is to destruct the waste without the emission of harmful pollutants. Recently, gasifying the waste has gained popularity, as the end product is gas, which is easy to burn. The objective here is to perform pyrolysis and combustion on the surrogate wastes and to determine how much energy is required or released from the waste destruction. This information can then be used in the design of advanced incinerators.

To obtain comprehensive understanding of the thermal destruction behavior of different kinds of wastes, the focus should be on exploring the basic chemical processes that occur during their thermal decomposition. This then requires an understanding of how the various parameters, such as temperature, heating rate time provided for combustion, and chemical composition, influence the thermal decomposition of specific wastes. Thermogravimetric analysis (TGA) has been used to study the pyrolysis of various waste materials. Raman et al. [47] used it to study the devolatilization of biomass at two heating rates of 5 and 80°C/min. Williams and Besler [48] investigated the thermal decomposition of municipal solid waste at temperatures of up to 560°C using TGA. In this study we provide kinetic information on the thermal destruction of cellulose, polypropylene, and polystyrene at different heating rates under conditions of pyrolysis and oxidative pyrolysis. We recognize that real wastes have many of these compounds present in some proportion that vary from day to day and from location to location. In order to decouple the complexity associated with the real wastes, one must then determine the waste thermal destruction behavior under some ideal conditions. This information can then be used to simulate the thermal destruction behavior for real wastes.

15.6.3. Experimental Procedure

Simultaneous thermogravimetric analysis (TGA) and differential scanning calorimetry (DSC) tests have been conducted to examine the decomposition of cellulose, polypropylene, polystyrene, polyethylene, and polyvinyl chloride in inert (nitrogen) and oxidative (air) environments. The tests were performed using an analyzer that allowed simultaneous measurements on calorimetry (DSC) and thermogravimetry (TG) in the temperature from ambient up to 1500°C (2732°F). The samples were subjected to temperature ramps in an inert gas (nitrogen) or an oxidative gas (air) in the temperature range of 25–1000°C (77–1832°F). Nitrogen and air, both at a flow rate of 40 mL/min, where used as the inert and oxidative atmosphere, respectively. The heating rate of the material and gas was kept at 5, 10, 30, or 50°C/min. The role of gas flow rate is significant in terms of heat transfer mechanisms (radiation, convection) from the hot gases to the material and mass transfer. In particular under oxidative conditions it is also the source of reaction gas (oxygen). The 4-mm pans made out of platinum were utilized throughout testing. The material sample sizes were in the range of 5–7 mg. The surrogate material were as follows: cellulose (Sigma Chemical Co., Sigmacell; ref. 9004-34-6), polypropylene (Aldrich Chemical Company, Inc.), polypropylene isotactic (25085-53-4), polystyrene (Aldrich Chemical Company, Inc.), standard polystyrene (9003-53-6).

The mass evolution of these samples was determined as a function of temperature. The fractional reaction, defined as $\alpha = (m_0 - m)/(m_0 - m_f)$ (reaction progress) [49], was therefore plotted versus sample temperature. TGA has been extensively used to determine the devolatilization characteristics and to determine kinetic parameters [50, 51]. It is to be noted that these ideal conditions as well as some simplifying assumptions [49, 52] made here may not necessarily correspond to the actual thermal chemical decomposition of the materials. Nevertheless the data provides useful comparison of reaction parameters such as temperature and heating rate.

15.6.4. Determination of Some Important Parameters

Determination of the Arrhenius Parameters The Arrhenius parameters (A and E) for the thermal decomposition of the sample [52–54] were determined assuming a first-order chemical reaction ($n = 1$). The rate constant is defined as:

$$k(T) = A \times \exp\left(-\frac{E}{RT}\right)$$

where A = preexponential factor, min^{-1}
E = activation energy, kJ/mol K
R = universal gas constant, J/mol K
T = temperature, K
t = time, s

This equation defines the temperature dependence of the specific rate constant. A second equation that relates the reaction progress to time through the rate constant is also required:

$$\frac{d\alpha}{dt} = A \times \exp\left(-\frac{E}{RT}\right) \times (1 - \alpha)$$

Integrating yields the following:

$$\int \frac{d\alpha}{1-\alpha} = A \times \int \exp\left(-\frac{E}{RT}\right) dt$$

$$\frac{dT}{dt} = \text{heating rate } (\beta)$$

$$-\ln(1-\alpha) = \frac{A}{\beta} \times \int \exp\left(-\frac{E}{RT}\right) dT$$

where α = reaction progress
β = heating rate, °C/min

In the last equation, the right-hand side has no exact solution, and several approximate solutions to this equation have been used to determine the order of reaction, activation energy, and preexponential factor:

$$-\ln(1-\alpha) \cong \frac{A}{\beta} \times \frac{RT^2}{E} \times \exp\left(-\frac{E}{RT}\right)$$

Taking the natural logarithm of this equation yields:

$$\ln[-\ln(1-\alpha)] = \ln\frac{ART_{max}^2}{\beta E} - \frac{E}{RT}$$

The value of E and $\ln(A)$ was evaluated from the slope and intercept of function $\ln[-\ln(1-\alpha)]$ versus $1/T$ [49, 52]. The data to construct this plot are taken from the TGA curve; see Figure 15.23.

Determination of the Maximum Decomposition Temperature The value of maximum decomposition temperature was calculated from the following two equations:

$$\left(\frac{dm}{m}\right) = \frac{m_i - m_{i-1}}{m_i}$$

$$f_{(t\,max)} = \left(\frac{dm}{m}\right)_{max}$$

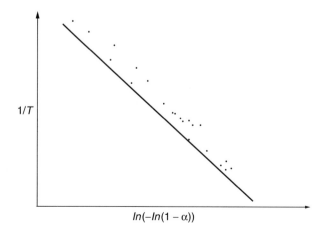

Figure 15.23. $1/T$ vs. $\ln[\ln(1-\alpha)]$ for determining Arrhenius parameters.

Notice that in the second of these two equations, the point of maximum decomposition can be obtained by plotting dm/m versus time or temperature, and is the minimum of this curve where m_i = initial mass.

Determination of the Heat of Pyrolysis The value of the heat of pyrolysis was calculated from the following two equations:

$$h = \frac{1}{m_i} \times \int_{t1}^{t2} P_{(t)}\, dt$$

$$h = \frac{T2-T1}{\beta} \times \frac{1}{m_i} \times \int_{T1}^{T2} P_{(T)}\, dT$$

where h = heat of pyrolysis
m_i = initial mass
$P_{(t)}$ = heat flow
t = time
T = temperature, (K)

Estimation of Required Heat of Pyrolysis for Different Compounds Polyethylene (PE) is usually referred as $(C_2H_4)_n$, or

$$\text{-[CH}_2\text{-CH}_2\text{]}_n\text{-} \longrightarrow H_2C=CH_2$$

The total molecular weight of polyethylene is:

$$M_w = 200{,}000 \sim 300{,}000$$

Polypropylene (PP) is usually referred as C_3H_6, or

$$\text{-[CH(CH}_3\text{)-CH}_2\text{]}_n\text{-} \longrightarrow H_3C\text{-}CH=CH_2$$

$$M_w = 250{,}000$$

Polystyrene (PS) is C_8H_8

$$\text{-[CH(C}_6H_5\text{)-CH}_2\text{]}_n\text{-} \longrightarrow C_6H_5\text{-}CH=CH_2$$

$$M_w = 400,000$$

$$h_{PE,PP,PS} = N_A/M_w \times M_w/M_{w(unit)} \times \Delta H_{lc}/N_A = \Delta H_{lc}/M_{w(unit)}$$

or

$$N_A/M_w \times M_w/M_{w(unit)} \times H_{(C-C)}/N_A = H_{(C-C)}/M_{w(unit)}$$

Polyvinyl chloride (PVC) is C_2H_3Cl, or

$$h_{PVC} = [(H_{(C-Cl)} + H_{(C-H)} + H_{(C-C)}) - (H_{(H-Cl)} + H_{(C=C)})]/M_{w(H-Cl)}$$

Cellulose is $C_6H_{10}O_5$, or

$$h_{Cel} = N_A/M_w \times M_w/M_{w(unit)} \times H_{(C-O)}/N_A = H_{(C-O)}/M_{w(unit)}$$

where $h_{PE, PP, PS, PV}$ = Heat of pyrolysis, kJ/g
N_A = Avogadro's number
M_w = molecular weight of polymer
$M_{w(unit)}$ = Unit molecular weight of substance
ΔH_{lc} = Heat of polymerization (kJ/mol)
H_0 = Enthalpy of atomization (kJ/mol)

The heat of polymerization [55], ΔH_{lc} (kJ/mol), for the various compounds is as follows:

$$PE = 88.7; \quad PP = 81.6; \quad \text{and } PS = 69.9 \text{ (kJ/mol)}$$

Enthalpy of atomization [56] H_0, B(A–B) (kJ/mol)

$$C-H = 412; \quad C-C = 348 \quad C=C = 612; \quad H-Cl = 431$$
$$C-O = 360 \text{ (kJ/mol)}$$

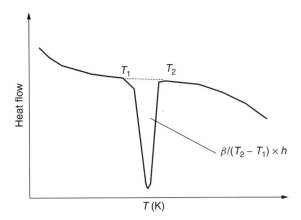

Figure 15.24. Heat flow vs. temperature for determining heat of pyrolysis.

Figure 15.24 shows the heat of pyrolysis (area under the curve between points T_1 and T_2)

Calculation of Required Amount of Air for Complete Combustion
The sample chemical reaction equations are:

Cellulose: $(C_6H_{10}O_5) + 6O_2 \rightarrow 6CO_2 + 5H_2O$
Polypropylene: $(CH_2CH(CH_3)) + 4.5O_2 \rightarrow 3CO_2 + 3H_2O$
Polystyrene: $(CH_2CH(C_6H_5)) + 10O_2 \rightarrow 8CO_2 + 4H_2O$

Similar equations can be written for other materials.
The required amount of air is calculated from

$$W_r = M_{w(\text{unit})}/m_s \times \text{Cr}/0.21 \times 22{,}400$$

where
$\quad W_r$ = required air amount
$\quad M_{w(\text{unit})}$ = unit molecular weight of substance
$\quad m_s$ = sample mass
$\quad \text{Cr}$ = coefficient of oxygen in perfect combustion (i.e., stoichiometric) conditions
$\quad 0.21$ = oxygen fraction in air
$\quad 22{,}400$ mL = 1 mol

15.6.5. Some Results

The results obtained on these three samples from TGA and DSC are shown in Figures 15.25–15.36. They show the mass left and heat flow into or out from the sample as a function of temperature. The TGA results on cellulose show the presence of combined moisture in the sample under both inert and oxidative

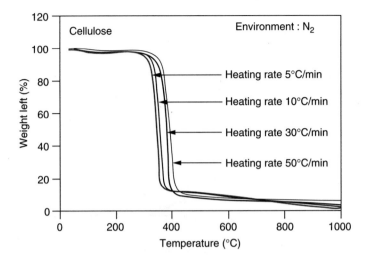

Figure 15.25. Thermal decomposition of cellulose in nitrogen.

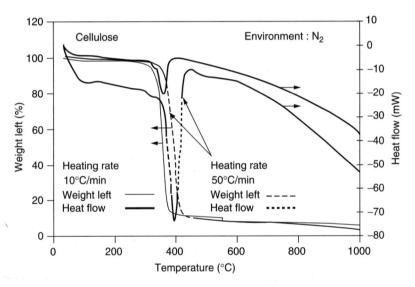

Figure 15.26. Thermal decomposition and heat of pyrolysis of cellulose in nitrogen.

conditions. The results under inert conditions (see TGA curves in Fig. 15.25) suggest that the decomposition occurs in a single-stage reaction while the results from oxidative condition (see TGA curves in Fig. 15.27) suggest that the decomposition occurs in multistages.

The DSC results of cellulose show endothermic behavior in inert conditions (see DSC curves in Fig. 15.26 showing negative values of heat flow indicating

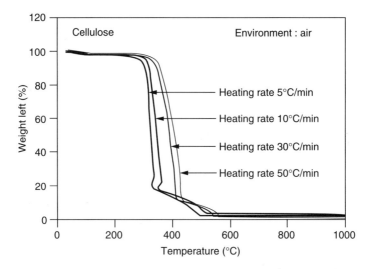

Figure 15.27. Thermal decomposition of cellulose in air.

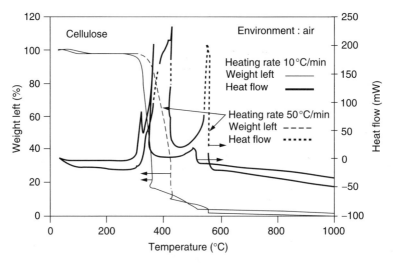

Figure 15.28. Thermal decomposition and heat of pyrolysis of cellulose in air.

heat absorbed by the sample during pyrolysis) and exothermic behavior in oxidative conditions (see TGA curves in Fig. 15.28 showing positive values of heat flow indicating heat evolved from the sample with combustion). This figure also shows the point of ignition (near 300°C) of the material. In addition DSC results in oxidative condition suggest that the decomposition is in multistages. The TGA results on polypropylene show that it has really few impurities in it (see TGA curves in Figs. 15.29 and 15.31).

THERMAL DESTRUCTION OF PP, PS, PE AND PVC   683

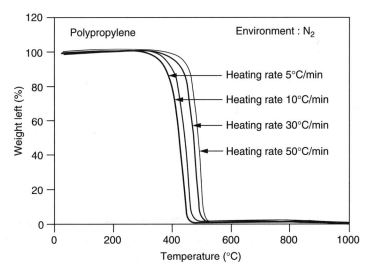

Figure 15.29. Thermal decomposition of polypropylene in nitrogen.

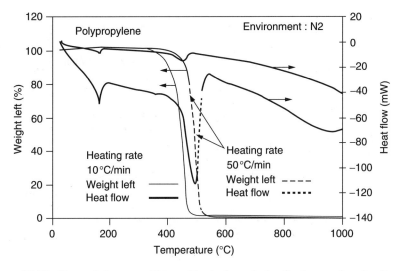

Figure 15.30. Thermal decomposition and heat of pyrolysis of polypropylene in nitrogen.

The DSC results on polypropylene show that at first it incurs a melting point and then decomposes. However, in inert conditions it has endothermic behavior and in oxidative condition it has exothermic behavior (see DSC curves in Figs. 15.30 and 15.32). The ignition occurs at a temperature of about 280°C (see Fig. 15.32). The values above zero heat flow correspond to exothermic region while those below the zero values correspond to endothermic region. So for the

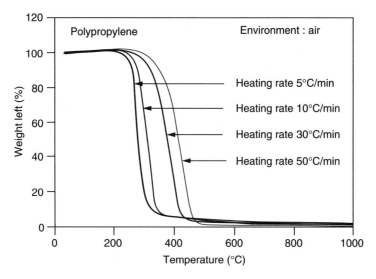

Figure 15.31. Thermal decomposition of polypropylene in air.

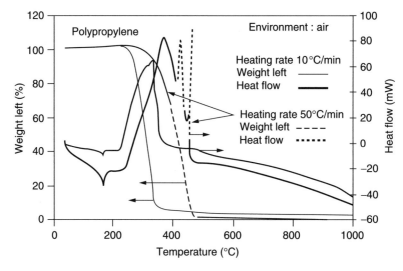

Figure 15.32. Thermal decomposition and heat of pyrolysis of polypropylene in air.

case of polypropylene the exothermic region follows endothermic region. The exothermic region evolves from the combustion of gases evolved from the initial pyrolysis of the material. The TGA results on polystyrene show that this material almost does not have any impurities, and at high heating rate in oxidative conditions the thermal decomposition does not follow the usual smooth decay (compare the TGA curves in Figs. 15.33 and 15.35).

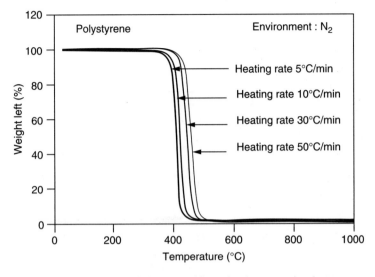

Figure 15.33. Thermal decomposition of polystyrene in nitrogen.

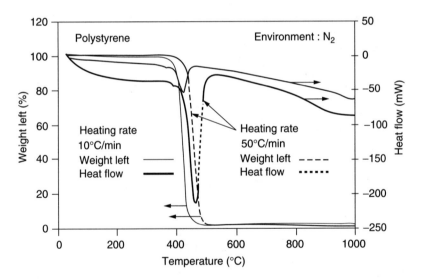

Figure 15.34. Thermal decomposition and heat of pyrolysis of polystyrene in nitrogen.

The DSC results suggest that polystyrene does not go through melting point before decomposition. Under inert conditions the results suggest decomposition in a single stage with endothermic heat of pyrolysis (see DSC curves in Fig. 15.34). Therefore in oxidative conditions the decomposition behavior varies between lower heating rate and higher heating rate. At lower heating rate it has a large amount of exothermic heat of pyrolysis, but at higher heating rate at first it shows

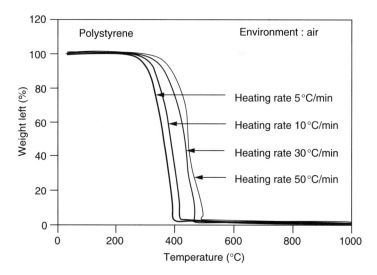

Figure 15.35. Thermal decomposition of polystyrene in air.

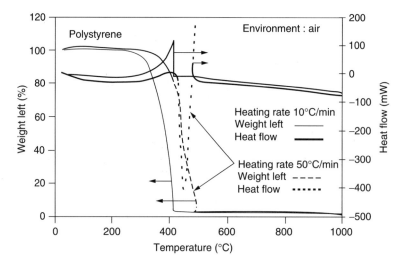

Figure 15.36. Thermal decomposition and heat of pyrolysis of polystyrene in air.

exothermic behavior and then it shows endothermic behavior (see DSC curves in Fig. 15.36). The results for the thermal decomposition of polyethylene and polyvinyl chloride in nitrogen are given in Figures 15.37 and 15.38, respectively.

Cellulose contained approximately 2% moisture and approximately 6% fixed carbon by weight. At the end of the TGA tests about 95% of weight loss was achieved. The remaining residue represents mainly the impurities present in the original sample. In contrast for polypropylene and polystyrene almost complete

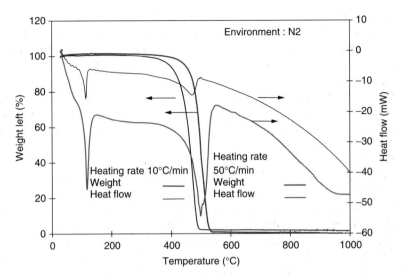

Figure 15.37. Polyethylene decomposition in nitrogen.

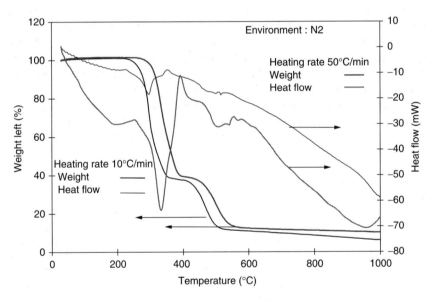

Figure 15.38. Polyvinyl chloride decomposition in nitrogen.

weight loss was achieved at the end of the TGA tests. The results show that under oxidative conditions all samples decompose in a single stage except for the drying process at the beginning of the test (up to 110°C for cellulose). The initial weight loss on each TGA curve for cellulose is due to the volatiles release during the pyrolysis.

The use of TGA was made in order to obtain information on the overall kinetics of the decomposition process for the surrogate waste materials. The decomposition rates for cellulose are shown in Figures 15.25–15.28, in Figures 15.29–15.32 for polypropylene, in Figures 15.33–15.36 for polystyrene, in Figure 15.37 for polyethylene, and in Figure 15.38 for polyvinyl chloride. The results provide the function $\ln[-\ln(1-\alpha)]$ versus $1/T$ from which the Arrhenius parameters A (the pre-exponential factor) and E (the activation energy) can be evaluated from the graphically determined slope and intercept of the function $\ln[-\ln(1-\alpha)]$ versus $1/T$.

The calculated results for $\ln(A)$, E, and maximum decomposition temperature are given in Tables 15.13–15.15. The results for cellulose and polystyrene show that the activation energy E decreases with increase of heat ramp rate. However, for polypropylene under inert conditions the results show an increase of activation energy E with an increase of heat ramp rate. The thermal decomposition temperature increases with increase in the ramp rate. In any case the amount of heat of pyrolysis is relatively larger under oxidative condition than in inert condition except for the high heating rate of polystyrene. The temperature dependence of heat of pyrolysis depends on the material properties. The reason for high endothermic heat of pyrolysis at high heating rate for polystyrene in oxidative condition is as follows. In the experiments we chose approximately constant mass and constant heat flow so that at different heating rate the sample receives different amounts of air per gram of sample. It is to be noted that the temperature of the air is at the sample temperature at the given condition. In real practice excess air levels may be 50–100% of the stoichiometric amount. Our results obtained can still be used under these conditions.

The results show that at higher heating rate the ratio of required amount of air to that provided to combust the polystyrene is not so high. So the assumption of constant partial pressure of oxygen becomes weak. Thus the correlation between

Table 15.13 Arrhenius Parameters and Maximum Decomposition Temperature for Cellulose in Nitrogen and Air

	Heating Rate (°C/min)	Sample Weight (mg)	E (kJ/mol °C)	log(A)	Temperature for Maximum Decomposition Rate (°C)
Nitrogen	5	5.080	261	42.7	353
Nitrogen	10	5.657	171	24.4	370
Nitrogen	30	6.367	163	22.9	394
Nitrogen	50	6.122	155	21.2	408
Air	5	5.277	261	44.1	332
Air	10	5.603	204	31.7	359
Air	30	5.206	147	19.5	411
Air	50	5.837	134	16.9	429

Table 15.14 Arrhenius Parameters and Maximum Decomposition Temperature for Polypropylene Decomposition in Nitrogen and Air

	Heating Rate (°C/min)	Sample Weight (mg)	E (kJ/mol °C)	log(A)	Temperature for Maximum Decomposition Rate (°C)
Nitrogen	5	5.369	148	16.2	458
Nitrogen	10	5.2022	163	18.7	481
Nitrogen	30	5.429	192	23.6	500
Nitrogen	50	5.248	239	31.1	522
Air	5	5.189	145	22.7	304
Air	10	5.083	129	18.3	336
Air	30	5.538	102	11.4	420
Air	50	5.759	104	13.4	463

Table 15.15 Arrhenius Parameters and Maximum Decomposition Temperature for Polystyrene Decomposition in Nitrogen and Air

	Heating Rate (°C/min)	Sample Weight (mg)	E (kJ/mol °C)	log(A)	Temperature for Maximum Decomposition Rate (°C)
Nitrogen	5	5.662	324	49.1	427
Nitrogen	10	5.525	321	48.3	440
Nitrogen	30	5.194	331	49.0	468
Nitrogen	50	5.744	319	46.5	483
Air	5	5.419	113	12.3	394
Air	10	5.717	113	12.1	415
Air	30	5.094	107	10.6	466
Air	50	5.112	125	13.8	493

heat and mass transfer becomes large, and this is thought to be reasonable. Results for heat of pyrolysis and temperature for maximum decomposition rate are given in Tables 15.16–15.21. The heat of pyrolysis depends on the differences in the chemical structure of the material. The weight loss due to the volatiles released during the pyrolysis exhibits itself as an endotherm, which goes though a well-defined minima. They correspond to the respective volatile loss that evolves from cellulose, polypropylene, or polystyrene. As for the temperature for maximum decomposition rate compare the results obtained with TGA with DSC results. The peak of the DSC curves evolves earlier than the peak from the TGA data. This may be due to the mass transfer limitations. The rate-controlling step (reaction

Table 15.16 Heat of Pyrolysis and Heat Evolved During Decomposition of Cellulose in N_2 and Air

	Heating Rate (°C/min)	Heat of Pyrolysis (J/g)	Beginning of Pyrolysis Temperature (°C)	Temperature for Maximum Decomposition Rate (°C)
Nitrogen	5	780 (endothermic)	314	345
Nitrogen	10	498 (endothermic)	337	360
Nitrogen	30	455 (endothermic)	350	383
Nitrogen	50	440 (endothermic)	362	396
Air	5	12,400 (exothermic)	267	
Air	10	9,460 (exothermic)	274	
Air	30	3,530 (exothermic)	288	
Air	50	3,580 (exothermic)	296	

Table 15.17 Heat of Pyrolysis/Heat Evolved During Decomposition of Polypropylene in N_2 and Air

	Heating Rate (°C/min)	Heat of Pyrolysis (J/g)	Beginning of Pyrolysis Temperature (°C)	Temperature for Maximum Decomposition Rate (°C)
Nitrogen	5	397 (endothermic)	340	439
Nitrogen	10	453 (endothermic)	350	458
Nitrogen	30	939 (endothermic)	361	483
Nitrogen	50	1,092 (endothermic)	369	496
Air	5	23,428 (exothermic)	209	
Air	10	9,542 (exothermic)	221	
Air	30	4,215 (exothermic)	233	420
Air	50	2,628 (exothermic)	235	

rate) in pyrolysis is dependent upon temperature, composition of the material, and its physical size. The products yield, composition, and their calorific value are highly dependent upon the material composition, heating rate, surrounding chemical environment in which the process takes place, and ultimate temperature of the material. The calculated amount of air required for complete combustion is given in Table 15.22.

The information presented here allows one to determine how much heat is required (or can obtain) during thermal decomposition of wastes. In the inert experiments, where this atmosphere was maintained throughout, no further weight loss was observed as seen by the constant DSC signal.

Table 15.18 Heat of Pyrolysis/Heat Evolved During Decomposition of Polystyrene in N_2 and Air

	Heating Rate (°C/min)	Heat of Pyrolysis (J/g)	Beginning of Pyrolysis Temperature (°C)	Temperature for Maximum Decomposition Rate (°C)
Nitrogen	5	1625 (endothermic)	362	412
Nitrogen	10	1617 (endothermic)	373	424
Nitrogen	30	1649 (endothermic)	395	383
Nitrogen	50	1742 (endothermic)	405	396
Air	5	7440 (exothermic)	250	391
Air	10	6656 (exothermic)	250	414
Air	30	479 (exothermic) [1290(exothermic), 811(endothermic)]	265	No data
Air	50	1980 (endothermic) [450(exothermic), 2430(endothermic)]	No data	No data

Table 15.19 Heat of Pyrolysis for Polyethylene Decomposition in Nitrogen[a]

Heating Rate (°C/min)	Heat of Pyrolysis (J/g)	Beginning Pyrolysis Temperature (°C)	Temperature for Maximum Decomposition Rate (°C)
5	254	332	455
10	354	417	470
30	358	424	491
50	351	448	500

[a]Estimated value; 3168 J/mol [3], 12429 J/mol [4].

Table 15.20 Heat of Pyrolysis for Polyvinyl Chloride Decomposition in Nitrogen[a]

Heating Rate (°C/min)	Heat of Pyrolysis (J/g)	Beginning Pyrolysis Temperature (°C)	Temperature for Maximum Decomposition Rate (°C)
5	757	232	281
10	736	232	298
30	695	253	321
50	727	259	335

[a]Estimated value; 1507 J/mol [5].

Table 15.21 Comparison Between Heat of Combustion and Experimental Heat of Pyrolysis for Different Materials[a]

Material	Heat of Combustion (J/g)	Heat of Pyrolysis (J/g)
PE	46,300	254
PP	44,000	423
PS	40,200	1499
PVC	18,100	757
Cellulose	17,500	584

[a] At a heating rate of 5°C/min.

Table 15.22 Ratio of Required Amount of Air to That Provided

Sample	Required Amount of Air for Perfect Combustion (mL)	Total Amount of Air Provided (mL) at 5°C/min	Total Amount of Provided Air (mL) at 10°C/min	Total Amount of Air Provided (mL) at 30°C/min	Total Amount of Air Provided (mL) at 50°C/min
Cellulose	21–29	330	280	120	80
Polypropylene	60–84	820	420	230	160
Polystyrene	54–75.5	1220	610	230	140

15.6.6. Closure

The results provided here show how thermogravimetry and differential scanning calorimetry analysis can provide information on the ignition, thermal decomposition characteristics of different surrogate materials, and amount of energy required to destruct a given material in controlled environment. The temperature dependence and loss characteristics of materials can be obtained from the thermogravimetry analysis. This information provided the desired Arrhenius kinetic parameters and therefore the decomposition rates under defined conditions of temperature, surrounding chemical environment, heating rate, and waste material type. The DSC also provides information on the amount of energy needed to thermally destruct the solid waste and how much energy can be recovered during the subsequent secondary combustion of the gaseous by-products. This helps one to understand the basic thermal decomposition of the surrogate materials from which information on the real waste materials can be formulated under distinct conditions.

The kinetic parameters of thermal decomposition were determined for the different samples as a function of the heating rates. The heating rate, surrounding environment, material composition, and temperature affects the thermal decomposition of materials. The rate-controlling step (reaction rate) in pyrolysis is the material and its physical size and ratio between the sample material and the surrounding chemical gas.

The information presented here helps to characterize and understand the thermal decomposition characteristics of waste materials. The information also assists in identifying materials of similar or different characteristics and for the design and development of advanced thermal destruction systems.

15.7. PROSPECTS, CHALLENGES, AND OPPORTUNITIES FOR ENERGY RECOVERY FROM WASTES

Despite the gaps in technology, America today disposes of nearly 5 million tons of hazardous wastes by burning them in more than 184 incinerators and 171 industrial furnaces, including 34 cement kilns. This volume is equivalent to full tank trucks stretched nose to tail from Washington, DC, to San Francisco. How we safely dispose of this great quantity of waste is an issue that affects the health and safety of all Americans. These words conjure up a picture of hundreds of thousands of tank trucks containing hazardous materials inundating our highways on their way to dump their loads at incineration facilities that pose dangers to public health. Thermal destruction of plastics in an environmentally benign fashion is a challenge. However, fundamental research combined with proven chemical engineering operations is a logical technology for reducing the volume of waste and for conversion to environmentally benign material. As a result, we have seen major growth in utilizing aerospace-related technology for waste incineration as well as in the combustion of hazardous waste in boilers and industrial furnaces. An interesting synergy exists in the use of cement kilns, which convert limestone to lime. Lime is an excellent trap for acid gases, so that burning of chlorinated organic wastes in a cement kiln provides a sound solution for both the ash disposal problem and effluent gas treatment problem in one simple solution.

Waste destruction by thermal decomposition makes good technical, environmental, and economic sense. Systems can be designed to meet almost any emissions standard: oxygen could be used instead of air (even in excess enthalpy combustion systems) to achieve higher temperatures and reduce the amount of effluent gas; a catalytic afterburner could be used to ensure complete combustion of trace compounds.

Solid-gas separation systems are available for removing trace particles that can then be packaged into an environmentally benign form. Solid adsorbents can be used for final gas cleanup. Water scrubbing streams can be processed through membrane separation systems to recover dissolved mineral matter and the recycled water. Despite the availability of technology for traditional thermal destruction of waste, a fundamental need exists for better understanding of the basics involved in thermal destruction and pollution control, cost, and size reduction in addition to the associated sociological issue. The aerospace experience obtained in the fundamentals of combustion, high-temperature materials engineering, plasma physics, mixing processes, high-temperature chemistry, and the like can make major contributions to manage the growing problem of wastes.

Indeed this is a particularly good opportunity for what is being termed *technology conversion*.

15.8. CLOSURE

It is anticipated that the amount of waste (plastic and nonplastic) generated will continue to grow, along with associated changes in its composition. Utilization of energy from waste can save millions of dollars. Thermal destruction of wastes to the molecular level allows one to more cleanly convert into usable energy. The challenges provide opportunities for combustion engineers, whose research is now becoming even more important with new emphasis. Research in combustion-related areas has expanded significantly in recent years, particularly in areas related to fossil fuels and wastes being used as fuels. Advanced combustion concepts for waste usage require detailed understanding of the physical and chemical processes that undergo during combustion. Advanced new technologies will require some trade-off between waste fuel properties, energy conservation, efficiency, and environmental pollution control. It is these challenges that are providing opportunities for combustion engineers to handle emerging environmental problems associated with incineration. Gasification and steam reforming of the waste offers good potential for converting low and varying heat content wastes into good clean gas energy.

REFERENCES

1. H. Chopra, A. K. Gupta, E. L. Keating, and E. B. White, Proc. 27th IECEC Conference, San Diego, CA, August 3–7, 1992, Paper No. 92–9224, Proc. IECEC, Vol. 1, 1992, pp. 377–381.
2. E. T. Oppelt, *JAPCA* **37**(5), 558–586 (1987).
3. A. K. Gupta, AIAA Aerospace Sciences Meeting, AIAA Paper No. 84–0444, Reno, NV, January 9–12, 1984.
4. A. K. Gupta, Invited Keynote Lecture at the RAN98 Conference, Nagoya University, Nagoya, Japan, 1998.
5. W. R. Niessen, *Combustion and Incineration Processes*, Marcel Dekker, New York, 1995.
6. C. R. Brunner, *Hazardous Waste Incineration*, McGraw-Hill, New York, 1993.
7. A. K. Gupta, and D. G. Lilley, *Flowfield Modeling and Diagnostics*, Abacus, Tunbridge Wells, UK, 1985.
8. A. C. Eckbreth, *Laser Diagnostics for Combustion Temperature and Species*, Abacus/Gordon and Breach, New York, 1988.
9. N. A. Chigier, *Combustion Measurements*, Hemisphere, New York, 1991.
10. W. Bartok, and A. F. Sarofim, eds., *Fossil Fuel Combustion—A Source Book*, Wiley, New York, 1991.
11. E. L. Keating, *Applied Combustion*, Marcel Dekker, New York, 1993.

12. T. Y. Toong, *Combustion Dynamics: Dynamics of Chemically Reacting Fluids*, McGraw-Hill, New York, 1983.
13. A. H. Lefebvre, *Gas Turbine Combustion*, McGraw-Hill, New York, 1983.
14. A. K. Gupta, D. G. Lilley, and N. Syred, *Swirl Flows*, Abacus, Tunbridge Wells, UK, 1984.
15. F. A. Williams, *Combustion Theory*, 2nd ed., Benjamin Cummings, Menlo Park, CA, 1985.
16. D. E. Rosner, *Transport Processes in Chemically Reacting Flow Systems*, Butterworths, Boston, 1986.
17. K. K. Kuo, *Principles of Combustion*, Wiley, New York, 1986.
18. E. S. Oran, and J. P. Boris, *Numerical Simulation of Reactive Flow*, Elsevier, New York, 1987.
19. J. Chomiak, *Combustion: A Study in Theory, Fact and Application*, Abacus/Gordon & Breach, New York, 1990.
20. R. W. Johnson, ed., *The Handbook of Fluid Dynamics*, CRC Press, Boca Raton, FL, 1998.
21. A. K. Gupta, E. Ilanchezhian, and E. L. Keating, Thermal Destruction Behavior of Plastic and Non-Plastic Wastes in a Laboratory Scale Facility, J. Energy Resources Technology, Vol. 118, 269–276 Dec. (1996).
22. T. Panagioutou, and Y. Levendis, *Combustion and Flame* **99**, 53–74 (1994).
23. O. C. Oppelt, *Air & Waste J.* **43**(Jan), 25–73 (1993).
24. D. T. Allen and N. Behmanesh, *Hazardous Waste and Mat.* **9**(1), 34–42 (1992).
25. A. K. Gupta, E. Ilanchezian, and E. L. Keating, ASME Design Technical Conference, Minneapolis, MN, Sept. 11–14, 1994.
26. D. W. Pershing, J. S. Lighty, and G. D. Silcox, *Combustion Energy and Sci.* **93**(1), 245–264 (1993).
27. H. Chopra, MS Thesis, Department of Mechanical Engineering, University of Maryland, May, 1993.
28. E. H. James and M. Narayani, *Pyrolysis Experiments with Municipal Solid Waste Components*, Publication No. 889513, ASME Waste Processing Conference, 1987, New York, NY.
29. A. K. Gupta, *J. Propulsion and Power* **15**(2), 187–194 (1999).
30. D. A. Tillman, *The Combustion of Solid Fuels and Wastes*, Academic, San Diego, CA 1991.
31. P. T. Williams, and S. Besler, *J. Institu. Energy*, (2), 192–200 (1992).
32. W. R. Seeker, *ASME* **15**, 57–62 (1992).
33. W. P. Linak and J. L. Wendt, *Progress in Energy and Combustion Sci.* **19**, 145–185 (1993).
34. C. R. Brunner, *Handbook of Incineration Systems*, McGraw-Hill, New York, 1991.
35. W. C. Reynolds, *STANJAN Chemical Equilibrium Solver*, V.3.89, IBM-PC, Stanford University, Stanford, CA, 1987.
36. T. M. Besman, *SOLGASMIX-PV for the PC*, Oak Ridge National Laboratory, Oak Ridge, TN, October 24, 1989.
37. *JANAF Thermo-chemical Tables*, 2nd ed., NSRDS-NBS 37, Washington DC, 1971.

38. F. Komatsu, A. Takusagawa, R. Wada, and K. Asahina, *Waste Manage.* **10**, 211–215 (1990).
39. P. J. Young and A. Parker, *Waste Manage. Res.* **1**, 213–226 (1992).
40. W. R. Seeker, Twenty Third Symposium (International) on Combustion, The Combustion Institute, Orleans, France, July 22–27, 1990, pp. 867–885.
41. U.S. Environmental Protection Agency, *Am. Lab.* Dec, 33–40 (1991).
42. P. T. Williams, *J. Instit. of Energy* **65**, 46–54 (1992).
43. L. Kun-Chieh, *JAPCA* **38**(12), 1542–1548 (1988).
44. Who special report, Assessment of the Health Risk of Dioxins: Re-evaluation of the Tolerable Daily Intake (TDI), WHO Consultation May, 25–29, 1998; *http://www.who.int/pcs/pubs/dioxin-exec-sum/exe-sum-final.html.*
45. Special Report on Environmental Endocrine Disruption: An Effects Assessment and Analysis, EPA/630/R-96/012, February 1997; *http://www.epa.gov/ORD/WebPubs/endocrine/.*
46. E. S. Domalski, T. L. Jobe, and T. A. Miline, *Thermodynamic Data for Biomass Materials and Waste Components*, American Society of Mechanical Engineers, United Engineering Center, New York, 1987.
47. P. Raman, W. P. Walawender, L. T. Fan, and J. A. Howell, *Ind. Engr. Chem. Process.* **20**, 630 (1981).
48. P. T. Williams and S. Besler, *J. Institut. of Energy* 192–200 Vol. LXV, No. 465, December (1992).
49. A. K. Gupta and P. Muller, *J. Propulsion and Power* **15**(2), 187–194 (1999).
50. R. K. Agrawal, ASTM-STP997, Symposium on Compositional Analysis by Thermogravimetry, Philadelphia, PA, March, 16–17, 1987.
51. A. A. Boateng, W. P. Walawender, and L. T. Fan, *Biomass for Energy and Industry*, Elsevier Applied Science, London, 1990.
52. A. Missoum, A. K. Gupta, and J. Chen, Proc. 1997 ASME Design Engineering Technical Conference and Computers in Engineering Conference, September 14–17, Paper No. DETC 97/CIE-4433, 1997.
53. A. K. Jain, S. K. Sharma, and D. Singh, IECEC Conference, Washington DC, August 11–16, Paper No. 96486, 1996.
54. D. Jinno, A. K. Gupta, and K. Yoshikawa, Proc. 26th International Technical Conference on Coal Utilization & Fuel Systems, Clearwater, FL, March 5–8, 2001, pp. 715–726.
55. H. Kanbe, *Baifu-kan*, (2), 217–251 (1974), (in Japanese).
56. P. W. Atkins, *Physical Chemistry*, Oxford University, Press, Oxford, 1990.

CHAPTER 16

RECYCLING OF CARPET AND TEXTILE FIBERS

YOUJIANG WANG,[1] YI ZHANG,[2] MALCOLM B. POLK,[1] SATISH KUMAR,[1] AND JOHN D. MUZZY[3]

[1]School of Textile and Fiber Engineering, [3]School of Chemical Engineering, Georgia Institute of Technology, [2]DuPont Teijin Films

A large amount of fibrous waste is generated each year in the United States. For economic and environmental reasons, industries and research organizations have been looking for various technologies to recycle fibrous waste. This chapter reviews the waste statistics, waste characteristics, recycling rates, and recycling technologies for fibrous waste.

16.1. INTRODUCTION

According to the U.S. Environmental Protection Agency (EPA) [1, 2] the municipal solid waste generated in the United States is about 210 million tons per year, among which about 40% is paper products, 9% is plastics, and 4% is carpet and textiles.

Most of the fibrous waste is composed of natural and synthetic polymeric materials such as cotton, polyester, nylon, polypropylene, among others. Table 16.1 presents the amount of fibrous waste including textiles and carpet. In 1996, approximately 6.8 million tons of textile waste was discarded, and 2.3 million tons of carpets and rugs (carpet contains fiber, rubber, and inorganic filler) were generated in the municipal solid waste. About 1% of carpet waste was recovered for recycling (Table 16.1). In comparison, 20 million tons of plastics were generated and 6% was recovered, and 4 million tons of rubber tire waste was generated and 19% was recovered in the United States.

There are several disadvantages associated with the current landfilling of fibrous waste. First, a tipping fee is required. Second, due to environmental concerns, there is increasing demand to ban polymers from landfills. Third, landfilling

Plastics and the Environment, Edited by Anthony L. Andrady.
ISBN 0-471-09520-6 © 2003 John Wiley & Sons, Inc.

Table 16.1 Selected Products in Municipal Solid Waste in the United States in 1996 (in thousands of tons) [1]

Products	MSW Generation	Recovered for Recycling
Carpet and rugs	2310	30 (1.3%)
Clothing and footwear	5340	700 (13.1%)
Towels, sheets, and pillowcases	750	130 (17.3%)

polymers is a waste of energy and materials. A variety of technologies have been developed in response to customer demands for recycled products and as alternatives to landfilling [3–8]. Except for the case of direct reuse, which is a common form of utilization for discarded textiles, some processing is involved to convert the waste into a product. Typically, recycling technologies are divided into primary, secondary, tertiary, and quaternary approaches. Primary approaches involve recycling a product into its original form; secondary recycling involves melt processing a plastic product into a new product that has a lower level of physical, mechanical, and/or chemical properties. Tertiary recycling involves processes such as pyrolysis and hydrolysis, which convert the plastic wastes into basic chemicals or fuels. Quaternary recycling refers to burning the fibrous solid waste and utilizing the heat generated. All these four approaches exist for fiber recycling.

16.2. TEXTILE WASTE AND RECYCLING

According to the Secondary Materials and Recycled Textiles Association (SMART) and the Council for Textile Recycling [9, 10], more than 1000 businesses and organizations employing many tens of thousands of workers divert some 2 million tons of textile waste from the solid waste stream. Textile waste can be classified as either preconsumer or postconsumer. Preconsumer textile waste consists of by-product materials from the textile, fiber, and cotton industries. Each year 750,000 tons of this waste is recycled into raw materials for the automotive, furniture, mattress, coarse yarn, home furnishings, paper, and other industries. Approximately 75% of the preconsumer textile waste is recycled.

Postconsumer textile waste consists of any type of garments or household article, made of some manufactured textile, that the owner no longer needs and decides to discard. These articles are discarded either because they are worn out, damaged, outgrown, or have gone out of fashion. They are sometimes given to charities but more typically are disposed of into the trash and end up in municipal landfills. Approximately 1,250,000 tons of postconsumer textile waste (4.5 kg per capita) is recycled annually. However, the recycled amount represents less than 25% of the total postconsumer textile waste that is generated. Almost half (48%) of the recovered postconsumer textile waste is recycled as secondhand clothing, which is typically sold to developing nations. Approximately 20%

of the material processed becomes wiping and polishing cloths. Finally, 26% of this postconsumer waste is converted into fiber to be used in products similar in nature to those manufactured from preconsumer textile waste.

16.3. CARPET WASTE AND COMPOSITION

During carpet manufacturing, the edges of a tufted carpet need to be trimmed and the face yarns sheared. This waste is approximately 60% edge trim and 40% shear lint. In the Dalton, Georgia, region where there is a high concentration of carpet manufacturing activities, over 20,000 tons of carpet manufacturing waste is generated every year. The carpet waste from the fitting process is concentrated in the automotive and prefabricated housing industries [11, 12]. During the fitting process, the carpet is formed and cut into various irregular shapes, and waste is generated as a result. The largest amount of carpet waste, however, is from the discarded postconsumer carpet. It is estimated that about 2.3 million tons of carpet and rugs were placed in the municipal solid waste stream in 1996, of which only 1% was recovered for recycling.

Carpet is a complex, multicomponent system. The tufted carpet, the most common type (90%) as shown in Figure 16.1, typically consists of two layers of backing (mostly polypropylene fabrics), joined by $CaCO_3$-filled styrene–butadiene latex rubber (SBR), and face fibers (majority being nylon 6 and nylon 6,6 textured yarns) tufted into the primary backing. The SBR adhesive is a thermoset material, which cannot be remelted or reshaped. The waste containing the SBR (postconsumer and some industrial waste) has not found suitable uses, and it forms the major part of the carpet waste going into the landfills. Figure 16.2 shows the typical masses for the various components [13].

Because about 70% of the carpet produced is for replacing old carpet, it is important to understand the amount and types of carpet produced. Using the typical life of a carpet of 5–10 years, one can estimate the amount of carpet being disposed of currently and to be disposed of in the next few years. According to carpet industry statistics [14], the total fiber consumption in 1999 was about 1.7 million tons: nylon 57%, olefin 36%, polyester 7%, and wool 0.4%. Among the nylon face fiber, about 40% is nylon 6 and 60% is nylon 6,6. In recent years, the use of polypropylene (labeled as polyolefin or olefin) in carpet is increasing, as tabulated in Table 16.2 [15]. Based on the aforementioned data, one expects the current rate of carpet disposal to be about 2 million tons

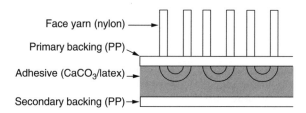

Figure 16.1. Typical carpet construction.

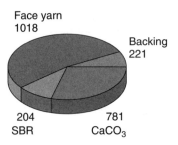

Figure 16.2. Component mass/area for a typical carpet (g/m^2). Total is 2223 g/m^2.

Table 16.2 Relative Market Share of Top Three Synthetic Fibers Used by the U.S. Carpet Industry [15]

Year	1992	1996	2000
Nylon	66%	62%	55%
Polyester	10%	6%	7%
Polypropylene	24%	32%	37%
Total (million tons)	1.4	1.6	1.7

per year, or about 70% of the shipment in 1996. The amount of the various materials used in carpet manufacturing in the United States for 1996 is given in Table 16.3.

Nylon generally performs the best among all synthetic fibers as carpet face yarn, but it is also the most expensive. Typical price per kilogram for the plastic resins are: nylon $2.50, polyester $1.20, and polypropylene $0.75. This price list provides a perspective on the economics of recycling as well. For example, if it takes the same processing effort to convert the fiber into resin, an operation on nylon would be most profitable. This also explains why most of the recycling effort is on nylon recovery.

Table 16.3 Estimate of Amount of Materials in Carpet Produced in 1996

	Mass (million tons)	Percent
Nylon 6	0.4	14
Nylon 6,6	0.6	21
Polyester	0.1	3
Polypropylene	0.5	18
Adhesive/filler	1.3	44
Total	2.9	100

16.4. FIBER RECYCLING TECHNOLOGIES

Many carpet manufacturers, fiber and chemical suppliers, recycling companies, and academic institutions are actively pursuing various methods to recycle fibrous waste. The approaches include chemical processes to depolymerize nylon and other polymers, recovery of plastic resins from carpet fibers, direct extrusion of mixed carpet waste, composites as wood substitutes, fibers for concrete and soil reinforcement, waste-to-energy conversion, and carpet as feedstock for cement kilns.

16.4.1. Depolymerization of Nylons

Because of the higher value of nylon resin in comparison with other polymers used in carpet, nylon carpet has been looked at as a resource for making virgin nylon via depolymerization. The majority of polyamides used commercially are nylon 6,6 or nylon 6, and the largest supply of waste for recycling of nylons is obtained from used carpets. The waste carpets are collected, sorted, and then subjected to a mechanical shredding process before depolymerization.

Hydrolysis of Nylon 6 A process for depolymerizing nylon 6 scrap using high-pressure steam was patented by Allied Chemical Corporation in 1965 [16], and subsequent patents by AlliedSignal, Inc. were obtained [17, 18]. Ground scrap was dissolved in high-pressure steam at 125–130 psig (963–997 kPa) and 175–180°C for 0.5 h in a batch process and then continuously hydrolyzed with superheated steam at 350°C and 100 psig (790 kPa) to form ε-caprolactam at an overall recovery efficiency of 98%. The recovered monomer could be repolymerized without additional purification. Braun et al. [13] reported the depolymerization of nylon 6 carpet in a small laboratory apparatus with steam at 340°C and 1500 kPa (200 psig) for 3 h to obtain a 95% yield of crude ε-caprolactam of purity 94.4%. Recently, patents were issued to AlliedSignal for the depolymerization of polyamide-containing carpet [19, 20].

Acid hydrolysis of nylon 6 wastes [21, 22] in the presence of superheated steam has been used to produce aminocaproic acid, which under acid conditions is converted to ε-caprolactam, and several patents have been obtained by BASF [23, 24]. Acids used for the depolymerization of nylon 6 include inorganic or organic acids such as nitric acid, formic acid, benzoic acid, and hydrochloric acid [23, 25]. Orthophosphoric acid [24] and boric acid are typically used as catalysts at temperatures of 250–350°C. In a typical process, superheated steam is passed through the molten nylon 6 waste at 250–300°C in the presence of phosphoric acid. The resulting solution underwent a multistage chemical purification before concentration to 70% liquor, which was fractionally distilled in the presence of base to recover pure ε-caprolactam. Boric acid (1%) may be used to depolymerize nylon 6 at 400°C under ambient pressure. A recovery of 93–95% ε-caprolactam was obtained by passing superheated steam through molten nylon 6 at 250–350°C [23].

Sodium hydroxide has been used successfully as a catalyst for the base-catalyzed depolymerization of nylon 6. At 250°C, a pressure of 400 Pa, and a sodium hydroxide content of 1%, the yield of ε-caprolactam was 90.5% [26].

Catalytic Pyrolysis Catalytic pyrolysis has been studied as a hybrid process for recovering caprolactam from nylon 6 followed by high-temperature pyrolysis of the polypropylene into a synthetic natural gas. Czernik et al. [27] investigated the catalysis of the thermal degradation of nylon 6 with an α-alumina supported KOH catalyst in a fluidized-bed reactor. In the temperature range of 330–360°C the yield of caprolactam exceeded 85%.

Bockhorn et al. [28] used a liquid catalyst composed of a eutectic mixture of 60 mol% NaOH and 40 mol% KOH, which melts at 185°C. At 290°C the caprolactam yield exceeded 95%. At this temperature the polypropylene is not degraded significantly. Based on a preliminary feasibility study, this process could be economically viable [29].

Recovery of Caprolactam Approximately 10–12% by weight of oligomers is formed in the synthesis of polycaprolactam (nylon 6). These oligomers are removed by extraction with water or by distillation under vacuum. In the process, two types of liquid wastes are formed: (1) a 4–5% aqueous solution of low-molecular-weight compounds, consisting of ca. 75% by weight of caprolactam and ca. 25% by weight of a mixture of cyclic and linear caprolactam oligomers and (2) a caprolactam–oligomer melt, containing up to 98% caprolactam and small amounts of dimer, water, and organic contaminants. The recycle of caprolactam involves two different stages: depolymerization of polymeric waste and purification of the caprolactam and oligomers obtained.

A general recovery of caprolactam from liquid waste generates 20–25% oligomers along with organic and inorganic compounds as impurities. The distillation of caprolactam under reduced pressure produces a residue that consists of inorganic substances such as permanganates, potassium hydrogen sulfate, potassium sulfate, sodium hydrogen phosphate, and sodium phosphate. The larger portion of the residue contains cyclic and linear chain oligomers plus 8–10% of caprolactam. The types and exact amounts of impurities depends on the method used for the purification and distillation of caprolactam.

The cyclic oligomers are only slightly soluble in water and dilute solutions of caprolactam. They tend to separate out from the extracted waste during the process of concentration and chemical purification of the caprolactam. The cyclic oligomers tend to form on the walls of the equipment used in the process equipment. 6-Aminocaproic acid or sodium 6-aminocaproate may also be found in the oligomeric waste, especially if sodium hydroxide is used to initiate the caprolactam polymerization.

Many impurities are present in commercial caprolactam that pass into the liquid wastes from polycaprolactam (PCA) manufacture from which caprolactam monomer may be recovered. Also, the products of the thermal degradation of PCA, dyes, lubricants, and other PCA fillers may be contained in the regenerated caprolactam. Identification of the contaminants by infrared (IR) spectroscopy has led to the detection of lower carboxylic acids, secondary amines, ketones, and esters. Aldehydes and hydroperoxides have been identified by polarography and thin-layer chromatography.

Caprolactam is a thermally unstable compound that on distillation may form methyl-, ethyl-, propyl- and n-amylamines. Also, at high temperatures, caprolactam reacts with oxygen to form hydroperoxides that in the presence of iron or cobalt ions are converted into adipimide. N-alkoxy compounds are also formed by the reaction of caprolactam with aldehydes during storage.

Therefore, caprolactam and the depolymerized product from which caprolactam is regenerated contain various impurities, which are present in widely fluctuating amounts depending on the processes involved. In particular, the presence of cyclohexanone, cyclohexanone oxime, octahydrophenazine, aniline, and other easily oxidized compounds affects the permanganate number. Also volatile bases such as aniline, cyclohexylamine, cyclohexanol, cyclohexanone, nitrocyclohexanone, and aliphatic amines may be present in the caprolactam.

Caprolactam must be very carefully purified to exclude small concentrations of (1) ferric ions, which would catalyze the thermal oxidative degradation of polycaprolactam and (2) aldehydes and ketones, which would markedly increase the oxidizability of caprolactam. The impurities in caprolactam may retard the rate of caprolactam polymerization as well as having a harmful effect on the properties of the polymer and fiber. In the vacuum depolymerization of nylon 6, a catalyst must be used because in the absence of a catalyst, by-products such as cyclic olefins and nitrides may form, which affects the quality of the caprolactam obtained [3].

The caprolactam obtained must meet the specifications of permanganate number, volatile bases, hazen color, ultraviolet (UV) transmittance, solidification point, and turbidity, in order to be used alone or in combination with virgin caprolactam [25]. Reported caprolactam purification methods include recrystallization, solvent extraction, and fractional distillation. One solvent extraction technique involves membrane solvent extraction. Ion-exchange resins have been shown to be effective in the purification of aqueous caprolactam solutions. In one such process, the oily impurities are removed by extraction with organic solvents, followed after treatment with carbon at 60–80°C. Cationic and anionic exchange resins are then used to complete the purification process. Ion-exchange resins remove all ionic impurities as well as colloidal and floating particles, that is, alkali metal salts formed in permanganate treatment are removed during ion-exchange treatment. Also the treatment of aqueous solutions of caprolactam with ion-exchange resins helps to remove the distillation residue. Treatment of caprolactam with activated carbon helps to remove anionic and cationic impurities.

Impurities in caprolactam have also been destroyed by oxidation with ozone followed by distillation. Ozonation treatment of waste caprolactam leaves no ionic impurities. However, the most commonly used oxidizing agents are potassium permanganate, perboric acid, perborate, and potassium bromate. Treatment of caprolactam with these oxidizing agents is carried out in a neutral medium at 40–60°C. Strongly alkaline or acidic conditions accelerate the oxidation of caprolactam to form isocyanates. The undesirable oxidation reaction is fast above pH 7 because of the reaction with isocyanate to form carbamic acid salts, which shifts the equilibrium to form additional isocyanate.

In a typical process, potassium permanganate is used to treat the cracked liquor exiting the depolymerization plant without any pH adjustment. The liquor is usually acidic because it contains some of the phosphoric acid depolymerization catalyst. The $KMnO_4$ treatment is followed by treatment of the caprolactam aqueous solution with carbon followed by filtration. Next the filtered 20–30% caprolactam aqueous solution is concentrated to 70% and the pH is adjusted to 9–10 by addition of sodium hydroxide. The caprolactam alkaline concentrate is treated with $KMnO_4$ followed by distillation under reduced pressure to remove water and low-boiling impurities.

Also the caprolactam aqueous solution may be hydrogenated at 60°C in the presence of 20% sodium hydroxide and 50% palladium absorbed on carbon to provide caprolactam of very high purity after distillation. Treatment with an ion-exchange resin before or after the oxidation or hydrogenation process also improves the quality of the caprolactam obtained after distillation. Caprolactam has also been purified by treatment with alkali and formaldehyde followed by fractional distillation to remove aromatic amines and other products.

Also, nylon 6 waste may be hydrolyzed in the presence of an aqueous alkali metal hydroxide or acid to produce an alkali metal or acid salt of 6-aminocaproic acid (ACA). The reaction of nylon 6 waste with dilute hydrochloric acid is very fast at 90–100°C. The reaction mixture is poured into water to form a dilute aqueous solution of the ACA salt. Filtration is used to remove undissolved impurities such as pigments, additives, and fillers followed by treatment of the acid solution with a strong cation-exchange resin. A sulfonic acid cationic-exchanger absorbs ACA, and pure ACA is eluted with ammonium hydroxide to form a dilute aqueous solution. Pure ACA is obtained by crystallization of the solution.

Alternatively, nylon 6 waste may be hydrolyzed with aqueous sodium hydroxide, and the sodium salt of ACA converted into pure ACA by passing the aqueous solution through an anion-exchange resin.

Applications of Depolymerized Nylon 6 Chemical recycling of nylon 6 carpet face fibers has been developed into a closed-loop recycling process for waste nylon carpet [25, 30–32]. The recovered nylon 6 face fibers are sent to a depolymerization reactor and treated with superheated steam in the presence of a catalyst to produce a distillate containing caprolactam. The crude caprolactam is distilled and repolymerized to form nylon 6. The caprolactam obtained is comparable to virgin caprolactam in purity. The repolymerized nylon 6 is converted into yarn and tufted into carpet. The carpets obtained from this process are very similar in physical properties to those obtained from virgin caprolactam.

The "6ix Again" program of the BASF Corp. has been in operation since 1994. Its process involves collection of used nylon 6 carpet, shredding and separation of face fibers, pelletizing face fiber for depolymerization and chemical distillation to obtain a purified caprolactam monomer, and repolymerization of caprolactam into nylon polymer [32].

Evergreen Nylon Recycling LLC, a joint venture between Honeywell International and DSM Chemicals, was in operation from 1999 to 2001. It used a two-stage selective pyrolysis process. The ground nylon scrap is dissolved with high-pressure steam and then continuously hydrolyzed with superheated steam to form caprolactam. The program has diverted over 100,000 tons of postconsumer carpet from the landfill to produced virgin-quality caprolactam [30, 31].

Hydrolysis of Nylon 6,6 and Nylon 4,6 The depolymerization of nylon 6,6 and nylon 4,6 involves hydrolysis of the amide linkages, which are vulnerable to both acid- and base-catalyzed hydrolysis. In a patent granted to the DuPont Company in 1946, Myers [33] described the hydrolysis of nylon 6,6 with concentrated sulfuric acid, which led to the crystallization of adipic acid from the solution. Hexamethylene diamine (HMDA) was recovered from the neutralized solution by distillation. In a later patent assigned to the DuPont Company by Miller [34], a process was described for hydrolyzing nylon 6,6 waste with aqueous sodium hydroxide in isopropanol at 180°C and 2.2 MPa pressure. After distillation of the residue, HMDA was isolated and on acidification of the aqueous phase, adipic acid was obtained in 92% yield. Thorburn [35] depolymerized nylon 6,6 fibers in an inert atmosphere at what was reported to be a superatmospheric pressure of up to 1.5 MPa and at a temperature in the range of 160–220°C in an aqueous solution containing at least 20% excess equivalents of sodium hydroxide.

Polk et al. [36] reported the depolymerization of nylon 6,6 and nylon 4,6 in aqueous sodium hydroxide solutions containing a phase-transfer catalyst. Benzyltrimethylammonium bromide was discovered to be an effective phase-transfer catalyst in 50% sodium hydroxide solution for the conversion of nylon 4,6 to oligomers. The depolymerization efficiency (percent weight loss) and the molecular weight of the reclaimed oligomers were dependent on the amount and concentration of the aqueous sodium hydroxide and the reaction time. Table 16.4 exhibits the effects of experimental conditions on the depolymerization efficiency and the average molecular weight of the oligomers. The viscosity-average molecular weight was calculated from the Mark–Houwink equation: $[\eta] = K M_v^a$, where M_v is the viscosity-average molecular weight, $K = 4.64 \times 10^{-2}$ dL/g and $a = 0.76$ at 25°C in 88% formic acid. Nylon 4,6 fibers ($M_v = 41,400$ g/mol) did not undergo depolymerization on exposure to 100 mL of 25 wt% sodium hydroxide solution at 165°C. Out of 6.0 g of nylon fibers fed for depolymerization, 5.95 g were unaffected. When the concentration of sodium hydroxide was increased to 50 wt%, the depolymerization process resulted in the formation of low-molecular-weight oligomers. Hence, even in the presence of a phase-transfer agent, a critical sodium hydroxide concentration exists between 25 and 50 wt% which is required to initiate depolymerization under the conditions used. Soluble amine salts were also obtained.

In order to establish the feasibility of alkaline hydrolysis with respect to recycling of nylon 4,6, it was necessary to determine whether the recovered oligomers could be repolymerized to form nylon 4,6. For this purpose, solid-state polymerization was performed on nylon 4,6 oligomers formed via alkaline hydrolysis with 50 wt% NaOH at 165°C for 24 h. The solid-state polymerization process was carried out in a round-bottom flask at 210°C for 16 h under vacuum. Solid-state polymerization of the nylon 4,6 oligomers resulted in an

Table 16.4 Effects of Experimental Conditions on Depolymerization Efficiency and Average Molecular Weight of Oligomers [36]

Experiment	1	2	3	4	5	6
Feed ratio of nylon 6,6/BTEMB (wt/wt)	5	10.3	20.6	29.4	59.8	no PTA
Decrease in weight of oligomers (%)	40.3	49.5	55.5	42.5	40.8	−15.9
M_v of oligomers	1556		1912	1697	2396	1644

increase in intrinsic viscosity from 0.141 to 0.740 dL/g. That corresponds to an increase in viscosity average molecular weight from 1846 to 16,343 g/mol. In theory, higher molecular weights would be obtained by heating for a longer time interval.

The product of the depolymerization of nylon 6,6 with 50% aqueous sodium hydroxide solution was relatively low molecular weight oligomers. A series of experiments were run in order to examine the applicability and efficiency of benzyltrimethylammonium bromide (BTEMB) as a phase-transfer catalyst in the depolymerization of nylon 6,6. Table 16.4 shows the effect of the feed ratio of the nylon 6,6 to BTEMB on the viscosity-average molecular weight of the depolymerized nylon. The product of the run with no phase-transfer agent showed a 15.9% increase in weight compared to the weight of the original nylon 6,6. The calculated percent increase in weight for a 19-fold decrease in molecular weight (due to the addition of water) would be ca. 1%. Therefore, a large part of the increase must be due to leaching of silicates of the glass container (resin reaction kettle) by the strong alkali (50 wt%) at the temperature of the reaction (130°C) over 24 h. The oligomer obtained had a viscosity-average molecular weight of 1644 g/mol (the original nylon 6,6 had a molecular weight of 30,944 g/mol). The runs with phase-transfer agent produced oligomers with decreases in weight of 40–50%. Although the occurrence of leaching of silicates from the glass container made quantitative assessment difficult, these results suggested that in the absence of phase-transfer agent only oligomers are formed; however, soluble low-molecular-weight products are formed in the presence of phase-transfer agent. The oligomers obtained were repolymerized in the solid-state by heating at 200°C in a vacuum. The viscosity-average molecular weight of the solid-state polymerized nylon 6,6 obtained was ca. 23,000 g/mol (the molecular weight of the oligomeric mixture was 1434 g/mol).

In order to isolate adipic acid, nylon 6,6 fibers were depolymerized under reflux with a 50% NaOH solution in the presence of catalytic amounts of benzyltrimethylammonium bromide. The oligomers formed in successive steps were depolymerized under similar conditions. The yields in steps 1, 2, and 3 were 57.8, 38.7, and 100% theoretical. However, hexamethylene diamine was not isolated. The overall yield of adipic acid was 59.6%.

Ammonolysis of Nylon 6,6 Ammonolysis is the preferred route currently in use at the DuPont Company for the depolymerization of nylon 6,6 carpet waste [7, 37]. McKinney [38] has described the reaction of nylon 6,6 and nylon 6,6/nylon 6 mixtures with ammonia at temperatures between 300 and 350°C and a pressure of about 68 atm in the presence of an ammonium phosphate catalyst to yield a mixture of the following monomeric products: HMDA, adiponitrile, and 5-cyanovaleramide from nylon 6,6 and ε-caprolactam, 6-aminocapronitrile, and 6-aminocaproamide from nylon 6. The equilibrium is shifted toward products by continuous removal of water formed. Most of the monomers may be transformed into HMDA by hydrogenation. Kalfas [39] has developed a mechanism for the depolymerization of nylon-6,6 and nylon-6 mixtures by the ammonolysis process. The mechanism includes the amide bond breakage and amide end dehydration (nitrilation) reactions, plus the ring addition and ring-opening reactions for cyclic lactams present in nylon 6. On the basis of the proposed mechanism, a kinetic model was developed for the ammonolysis of nylon mixtures.

Bordrero et al. [40] utilized a two-step ami/ammonolysis process to depolymerize nylon 6,6. The first step is based on an aminolysis treatment of nylon 6.6 by n-butylamine at a temperature of 300°C and a pressure of 45 atm. Free HMDA and NN'-dibutyladipamide are generated. The second step is ammonolysis of NN'-dibutyladipamide at a temperature of 285°C and a pressure of 50 atm. The end product is adiponitrile (ADN). It is estimated that the yields could be about 48% for ADN and about 100% for HMDA at optimized reaction conditions.

Recovery of Nylon 6,6 Monomers Adipic acid and HMDA are obtained from nylon 6,6 by the hydrolysis of the polymer in concentrated sulfuric acid (Fig. 16.3). The adipic acid is purified by recrystallization, and the HMDA is recovered by distillation after neutralizing the acid. This process is inefficient for treating large amounts of waste because of the required recrystallization of adipic acid after repeated batch hydrolyses of nylon 6,6 waste. In a continuous process [25], nylon 6,6 waste is hydrolyzed with an aqueous mineral acid of 30–70% concentration and the resulting hydrolysate is fed to a crystallization zone. The adipic acid crystallizes and the crystals are continuously removed from the hydrolysate. Calcium hydroxide is added to neutralize the mother liquor and liberate the HMDA for subsequent distillation.

Continuous recovery requires adipic acid crystals having an average diameter of ca. 40–50 nm. Such crystals are obtained by continuously introducing the hot hydrolysate containing 10–20% adipic acid into an agitated crystallization vessel while maintaining an average temperature of 20–30°C. The slurry obtained from

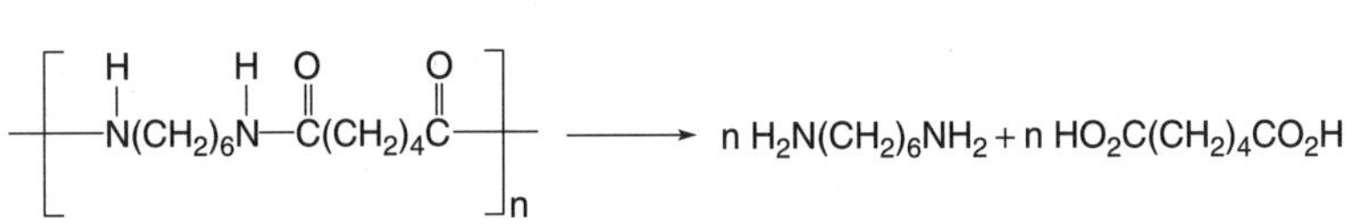

$$[-N(CH_2)_6N(H)-C(O)(CH_2)_4C(O)-]_n \longrightarrow n\ H_2N(CH_2)_6NH_2 + n\ HO_2C(CH_2)_4CO_2H$$

Figure 16.3. Depolymerization of nylon 6,6 by hydrolysis.

the crystallization vessel is filtered to collect the adipic acid crystals and the filtrate, which contains the HMDA acid salt, is continuously neutralized with calcium hydroxide. The calcium salt formed is removed by filtration and the HMDA in the filtrate is isolated by distillation.

In the case of nylon 6,6 waste recycled by ammonolysis, nylon is treated with ammonia in the presence of a phosphate catalyst. Reaction occurs at 330°C and 7 MPa. Distillation of the reaction mixture produces ammonia that is recycled and three fractions containing (a) caprolactam, (b) HMDA and aminocapronitrile, and (c) adiponitrile. Aminocapronitrile and adiponitrile are hydrogenated to yield pure HMDA, and the caprolactam is either converted to aminocapronitrile by further ammonolysis or distilled to produce pure caprolactam. The HMDA produced by this process is extremely pure ($>$ 99.8) [3]. The main impurities are aminomethylcyclopentylamine and tetrahydroazepine, which are expected to be removed more effectively in the larger distillation columns employed in larger plants.

16.4.2. Fiber Identification and Sorting

For many recycling processes such as nylon depolymerization and polymer resin recovery, it is desirable or required to sort the feedstock according to the fiber type. For carpet, the sorting is according to the type of the face fiber. Melt point indicator is an inexpensive instrument that can identify most fiber types, but it is generally slow and cannot distinguish between nylon 6,6 and polyester. Infrared spectroscopy is a much faster and accurate technology. A typical instrument consists of an alternating current (ac)-powered base unit for data acquisition, analysis, and display, and a probe connected to the based unit via a fiber-optic cable. Such units are suitable for carpet sorting in a central warehouse. A portable infrared spectrometer has been developed by Kip et al. [41], which is a lightweight, battery-operated unit. It is designed to identify the common carpet face fibers: nylon 6, nylon 6,6, polypropylene, polyester, and wool. Unidentifiable fibers, either due to operating conditions or fiber types other than those in the above list, would be shown as "unknown."

16.4.3. Mechanical Separation of Polymers from Carpet

Mechanical methods have been utilized to separate carpet components. One or more segregated components then are recycled into products that generally compete with products produced from virgin polymers.

In a process developed by DuPont [7, 42], nylon 6,6 carpet first is passed through dry processes consisting of a series of size reduction and separation steps. This provides a dry mix of 50–70% nylon, 15–25% polypropylene, and 15–20% latex, fillers, and dirt. Water is added in the second step where the shredded fiber is washed and separated using the density differences between the fillers, nylon, and polypropylene. Two product streams are obtained: one 98% pure nylon and the other is 98% pure polypropylene. The recycled nylon is compounded with the virgin nylon at a ratio of 1:3 for making automotive parts.

The United Recycling process [43, 44] starts with clipping the face fibers on loop carpet to open the loops. The next step is debonding in which the carpet is bombarded with a combination of air and steam to loosen the calcium-carbonate-filled latex backing. The secondary backing then is peeled off mechanically, exposing the primary polypropylene backing. Next, mechanical picks pluck the face fibers. It is claimed that the cost of this process is low, and that it yields a product stream with 93–95% pure face fibers. Other devices employing water jet [45] or mechanical actions [46–53] for size reduction and separation of carpet have also been reported. Many types of equipment are commercially available for processing textile and carpet waste [54].

A process has been developed to separate carpet waste using a cold, dry abrasive step [55]. Dry ice pellets are shot into an abrasive zone as a segment of discarded carpet on a conveyor system is stripped apart and disassembled. The dry ice pellets freeze the binder material (usually latex), lowering it to a temperature that makes the binder brittle and easy to break apart. The dry ice pellets sublimate directly into gas without any liquid residues. This process eliminates the need for a drying operation, which saves energy and avoids potential chemical pollution.

16.4.4. Physical/Chemical Separation of Polymers from Carpet

Solvent extraction has also been used to separate the high value nylon from carpet waste. The solvents used are aliphatic alcohol [56], alkyl phenols [57], and hydrochloric acid [58]. In the process developed by Booij et al. [56], the carpet waste is shredded into 0.5 to 20-cm^2 pieces. Then the carpet pieces are mixed with the extraction agent, such as methanol. The weight ratio of solvent-to-carpet waste is generally 5–20. An extraction time of 60 min is found to be sufficient to dissolve nylon 6 at a temperature of 135–140°C and a pressure of 0.2–2 MPa. Solids are filtered out, the solution cooled down, and the nylon 6 precipitated. The nylon obtained has at least 90% of the relative viscosity of the nylon present in the carpet waste, indicating that no serious degradation takes place in the extraction process. In addition, the yield of nylon is high (above 90%). The drawbacks of solvent extraction are the chemicals involved, modest temperature and pressures required, and time required. In comparison, the use of hydrochloric acid [58] as a solvent requires lower temperature (20–100°C) and shorter dissolution time (2–30 min). Based on relative viscosity data, no degradation of the recycled nylon is observed. However, hydrochloric acid solvent is not recyclable due to its reaction with the calcium carbonate filler in the carpet waste.

Another approach to separate carpet components is to use a supercritical fluid (SCF) method [59, 60]. The solubility of the polymer changes with the variation in pressure and temperature of the SCF. Sikorski [59] disclosed that the individual polymers in carpet could be extracted sequentially using an SCF such as CO_2 by increasing temperature and pressure. However, high temperatures (170–210°C) and pressures (50–1000 atm) are required to dissolve the various polymers in the SCF solvent. A recent development enables the separation of carpet waste at close to room temperature and moderate pressure [60]. Up to 2.3 wt% nylon was dissolved in an 88 wt% formic acid solution. Then supercritical CO_2 as an antisolvent was added to precipitate the nylon out of solution at a temperature of 40°C and a pressure between 84 and 125 atm. Both the solvent and the antisolvent can be recycled. The whole process is very controllable and the resultant nylon is of high quality.

16.4.5. Plastic Resin Compounds from Waste

Most carpet waste contains two immiscible plastics: nylon and polypropylene. The immiscibility of these two components leads to poor mechanical properties. When carpet is recycled using melt blending, compatibilizers are used to improve the properties of the blends [61–67].

United Recycling Inc. (URI), which was in operation from the early 1990s to 1999, introduced two extruded blends (URI 20-001 and URI 10-001) from postconsumer carpet waste for injection molding in 1993 [61]. These were the first commercial recycled carpet compounds. The process used both polypropylene and nylon carpet. Their products were described as proprietary blends containing polypropylene. The composition and properties of the two compounds are listed in Table 16.5. The molding compounds developed by URI were used to extrude carpet tack strip.

In 1994, Monsanto patented a process to recycle all the components of postconsumer nylon 6,6 carpet, without separation, into a filled thermoplastic product suitable for injection molding [62, 63]. It used a twin-screw extruder to accomplish high-intensity mixing of the thermoplastic from carpet samples. The recycled material contained 35–67 wt% nylon, 8–21 wt% polypropylene, 5–29 wt% SBR, and 10–40 wt% inorganic filler. In one study, no compatibilizer was used [62]. The carpet samples were fed directly into a twin-screw extruder operating at about 250–260°C and at a shear rate of 200–400 s^{-1}. The tensile

Table 16.5 Composition and Properties of Molding Compounds Produced by United Recycling, Inc. from Carpet [61]

	URI 20-001	URI 10-001
Composition		
Nylon 6 (min. %)	60	75
PP (max. %)	15	10
Other polymers (max. %)	10	10
Inorganic filler (max. %)	15	15
Moisture (max. %)	2	2
Properties		
Tensile strength (MPa)	38	23
Elongation at break (%)	10	10
Flexural modulus (MPa)	790	760
Melt flow index (g/10 min)	6.0	8.6
Extrusion/injection temp. (°C)	260–288	232–288

Table 16.6 Comparison Between Extruded Carpets and Virgin Plastics [62]

Samples	Tensile Properties			Impact Properties
	Tensile Strength (MPa)	Tensile Modulus (GPa)	Elongation at Break (%)	Izod Impact Strength (J/m)
Extruded carpets	26–31	2.1–2.2	4.5–6.7	41–57
Control sample, polystyrene	34	2.8	2–3	22
Nylon 6,6[a]	77	1.7	>200	267

[a] Data for nylon 6,6 are from *'Nylon Plastics Handbook'* [68]. All the data are collected at room temperature and 50% RH.

and Izod strengths of the extruded carpet were compared with those of the control sample, virgin polystyrene (Table 16.6). The properties of the directly extruded thermoplastic were comparable to those of the virgin polystyrene. However, the properties of the extruded carpet were lower than those of virgin nylon 6,6 [68]. In their subsequent work [63], a maleic anhydride grafted polypropylene, PolyBond 3150, was added to compatibilize the nylon and polypropylene in the carpet waste. Addition of 3 wt% PolyBond 3150 resulted in a tensile strength of about 40 MPa, tensile modulus of 3.1 GPa, and Izod impact strength of 35 J/m. The elongation at break was low, about 2.3%. These properties appeared promising enough for the product to compete with some virgin polymers.

Studies have been carried out on the compatibilization of polypropylene and nylon using maleated polypropylene, PolyBond 3002 [64]. An addition of 3 wt% of PolyBond 3002 to the nylon–polypropylene carpet increased the blend tensile strength to the level of virgin nylon 6,6. In another study [65], a compatibilizer, Kraton (Kraton is a trademark of Shell Company), and a toughening agent, styrene–ethylene/butylene–styrene block copolymer (SEBS) were also used in the extrusion process. It was found that the addition of compatibilizer or toughening agent improved the strain to failure. The addition of Kraton increased the strain and the work of rupture significantly while the tensile strength was decreased slightly (Table 16.7). This decrease in tensile strength was attributed to the elastomeric character of Kraton.

Young et al. [66] patented a recycling method of preparing a polymeric blend formed from carpet scrap through the use of selected compatibilizing agents — PolyBond, Kraton, or polyethylene-*co*-vinyl acetate (EVA). The carpet scrap was from automotive carpet, and the composition of the waste was significantly different from that of the residential carpet waste. The automotive carpet scrap contained 14 wt% nylon 6,6, 4 wt% PP, 11 wt% EVA, and 71 wt% $BaSO_4$-filled EVA. The addition of the compatibilizing agents significantly improved the mechanical properties. For example, PolyBond 1001, even at

Table 16.7 Effect of Weight Fractions of Kraton on the Mechanical Properties of Extruded Carpets.[a]

Kraton Content (wt%)	Tensile Strength (MPa)	Strain at Failure (%)	Work of Rupture (J/cm^3)
0	46	7	1.6
5	43	9	2.3
10	38	15	3.5
15	37	31	8.0

[a]The carpet was composed of approximately 70% nylon, 20% PP, and 10% SBR and calcium carbonate [65].

2 wt%, increased the tensile strength from 4.7 to 9.1 MPa, elongation at break from 7.6 to 17.5%, and Izod impact strength (un-notched) from 254 to 471 J/m.

Zegler and Weinle [69] developed a process to make the secondary backing for new carpet from the shredded carpet waste. The carpet waste could contain nylon 6, nylon 6,6, polyvinyl chloride, vinyl copolymer, polypropylene, and polyethylene. The content of nylon was about 15–50 wt%. The chopped carpet waste was extruded at a temperature of 215°C, then calendared and bonded to a primary backing of the new carpet through an adhesive coating.

16.4.6. Use of Recycled Polymers in Glass-Fiber-Reinforced Composites

Glass fiber reinforcement can be used to enhance the properties of melt-processed carpet. In the study by Hagberg et al. [63], carpet waste was first compatibilized with 10% maleic anhydride grafted polypropylene (PP-*g*-MA), and then compounded with 15 or 30% glass in an extrusion process as shown in Figure 16.4. The average composition of the carpet waste is 50/30/15/5:nylon/CaCO$_3$/PP/SBR. Mechanical properties of the glass-filled and unfilled carpet waste are compared in Table 16.8. The addition of 30% glass fibers improved the tensile strength by about 180%, tensile modulus by about 190%, and Izod impact strength by about 130%. The properties of the glass-filled carpet waste are competitive with several commercial resins.

Mantia et al. [70] used short glass fibers to reinforce a PET/HDPE mixture (Table 16.9), which is one of the typical compositions in packaging waste. The mechanical properties of the blends, except for the elongation at break, are enhanced with the increase of glass fiber content. The addition of 20 wt% glass in PET/HDPE blends increased the tensile modulus by 50%, tensile strength by 110%, and impact strength by 70%. In addition, the heat distortion temperature of the blend was remarkably improved.

One of the important applications of polypropylene is to make glass-mat-reinforced thermoplastic (GMT) composites. Such composites are extensively

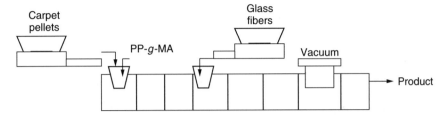

Figure 16.4. Extrusion process developed by Monsanto for recycling carpet waste [63].

Table 16.8 Effect of the Addition of Glass Fibers on the Carpet Waste Compatibilized with 10% Maleic Anhydride Grafted PP (PP-g-MA)[a]

Samples	Tensile Strength (MPa)	Tensile Modulus (GPa)	Izod Impact Strength (J/m)
Unfilled carpet waste	42	3.0	37
Carpet waste with 15% glass	77	5.2	69
Carpet waste with 30% glass	117	8.8	87

[a] Estimates based on the charts presented in Ref. [63].

Table 16.9 Mechanical Properties of Glass Fiber (GF) Filled PET/HDPE Blends [70]

Blend	Tensile Modulus (GPa)	Tensile Strength (MPa)	Elongation at break (%)	Impact Strength (J/m)	Heat Distortion Temperature (°C)
PET/HDPE	1.6	12	1.3	19	120
PET/HDPE + 10% GF	2.0	21	1.4	28	158
PET/HDPE + 20% GF	2.4	26	1.5	33	233
PET/HDPE + 40% GF	3.1	30	1.4	39	239

used in the automotive industry. Both carpet edge trim and polypropylene separated from carpet waste are effective substitutes for virgin polypropylene in GMT (Table 16.10) [71]. The Azdel PM 10400 is a commercial GMT based on virgin polypropylene. Compared to short-fiber-reinforced thermoplastics, these GMT materials are both rigid and tough. Their primary application is in automotive bumper beams. This approach to recycling carpet waste has been patented [72]. Also, the material is produced commercially by Georgia Composites, Inc. [73].

Table 16.10 Density and Mechanical Properties of Compression Molded GMT Samples Containing 40 wt% glass [71]

Materials	Edge Trim	Separated PP	Azdel PM 10400
Density (g/cm^3)	1.52	1.18	1.18
Tensile strength (MPa)	109	95.8	95.2
Tensile modulus (GPa)	5.72^a	6.83	6.27
Elongation at break (%)	7.9	8.8	3.0
Total energy absorbed (J)^b	21.4	28.3	22.4
Sample thickness (mm)^b	3.0	4.0	3.6

^{a}Estimated without the use of an extensometer.
^{b}ASTM D3793 drop weight impact test at 3.4 m/s.

16.4.7. Polymer Composites Utilizing Waste Fibers as Reinforcement

Kotliar et al. [74–76] have explored the use of carpet face yarn and textiles as a fibrous filler for a composite or laminate. Because of the fine diameter of the fibers involved, a low-viscosity prepolymer in a water base was used to ensure complete coverage of the fibers. Adhesives were selected to result in a high-modulus and creep-resistant material with good weathering characteristics.

The work emphasized shredded carpet selvage to which various amounts of cut waste fibers such as nylon 6, nylon 6,6, polyester, and cotton were added. Fabric bits of waste denims and cotton–polyester fabrics were also used. The waste carpet blend was then coated with phenolic or urea formaldehyde resins that were dispersed in a water base. The composites contained various amounts of different fibers or fabrics and 7.5–20 wt% adhesive solids with respect to the fiber content. The fibers were spray coated and molded in a heated press at 150–200°C and 3.4 MPa. Test results show that one can achieve high flexural moduli of 2.4–2.8 GPa with face yarn, that is, fibers that bind to the matrix such as nylon, polyester, and cotton. These values together with flexural strengths of 34–48 MPa make the products suitable for many outdoor and transportation applications.

Laminates directly from waste carpet pieces were also made by coating the face yarn with a phenol formaldehyde resin and molding the carpet pieces back to back with the face yarn on the outside to achieve a high flexural modulus [75]. Holes were punched into the carpet prior to spray coating the face yarn so that protrusions of the matrix material could flow into the backing during the molding process to avoid shear delamination. Additional work has been done to make honeycomb sandwich structures for high flexural stiffness and light weight.

Gowayed et al. [77] explored the use of edge trim of polypropylene (PP) fabric waste (from the carpet backing) to reinforce a polyethylene (PE) matrix. Four layers of 0.1-mm-thick PE film were laid with a single layer of washed PP fabric waste. Then they were molded at a temperature of 150°C and a pressure of 290 kPa. It was found that the resulting PE/PP composite, with 25% PP volume fraction, exhibited a three-time increase in tensile strength and a 60% increase in flexural modulus when compared to the properties of pure PE. The composite developed can be used as a lining material for corrosive mixtures.

16.4.8. Waste Fibers as Reinforcement in Concrete and Soil

Fiber-Reinforced Concrete Concrete is the most heavily used construction material in the world. Adding a small fraction (usually 0.5–2% by volume) of short fibers to the concrete mix can increase the toughness (energy absorption) of concrete by orders of magnitude. Reduced shrinkage cracking has been observed even with fiber volume fractions as low as 0.1% of polypropylene fibers. Besides reducing the need for landfilling, the use of waste fiber for concrete reinforcement could lead to improved infrastructure with better durability and reliability. Potential applications could include pavements, columns, bridge decks and barriers, and for airport construction as runways and taxiways.

In a study on concrete reinforcement with carpet waste fibers [78, 79], recycled carpet waste fibers about 12–25 mm in length and fiber volume fractions of 1 and 2% were used. FiberMesh, a virgin polypropylene (PP) fiber (19 mm long), at 0.5 and 1% volume fractions was included for comparison.

Four-point flexural test and cylinder compressive test were conducted on a hydraulic testing machine, and the results are given in Table 16.11. Six or seven specimens were tested for each setup. In the 1-day compressive test, similar strength values were observed for plain concrete and various fiber-reinforced concrete (FRCs). It appeared that the 28-day compressive strengths of FRCs with 2% carpet waste fibers were lower than that of plain concrete. The plain concrete specimens failed in a brittle manner and shattered into pieces. In contrast, all the FRC samples after reaching the peak load could still remain as an integral piece, with fibers holding the concrete matrices tightly together.

The flexural strengths of all mixes tested were essentially the same and the standard deviations were low. The plain concrete samples broke into two pieces once the peak load was reached, with very little energy absorption. The FRC specimens, on the other hand, exhibited a pseudo-ductile behavior (Fig. 16.5), and fibers bridging the beam crack can be seen. Because of the fiber-bridging

Table 16.11 Compressive and Flexural Tests for FRC [86]

Fiber	V_f (%)	Comp. Strength (MPa)	Flex. Strength (MPa)	I_5	I_{20}
None	0	52.6	4.65	1.0	1.0
Fiber Meshpp	0.5	52.2	4.58	2.6	6.9
Fiber Meshpp	1.0	51.5	4.99	3.3	12.05
Waste fiber	1.0	61.8	4.09	2.1	5.4
Waste fiber	2.0	40.7	4.35	3.5	9.8

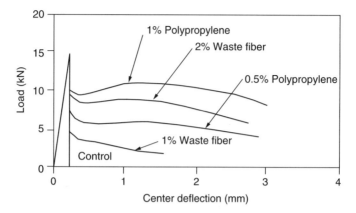

Figure 16.5. Typical flexural test curves of fiber-reinforced concrete [78].

mechanism, the energy absorption during flexural failure was significantly higher than that for plain concrete. The flexural toughness indices (I_5 and I_{20}) were calculated according to ASTM C1018.

Shaw Industries, Inc., the largest carpet manufacturer in the world, constructed a 11,000 m² R&D Center in Dalton, Georgia, which used concrete reinforced with carpet waste fibers in the construction project [80]. The amount of waste fiber included was 5.95 kg/m³, and about 20 tons of carpet production waste was consumed in the project. Mixing was done by adding fibers to the mixing truck directly, after which the fibers were found to be uniformly dispersed in the concrete without balling or clumping. Mixing, pouring, and finishing followed standard procedures, used conventional equipment, and went smoothly. The compressive and flexural strengths exceeded specifications, and reduced shrinkage cracking was observed. Such concrete containing waste fibers was used for floor slabs, driveways, and walls of the building. The project demonstrated the feasibility of using large amount of carpet waste for concrete reinforcement in a full-scale construction project.

Other studies on the use of polypropylene fibers from carpet waste in concrete [81], used tire cords in concrete [82, 83], and using recycled nylon fibers to reduce plastic shrinkage cracking in concrete [84] have also been reported and reviewed [85]. Gordon et al. [86] used the waste nylon fibers and ground carpet to stabilize asphalt concrete. Increase of asphalt content in asphalt concrete is favorable because it leads to more durable roads. But it is limited by the resultant flushing and bleeding of pavements and possible permanent deformation of the pavement. Addition of 0.3 wt% waste fibers increased the allowed asphalt content by 0.3–0.4 wt%.

Fiber-Reinforced Soil Studies reported in the literature have shown that fiber reinforcement can improve the properties of soil, including the shear strength, compressive strength, bearing capacity, postpeak load strength retention, and the elastic modulus. Recycled-fiber-reinforced soil was studied by laboratory evaluation, field trials, and ranking of potential applications [86–88]. Carpet waste, apparel waste, and virgin fibers were used for this study at different dosage rates. Additionally, the effect confining pressure and saturation were investigated. Compaction tests were performed to determine the moisture density relationships of the fiber-reinforced soil.

The tests were conducted on a silty to clayey sand. The soil–fiber mixture was allowed to hydrate for 24 h prior to compaction and triaxial testing. The triaxial test specimens were compressed hydraulically with a static-loading machine in a metal split mold. To determine the effect of reinforcing fibers on the moisture density relationship, standard compaction tests (ASTM D698) were conducted for soil reinforced with carpet fibers and polypropylene fibers at fiber contents of 1, 2, and 3%. It was observed that the reinforcing fibers impeded the compaction process, and thus increasing the fiber content had the same effect as reducing the compactive effort (energy). With the addition of fibers, more water was required to lubricate the soil grains during compaction, resulting in a higher optimal moisture content.

The triaxial compression tests were performed on as-compacted and soaked samples. Fiber type, fiber content, and confining pressure were varied, and for each specimen the dry density and moisture content was maintained at 1597 kg/m^3 (100 lb/ft^3) and 19.0%, respectively. In the unconfined compression tests performed on soil reinforced with carpet fiber, polypropylene fiber, and apparel fibers, the peak compressive stress increases with increasing fiber content for all three fiber types with the exception of the 0.3% polypropylene-reinforced specimen, which showed a decrease in peak compressive stress. Fiber reinforcement also resulted in a reduction of postpeak strength loss with increasing fiber content for all fiber types. The control, 0% fiber content specimen exhibits strain-softening behavior whereas strain-hardening behavior is exhibited at higher fiber contents. Triaxial compression tests on soil reinforced with carpet fibers under confining pressures of 34.5 and 69.0 kPa were also performed (Fig. 16.6, Table 16.12). The fiber-reinforced specimens showed significant increases in peak stress ranging from over 121 to 303%. More importantly, with increasing fiber content, the soil behavior changed from strain softening to strain hardening. The compressive stress at an axial strain of 10% is given in Table 16.12. From these tests, it was generally observed that the enhancement of soil properties generally occurs at large deformation levels. At very small strains, the stiffness of the soil is actually decreased due to a reduction in soil compaction.

The specimens in the soaked triaxial compression tests were allowed to absorb water for 48 h prior to testing. These tests were used to simulate in-service saturation that can occur during periods of heavy rainfall or due to other natural or man-made events. The soaked tests showed reduced strength at all strain levels as compared to the as-compacted condition. However, the 1 and 2% fiber content soaked specimens confined at 34.5 and 69.0 kPa exhibited increases in strength over the unreinforced soaked specimens. Thus, the use of fiber reinforcement can greatly reduce the strength losses associated with in-service saturation.

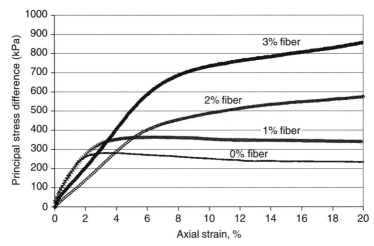

Figure 16.6. Stress–strain relationships for triaxial compression test of carpet-fiber-reinforced soil confined at 34.5 kPa [87].

Table 16.12 Compressive Stress (kPa) at 10% Axial Strain for Soil with Carpet, Polypropylene, and Apparel Fibers [87, 88]

Fiber Type	Fiber %	Comp. Stress at 0 kPa Confinement	Comp. Stress at 34.5 kPa Confinement	Comp. Stress at 69.0 kPa confinement
Carpet	0	0	253	337
	1	213	355	342
	2	318	489	577
	3	501	736	859
Polypropylene	0	0	253	337
	0.3	111	474	335
	0.5	192	553	494
	1.0	335	635	722
Apparel	0	0		
	1	163		
	2	341		
	3	480		

A field study was carried out to demonstrate the feasibility of incorporating carpet waste fibers in road construction [80]. The field trial sites for unpaved county roads were selected to represent typical types of soils found in Georgia. Trial sections with carpet waste fibers and virgin fibers were installed in several Georgia counties. The installation involves ripping the soil to a 150-mm depth, spreading the shredded carpet fibers, blend the fibers into the soil, and smooth and compact the soil. The fibers used were shredded into a length up to 70 mm long, and about 0.33% by weight of fibers was added. It was observed that the fibers could be mixed into soil with reasonable consistency in the field. Assessment of the unpaved roads by visual inspections confirmed that fibers in soil could improve the durability for certain types of soils, and thus reducing the need for frequent regrading. Promising applications for fiber-reinforced soil include: levees, landfills, retaining structures backfill, roadway slopes, and sports surfaces.

16.4.9. Waste to Energy Conversion

The energy content of the waste materials may be recovered, at least in part, by burning the waste materials in air (incineration) [3, 89–92]. Together, about 100 municipal solid waste (MSW) combustion facilities incinerated about 17% (35 million tons) of the municipal solid waste in the United States in 1996 [1, 2]. Most of these facilitates have a waste-to-energy conversion process. Waste containing used paper/wood products, contaminated packaging, and discarded tires has been combusted. The volume of these MSW is reduced by about 75% after incineration. The postcombustion ash still needs to be treated separately and then landfilled. Public concerns exist for the incineration of polymer waste. However, with advanced technologies and proper management, waste-to-energy conversion can be a viable alternative to landfilling. It is estimated that, if all the MSW currently generated in the United States were incinerated, the resultant carbon dioxide would be only 2% of that produced from the combustion of all other fossil fuels [90]. The current challenges for the incineration of polymer waste include further improving the incineration efficiency and reducing the harmful end products in the form of ash and noxious gases. The high combustion energy of polymer waste leads to decreased incineration capacity if the incinerator is heat limited [3].

Carpet waste has a simple composition compared to some other plastic waste streams. The major component in current carpet design is nylon, which requires up to 155 MJ/kg to manufacture, but gives off 29 MJ/kg when burned (see Table 16.13 [89]). The criterion for energy efficiency for incineration is that if it takes more than twice as much energy to make a plastic than what is recovered by burning, it is better to recycle the plastic than to burn it [61, 66]. This criterion clearly favors recycling nylon carpet waste over incineration. In general, the following factors have made incineration a viable option for carpet waste:

Table 16.13 Energy Content and Combustion Energy of Various Polymers Used in Carpets [89]

Plastic	Combustion Energy (MJ/kg)	Energy Content (MJ/kg)
Nylon	29	155
Polypropylene	44	73
Polyester	31	84

1. Incineration is becoming more acceptable for handling the waste due to more and more restrictions and regulations being put on landfilling waste [93].
2. The combustion energy values of typical components in carpet are listed in Table 16.13. These values are comparable to, or higher than, coal (29 MJ/kg). Thus carpet waste is a good candidate for heat generation.
3. The use of polypropylene in carpet manufacturing is increasing continuously. Polypropylene in carpet manufacturing is about 40% of total fiber used [14, 15]. Incineration may be a better option for carpet waste with polypropylene as the major component. Polypropylene requires 73 MJ/kg to make and releases 44 MJ/kg when burned.
4. The incineration option also can fit with other recycling routes. For example, after the valuable face yarns, such as nylon, have been separated, the remaining components may be gainfully incinerated.

Solid waste such as tires has been use in cement kilns as fuel supplement for making Portland cement. In an Atlanta, Georgia, plant, the use of tires has decrease the plant's air emissions by up to 30% and allows the company to meet tighter nitrogen oxides (NO_x) guidelines [94]. The use of carpet waste in cement kilns is also quite attractive and an effort is being made in this direction [95]. The relatively high fuel value of carpet polymers can reduce the need for fuels, and the calcium carbonate in carpet becomes raw material for cement.

16.4.10. Carpet Redesign and Reuse

As the search of technology for carpet recycling continues, carpet also is being redesigned for better recyclability. Reuse of reconditioned carpet is being considered for extending its life.

One approach is to make a carpet from one type of material to eliminate the need for separation in recycling. Hoechst-Celanese Corporation manufactures a one-component recyclable carpet from polyester fibers [96, 97]. The carpet used undyed polyester as the face yarns in the carpet. These face yarns were tufted on a primary polyester nonwoven backing fabric. A secondary backing fabric, also a polyester nonwoven, was sewn to the back of the tufted primary backing to lock the tufted face yarns in place. At least one of the two backings contained low-melt polyester fibers having a melt temperature of approximately 105–110°C. When the carpet was heated in the subsequent dyeing and drying steps, the two backings melted together thereby keeping the face yarns from being removed easily. The Hoechst process is suitable for polyester fiber, which is not predyed in the fiber-forming stage.

A recyclable carpet based on polypropylene fibers also has been developed [98, 99]. Unlike polyester fiber, polypropylene fiber is pigmented during the fiber-forming stage. The tufted polypropylene fiber is manufactured by the use of a needle-bonding method to interlock the primary and secondary backings. The polypropylene face yarns were tufted into the primary backing first. Then the secondary backing was needle punched into the tufted primary backing. The secondary backing then was heated using infrared radiation to strengthen the bond between the two backings.

Carpet is removed from homes or businesses due to a variety of reasons, including being dirty, stained, worn out, or requiring a style change. In many cases, used carpets can still be reused. The Earth Square program by the Milliken Company is to promote reuse through reconditioning old carpet tiles [100]. Milliken's process includes four steps: collecting the old carpet, cleaning, restoring the texture, and adding new color/design. This process helps to extend the service life of the carpet.

16.5. SUMMARY

A large amount of textile waste is disposed of in landfills each year. This not only poses economical and environmental problems to the society, it also represents a severe waste of resources. Waste statistics, waste characteristics, recycling rates, and recycling technologies for fibrous waste have been reviewed in this chapter. There have been several commercial carpet recycling operations in the United States, and the products range from virgin-equivalent nylon 6 resin (or fiber), resin for automotive parts, fibrous mats, and vinyl carpet backing, among others. While many commercial operations are still active, many of the carpet recycling facilities have been discontinued due to economical reasons. Although a recycling operation handling one type of carpet waste can be feasible, the overall economical viability is compromised if only part of the carpet collected can be recycled. This signifies the need for further research to develop diversified approaches that can recycled all types of fibrous waste collected.

REFERENCES

1. U.S. Environmental Protection Agency, Report EPA 530-R-98-007, Franklin Associates, Prairie Village, KS, 1998.
2. S. Kumar, *Polym.-Plastics Tech. Engr.* **38**(3), 401–410 (1999).
3. J. Scheirs, *Polymer Recycling, Science, Technology and Applications*, Wiley, New York, 1998.
4. C. Hawn, *Int. Fiber J.* June, 68–72 (1999).
5. V. Nadkarni, *Int. Fiber J.* June, 18–24 (1999).
6. D. O. Taurat, *Int. Fiber J.* June, 58–62 (1999).
7. H. P. Kasserra, in P. N. Prasad et al., eds., *Science and Technology of Polymers and Advanced Materials*, Plenum, New York, 1998, pp. 629–635.
8. H. C. Gardner, *Int. Fiber J.* August, 36–49 (1995).
9. B. Brill, 4th Annual Conference on Recycling of Fibrous Textile and Carpet Waste, Dalton, GA, May 17–18, 1999.
10. Council for Textile Recycling, *http://www.textilerecycle.org/*, 2001.
11. A. Hoyle, *Carpet & Rug Industry*, May, 36–40 (1995).

12. A. Hoyle, *Carpet & Rug Industry*, July, 39–46 (1995).
13. M. Braun, A. B. Levy, and S. Sifniades, *Polym.-Plastics Tech. & Engr.* **38**(3), 471–484 (1999).
14. Carpet and Rug Institute, Carpet Industry Statistics, *http://www.carpet-rug.com*, Dalton, GA, 2001.
15. D. Dennett, Polypropylene Fiber Technology Conference, Clemson University, Clemson, SC, September 11–13, 2001.
16. J. H. Bonfield, R. C. Hecker, O. E. Snider, and B. G. Apostle, U.S. Patent. 3,182,055 (1965).
17. S. Sifniades, A. Levy, and J. Hendrix, U.S. Patent 5,932,724 (August 3, 1999).
18. T. Jenczewski, L. Crescentini, and R. Mayer,U.S. Patent 5,656,757 (August 12, 1997).
19. R. Mayer, L. Crescentini, and T. Jenczewski, U.S. Patent 6,187,917 (February 13, 2001).
20. S. Sifniades, A. Levy, and J. Hendrix, U.S. Patent 5,929,234 (July 27, 1999).
21. H. V. Datye, *Indian Fibre Textile Res.* **16**(1), 46 (1991).
22. N. Chaupart, G. Serpe, and J. Verdu, *Polymer*, **39** (6–7), 1375 (1998).
23. P. Bassler and M. Kopietz, U.S. Patent 5,495,015 (February 27, 1996).
24. T. Corbin, A. Handermann, R. Kotek, W. Porter, J. Dellinger, and E. Davis, U.S. Patent 5,977,193 (November 2, 1999).
25. P. Bajaj and N. D. Sharma, in V. B. Gupta and V. K. Kothari, eds., *Manufactured Fibre Technology*, Chapman & Hall, New York, p. 615 (1997).
26. A. R. Mukherjee and D. K. Goel, *J. Appl. Polym. Sci.* **22**(2), 361 (1978).
27. S. Czernik, C. Elam, R. Evans, R. Meglen, L. Moens, and K. Tatsumoto, *J. Anal. Appl. Pyrolysis* **56**, 51–64 (1998).
28. H. Bockhorn, S. Donner, M. Gernsbeck, A. Hornung, and U. Hornung, *J. Anal. Appl. Pyrolysis* **58–59**, 79–84 (2001).
29. J. Muzzy, research in progress.
30. T. Brown, 6th Annual Conference on Recycling of Fibrous Textile and Carpet Waste, Dalton, GA, April 30-May 1, 2001.
31. C. C. Elam, R. J. Evan, and S. Czernik, Preprint papers—*Am. Chem. Soc. Div. Fuel Chem.* **42**(4), p 993–997 (1997).
32. BASF Corp., BASF 6ix Again Program, *www.nylon6ix.com*, 2001.
33. C. D. Myers, U.S. Patent 2,407,896 (1946).
34. B. Miller, U.S. Patent 2,840,606 (1958).
35. J. Thorburn, U.S. Patent 3,223,731 (1965).
36. M. B. Polk, L. L. LeBoeuf, M. Shah, C.-Y. Won, X. Hu, and Y. Ding, *Polym.-Plast. Technol. Eng.* **38**(3), 459 (1999).
37. H. P. Kasserra and D. Trickett, Presentation at 2nd Annual Conference on Recycling of Fibrous Textile and Carpet Waste, Atlanta, GA, May 1997.
38. R. J. McKinney, U.S. Patent 5,302,756 (1994).
39. G. A. Kalfas, *Polym. React. Engr.* **6**(1), 41 (1998).
40. S. Bodrero, E. Canivenc, and F. Cansell, 4th Annual Conference on Recycling of Fibrous Textile and Carpet Waste, Dalton, Georgia, May 17–18, 1999.

41. B. J. Kip, E. A. T. Peters, J. Happel, T. Huth-Fehre, and F. Kowol, U.S. Patent 5,952,660 (September 14, 1999).
42. J. Herlihy, *Carpet and Rug Industry*, Nov./Dec. 17–25 (1997).
43. J. A. E. Hagguist and R. M. Hume, U.S. Patent 5,230,473 (July 1993).
44. J. H. Schut, *Plastics World*, December, 25 (1995).
45. M. A. Howe, S. H. White, and S. G. Locklear, U.S. Patent 6,182,913 (February 6, 2001).
46. F. L. Robinson and W. R. Campbell, U.S. Patent 6,126,096 (October 3, 2000).
47. R. G. Rowe, U.S. Patent 6,061,876 (May 16, 2000).
48. D. W. White, U.S. Patent 6,029,916 (February 29, 2000).
49. F. C. Bacon, W. R. Holland, and L. H. Holland, U.S. Patent 5,897,066 (April 27, 1999).
50. M. Deschamps, U.S. Patent 5,829,690 (November 3, 1998.)
51. R. A. Sferrazza, A. C. Handermann, C. H. Atwell, and D. K. Yamamoto, U.S. Patent 5,535,945 (July 16, 1996).
52. D. K. Yamamoto and P. Viveen, U.S. Patent 5,516,050 (May 14, 1996).
53. P. C. Sharer, U.S. Patent 5,518,188 (May 21, 1996).
54. K. Hawn, 6th Annual Conference on Recycling of Polymer, Textile and Carpet Waste, Dalton, GA, April 30–May 1, 2001.
55. F. C. Bacon, W. R. Holland, and L. H. Holland, U. S. Patent 5,704,104 (January 6, 1998).
56. M. Booij, J. A. J. Hendrix, and Y. H. Frentzen, European Patent 759,456 (February 1997).
57. Y. H. Frentzen, M. P. Thijert, and R. L. Zwart, World Patent 97,03,04 (1997).
58. A. K. Sarian, A. A. Handerman, S. Jones, E. A. Davis, and A. Adbye, U.S. Patent 5,849,804 (December, 1998).
59. M. E. Sikorski, U. S. Patent 5,233,021 (August 1993).
60. A. T. Griffith, Y. Park, and C. B. Roberts, *Polym.-Plastics Tech. Engr.* **38**(3), p 411–432 (1999).
61. J. H. Schut, *Plastic Tech.* April, 22–25 (1993).
62. D. J. David, J. L. Dickerson, and T. F. Sincock, U.S. Patent 5,294,384 (March 1994).
63. C. G. Hagberg and J. L. Dickerson, *Plastics Engr.* April, 41–43 (1997).
64. R. J. Datta, M. B. Polk, and S. Kumar, *Polym.-Plastics Tech. Engr.* **34**(4), 551–560 (1995).
65. T. Chen, M.S. Thesis, School of Textile and Fiber Engineering, Georgia Institute of Technology, March 1996.
66. D. Young, S. Chlystek, R. Malloy, and I. Rios, U.S. Patent 5,719,198 (February 1998).
67. S. S. Dagli, M. Xanthos, and J. A. Biesenberger, *Am. Chem. Soc. Symp.* **513**, 241–257 (1992).
68. J. C. L. Williams, S. J. Watson, and P. Boydell, in M.I. Kohan ed., *Nylon Plastics Handbook*, Hanser/Gardner, Cincinnati, 1995, pp. 291–358.

69. S. A. Zegler and P. L. Weinle, U.S. Patent 5,728,741 (March 1998) and U.S. Patent 5,855,981 (January 1999).
70. F. P. La Mantia, *Macromol. Symp.* **135**, 157–165 (1998).
71. Y. Zhang, J. Muzzy, and S. Kumar, *Polym.-Plast. Tech. Engr.* **38**(3), 485–498 (1999).
72. J. Muzzy, D. Holty, D. Eckman, and J. Stoll, U.S. Patent 6,271,270 (2001).
73. Georgia Composites, Inc., *www.gacomposites.com*.
74. A. Kotliar and D. P. Fountain, U.S. Patent 5,626,939 (May 1997).
75. A. M. Kotliar and S. Michielsen, U.S. Patent 5,912,062 (June 15, 1999).
76. A. Kotliar, *Polym.-Plast. Tech. Engr.* **38**(3), 513–531 (1999).
77. Y. A. Gowayed, R. Vaidyanathan, and M. El-Halwagi, *J. Elastomers Plastics* **27**, 79–90 (1995).
78. Y. Wang, A. H. Zureick, B. S. Cho, and D. E. Scott, *J. Mat. Sci.* **29**(16), 4191–4199 (1994).
79. Y. Wang, in Y. Ohama, ed., *Disposal & Recycling of Organic and Polymeric Construction Materials*, E & FN Spon, London, 1995, pp. 297–305.
80. Y. Wang, *J. Polym.-Plast. Tech. & Engr.* **38**(3), 533–546 (1999).
81. A. E. Naaman, S. Garcia, M. Korkmaz, and V. C. Li, In K. Chong ed., *ASCE Proc. Materials Eng. Conf.*, 1996, pp. 782–791.
82. H. C. Wu, Y. M. Lim, V. C. Li, and D. J. Foremsky in K. Chong, ed., *ASCE Proc. Materials Eng. Conf.*, 1996, pp. 799–808.
83. H. C. Wu, Y. M. Lim, and V. C. Li, *J. Mater. Sci. Lett* **15**, 1828–1831 (1996).
84. J. L. Groom, D. V. Holmquist, and K. Y. Yarbrough, in *Proc. Recovery & Effective Reuse of Discarded Materials and By-Products for Construction of Highway Facilities,* Denver, CO, Oct. 19–22, 1993, pp. 179–183.
85. Y. Wang, H. C. Wu, and V. C. Li, *J. Mat. Civil Eng.*, **12**(4), 314–319 (2000).
86. G. S. Gordon, Darrel V. Holmquist, and T. W. Kennedy, 2nd Annual Conference on Recycling of Fibrous Textile and Carpet Waste, Atlanta, GA, May 19–21, 1997.
87. J. Murray, J. D. Frost, and Y. Wang. *Transport. Res. Rec.*, No. 1714, 9–17 (2000).
88. Y. Wang, in Proceedings of the Seventh International Conference on Composites Engineering, July 2–9, 2000, Denver, CO., B19–B22.
89. H. Kindler and A. Nikles, *Kunststoffe* **70**(12), 802–807 (1980).
90. M. M. Nir, *Plast. Eng.* October, 21–28, (1990).
91. J. M. Henshaw and W. Han, *J. Thermoplast. Compos. Mat.* **9**, 4–20, (1996).
92. S. J. Huang, *Plast. Engr.* No. 29, 1–6 (1996).
93. D. Curto and Y. Basar, in G. Akovali, ed., *Frontiers in the Science and Technology of Polymer Recycling*, Kluwer Academic, Netherlands, 1998, pp. 17–28.
94. Georgia Department of Community Affairs, Solid Waste Management Report 1999–2000, *http://www.dca.state.ga.us/solidwaste/swar.html*, 2001.
95. M. Realff, J. Clark, and F. Cook, 6th Annual Conference on Recycling of Polymer, Textile and Carpet Waste, Dalton, GA, April 30–May 1, 2001.
96. J. A. Corbin, R. D. Johnson, W. G. Neely, I. S. Slack, and B. L. Davis, U.S. Patent 5,532,035 (July 1996).
97. J. A. Corbin and I. S. Slack, U.S. Patent 5,538,776 (July 1996).

98. Recyclable Fair Carpets Made from PP and Other Polyolefins, *Chem. Fibers Int.*, **46** (October 1996).
99. J. M. Long and K. A. Snyder, U.S. Patent 5,604,009 (February 1997).
100. W. Blackstock, 4th Annual Conference on Recycling of Fibrous Textile and Carpet Waste, Dalton, GA, May 17–18, 1999.

CHAPTER 17

POLYMERS IN AUTOMOBILE APPLICATIONS

WENDY LANGE
General Motors Corporation

17.1. INTRODUCTION

The use of polymers in automobiles has increased dramatically over the last 100 years. Particularly significant gains were made within the last two decades. Figure 17.1 shows the dramatic increase in the polymer content in a typical vehicle over recent years. These materials provide increased comfort, better aesthetics, safety enhancements, and environmental benefits over the conventional materials they replaced. This chapter will discuss where these materials are used and the environmental benefits they provide in such applications.

The use of recycled polymers in automobiles has also grown in recent years. There are many applications where recycled materials work particularly well. Components in the automobile that are generally hidden from view, covered, or wrapped offer the best potential for using recycled plastics. Recycled materials can often provide equivalent performance to virgin materials at a lower cost and reduce the use of limited natural resources. Specifying and sourcing recycled materials in high-volume automotive applications also helps to create a stable market for such materials. This helps to develop and grow a commercially viable infrastructure for recycling polymers. Typical applications of recycled materials in General Motors vehicles are shown below in Figure 17.2.

The extent to which polymers are used in its construction affects the recyclability of the vehicle at the end of its useful life. The current infrastructure for vehicle recycling is particularly successful. Over 94% of vehicles that reach the end of their service life enter the recycling infrastructure [2]. This is higher than that for any other nondurable or durable consumer good. In comparison,

---

Plastics and the Environment, Edited by Anthony L. Andrady.
ISBN 0-471-09520-6 © 2003 John Wiley & Sons, Inc.

728 POLYMERS IN AUTOMOBILE APPLICATIONS

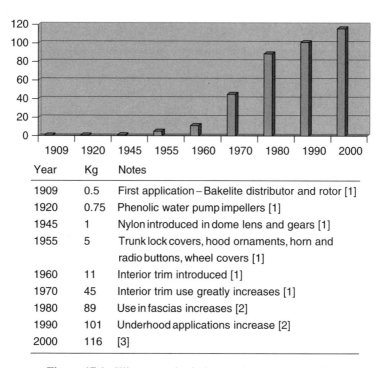

Year	Kg	Notes
1909	0.5	First application – Bakelite distributor and rotor [1]
1920	0.75	Phenolic water pump impellers [1]
1945	1	Nylon introduced in dome lens and gears [1]
1955	5	Trunk lock covers, hood ornaments, horn and radio buttons, wheel covers [1]
1960	11	Interior trim introduced [1]
1970	45	Interior trim use greatly increases [1]
1980	89	Use in fascias increases [2]
1990	101	Underhood applications increase [2]
2000	116	[3]

Figure 17.1. Kilograms plastic in a typical vehicle [1–3].

aluminum cans are recycled only at a rate of 62% [6], PET bottles a rate of 36% [4], and paper at a rate of 45% [5].

Most vehicles enter the recycling infrastructure through a vehicle dismantler (see Fig. 17.3). All fluids and the battery are first removed from the vehicle. These items are reused, recycled, or incinerated to generate energy. Any components that can be sold for reuse (e.g., bumpers, windshields, and trim), remanufacture (e.g., engines and electric motors), and for material recovery (e.g., aluminum wheels) are next removed. Once all parts of value have been removed from the vehicle, the remainder, or the hulk, is usually crushed and sold to a shredder. A small number of vehicles, generally older vehicles, enter the recycling infrastructure directly at the shredder skipping the dismantling step.

The shredder shreds the vehicle into approximately fist-sized chunks. The ferrous metals are separated using a magnet and recycled in electric arc furnaces. The nonferrous metals are separated using a variety of techniques and recycled as individual materials. The remainder, including most of the polymers, is sent to landfill. Currently about 75% of a vehicle, by mass, (more for a truck) is metal and gets recycled.

The drive toward lightweight vehicles is leading to a change in their material composition. While the amount of ferrous metal used in vehicles is slowly decreasing, that of nonferrous metals and polymers is increasing in more recent models. The nonferrous metals are generally high-value materials. Thus, the

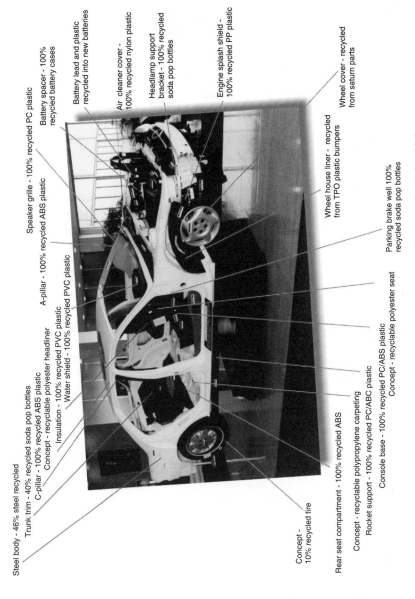

Figure 17.2. Typical recycled content applications in GM vehicles.

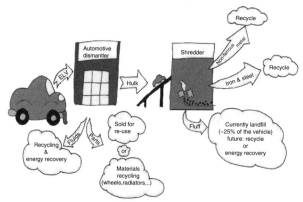

Figure 17.3. Vehicle recycling infrastructure.

increase in the use of nonferrous metals improves the economics of recycling. The polymers are consistently replacing heavier metals that are generally recovered by recycling. This leads to an overall reduction in the percentage of metal (by mass) in the vehicle. As the percentage of polymer content in vehicles increases and that of metals decreases, the capability of the existing infrastructure to recycle them must decrease. Efforts are being made to increase the rate of recycling of automotive plastics through improved technology and infrastructure expansion.

17.2. GENERAL ENVIRONMENTAL IMPACTS

There are several key areas where polymers improve the environmental performance of the vehicle. These include the reduction in vehicle mass, process improvements, elimination of corrosion, better durability, and elimination of the use of paint. Lighter vehicles provide better fuel economy during operation. They also provide significant reductions in the use of energy and raw materials as well as reduced atmospheric emissions from the manufacturing process. Polymers are moldable, more economically, into a greater variety of shapes than metal. This allows the reduction in the number of parts needed in complex designs (recycled polymers can be used in many such designs). Reducing the number of design components and using recycled materials reduces the demand on virgin materials with corresponding reductions in energy, materials, and emissions. Increased durability has a similar effect since less replacement parts are required during the life of the vehicle. The energy, CO_2 emissions, and the total inorganic air emissions created when producing 1 kg of some common materials are shown in Table 17.1.

Using molded plastics in requisite colors can minimize the use of paint in automobiles. The environmental benefits accrued from not having to produce and apply the automotive paint because of plastics use is quite significant. These environmental impacts include the energy to paint a component and the emissions

Table 17.1 Environmental Impacts of Material Production [7]

Material-impacts per 1 kg produced	Energy to Produce (MJ)	CO_2 Emissions (kg)	Inorganic Air Emissions (kg)
Polypropylene (PP)	1.68	1.10	1.12
Nylon (PA)	3.11	7.00	7.08
Acrylonitrile Butadiene Styrene (ABS)	2.40	3.14	6.00
Polymethyl Methacrylate (PMMA)	2.31	3.13	6.09
Polyester—amorphous (PET)	1.73	2.22	2.28
Polyester—semi-crystalline	1.77	2.33	5.00
Polycarbonate (PC)	2.55	5.00	5.04
Polyurethane (PU)	2.23	3.74	3.78
Polyethylene (PE)	1.73	1.10	1.12
Aluminum—primary mix ingot	6.16	12.74	37.24
Steel—billet	0.78	1.31	1.98
Steel—primary sheet	1.50	2.95	2.98

of volatile chemicals and aerosol particles created during the process. Table 17.2 shows some typical environmental impacts associated with the painting process for some common coating systems.

17.3. USING POLYMERS IN VEHICLES

Different components in the vehicle have unique performance requirements and therefore require different types of materials. For the purpose of the present discussion the vehicle can be split into four sections: interior, exterior, underhood, and chassis. Each of these areas has a unique set of performance criteria.

17.3.1. Plastics Used in the Interior of Vehicles

There are six main subsystems making up the interior of a vehicle. They are the instrument panel, console, seats, carpet, headliner, door, and pillar trim. The most complex of these is the instrument panel.

Instrument Panel The instrument panel (IP) consists of the carrier, the structure, skin and padding, cluster, air bags, air ducts, glove box, brackets, sound-absorbing materials, and various other small parts. The heating/cooling controls and radio are also typically attached to the instrument panel. This often requires a wide variety of materials. From an end-of-life recycling perspective, all the different materials must be separated to be recycled effectively. Therefore, using the least number of compatible materials wherever possible simplifies end-of-life recycling.

Table 17.2 Environmental Impacts of Painting [8]

5a. Environmental Impacts of Producing Paint Materials			
Paint Material Production	Energy to Produce 1 kg (MJ)	CO_2 Equivalent Emissions Per 1 kg coating system (kg)	Inorganic Air Emissions Per 1 kg coating system (g)
A. Acrylic Powder Primer	275	14.2	0.06
B. Polyester Powder Primer	107	5.5	0.02
C. Polyester Solventborne Primer	90	2.5	0.006
D. Waterborne Basecoat	50	1.8	0.03
E. Acrylic Powder Clearcoat	290	14.9	0.06
F. Acrylic Solventborne Clearcoat	126	4.2	0.03

5b. Environmental Impacts of the Painting Process			
Painting Process—values to paint a mid-sized sport utility vehicle	Energy (MJ)	VOC Emissions (kg)	Solid Wastes (kg)
C + D + F	3400	2.1	26
A + D + F	3100	1.8	24.5
B + D + F	3100	1.8	24.5
B + D + E	2500	1.3	18.5

IP: Structure The use of polymers in the design of the instrument panel has helped to reduce its mass considerably. Polymers also allow integration of parts. For example, the 1992 Pontiac Grand Am was the first model where a blow-molded engineering thermoplastic IP structure was used. The material and process used allowed the air duct to be molded directly into the instrument panel. This reduced the number of different materials used, facilitating easier recycling, and also eliminating the need to separately mold the duct. This modification also eliminated the environmental impacts of that process such as energy use and emissions.

The most common polymers used for instrument panel structures are acrylonitrile–butadiene–styrene (ABS), acrylonitrile–butadiene–styrene blended with polycarbonate (ABS/PC), polycarbonate (PC), poly(phenylene oxide) blended with nylon (PPO/nylon), poly(phenylene oxide) blended with styrene (PPO/styrene), polypropylene (PP), and styrene–maleic anhydride copolymer (SMA) [3]. The percentage of each of these polymers typically used in year 2000 models is shown below in Table 17.3 [3]. All of these materials by themselves

Table 17.3 Distribution of IP Structure Materials Used in 2000

Material	Percent Usage
ABS	15
ABS/PC	32
PC	15
PPO/Nylon	2
PPO/Styrene	5
PP	9
SMA	22

are recyclable. However, the structure generally has a protective skin applied to it, in many cases fabricated of polyurethane foam and polyvinyl chloride (PVC). These materials are generally not compatible from a recycling standpoint with the materials used in the structure itself. Technology has been developed to separate the urethane foam and PVC skin from the structure substrate. This, however, constitutes an extra step in the recycling process that adds cost. The separation process also has associated environmental impacts, energy use, water and/or chemical use, and emissions.

Recently there has been a trend toward the use of either thermoplastic polyolefin (TPO) or urethane polymers for the skin. Where TPO is used for the skin, with expanded PP foam and injection-molded PP used as the panel structure material, recycling (both end-of-life recycling and the recycling of the manufacturing scrap) becomes particularly easy because these materials are compatible substrates. Another design option is to use a soft-touch paint (paint that is flexible and feels similar to a foam) instead of the skin with a foam cover on the structure. In some instances, the paint can be chosen to be compatible with the substrate. Current processes to remove paint from plastics, however, are costly and have environmental burdens such as release and disposal of waste solvents associated with them. A third option is to use only a grained substrate as used in the 1994 Dodge Neon. The instrument panel structure was a grained PP eliminating the need for painting. The same material was also used in many of the other interior parts, allowing many of the interior parts to be recycled together.

Another interesting application was the top trim piece on the 1998 Saturn. The part was made using TPO, reducing the mass by approximately 0.45 kg (1 lb) over the previously used painted polycarbonate trim. The need for paint and its associated environmental impacts were also eliminated.

IP: Carrier The instrument panel carrier has typically been steel, but the trend has been toward using lighter weight metals and plastics. The 1997 GM APV minivan used an SMA polymer instead of steel for the support element with a mass savings of over 1.6 kg (3.5 lb).

IP: Ducts Heaters with ducts did not appear in the instrument panel in automobiles until the late 1940s. The original ducts were made of metal. The trend has since favored polymeric materials that can be molded into a variety of shapes for easier packaging. Currently about 55% of automotive IP ducts are made of polyethylene (PE) and 40% are polypropylene PP [3]. Both these materials are much lighter in weight compared to steel. Choosing the material that is most compatible with the structure also facilitates easier recycling. IP ducts are also candidates for use of recycled materials, such as recycled PP.

IP: Glove Box Glove box bins, first seen in vehicles in the early 1930s [1], were also initially made of steel. As the use of plastics in automobiles caught on, these bins changed over to the lighter plastic materials. The glove box bin is a good candidate for use of recycled material because it is typically covered or flocked. Recycled ABS and recycled PP have been successfully used in this application.

IP: Air Bags General Motors offered the first air bags in the mid-1970s as an option on selected models. Air bags began appearing in high volume in vehicles only in the late 1980s. The original systems consisted of an aluminum or steel canister, a nylon bag, and a steel door covered with skin and foam or a reaction injection molded (RIM)/scrim/urethane door with a metal insert. The mostly metal system was recyclable but did not have optimum environmental characteristics.

The door of the airbag module presented several areas that could be improved from an environmental standpoint. For instance, the skin and foam covering over the steel door was environmentally undesirable because of the potential for significant emissions during manufacturing (from the foaming process and that employed to apply the skin). The present trend is toward the use of thermoplastic elastomers (TPEs) in doors typically painted to achieve proper color match with the rest of the interior panel. The polymeric door is lighter than the steel door and the environmental impacts of the skin and foam are eliminated. Although the emissions associated with painting are now added, the overall environmental characteristics of the product improved very significantly with the use of plastics.

The passenger-side air bag door of the 1992 Lincoln Town Car switched from RIM/scrim/urethane structure to an injection-molded polyester elastomer that was painted. This resulted in a mass saving of about 50% [3] and was also simpler to manufacture, using less energy and with fewer associated emissions. The first application of an unpainted air bag door in North America was on the 1995 Geo Metro and Suzuki Swift [3] where the passenger-side air bag door was molded in color from thermoplastic polyolefin. Another novel approach was used in the 1997 Cadillac DeVille where instead of having a separate door for the passenger-side air bag, the top IP trim piece functioned as the door. The trim piece was fastened with permanent fasteners on the side toward the driver and with releasable fasteners on the passenger side. When the air bag deployed, the part bent up and out of the away. By eliminating the separate door, all of the manufacturing processes along with their environmental burdens to make the door were eliminated.

Another area with a potential for a significant reduction in mass is the canister, generally constructed of metal. Nylon is being evaluated as a replacement for the metal in the canister. The first passenger system with a molded nylon canister housing was on GM's 1995 Opel Vectra B and Omega models. The nylon molded part allowed the ducts for cable and plug attachments to be molded in and resulted in a system that had six fewer component parts. This simplified the manufacturing process and assembly as well as reduced the environmental burdens.

IP: Other Parts The brackets, sound-absorbing materials, electronics, and other parts attached to the instrument panel are made from a variety of materials. For items such as brackets and sound-absorbing materials, it is best to employ materials that are compatible with the IP substrate. For electronics and other complicated small parts, it is best to design them to be easily removed during recycling.

Console Vehicles of the early 1900s had bench seats and no consoles. As the amount of plastic trim in the vehicle interior increased, consoles gradually came into use [1]. The original consoles were made from ABS or ABS/PC and painted. In the early 1990s automotive companies started considering PP as a possible replacement for these resins to decrease cost.

The floor console of the 1994 Chrysler Neon was the first unpainted polypropylene floor console used by a U.S. manufacturer [3]. Besides achieving cost reduction, the significantly lighter part eliminated the need for paint and its associated environmental burdens. The part was also of the same material as the IP and pillar trim, allowing for a large amount of a single material to be harvested from the interior of the vehicle. This helped to make end-of-life recycling more economical. The 1996 Ford F-150 also used a molded in color PP console with all of the parts made from PP [3]. The console was attached to the vehicle with two bolts that were easily accessible.

Seats Seats in automobiles have gone from basic metal and wood structures with leather covering to complex structures with everything from steel and magnesium to plastics, often with embedded electronics. Seat frames have remained mostly metal, although lighter weight metals are now routinely used. Components such as heaters, air bag sensors, and electronic positioning devices have added parts made of different materials that complicate end-of-life recycling.

In 1896, in the beginning years of the auto industry, vehicles were open to the environment. Leather was the material of choice because of its ability to withstand the elements. In 1923 the first closed car came on the scene, allowing greater flexibility in the choice of trim materials. In the 1920s and 1930s soft woolen broadcloths coverings were popular [9]. A seat covering material that soon came into use was vinyl. Vinyl (PVC) was invented in 1926 by Waldo Semon, an organic chemist at B.F. Goodrich [10]. The material was durable, resisted stains, and when properly compounded had the appearance of leather but was considerably less expensive. Over the next 40 years much work was done to improve the quality of the vinyl and to experiment with competing materials.

In the mid-1970s nylon velour, a material with a soft feel and a variety of possible color combinations, came into use. Today's seat coverings are mainly nylon or polyester cloth. Though leather is a "renewable resource," the process to make it ready for use in the automobile is not a "clean" process, requiring large amounts of chemicals. The nylon and polyester materials have a relatively smaller processing impact on the environment and are easily recycled.

Soft padding usually made from urethane foam is used under the seat covering for comfort. The foam is recyclable with a stable market available for the recycled foam, which is primarily used in carpet underlay. The demand is so great for this material that scrap urethane foam is imported from Europe into the United States. There are approximately 31 lb of urethane foam in an average vehicle [3]. Provided the foam is easy to remove, it could be a viable recyclable product. Design choices that imbed metal rods, heaters, sensors, and the like into the foam hinder recycling. The method used for attaching the seat cover can also affect recycling.

Carpet Original floor coverings were durable rubber mats. Once the closed cars were developed, other materials were evaluated for use. Initially natural fiber matting was used, but as synthetic materials came onto the market, they replaced the natural materials. This started in the 1950s with tufted rayon [9]. Soon rayon was replaced by nylon, a particularly durable plastic that has remained the material of choice in modern automotive carpet systems.

Over time, the acoustic properties of floor coverings became an important consideration. To accomplish better acoustic properties and to shape the product to fit more complex floor pans, a backing layer is often used. A variety of materials such as rubber foams, and shoddy pad (an acoustic padding made from recycled cotton products) can be used for the purpose. These materials are not compatible with nylon in recycling operations. Technologies have been developed to separate these materials, but these processes add both cost and environmental burden to the operation. There has been some effort to develop all polypropylene systems that are more easily recyclable, though durability continues to be a challenge for these systems. Carpeting is also used in the trunk of many vehicles. The material typically used is recycled polyester. This large volume use of a recycled material also helps to support the market for recycled resins encouraging resource conservation.

Headliner The first one-piece plastic headliner was released for use in station wagons in 1956, and the first application of a fiberglass substrate with a cloth covering was introduced in 1967 [9]. Today a variety of materials are used for the purpose. These include fiberglass systems with a phenolic resin and a polyester cloth, polyurethane, fiberglass, corrugated cardboard, and resinated cotton. The headliner construction contains a variety of materials that are not compatible for recycling. Much effort has been devoted in the past 5 years to develop all-polyester systems. The system consists of polyester batting, foam, and cloth. These systems could be ground up and recycled at any stage of the product life cycle (i.e., during manufacturing, parts scrapped during assembly, and at the end of life). The first use of postconsumer recycled polyester was in the 1992 Cadillac Seville and Eldorado headliners. These parts contained polyester recycled from post-consumer soda bottles.

Door and Pillar Trim Door and pillar trim did not come widely into use in automobiles until the 1960s [1]. A variety of materials, particularly thermoplastic substrates such as PP and ABS; PP or ABS covered with vinyl skin, cloth, or carpet; covered thermoset RIM substrates; and covered natural fiber substrates are used in their construction. The trend is toward lighter materials that flow easily and are easier to mold during processing, to reduce the mass and therefore the material use [3]. The thermoplastic substrates are easily recycled. If a noncompatible covering material is added, the recycling becomes more difficult. There has been an effort to develop covered systems that are more recyclable; for instance, PP substrate with a TPO skin.

17.3.2. Plastics Used on the Exterior of Vehicle

Polymers are used in many areas on the exterior of the vehicle. These include a variety of components such as moldings, trim, appliqués, body panels, doors and exterior mirror housings, front and rear fascia, and lighting systems and pickup beds. Door seals can be made from thermoplastic elastomers. These materials must meet performance, durability, weatherability, and styling criteria.

Moldings and Trim Exterior plastic moldings are a fairly recent addition to the automobile. The move toward reduced mass has driven the trend for the original all-steel body panels to be made increasingly lighter. Plastics offer flexibility of shape, relatively lower energy to form, and are significantly lighter in weight. Plastic trim started replacing metal trim in the early 1980s because it eliminated the corrosion problem and reduced weight [3]. The addition of plastic moldings also helps to maintain dent resistance on the body panels, which increased customer appeal while allowing the panels to be lighter weight. Typical materials for this application are PVC, ionomer, and TPO.

Plastics allow innovative styling while minimizing mass and also allow covered joints and attachments. Styling trends have led to shapes in modern vehicles that are difficult or even impossible to create in metal.

Body Panels Plastics can help considerably reduce the mass of body panels while providing better durability, damage resistance, and corrosion resistance. Their first application in body panels was in 1953 [1]. General Motors introduced the Corvette with a fiber-reinforced thermoset plastic body (see Fig. 17.4). At almost the same time, Kaiser introduced a reinforced-plastic body on a sports car. Since then polymers have been used for fenders, hoods, trunk lids, roofs, doors, quarter panels, and lift gates.

Fenders Fenders are particularly good candidates for polymer use. A PPO/nylon blend (presently used in Saturn models) is one material that has been shown to

738 POLYMERS IN AUTOMOBILE APPLICATIONS

Figure 17.4. 1953 Corvette.

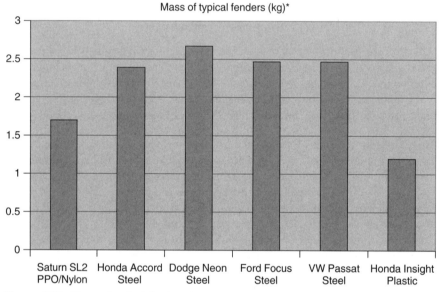

Figure 17.5. Mass of typical fenders (based on General Motors benchmarking disassembly data).

work well in this application. RIM urethane also works for fenders and was used on the 2000 Corvette and the 1996 GM APV minivan. The Honda Insight also used a polymeric fender. Figure 17.5 shows the comparative masses of some steel and polymeric fenders.

Doors Doors are an area where efforts to reduce the mass is limited by structural load criteria. Several vehicles have been designed with doors made of structural polymers to reduce the overall mass. The doors in a Saturn are a PC/ABS blend while the GM APV minivans used SMC doors. The GM EV1 design was optimized for mass and performance. The doors in the EV1 are a

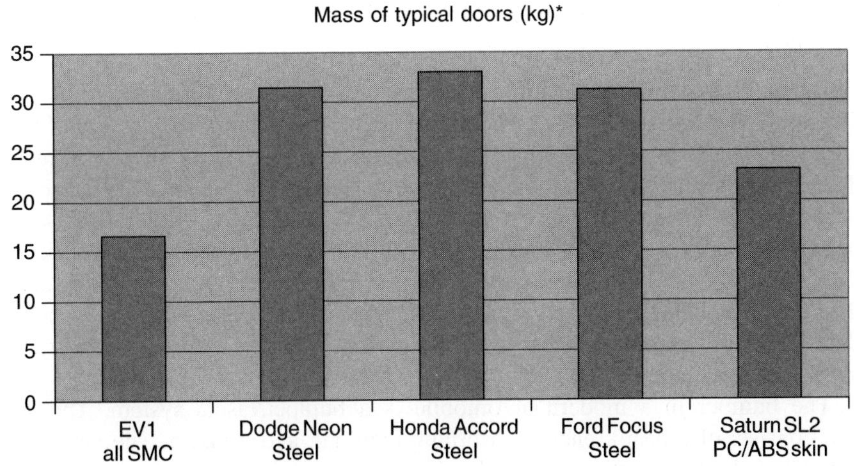

Mass of typical doors (kg)*

[Bar chart showing:
- EV1 all SMC: ~17
- Dodge Neon Steel: ~32
- Honda Accord Steel: ~33
- Ford Focus Steel: ~32
- Saturn SL2 PC/ABS skin: ~24]

Figure 17.6. Mass of typical doors (based on General Motors benchmarking disassembly data).

two-piece SMC design that is bonded together. A steel beam runs through the middle of the door for stiffness. This design is approximately 50% lighter than a typical small-car door made of steel. Figure 17.6 shows the mass of several small-car doors for comparison.

Door seals have typically been made of rubber materials with metal inserts. The metal inserts allow the rubber part to snap-fit onto the frame. There have been efforts to develop a thermoplastic elastomer seal that would be relatively easy to recycle and would meet the same performance requirements. These can also be molded in different densities or fitted over compatible plastics to eliminate the need the metal inserts. A good example is the Mitsubishi Motors Mini-Cab (in Japan), which is a SUV-like commercial vehicle with a large rear cargo door. The rubber seal for the door was replaced with a TPE seal with the new part weighing 30% less [3].

Mirror Housings In the early 1980s auto companies started experimenting with several types of plastics to replace zinc die-cast parts originally used in housing applications. Nylon, ABS, and ABS/PC were among the first plastics used for the purpose. In the late 1980s the trend moved to mineral-filled nylon and nylon blended with PPO. These plastics reduced the mass but did not eliminate the need for paint. The magnitude of mass savings was calculated for Chrysler's 1985 Omni and Horizon. The mineral-filled nylon mirror housing weighed 0.68 kg (1.5 lb) while the same housing in thin zinc weighed 1.3 kg (2.87 lb) [3] yielding a 48% mass savings.

In the early 1990s unpainted plastic mirror housings started to appear in automobiles. The optional electric mirror on the 1991 Chevrolet Blazer and GMC Jimmy was molded in color with polycarbonate and poly(butylene terephthalate) blend (PC/PBT). The technology was a joint development between GE Australia and GM Holden. GM introduced an unpainted nylon mirror housing on the 1990 S-10 pickup.

Bumper/Fascia Bumpers first appeared on cars as an option in 1919 [11]. These first bumpers were simple steel bars. The use of bumpers from that time until 1973 was mainly for styling and therefore had a variety of shapes. Some of them were chromed, and some were painted. New government safety standards introduced in 1973, however, required bumpers to become a functional safety item. Bumpers that could meet the standard at the time were large steel beams that were very heavy and not particularly appealing. The 1973 Corvette solved the styling problem by covering the steel beam with a urethane decorative cover. Since that time improved technologies to absorb energy and offer flexible styling have been developed.

The bumper in a modern automobile is a bumper/fascia system. The system consists of a beam, energy-absorbing material, and a fascia. The beams are typically a high-strength steel or aluminum. The fascias are made from several materials including PC/PBT, PP, ionomer, polyester elastomer, TPO, and RIM. Today the vast majority is made from TPO or RIM. These fascias are much lighter than the steel bumpers of the past. Table 17.4 shows the approximate percentage of each material used in 2000 [3]. Energy-absorbing options include metal shock absorbers, polyethylene honeycomb, polyurethane foam, and polypropylene foam. The polypropylene foam offers good energy absorption at a lower mass and is also compatible with TPO and PP fascia materials.

An excellent example of a compatible system is the 1997 Buick Park Avenue bumper/fascia system. The fascia is a TPO material and the energy absorber is PP foam snap fit to the aluminum bumper beam. Signal markers and trim strips are also snap-fit for easy removal. Therefore, a large amount of compatible material can be quickly removed from it for recycling. There has been a trend toward molded-in-color fascias on certain types of vehicles. These not only reduce mass but also eliminate the environmental burdens associated with painting. Molded-in-color fascias have been used on a variety of small cars and some sport utility vehicles.

Exterior Lighting Early automotive lighting systems used mainly glass and metal components in their design. As plastics technology evolved and the need for

Table 17.4 Distribution of Fascia Materials Used in 2000

Material	Percent of Total Usage
PC/PBT	6.5
Ionomer	2
Polyester Elastomer	1.5
TPO	61
RIM	29

mass reductions grew, plastics were used in automobile lighting fixtures. Changes in the design of fixtures enabled by the flexibility in molding plastic shapes have led to significant reductions in component mass in recent models. Ford initiated work on plastic headlamp lenses and parabolic housings. The 1984 Mark IV headlamp system was a breakthrough in this area. It had five plastic parts; the lens and body were PC; the replaceable bulb base, the electric connector, and locking retainer ring were also made of plastic. This system saved 0.9 kg (2 lb) over a conventional two-lamp system and 1.4 kg (3 lb) over a four-lamp system done in glass [3]. Additional mass savings resulted from the elimination of a separate trim piece.

Plastics are now used extensively in exterior lighting to reduce mass. Lenses are made from acrylic and polycarbonate. Housings are made from many materials including PC, PP, polyetherimide (PEI), polyetherimide plus polyethylene terephthalate (PEI/PETP), bulk molding compound (BMC), and nylon. The PC and acrylic are much lighter weight than glass and are also tougher.

The brackets that hold the lighting systems in place have also gone from steel to plastic. Typical materials for this application include glass-filled nylon, PC, glass-filled PET, and sheet molding compound (SMC). The plastic brackets are much lighter in weight than the metal part they replaced.

Pickup Beds Both SMC and RIM have been used for pickup beds and demonstrated to yield at least a 20% mass reduction over steel. They also eliminate the need for a separate bedliner, thus reducing raw material needs. The 2000 Chevrolet Silverado offered an optional composite pickup box (Fig. 17.7). The bed and tailgate were structural RIM while the fenders and tailgate outer were reinforced RIM. The tailgate was 6.8 kg (15 lb) lighter, which not only helps fuel economy but also requires less effort to close. The overall system was 23 kg (50 lb) lighter than the original steel system.

17.3.3. Underhood Polymers

Polymers used under the hood have to be able to withstand very high temperatures. In a typical vehicle temperatures can get as high as 125°C under the hood

Figure 17.7. Silverado composite pickup box.

during operation. In a turbo diesel they can get as high as 150°C. Therefore, the choice of available polymers and the extent of their use under the hood is more limited. Rubber materials are typically used for hoses. Polymers are also used for brackets, splash shields, fluid reservoirs, cases, and fans. They all offer mass savings and better corrosion resistance over metal parts as well as afford increased design flexibility.

Hoses There are a variety of rubber hoses found under the hood, made from various types of elastomers. The most common elastomer used is ethylene-propylene-diene (EPDM) with a reinforcement material, mostly rayon yarn and aramid yarn. The rubber hoses are lighter in weight than metal and can better endure the harsh vibrations from engine–vehicle and vehicle–road interactions. They can also be formed into a variety of shapes. With the ever-shrinking size of the engine compartment this is very important.

Radiator System There are several applications that have gone from metal to a polymer in the radiator with considerable reduction in vehicle mass. These applications include SMC and glass-filled PP supports, nylon and PP fans and fan shrouds, nylon end tanks, PE overflow, and nylon and polyphenylene sulfide (PPS) water pumps. The nylon end tanks have been studied for recycling at end of life. The radiators along with the end tanks are removed from the vehicle for recycling. This eliminates the cost of dismantling the end tanks from the radiation during recycling of the nylon. A part in the radiator that contains recycled material is the seal. It is generally made from a PP modified with recycled rubber.

Intake Manifold Nylon has been extensively used in the intake manifold. The applications include the 1993 4.0-liter V-8, 3800 engine in the Aurora, the 1994 Dodge Neon, the 1995 GM Quad IV, and the 1996 4.6-liter Mustang.

Engine Throttle Body Nylon has also found its way into an engine throttle body application. The 2000 Dodge Neon provides an example of this application. The nylon part was 60% lighter than aluminum [3], and the machining generally required with aluminum was eliminated with the nylon part. Thus the scrap, and energy used in manufacturing the part were reduced.

Battery Case and Tray The automotive battery is a great recycling success story. Today, about 99% of automobile batteries are collected for recycling. The lead and acid as well as the PP cases are recovered. The recycled PP from the case can then be used in other cases or used to produce the battery tray. Approximately 40% of the recycled case material is used in manufacturing cases [3]. An example of a unique battery tray is seen in the GM EV1 electric vehicle with a particularly large set of batteries. The batteries in the EV1 weigh over 500 kg. The structural load requirements included 5 G vertical load when bouncing over potholes, 25 G in a frontal collision, and 18 G in side and rear impact deceleration. The original design in aluminum had 16 parts and weighed 22.7 kg (50 lb). To minimize mass

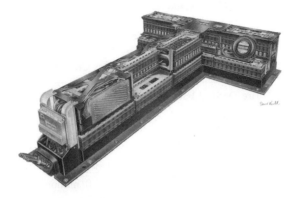

Figure 17.8. EV1 glass-reinforced PP battery tray.

in this electric vehicle, plastics were considered for the tray application. The material chosen was a long glass-fiber random-reinforced polypropylene sheet. The final design was a single part that weighed only 15 kg (33 lb) (shown in Figure 17.8).

Fender Liner and Splash Shields PP is also the material of choice for fender liners and splash shields. The material is light weight and can be molded into a variety of shapes in different colors. Recycled PP can be used in many of these applications.

Fluid Reservoirs Newer vehicles are being continuously studied for mass savings opportunities. Some of the mass savings initiatives along with other design parameters have caused the engine compartment to become progressively smaller. Therefore, packaging the components within them has become increasingly difficult. Polymers enable packaging of fluid reservoirs in tight spaces. Power steering and brake fluid reservoirs are generally made from nylon. Coolant and windshield washer fluid reservoirs are made from PE and PP, respectively. Some pressurized coolant reservoirs, however, are made from nylon.

17.3.4. Chassis Polymers

Tires The early automobile wheels were constructed of wood. In time, to increase durability and the quality of the ride, coatings of harder wood, metal, or leather were added. ("The wheel, therefore, had an attire to enhance its performance. Through the years (or centuries), the term was shortened to tire" [1].)

Over time it was discovered that a coating inflated with air would greatly improve its performance. The tires of today are complex designs with rubber, steel belts, and fabric reinforcements. The modern tire is optimized to give the least amount of rolling resistance while still meeting its other performance requirements. This helps to increase fuel economy.

Table 17.5 Markets for End-of-Life Tires

1996 Scrap Tire Markets	
Scrap Market	Tires Consumed
Tire-Derived Fuel	152.0 million
Export	15.0 million
Ground Rubber	12.5 million
Civil Engineering	10.0 million
Stamped Products	8.0 million
Agricultural Uses	2.5 million
Miscellaneous Uses	1.0 million

The modern tire is also much more durable compared to the early designs. This has reduced the need for frequent replacement tires, and thus the raw materials and manufacturing effort needed to produce them. Improvements in design and rubber technology have made this possible.

The design of a tire with its several component materials yields its unique performance characteristics. It also makes the tire a particularly challenging product to recycle. The recycling infrastructure has developed technologies to grind tires and separate the materials in them. There are several markets for end-of-life tires. Some ground tire rubber finds its way back into automotive parts such as splash shields and brake pedal pads. A list of scrap tire markets is given in Table 17.5 [12].

Fuel Tanks The material of choice for fuel tanks was steel for many years. The main challenge for polymers in this area was fuel permeation. In the late 1980s and early 1990s the technology to co-extrude a multilayer tank using high-density polyethylene (HDPE), adhesives, and ethylene vinyl alcohol (EVOH) copolymer or nylon (as the fuel barrier) was developed. These polymeric fuel tanks were lighter in weight and could be molded into complex shapes for easier packaging.

A life-cycle study was recently carried out using the fuel tank on a 1996 GMC full-size van. The van was available in two versions, one with a steel tank and one with an HDPE tank. The mass and design information was accurate and directly comparable. The van with the HDPE tank was estimated to consume 31 less liters of fuel over its useful life (177,000 km or 110,000 miles). Usage figures were 25,390 liters when equipped with the steel tank versus 25,359 when equipped with the HDPE tank [13]. The HDPE tank weighed 7.85 kg less, a 36% mass savings, and had equivalent performance to the steel tank.

17.4. SUMMARY

Polymer materials began appearing in vehicles as early as 1909. Their use has continued to grow dramatically because they add value to the vehicle. Polymers make the vehicle more comfortable, safer, more durable, more resistant to corrosion and collision damage, and lighter weight. Many of these improvements also benefits the environment. The primary environmental benefits include very substantial reductions in the mass of vehicles, the energy and materials used in their manufacture, and consequent reductions in the emissions related to their manufacturing. Key impacts of the use of polymers in automobiles might be summarized as follows.

- Plastics use in automobiles can offer very significant mass reductions, sometimes as large as 60% or more when the whole vehicle is considered, that invariably improves the fuel economy. Improving fuel economy lowers the rate at which fossil fuel resources are used.
- Polymers also deliver increased durability to the vehicles, reducing the demand on scarce raw materials to manufacture replacement parts and vehicles. For instance, when the service life of a tire is doubled, four to eight fewer tires are required during the useful lifetime of a vehicle. Considering the number of vehicles produced worldwide, this amounts to a significant amount of synthetic rubber (based on scarce petroleum reserves) that is conserved.
- Polymers can be molded in color, which can eliminate the need for painted parts. Eliminating paint is particularly desirable as it reduces air emissions of volatiles and aerosols associated with painting. Eliminating paint also makes the parts easier to recycle. Also, molding plastics require less energy compared to metal forming operations.
- Polymers used in automobiles are somewhat difficult to recycle from end-of-life vehicles. Smart design practices and improvements in recycling technology should continue to close this gap. Automotive design engineers can help to grow the markets for polymers by adopting recycling-friendly designs and by specifying recycled materials for parts. Growth of the infrastructure that recycles end-of-life vehicles will invariably be best motivated by economics.

REFERENCES

1. *The Automobile: A Century of Progress*, Society of Automotive Engineers, August, 1997.
2. *AAMA Motor Vehicle Facts & Figures '95*, American Automobile Manufacturers Association, September, 1995.
3. *Automotive Plastics Report—2000, 21st Annual Multiclient Study of Automotive Plastics Opportunities*, Market Search, 2000.
4. R. W. Beck, *1998 National Post-consumer Plastics Recycling Rate Study*, July 1998.
5. American Forest and Paper Association.
6. Aluminum Association Inc.

7. Compiled by Dr. Candace Wheeler (Staff Research Scientist, Vehicle Emissions and Life Cycle Analysis, General Motors Corporation) using GaBi Software System for Life Cycle Engineering (University of Stuttgart and PE Europe GmbH).
8. S. Kia, J. Claya, R. Gunther, and S. Papasavva, *J. Coatings Tech.* **74**(925), 65–76, 2002.
9. G. Neale, Design in a Textile Revolution, SAE 750338, 1975.
10. *The Resin Review*, 2000 edition, American Plastics Council, November 2000.
11. P. C. Wilson, *Chrome Dreams; Automobile Styling Since 1893*, Chilton, 1976.
12. Scrap Tire Management Council, 1996.
13. G. A. Keoleian, S. Spatari, R. T. Beal, R. D. Stephens, and R. L. Williams, *Int. J. Life Cycle Assessment* **3**(1), p. 25, (1998).

INDEX

Accumulation rate approach, marine environment sampling and surveys, 383
Acetate, textile processing wastes, 293
Acid dyes, textile processing wastes, 290–291
Acidification and acid rain:
  polymeric materials, 320
  polymer industry, 34, 51–52
  recycling, life-cycle analysis, 547–548
  volatile organic compounds regulation, 223–224
Acrylics, textile processing wastes, 291–292
Acrylonitrile-butadiene-styrene, degradation mechanisms, 322–323
Additives:
  coatings industry regulation, 226–228
  identification technologies, recycling, 587–588
  plastics manufacturing, 106
Adhesives, polymer industry pollution potential, 47–48
Aerobic environments, biodegradable water-soluble polymer test methods, 497
Agricultural applications, 185–209
  drip-irrigation systems, 201–204
  greenhouse films, 187–190
  mulch films, 197–199
  overview, 185–187
  silage, 199
  stabilization procedures, 192–197
  waste disposal, 199–201
  weather parameters, 190–192
Agriculture:
  fertilizers, 9–10
  global warming, 67–68
  ozone depletion, 73–74
  renewable sources, biodegradable polymers, 362–364
Air bags, automotive polymers, 734–735
Air classification methods, plastic recycling, 596–597
Air pollution. *See also* Environment; Pollution
  polymeric materials, 320
  polymer industry, 44–50
  textile processing wastes, 253–261
Alcaligenes eutrophus, biodegradable polymers, 365
Aliphatic isocyanate oligomers, 215
Aliphatic polyamides. *See also* Nylons
  degradation mechanisms, 326
  stabilization methods, 337

747

Aliphatic polymers, degradation mechanisms, 321–322
Aluminum, recycling, 133–134
Ammonolysis, of nylons, textile fiber recycling, 707
Amorphophallus konjac, biodegradable polymers, 364
Anaerobic environments, biodegradable water-soluble polymer test methods, 498–499
Anionic materials, textile processing wastes, 292
Appliances:
   flammability, UL 94 Standard Test, 464–465
   recycling collection, 579–581
Aromatic polyamides:
   degradation mechanisms, 326–327
   stabilization methods, 337
Arrhenius parameters, thermal destruction, 676–677
ASTM D 2863-70 Test, flammability, 465–467
ASTM E84-00a Test, flammability, 468–472
ASTM E1321 Standard Test, flammability, 473–474
ASTM E162-98 Test, flammability, 467–468
ASTM E648-99 Test, flammability, 472–473
Aureobasidium pullulans, biodegradable polymers, 365
Automobile industry, 727–746
   end-of-life vehicles:
      recycling collection, 581–583
      recycling process, 727–728
      environmental impacts, 730–731
      flammability, FMVSS 302 test, 464
      overview, 727–730
      polymer applications:
         chassis, 743–744
         exteriors, 737–741
         interiors, 731–737
         underhood, 741–743
      resource depletion, 6
      sustainability, 526–527
      waste production, 634, 635
Automotive coatings. *See* Coatings

Azoic (nephthol) dyes, textile processing wastes, 295

Bacterial polyesters (polyhydroxyalkonoates), biodegradable polymers, 365
Batch dyeing machines, textile processing wastes, 286–287
Batch processes, coloration (textile processing wastes), 285–286
Batch ratio, textile processing wastes, 287–288
Battery case and tray, automotive polymers, 742–743
Binders, coatings, 215–220
Biodegradable polymers, 359–377. *See also* Biodegradable water-soluble polymers; Photodegradable plastic; Polymeric materials
   commercially available, summary table, 369
   defined, 494
   design and manufacture requirements, 361
   future prospects, 368, 370
   justification, 360–361
   overview, 359–360
   packaging, 158–160
   petroleum-derived products, 367–368
   renewable sources, 361–367
      agricultural, 362–364
      biosynthesis, 365–367
Biodegradable water-soluble polymers, 491–519. *See also* Biodegradable polymers; Photodegradable plastic; Polymeric materials
   definitions, 493–495
   modified natural polymers, 509–512
   opportunities, 495–497
   overview, 491–493
   synthesis of, 499–509
      carbon chain polymers, 499–505
      heteroatom chain polymers, 505–509
   test methods, 497–499
Biomass-derived power, possibilities of, 15–16
Biosynthesis, biodegradable polymers, 365–367
Biota, marine environment, 386–388

Bleaching, textile processing wastes, 282
Blow-molded containers, packaging, 146–147
Blow molding, plastics manufacturing, 110–111
Body panels, automotive polymers, 737
Bottle deposits, recycling, 567
Bottom ash, metals in, thermal destruction, 671–672
Building materials, flammability, ASTM E84-00a Test, 468–472
Bulk systems/auto dispensing, textile processing wastes, 298–299
Bumper/fascia system, automotive polymers, 740
Business enterprises. *See* Industry

Caprolactam recovery, textile fiber recycling, 702–704
Carbon arcs, polymeric materials weathering tests, 346–347
Carbon chain polymers, biodegradable water-soluble polymer synthesis, 499–505
Carbon dioxide emissions, public policy, 4
Carbon monoxide emission, thermal destruction, 665–666
Carpet fiber recycling. *See* Textile fiber recycling
Carpets, automotive polymers, 736
Catalytic pyrolysis, of nylons, textile fiber recycling, 702
Cationic materials, textile processing wastes, 291–292
Cellulose:
  biodegradable polymers, 362–363
  thermal destruction:
    experimentation methods, 643–646
    experimentation results, 646–655
    generally, 641
    problem statement, 641–643
Centrifuges, recycling technology, plastic-plastic separation, 603–604
Chemical handling systems, textile processing wastes, 288
Chemical markings, recycled products, 577
Chemical recycling, technology, 611–615
Chitin, biodegradable polymers, 366–367

Chitosan, biodegradable polymers, 366–367
Chlorofluorocarbons:
  ozone depletion, 24, 42–44, 71, 72
  polystyrene manufacture, 102–103
  polystyrene packaging, 156
  polyurethane foam manufacture, 117, 118
Chromium compounds, coatings industry regulation, 226–228
Clean Air Act Amendments of 1977, 220
Clean Air Act Amendments of 1990, 42, 220, 223, 256–260
Clean Air Act of 1970, 52, 220
Clean Room Flammability Standard, 474–481
Coarse manual sorting, recycling technology, 583–584
Coatings, 211–240
  binders and solvents in, 215–220
  cost benefit analysis, 211–212
  energy crisis, 212
  historical perspective, 213–215
  materials crisis, 212–213
  polymer industry pollution potential, 47–48
  regulation of industry, 220–228
    hazardous air pollutants, 224–226
    pigments and additives, 226–228
    volatile organic compounds, 220–224
  regulation response of industry, 228–239
    application methods, 229–230
    conversion to high-solids coatings, 234–236
    conversion to waterborne, 231–234
    powder and radiation cure coatings, 236–238
    solvent abatement, 238–239
Coil coatings, binders in, 215–216
Coloration (textile processing wastes), 283–299
  acid dyes, 290–291
  anionic materials, 292
  batch dyeing machines, 286–287
  batch ratio, 287–288
  bulk systems/auto dispensing, 298–299
  cationic materials, 291–292
  chemical handling systems, 288

750 INDEX

Coloration (textile processing wastes) (*Continued*)
  disperse dies, 293
  dyebath reuse, 298
  fiber-reactive dies, 293–295
  metals, 285
  nephthol (azoic) dyes, 295
  pigments, 296
  pollutants for dye classes, 288–289
  processes in, 285–286
  product design, 283, 285
  quality, 286
  sulfur dyes, 296–297
  vat dyes, 297–298
  waterless coloration technologies, 298
Combustion aerodynamic problem, thermal destruction, 639–640
Composite manufacture, polymer industry pollution potential, 49
Composite reinforcement, textile fiber recycling applications, 714–715
Concrete reinforcement, textile fiber recycling applications, 715–716
Console, automotive polymers, 735
Consumer electronics, recycling collection, 579–581
Consumers:
  environmental stewardship responsibility, 30–31
  media and, 32
Contaminants, in fiber, textile processing wastes, 244
Contingent valuation method, 26
Continuous processes, coloration (textile processing wastes), 285–286
Conversion processes, energy requirements, 132–133
Cost-benefit analysis, 26, 211–212
Cost-benefit analysis, sustainability, recycling, 552–560
Cotton:
  coloration, textile processing wastes, 295, 296–297
  contaminants in, 276, 277–279
Cutting room waste, textile processing wastes, 303–304
Cyclone separators, plastic recycling, 596–597

Decomposition, flammability, 417

Degradable plastic. *See also* Biodegradable polymers; Biodegradable water-soluble polymers; Photodegradable plastic
  defined, 494
  marine environment, 389–394. *See also* Marine environment
  polymers, 403
Density-based dry methods, recycling technology, plastic-plastic separation, 604
Density-based wet methods, recycling technology, plastic-plastic separation, 598–604
Depolymerization, of nylons, textile fiber recycling, 701–708
Desizing, textile processing wastes, 281
Developing countries, renewable resources, 13
Devices, flammability, UL 94 Standard Test, 464–465
Dioxins:
  emission of, thermal destruction, 668–670
  polyvinyl chloride, environmental and safety issues, 100–101
Disperse dies, textile processing wastes, 293
Doors, automotive polymers, 738–739
Door trim, automotive polymers, 737
Drip-irrigation systems, agricultural applications, 201–204
Ducts, automotive polymers, 734
Dyebath reuse, textile processing wastes, 298
Dyeing, textile processing wastes, 283–299. *See also* Coloration (textile processing wastes)

Ecosystems:
  described, 20–22
  global warming, 67
Elastomers:
  flammability, 407
  stabilization methods, 336
Electronics, recycling collection, 579–581
Emergency Planning and Community Right to Know Act (EPCRA), 224–226

End-of-life vehicles:
  recycling collection, 581–583
  recycling process, 727–728
Energy crisis:
  coatings, 212
  environment, 11–16
Energy recovery, recycling technology, 565, 615–617
Energy requirements, 123–135
  contributions to, 124
  conversion processes, 132–133
  feedstock, 125
  gross and net calorific values, 127–128
  gross energies, typical, 128–132
  historical perspective, 123–124
  packaging, 166–177
  recycling, 133–135
  representations, 126–127
  transportation, 125–126
Engine throttle body, automotive polymers, 742
Environment, 3–75
  automobile applications, 730–731
  business enterprises, 18–25
  energy crisis, 11–16
  global warming, 62–68
  marine environment, 385–389. *See also* Marine environment
  materials crisis, 17–18
  overview, 3–11
  ozone depletion, 69–75
  packaging, 166–177
  perspectives on, xiii–xv
  polyethylene manufacture, 92–93
  polymer industry, 32–50. *See also* Polymer industry
  polyvinyl chloride, 98–101
  proactive treatment, 56–57
  stewardship responsibility, 27–32, 56
    consumers, 30–31
    government, 31–32
    industry, 27–29
    valuation methods, 25–27
Environmental weathering tests:
  described, 339–342
  laboratory-accelerated weathering tests compared, 351–352
Epibionts, marine environment, 386–388
Epoxy binders, 216

Ethylene vinyl alcohol, packaging applications, 157
Eutrophication, recycling, life-cycle analysis, 547–548
Excited-state quenchers, polymeric material stabilization, 330–331, 333
Exterior lighting, automotive polymers, 740–741
Extrusion blowing, of film, plastics manufacturing, 111–113
Extrusion processing, plastics manufacturing, 107–108

Fabric formation, textile processing wastes, 269–271
Feedstock, energy requirements, 125
Feedstock recycling:
  life-cycle analysis, sustainability, 535
  technology, 611–615
Fender liner, automotive polymers, 743
Fenders, automotive polymers, 737–738
Fertilizers, agriculture, 9–10
Fiber, contaminants in, 244, 276. *See also* Textile fiber recycling; Textile processing wastes
Fiber-reactive dies, textile processing wastes, 293–295
Fiber spinning, polymer industry pollution potential, 48–49
Films:
  greenhouse, agricultural applications, 187–190
  mulch, agricultural applications, 197–199
Filtered xenon arcs, polymeric materials weathering tests, 343–346
Financing, recycling, sustainability, 552
Finishing operations, textile processing wastes, 299–303
Fire. *See* Flammability
Flame spread, flammability, 407
Flaming combustion, flammability, 406
Flammability, 403–489
  fire propagation, 461–481
    ASTM D 2863-70 Test, 465–467
    ASTM E84-00a Test, 468–472
    ASTM E1321 Standard Test, 473–474
    ASTM E162-98 Test, 467–468

752 INDEX

Flammability (*Continued*)
    ASTM E648-99 Test, 472–473
    Clean Room Flammability Standard, 474–481
    FMVSS 302 test, 464
    generally, 461–464
    UL 94 Standard Test, 464–465
    heat exposure, 406–407
    melting, 408–416
    overview, 403–406
    vapor combustion, 435–461
        fire properties, 437–457
        generally, 435–436
        heat flux, 458–461
        oxygen availability, 437
    vapor ignition, 417–435
    vaporization/decomposition, 417
    vapor release, 407–408
Flexible packaging, 145–146
Floor coverings, flammability, ASTM E648-99 Test, 472–473
Fluid reservoirs, automotive polymers, 743
Fluorescent UV lamps, polymeric materials weathering tests, 347–348
Fly ash, metals in, thermal destruction, 670–671
FMVSS 302 test, flammability, 464
Food production, population growth, 9
Fossil fuels:
    consumption statistics, 212
    depletion of, 3, 13–14, 33, 38–40
    energy crisis, 11–16
    population growth, 8–9
    transportation, 6
Fourier transform infrared, identification technologies, recycling, 587
Free-radical scavengers, polymeric material stabilization, 333–334
Froth flotation, recycling technology, plastic-plastic separation, 604–608
Fuel recovery, recycling technology, 615–617
Fuel tanks, automotive polymers, 744
Furan emission, thermal destruction, 668–670

Gasification, thermal destruction, 637
Glass-fiber-reinforced composites, textile fiber recycling applications, 712–714
Glass-reinforced plastics, agricultural applications, 186
Global perspective:
    environmental issues, polymer industry, 33–34, 37–38
    recycling, 566–567, 569, 571–573
Global warming:
    described, 62–68
    polymer industry, 33, 41–42
Glove box, automotive polymers, 734
Government:
    coatings industry, 220–228
        hazardous air pollutants, 224–226
        pigments and additives, 226–228
        response of industry, 228–239
            application methods, 229–230
            conversion to high-solids coatings, 234–236
            conversion to waterborne coatings, 231–234
            powder and radiation cure coatings, 236–238
            solvent abatement, 238–239
        volatile organic compounds, 220–224
    environmental stewardship responsibility, 4, 29, 31–32, 56
    packaging applications, 177–181
Graft polymers, biodegradable water-soluble polymers, 510
Greenhouse films, agricultural applications, 187–190
Greenhouse gas emissions, polymer industry, 41–42. *See also* Global warming
Gross calorific values, energy requirements, 127–128. *See also* Energy requirements

Hazardous air pollutants:
    coatings industry regulation, 224–226
    textile processing wastes, 256–261
Hazardous wastes:
    recycling, life-cycle analysis, 548–549
    textile processing wastes, 265
Headliner, automotive polymers, 736–737
Health. *See* Public health
Heat flux, flammability, 458–461
Heating ducts, automotive polymers, 734
Heat of pyrolysis, thermal destruction, 678–680

Heat setting, textile processing wastes, 282
Herbicides, 24
Heteroatom chain polymers, biodegradable water-soluble polymer synthesis, 505–509
High-density polyethylenes. *See* Polyethylene(s)
High-impact polystyrene, degradation mechanisms, 322
High-solids solvent-borne (HSSB) coatings:
   conversion to, 234–236
   described, 215, 218, 220
Hindered amine light stabilizers, polymeric materials, 331, 334–339
Home insulation, sustainability, 525–526
Hoses, automotive polymers, 742
Household packaging materials, recycling, sustainability, 531–535
Humidity, agricultural applications, 192
Hydrocyclones, recycling technology, plastic-plastic separation, 600–603
Hydrolysis, of nylons, textile fiber recycling, 701, 705–706
Hydrolytically degradable plastic, defined, 494
Hydroperoxide decomposers, polymeric material stabilization, 333
Hydropower, possibilities of, 14–15

Identification technologies:
   additives, recycling technology, 587–588
   plastics, recycling technology, 584–587
Ignition. *See also* Flammability
   flammability, 406
   polymer vapors, 417–435
Incineration, coatings industry, solvent abatement, 238–239. *See also* Thermal destruction
Incineration hazards, solid waste management, 55–56
Indoor air pollution, textile processing wastes, 254–256
Industrial ecology, term of, 20
Industry. *See also* Automobile industry; Polymer industry; specific industries and industrial applications
   environment, 18–25
   environmental stewardship responsibility, 27–29, 56
Injection-molded packaging, 147–148
Injection molding, plastics manufacturing, 108–110
Instrument panel, automotive polymers, 731–735
Insulation, home, sustainability, 525–526
Intake manifold, automotive polymers, 742
Integrated resource management, defined, 567
International test standards, flammability, 404–405
Irradiance, laboratory-accelerated weathering tests, polymeric materials, 349
Irrigation, drip-irrigation systems, 201–204

Kice multiaspirators, plastic recycling, 597
Kongskilde aspirator, plastic recycling, 596
Konjac, biodegradable polymers, 364

Laboratory-accelerated weathering tests, 342–351
   environmental weathering tests compared, 351–352
   irradiance, 349
   light sources, 343–348
   moisture, 350–351
   spectral power distribution effect, 348–349
   temperature, 349–350
Lactones, polymer industry pollution potential, 49–50
Laminate reinforcement, textile fiber recycling applications, 714–715
Landfills. *See also* Solid wastes
   biodegradability, 158–160, 360
   solid waste management, 52–53, 140
Laser velocimetry, thermal destruction, diagnostic tools, 639
Lateral Ignition and Flame Spread Test (LIFT), ASTM E1321 Standard Test, 473–474
Lead compounds, coatings industry regulation, 226–228

Legislation. *See* Government
Life-cycle analysis, 26
  packaging, 166–167
  recycling, sustainability, 535–551
Lighting, automotive polymers, 740–741
Light screeners, polymeric material stabilization, 331–333
Limited Oxygen Index of Materials, ASTM D 2863-70 Test, 465–467
Linear low-density polyethylenes. *See* Polyethylene(s)
Litter, solid waste management, 53–55
Local issues. *See* Regional/local environmental issues
Lopha cristagalli, 388
Love Canal, 23, 100
Low-density polyethylenes. *See* Polyethylene(s)
Low-temperature oxygen/air-enriched systems, thermal destruction, 636
Low-temperature starved air systems, thermal destruction, 636

Macrolitter, marine environment, 382
Major appliances, recycling collection, 581
Malthus, Thomas, xiii, xiv, xv, 9
Manual sorting, coarse, recycling technology, 583–584
Marine environment, 379–401
  biological and environmental impacts, 385–389
    ecotoxicology, 388–389
    entanglement and ingestion, 385–386, 387
    epibionts and biota, 386–388
  degradation, 389–394
    defined, 390–391
    deterioration, 391–393
    floating plastic debris, 393–394
    generally, 389–390
  overview, 379–380
  ozone depletion, 74
  photodegradable plastics as mitigation strategy, 394–397
  plastic litter, 381–385
    categories and sources, 381–382
    quantities and distribution, 383–385
    sampling and surveys, 383
Markings, recycled products, 574–577
Materials. *See* Raw materials

Maximum decomposition temperature, thermal destruction, 677–678
Mechanical recycling, life-cycle analysis, sustainability, 535–547
Media, consumers and, 32
Megalitter, marine environment, 382
Melamine formaldehyde (MF) resins, 215
Melting, flammability, 408–416
Mercerization, textile processing wastes, 282
Mercury, 24
  coatings industry regulation, 226–228
  coloration (textile processing wastes), 285
Mesolitter, marine environment, 381–382
Metal halide lamp, polymeric materials weathering tests, 348
Metals:
  in bottom ash, thermal destruction, 671–672
  coloration (textile processing wastes), 285
  in fly ash, thermal destruction, 670–671
  recycling of, 564
  removal of, plastic recycling, 595–596
Mid-infrared (MIR) spectroscopy, recycling technology, 586
Mineralization, degradation, 390–391
Minerals. *See* Raw materials
Mirror housings, automotive polymers, 739–740
Modified natural polymers, biodegradable water-soluble polymers, 509–512
Moisture:
  laboratory-accelerated weathering tests, polymeric materials, 350–351
  polymeric materials, 318–319
Moldings, automotive polymers, 737
Montreal Protocol of 1987, 4, 24, 28, 42, 44, 72, 113
Mulch films, agricultural applications, 197–199
Multiaspirators, plastic recycling, 596–597
Municipal solid waste. *See* Solid wastes

Natural resources. *See also* Fossil fuels; Raw materials
  availability of, xiv
  conservation of, xv

environmental resources valuation
   methods, 25–27
  transportation, 6
Near-infrared (NIR) spectroscopy,
   recycling technology, 586, 588–590
Nephthol (azoic) dyes, textile processing
   wastes, 295
Net calorific values, energy requirements,
   127–128. *See also* Energy
   requirements
NFPA 318[77] FMR 4910 Test Standard,
   474–481
Nonflaming combustion, flammability, 406
Nuclear power, possibilities of, 16
Nurdles, marine environment, 382
Nylons. *See also* Aliphatic polyamides
  depolymerization of, textile fiber
     recycling, 701–708
  packaging applications, 157

Oil crisis, recycling, 567
Organization of Petroleum Exporting
   Countries (OPEC), 567
Oxidative degradable plastic, defined, 494
Oxygen:
  flammability, 437
  Limited Oxygen Index of Materials,
     ASTM D 2863-70 Test, 465–467
  polymeric materials, 319–320
Ozone, polymeric materials, 320
Ozone depletion:
  chlorofluorocarbons, 24
  described, 69–75
  Montreal Protocol, 4
  polymer industry, 34, 42–44
  volatile organic compounds regulation,
     221–223

Packaging:
  recycling collection, 577–579
  sustainability, 527–530, 531–535
Packaging applications, 139–183
  advantages, 143–145
  biodegradability, 158–160
  energy and environmental assessments,
     166–177
  functions, 142–143
  government regulation, 177–181
  overview, 139–142

plastics in, 148–158
  generally, 148–149
  polyethylenes, 149–152
  polyethylene terephthalate, 152–154
  polypropylene, 154
  polystyrene, 155–156
  polyvinyl chloride, 156–157
  specialty polymers, 157–158
recycling, 163–166
reuse, 162–163
source reduction, 160–162
types, 145–148
  blow-molded containers, 146–147
  flexible, 145–146
  injection-molded components,
     147–148
  thermoformed shapes, 148
Paints. *See* Coatings
Particle image velocimetry, thermal
   destruction, diagnostic tools, 639
Particulate matter emission, thermal
   destruction, 666–668
Pesticides, textile processing wastes, 276,
   277–279
Petroleum-derived products, biodegradable
   polymers, 367–368
Phenolic foams, described, 117
Photodegradable plastic. *See also*
   Biodegradable polymers;
   Biodegradable water-soluble
   polymers
  defined, 494
  as mitigation strategy, marine
     environment, 394–397
Physical markings, recycled products,
   574–577
Pickup beds, automotive polymers, 741
Pigments:
  coatings industry regulation, 226–228
  textile processing wastes, 296
Pillar trim, automotive polymers, 737
Plasticizers, polyvinyl chloride,
  environmental and safety issues,
     99–100
Plastics, 77–121. *See also* Polymeric
   materials; Polymer industry; specific
   plastics
  consumption of, 4–6, 79–82
  environmental concerns, 6–7, 82–83

756 INDEX

Plastics (*Continued*)
  historical perspective, 77–79
  manufacturing process, 105–113
    blow molding, 110–111
    energy requirements in, 123–135.
      See also Energy requirements
    extrusion blowing of film, 111–113
    extrusion processing, 107–108
    generally, 105–107
    injection molding, 108–110
  perspectives on, xvi
  polyethylenes, 83–93
    consumption of, 83–88
    environmental aspects of manufacture, 92–93
    manufacture of, 88–92
  polyethylene terephthalate, 103–105
    consumption of, 103–104
    manufacture of, 104–105
  polypropylene, 93–95
    consumption of, 93–94
    manufacture of, 94–95
  polystyrene, 101–103
    characteristics of, 101
    manufacture of, 101–103
  polyurethane and polymeric foams, 113–119
    generally, 113–114
    phenolic foams, 117
    polyurethane foam, 114–117
    uses of, 117, 119
  polyvinyl chloride, 96–101
    characteristics of, 96–97
    environmental and safety issues, 98–101
    manufacture of, 97–98
Plastics industry. *See* Polymer industry
Pollution:
  governmental stewardship responsibility, 31–32
  marine, 379, 385–389. *See also* Marine environment
  polymer industry, 44–50
  solid waste recycling, 22–25
  textile processing wastes, coloration, 288–289
  threat of, 3–4
Pollution Prevention Act of 1990, 32

Polyacrylonitrile, packaging applications, 157
Polyactic acid, biodegradability, 160
Polyamides:
  biodegradable water-soluble polymer synthesis, 505–507
  stabilization methods, 337
Polybutylene/ethylene succinate/adiphate/terepthalate, biodegradable polymers, 368
Polycaprolactone, biodegradable polymers, 367–368
Polycarbonate:
  degradation mechanisms, 328–329
  packaging applications, 157
  stabilization methods, 338
Polycarboxylates, biodegradable water-soluble polymer synthesis, 502–505
Polychlorinated biphenyls, 24, 386
Polyesters:
  biodegradable water-soluble polymer synthesis, 505–507
  degradation mechanisms, 329–330
  stabilization methods, 338–339
  textile processing wastes, 291–292, 293
Polyethers, biodegradable water-soluble polymer synthesis, 507–509
Polyethylene(s), 83–93
  agricultural applications, 185, 186, 187–188, 189, 202–203
  biodegradability, 158–159
  consumption of, 83–88
  environmental aspects of manufacture, 92–93
  historical perspective, 78
  manufacture of, 88–92
  marine environment, 381
  packaging applications, 145, 147, 149–152
  recycling, 164–166, 570
  thermal destruction, 673–693
Polyethylene naphthalate, packaging applications, 157
Polyethylene terephthalate, 103–105
  agricultural applications, 186
  consumption of, 103–104
  manufacture of, 104–105

marine environment, 381
packaging applications, 152–154
recycling, 567, 569–571. *See also* Recycling
Polyhydroxyalkonoates (bacterial polyesters), biodegradable polymers, 365
Polyhydroxybutyrate-valerate, biodegradability, 159
Polylactic acid, biodegradable polymers, 367
Polymer foams, ozone depletion, 43–44. *See also* Polyurethane and polymeric foams
Polymeric materials, 313–358. *See also* Biodegradable polymers; Biodegradable water-soluble polymers; Plastics; specific polymeric materials
  degradation mechanisms, 320–330
    aliphatic polyamides (nylons), 326
    aromatic polyamides (polyaramids), 326–327
    generally, 320–323
    polycarbonates, 328–329
    polyesters, 329–330
    polyolefins, 323–324
    polystyrene, 324–325
    polyvinyl chloride, 327–328
  environmental weathering tests, 339–342
  laboratory-accelerated weathering tests, 342–351
    environmental weathering tests compared, 351–352
    irradiance, 349
    light sources, 343–348
    moisture, 350–351
    spectral power distribution effect, 348–349
    temperature, 349–350
  overview, 313–314
  stabilization methods, 330–339
    elastomers, 336
    polyamides, 337
    polycarbonate, 338
    polyesters, 338–339
    polyolefins, 335–336
    polyvinyl chloride, 337–338

stabilizer types, 330–335
styrenic polymers, 336–337
weather factors, 314–320
  atmospheric pollutants, 320
  moisture, 318–319
  oxygen, 319–320
  solar radiation, 314–317
  temperature, 317–318
Polymer industry, 32–56
  acidification, 51–52
  fossil fuels depletion, 38–40
  global environmental issues, 33–34, 37–38
  greenhouse gas emissions, 41–42
  ozone depletion, 42–44
  pollution potential, 44–50
  regional/local environmental issues, 35–38
  segmental perspective on, 7
  solid waste management, 52–56
Polymethylmethacrylate, agricultural applications, 186
Polyolefin(s):
  degradation mechanisms, 323–324
  stabilization methods, 335–336
Polyolefin pipes, drip-irrigation systems, 202–203
Polyols, polymer industry pollution potential, 49–50
Polypropylene, 93–95
  consumption of, 93–94
  manufacture of, 94–95
  marine environment, 381
  packaging applications, 154
  thermal destruction, 673–693
Polysaccharides, modification of, biodegradable water-soluble polymers, 510–512
Polystyrene, 101–103
  characteristics of, 101
  degradation mechanisms, 324–325
  historical perspective, 78
  manufacture of, 101–103
  marine environment, 381
  packaging applications, 144, 155–156
  thermal destruction, 673–693
  uses of, 117, 119

Polyurethane and polymeric foams, 113–119
  generally, 113–114
  phenolic foams, 117
  polyurethane foam, 114–117
  uses of, 117, 119
Polyvinyl acetate, packaging applications, 157
Polyvinyl alcohol:
  biodegradable polymers, 368
  biodegradable water-soluble polymer synthesis, 500–502
Polyvinyl alcohol, biodegradability, 160
Polyvinyl chloride, 96–101
  agricultural applications, 185, 186, 203–204
  characteristics of, 96–97
  degradation mechanisms, 327–328
  environmental and safety issues, 98–101
  historical perspective, 78
  incineration hazards, 55–56
  manufacture of, 97–98
  marine environment, 381
  packaging applications, 144, 156–157
  stabilization methods, 337–338
  thermal destruction, 673–693
Population growth, challenge of, 8–10
Postconsumer plastics, defined, 565–566
Powder coatings, shift to, 236–238
Preconsumer plastics, defined, 565–566
Primary recycling, defined, 565
Printing, textile processing wastes, 283–299. *See also* Coloration (textile processing wastes)
Public health:
  global warming, 67
  ozone depletion, 72–73
  polyvinyl chloride, 98–101
Public policy. *See* Government
Pullulan, biodegradable polymers, 365–366
Pullularia pullulans, biodegradable polymers, 365
Pyrolysis:
  catalytic, of nylons, textile fiber recycling, 702
  heat of, thermal destruction, 678–680
Pyrolysis, thermal destruction, 637

Quality control, recycling technology, 610–611
Quaternary recycling, defined, 565

Radiation cure coatings, shift to, 236–238
Radiator, automotive polymers, 742
Raw materials. *See also* Fossil fuels; Natural resources
  coatings, 212–213
  depletion of, 38–40
  environment, 17–18
Recovery, coatings industry, solvent abatement, 238–239
Recycling, 563–627. *See also* Sustainability; Textile fiber recycling
  collection, 577–583
    consumer electronics, 579–581
    consumer packaging, 577–579
    end-of-life vehicles, 581–583
    major appliances, 581
  definitions and nomenclature, 564–566
  energy requirements, 133–135
  future prospects, 617
  global perspective, 566–567
  historical perspective, 567–569
  markings, 574–577
  overview, 563–564
  packaging applications, 163–166
  polyethylene packaging, 149–150
  polyethylene terephthalate, 152–154
  polypropylene packaging, 154
  polystyrene packaging, 155–156
  solid wastes, 22–25
  statistics, 569–574
    Europe, 571–572
    Japan, 572–573
    rubber, 573–574
    United States, 569–571
  sustainability, 530–560
    financing, 552
    generally, 530–531
    household packaging materials, 531–535
    life-cycle analysis, 535–551
  technology, 583–617, 590–608
    coarse manual sorting, 583–584
    color sorting, 610
    energy/fuel recovery, 615–617
    feedstock/chemical recycling, 611–615

identification technologies (additives), 587–588
identification technologies (plastics), 584–587
near-infrared spectroscopy, 588–590
nonplastic materials removal, 595–597
plastic-plastic separation, 598–608
quality control, 610–611
selective dissolution, 608–609
size reduction, 590–595
steel blast furnaces, 615
thermosets, 610
Regenerated fibers, contaminants in, 276
Regional/local environmental issues, polymer industry, 35–38
Regulation. *See* Government; specific legislation
Renewable resources:
biodegradable polymers, 361–367
agricultural, 362–364
biosynthesis, 365–367
developing countries, 13
sources of, 14–16
Resource management, recycling, global perspective, 566–567
Resource recovery, recycling, 565
Resources. *See* Natural resources
Reuse, packaging applications, 162–163
Right to know, coatings industry regulation, 224–226
Rio Declaration of 1992, 25
Rotary grinders, size reduction, recycling technology (mechanical), 592–595
Rubber, recycling statistics, 573–574

Scouring, textile processing wastes, 281–282
Sea levels, global warming, 66–67
Seats, automotive polymers, 735–736
Secondary recycling, defined, 565
Selective dissolution, recycling technology, 608–609
Semiconductor industry, Clean Room Flammability Standard, 474–481
Silage, agricultural applications, 199
Singeing, textile processing wastes, 281

Sink-float tanks, recycling technology, plastic-plastic separation, 599–600
Size reduction, recycling technology (mechanical), 590–595
Small appliances, recycling collection, 579–581
Softening point, recycling technology, plastic-plastic separation, 608
Soil pollution, polymer industry, 44–50
Soil reinforcement, textile fiber recycling applications, 716–719
Solar power, possibilities of, 15
Solar radiation, polymeric materials, 314–317. *See also* Polymeric materials
Solid wastes. *See also* Thermal destruction
agricultural applications, 199–201
biodegradability, 158–160
biodegradable polymers, 360–361. *See also* Biodegradable polymers
biodegradable water-soluble polymers, 495–497. *See also* Biodegradable water-soluble polymers
industry stewardship, 27–29
marine environment, 379. *See also* Marine environment
packaging applications, 139–142
plastics, 6–7
polymer industry, 52–56
polypropylene packaging, 154
problem of, 630–634
recycling, 22–25, 548–549, 568–569
textile processing wastes, 252–253. *See also* Textile processing wastes
Solvents, coatings, 215–220
Sorting. *See also* Recycling
coarse manual, recycling technology, 583–584
near-infrared spectroscopy, recycling technology, 588–590
Source reduction, packaging applications, 160–162
Spectral power distribution (SPD) effect, laboratory-accelerated weathering tests, 348–349
Splash shield, automotive polymers, 743
Stabilization methods:
polyvinyl chloride, environmental and safety issues, 100

Stabilization methods (polymeric materials), 330–339
  elastomers, 336
  polyamides, 337
  polycarbonate, 338
  polyesters, 338–339
  polyolefins, 335–336
  polyvinyl chloride (PVC), 337–338
  stabilizer types, 330–335
  styrenic polymers, 336–337
Standing crop approach, marine environment, sampling and surveys, 383
Starch, biodegradable polymers, 363–364
Steam reforming, thermal destruction, 637
Steel blast furnaces, recycling technology, 615
Styrene emissions, polymer industry pollution potential, 49
Styrenic polymers, stabilization methods, 336–337
Styrofoam, environment, 7
Substrate preparation, textile processing wastes, 276–283
Sulfur dyes, textile processing wastes, 296–297
Surface-property-based methods, recycling technology, plastic-plastic separation, 604–608
Surrogate solid waste. *See* Cellulose
Sustainability, 523–562. *See also* Recycling; Textile fiber recycling
  defined, 523–524
  plastics production, 524–525
  plastics use, 525–530
    automobile industry, 526–527
    home insulation, 525–526
    packaging, 527–530
    technologies, 530
  recycling, 530–560
    correlations, 552–560
    financing, 552
    generally, 530–531
    household packaging materials, 531–535
    life-cycle analysis, 535–551
Sustainable growth, shift to, 12–13
Synthetic fibers:
  coloration, textile processing wastes, 293
  contaminants in, 276, 279–280

Temperature:
  agricultural applications, 191–192
  laboratory-accelerated weathering tests, polymeric materials, 349–350
  polymeric materials, 317–318
Tertiary recycling, defined, 565
Test standards, flammability, 404–405
Textile fiber recycling, 697–725. *See also* Recycling; Sustainability; Textile processing wastes
  carpet waste, 699–700
  overview, 697–699
  technology, 701–721
    depolymerization of nylons, 701–708
    fiber identification and storage, 708
    mechanical separation of polymers, 708–709
    physical/chemical separation of polymers, 709–710
    plastic resin compounds, 710–712
    redesign and reuse, 720–721
    use in composite or laminate reinforcement, 714–715
    use in concrete reinforcement, 715–716
    use in glass-fiber-reinforced composites, 712–714
    use in soil reinforcement, 716–719
    waste to energy conversion, 719–720
Textile processing wastes, 243–309. *See also* Textile fiber recycling
  airborne waste, 253–254
  coloration, 283–299
    acid dyes, 290–291
    anionic materials, 292
    batch dyeing machines, 286–287
    batch ratio, 287–288
    bulk systems/auto dispensing, 298–299
    cationic materials, 291–292
    chemical handling systems, 288
    disperse dies, 293
    dyebath reuse, 298
    fiber-reactive dies, 293–295
    metals, 285

nephthol (azoic) dyes, 295
pigments, 296
pollutants for dye classes, 288–289
processes in, 285–286
product design, 283, 285
quality, 286
sulfur dyes, 296–297
vat dyes, 297–298
waterless coloration technologies, 298
contaminants in fiber, 244
fabric formation, 269–271
finishing, 299–303
global view examples, 245–251
hazardous wastes, 265
indoor air pollution, 254–256
overview, 243–244
product fabrication, 303–304
solid waste, 252–253
substrate preparation, 276–283
toxic air emissions, 256–261
wastewater, 261–265
wet processing chemicals, 271–275
yarn formation, 266–267
yarn preparation, 267–269
Thalamoporella evelinae, 388
Thermal destruction, 629–696
cellulose, 641–655
conclusions, 655
experimentation methods, 643–646
experimentation results, 646–655
generally, 641
problem statement, 641–643
future prospects, 693–694
overview, 629–630
plastic and nonplastic solid waste, 655–673
conclusions, 672–673
experimentation methods, 657–659
experimentation results, 659–672
generally, 655–656
problem statement, 656–657
polypropylene, polystyrene, polyethylene, and polyvinyl chloride, 673–693
conclusions, 692–693
experimentation methods, 675–676
experimentation results, 680–692
generally, 673–674
parameter determination, 676–680

problem statement, 674–675
problem magnitude, 630–634
sources, disposal, and recycling, 634–641
combustion aerodynamic problem, 639–640
diagnostic tools, 638–639
disposal options, 635–637
generally, 634–635
process of, 637–638
turbulent reacting flows problem, 640–641
Thermoformed packaging, 148
Thermoplastics. *See also* Plastics; Polymeric materials; specific thermoplastics
flammability, 407
term of, 78n2
Thermosets. *See also* Plastics; Polymeric materials; specific thermosets
flammability, 407
recycling technology, 610
term of, 78n2
Tires, automotive polymers, 743–744
Titanium dioxide, 23, 213
Toxic air pollutants. *See also* Air pollution; Pollution
textile processing wastes, 256–261
thermal destruction, 672
Transportation:
energy requirements, 125–126
resources depletion, 6
Trim, automotive polymers, 737
Turbulent reacting flows problem, thermal destruction, 640–641

UL 94 Standard Test, 464–465
UL 2360 Test Standard, 474–481
Ultraviolet absorbers, polymeric material stabilization, 331–333, 334–339
Ultraviolet radiation:
agricultural applications, 190–191
polymeric materials, 314–317. *See also* Polymeric materials
Urbanization, increase in, 10
Urethane binders, 216

Valuation methods, environmental resources, 25–27

Vapor combustion, flammability, 435–461. *See also* Flammability
Vapor ignition, flammability, 417–435. *See also* Flammability
Vaporization, flammability, 417
Vapor release, flammability, 407–408
Vat dyes, textile processing wastes, 297–298
Vehicles, flammability, FMVSS 302 test, 464. *See also* Automobile industry
Ventilation ducts, automotive polymers, 734
Vinyl chloride, 96. *See also* Polyvinyl chloride
Volatile organic compounds:
　coatings industry regulation, 220–224
　limits on, 29, 31
　polymer industry pollution potential, 47–48
　textile processing wastes, 256–261

Wastes. *See* Solid wastes; Textile processing wastes
Waste to energy conversion, textile fiber recycling, 719–720
Wastewater, textile processing wastes, 261–265
Waterborne coatings:
　binders in, 216–220
　conversion to, 231–234
Waterless coloration, textile processing wastes, 298
Water pollution, polymer industry, 44–50
Water-soluble polymers. *See* Biodegradable water-soluble polymers; Polymeric materials
Weathering, polymeric materials, 313
Weathering tests. *See* Environmental weathering tests; Laboratory-accelerated weathering tests
Weather parameters, agricultural applications, 190–192
Wet processing chemicals, textile processing wastes, 271–275. *See also* Textile processing wastes
White goods (major appliances), recycling collection, 581
Wind power, possibilities of, 15
Wool:
　contaminants in, 276
　finishing operations, textile processing wastes, 302–303

X-ray fluorescence, identification technologies, recycling, 587–588

Yarn. *See* Textile processing wastes